SEPARATION TECHNIQUES I

SEPARATION TECHNIQUES I
LIQUID-LIQUID SYSTEMS

Edited by
Larry Ricci
and
the Staff of
Chemical Engineering

McGraw-Hill Publications Co., New York, N.Y.

Printed in the United States of America.

Library of Congress Cataloging in Publication Data
Main entry under title:

Liquid-liquid systems:

(Separation techniques; 1)
1. Separation (Technology) 2. Liquids I. Chemical engineering. II. Series.
TP156.S45S45 vol. 1 660.2′842s [660.2′842] 79-27362
ISBN 0-07-606652-5 (paper)
ISBN 0-07-010711-4 (case)

CONTENTS

Introduction

Effectively separating mixtures of liquids is one of the more challenging tasks faced by chemical engineers. Indeed, in any chemical process industry—whether it be manufacture of pharmaceuticals, organics, petrochemicals, cosmetics, foods, wood chemicals, or what have you—the chemical engineer will, at some point, be called on to use various principles of liquid-liquid separation.

The aim of this book, therefore, is to provide largely non-theoretical information that the engineer needs to make intelligent decisions concerning separations of liquids. Instead of an academic treatment, a practical "how to" approach has been taken.

Because distillation is so widespread in the CPI, roughly the first half of the book covers that important operation extensively. The second part takes a look at alternatives to distillation, especially less-energy-hungry options.

In Section One, the basics of batch and steam distillation are presented, focusing on practical aspects. Data analysis—predicting liquid activity coefficients and equilibrium relationships, for example—is detailed in Section Two. A long section on design follows, basically providing: (1) an overview of the essentials of distillation design—energy considerations, equipment layout problems, cost estimating and the like; (2) a guide to specifying particular equipment; and (3) an overview of how to design by computer, by hand-held calculator, or by other methods. Section Four wraps up the section on distillation, with articles covering various aspects of operation and maintenance—for instance, troubleshooting, minimizing foaming, and evaluating packings.

Non-distillation liquid-liquid separation methods follow the distillation sections. One of the most widely used options is liquid-liquid extraction, which is covered in Section Five; for both metals and nonmetallic materials, process aspects and equipment selection and operation are covered. Section Six focuses on another much-used separation technique—adsorption. System selection, specification, design and operation are detailed in a series of articles. Finally Section Seven covers other separation methods, such as ultrafiltration and reverse osmosis.

Thus, the editors believe this book will be a valuable, practical tool for chemical engineers—homing in on the type of information needed to do the job well, whether it be in design, process engineering, operations or maintenance.

Part I
DISTILLATION

Section 1 Basics of Distillation

Batch distillation basics
Steam distillation basics

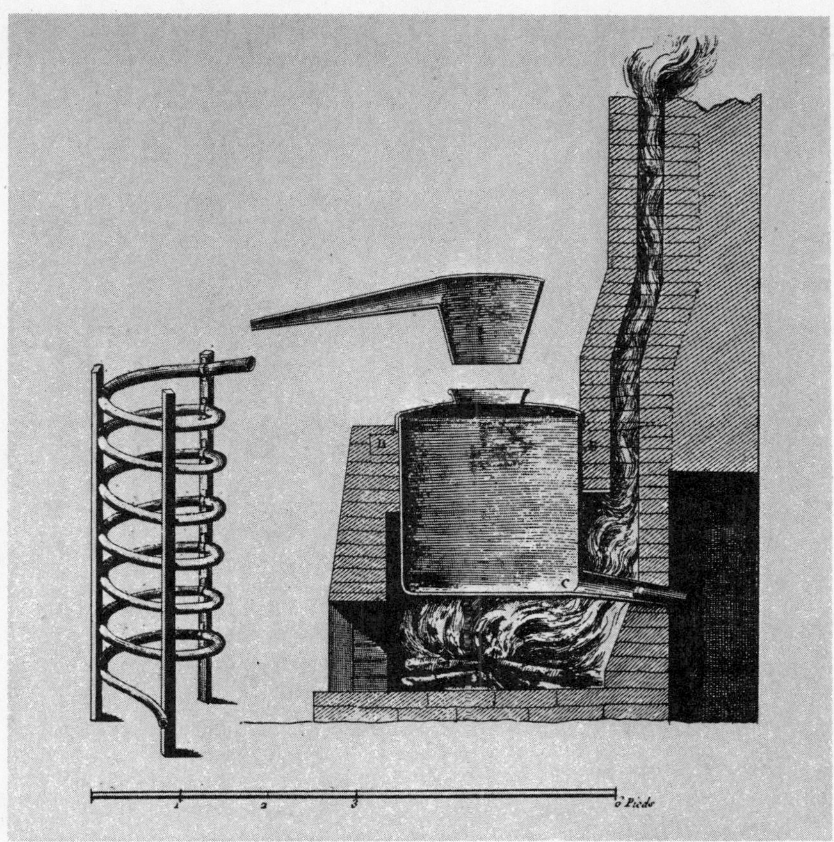

BATCH STILL of a bygone era. This still is of the type used in France during the late 18th century for the manufacture of brandy. The illustration is from Diderot's famous encyclopedia of industrial processes.

Batch Distillation Basics

Although continuous distillation columns are everywhere in the chemical process industries, the batch-still remains useful in situations where feeds may change from batch to batch, and where distillation is required at irregular intervals.

R. W. ELLERBE, The Rust Engineering Co.

In batch distillation, a liquid mixture is charged to a vessel, and the components are separated by boiling the liquid, condensing the vapors, and collecting the components according to their boiling points. The most outstanding attribute of this operation is flexibility.

Batch distillation is often preferable to continuous distillation when relatively small amounts of material are to be handled at irregularly scheduled periods. In many cases, the composition of the material to be distilled may vary widely from period to period. In batch distillation, very little change is required when switching from one fluid mixture to another; reflux rate can be varied easily without balancing or adjusting several instruments; no balance of feed and drawoff need be maintained. In a situation where different mixtures must be handled from day to day, the versatility of the batch still is unexcelled.

Originally published May 28, 1973

The simplest case of batch distillation is one in which the mixture to be separated is charged to a heated kettle fitted with a total condenser and a product receiver. The mixture is distilled without reflux until a definite quantity of one of the components of the mixture has been recovered, or until a definite change in composition has been effected.

Volatility and Relative Volatility

Generally, the term "volatility" is used for comparing vapor pressures of substances. When comparing one with another, the substance with the higher vapor pressure is considered the more volatile. Vapor pressure alone is not a valid criterion for determining the ease of separation of components from liquid mixtures, since the vapor pressure of each is lowered by the presence of the others.

For a solution containing components A and B:

$$v_A = \pi(y_A/x_A) \qquad v_B = \pi(y_B/x_B) \tag{1}$$

where v_i is the activity of each component, where y_i and x_i are the mole fractions of each component in the vapor and liquid phases, and where π is the total system pressure. Relative volatility is defined as:

$$\alpha = v_A/v_B = (y_A \cdot x_B)/(y_B \cdot x_A) \tag{2}$$

The activity coefficient γ is v_i/P_i for each component; therefore:

$$y_A = \gamma_A P_A x_A \qquad y_B = \gamma_B P_B x_B \tag{3}$$

where P_i is the pure-component vapor pressure. Accordingly:

$$\alpha = \frac{(y_A \cdot x_B)}{(y_B \cdot x_A)} = \frac{\gamma_A P_A}{\gamma_B P_B} = \text{relative volatility} = \frac{K_A}{K_B} \tag{4}$$

K_A and K_B are equilibrium vaporization constants for A and B, respectively.

If $\gamma_A = \gamma_B$,

$$\alpha = P_A/P_B \tag{5}$$

Relative volatility is a direct measure of the ease of separation of components by a distillation process. Therefore, substances readily separable by distillation have large values of relative volatility. A relative volatility of unity means no separation is possible, and relative volatility less than unity means that the molar ratio of the constituents in the vapor will be α times their molar ratio in the liquid. Where a difference in volatilities exists, the value of α will always exceed 1 if the volatility of the more volatile component be made the numerator in the above expression for α. The relative volatility of two components changes with temperature. Viewed from the standpoint of ease of separation alone, the optimum temperature range in which to conduct a distillation is that in which the value of α is a maximum. The following relationship derived from the expression for α gives, for a binary mixture, the relationship between x and y for a constant α.[7]

$$y = \frac{\alpha \cdot x}{1 + x(\alpha - 1)} \tag{6}$$

At left is a diagram of Eq. (6) for several values of relative volatility (Fig. 1).

During a batch distillation at constant pressure, the temperature rises as the volatiles concentration diminishes in the residual liquid. For mixtures of similar substances having low values of α, there will be a very slow change in α with temperature; hence, the above equation should give reliable results. These conditions are met, for example, by adjacent members of a homologous series such as the paraffin hydrocarbons. Variations of α, however, should be investigated when using the above equation.[7]

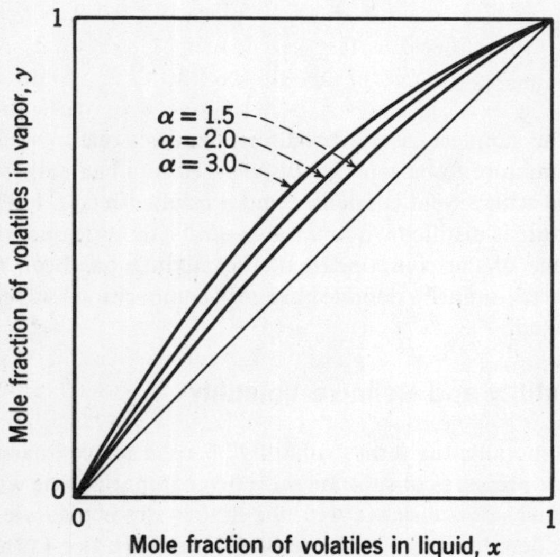

BINARY system with constant relative volatility—Fig. 1

Nomenclature

D	Moles of distillate per unit time, in Eq. (13); moles of distillate at time θ, in Eq. (15); distillate drawoff, in Eq. (23)
K_A, K_B	Equilibrium vaporization constants for A and B, respectively
L	Moles of liquid per unit time, in Eq. (11); liquid returned to column, in Eq. (23)
P	Distillate drawoff percentage = $100/(R + 1)$
P_i	Pure-component vapor pressure, mm. Hg.
R	Reflux ratio—(liquid returned to column)/(distillate drawoff); subscripts indicate number of plates; $R_{min.}$ indicates an infinite number of plates
V	Vapor load, moles of vapor per unit time
v_i	Activity of each component
W	Moles of liquid being distilled, in Eq. (7) to (10); moles of liquid in still at time θ, in Eq. (14), (15)
x	Mole fraction of more-volatile component in liquid, in Eq. (13)
x_i	Mole fraction of component in liquid phase
x_D	Instantaneous mole fraction of the component in the distillate that is leaving the condenser at time θ
x_{D_i}	Initial distillate composition, mole fraction
x_W	Composition of liquid in still, mole fraction
x_{W_o}	Initial mole fraction of the more volatile component in the mixture
y_i	Mole fraction of component in vapor phase
Subscripts	
$n, n + 1$	Plate numbers in rectification column, from top down
o	Initial conditions
Greek letters	
α	Relative volatility—defined in Eq. (2)
γ	Activity coefficient, v_i/P_i, for a component
θ	Time
π	Total system pressure

Differential Distillation

The simplest case of batch distillation is commonly called differential distillation. The vapor generated by boiling the liquid is withdrawn from contact with the

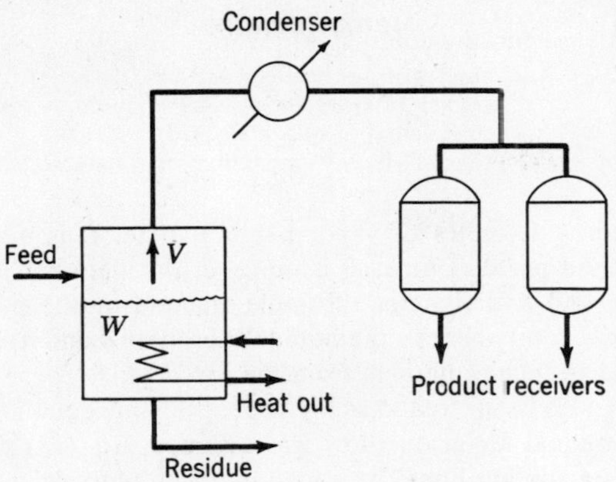

PROCESS SCHEMATIC for differential distillation—Fig. 2

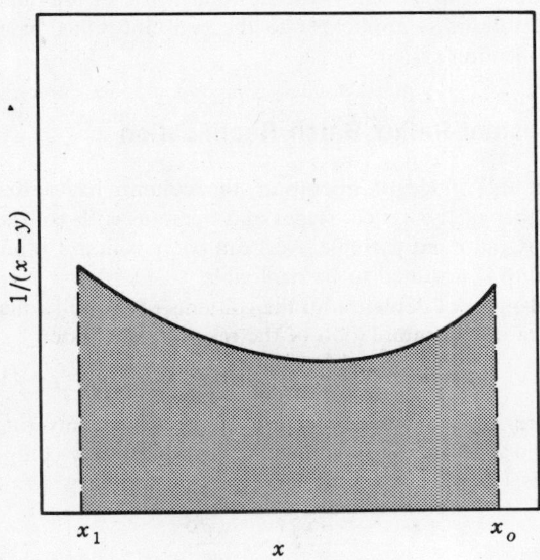

GRAPHICAL integration of Rayleigh equation—Fig. 3

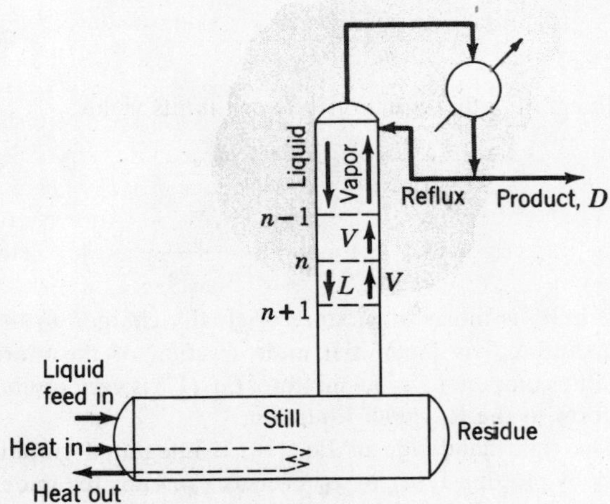

BATCH rectification, process schematic—Fig. 4

liquid and condensed as fast as it is formed. Rayleigh developed an equation for differential distillation. Fig. 2 is a schematic diagram of the differential distillation process.

Consider a binary batch of W_o moles of liquid. At any instant during the distillation, there are W moles of liquid left in the still. For a given instant, when the mole fraction of the more volatile component in the liquid is x_A, and its corresponding vapor composition is y_A, the total amount of component A in the liquid will be $x_A W$. If a differential amount of liquid, dW, is vaporized, the liquid composition will diminish from x_A to $x_A - dx_A$, and the amount of liquid will diminish from W to $W - dW$ moles. There will be left in the still $(x_A - dx_A)(W - dW)$ moles of A, while the amount $y_A dW$ has been removed from the still. By material balance with respect to component A:

$$xW = (x - dx)(W - dW) + y\,dW \tag{7}$$

Expanding Eq. (7):

$$xW = xW - x\,dW + dx\,dW - W\,dx + y\,dW \tag{8}$$

The second-order differential in Eq. (8) may be neglected, and the equation may be written as:

$$dW/W = dx/(y - x) \tag{9}$$

By applying the appropriate limits to Eq. (9), the equation may be written as:

$$\int_{W_1}^{W_o} \frac{dW}{W} = \ln(W_o/W_1) = \int_{x_1}^{x_o} \frac{dx}{(y - x)} \tag{10}$$

Eq. (10) is the well-known Rayleigh Equation.[2,3,7]

Assuming equilibrium exists between phases, the relation between x and y for a binary system can be obtained from the equilibrium curve for the mixture. Values of x are chosen, and the corresponding values of y are taken from the equilibrium curve; then the values of $1/(y - x)$ are tabulated. A plot of x versus $1/(y - x)$ permits graphical integration of the function $dx/(y - x)$; Fig. 3 shows the typical shape of such a plot. The area under the curve between the appropriate limits equals ln W_o/W_1, where W_1 is the final content in moles.

Simple or differential distillation is approached by commercial batch-distillation processes, in which the vapor is removed as fast as it is formed. Although, as a method of separation, this process is not effective, many such stills are used, especially where the components to be separated have widely different boiling points, so that methods providing sharp separations are unnecessary. The older design of the batch type of petroleum still, known as the "topping still," is an example.[6]

It is a general characteristic of this type of distillation that for a high volatiles-concentration in the distillate, a low recovery must be accepted. High recovery is achieved at the expense of a lowered distillate concentration. Of course, if a relatively small amount of a highly volatile component were to be removed from a large amount of a component of low volatility, a satisfactory recovery with high volatiles-concentration could be obtained.

One way to obtain a high degree of separation, along with high concentration, by simple distillation would be to redistill both the residual liquid and the overhead products. This can be accomplished by a repetition of simple distillations. A better way to accomplish separation by batch distillation would be through the use of a batch still with a rectifying column, as shown in Fig. 4.

Batch Rectification

A batch rectifying unit consists primarily of: (1) a still in which vapor is generated, (2) a rectifying column through which this vapor rises in countercurrent contact with a descending stream of liquid, and (3) a condenser that condenses all of the vapor leaving the top of the column, sending part of the liquid (reflux) back to the column to descend counter to its rising vapors, and delivering the rest of the condensed vapors as product.

As the liquid stream descends in the column, it is progressively enriched with the high-boiling component; and as the vapor stream ascends, it is progressively enriched with the low-boiling component. The column then becomes an apparatus for bringing these streams into intimate contact, so that the vapor stream tends to vaporize the low-boiling constituent from the liquid, and the liquid stream tends to condense the high-boiling constituent from the vapor. The top of the column is cooler than the bottom, so that the liquid stream becomes progressively hotter as it descends, and the vapor stream becomes progressively cooler as it rises. This heat transfer is accomplished by actual contact of liquid and vapor; and, for this purpose, effective contacting is desirable.[6] Contacting may be brought about by means of trays or plates, or packing.

For this development, it is sufficient to remember that a "theoretical tray" is one on which the vapor leaving the tray is in equilibrium with the liquid leaving the tray. As the vapor moves up the column, it passes through layers of liquid held on trays; this results in transfer of heat and mass on each tray. Some of the vapor condenses by direct mixing with the liquid, and this, in turn, results in the vaporization of a portion of the liquid. In many cases, the sensible heat lost by the vapor in cooling to its dewpoint is balanced by the sensible heat gained by the liquid in being heated to its boiling point.[2]

For this situation, most of the heat transferred between the two phases, and effective for vaporization, is due to the latent heat of the two components. If the molal heats of vaporization of the two components are very nearly the same, then the moles of vapor condensed are approximately equal to the moles of liquid vaporized. As a result, the molal upflow of vapor from tray to tray remains constant at any instant during the process. Also, the liquid downflow from tray to tray remains constant. This development assumes equal molal overflow; it also assumes adiabatic operation.

Mass is transferred on each tray in such a fashion that the more volatile component passes from the liquid to the vapor phase, and the less volatile component passes from the vapor into the liquid phase. Thus, there is an enrichment of the vapors as they pass up the column from tray to tray. There is also an increase in the concentration of the less volatile component in the liquid as it passes down the column from tray to tray.

Referring to Fig. 4, a material balance for the section within the envelope gives:

$$V_{n+1} = L_n + D \tag{11}$$

A balance for any component gives:

$$(V_{n+1})(y_{n+1}) = L_n x_n + D x_D \tag{12}$$

$$y_{n+1} = \frac{L_n x_n}{V_{n+1}} + \frac{D x_D}{V_{n+1}} \tag{13}$$

Where V is moles of vapor per unit time, L is moles of liquid per unit time, D is moles of distillate per unit time, and x designates the mole fraction of the more volatile component in the liquid. Subscripts n and $n + 1$ designate plate numbers from the top down.[2]

Eq. (13) is referred to as the "operating line equation." Since equal molal overflow was assumed, Eq. (13) produces a straight line. The equation, when plotted on the same graph as the equilibrium curve for the system under consideration, permits the classical McCabe-Thiele method of solving a batch rectification problem.

Batch rectification problems generally fall into two groups—constant and variable reflux. The development that follows is applicable to the constant-reflux batch-rectification system.

Constant-Reflux Batch Rectification

In this mode of operation, the column has a fixed number of theoretical stages and operates with constant reflux ratio and variable overhead composition. Column holdup is assumed to be negligible.

The rate of depletion of the contents of the still equals the rate of accumulation of the resulting distillate:

$$-dW/d\theta = dD/d\theta \tag{14}$$

where W is moles of mixture in the still at any time, θ, and D is moles of distillate at time θ. For any component:

$$\frac{-d(Wx_W)}{d\theta} = \frac{x_D\,dD}{d\theta} \tag{15}$$

where x_W is mole fraction of the more volatile component in the still at time θ, and x_D is instantaneous mole fraction of the component in the distillate that is leaving the condenser at time θ.

Combining Eq. (14) and (15) gives:

$$\frac{dW}{W} = \frac{dx_W}{x_D - x_W} \tag{16}$$

Integrating this equation between limits yields:

$$\int_{W_o}^{W_1} \frac{dW}{W} = \int_{x_{W_o}}^{x_W} \frac{dx_W}{x_D - x_W} \tag{17}$$

$$\ln (W_1/W_o) = \int_{x_{W_o}}^{x_W} \frac{dx_W}{x_D - x_W} \tag{18}$$

where W_o is moles of mixture originally charged to the still, and x_{W_o} is the initial mole fraction of the more volatile component in the mixture. Eq. (17) is very similar in form to the Rayleigh Equation.

The right-hand side of Eq. (18) is integrated graphically by plotting $1/(x_D - x_W)$ versus x_W. The area under such a curve, taken between the limits x_{W_o} and x_W, is the value of the integral.[2]

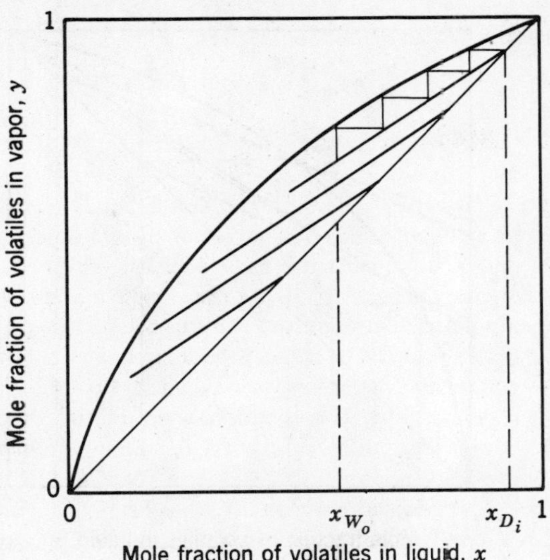

CONSTANT-REFLUX batch-process solution—Fig. 5

To establish the relationship between x_D and x_W, the equilibrium diagram is first plotted. Several values of x_D are selected, and operating lines having the same slope (L/V is constant) are drawn through the intersection of x_D and the diagonal; Fig. 5 is a diagram representing the procedure. The diagonal has the equation $y = x$, and operating lines intersect this diagonal at x_D.

Having completed these lines, steps are drawn between the operating line and equilibrium curve, as in the well-known McCabe-Thiele method. Such a procedure is a graphical approach to simultaneously solving the operating and equilibrium equations. The correct line is that which requires the specified number of theoretical stages in going from the initial distillation composition to the composition of the mixture initially charged to the still.

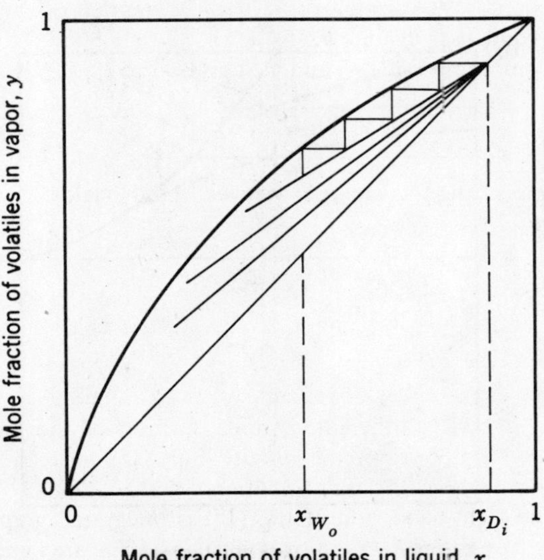

VARIABLE-REFLUX batch-process solution—Fig. 6

It must be remembered that the still plays the same role as a theoretical stage. The intersection of the last horizontal step (going down from x_D) with the equilibrium curve is the composition x_W of the liquid in the still.

When the internal reflux ratio is specified, the initial distillate composition, x_{D_i}, is the composition of the first drop of distillate. In such a system, the procedure for determining the *initial* distillate composition is one of trial-and-error. However, once this composition has been determined, successively lower distillate compositions are selected and tabulated beside the corresponding composition of the mixture in the still. Having reduced the still concentration to the desired value, the distillation is stopped.

From the overall material balance and the fact that a constant reflux ratio is used, the total vapor generated by the still for the distillation process can be computed. Since the heat requirement consists primarily of latent heat, the quantity of vapor calculated above can be used to estimate the heat supplied to the still for the process.[2,3] The material balance can also be extended to calculate the percent of the more volatile component that was recovered in the distillate. In addition, the purity of the distillate product can be calculated.

Variable-Reflux Batch Rectification

Operation of a batch column under variable-reflux conditions requires the continuous adjustment of reflux to maintain a distillate-purity specification. Initially, the still liquid is rich in the more volatile component, and a low reflux-ratio will produce the desired distillate purity. As the distillation proceeds, the reflux ratio must be continuously increased until a practical maximum is reached. At this time, the receivers could be switched, reflux reduced, and a tails-cut taken for reclaiming later.[3] Overall material balance at any time is:

$$W = W_o\left(\frac{x_D - x_{W_o}}{x_D - x_W}\right) \tag{19}$$

Differentiating with respect to time gives:

$$\frac{dW}{d\theta} = \frac{W_o\,(x_D - x_{W_o})\,dx_W}{(x_D - x_W)^2\,d\theta} \tag{20}$$

Assuming constant molal overflow for any instant, the rate of distillation may be expressed as:

$$\frac{dW}{d\theta} = \frac{(L - V)}{d\theta} \tag{21}$$

Substituting in Eq. (20) and solving for time gives:

$$\theta = \frac{W_o(x_D - x_W)}{V}\int_{x_W}^{x_{W_o}} \frac{dx_W}{(1 - L/V)(x_D - x_W)^2} \tag{22}$$

This expression is the time required for the distillation itself (exclusive of that required for charging the still, heating up, shutting down, and cleaning). Eq. (22) is the same as that of Bogart.[2,3]

The vapor load, V, can be calculated if the diameter of the column, estimated allowable vapor rate, operating pressure, and temperature are known. Conversely, if the time were fixed, then the diameter of the column could be calculated using V obtained from Eq. (22).

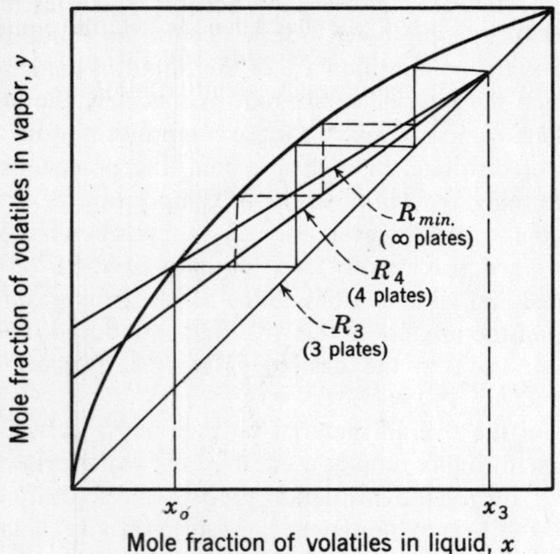

MINIMUM reflux, graphical illustration—Fig. 7

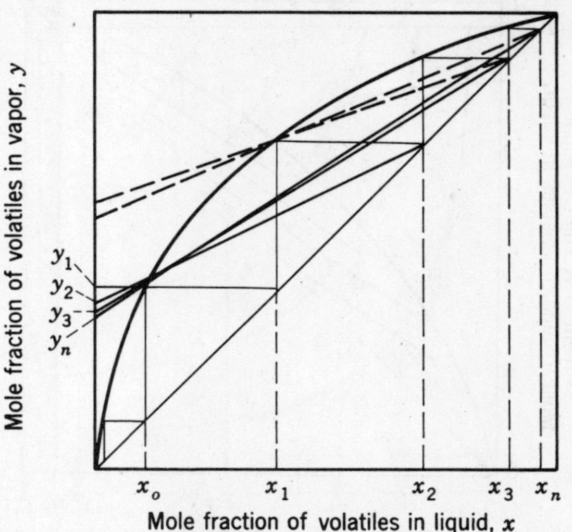

MINIMUM drawoff percentage calculation—Fig. 8

To evaluate the integral term on the right-hand side of Eq. (22), the equilibrium diagram is drawn as before. Several operating lines with different slopes ($1/V$) are drawn, each passing through x_D on the diagonal; Fig. 6 is a diagram representing the procedure.

The steps equivalent to the number of theoretical stages are drawn in the usual manner for each line. Again, the intersection of the last horizontal step with the equilibrium line gives x_W. This is continued until x_W reaches the desired final value. The integral is then evaluated graphically by plotting $1/(1 - (L/V))(x_D - x_W)^2$ versus x_W and taking the area under the curve between x_{W_o} and x_W.

Following is a simple method for construction of a chart that can be used to program a variable-reflux batch distillation. The chart is derived from the McCabe-Thiele diagram for a binary mixture, using a modification of the well-known minimum-reflux-ratio concept.[5]

Maximum Drawoff Percentage

As shown in Fig. 7, a separation, $x_3 - x_o$, requiring three theoretical stages at total reflux can be obtained with four theoretical stages and the finite reflux-ratio, R_4, as well as with infinite stages and the minimum reflux-ratio, R_{min}. The value of R can be determined from the graph, using the operating-line equation, the y-intercept for which:

$$y = x_D/(R + 1) \tag{23}$$

where x_D is the distillate volatiles concentration in mole percent; R is the reflux ratio L/D (L is the liquid returned to the column, and D is the distillate drawoff).

It is impractical to represent reflux-ratio graphically because it varies from zero to infinity. Accordingly, another parameter, P, is substituted for R. P, the distillate drawoff percentage, equals $100/(R + 1)$ and ranges from 0% (total reflux) to 100% (zero reflux). The expression can be rearranged so that $P = 100\ y/x_D$.

The situation shown in Fig. 7, where an overhead composition, x_3, is obtained from a bottoms composition, x_o, with infinite stages thus corresponds to the maximum drawoff percentage.

Starting from the same bottoms composition, x_o, and with infinite stages, overhead compositions other than x_3 will be obtained with different values for the maximum drawoff percentage, as Fig. 8 illustrates. For convenience of drawing, overhead compositions x_1, x_2, and so forth have been chosen so that they differ by one theoretical stage.

Fig. 9 is obtained by plotting the drawoff percentages obtained by the method described above—for example, percent volatiles in the vapor corresponding to drawoff percentages $100\ y_1/x_1$, $100\ y_2/x_2$, and so forth. The drawoff percentages are presented by points P_1, P_2, and

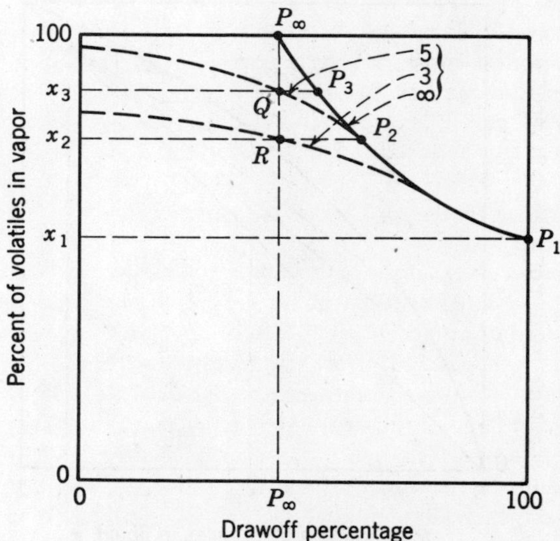

OVERHEAD composition, effect of stages, drawoff—Fig. 9

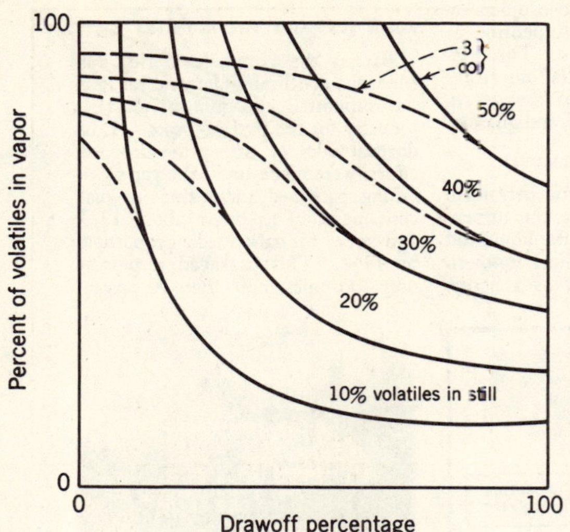

PROGRAMMING correlation for variable-reflux batch distillation—Fig. 10

so forth that appear to lie on a hyperbola-like curve.[5]

Higher drawoff percentages correspond to lower overhead compositions. At 100% drawoff, an overhead composition x_1 (point P_1 on Fig. 9) is obtained with one theoretical stage (the kettle). In the McCabe-Thiele diagram of Fig. 8, this situation is represented by a horizontal operating line, $y_1 = (x_1)^{0.5}$

Finite Number of Theoretical Stages

The same drawoff percentage, P_∞, with a finite number of stages, gives lower overhead-compositions. Operating lines having the same slope can be drawn so that upon stepping off stages as usual, three or five stages will fit between the equilibrium curve and the operating lines. Points Q and R of Fig. 9 were obtained by stepping off a finite number of stages, using an operating line of slope P_∞. Note that the distillate-volatiles concentrations decrease as the number of stages decreases.

Drawoff percentages lower than P_∞ give higher overhead compositions, but the maximum overhead composition that may be obtained is at zero drawoff. With drawoff percentages higher than P_∞, lower overhead percentages are obtained.

Effect of Bottoms Composition

Many curves similar to those in Fig. 9 will appear when the procedure described above is repeated for bottoms compositions other than x_0. Fig. 10 illustrates the effect of still composition. This graph illustrates that as the bottoms volatiles-concentration decreases in the progress of a distillation, the separation becomes more difficult.

A glance at the graph shows that the drawoff percentage must be decreased to maintain a distillate specification as the distillation proceeds. The graph further illustrates that too low a reflux ratio cannot be compensated for by a large number of theoretical stages; this is emphasized by the fact that the curves for finite and infinite plates converge to a single value at high drawoff percentages. Also, it is easy to see that when the volatiles content of the still is low, the separating effect of a column, especially one of many stages, greatly diminishes as the drawoff percentage is increased.

Programming Batch Distillation

A glance at a plot similar to that of Fig. 10 is enough to decide what range of drawoff percentages is needed to meet a specified distillate composition. A horizontal line of constant overhead-composition intersects a number of sequential curves of different volatiles concentrations in the kettle. The batch distillation is programmed along a horizontal line corresponding to a distillate specification.

References

1. Dechman, D. A., Correcting the McCabe-Thiele Method for Unequal Molal Overflow, *Chem. Eng.*, Dec. 21, 1964, p. 79.
2. Coates, J., Pressburg, B. S., How to Analyze the Calculations for Batch Rectification in Tray Columns, *Chem. Eng.*, Jan. 23, 1961, p. 131.
3. Block, B., Batch Distillation of Binary Mixtures Provides Versatile Process Operations, *Chem. Eng.*, Feb. 6, 1961, p. 87.
4. Alleva, R. Q., Improving McCabe-Thiele Diagrams, *Chem. Eng.*, Aug. 6, 1962, p. 111.
5. Haring, H. G., others, Programming Batch Distillation, *Chem. Eng.*, Mar. 16, 1964, p. 157.
6. Badger, W. L., Banchero, J. T., "Introduction to Chemical Engineering," McGraw-Hill, New York, 1955.
7. Perry, J., ed., "Chemical Engineers' Handbook," 3rd ed., McGraw-Hill, New York, 1950.
8. Brown, G. G., others, "Unit Operations," Wiley, New York, 1950.
9. Robinson, C. S., Gilliland, E. R., "Elements of Fractional Distillation," 4th ed., McGraw-Hill, New York, 1950.
10. McCabe, W. L., Smith, J. C., "Unit Operations of Chemical Engineering," McGraw-Hill, New York, 1956.
11. Chen, N. H., Calculating Theoretical Plates in Absorbers or Strippers, *Chem. Eng.*, May 11, 1964, p. 159.
12. Chidambaram, S., Narsimhan, G., Generalized Chart Gives Activity Coefficient Correction Factors, *Chem. Eng.*, Nov. 23, 1964, p. 135.
13. Glotzer, H. L., Generalized Equation Speeds Batch Distillation Design, *Chem. Eng.*, May 14, 1962, p. 202.
14. Coates, J., Pressburg, B. S., How to Make Distillation Calculations, *Chem. Eng.*, Mar. 20, 1961, p. 155.
15. Coates, J., Pressburg, B. S., Review Mass Transfer Concepts, *Chem. Eng.*, Sept. 5, 1960, p. 137.
16. Smoker, E. H., Rose, A., Graphic Determination of Batch Distillation Curves for Binary Mixtures, *Trans. AIChE*, **36,** p. 285 (1940).
17. Smoker, E. H., Analytic Determination of Plates in Fractionating Columns, *Trans. AIChE*, **34,** p. 165 (1937).
18. Bogart, M. J. P., The Design of Equipment for Fractional Batch Distillations, *Trans. AIChE*, **33,** p. 139, (1937).
19. Rose, A., Johnson, R. C., The Theory of Steady-State Distillation: Batch Distillation Calculations, *Chem. Eng. Progr.*, **49,** p. 15 (1953).
20. Converse, A. O., Gross, G. D., Optimal Distillation-Rate Policy in Batch Distillation, *Ind. Eng. Chem. Fund.*, **2,** p. 217 (1963).
21. Chen, N. H., New Equations for McCabe-Thiele Diagram, *Chem. Proc. Eng.*, **44,** p. 302 (1963).
22. McCabe, W. L., Thiele, E. W., Graphical Design of Fractionating Columns, *Ind. Eng. Chem.*, **17,** p. 605 (1925).

Meet the Author

R. W. Ellerbe is a staff engineer for The Rust Engineering Co., 1130 South 22nd St, P. O. Box 101, Birmingham, AL 35201. He has held process engineering and administrative assignments in petroleum refining, natural gas processing and pulp and paper. He holds an M.S.Ch.E. from Louisiana Tech University and has taken other graduate courses in chemical engineering and business administration. His professional affiliations have included AIChE, Technical Assn. of the Pulp and Paper Industry and the Arkansas Soc. of Professional Engineers—he is a registered professional engineer in Arkansas.

CRUDE PROCESSING

This turpentiner—call him Grandpa Jones—was a man who didn't know the meaning of chemical processing. His distilling equipment was wood-

DAVID N. COLLINS and E. L. PATTON are members of the Naval Stores Research Division of the Bureau of Agricultural and Industrial Chemistry, United States Department of Agriculture. Their location: Olustee, Florida.

cation from processor to factor to dealer to sales agent to consumer, individual shipments of turpentine or rosin lost all identification. The soapmaker and paint maker had no means of tracing the source of the deceit, however righteous their indignation.

PETROLEUM SOLVENTS COMPETE

Then the salesman for petroleum solvents began to call on the turpentine user. The uses for the liquid did not depend on its chemical composition but only its power as a hydro-

uniform product.

ROSIN SUPPORTS THE INDUSTRY

Barring the extra sales during wartime, the continuing demand for rosin has supported the gum naval stores industry for the past 30 years. As industrial sales of turpentine dropped, efforts were made to fill the gap by retailing packaged turpentine in small containers. This began about 1925, and grew to nationwide proportions by 1940. This packaged turpentine now consumes more than 90 percent

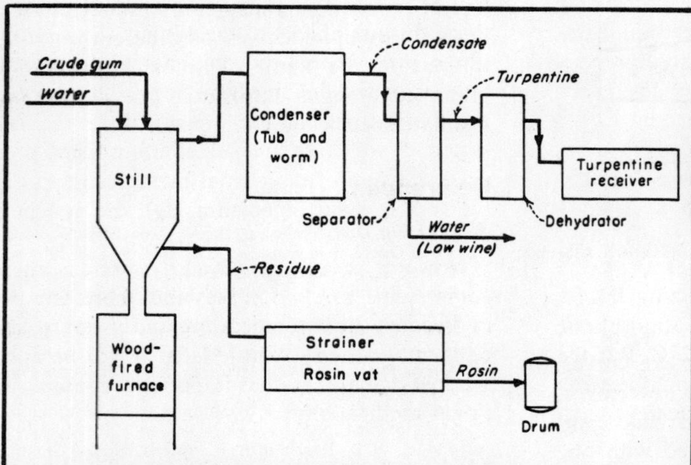

YESTERDAY the old fire stills usually consisted of large copper kettles heated directly by a wood fire. The kettle was connected to a coil of copper pipe in a wooden tub. The turpentine was distilled off with water and separated by gravity.

Rosin, remaining in the kettle was removed and strained, while still in the molten condition, at the end of each run. Cut (above) and flowsheet (left) show the fire still as it operated in 1935 before the Olustee Process.

154 *September 1951*—CHEMICAL ENGINEERING

STEAM DISTILLATION of naval stores as it was done in 1935 (from a 1951 issue of *Chem. Eng.*).

Steam-Distillation Basics

Here is a convenient technique for distilling certain difficult-to-handle materials. A summary of both theory and practice is included.

R. W. ELLERBE, The Rust Engineering Co.

Steam distillation offers a simple solution to some very knotty chemical-processing problems. In general, the chemical engineer should consider steam distillation when he has separation or purification problems involving:

- Chemicals with relatively high atmospheric boiling points, e.g., glycerin.
- Chemicals that may decompose at their atmospheric boiling points, e.g., fatty acids.
- Chemicals presenting hazardous conditions when distilled by the only other means available, namely, direct heat, e.g., turpentine.

It is doubtful that Avicenna [*1*], the Arabian physician who discovered steam distillation of volatile oils, understood the principles underlying his discovery. Many of the essential oils presently used in perfumery are obtained by steam distillation of flowers, leaves, bark, etc.; but even today, the practice of steam distillation in the cosmetic industry remains more an art than a science (because the essential oils are usually mixtures of complex organic chemicals).

Steam distillation is also used in the petroleum, coal-

Originally published March 4, 1974

chemicals, wood-chemicals, fats and oils, petrochemicals, and paper industries, as well as many others.

STEAM-DISTILLATION THEORY

Conventional distillations involve separations of miscible liquids, but steam distillations rely on the unique vapor-pressure behavior of *immiscible* liquids. The vapor-pressure relations of two immiscible liquids are shown in Fig. 1. Each liquid exerts its own vapor pressure independently of the other. In such a system, at a given temperature, the total vapor pressure is the sum of the vapor pressure of the two liquids (if neither liquid dissolves in the other) [*2*]. Furthermore, the partial pressure for each component and the composition of the vapor phase, at constant temperature, are independent of the moles of liquid water or hydrocarbon present.

In the example shown, the total vapor pressure reaches 760 mm Hg (atmospheric pressure) at 95°C. Boiling begins at that temperature, and both liquids distill together. When either one of the liquids distills away, the vapor pressure drops to that of the remaining liquid. Ordinarily, bromobenzene would not boil at atmospheric pressure until its temperature reached 156.2°C. But the presence of a little water (or steam) makes bromobenzene boil at 95°C, a temperature that is easy to achieve even with low-pressure steam.

The Phase Rule

Before going into the equations ordinarily used in steam distillation, the unique characteristic of immiscible liquids presented in the preceding two paragraphs will be briefly discussed using Gibbs' phase rule. This rule, formulated by Willard Gibbs, is a criterion for phase equilibrium. It states how many phases may exist at equilibrium or how many degrees of freedom exist for a given system [*3*]. It provides no further information.

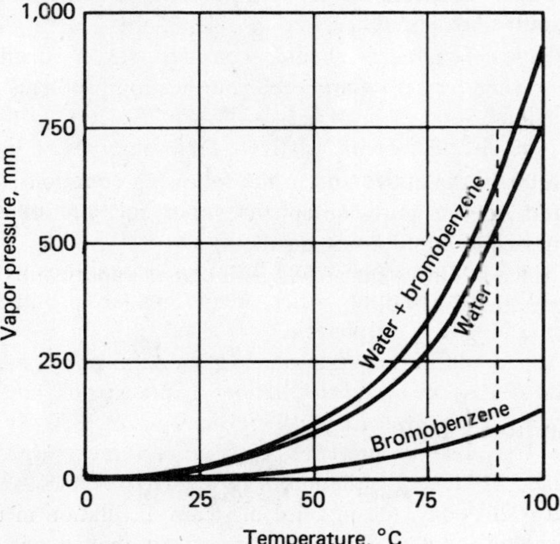

VAPOR-PRESSURE vs. temperature plot for system made up of water and bromobenzene—Fig. 1

The rule is given as follows:

$$F = 2 + C - P \tag{1}$$

The *degrees of freedom* simply tells how many independent variables such as temperature, pressure, and concentrations of the components must be fixed in order to define a system completely. There are two cases to consider in steam-distillation problems:

1. Water is present as both a liquid and vapor.
2. Water is present as a vapor only.

Returning to the bromobenzene-water example used earlier, if water is present as a liquid phase, there are three distinct phases to the system. Therefore, from the phase rule, when two components are present there is only one degree of freedom. So, in this case, fixing either the temperature or the pressure fixes the state of the system. If we chose to distill at atmospheric pressure, and since the sum of the partial pressures necessarily equals the total pressure (760 mm Hg), the system must boil at 95°C.

However, if no liquid water is present, there are two components and two phases and, therefore, two degrees of freedom. Both temperature and pressure can be independently varied. (A little later, it will be apparent that a liquid-water phase has important economic and operational implications.)

Law of Partial Pressures

It has been shown that when two liquid phases are present, each exerts the same vapor pressure as if it existed alone. Thus, if a water and a bromobenzene phase exist together in a system, the vapor pressure within the system is the sum of the vapor pressure of the water and the bromobenzene phases. For example, if water and bromobenzene liquid phases are in equilibrium with a vapor phase of these components at 110°C, the vapor pressure of the water phase will be 1,075 mm Hg, and the vapor pressure of the bromobenzene will be 200 mm Hg. The total pressure on the system will be 1,275 mm Hg. This is true provided that the total pressure is not high enough to cause appreciable deviations from the ideal-gas law. This law is considered applicable for pressures below about 3 atm.

Strictly speaking, the fugacity of the vapor phase equals the fugacity of the liquid phases. But since steam distillations are carried out so that low temperatures can be maintained, it is most economical to distill at low pressure, where the ideal-gas law applies. Thus, the solving of steam-distillation problems may be based on partial pressures [*3*]. In many steam distillations, no water phase is present, but the total pressure on the system is made up of the steam partial pressure plus that of the immiscible hydrocarbon. Kirkbride [*3*] developed the following equations:

$$y_s = p_s/\pi \tag{2}$$

and

$$y_{im} = p_{im}/\pi \tag{3}$$

Since the partial pressure of the immiscible liquid is a function of its temperature, the amount of steam required for its distillation can be calculated by rearranging Eq. (2). Since:

$$\pi = p_s + p_{im} \tag{4}$$

$$y_s = \frac{\pi - p_{im}}{\pi} \tag{5}$$

The moles of steam required per mole of immiscible liquid vaporized is therefore:

$$\frac{y_s}{y_{im}} = \frac{\pi - p_{im}}{p_{im}} = \frac{p_s}{p_{im}} \tag{6}$$

or

$$N_s = \frac{N_{im}(\pi - p_{im})}{p_{im}} \tag{7}$$

These equations show that when the sum of the partial pressures of the steam and the material distilled reach the system pressure, boiling begins and both materials pass over in the mole ratio of their partial pressures. If these mixed vapors are condensed, two layers form in the condensate receiver, and the materials can be separated by gravity. Any nonvolatile impurity will be left behind in the still. Regardless of whether or not there is a liquid phase, Badger and McCabe [*4*] have shown that the weight ratio of steam to the immiscible liquid in the vapor is:

$$\frac{W_s}{W_{im}} = \frac{p_s M_s}{p_{im} M_{im}} \tag{8}$$

The Hausbrand Vapor-Pressure Diagram

In 1918, Hausbrand [*4*] published a vapor-pressure diagram that proved to be very useful in the calculation of steam distillations. Fig. 2 shows that diagram. It plots $\pi - p_s$ at three system pressures (760, 300, and 70 mm Hg) versus temperature. These curves cut across the ordinary vapor-pressure curves of the materials to be distilled. The intersection of the water curve with the curves of the other materials gives the temperature at which steam distillations can take place.

Suppose an atmospheric still contains some impure toluene, and steam is blown into this still. Suppose also that the impurities are very-high-boiling compounds with negligible vapor pressure. Further, consider the case where the liquid in the still is heated solely by condensation of the steam. A water layer will therefore accumulate. As the temperature rises, the vapor pressures of the toluene and water layers rise.

When the sum of the two vapor pressures equals 760 mm Hg, the mixture begins to distill. At this point, the vapor pressure of the toluene is p_{im} mm, the vapor pressure of water is $\pi - p_{im}$ mm; Fig. 2 shows that this occurs at a temperature of about 84°C. In this case, the vapor passing over consists of toluene with a partial pressure of 350 mm, and water vapor with a partial pressure of 410 mm. The molar ratio of toluene to water would therefore be 350/410, or 85.4 parts toluene to 100 parts water [*5*].

The Hausbrand diagram makes possible a quick determination of the temperature at which a steam distillation takes place. It also presents graphically the molar richness of the vapor. Since most substances on the chart have a molecular weight considerably greater than that of water, the composition of the distillate by weight is much richer than would appear from the diagram.

Nomenclature

C	Number of components
E	Vaporization efficiency of steam distillation
F	Degrees of freedom
M	Molecular weight of material
N	Number of moles
N_o	Number of moles of nonvolatile material present
P	Number of phases
p	Partial pressure, mm Hg
W	Weight of material in vapor
y	Mole fraction of material in vapor
π	Total system pressure, mm Hg
	Subscripts
im	Immiscible liquid
s	Steam
1	Initial
2	Remaining

Effect of Nonvolatile Material

In steam-stripping operations, the partial pressure of the distilling component will generally be low. Therefore, it will be sensitive to slight changes in pressure or to changes of the quantity of the volatile component in the vapor [*6*]. If the volatile hydrocarbon is being distilled from a mixture containing *traces* of nonvolatile impurities, steady-state conditions may be assumed to exist.

However, if the composition of the material in the still changes appreciably as the batch is distilled, unsteady-state relationships, similar to those encountered in differential distillation, must be integrated. Using unpublished notes of McAdams (Massachusetts Institute of Technology), Carey [*7*] shows how the concentration of nonvolatile impurities affects steam usage during distillations.

Nonvolatile-Material Concentration Is Significant— This development assumes a batch distillation for which no liquid water is present, so p_s is less than the saturation pressure of steam at the temperature of the still. If Raoult's law applies,

$$E = \frac{p_{im}}{p_{im}\left(\dfrac{N_{im}}{N_{im} + N_o}\right)} \tag{9}$$

(Note: In the above and in the following equations, p_{im} refers to the pure-component vapor pressure of the immiscible liquid being distilled.)

Since adding steam to the still causes vaporization of the immiscible liquid,

$$\frac{+dN_s}{-dN_{im}} = \frac{p_s}{p_{im}} = \frac{\pi - p_{im}}{p_{im}} = \frac{\pi}{p_{im}} - 1 \tag{10}$$

But from Eq. (9):

$$p_{im} = Ep_{im}[N_{im}/(N_{im} + N_o)] \tag{11}$$

Therefore:

$$dN_s = -\left(\frac{\pi}{Ep_{im}} = 1\right) dN_{im} - \frac{\pi N_o}{Ep_{im}}\left(\frac{dN_{im}}{N_{im}}\right) \tag{12}$$

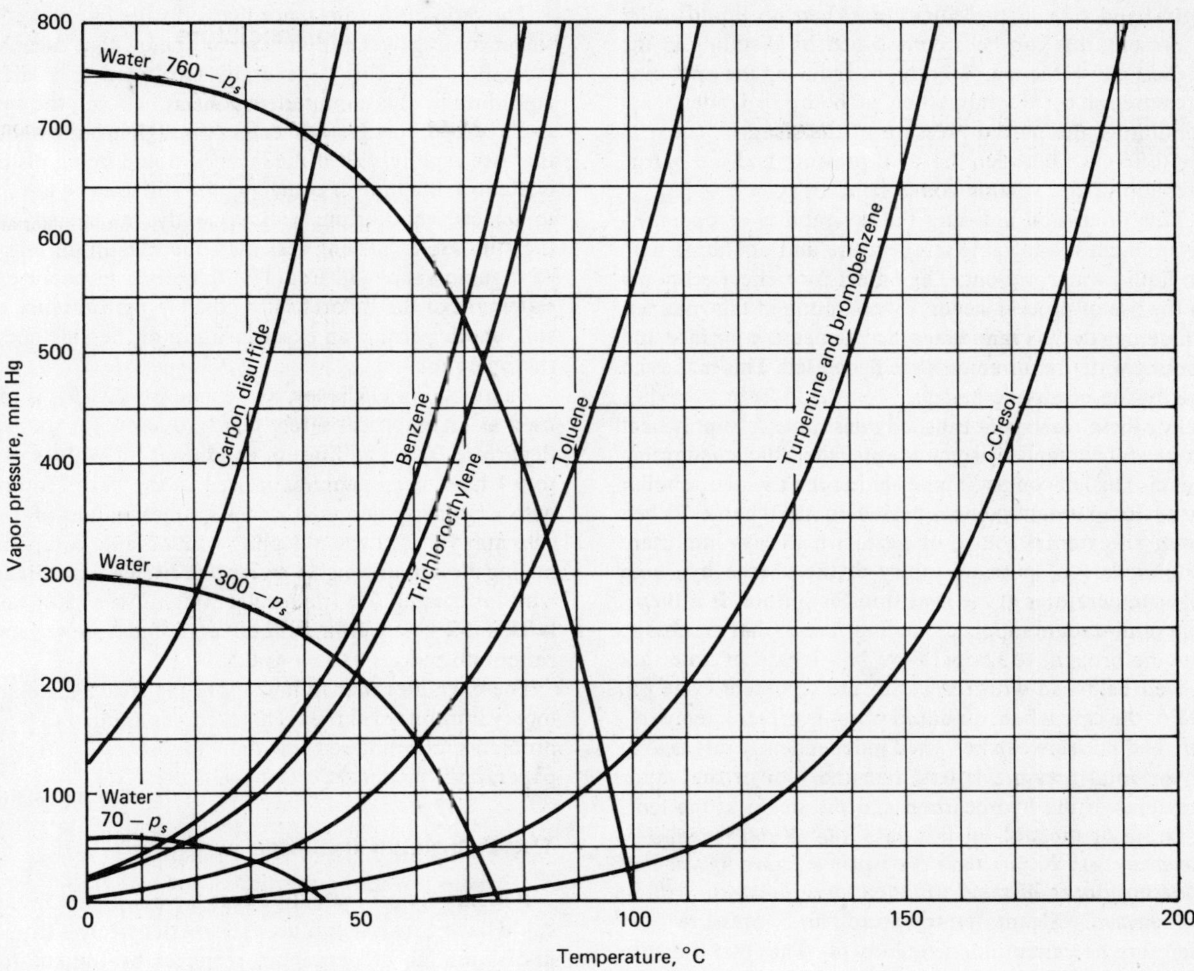

HAUSBRAND vapor-pressure diagram for various liquids, and at three system pressures—Fig. 2

If the distillation temperature remains constant, p_{im} is constant. For constant total pressure:

$$N_s = \left(\frac{\pi}{Ep_{im}} - 1\right)\left(N_{im_1} - N_{im_2}\right) + \frac{\pi N_o}{Ep_{im}} \ln \frac{N_{im_1}}{N_{im_2}} \quad (13)$$

If the nonvolatile material is present in large amount, and a batch distillation is used, the partial pressure of steam rises as the distillation progresses. This is, of course, due to the depletion of the volatile component in the charge, and the consequent decrease in p_{im}. Since the volatile hydrocarbon is usually stripped down to a low residual concentration, p_s closely approaches π toward the end of the batch process. Therefore, a distillation temperature for which the saturation pressure of steam is greater than π, ensures no condensation of open steam during the process.

Nonvolatile-Material Concentration Is Insignificant— Under these conditions, N_o, as used in the preceding derivation, is small relative to N_{im}. Hence p_{im} can be considered constant, and by Eq. (9):

$$p_{im} = Ep_{im} \quad (14)$$

Therefore:

$$\frac{+dN_s}{-dN_{im}} = \frac{p_s}{p_{im}} \quad (15)$$

Integrating this equation gives:

$$N_s = \frac{p_s}{p_{im}}(N_{im_1} - N_{im_2}) \quad (16)$$

$$\frac{N_s}{N_{im_1} - N_{im_2}} = \frac{p_s}{p_{im}} \quad (17)$$

If the weight ratio of the distillate is desired:

$$\frac{W_s}{W_{im}} = \frac{(\pi - p_{im})\, M_s}{p_{im}\, M_{im}} \quad (18)$$

Effect of Liquid Water in the Still

As shown earlier, the effect of liquid water in the still can be analyzed using the principles of Gibbs' phase rule. If no liquid water is present, there are two components and two phases, and, therefore, two degrees of freedom. Both temperature and pressure can be varied independently. Referring to Eq. (18), one can see that steam consumption for the process can be decreased by increasing the denominator and decreasing the numerator of the right-hand side of the equation. When no liquid water is present, this can be accomplished by distilling at the highest allowable temperature, by reducing the operating pressure, or by a combination of both [7]. Under these conditions, the partial pressure of the steam is equal to the difference between the total pressure and the partial pressure of the volatile component, or $p_s = \pi - p_{im}$.

This discussion indicates the desirability of operating at the highest allowable temperature and operating with no liquid water present. The object is to effect economy in the use of process steam. By operating in this manner, condenser duty is minimized, so condensing surface and cooling water requirements are decreased. This makes an interesting economic balance.

For these reasons, it is usually desirable to supply heat to the still contents by some means other than condensing steam. This, of course, implies either the use of a reboiler or a closed steam-coil immersed in the charge. When using an external source of heat, it is always important to be aware of possible injury to the charge by high film-temperatures at the heat-transfer surface. It is therefore important to supply enough surface so that the charge may be brought to temperature in a length of time that is well balanced with respect to the complete cycle [7].

For the case where no liquid water is present, temperature and pressure can be varied independently just as long as the total pressure is less than the sum of the vapor pressures of the hydrocarbon and the steam at the temperature of the still. In this case, *the partial pressure of the steam will be less than the vapor pressure of water at the temperature of the distillation, and the steam will be superheated.* Steam temperature and pressure can therefore be varied independently [4]. The total pressure is determined by the cooling water and the condenser employed [8].

STEAM-DISTILLATION PRACTICE

The use of open steam to reduce operating temperatures is practiced in a great many industries. Examples of problems encountered in the petroleum, paper, coal-chemicals, and naval-stores industries will be given in the paragraphs that follow. But first, a brief discussion of the equipment and typical operating procedures.

The design and operation of a *continuous* steam-distillation or stripping column may be identical with that of fractionating towers used for miscible liquids. Heat may be supplied by a reboiler or by condensation of some of the open steam. It may also be supplied by the sensible heat of the liquid to be stripped. This liquid frequently enters the column in a superheated state [6].

Batch steam distillation is generally carried out in a cylindrical still. The still is sometimes externally heated, either by a steam jacket or by direct fire. More often, however, steam is injected through spargers placed in the bottom of the still. These spargers have numerous perforations to ensure uniform distribution of the steam. In some cases, a closed steam-coil is used to heat up the batch and to replace heat losses as the distillation progresses. The still must, of course, be fitted with a condenser and with a distillate accumulator designed for gravity separation of the two phases being distilled.

The ratio of the substance being distilled to that of the steam coming over with it is always below that calculated from Eq. (17). This is generally attributed to a lack of equilibrium, due to imperfect contact between the steam and hydrocarbon. If steam comes through in large bubbles that rise rapidly, and if the layer of liquid being distilled is shallow, the time of contact in the still may be too short to achieve equilibrium. Consequently, the steam leaves the still while carrying less than the maximum amount of organic vapor [8]. Eq. (17), therefore, gives an exact statement of the theoretically possible performance of a still, and it permits an exact measure of the efficiency of the operation.

Vaporization efficiency, as defined by Eq. (9), is often used as an empirical safety factor to allow for the usual departure from equilibrium conditions [7]. Values of 0.6 to 0.7 have been commonly used in the past. However, with a properly designed steam sparger, values of 0.9 to 0.95 may be used when dealing with organic compounds having molecular weights under 100. In steam distillations with lubricating oils (molecular weight over 250), calculations using data from refinery equipment show vaporization efficiencies as low as 0.5.

The examples that follow apply the steam-distillation theory introduced earlier. This theory is applied to typical problems encountered in four industries: petroleum, paper, coal-chemicals, and naval-stores.

The Petroleum Industry

It seems almost incredible that the petroleum industry could have grown so rapidly and so extensively—from the first skimming of petroleum seepages by Indians for a medicinal or rubbing oil, or the first modern petroleum well drilled near Titusville, Pa., by Colonel Drake in 1859, to the tremendous industry now flourishing not only in the U.S. but throughout the world [1]. Chemical engineering owes a great deal to the discovery and application of engineering principles in distillation, heat transfer, fluid flow, etc., by the refinery engineers, in the designing, building, and operating of their refineries. The example that follows is typical of refinery-engineering ingenuity; it uses steam-distillation principles to analyze and to troubleshoot the operation of a stripping column (Fig. 3).

Example: Light-hydrocarbon-rich absorber oil is charged to an open steam-stripping column at a rate of 10,000 mol/h. The overhead hydrocarbon product of 1,000 mol/h is withdrawn as 60% vapor and 40% liquid at the accumulator conditions of 50 psia and 140°F. In addition, 800 mol/h of reflux is in the top vapor, where the temperature is 180°F. Superheated steam is admitted beneath the bottom tray at a rate of 13,500 lb/h. The mixed vapor at this point indicates a water partial-pressure of 20 psi, and the temperature is 300°F. A dehydrator tray up in the tower will remove any liquid water that accumulates in the tower. There is also a liquid-water drawoff in the accumulator. All pressure drops and water solubility in hydrocarbon are neglected. (1) How much water is removed from each of the possible places? (2) If the dehydrator tray becomes plugged, additional heat is added either to the feed or the bottom of the column (none by open steam) to remove all of the water overhead. Under these conditions, all other things remaining constant, what must the reflux rate be? (Assume top-vapor-temperature constant.) The answer follows:

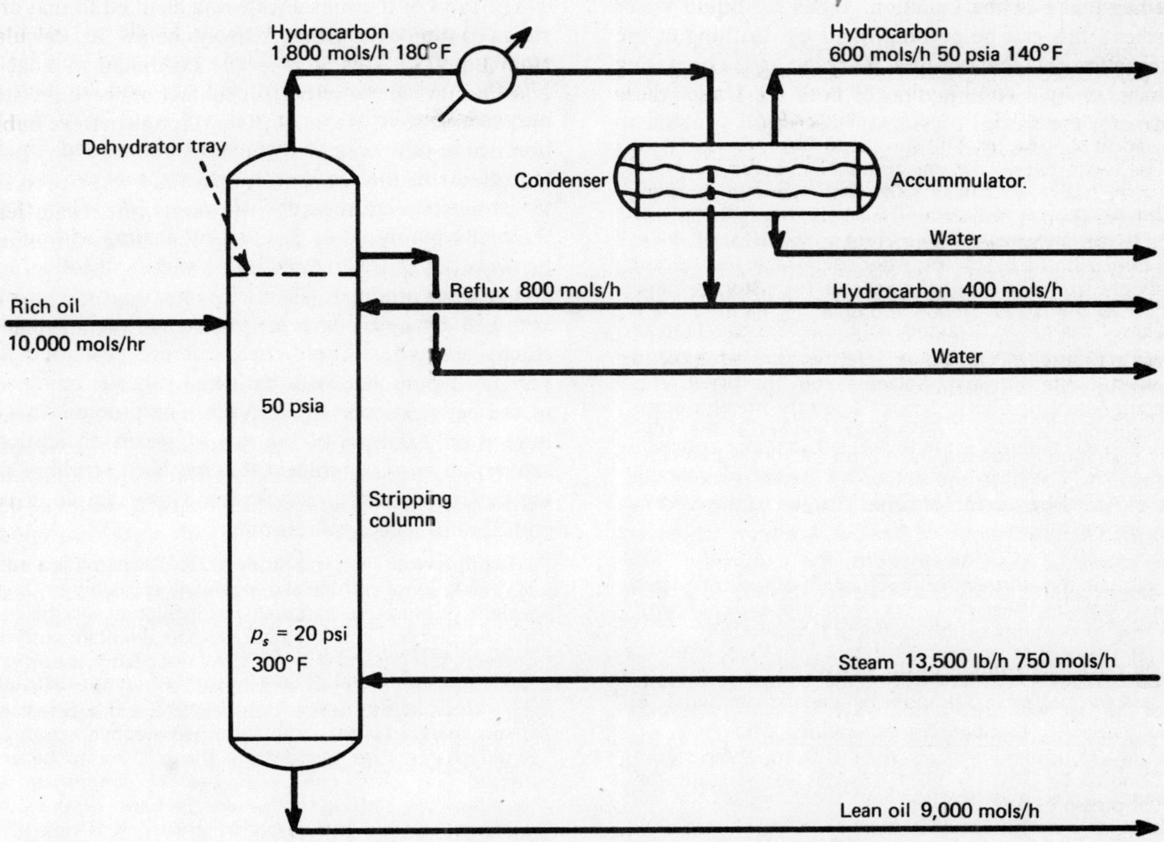

STEAM DISTILLATION principles are used to troubleshoot this petroleum-industry stripping column—Fig. 3

Basis: 1 hour of operation

Temperature, °F	Vapor Pressure of Water, Psi
140	2.89
180	7.51
300	67.01

Mol fraction of water vapor at top of tower = 7.51/50 = 0.15

Mol fraction of hydrocarbon at top of tower = 1 − 0.15 = 0.85

Total mols vapor at top of tower = 1,800/0.85 = 2,120

Mols water vapor = 2,120 − 1,800 = 320

Mol fraction of water in accumulator vapor = 2.89/50 = 0.0578

Mol fraction of hydrocarbon in accumulator vapor = 0.9422

Total mols vapor leaving accumulator = 600/0.9422 = 636

Mols water vapor leaving accumulator = 636 − 600 = 36

Mols liquid water withdrawn from accumulator = 320 − 36 = 284

Mols liquid water collected on dehydration tray = 750 − 320 = 430

Now, if the dehydrator tray becomes plugged, all water is removed overhead. To maintain p_s at the top of the tower, hydrocarbon reflux must be increased.

Since the molar ratios at the top of the tower remain the same as before, by material balance (where R is reflux in Mol/h), we get:

$$(1{,}000 + R)/0.85 = 1{,}000 + R + 750$$

$$R = 498/0.15 = 3{,}320 \text{ Mol/h}$$

The partial pressure of water at the bottom of the tower is less than the saturation pressure of steam at 300°F; therefore, no condensation occurs in the bottom of the tower.

The Paper Industry

Between 2500 and 2000 B.C., the manufacture of writing paper was begun from a tall reed called papyrus growing along the Nile; hence the word "paper." The industry did not obtain a firm foothold in England until the seventeenth century, and in 1690 America's first paper mill was established [*1*].

Technological advancements in the field of packaging, using paper and paperboard, have created waste-paper contaminants such as plastic coatings, waxes, and wet-strength resins [*9*]. Successful methods of removing these contaminants are important to this nation's recycling effort. To this end, a patented, solvent-type, fiber-recovery system known as the "Polysolv" process was developed. The process uses a nonflammable solvent to remove various contaminants from waste paper. The contaminated solvent (trichloroethylene) used in the extraction step of the process is steam distilled for reuse as clean solvent in subsequent batches.

Example: A rotary reactor is loaded with 3,200 lb of wax- and poly-coated waste paper. The reactor is closed and the extraction cycles are carried out in three stages using trichloroethylene (TCE). In each cycle, 1,400 gal of solvent are fed to the reactor. The solvent to fiber ratio is 4½ to 1. The waste paper contains about 10% contaminants on a weight basis—essentially 100% of the contaminants are dissolved. The first extraction is made with solvent already used twice; the second with solvent previously used once; the third with fresh solvent. After each stage, the solvent is simply siphoned from the reactor to a storage tank. Following the last extraction and draining, 1.4 lb of 200°F solvent per lb of paper remain in the reactor. Superheated steam (110 psig, 380°F) is then admitted to the reactor to steam-strip the remaining solvents from the paper. This operation is carried out at 9 psig. When the pressure starts to drop, there is no more recoverable solvent, so the stripping operation is stopped and a 3-psi vacuum is pulled on the reactor. Calculate the pounds of steam required for solvent-stripping per ton of paper charged to the reactor. The answer follows:

Assumptions

(1) Assume that all heat for vaporization of solvent comes from steam superheat, steam condensation, and cooling of the reactor contents to the stripping temperature.

(2) Neglect heat content of the reactor shell.

(3) Assume perfect equilibrium between steam and TCE.

Distillation temperature

From a Hausbrand diagram, the distillation temperature is found to be 86°C (or 187°F). At this temperature, the partial pressures are:

Trichloroethylene	750 mm Hg
Water	470 mm Hg
Reactor pressure	1,220 mm Hg
or	23.7 psia

Steam Requirement

Basis = 1 batch of paper (3,200 lb)
Contaminant-free paper = 0.9 (3,200) = 2,880
TCE to be vaporized = 2,880 (1.4) = 4,040 lb

Heat required for vaporization
103 (4,040) = 415,000 Btu

Sensible heat given up by still contents
Paper 0.50 (2,880) (200 − 187) = 19,000 Btu
TCE 0.266 (4,040) (200-187) = 14,000 Btu
Total sensible heat change = 33,000 Btu

Steam required for distillation

$$W_s = \frac{4{,}040\ (750)\ (18)}{470\ (131.5)} = 880 \text{ lb}$$

Sensible-heat-change of distillation steam
0.48 (880) (380 − 187) = 82,000 Btu

Heat input of condensing steam per pound
Latent heat (1,265 − 344) = 921 Btu
Sensible heat (344 − 154 = 190 Btu
Total (per pound) = 1,111 Btu

Condensing Steam Required

Heat to vaporize TCE	415,000 Btu
Sensible heat from still contents	33,000 Btu
Sensible heat of distillation steam	82,000 Btu
Heat required from condensing steam	300,000 Btu

$$\text{Weight of condensing steam} = \frac{300{,}000}{1{,}111} = 270 \text{ lb}$$

Total steam per batch = 880 + 270 = 1,150 lb

$$\text{Steam per ton of paper} = \frac{2{,}000}{2{,}880}(1{,}150) = 800 \text{ lb}$$

Coal-Chemicals Industry

It is known that coke was an article of commerce among the Chinese over 2,000 years ago, and in the Middle Ages it was used in the arts and for domestic purposes. Nevertheless, it was not until 1620 that the production of coke in an oven was first recorded. Up until the middle of the nineteenth century, coal tar and coal-tar products were regarded as wastes [*1*]. When coal is thermally pyrolyzed or distilled by heating without contact with air, it is converted into a variety of solid, liquid, and gaseous products. Chemicals from coal were initially obtained by destructive distillation of coal, furnishing chiefly aromatics. These chemicals are now used for a variety of purposes, with dyes and solvents being some of the better-known applications. The problem that follows is an example of the use of steam distillation to recover an aromatic solvent that has been scrubbed from the destructive distillation fumes. This example involves equilibrium consideration only.

Example (Ref. [*7*]): It is proposed to debenzolize a batch of 10,000 lb of benzolized wash oil containing 10% by weight of benzene. Batch steam-distillation will be used for the process, and benzene will be simultaneously recovered. The proposed process will use a batch steam-still, provided with a closed steam coil to heat and maintain the charge at the desired temperature and a perforated steam sparger to distribute the open steam through the liquid charge. The vapors from the still are to be condensed in a worm condenser, and the condensate will then flow to a settling tank where the benzene and water will separate into two layers by gravity. It is desired to calculate only the pounds of *open steam* that must be blown through the charge to reduce the benzene content to 50 lb, under the following conditions:

(1) The contents of the still will be maintained at 350°F during the distillation by means of high-pressure steam condensing in a closed coil. Atmospheric steam, superheated to 350°F, will be used in the steam spray. Heat losses should be neglected since the still is well insulated. Raoult's law applies to the solution of benzene in oil when the molecular weight of the oil is taken as 220. The vapor pressure of the oil is so low at distillation temperatures that its volatility may be neglected. Vapor pressure of benzene at 350°F is 7,100 mm. The molecular weight of benzene is 78 and the vaporization efficiency of the process may be taken as 0.90. The answer follows:

Component	Lb	Mol. Wt.	Moles	Mol. Fraction	Vapor Pressure, mm Hg	Partial Vapor Pressure, mm Hg
Benzene	1,000	78	12.81	0.239	7,100	1,700
Oil	9,000	220	40.85	0.761		
Total	10,000		53.66	1.000		

The partial pressure of benzene in this solution is noted to be greater than 760 mm at 350°F. This indicates that a portion of benzene will vaporize during the heating of the charge to 350°F, and before introduction of the process steam. The amount of benzene removed by this preliminary vaporization may be calculated as follows:

Let F = mols of benzene vaporized. Mole-fraction benzene in residual solution =

$$\frac{12.81 - F}{53.66 - F}$$

Partial pressure of benzene from residual solution =

$$\left(\frac{12.81 - F}{53.66 - F}\right) 7{,}100 = 760 \text{ mm}$$

Solving, F = 7.91 moles of benzene vaporized, or almost

62% of that in the charge to the still. Moles of benzene remaining = 12.81 − 7.91 = 4.90. Using Eq. (13):

$$N_s = \left(\frac{760}{(0.9)\,(7{,}100)} - 1\right)(4.90 - 0.64) + \left(\frac{(760)\,(40.85)}{(0.9)\,(7{,}100)}\right)(2.3)\log\frac{4.90}{0.64}$$

N_s = 6.12 mols steam

W_s = 6.12 (18) = 110 lb steam

Pounds steam per pound benzene stripped =

$$\frac{110}{(4.90 - 0.64)\,(78)} = 0.329$$

Pounds steam per pound benzene recovered =

$$\frac{110}{(12.81 - 0.64)\,(78)} = 0.10$$

Naval-Stores Industry

Turpentine and rosin are produced from the longleaf pine in the southeast and southern parts of the U.S. In fact, in the early days of America, the forests in this region were said to extend in a belt approximately 200 miles wide and 1,000 miles long. In 1608, what was probably the first cargo of American naval stores was shipped to England. Florida and Georgia now produce 95% of today's gum turpentine and rosin. Gum naval stores are made from crude gum (oleoresin) from living longleaf and slash pine trees [*1*]. The example that follows illustrates the application of steam distillation to problems in the naval-stores industry.

Example (Ref. [*4*]): One hundred gal/h of turpentine are to be steam-distilled at atmospheric pressure. The latent heat of vaporization of turpentine is 74 cal/g, its molecular weight is 136, and its specific gravity is 0.865. No external heat is applied. (1) How much steam at 5 lb gage will be used per hour? (2) What would be the effect of operating at 300-mm Hg absolute pressure instead of atmospheric? (Neglect heat required to heat up the charge and heat lost by radiation.) The answer follows:

From the Hausbrand diagram, it is seen that the curves for turpentine and water at atmospheric pressure cross at 95.5°C. At this temperature the vapor pressure of turpentine is 114 mm, and that of water is 646 mm. From Eq. (8) the ratio of turpentine to water in the product is

$$\frac{W_{im}}{W_s} = \frac{(114)\,(136)}{(646)\,(18)} = 1.333$$

One hundred gal of turpentine is 100 × 8.33 × 0.865 = 720.5 lb. Since the ratio of turpentine to water by weight is 1.333, the steam passing over into the product will be 720.5/1.333 or 541 lb.

The total heat needed is

$$720 \times 74 \times 1.8 = 96{,}000 \text{ Btu/h}$$

Steam at 5 lb gage has a temperature of 108.3°C. The steam that goes over with the product cools from 108.3 to 95.5°C as superheated steam (sp. ht. = 0.48) and therefore furnishes

$$720 \times 0.48 \times (108.3 - 95.5) \times 1.8 = 6{,}000 \text{ Btu/h}$$

The rest of the heat needed, or 90,000 Btu, must be supplied by steam that condenses and then cools to 95.5°C. This steam furnishes, per pound,

As latent heat	960.4 Btu
As sensible heat (108.3 − 95.5) × 1.8	23.0 Btu
Total	983.4 Btu

The weight of steam condensed is

$$\frac{90{,}000}{983.4} = 91.5 \text{ lb}$$

The total steam used per hour is 541 + 91 = 632 lb.

At 300-mm absolute pressure the results are:

Temperature of distillation	72°C
Vapor pressure of water	255 mm
Vapor pressure of turpentine	45 mm
Ratio of turpentine to water	1.334
Weight of steam in product	541 lb
Total heat needed	96,000 Btu/h
Heat from steam in product	17,070 Btu/h
Steam condensed	77 lb
Total steam used	618 lb
Steam savings over atmospheric distillation	2.2%

This example deals with equilibrium behavior and the distillation heat balance.

Conclusion

Steam distillation affords the chemical engineer a very handy tool for solving some difficult separations problems. The unique vapor-pressure behavior of immiscible fluids is at the heart of the theory developed for this unit operation, and an understanding of the theory then enables the chemical engineer to establish the equilibrium relationships needed for distillation calculations. From this point, the method for calculating most of the factors involved in the design of stills and auxiliaries for steam distillations are but the application of other chemical engineering principles. Using them, such things as the heat that must be supplied in vaporization and removed in condensation, the volumes of vapors and liquids to be handled, and the heating and cooling surfaces necessary, can be readily calculated.

References

1. Shreve, R. N., "Chemical Process Industries," McGraw-Hill, 1967, pp. 66, 501, 617, 738.
2. Daniels, F., Alberty, R.A., "Physical Chemistry," Wiley, New York, 1956, pp. 184, 197–199.
3. Kirkbride, C. G., "Chemical Engineering Fundamentals," McGraw-Hill, New York, 1947, p. 67.
4. Badger, W. L., McCabe, W. L., "Elements of Chemical Engineering," McGraw-Hill, New York, 1936, p. 368.
5. Badger, W. L., Banchero, J. T., "Introduction to Chemical Engineering," McGraw-Hill, New York, 1955, p. 314.
6. Brown, G. G. others, "Unit Operations," Wiley, New York, 1953, pp. 391–393.
7. Perry, J. H., "Chemical Engineers' Handbook," 3rd ed., McGraw-Hill, New York, 1950, pp. 582–585.
8. Walker, W. H., others, "Principles of Chemical Engineering," McGraw-Hill, New York, 1937, pp. 537–543.
9. Johnson, A. E., What the Polysolv Process Is and How it Works, *Paper Trade J.*, Oct. 6, 1969, p. 55.

Meet the Author

R. W. Ellerbe is project engineer for The Rust Engineering Co., 1130 South 22nd St., P. O. Box 101, Birmingham, AL 35201. He has held process engineering and administrative assignments in petroleum refining, natural-gas processing and pulp and paper. He holds an M.S.Ch.E. from Louisiana Technological University and has taken other graduate courses in chemical engineering and business administration. His professional affiliations have included AIChE, Technical Assn. of the Pulp and Paper Industry and the Arkansas Soc. of Professional Engineers—he is a registered professional engineer in Arkansas.

Section 2 Data Analysis

An approach to multiphase vapor-liquid equilibria

Improved estimation techniques using the NRTL (nonrandom two-liquid) method handle nonideal vapor-liquid equilibria having two or more coexisting liquid phases.

M. J. Leach, *The British Petroleum Co. Ltd.*

☐ Effective design for complex distillations demands an accurate knowledge of the phase equilibria of the system being processed. Many systems prove to be multicomponent and highly nonideal—especially in the petrochemical industry. Nonideality usually predominates in the liquid phase.

Experimental vapor-liquid equilibrium (VLE) data are often available for binary mixtures, but are less so for multicomponent systems. For the design engineer, it is essential to have techniques that can predict multicomponent behavior using only binary data. A significant portion of the systems encountered contain one or more partially miscible binary pairs, so a flexible technique should be capable of handling these.

Phase-equilibrium thermodynamics

For equilibrium between phases, the fugacity of a component in one phase must be equal to that of the same component in the other phase. The vapor fugacity of component i may be expressed as:

$$f_i^v = \phi_i y_i^P \tag{1}$$

and the liquid fugacity as:

$$f_i^L = \gamma_i x_i f_i^{OL} \tag{2}$$

Five different data types can be used to find NRTL interaction parameters

Type of data	Parameter to be fitted
VLE: $x-y-P$ VLE: $x-y-T$	y
VLE: $x-P-T$ VLE: $y-P-T$	P
VLE: $x-y-P-T$ Activity data	γ
Mutual solubility data	$x\gamma$
Heat of mixing data	ΔH^M

If we equate Eq. (1) and (2) and rearrange, a relationship for the equilibrium K-value results:

$$K_i = y_i/x_i = \frac{\gamma_i f_i^{OL}}{\phi_i P} \tag{3}$$

Prausnitz et al. [1] give suitable expressions for the fugacity coefficient, ϕ_i, and the standard-state liquid-phase fugacity, f_i^{OL}. The activity coefficient, γ_i, is normally calculated using a liquid-phase interaction model; the Wilson equation [2] is probably the best known of these models, but it has the disadvantage of being unable to describe partial miscibility.

The NRTL equation

The nonrandom two-liquid (NRTL) equation, proposed by Renon and Prausnitz [3], is capable of describing partial miscibility. The NRTL equation has the form:

$$\ln \gamma_i = \frac{\sum_j \tau_{ji} G_{ji} x_j}{\sum_k G_{ki} x_k} + \sum_j \frac{x_j G_{ij}}{\sum_k G_{kj} x_k}\left[\tau_{ij} - \frac{\sum_p x_p \tau_{pj} G_{pj}}{\sum_k G_{kj} x_k}\right] \tag{4}$$

where $\tau_{ji} = (g_{ji} - g_{ii})/RT$; $G_{ji} = \exp(-\alpha_{ji}\tau_{ji})$; and $\alpha_{ji} = \alpha_{ij}$.

For any two components, 1 and 2, this equation contains three adjustable interaction parameters: $(g_{21} - g_{11})$, $(g_{12} - g_{22})$ and α_{12}. Hence, for N components, $3N(N-1)/2$ parameters are required to describe the system. The temperature interval over which predictions are needed may often be large, or the available experimental data may not be at the required temperature. To overcome such problems, the $(g_{ji} - g_{ii})$ terms can be estimated as linear functions of temperature. However, this technique increases the number of parameters to five per binary pair.

Calculation of interaction parameters

Data needed to determine the NRTL interaction parameters are shown in the table, together with the ap-

Originally published May 23, 1977

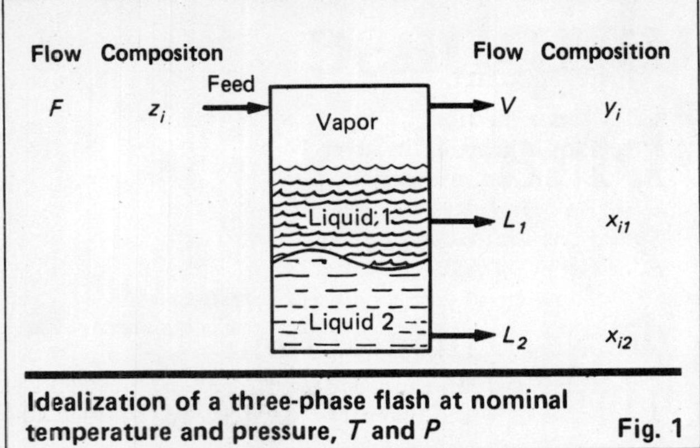

Idealization of a three-phase flash at nominal temperature and pressure, T and P — Fig. 1

propriate fitting parameters. The calculated value of the fitting parameter should be compared with the experimental value, and the difference between the two should be minimized. If *x-y-P-T* data are available, it is possible to calculate and fit directly the activity coefficient, γ. With other types of VLE data, it is necessary to use conventional bubble- and dew-point routines such as those described by Judson-King [19], to produce the calculated value of the fitting parameter.

All of the different types of data listed in the table may be used simultaneously to calculate the interaction parameters. Details on the objective function to be used in each case are given elsewhere [4].

It is most important to include in the calculation as many different types of data as possible. Both binary and multicomponent data may be used, as well as azeotropic data. However, the data *must* be checked for compatibility and internal consistency, and any doubtful points should be discarded. If a conventional equilibrium prediction-routine is used, the VLE data should derive from systems having a homogeneous liquid phase. If the data are very limited, it may be necessary to fix the value of α_{ij}, using the rules suggested by Renon and Prausnitz [3].

No data available

This situation arises fairly frequently. It becomes

Comparing conventional prediction routine

***Example 1**—Ethyl acetate/water*

Ethyl acetate and water form a partially miscible mixture. The NRTL interaction parameters for the system were calculated using mutual solubility data [13]; *x-y-P-T* VLE data at 70°C, 55°C and 40°C [14]; and *x-P-T* and *y-P-T* data at 760 mm Hg [15]. Care was taken to use only VLE data with a homogeneous liquid phase, since a conventional VLE prediction routine was used in the fitting package [19].

The calculated NRTL interaction parameters were:

$$(g_{21} - g_{11}) = 2{,}599.45 \text{ kcal/kg-mole}$$
$$(g_{12} - g_{22}) = -455.96 \text{ kcal/kg-mole}$$
$$\alpha_{12} = \alpha_{21} = 0.082$$

Horsley [16] presents experimental values for the heterogeneous azeotropic composition of the system at various pressures. These data are shown here, together with azeotropic compositions predicted using NRTL interaction parameters in the VLLE algorithm shown in Fig. 3, and with misleading data derived from a conventional VLE algorithm [19] that ignores the existence of two liquid phases:

Azeotropic composition (mole fraction water)

Pressure, mm Hg	Experimental	VLLE algorithm	Conventional VLE algorithm*
25	0.1544	0.1695	0.2189
250	0.2468	0.2540	0.3044
760	0.3116	0.3058	0.3466

The VLLE algorithm predicts azeotropic compositions fairly close to the experimental values, whereas the predictions of the conventional VLE routine differ considerably from the experimental values.

***Example 2**—Benzene/water*

Seidell [17] gives the mutual solubilities of benzene and water at 80°C. The resulting NRTL interaction parameters are:

$$(g_{21} - g_{11}) = 4{,}001.43 \text{ kcal/kg-mole}$$
$$(g_{12} - g_{22}) = 1{,}736.38 \text{ kcal/kg-mole}$$
$$\alpha_{12} = \alpha_{21} = 0.2 \text{ (fixed)}$$

These parameters were used in the Fig. 3 VLLE algorithm, and in a conventional VLE algorithm to predict the VLE data of the system at atmospheric pressure. The resulting curves point up the inadequacies of the latter:

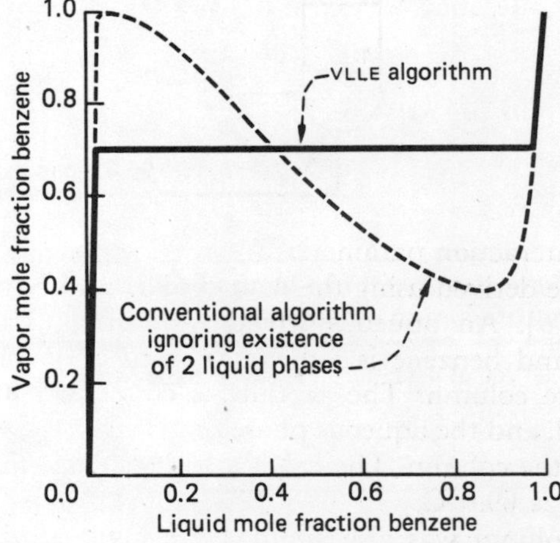

The VLLE routine predicts a constant vapor composition over a large range of liquid composition, as would be expected for a system with very limited miscibility. In contrast, the conventional routine [19] predicts a vapor composition for which $(\partial y_i/\partial x_i)$ is negative over most of the range.

necessary to predict activity coefficients using the "solution of groups" approach of Derr and Deal [5]—recently extended by Fredenslund et al. [6]—or, alternatively, the correlation of Pierotti et al. [7]. or that of Helpinstill and Van Winkle [8].

Vapor-liquid-liquid equilibrium (VLLE)

Equilibrium between a vapor and two coexistent liquid phases is shown schematically in Fig. 1. From phase-equilibrium and material-balance considerations [4], we can show that:

$$z_i = x_{i1}J_i = x_{i2}J_iK_{i2}/K_{i1} = y_iJ_i/K_{i1} \quad (5)$$

where $J_i = b(1-a) + (1-b)(1-a)K_{i1}/K_{i2} + aK_{i1}$

$$a = V/F \text{ and } b = L_1/(L_1+L_2)$$

$$K_{i1} = y_i/x_{i1} \text{ and } K_{i2} = y_i/x_{i2}$$

and

$$F_1 = \Sigma[z_i(1-K_{i1})/J_i] = 0 \quad (6)$$

$$F_2 = \Sigma[z_i(1-K_{i1}/K_{i2})/J_i] = 0 \quad (7)$$

Nomenclature

a	Vapor fraction
b	Liquid fraction in layer 1
E	Modification function
F	Feed flowrate (kg-moles/h)
F_1	Function defined by Eq. (6)
F_2	Function defined by Eq. (7)
f_i^L	Fugacity of component i in liquid, bar
f_i^{OL}	Standard-state liquid fugacity of component i, bar
f_i^V	Fugacity of component i in vapor, bar
G_{ji}	Coefficient in NRTL equation.
g_{ji}	NRTL interaction energy between an i-j pair of molecules (kcal/kg-mole)
K_i	K-value of component i
K_{i1}	K_i in layer 1
K_{i2}	K_i in layer 2
L_1	Flowrate of layer 1 (kg-moles/h)
L_2	Flowrate of layer 2 (kg-moles/h)
P	Pressure, bar
P_n	nth estimate of calculated pressure, bar
T	Temperature, °K
V	Vapor flowrate, kg-moles/h
x_i	Liquid mole fraction of component i
x_{i1}	x_i in layer 1
x_{i2}	x_i in layer 2
y_i	Vapor mole fraction of component i
z_i	Feed mole fraction of component i
α_{ji}	NRTL parameter for i-j interactions
γ_i	Activity coefficient of component i
ϕ_i	Fugacity coefficient of component i
τ_{ji}	Coefficient in NRTL equation

with the VLLE method

Example 3—Ethyl acetate/benzene/water

The VLLE routines have been incorporated into a distillation design program, which was used to simulate a simple distillation column:

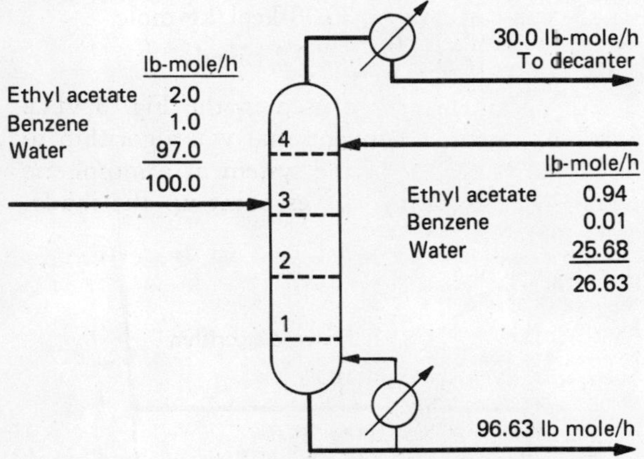

The interaction parameters for ethyl acetate/benzene were derived using the data of Carr and Kropholler [18]. An aqueous stream containing ethyl acetate and benzene is fed to the third plate of a four-plate column. The product is condensed and decanted, and the aqueous phase returned to the top plate of the column. The calculated condenser temperature is 69.6°C.

The column was also simulated using a similar program with a conventional VLE routine that ignores the existence of two liquid phases [19]. The predicted condenser temperature in this case was 35.9°C—an error of over 30°C. This difference in temperature would significantly affect the design of the condenser and, possibly, the choice of cooling medium for the duty.

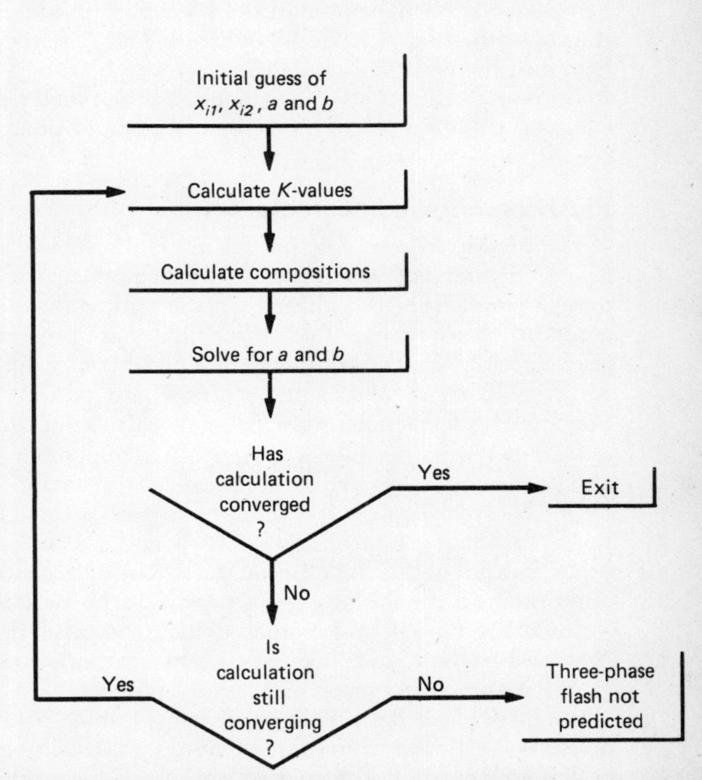

Algorithm for flash calculations in which T and P are fixed, and a is calculated **Fig. 2**

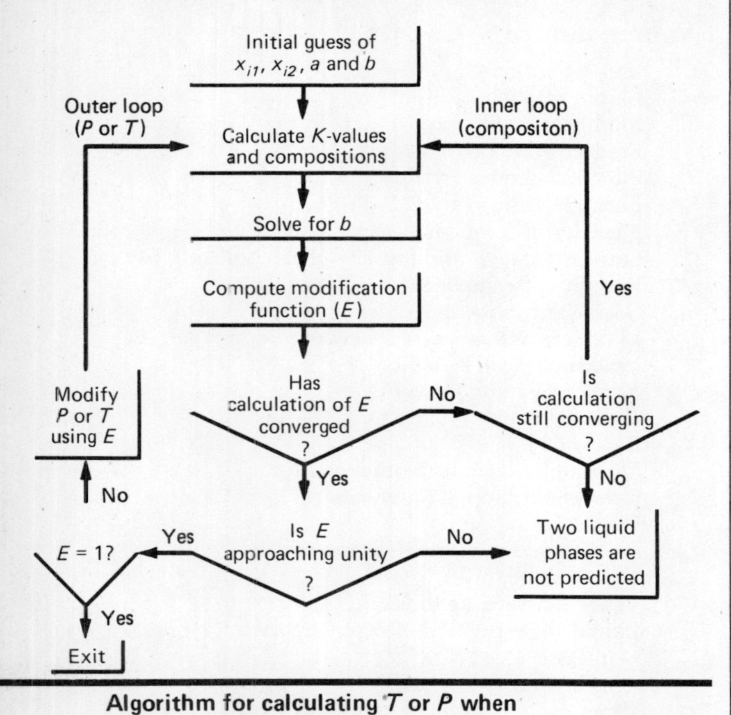

Algorithm for calculating T or P when vapor fraction a is specified **Fig. 3**

The variety of calculations that might be performed on the system shown in Fig. 1 may be classified into two basic types:

- Flash equilibrium calculations in which T and P are both specified and a, the vapor fraction, is calculated.
- Calculation of temperature or pressure in which the vapor fraction, a, is specified. Either T or P is specified, and the other is calculated.

(This analysis can also be extended to the much less common situation where more than two liquid phases coexist.)

Flash equilibrium calculation

The algorithm used for this calculation is shown in Fig. 2. It does not work if only two components are present since, from the phase rule, the temperature is constant for any value of a if three phases are present. (Hence, with only two components, a number of values for a could result at any temperature and pressure.) Fortunately, the temperature/pressure calculation described below is possible with only two components, and results can normally be obtained by this method.

The Newton-Raphson procedure has been suggested [9,10] for solving Eq. (6) and (7) for a and b. Convergence is rapid with a good initial guess, but divergence often results if the initial guess is poor. For this reason, we used the Powell [11] method, which combines the Newton-Raphson and steepest-descent methods and gives a rapid convergence with steady progress.

If a three-phase flash is not predicted, a bubble-point pressure calculation assuming two liquid phases should be undertaken. If the calculated pressure is below the specified pressure, the system lies below its bubble point. If not, a conventional flash calculation is begun assuming a homogeneous liquid phase.

Temperature/pressure calculation

The algorithm used for this calculation is shown in Fig. 3. The composition loop resembles the corresponding loop in the preceding flash calculation. In the outer P or T loop, the temperature or pressure is adjusted using the function E, where

$$E = \sum_i y_i \Big/ \left(b \sum_i x_{i1} + (1 - b) \sum_i x_{i2} \right) \tag{8}$$

The pressure is adjusted using

$$P_{n+1} = P_n E$$

whereas the temperature is adjusted assuming that log E is a linear function of $(1/T)$.

Use of the algorithms

The two algorithms have been widely used for equilibrium calculations and have performed very well on systems containing up to eight components. They have also been used within a distillation-column design program. Details on a design program that permits two coexistent liquid phases are given by Block and Hegner [12]. (See above for examples.)

References

1. Prausnitz, J. M., Eckert, C. A., Orye, R. V. and O'Connell, J. P., "Computer Calculations for Multicomponent Vapor-Liquid Equilibria," Prentice-Hall, Englewood Cliffs, N.J., 1967.
2. Wilson, G. M., *J. Am. Chem. Soc.*, Vol. 86, 1964 p. 127.
3. Renon, H., and Prausnitz, J. M., *A.I.Ch.E.J.*, Vol. 14, 1968 p. 135.
4. Leach, M. J., paper presented at I. Chem. E. *Design Congress—'76*, Birmingham, England, Sept. 1976.
5. Derr, E. L., and Deal, C. H., "Distillation," *I. Chem. E.*, London, 1969.
6. Fredenslund, A., Jones, R. L., and Prausnitz, J. M., *A.I.Ch.E.J.*, Vol. 21, 1975, p. 1086.
7. Pierotti, G. J., Deal, C. H., and Derr, E. L., *Ind. Eng. Chem.*, Vol. 51, 1959, p. 95.
8. Helpinstill, J. G., and Van Winkle, M., *I.E.C. Proc. Des. and Dev.*, Vol. 7, p. 213.
9. Henley, E. J., and Rosen, E. M., "Material and Energy Balance Computations," Wiley, New York, 1969.
10. Deam, J. R., and Maddox, R. N., *Hydroc. Proc.*, Vol. 48, No. 7, 1969, p. 163.
11. Powell, M. J. D., Atomic Energy Research Establishment (U.K.) Report No. R 5947, 1968.
12. Block, U., and Hegner, B., *A.I.Ch.E.J.*, Vol. 22, 1976, p. 582.
13. Stephen, M., and Stephen, T., "Solubilities of Inorganic and Organic Compounds," Pergamon Press, Oxford, 1963.
14. Mertl, I., *Coll. Czech. Chem. Comm.*, Vol. 37, No. 2, 1972, p. 366.
15. Kato, M., Konishi, H., and Hirata, M., *J.C.E.D.*, Vol. 15, 1970, p. 435.
16. Horsley, L. H., "Azeotropic Data—III," ACS No. 116, Am. Chem. Soc., 1973.
17. Seidell, A., "Solubilities of Organic Compounds," 3rd ed., Van Nostrand, New York, 1941.
18. Carr, A. D., and Kropholler, H. W., *J.C.E.D.*, Vol. 7, 1962, p. 26.
19. Judson-King, C., "Separation Processes," McGraw-Hill, New York, 1971.

The author

Malcolm Leach is a technologist in the Group Research and Development Dept. of The British Petroleum Co. Ltd., Chertsey Rd., Sunbury-on-Thames, Middlesex, England. He is engaged in process development and economic evaluation, and has been particularly concerned with nonideal phase equilibrium. A member of the Institution of Chemical Engineers' subcommittee on phase equilibrium, he obtained his B.Sc. and Ph.D. in chemical engineering from the University of Leeds, England.

Prediction of liquid activity coefficients

If you need equilibrium data for a preliminary distillation design but cannot wait for costly experimental results, here is a guide that will help you locate an appropriate estimation technique.

A. K. Rao, Catalytic, Inc.

☐ Differences between the equilibrium behavior of real vapor-liquid mixtures and that of ideal ones can be accounted for by two arbitrarily defined correction factors, the vapor-fugacity coefficient (ϕ_i) and the liquid activity coefficient (γ_i). When both are equal to 1, the system is ideal (i.e., Raoult's law applies to the liquid; the ideal-gas law and Dalton's law to the vapor).

The vapor fugacity coefficient accounts for the properties of the vapor phase as a real gas. As temperature decreases and pressure increases, the behavior of a real gas tends to deviate more from that of an ideal gas. At pressures below 1 atm, this deviation can be neglected; at higher pressures, it can be accurately predicted from critical properties of pure components [*17,18*].

The liquid activity coefficient represents the correction factor for liquid-phase nonideality. It depends mainly on the composition and, to a certain extent, on the temperature. The effect of pressure on this coefficient can usually be neglected.

Reliable prediction of activity coefficients requires experimental data on the mixture. In the absence of experimental data at different operating conditions, we can resort to estimating the activity coefficient by using various theories of solution (Table I). However, none of the theories currently available are totally reliable.

Functional dependence of the coefficient on mole fractions is fairly well understood. Several equations have been proposed to show this dependence, the more commonly used ones having been advanced by Wohl, Scatchard and Hamer, Van Laar, Margules, and Wilson [*7*]. More recently, the Wilson equation has gained added popularity, especially because it eases the use of binary data for characterizing multicomponent systems, and contains a built-in temperature adjustment [*17*].

The most recent development in this area is a semitheoretical equation for activity coefficients in liquid mixtures, generalized from Guggenheim's quasichemical model [*1*]. Called UNIQUAC (for universal quasichemical equation), this method uses only two adjustable parameters per binary pair in a multi-component mixture. It works for a wide range of mixtures of polar and nonpolar liquids, such as hydrocarbons, ketones, esters, amines, alcohols, nitriles and water. It can, unlike Guggenheim's model, handle mixtures whose molecules differ appreciably in size and shape; and, unlike Wilson's equation, it accounts for partial miscibility when liquid-liquid equilibria become important.

Due to the very generalized nature of the UNIQUAC model, it is fairly complex, favoring the use of a digital computer for implementation. With well-defined assumptions, it reduces to more-common models, such as the Van Laar, Wilson, NRTL (nonrandom two-liquid) and Scatchard-Hamer ones.

All the models mentioned above have been derived so that they satisfy the Gibbs-Duhem test for thermodynamic consistency. The empirical constants or "parameters" of the models are unique to each mixture. For reliable prediction of liquid-activity coefficients, at least one of the following types of experimental binary data is required [*17,18*].

1. Isothermal dependence of total pressure on liquid composition.
2. Isobaric dependence of boiling point on liquid composition.
3. Activity coefficients of each component when it is infinitely dilute.
4. Known equilibrium composition of vapor and liquid at one point.
5. Azeotropic composition.

Prediction from pure-component properties

The first step in estimating liquid activity coefficients is to examine qualitatively the behavior of the system components.

Originally published May 9, 1977

Some methods for estimating liquid activity coefficients — Table I

Basis	Method	Class of compounds	Raoult's law deviation	Temperature range	Expected error	Other restrictions
Equation of state	Van Laar	Limited to fluids for which a two-constant equation of state is applicable	Only slightly positive		Can be very high, and in some cases, worse than if ideality had been assumed	Volume change on r ing must be neglig
Classification of pure compounds based upon hydrogen bonding	Ewell	Mixtures of compounds whose structures and hydrogen-bonding tendencies are well-defined	Positive and negative		Gives rough estimate of direction and magnitude of Raoult's law deviation	Only qualitative
	Gilmont et al.	All except mixtures of strongly associating compounds, or ones with radically different molecular sizes			For most mixtures, differences between predicted and experimental vapor compositions average ±1.7%; max. is ±4.5%. With ketones, nitriles or picolines, differences run 5-15%	
Regular solution theory and its extensions	Scatchard-Hildebrand	Nonpolar, especially hydrocarbons that have nearly equal molar volumes	Only positive		In naphthenic, paraffinic and aromatic oils, av'g deviation of K values is about 15%	Molar liquid volume solubility parameter always available; hy thetical liquid state be needed at s operating condition
	Van Arkel modification	Slight correction for weakly polar pure components	Only positive		Overall accuracy comparable to Scatchard-Hildebrand	
	Weimer-Prausnitz modification	Highly polar solvents. Good for polar-nonpolar mixtures, but not polar-polar ones	Positive and negative	0-125°C	For many polar-nonpolar mixtures, av'g difference between predicted and experimental coefficients at infinite dilution is ±10%	Only terminal co cient (infinite dilut can be easily estim
	Helpinstill-Van Winkle	Good for all combinations of polar and nonpolar compounds. Saturated, unsaturated and aromatic hydrocarbons	Postitive and negative	0-125°C	For many mixtures of sat. hydrocarbons and polar solvents, absolute error in terminal coefficients is ±11.6%. In unsat. and aromatic mixtures it runs 8.5% and 13.5%, respectively	
	Null modification	General, including systems containing strongly polar or strongly associating compounds such as alcohols	Positive and negative		Average error in terminal coefficient for over 800 points was 25%—equivalent to 41% standard deviation. For many solvents containing saturated hydrocarbons, av'g error less than 9% (standard deviation, 14%). Errors greater when polar or associating component is solute rather than solvent	Introduces added p meters, complica prediction. Gives terminal coefficient
Group contribution	Pierrotti, Deal and Derr	Structurally related components (e.g., R_1-X_1 methylene homologue in R_2-X_2 methylene homologue)		25-100°C	Average deviation in terminal coefficient runs about 8%. Error is slightly greater for cyclic and alkyl cyclic compounds	Gives only terminal efficients.
	ASOG	General	Positive and negative	0-125°C	With few exceptions, agreement between predicted and measured values of the log of the activity coefficient is reported to be ±10%	Complex molecules give misleading res
	UNIFAC	General. Predicts two-phase region for partially miscible liquids	Positive and negative	0-125°C	In most typical cases, predicted activity coefficients at infinite dilution deviate less then 20% from measured result	

Ewell, Harrison and Berg [4] grouped all chemicals into five classes, according to their ability to form hydrogen bonds (Table II). With this characterization, the direction of the deviation from ideal behavior, and the tendency to form azeotropes, can be predicted. Gilmont et al. [6] subdivided these classes into smaller subgroups and used these for quantitative prediction of activity coefficients. The formation of an azeotrope depends on both the difference in boiling points of the pure components, and the extent of deviation (positive or negative) from Raoult's law.

In general, the smaller the deviation from Raoult's law for a pair of substances, the smaller the difference in boiling point must be before an azeotrope will form. If an azeotrope forms, then positive deviation yields a minimum-boiling azeotrope and, conversely, a negative deviation yields a maximum boiling one. The following generalizations are useful:

- Mixtures of two chemicals that are homologs, or isomers, are generally ideal.
- Mixtures that form these combinations of Ewell classes tend to be ideal: II + III; III + IV; IV + IV; IV + V; V + V.

Table III gives some examples of known systems whose behavior can be predicted from Ewell's classification.

Methods based on equations of state

The Van Laar technique correlates activity coefficients with temperature, liquid mole fractions, and two

Ewell classification of chemicals according to tendency toward hydrogen bonding — Table II

Group combinations	Deviations	Remarks
I + V	Always positive	Frequently of limited miscibility
II + V	Always positive	
III + IV	Always negative	
I + IV	Always positive	Frequently of limited miscibility
II + IV	Always positive	
I + I I + II I + III II + II II + III	Usually positive; exceptions can be negative and give maximum-boiling azeotropes	
III + III III + V IV + IV IV + V V + V	Most of these systems will be close to ideal, or show slightly positive deviation. Azeotropes not common, and will be minimum-boiling.	

Behavior of some common binary systems — Table III

System	Ewell-class combination	Known behavior
Ethanol-benzene	II + V	Positive deviation; minimum-boiling azeotrope
Acetone-chloroform	III + IV	Negative deviation; maximum boiling azeotrope
Water-ethanol	I + II	Positive deviation; minimum-boiling azeotrope
Benzene-toluene	V + V	Almost ideal
Dichloromethane-chloroform	IV + V	Almost ideal; slight positive deviation

Class I—These liquids form a strong three-dimensional network of hydrogen bonds: e.g., water, glycol, glycerol, amino alcohols, hydroxylamines, hydroxyacids, polyphenols, amids. Compounds such as nitromethane and acetonitrile (Class II) also form a network, but bonds are weaker than those that involve –OH or –NH groups. Class I liquids have high dielectric constants and are invariably water soluble.

Class II—Molecules have both active hydrogen and electron-donor atoms (e.g. O, N, F), but do not form strong network of hydrogen bonds: alcohols, acids, phenols, primary and secondary amines, oximes, nitro compounds with α-hydrogen, nitriles with α-hydrogen, ammonia, hydrazine, hydrogen halides. These liquids also have high dielectric constants and are generally water soluble.

Class III—These liquids have an electron donor atom, but not an active hydrogen. Dielectric constants may not be high, but they can dissolve in water, since they are capable of forming hydrogen bonds with water molecules—e.g., ethers, esters, ketones, aldehydes, tertiary amines, pyridine, nitro and nitrile groups with no α-hydrogen, etc.

Class IV—Molecules have an active hydrogen but no electron donor atom. Typically, molecules have two or three chlorine atoms on the same carbon atom as a hydrogen atom, or one chlorine atom on the same carbon atom and one or more chlorine atoms on an adjacent carbon atom. Examples: $CHCl_3$; CH_3CHCl_2. These compounds are frequently insoluble in water or have partial miscibility, since hydrogen bonds formed between water and these liquids are not strong enough to break water-water hydrogen bonds.

Class V—All other liquids incapable of forming hydrogen bonds, such as hydrocarbons, carbon disulfide, mercaptans, some halogenated hydrocarbons, etc. Cannot dissolve in water, since no compensating hydrogen bonds can be formed to account for water hydrogen bonds that need to be broken in order for dissolution to take place.

parameters. These parameters are related to constants in the van der Waals equation of state, and can be evaluated from the critical properties of the pure components. This method is very approximate and should only be used when no information other than critical constants is available [*19*]. It assumes that volume change on mixing is negligible; excess partial molal entropy is zero; and that pure components, as well as their mixtures, approximate van der Waals fluids.

The Benedict-Webb-Rubin and Wilson-Redlich-Kwong equations have been increasingly used to describe both liquid and vapor phases quite accurately [*9,24*]; a modification of the latter has been used for three-phase systems of water and hydrocarbons. Unfortunately, these equations all require interaction parameters that must be evaluated from binary data. Until a generalized method of predicting these interaction parameters is developed, the usefulness of these equations will remain limited.

Nomenclature

- ΔH_{vap} Molar heat of vaporization, cal/mole
- R Ideal gas law constant, 1.987 cal/(mole)(°K)
- T_B Boiling-point temperature, °K
- v Molar volume, cm^3/mole
- x Mole fraction in liquid phase
- z Volume fraction in liquid phase
- γ Liquid activity coefficient
- δ Solubility parameter
- λ Nonpolar component of solubility parameter
- μ Dipole moment, Debye units
- τ Polar component of solubility parameter
- ψ Energy term for induction effect
- ω Orientation parameter

Subscripts

- ∞ Measured at infinite dilution
- i,j Pure-component designations

Methods based on regular-solution theory

The Scatchard-Hildebrand equation [*11*] is applicable to regular solutions whose nonideality is entirely due to heat of mixing. Nonpolar or weakly polar components having nearly equal molar volumes form solutions that can be approximated by this model.

Terminal coefficients for mixture having positive deviation

Example—Calculate the infinite-dilution activity coefficients at 36°C for a benzene and 2,4-dimethylpentene [1,2] mixture, using Scatchard-Hildebrand regular-solution theory.

Physical properties at 36°C:

	Benzene	2,4-dimethylpentene
Molecular wt	78.11	100.20
Density, g/cm^3	0.868	0.663
Boiling point, °C	80.1	80.5
Dipole moments, Debyes	0.0	0.0

Using Eq. (5) for the heat of vaporization $(\Delta H_{vap})_{298°K}$ we get:

$$\Delta H_{vap,1} = -2{,}950 + 23.7\,(273.0 + 80.1) + 0.02\,(273 + 80.1)^2 = 7{,}912 \text{ cal/mole}$$

and

$$\Delta H_{vap,2} = 2{,}950 + 23.7\,(273.0 + 80.5) + 0.02\,(273 + 80.5)^2 = 7{,}918 \text{ cal/mole}$$

We have neglected the effect of specific heat on ΔH_{vap}. Using Eq. (4) to determine solubility parameters, we calculate:

$$\delta_1 = \left(\frac{7912 - 615}{78.11/0.868}\right)^{1/2} = 9.00$$

$$\delta_2 = \left(\frac{7918 - 615}{100.2/0.663}\right)^{1/2} = 6.95$$

and $(\delta_1 - \delta_2)^2 = 4.2$

At infinite dilution, the volume fraction becomes 1 for the solvent and 0 for the solute. Hence, $z_1 = 1$ and $z_2 = 0$, or vice-versa. Substituting into Eq. (1) and (2), we obtain:

$$\ln \gamma_1^{\infty} = \frac{78.11/0.868}{615} \times 4.2 = 0.615$$

$$\ln \gamma_2^{\infty} = \frac{100.2/0.663}{615} \times 4.2 = 1.032$$

Activity coefficients at infinite dilution are $\gamma^{\infty}_{benzene} = 1.85$ and $\gamma^{\infty}_{dimethylpentene} = 2.81$

Note: The above mixture is fairly nonideal, with a positive Raoult's-law deviation. At the same time, the boiling points are fairly close. Hence, one can expect a minimum-boiling azeotrope for this mixture.

Activity coefficients of two components in their binary mixtures can be expressed as:

$$\ln \gamma_1 = \frac{v_1 z_2}{RT}(\delta_1 - \delta_2)^2 \tag{1}$$

and

$$\ln \gamma_2 = \frac{v_2 z_1}{RT}(\delta_1 - \delta_2)^2 \tag{2}$$

where v_1, v_2 are molar volumes in the liquid phase; z_1, z_2 are volume fractions in the liquid phase; and δ_1, δ_2 are solubility parameters. Volume fractions and solubility parameters are defined as:

$$z_i = \frac{x_i v_i}{x_i v_i + x_j v_j} \tag{3}$$

and

$$\delta_i = \left[\frac{\Delta H_{vap,i} - RT}{v_i}\right]^{1/2} \tag{4}$$

where $\Delta H_{vap,i}$ is the molar heat of vaporization (cal/mole) of the pure component i, and x_i is its mole fraction in the liquid phase.

Heat of vaporization can be estimated from the boiling point, using the approximation [11]:

$$(\Delta H_{vap})_{298°K} = -2{,}950 + 23.7\,T_B + 0.02\,T_B^2 \tag{5}$$

$(\Delta H_{vap})_{298°K}$ can be corrected to any temperature (°K) by using heat-capacity data.

The original Scatchard-Hildebrand equation has since been generalized by several authors.

Van Arkel [20] introduced an additional term in Eq. (1) and (2) to correct for orientation forces:

$$\ln \gamma_i = \frac{v_i z_j}{RT}\left[(\delta_j - \delta_i)^2 + (\omega_j - \omega_i)^2\right] \tag{6}$$

where $(\omega_j - \omega_i)^2$ was defined as:

$$(\omega_j - \omega_i)^2 = \frac{3.25 \times 10^7}{RTv_j^2}\left[\frac{\mu_j^2}{\sqrt{v_j}} - \frac{\mu_i^2}{\sqrt{v_i}}\right]^2 \tag{7}$$

and μ_j, μ_i are the dipole moments in Debye units for the pure components.

The most significant improvement to Eq. (6) has been the introduction of a third residual term, $-c\delta_j v_i/\delta_i v_j$, inside the outer parentheses. This negative term accounts for negative Raoult's law deviations. The constant, c, has been measured experimentally for a number of systems [12], and a simple determination of retention time in a gas chromatograph is sufficient to estimate it.

More recent extensions of the regular-solution theory model, given by Weimer and Prausnitz [21], and subsequently modified by Helpinstill and Van Winkle [10], apply to significant nonpolar-polar and polar-polar interactions. For polar-polar systems, the infinite-dilution activity coefficient, γ^∞, is:

$$\ln \gamma_2^\infty = \frac{v_2}{RT}[(\lambda_1 - \lambda_2)^2 + (\tau_1 - \tau_2)^2 - 2\psi_{1,2}] + \left[\ln\left(\frac{v_2}{v_1}\right) + \left(1 - \frac{v_2}{v_1}\right)\right] \tag{8}$$

where λ and τ are nonpolar and polar components of the solubility parameter δ. Plots of λ^2 versus molar volume for different reduced temperatures (homomorphs) are given for aliphatic, naphthenic and aromatic hydrocarbons. The polar solubility parameter, τ, is estimated from λ and Hildebrand's parameter, δ, by the following relationship:

$$\tau_i = (\delta_i^2 - \lambda_i^2)^{1/2} \tag{9}$$

The interaction parameter, $\psi_{1,2}$, accounts for the induction effect due to molecular polarity, and varies linearly with $v_2(\tau_2 - \tau_1)^2$ for saturated, unsaturated and aromatic compounds.

The most recent extension of the Helpinstill and Van Winkle correlation is that of Null [13]. The liquid activity coefficient at infinite dilution (terminal coefficient, γ^∞) is estimated by adding together various contributions to $\ln \gamma^\infty$.

$$\ln \gamma_i^\infty = (\ln \gamma_i^\infty)_a + (\ln \gamma_i^\infty)_b + (\ln \gamma_i^\infty)_c + (\ln \gamma_i^\infty)_d \tag{10}$$

These contributions have been broken down as: (a)

regular solution heat of mixing; (b) polar energy and induced dipole effect; (c) entropy effect attributed to an effective polymerization equilibrium; and (d) free energy of the broken polymer bonds as the associating component goes into solution.

The preceding equations are thus reduced to special cases of Null's equation. The main disadvantage of this correlation is its introduction of six additional parameters to characterize the pure components; these are given for nearly 100 commonly encountered compounds.

Group contribution methods

These methods assume that the physical properties of a fluid can be gleaned from the sum of contributions made by various functional groups (e.g., $-CH_3$; $-OH$; $-CH_2-$; $-CO-$; etc.) that comprise the molecule. This assumption drastically cuts experimental data requirements.

Pierrotti et al. [*16*] derived an expression for the terminal activity coefficient of a methylene homologue R_1-X_1 in the methylene homologue R_2-X_2:

$$\ln \gamma_1^\infty = A_{1,2} + B_2\frac{n_1}{n_2} + \frac{C_1}{n_1} + D(n_1 - n_2)^2 + \frac{F_2}{n_2} \quad (11)$$

where n_1, n_2 = number of carbon atoms in R_1 and R_2 radicals; $A_{1,2}$ = coefficient that depends on nature of functional groups, X_1 and X_2; B_2 = coefficient which depends only upon nature of functional group X_2; C_1 = coefficient which depends only on functional group X_1; D = coefficient independent of both X_1 and X_2; and F_2 = coefficient which essentially depends only on nature of X_2.

Constants in the above equation have been tabulated by the authors for individual groups. Although Eq. (11) was developed for straight-chain R_1 and R_2, estimates for components with moderate branching of Rs not in the immediate vicinity of the functional group X still yielded good results. The equation has been generalized to systems of polar components [*2*].

Based upon previous group-contribution work, two important correlation methods are now available.

The ASOG (analytical solution of groups) method is considered fairly accurate and has recently gained popularity in industry. The method predicts activity coefficients (at different compositions and temperatures) from the sums of interactions between groups in the mixture, adjusted for interactions between groups of the pure components. Each interaction is correlated using a two-parameter equation developed by Wilson [*22*], which yields a group activity coefficient.

Differences in molecular size that affect entropy are accounted for by counting "size" groups, which may be $-CH_2-$; $-CH_3$; $=CH-$; $-CO-$; $-O-$; or $-OH$. The size contribution is added to the interaction contribution to predict the activity coefficient.

The most recent variant of ASOG is the UNIFAC (universal quasichemical functional activity coefficient), method of Fredenslund, Jones and Prausnitz [*5*]. This correlation technique combines the basic concept of ASOG with the universal quasichemical theory, enabling it to predict two-phase phenomena, as well as the dependence of the activity coefficient on composition and temperature. The model features adjustments for both molecular size and shape (combinatorial) contributions, and energy interactions.

References

1. Abrams, D. S., and Prausnitz, J. M., *AIChE J.,* Vol. 21, no. 1, 1975, p. 116.
2. Deal, C. H., Derr, E. L., and Papadopoulos, M. N., *Ind. Eng. Chem. Fund.,* Vol. 1, no. 1, 1962, p. 17.
3. Derr, E. L., and Deal, C. H., *I. Chem. E. Symposium Series,* Vol. 32, no. 3, "International Symposium on Distillation, 1969, Brighton, England, Proceeding," Inst. of Chem. Engr., London, 1969, p. 40.
4. Ewell, R. H., Harrison, J. M., and Berg, L., *Ind. Eng. Chem.,* Vol. 36, 1944, p. 871.
5. Fredenslund, A., Jones R. L., and Prausnitz, J. M., *AIChE J.,* Vol. 21, no. 6, 1975, p. 1086.
6. Gilmont, R., Zudkevitch, D., and Othmer, D. F., *Ind. Eng. Chem.,* Vol. 53, 1961, p. 223.
7. Hála, E., Pick, J., Fried, V., and Vilím, O., "Vapour-Liquid Equilibrium," 2nd ed., Pergamon Press, 1967.
8. Hála, E., Wichterle, I., Polák, J., and Boublík, T., "Vapour-Liquid Equilibrium Data at Normal Pressures," 1st ed., Pergamon Press, 1968.
9. Heidemann, R. A., *AIChE J.,* Vol. 20, no. 5, 1974, p. 847.
10. Helpinstill, J. G., and Van Winkle, M., *Ind. Eng. Chem. Proc Des. and Develop.,* Vol. 7, 1968, p. 213.
11. Hildebrand, J. H., and Scott, R. B., "Solubility of Non-Electrolytes," Reinhold, New York, 1950.
12. Martire, D. E., *Anal. Chem.,* Vol. 33, 1961, p. 1143.
13. Null, H. R., "Phase Equilibrium in Process Design," Wiley, New York, 1970.
14. Orye, R. V., and Prausnitz, J. M., *Ind. Eng. Chem.,* Vol. 57, no. 5, 1965, p. 18.
15. Palmer, D. A., *Chem. Eng.,* June 9, 1975, p. 80.
16. Pierrotti, G. J., Deal, C. H., and Derr, E. L., *Ind. Eng. Chem.,* Vol. 51, no. 1, 1959, p. 95.
17. Prausnitz, J. M., Eckert, C. A., Orye, R. V., and O'Connell, J. P., "Computer Calculations for Multicomponent Vapor-Liquid Equilibria," Prentice-Hall, 1967.
18. Prausnitz, J. M., "Molecular Thermodynamics of Fluid-Phase Equilibria," Prentice-Hall, 1969.
19. Robinson, C. S., and Gilliland, R. E., "Elements of Fractional Distillation," 4th ed., McGraw-Hill, 1950.
20. Van Arkel, A. E., *Trans. Farad. Soc.,* Vol. 42, no. 2, 1946, p. 81.
21. Weimer, R. F., and Prausnitz, J. M., *Hydro. Proc.,* Vol. 44, 1965, p. 237.
22. Wilson, G. M. and Deal, C. H., *Ind. Eng. Chem. Fund.,* Vol. 1, no. 1, 1962, p. 20.
23. Wilson, G. M., *J. Am. Chem. Soc.,* Vol. 86, 1964, p. 127.
24. Wilson, G. M., "A Modified Redlich-Kwong Equation of State, Application to General Physical Data Calculations," 65th National AIChE Meeting, Cleveland, Ohio, 1969.

The author

A. K. Rao, a Senior Process Engineer for Catalytic, Inc., 1500 Market St., Phila., PA 19102, is involved in process design of chemical and petrochemical plants, and in development work related to coal conversion. A member of the ACS and The California Academy of Science, he holds an M.S. in chemical engineering from Syracuse University, and a B. Tech. in chemical engineering from Indian Institute of Technology in Bombay. He has done graduate work in industrial engineering and systems sciences.

Predicting Equilibrium Relationships for Maverick Mixtures

This method, used in the design of distillation and extraction columns, predicts activity coefficients for widely varying mixtures from a relatively small number of functional groups.

DAVID A. PALMER, Amoco Chemicals Corp.

The expanding use of polymers has presented chemical engineers with an ever-increasing list of component mixtures that are best separated by distillation, extractive distillation, and liquid-liquid extraction. However these mixtures often exhibit equilibrium relationships that vary considerably from those predicted according to ideal behavior.

A small sampling of nonideal distillations encountered in modern processing includes: extractive distillation of butadiene from butenes, using acetonitrile and water; distillation of maleic anhydride from *o*-xylene; and azeotropic distillation of formic acid from acetic acid, using benzene as an entrainer.

It has often been necessary to determine the vapor-liquid equilibrium of such mixtures experimentally. Such measurement of equilibrium data cannot usually be justified during screening of alternatives for a new chemical process, because of the high cost. However, the entire flowsheet is sometimes changed by such data, depending on the degree of nonideality of the mixtures.

While the cost of data measurement can sometimes be justified during process development, the necessary time delays are often not tolerable. Nevertheless, applying a large excess capacity but little consideration of the true phase equilibrium can lead to very costly errors [*4*] when designing distillation columns or other separation units.

Finally, optimization of the design of a distillation column is usually less costly when done on the computer with phase-equilibrium correlations than when performed in the pilot plant.

Consequently, reliable phase-equilibrium correlations and prediction methods are very valuable.

Nonidealities in the vapor phase can usually be predicted quite accurately, but prediction of liquid-phase nonidealities (represented by activity coefficients) is a much more difficult problem. Reliable correlations can be made for moderately nonideal mixtures of nonpolar liquids by means of the widely used Scatchard-Hildebrand activity coefficient equation [*5*]. However, the situation is quite different for mixtures containing polar liquids. The Scatchard-Hildebrand equation can lead to predictions that are worse than those made by assuming ideal liquid mixtures [*6*]. Correlations based on empirical extensions of regular solution theory to polar compounds have been developed by Weimer and Prausnitz [*7*], Helpinstill and Van Winkle [*8*], and Null and Palmer [*9*]. While these empirical correlations have been useful, they are limited in scope.

Consequently, the prediction of vapor-liquid equilibrium of nonideal mixtures has remained one of the major problems of modern process technology.

The Evolution of ASOG

We have concluded that the "Analytical Solutions of Groups" (ASOG) method is sufficiently accurate for exploratory and preliminary design calculations; and we have used this method for a variety of systems at Amoco Chemicals, including not only the design of distillation and extraction columns but also the preliminary design of liquid-phase reactors.

The ASOG approach was developed by Derr and Deal [*1*] from previous work on group-contribution theory by Wilson, Pierotti, Deal and Derr [*2,3*]. Although Derr and Deal have presented a large number of examples in support of their approach, the computation method was published in a relatively obscure place and was not ex-

Originally published June 9, 1975

plained with sufficient clarity to permit ready use. Therefore this paper.

The ASOG method correlates the interaction of functional groups—such as methylene ($—CH_2—$), hydroxyl ($—OH$), benzene ($C_6H_5—$), acid ($—COOH$), and anhydride ($—CO—O—CO—$). Any given combination of such groups will have two characteristic "interaction parameters," which are independent of composition. Accordingly, the interaction parameter for a combination of groups can be calculated (by regression analysis) and then applied to almost any other mixture containing the same combination. A temperature-dependence of the parameters can be included.

Activity coefficients are predicted as the sums of the interactions between groups in the mixture, less the interactions between groups of the pure components. Each interaction between groups is correlated as if it were a mixture of two pure components, using a two-parameter equation developed by G. M. Wilson.

To this is added the effect of differences in molecular size that affect the entropy. The relative molecular sizes are obtained by counting "size groups," which are the hydroxyl group and the individual carbon-containing groups of the organic molecule. These may differ in number from the interaction groups; for example, the anhydride interaction group contains two size groups. The size contribution is added to the interaction contribution to predict the activity coefficient.

A major advantage of the ASOG equations is that they can be used to predict the effect of composition on the activity coefficients. Most other approaches provide activity coefficients at infinite dilution only, and the constants for equations that depend on composition must be generated in a separate step.

Calculation Method

The following nine steps outline the method for applying the working equation. The first three steps give the contribution of the size difference, the next five the contribution of the interaction groups, and the ninth step the sum of the two contributions.

Size-difference contribution:

1. Identify the number of size groups, as k_1, k_2, etc., that are contained in the molecule of each component in the mixture. These size groups may be $—CH_2—$, $—CH_3$, $=CH—$, $—CO—$, $—O—$ and $—OH$ (see Table II).

2. Using the mole fraction of the components in the mixture, along with the number of size groups for each component, calculate the size term, R_i, as:

$$R_i = \frac{S_i}{\sum_j S_j x_j} \tag{1}$$

where: S_i = number of size groups in molecule i.
S_j = number of size groups in molecule j.
x_j = mole fraction of molecule j in the mixture.

3. Using the size term in step 2, calculate to the activity coefficient, $\gamma_i{}^S$, for each molecule, as:

$$\ln \gamma_i{}^S = 1 - R_i + \ln R_i \tag{2}$$

Group interaction contribution:

4. Calculate the mole fraction, X_k, of each interaction group from the mole fraction of the components, x_i, and

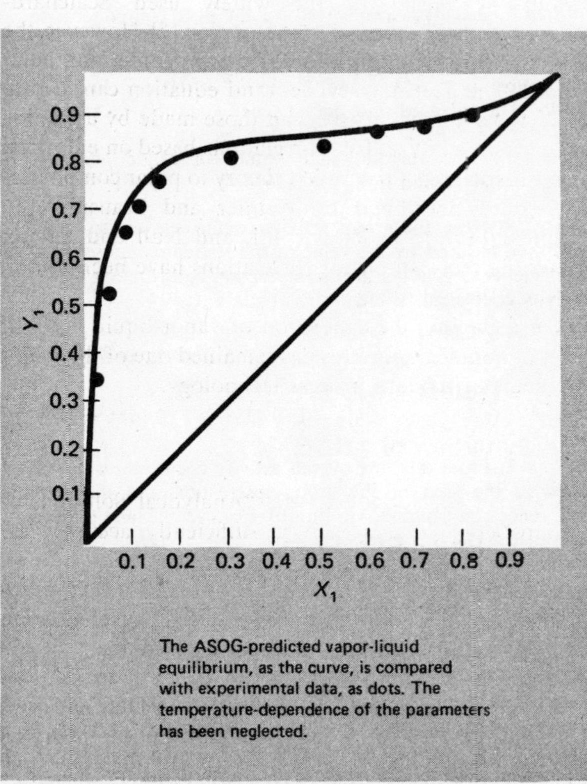

The ASOG-predicted vapor-liquid equilibrium, as the curve, is compared with experimental data, as dots. The temperature-dependence of the parameters has been neglected.

ASOG PREDICTION of vapor-liquid equilibrium for ethanol (1) and ethylbenzene (2) at 1 atm—Fig. 1

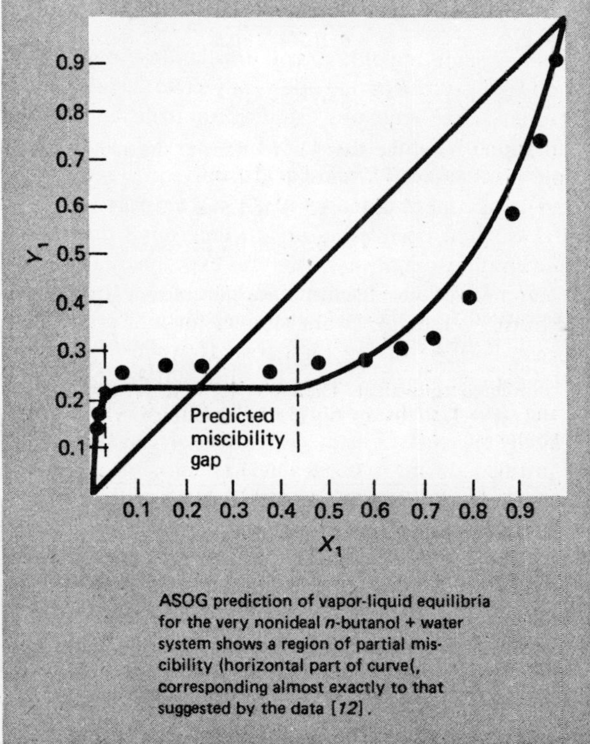

ASOG prediction of vapor-liquid equilibria for the very nonideal *n*-butanol + water system shows a region of partial miscibility (horizontal part of curve(, corresponding almost exactly to that suggested by the data [*12*].

ASOG PREDICTION of vapor-liquid-liquid equilibrium for butanol (1) and water (2) at 1 atm—Fig. 2

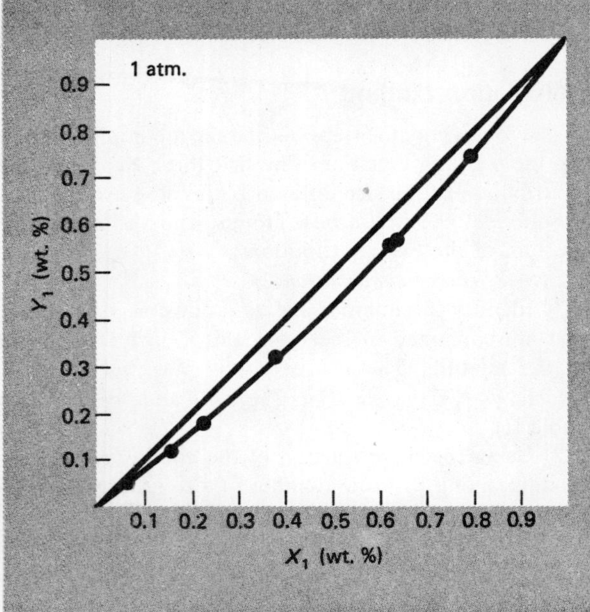

COMPARING ASOG correlation with data for the system water (1) + phenol (2) at 1 atm—Fig. 3

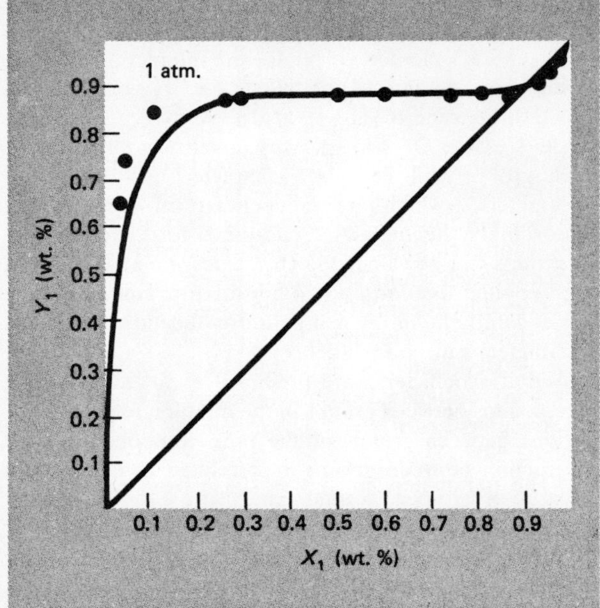

ASOG PREDICTION of vapor-liquid equilibrium for o-cresol (1) and phenol (2) at 1 atm—Fig. 4

ASOG in Action

What are the activity coefficients for a 50/50 mole % mixture of cyclohexane and acetic acid?

Step 1: Cyclohexane has six size groups of $—CH_2—$, while acetic acid has two size groups, one —COOH and one $—CH_3$.

Step 2: At $x_1 = 0.50$, the value of R_i from Eq. (1) is:

$$R_1 = \frac{6}{6 \cdot 0.5 + 2 \cdot 0.5} = 1.5$$

$$R_2 = \frac{2}{6 \cdot 0.5 + 2 \cdot 0.5} = 0.5$$

Step 3: Now, the size contributions to the activity coefficients are calculated from Eq. (2), as:

$$\ln \gamma_1{}^S = 1 - 1.5 + \ln 1.5 = -0.0945$$

$$\ln \gamma_2{}^S = 1 - 0.5 + \ln 0.5 = -0.1931$$

Step 4: The mole fraction of each interaction group is now calculated from the molecular composition. The $—CH_2—$ group is designated arbitrarily as group 1, and the —COOH group as group 2. The $—CH_3$ and $—CH_2—$ groups are considered equivalent. Therefore, $\nu_{11} = 6$, $\nu_{12} = 1$, $\nu_{21} = 0$ and $\nu_{22} = 1$, so that by Eq. (3) the mole fraction of $—CH_2—$ groups is:

$$X_1 = \frac{0.5 \cdot 6 + 0.5 \cdot 1}{0.5 \cdot 6 + 0.5 \cdot 1 + 0.5 \cdot 0 + 0.5 \cdot 1} = 0.875$$

And the mole fraction of —COOH groups is:

$$X_2 = \frac{0.5 \cdot 0 + 0.5 \cdot 1}{0.5 \cdot 6 + 0.5 \cdot 1 + 0.5 \cdot 0 + 0.5 \cdot 1} = 0.125$$

Step 5: Now that the group mole fractions have been calculated, the interaction parameters for the $—CH_2/—COOH$ interaction are taken from Table I as:

$$A_{11} = 1.0 \qquad A_{12} = 1.1333$$

$$A_{12} = 0.0340 \qquad A_{22} = 1.0$$

Step 6: From these, the Wilson equation, Eq. (4), is used to calculate the group activity coefficients, Γ_i, as:

$$\Gamma_1 = 0.0954$$

$$\Gamma_2 = 1.0828$$

Step 7: Eq. (4) is also solved for the pure components to give $\ln \Gamma_k{}^*$, the standard-state group activity coefficients. The standard-state activity coefficients for pure cyclohexane are the values at $x_1 = 1.0$, and are $\ln \Gamma_1{}^* = 0.0$ and $\ln \Gamma_2{}^* = 3.248$. (The standard-state activity coefficient for $—CH_2—$ in cyclohexane had to be 0, because in pure cyclohexane, $X_1 = 1.0$.) The standard-state values in acetic acid are $\ln \Gamma_1{}^* = 0.4338$ and $\ln \Gamma_2{}^* = 0.1614$.

Step 8: Next, the group activity coefficients are summed and then reduced by the value of the standard-state activity coefficients to give the group interaction contribution to the activity coefficient, $\gamma_i{}^G$, from Eq. (5) as:

$$\ln \gamma_1{}^G = 6 \cdot 0.0954 + 0 \cdot 1.0828 - 6 \cdot 0.0 - 0 \cdot 3.248 = 0.5724$$

$$\ln \gamma_2{}^G = 1 \cdot 0.0954 + 1 \cdot 1.0828 - 1 \cdot 0.4338 - 1 \cdot 0.1614 = 0.583$$

Step 9: From this, the overall activity coefficient is calculated as the sum of the group interaction and the size-difference contribution, via Eq. (6). For cyclohexane (1) and acetic acid (2), this gives:

$$\gamma_1 = \exp(0.5724 - 0.0945) = 1.61$$

$$\gamma_2 = \exp(0.583 - 0.1931) = 1.48$$

These activity coefficients are used, together with vapor pressures and vapor-nonideality corrections, to calculate the vapor-liquid equilibrium. The equations can also be used in iterative form to predict multicomponent liquid-liquid equilibria. These methods for converting activity coefficients to equilibrium values are given by Null [*10*], Prausnitz [*6*], and other standard references.

the number of each type of interaction functional group, ν_{ki}, as:

$$X_k = \frac{\sum_i x_i \nu_{ki}}{\sum_k \sum_i x_i \nu_{ki}} \quad (3)$$

where ν_{ki} = number of interaction functional groups, k, in molecule i.

5. Using Table I, determine the interaction parameter, $A_{k,l}$, for each function group in the mixture. The parameter of each group's interaction with all the others is obtained by reading from left to right, and then up. For example, in a mixture containing $-CH_2/{=}CH-$, with CH_2- designated as component 1 and $=CH-$ (benzene carbon) designated as component 2, the interaction parameters are $A_{11} = 1.0$, $A_{12} = 0.734$, $A_{21} = 1.24$ and $A_{22} = 1.0$.

The parameters in Table I have been calculated by nonlinear regression analysis from experimental data. Those bracketed by the dotted line in the upper left-hand corner of the table were given by Derr and Deal [1], while the others have been calculated as part of this study. Derr and Deal have reported many other parameters, some of which are presented in the reference [1].

6. Calculate the group activity coefficients, Γ_i, from the Wilson equation as:

$$\ln \Gamma_k = -\ln \sum_l X_l A_{kl} + \left[1 - \sum_l \frac{X_l A_{lk}}{\sum_m X_m A_{lm}}\right] \quad (4)$$

7. Using the Wilson equation, also calculate the standard-state group activity coefficients, Γ_k^*, by repeating step 6 for a 100% solution of each component in the mixture. For components such as water, hexane or benzene, which contain only one kind of interaction group, $\ln \Gamma_k^*$ is zero. For a molecule such as ethanol, $\ln \Gamma_k^*$ will have a finite value for both hydroxyl and methyl groups, which make up the molecule.

8. Add the group activity coefficients, and then subtract the value of the standard-state group activity coefficients, to obtain the group-interaction contribution to the activity coefficient, γ_i^G, as:

$$\ln \gamma_i^G = \sum_k \nu_{ki} \ln \Gamma_k - \sum_k \nu_{ki} \ln \Gamma_k^* \quad (5)$$

9. Finally, add the overall group interaction coefficient, γ_i^G, and the contribution of the size term, γ_i^S, from step 3, to get the overall activity coefficient γ_i, as:

$$\ln \gamma_i = \ln \gamma_i^G + \ln \gamma_i^S \quad (6)$$

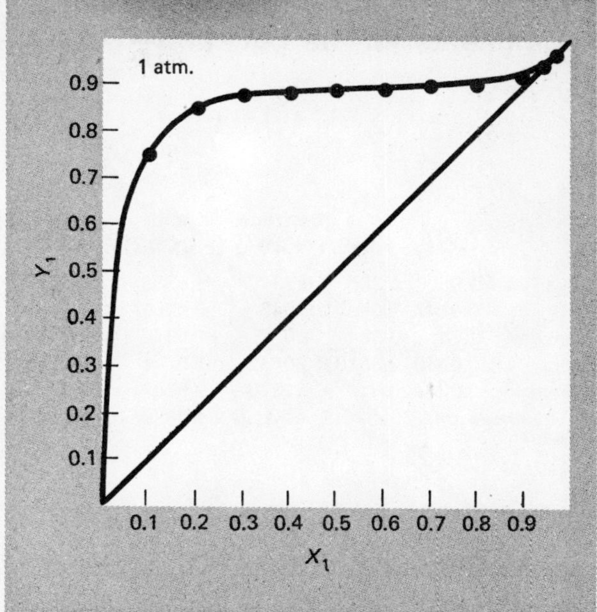

COMPARING ASOG correlation with data for the system cyclohexane (1) + acetic anhydride (2) at 1 atm—Fig. 5

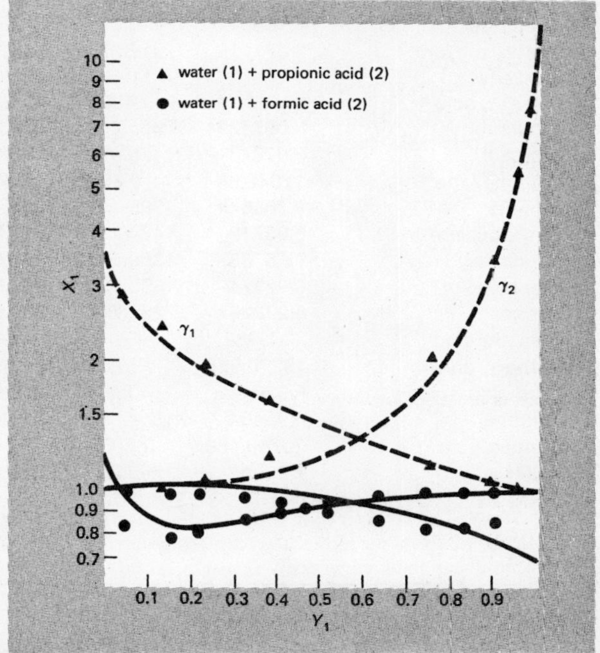

ASOG ACTIVITY COEFFICIENTS for propionic acid (2) + water (1) and formic acid (2) + water (1)—Fig. 6

Testing ASOG

Using interaction parameters calculated by Derr and Deal [1], we have applied ASOG to four mixtures containing hydroxyl compounds: ethanol plus ethylbenzene (Fig. 1), 1-butanol plus water (Fig. 2), phenol plus water (Fig. 3), and phenol plus *o*-cresol (Fig. 4). The accuracy of the results in these polar systems encouraged us to calculate interaction parameters for carboxylic acid and anhydride groups (Table I).

The mixtures from which the carboxylic acid and anhydride interaction parameters were calculated are given in Table III. Activity coefficients for the mixtures containing acetic acid were calculated (or recalculated) to allow for dimerization and trimerization of acetic acid vapor. Activity coefficients for the other two mixtures were used as published. Interaction parameters for carboxylic acid and anhydride with other groups were then calculated by nonlinear regression analysis. The correla-

Constants for the Equation—Table I

$$\ln \Gamma_k = -\ln \sum_l X_l A_{kl} + \left[1 - \sum_l \frac{X_l A_{lk}}{\sum_m X_m A_{lm}}\right]$$

	—CH_2	—OH	Benzene (=CH—)	Acid (—COOH)	Anhydride (—COO—)
CH_2	1.0	0.305	0.734	1.1333	0.7703
OH	0.0147	1.0	0.045	0.1960	
Benzene	1.24	0.534	1.0	1.1372	0.9841
Acid	0.0340	2.3518	0.0442	1.0	0.6205
Anhydride	0.0249		0.1876	0.5369	1.0

Example: Where CH_2 = 1, OH = 2, BENZENE = 3, ACID = 4, and ANHYDRIDE = 5

$A_{2,4} = 0.1960$
$A_{4,1} = 0.0340$

tion of data for one of these systems, cyclohexane plus acetic anhydride, is illustrated in Fig. 5.

The parameters obtained from these data regressions were then used to predict the vapor-liquid equilibrium for other systems, using no data other than the pure component vapor pressures. As a first example, the activity coefficients were predicted for water plus propionic acid and water plus formic acid (Fig. 6).

As shown in the figure, the predictions match the data [*19*] quite well. A perfect fit for the water-plus-formic acid system would be impossible, because the data of this system are not consistent with respect to the Gibbs-Duhe Equation.

Graphically presented data are available for aromatic systems containing acid and anhydride groups [*20*], including mixtures of *o*-toluic acid plus phthalic anhydride and banzoic acid plus phthalic anhydride. These data are compared with ASOG predictions for mixtures of ben-

Pure Component Parameters—Table II

	Vapor-Pressure Constants				Interaction Groups				
	A	B	C	Size Groups	CH_2	OH	Benz	Acid	Anhydride
Acetic acid	7.18807	1416.7	211.	2	1			1	
Acetic anhydride	8.071176	2012.3234	248.6623	4	2				1
Benzene*	11578.36	−11.4833	50.3957	6			6		
Benzoic acid*	19426.496	−8.4536	45.336	7			6	1	
n-butanol	9.067429	2295.0611	254.2902	5	4	1			
o-cresol	6.97943	1479.400	170.00	8	1	1	6		
Cyclohexane*	11042.68	−10.7089	47.3076	6	6				
Ethanol	8.708844	1935.1248	253.426	3	2	1			
Ethylbenzene	6.95719	1424.2550	213.206	8	2		6		
Formic acid	7.65788	1750.5992	265.5058	2				1	
n-octane	6.92374	1355.126	209.517	8	8				
Phenol	8.232267	2251.7779	239.831	7		1	6		
Phthalic acid	—	—	—	8			6	2	
Phthalic anhydride*	19270.926	−10.7695	51.6957	8			6		1
Propionic acid	8.320101	2140.4889	252.8768	3	2			1	
Toluene	6.95464	1344.800	219.482	7	1		6		
o-toluic acid*	20789.480	−9.2081	48.8644	8	1		6	1	
Water	7.96681	1668.21	228.0	1		1.4			

* These are constants for the equation

$$\ln P(\text{atm}) = \frac{-A}{RT} + \frac{B}{R} \ln T + C$$

All other constants are for the Antoine Equation

$$\log P(\text{mm Hg}) = A + \frac{B}{C + t(°C)}$$

Sources of Data for Calculation of Interaction Parameters for Carboxylic Acid or Anhydride—Table III

Group Interaction	Mixture	Reference
Carboxylic acid – methylene	Cyclohexane + acetic acid	14
Carboxylic acid – hydroxyl	Acetic acid + water	15
Carboxylic acid – benzene	Toluene + acetic acid	16
Carboxylic anhydride – methylene	Cyclohexane + acetic anhydride	17
Carboxylic anhydride – benzene	Acetic acid + benzene	18
Carboxylic acid – carboxylic anhydride	Acetic acid + acetic anhydride	17

Note: No data could be found that would give hydroxyl + anhydride interactions.

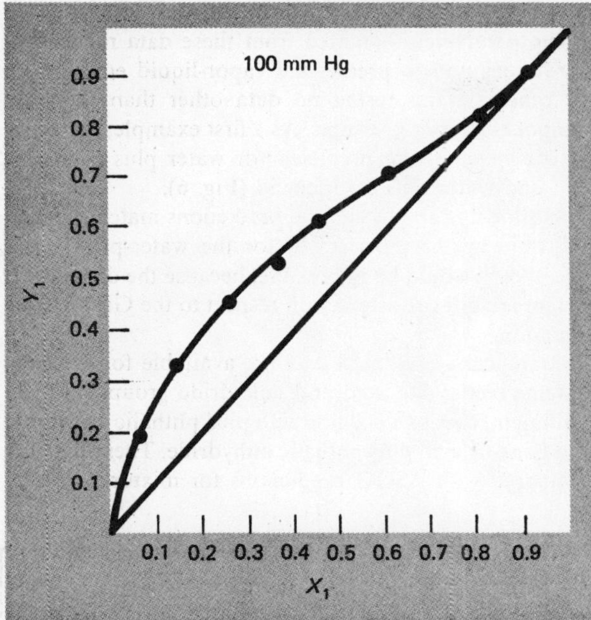

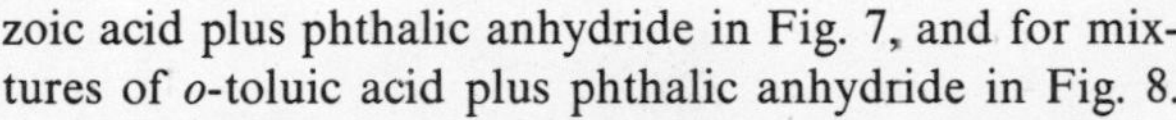
ASOG PREDICTION of vapor-liquid equilibrium for benzoic acid (1) + phthalic anhydride (2) at 100 mm Hg—Fig. 7

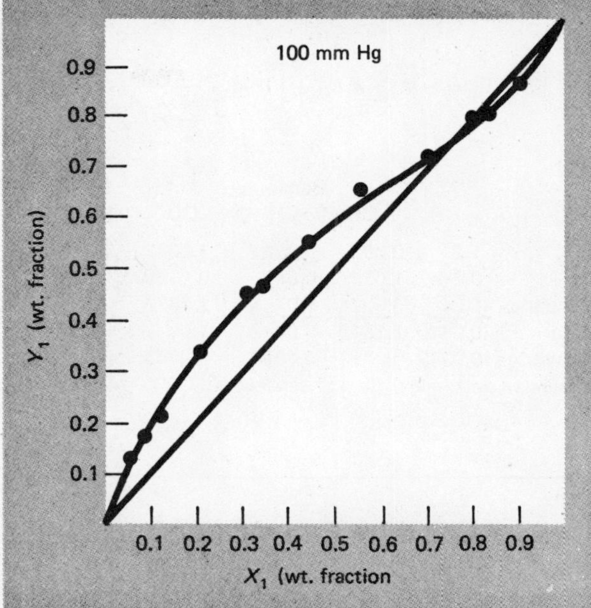

ASOG PREDICTION of vapor-liquid equilibrium for o-toluic acid (1) and phthalic anhydride (2) at 100 mm Hg—Fig. 8

zoic acid plus phthalic anhydride in Fig. 7, and for mixtures of *o*-toluic acid plus phthalic anhydride in Fig. 8.

Although the computations represent a sizable excursion beyond the temperature range of the data base, and although no aromatic acids or anhydrides were used in developing the correlation parameters, the results are very good.

Summary

The numbers of interaction groups and size groups for each molecule of the foregoing studies are given in Table II. As indicated in the table, the procedure, suggested by Derr and Deal, for counting water as equal to 1 size group but 1.4 interaction groups was followed. The acid group was counted as one size group and one interaction group, while the anhydride group was counted as two size groups and one interaction group—as indicated, for example, for acetic anhydride.

Table II also presents the vapor-pressure-equation constants used in this study in conjunction with activity coefficients for calculating the vapor-liquid equilibrium. Vapor-phase nonidealities were neglected except for acid systems having associating vapors, where dimerization was allowed for.

The ASOG capabilities here presented can be further expanded by additional data representing different sets of interaction groups. These data should be thermodynamically consistent and available at a variety of temperatures. Ideally, the data should be assembled by a single, non-commercial individual or group that could act as a reliable source to all users. As the data are compiled, evaluated and presented through the published literature, they will substantially improve the capabilities of the Derr and Deal equations, or provide a foundation for modification of the approach.

References

1. Derr, E. L. and Deal, C. H., *I. Chem. E. Symposium Series* **32,** *3:40* (1969), in "International Symposium on Distillation 1969, Brighton England Proceedings," Inst. of Chem. Engr., London.
2. Pierotti, G. J., Deal, C. H., and Derr, E. L., *Ind. Eng. Chem.*, **51,** 95 (1959).
3. Wilson, G. M. and Deal, C. H., *Ind. Eng. Chem. Fund.*, **1,** 20 (1962).
4. Martin, H. W., *Chem. Eng. Prog.* **60(10),** 50 (1964).
5. Hildebrand, J. H., and Scott, R. L., "The Solubility of Non-Electrolytes," 3rd ed., Reinhold, New York (1953).
6. Prausnitz, J. M., "Molecular Thermodynamics of Fluid Phase Equilibria," Prentice-Hall, New Jersey (1969).
7. Weimer, R. F. and Prausnitz, J. M., *Hydro. Proc.*, **44,** 237 (1955).
8. Helpinstill, J. G., and Van Winkle, M., *Ind. Eng. Chem. Proc. Des. and Develop.*, **7,** 213 (1968).
9. Null, H. R. and Palmer, D. A., *Chem. Eng. Prog.*, **65(9),** 47 (1969).
10. Null, H. R., "Phase Equilibrium in Process Design," Wiley, New York (1970).
11. Ellis, S. R. M. and Spurr, M. J., *Brit. Chem. Eng.*, **92** (Feb., 1961).
12. Ellis, S. R. M. and Garbett, R. D., *Ind. Eng. Chem.* **52,** 385 (1960).
13. Rhodes, F. H., Wells, J. H., and Murray, G. W., *Ind. Eng. Chem.*, **17,** 1199 (1925).
14. Baradarajan, A. and Satyanarayana, M., *Indian J. Tech.*, **5,** 264 (1967).
15. Ito, T. and Yoshida, F., *J. Chem. Eng. Data,* **8,** 315 (1963).
16. Haughton, C. O., *Chem. Eng. Sci.*, **16,** 82 (1961).
17. Jones, H. E., *J. Chem. Eng. Data,* **7,** 13 (1962).
18. Molochnikov, M. L., Markova, L. P., and Kogan, V. B., *Zhur-Prikl. Khim.*, **40(9),** 2083 (1967).
19. Ito, T. and Yoshida, F., *J. Chem. Eng. Data,* **8,** 315 (1963).
20. Suter, H., "Phthalsaureanhydrid und seine Verwendung," BASF: Ludwigshafen, Steinkopff Verlag, Darmstadt (1972).

Meet The Author

David A. Palmer is a research engineer at the Amoco Research Center, Amoco Chemicals Corp., P.O. Box 400 Naperville, IL 60540, where his responsibilities include application of engineering fundamentals to the development and design of new chemical processes. He holds a B.E.S. in chemical engineering from Brigham Young University and a D.Sc. in chemical engineering from Washington University, where he did research on vapor-liquid equilibrium in the Thermodynamics Research Laboratory. His hobby is ancient Mesoamerican history and early migrations to America.

Section 3 Design

A. Overview
Shortcuts for distillation design
Energy-saving schemes in distillation
Layout arrangements for distillation columns
Installed cost of a distillation column
Minimizing distillation costs via graphical techniques

B. Specific equipment
How to design overhead condensing systems
How to design reflux drums
More on design of reflux drums
Demonstrated data for tower design

C. Computer design
A fast computer method for distillation calculations
Computer design of sieve trays and tray columns
Heavy-oil distillation via computer simulation

D. Calculator programs
Streamline flash computations with calculator program
Calculator program for sour-water-stripper design
Determining ideal stages on a pocket calculator

E. Other calculation methods
Estimating recoveries in multicomponent distillation
Calculating actual plates in absorbers and strippers
Optimum-feed-stage location in multicomponent distillations

A. Overview

Shortcuts for Distillation Design

For both batch and continuous distillations, the author offers shortcut techniques, as well as practical advice, for fixing the number of equilibrium stages, setting the operating conditions, sizing the column, and choosing the internals.

Otto Frank, *Allied Chemical*

☐ Design of a distillation column progresses through four major steps:

- Evaluation and regression of vapor-liquid equilibrium (VLE) data.
- Calculation of the number of equilibrium stages.
- Determination of tray hydraulics.
- Selection of tray or packing efficiencies.

Of these, only the second lends itself to a strictly theoretical approach. In contrast, thermodynamically consistent equilibrium data are frequently difficult to come by, and their application to multicomponent systems can yield results of indifferent accuracy. Efficiency correlations disclosed in the open literature are still of a very rudimentary nature, with little advancement recorded over the past 20 years. More often than not, tray or column efficiencies reflect the uncertainties of the equilibrium data rather than the actual system characteristics or column configuration.

Column hydraulics have been extensively investigated. Reasonably good empirical data are available to describe the performance of the more popular types of trays and packing. These have been redefined in terms of acceptable operating ranges that are often no less accurate than the published correlations.

Presented below are selection guidelines and shortcut design procedures that will simplify the evaluation of many standard distillation columns. The author has earmarked cases where it is advisable not to cut corners, and where the development of a detailed, computerized solution is justified.

Where to use shortcut procedures

The use of *shortcut,* hand calculation procedures should be considered for several purposes:

1. Scoping studies suitable for preliminary costs.
2. Parametric evaluation of operating variables.
3. Separations having coarse purity requirements (i.e., contaminants >0.5 wt%).
4. Detailed designs for ideal and close-to-ideal systems.
5. Designs for systems for which equilibrium data are unavailable.

On the other hand, *rigorous* design procedures should be applied if the following are true:

1. The separation is a multicomponent one, requiring high product purity.
2. The system is highly nonideal but good equilibrium data are available.
3. The relative volatility between key components is less than 1.3.
4. One or more of the components is near the critical pressure.

A number of rules-of-thumb are helpful in judging whether or not mixture characteristics favor the use of simple VLE relationships to design the column. These rules are listed in Table I according to the ideality of the mixture to which they refer.

Selecting operating parameters

Before any detailed distillation column design or evaluation can be undertaken, it is essential to define a number of operating parameters:

Feed composition—When the component ratio is large (>5 : 1), variation in the feed composition can noticeably influence reflux ratio and number of trays. It is less significant when the concentrations of key components in the feed are of about equal magnitude.

Product purity—The specified concentration levels of high-boiling components in the distillate and low-boiling ones in the bottoms are the sole criteria for establishing the number of trays and reflux ratio. The finer the split, the greater will be the number of stages or the required reflux ratio.

Feed equilibrium—Whenever possible, the equilibrium relationship among system components should be established experimentally. The blind assumption of an ideal system is certainly not valid. In nonideal systems, deviation of activity coefficient from ideality is mostly positive. The relative volatility between components is thus curtailed at the top of the column, making it harder to obtain a pure overhead than assumption of an ideal system would have predicted.

Thermal state of feed and reflux—Feed quality can have a noticeable effect on tray requirements. If the feed is subcooled below its bubble point, the number of trays in the rectifying section will decrease, whereas those in the stripping section will increase. More heat will be required in the reboiler and less cooling in the condenser. The reverse is true for a feed containing vapor.

Subcooled reflux will increase the molar ratio of liquid and vapor flows, and thereby increase internal reflux. The top tray will act as a partial condenser which, at the expense of efficiency, condenses vapor to reheat the external reflux to its equilibrium temperature. Usually there is no justification to subcool reflux before returning it to the column.

Column pressure—Raising the column operating pressure will increase reboiler and condenser temperatures.

Originally published March 14, 1977

Rules-of-thumb regarding equilibrium properties of vapor-liquid mixtures **Table I**

(Declining ideality ↓)

Mixtures of isomers usually form ideal solutions.

Mixtures of close-boiling aliphatic hydrocarbons are nearly ideal below a pressure of 10 atm.

Mixtures of compounds close in molecular weight and structure frequently do not deviate greatly from ideality (e.g. ring compounds, unsaturated compounds, naphthenes, etc.).

Mixtures of simple aliphatics with aromatic compounds deviate modestly from ideality.

"Inerts" such as CO_2, H_2S, H_2, N_2, etc., that are present in mixtures of heavier components tend to behave nonideally with respect to the other compounds.

Mixtures of polar and nonpolar compounds are always strongly nonideal. (Look for polarity in molecules containing oxygen, chlorine, fluorine or nitrogen, in which electrons in bonds between these atoms and hydrogen are not equally shared).

Azeotropes and phase separation represent the ultimate in nonideality, and their occurrence should always be confirmed before detailed distillation studies are undertaken.

On the other hand, this change will also decrease vapor velocity, since the increase in vapor density more than offsets a corresponding temperature effect on volume.

Often, relative volatility improves at lower pressures, making separations easier. In vacuum service, this effect may be quite pronounced.

Sometimes the operating pressure may be dictated by separation requirements, making it necessary to operate outside the region of an azeotrope or below the critical pressure of a component.

Column temperature—Column operating conditions are frequently selected to allow available plant utilities to be used as a heat source and heat sink. It is also important to select a reboiler temperature that is not so high that it will degrade the products. Given here are guidelines regarding the approach temperatures of the heat sink and the heat source for different reboiler and condenser services:

Heat-sink approach, °C	
Refrigeration	3–10
Cooling water	6–20
Pressurized fluid	10–20
Boiling water	20–40
Air	20–50
Heat-source approach, °C	
Process fluid	10–20
Steam	10–60
Hot oil	20–60

Energy utilization

The customary practice of adding heat into the bottom of a column and then abstracting an almost equal quantity from the condenser at the top makes distillation one of the highest energy consumers in a chemical or petrochemical plant. Energy conservation measures should be considered during the early stages of column design and layout. These measures include:

- Recuperative heat exchange between a cold feed and a hot bottoms stream. (The reboiler effluent must be a large fraction of the feed.)
- Columns cascaded so that vapor from one is condensed in the reboiler of another.
- Insulation. An economic evaluation may justify insulation even if process temperatures are as low as 50–60°C.
- Generation of low-pressure steam in condensers.
- Reduction of refrigeration levels in low-temperature separations by operating at the highest possible pressure.
- Recompression of overhead vapor to raise its energy to a level where it can be used as a heat source for the reboiler. Worthwhile when the temperature drop across a column is small and a single compression stage is sufficient to raise the vapor temperature above that of the reboiler.

Shortcut stage calculations

Two or more components in a liquid system can be separated by taking advantage of their different boiling points. In a boiling mixture, the coexisting vapor and liquid phases have different compositions; lower-boiling constituents predominate in the vapor, whereas higher-boiling ones concentrate in the liquid.

The maximum concentration difference between the two phases is reached at equilibrium—assumed for an ideal stage. Since it is possible to calculate rationally the number of successive equilibrium stages required for a specified separation, it has become the practice to establish first what degree of separation is theoretically possible and then to estimate how closely commercial equipment will approach this goal.

Graphical design procedures, as well as most shortcut algebraic correlations [*5, 10, 13, 14*], usually deal with separations of binary mixtures. However, since few true binary systems are encountered in industry, calculation procedures are usually applied to pseudo-binary mixtures in which two principal components are designated as the light key and the heavy key.

When it is necessary to produce high-purity product streams, compounds boiling adjacent to each other on the temperature scale become the key components. The light key is the lowest boiling component present in the bottoms stream, whereas the heavy key is the highest boiling component present in the overhead.

When a rough separation will suffice, the keys are selected to yield a specified product mix. It is then probable that the light and heavy keys do not fall adjacent to each other, but have an intermediate boiling component between them—usually referred to as a distributed key.

Mixed-products are only occasionally specified in the chemical processing industries; consequently, the design procedures outlined here will deal only with adjacent keys.

Graphical stage calculations

The simplest and most direct approach for analyzing binary distillations is still the graphical technique devised by McCabe and Thiele in 1925. Despite its age, the *x-y* diagram remains a highly useful tool for quick evaluation of a column (Fig. 1). The influence of reflux ratios, product concentrations, and feed conditions can be readily spotted by adjusting the slope and point of origin of the operating and q lines.

The thermal condition of the feed is defined by the slope of the q line:

$$q = \frac{\text{Heat to convert 1 mole of feed to saturated vapor}}{\text{Molar heat of vaporization}} \quad (1)$$

Values of q that characterize a range of feed qualities are given here:

Feed quality	q	q-line slope
Saturated liquid	1	infinite
Saturated vapor	0	zero
Cold liquid feed	>1	positive
Superheated vapor	<0	positive
Two phase	1.0 to 0	negative

The VLE relationship expressed by $y = \alpha x/(1 + (\alpha - 1)x)$ assumes a constant relative volatility (α) across the entire pressure range found in the column. If variations are small, an algebraic or geometric average may suffice. However, if the relative volatilities at the top and the bottom are more than 15% apart, the designer may resort to constructing the equilibrium curve incrementally by calculating the relative volatility at several points along the column.

The McCabe-Thiele diagram assumes a constant molal overflow—the two key components must have identical molal heats of vaporization; and all other heat effects, such as heats of mixing and sensible heat losses, are zero. Thus, if no material or energy is added or withdrawn, the moles of liquid overflowing from stage to stage must be constant, as must the moles of vapor rising from reboiler to condenser.

In only a few systems will the molal heats of vaporization vary by more than 10%. Hence, for most cases, the assumption of constant molal overflow is quite reasonable and the McCabe-Thiele procedure can be applied whenever shortcut solutions are acceptable.

The McCabe-Thiele diagram can accommodate other than ideal, 100% efficient, single-feed separations. It can be adapted to handle systems in which molal overflow is not constant, as well as Murphree tray efficiencies, multiple feed and withdrawal points, subcooled reflux, partial condensation, and open steam addition. These variants are well covered in standard texts [*5*, *10*, *11*, *13*, *14*], and therefore will not be discussed here. It must be noted, however, that modifications of the basic graphical procedure can be quite time-consuming and are usually not justified. It is best suited for quick parametric studies and design estimates; if more-detailed evaluations are desired, the engineer should opt for a computer solution.

Because of its inherent limitations, the McCabe-Thiele diagram should not be used when more than 25 theoretical trays are required for the specified separation; when the relative volatility is less than 1.3, and even a small deviation from ideality can appreciably affect the separation; when the relative volatility is greater than 5 and most systems are highly nonideal; when the reflux ratio is exceptionally small (<1.1 times minimum); and when there are large differences among the molal heats of vaporization. A more rigorous graphical solution is the Ponchon-Savarit method, which incorporates an enthalpy balance. It is appropriate when component latent heats per mole differ drastically, or when the temperature spread between the top and bottom of a column is large.

To use this method, one must construct an enthalpy vs. concentration diagram for the two-phase, vapor-liquid region pertinent to the column temperature profile.

Since adequate enthalpy diagrams are seldom available in the literature, it is generally necessary to construct them for each system under consideration. With the availability of comprehensive computer programs there is little incentive to spend time constructing a graph which, at best, can only represent a binary distillation having a limited number of trays.

The Ponchon-Savarit technique has found use pri-

Nomenclature

A_c	Cross-sectional area of column, ft^2
A_h	Open hole area, ft^2
C_0	Orifice coefficient
d	Packed bed dia., ft
D	Distillate flowrate, moles/h
G	Vapor flowrate, lb/(s)(ft^2)
h_h	Dry pressure drop across perforations, in. liquid
h_l	Aerated liquid head on tray, in. liquid
h_{ow}	Head of liquid over weir, in. liquid
h_{ss}	Static seal, in. liquid (bubble cap)
h_t	Tray pressure drop, in. liquid
h_w	Liquid head equivalent to weir height, in. liquid
k	Pressure loss factor (downcomerless tray)
L	Reflux to top of column, moles/h
L_1; L_2	Initial and final liquid in reboiler (batch), moles
L_w	Weir length, in.
N_{min}	Theoretical trays at total reflux
ΔP	Column pressure drop, ft H_2O
q	Thermal condition of feed
Q	Liquid flowrate onto tray, gpm
R_{min}	Minimum reflux ratio (L/D)
U_h	Hole velocity, ft/s
v	Superficial vapor velocity, ft/s
W	Vapor rate, lb/s
x	Liquid-phase mole fraction
y	Vapor-phase mole fraction
α	Relative volatility
β	Aeration factor
$\Delta/2$	Average liquid gradient on tray
ρ	Density, lb/ft^3
τ	Pressure-drop loss term (downcomerless trays)
θ	Underwood constant

Subscripts

B	Bottoms stream
D	Distillate stream
F	Feed stream
HK	Heavy key
LK	Light Key

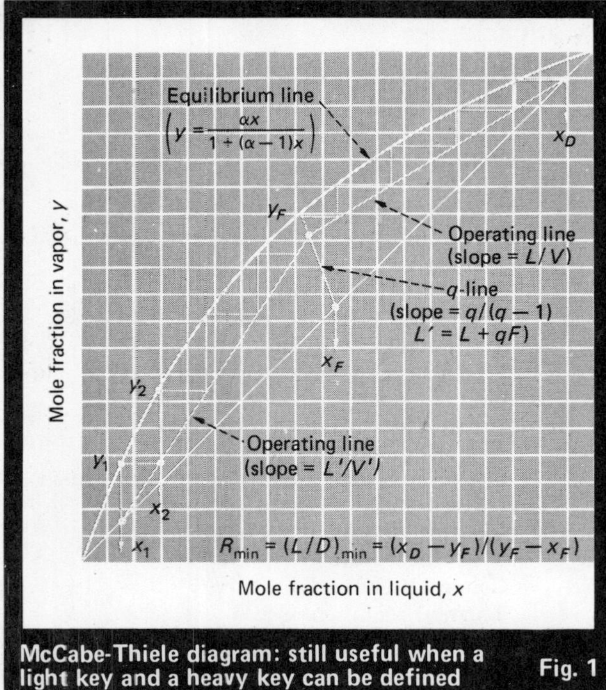

McCabe-Thiele diagram: still useful when a light key and a heavy key can be defined **Fig. 1**

marily for analyzing mixtures that exhibit a large heat of mixing—especially aqueous solutions of ionic compounds such as HCl, HF and NH_3, for which extensive enthalpy data happen to be available.

Algebraic routines

A number of algebraic procedures (and many variations on these) have been developed for calculating the number of trays and reflux ratios for distillation systems. No attempt will be made here to offer a compilation of methods. In fact, because little gain in accuracy can be realized by resorting to the more complex algebraic procedures, the engineer is advised to stick with the basic Fenske, Underwood and Gilliland calculation techniques.

The Fenske equation relates the minimum number of equilibrium stages (at total reflux) to the distillate and bottoms compositions and the average relative volatility:

$$N_{min} = \frac{\log\left[\left(\frac{x_{LK}}{x_{HK}}\right)_D\left(\frac{x_{HK}}{x_{LK}}\right)_B\right]}{\log(\alpha_{LK-HK})} \quad (2)$$

To solve for the component splits, the equation can be arranged as:

$$\left(\frac{x_{LK}}{x_{HK}}\right)_D = \left(\frac{x_{LK}}{x_{HK}}\right)_B (\alpha_{LK-HK})^{N_{min}} \quad (3)$$

where N_{min} = number of theoretical trays at total reflux; x_{LK} = liquid mole fraction of light key; x_{HK} = liquid mole fraction of heavy key; and α_{LK-HK} = average relative volatility.

A reasonably good estimate of equilibrium-stage requirements can be obtained by solving for N_{min} (number of equilibrium stages at total reflux), and then doubling this value. The result is equivalent to a reflux ratio of about 1.3 times the minimum.

The minimum reflux ratio required for a given separation can be conveniently calculated by the Underwood method. Like most shortcut methods, this procedure is limited by its assumption of constant molal overflow, constant (or averaged) αs, and optimum feed-tray location. It is not limited to a binary system but can be applied to any number (n) of components:

$$\left(\frac{L}{D}\right)_{min} + 1 = \sum \frac{\alpha x_D}{\alpha - \theta} = \frac{(\alpha_a x_a)_D}{\alpha_a - \theta} + \frac{(\alpha_b x_b)_D}{\alpha_b - \theta} \cdots + \frac{(\alpha_n x_n)_D}{\alpha_n - \theta} \quad (4)$$

To evaluate θ (the Underwood constant, which ranges between α_{LK} and α_{HK}), a second equation is required:

$$1 - q = \sum \frac{\alpha x_F}{\alpha - \theta} = \frac{(\alpha_a x_a)_F}{\alpha_a - \theta} + \frac{(\alpha_b x_b)_F}{\alpha_b - \theta} + \cdots + \frac{(\alpha_n x_n)_F}{\alpha_n - \theta} \quad (5)$$

where: q is 1 for a feed at its bubble point, and zero at its dewpoint. Subscript D refers to the distillate; subscript F to the feed.

This procedure applies when there are no distributed keys. If one component has a relative volatility falling between those of the light and heavy keys, it becomes necessary to solve for two values of θ [*14*].

Eq. (4) requires estimation of an overhead composition consistent with minimum reflux conditions. Unfortunately it is not easy to predict closely this composition. In practice, the same overhead concentrations that are substituted into or calculated by the Fenske equation are also used in the Underwood equation. A more

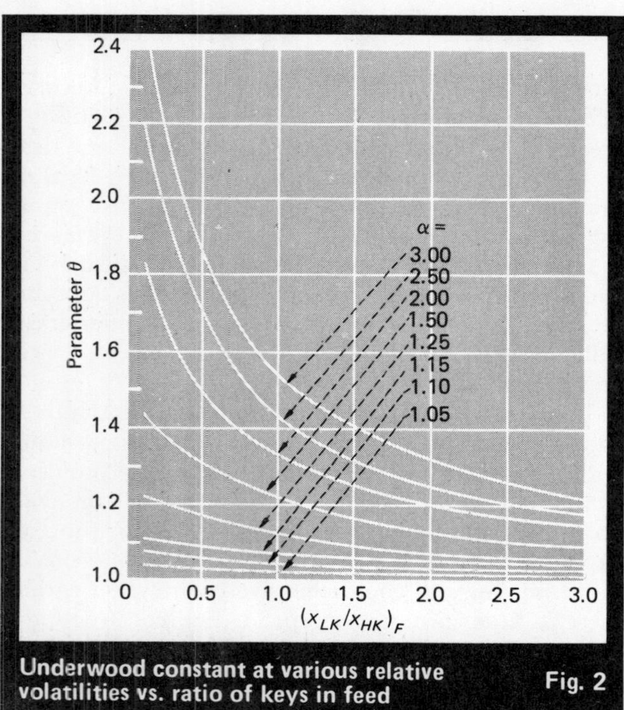

Underwood constant at various relative volatilities vs. ratio of keys in feed **Fig. 2**

1.0
0.8
0.6
0.4
0.2
0
0.2
0.4
0.6
0.8
1.0
$\frac{N - N_{min}}{N+1}$
$\frac{R - R_{min}}{R+1}$
0.3
0.1
0.08
0.06
0.04
0.03
0.02
0.01
0.02
0.04
0.06
0.10

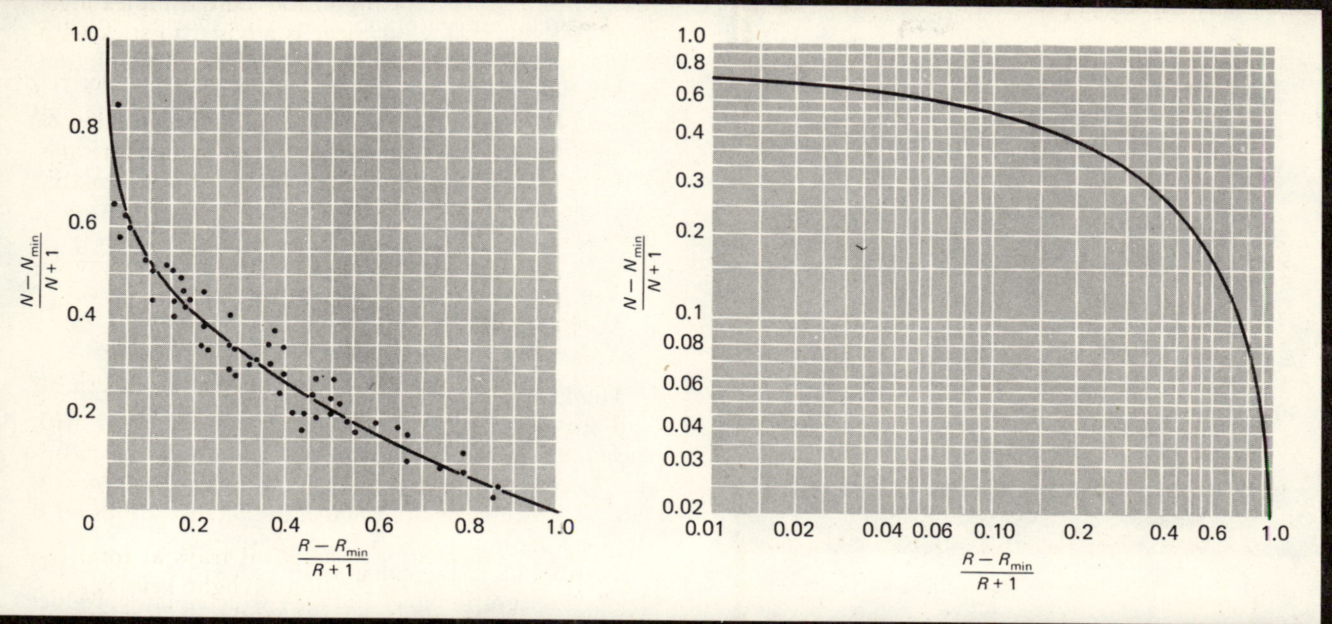

Gilliland empirical solution for the number of theoretical equilibrium stages

Fig. 3

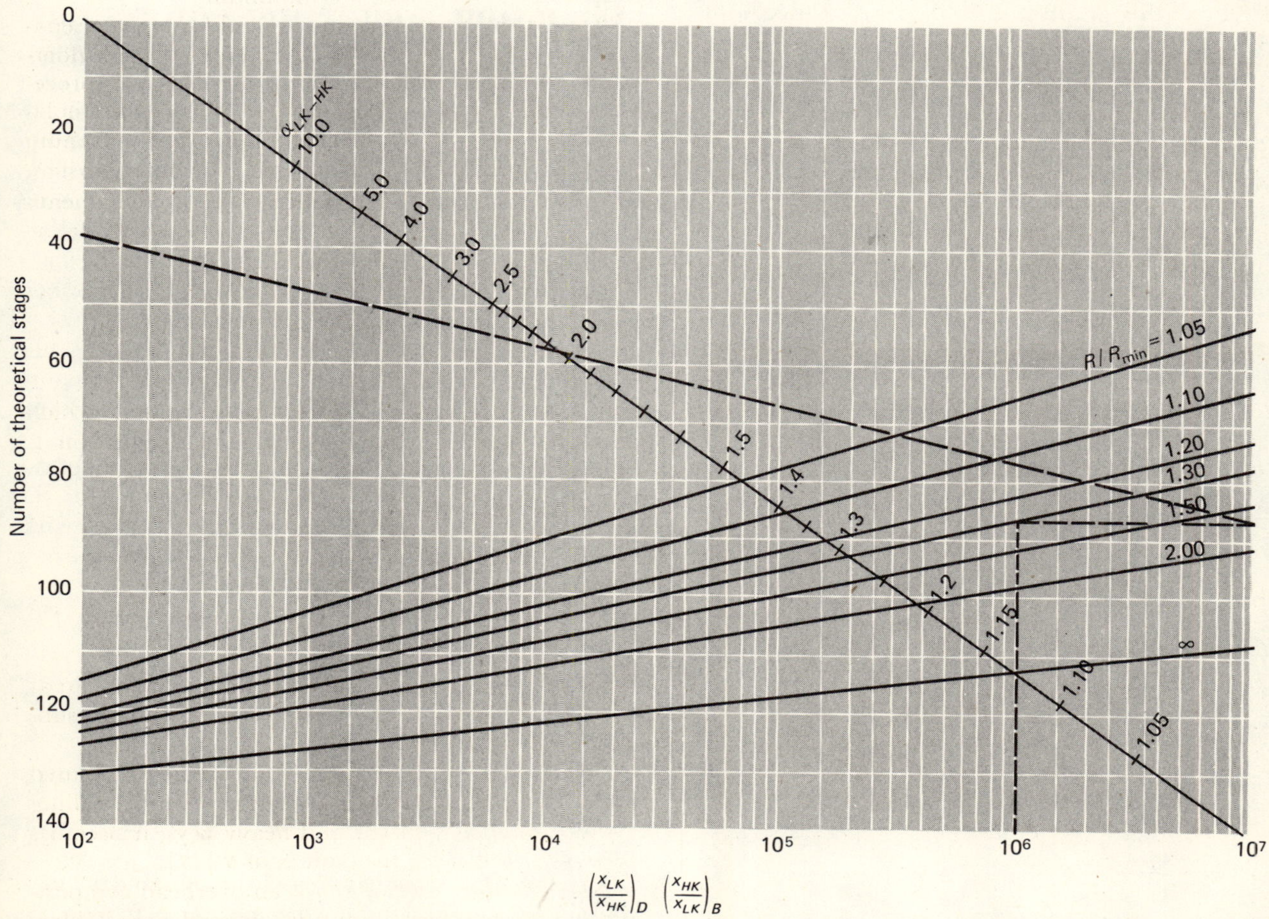

Number of theoretical stages
0
20
40
60
80
100
120
140
α_{LK-HK}
10.0
5.0
4.0
3.0
2.5
2.0
1.5
1.4
1.3
1.2
1.15
1.10
1.05
R/R_{min} = 1.05
1.10
1.20
1.30
1.50
2.00
∞
10^2
10^3
10^4
10^5
10^6
10^7
$\left(\frac{x_{LK}}{x_{HK}}\right)_D \left(\frac{x_{HK}}{x_{LK}}\right)_B$

Graphical solution for number of trays uses Fenske-Underwood-Gilliland correlations

Fig. 4

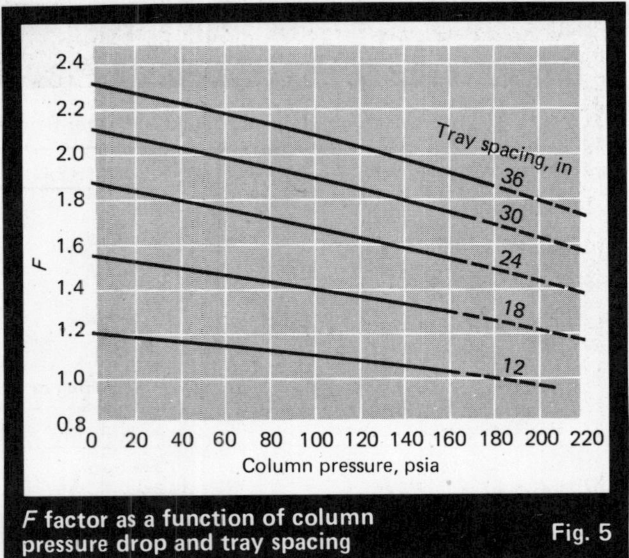

F factor as a function of column pressure drop and tray spacing — Fig. 5

rigorous approach for calculating the actual minimum reflux composition is not usually justified.

To eliminate the trial-and-error calculation necessary to solve Eq. (4) and (5), Van Winkle and Todd [*15*] developed a graphical solution (Fig. 2) for a liquid feed at its bubble point. The Underwood constant is read directly from this graph and is then substituted into Eq. (4).

The Gilliland empirical solution for the number of stages (calculated for the limits of total and minimum reflux) is shown in Fig. 3. Although it was prepared from widely scattered data, and probably could be bettered by using more than a single line, such refinements would probably not improve the estimates of actual stages and reflux, given the inherent limitations present in the values of N_{min} and R_{min}. In combination with the Fenske and Underwood equations, the Gilliland correlation is a practical tool for shortcut calculations of theoretical equilibrium stages.

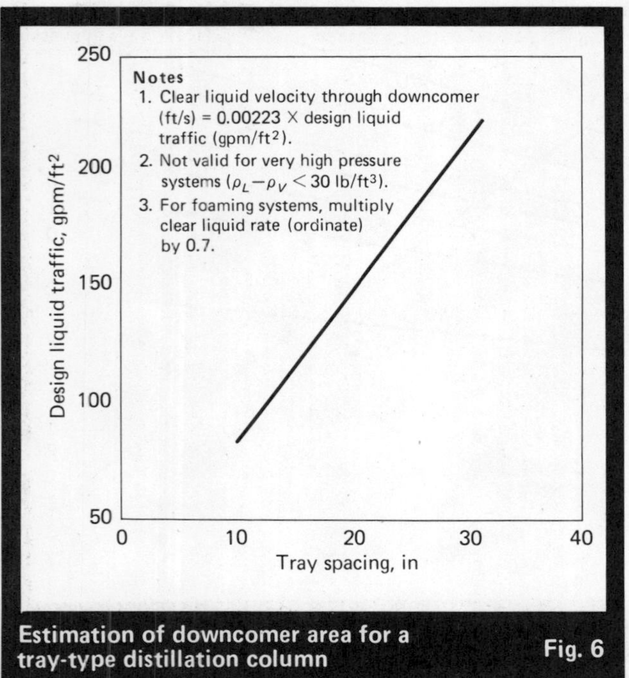

Estimation of downcomer area for a tray-type distillation column — Fig. 6

The optimum reflux ratio for a system can be derived by comparing operating (energy) costs with capital costs and then optimizing the two. The following are reasonably good estimates of R/R_{min} for several services, suitable for use with Fig. 3:

	R/R_{min}	N/N_{min}
Low-level refrigeration ($<-150°F$)	1.05–1.1	2.5–3.5
High-level refrigeration	1.1–1.2	2.0–3.0
Water- and air-cooled condensers	1.2–1.3	1.8–2.5

The Fenske-Underwood-Gilliland method is convenient for designing new columns since it begins with specifications of component splits. However, its usefulness for checking product purity in existing columns is limited because the accuracy of the method drops off if the feed contains close-boiling components.

To streamline the calculation procedure, I have developed the graph shown in Fig. 4, which relates product purity, reflux ratio and relative volatility to the number of equilibrium stages. Values obtained from the graph vary less than 10% from those calculated directly from the Fenske-Underwood-Gilliland correlations.

Choosing contacting equipment

Over the past few decades, a large number of different fractionating devices have been developed. Some of the better-known ones are listed in Table II.

Because manufacturers' claims for their fractionating devices are sometimes overly optimistic, it is essential to pinpoint the conditions under which a device functions most efficiently. Table III offers some general guidelines regarding the suitability of different internals for a number of services and operating conditions. Obviously, there are no rigid lines of demarcation between column applications. Furthermore, if more than one special condition is present, the designer may have to compromise in his demands for efficiency or performance.

Most distillation columns now in operation are of the crossflow-tray type, with perforated trays being the most popular. Packed columns are used to a somewhat lesser extent. There are relatively few countercurrent trays in operation.

Shortcut designs for tray columns

Column diameter—The active cross-section of the column can be quickly estimated from the simple "F" factor (Fig. 5):

$$F = v\sqrt{\rho_V} \tag{6}$$

where v = superficial vapor velocity, ft/s, and ρ_V = vapor density, lb/ft³.

The free cross-sectional area of the column (the total column area minus the downcomer area, ft²) can be obtained directly.

$$A_c = \frac{W}{F\sqrt{\rho_V}} \tag{7}$$

where W = vapor rate in lb/s. (For foaming systems, the F factor obtained from Fig. 5 should be multiplied by 0.75.)

At low and moderate pressures (<10 atm), the accuracy of this expression is surprisingly good, probably falling within ±15%.

Downcomer and weir design—Downcomer area can be estimated from the correlation [6] shown in Fig. 6. For foaming liquids, the calculated downcomer area should be multiplied by 1.5.

Regardless of the area calculated from Fig. 6, segmental downcomers should never be smaller than 5% of the total column cross-sectional area. If liquid rates are extremely small, tubular downcomers may be installed within the area subtended by a segmental weir (Fig. 8a).

The maximum recommended liquid flow over a straight, segmental weir is 70 gpm/ft. This range can be extended to 80 gpm/ft of projected weir if relief weirs are installed. For rates above 80 gpm/ft, a multiple-downcomer arrangement should be considered.

At low weir-loadings (frequently encountered in vacuum service), notched or castellated weirs can aid liquid distribution across the tray. Their use is recommended when the calculated flow over the weir is less than 3 gpm/ft.

Installation of adjustable weirs is seldom justified. At elevated pressure, weir height is only limited by tray spacing, since pressure drop is not critical. However, it is recommended that the weir height not exceed 15% of tray spacing in order to hold down jetting and entrain-

Some common types of contacting devices **Table II**

Staged tray columns (separate liquid and vapor flowpaths)

Common types	Proprietary types
Bubble cap	Angle
Sieve	Uniflux
Valve	Montz
	Linde
	Thorman
	Jet

Differential columns (countercurrent flow on packing or tray-surface)

Randomly packed towers	Systemically packed towers
Raschig or partition rings	Flexipac
Saddles	Goodloe
Slotted rings	Hyperfil
Tellerettes	Sulzer
Maspac	Glitsch Grid
	Leva film trays

Pseudoequilibrium stages (countercurrent flow across discrete trays)

Downcomerless trays	Low pressure drop trays
Perforated	Disk and doughnut (shower deck)
Turbogrid	
Ripple	

Special devices (low pressure drop)

Kloss (vertical springs)	Horizontal, agitated columns
Neo-Kloss (concentric vertical cylinders)	

Selection guide for distillation column internals **Table III**

	Staged columns		Differential columns		Pseudo-equilibrium	
	Perforated, or valve trays	Bubble cap or tunnel trays	Randomly packed	Systematically packed	Downcomer-less	Disk and doughnut
Low pressure (<100mm Hg)	2	1	2	3	0	1
Moderate pressure	3	2	2	1	1	1
High pressure (>50% of critical)	3	2	2	0	2	0
High turndown ratio	2	3	1	2	0	1
Low liquid rates	1	3	1	2	0	0
Foaming systems	2	1	3	0	2	1
Internal tower cooling	2	3	1	0	1	0
Solids present	2	1	1	0	3	1
Dirty or polymerizing solution	2	1	1	0	3	2
Multiple feeds and sidestreams	3	3	1	0	2	1
High liquid rates (scrubbing)	2	1	3	0	3	2
Small dia. columns	1	1	3	2	1	1
Columns with dia. 3-10 ft.	3	2	2	2	2	1
Large dia. columns	3	1	2	1	2	1
Corrosive fluids	2	1	3	1	2	2
Viscous fluids (at boiling pt.)	2	1	3	0	1	0
Low ΔP (efficiency no concern)	1	0	2	2	0	3
Expanded column capacity	2	0	2	3	2	0
Low cost (performance no concern)	2	1	2	1	3	3
Available design procedures	3	2	2	1	1	1

Key
0-Don't use 1-Evaluate carefully 2-Usually applicable 3-Best selection

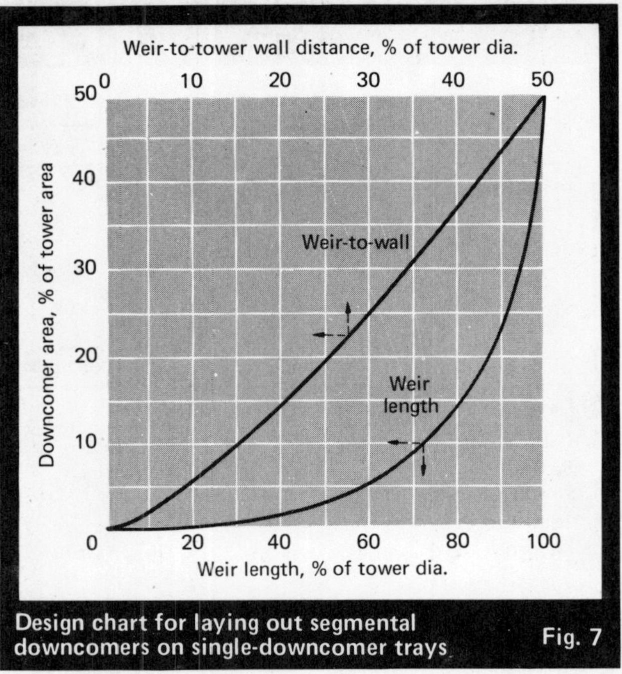

Design chart for laying out segmental downcomers on single-downcomer trays Fig. 7

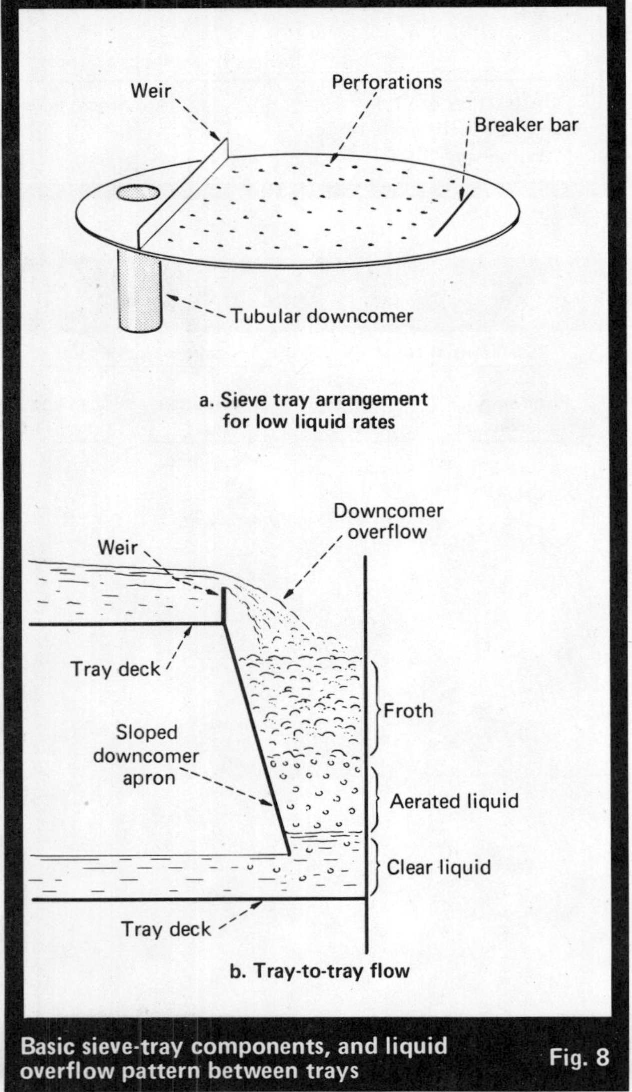

Basic sieve-tray components, and liquid overflow pattern between trays Fig. 8

ment. For vacuum operations, a 1-in weir seems to be satisfactory for most sieve- and valve-tray columns.

If the system is known to foam, or if the separation is to be conducted at high pressures, installing sloped downcomers may prove helpful (Fig. 8b). These provide enough volume for vapor-liquid disengaging at the top of the downcomer while leaving the maximum active area on the tray below. The ratio of top to bottom area is usually 1.7.

To ensure good flow distribution on trays, weir length for segmental downcomers should not be less than 50% of the column diameter. Maximum length is governed by economics: too big a downcomer will take away area more profitably used for mass transfer. A weir length 80% of column diameter is considered reasonable, although for columns handling extremely high liquid loads a 95% weir is not unheard of.

The curves in Fig. 7, reproduced from the Glitsch Design Manual [4], are helpful for laying out segmental downcomers for single-downcomer trays.

To prevent vapor flow up the downcomer, a seal must be provided at the outlet where the liquid flows onto the tray (Fig. 8b). This is accomplished by keeping the distance between tray-deck and downcomer-apron at less than the weir height and maintaining a $\frac{1}{2}$-in static seal if at all possible. As a rule, it is recommended that the distance to the downcomer be half of the weir height or $\frac{3}{4}$ in, whichever is greater. Under no circumstance should the clearance be less than $\frac{1}{2}$ in. The liquid flowrate at the opening below the downcomer skirt should be held to less than 1 ft/s. Inlet weirs, which are sometimes recommended to prevent vapor bypassing, are seldom, if ever, justified.

Bubble caps

Although bubble caps were the workhorse of the chemical and petroleum industries until the mid-1950s, they are installed only infrequently today. Sieve trays and valve trays have almost completely replaced them. The primary reasons for this shift are:

- High cost (2 to 3 times that of the equivalent sieve tray).
- Tendency to foul and collect solids.
- High pressure drop due to a complex vapor flowpath.
- High liquid gradients, which hinder design of large columns.
- Limitations on cap flow at low pressures, which increases column size.
- Corrosion effects—more severe than for sieve trays.

There are still a few instances when the installation of bubble-cap trays can be justified; these include columns subject to extremely low liquid rate ($\leq$2 gal/ft of average flow width); and columns requiring extremely high turndown ($>$5:1). Although there will be few occasions for installing new bubble-cap columns, *there are still a large number of existing facilities that may require analysis.**

A wide variety of bubble-cap shapes have been employed in columns, the majority of them being of the round, bell-shaped type having vertical slots. Other

*The chapter by W. Bolles in [11] contains a procedure recommended for bubble-cap design.

popular types are inverted, rectangular boxes (tunnel trays), and inverted "teacups" without slots.

Round caps are usually supplied in three sizes and are selected according to column diameter:

Tower dia., ft	Cap dia., in
2.5–5	3
4–12	4
10 and up	6

1- and 2-in caps have sometimes been employed in small, low-temperature stills.

Normal slot height varies between 0.25 and 1.0 in for 3- and 4-in caps, but may go as high as 1.5 in for 6-in caps. It is recommended that slots not be left open at the bottom, and that caps be equipped with a strength-giving shroud ring to forestall deformation of the teeth. Experience indicates that cap configuration has little effect on performance. Furthermore, there appears to be no loss of efficiency if slots are omitted entirely, and all vapor passes under the lip of the cap.

The critical dimension in bubble-cap design is the liquid seal—i.e., the depth of liquid through which the vapor must travel. Large slot seals enhance plate efficiency at the expense of pressure drop. For design purposes, a dynamic slot seal has been defined:

$$\text{Dynamic slot seal} = h_{ss} + h_{ow} + \Delta/2 \qquad (8)$$

where h_{ss} is the static seal, in. liquid; h_{ow} is the height of liquid over the weir, in; and $\Delta/2$ is the average liquid gradient, in (Fig. 9a). Practical limits for the dynamic slot seal are given here:

Pressure, psig	Dynamic slot seal, in
<0	0.5–1.5
0–50	1.0–2.5
50–200 psig	1.5–3.0
200–400 psig	2.0–4.0

The weir height for bubble-cap trays is set by adding the static seal to the slot height. In low-pressure and atmospheric columns, the static seal usually runs between ¼ and ½ in. The pressure drop across a bubble-cap tray usually ranges between 2.5 and 5 in of liquid.

Caps are generally arranged in an equilateral triangular pattern, with rows normal to liquid flow. The caps are usually spaced to have a pitch between 1.25 and 1.5, but never with less than 1-in clearance between adjoining caps.

Not all of the active area can be used for cap placement. Stilling sections are customarily provided next to the inlet downcomer and next to the overflow weir. There is also considerable wasted area where, for structural reasons, it is impossible to place caps (Fig. 9b).

A liquid gradient exists across trays between the point of liquid inflow and the overflow weir. This gradient provides the necessary driving force to overcome frictional resistance due to vapor flow, tray surface and bubble caps. Because rows of caps offer a flow barrier very much like rows of baffles, the gradient across a bubble-cap tray can be quite pronounced. This is especially a concern in large-diameter columns (>5 ft for moderate and high-pressure systems; >7 ft for vacuum systems) and at high liquid rates. Too high a gradient

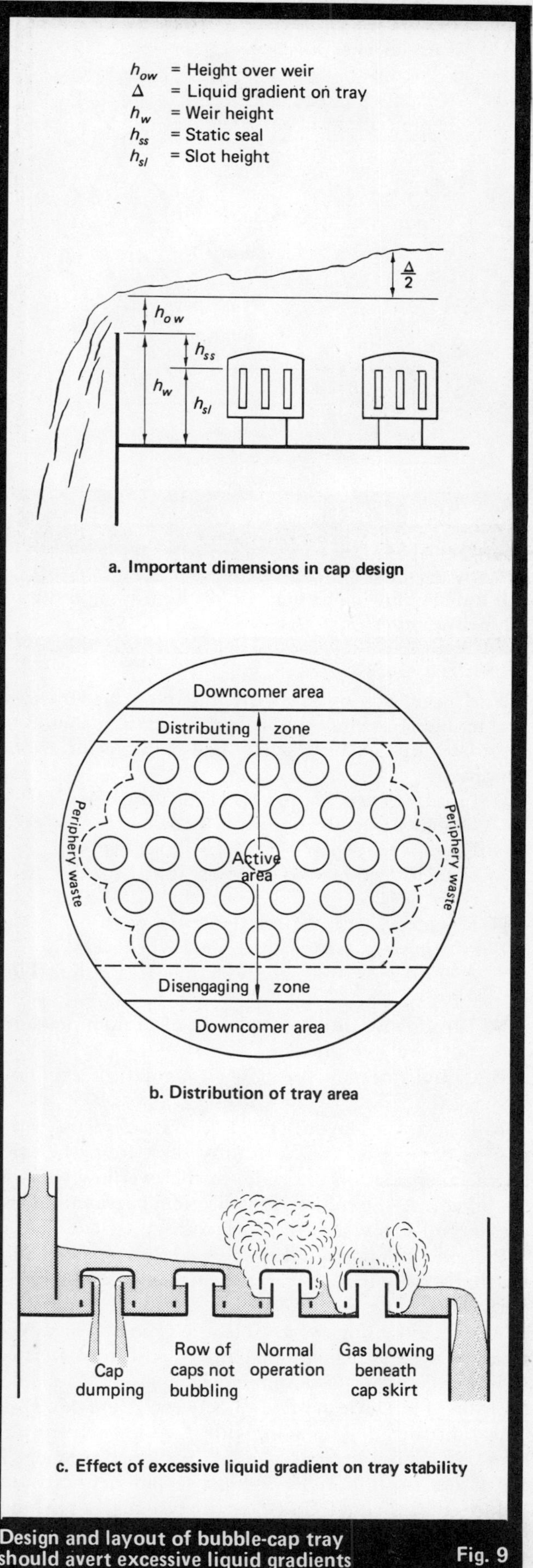

Design and layout of bubble-cap tray should avert excessive liquid gradients — Fig. 9

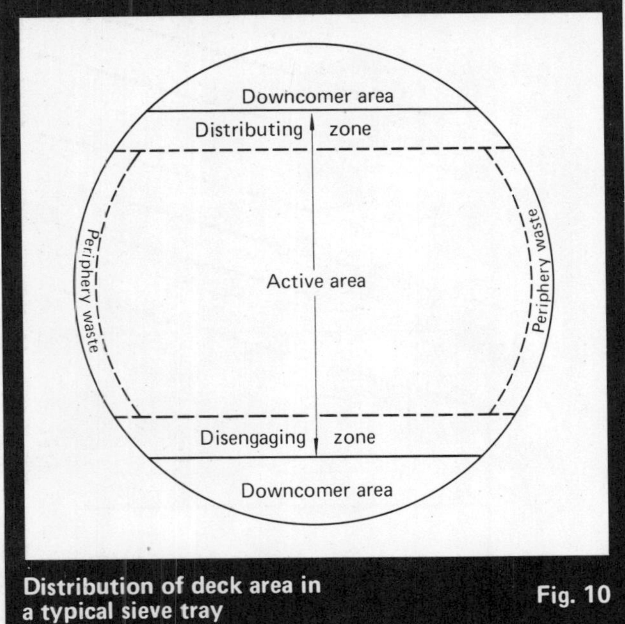

Distribution of deck area in a typical sieve tray **Fig. 10**

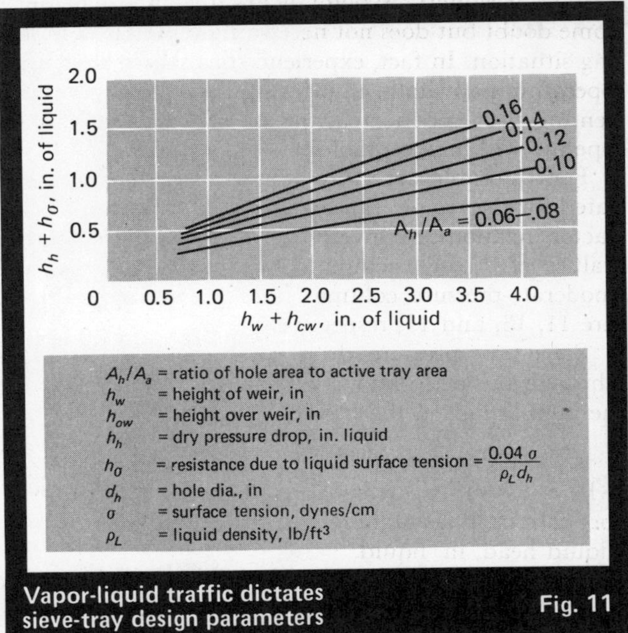

Vapor-liquid traffic dictates sieve-tray design parameters **Fig. 11**

will induce blowing at one end of the tray, and dumping at the other (Fig. 9c).

Sieve trays

The sieve tray is probably the most widely used contacting device found in columns today. It should be considered first in the design of new columns for several reasons:

- □ Installed costs are lowest of all tray-type devices.
- □ Design procedures are well known.
- □ Fouling tendency (with large holes) is low.
- □ Capacity equals or exceeds that of other tray types.
- □ Efficiency, with proper design, is good.

Sieve trays are *not* recommended whenever:

- ■ Pressure drop must be very low (<2.5 mm Hg/tray).
- ■ Turndown ratios are high (>3:1 at high pressure; >2:1 at low pressure).
- ■ Liquid rates are very low (<2 gal/ft of avg. flow width).

In a sieve tray layout (Fig. 10), the active tray area can be defined as the entire tray deck from the inlet downcomer-skirt on one side, to the overflow weir on the other (i.e., the column cross-sectional area minus the area occupied by the two downcomers).

Placement of holes in the active area is restricted only by the positioning of tray support rings and beams. Perforations can be made within 2 or 3 in. of the inlet downcomer or outlet weir. Holes are generally arranged in an equilateral triangular pattern, with rows normal to liquid flow. For optimum efficiency, pitch-to-diameter ratio of the hole should fall between 2 and 4.5. Hole sizes range from ¼ to 1 in, with ½-in holes being the standard. Small holes (¼ and ⅜ in) lessen weeping if the liquid downflow has a high surface tension, and reduce entrainment in low-pressure systems. Large holes (¾ and 1 in) should be used in fouling service.

Pressure drop is lower in trays that are installed with perforations punched upwards to create a nozzle effect in the direction of vapor flow. Despite this advantage, tray panels are usually installed with downpunched holes in order to reduce the risks presented by jagged edges to personnel installing or inspecting tower internals.

Open hole area depends on vapor rate and the specified tray pressure-drop. In most cases, it ranges between 4% and 16% of the active tray area. Because the rate of vapor flow is inversely proportional to the square root of pressure, large open areas are found in vacuum towers, whereas in pressurized towers open areas are smaller.

Outlet weir height on a sieve tray ranges between 0 and 4 in. For low- and atmospheric-pressure systems, the height is usually 1 to 2 in. In high-vacuum systems, where the mass flowing over the tray is more a vapor continuum than a liquid continuum, a weir actually serves little purpose. It could be eliminated entirely, although most engineers prefer to play safe by providing at least a 1-in. weir.

Liquid gradients on sieve trays are considerably less pronounced than on bubble-cap trays. They can be ignored entirely in pressurized columns smaller than 8 ft dia., and in low-pressure columns less than 10 ft dia.

The most critical variable in sieve tray design is the open hole area. Too small an area leads to a high pressure drop and, in extreme cases, jetting. An excessively large open hole area encourages weeping or even dumping, where no liquid passes over the sieve-tray weir.

There is considerable uncertainty regarding how much weepage is detrimental to column performance. It has often been recommended that liquid flow through the perforations be kept at less than 25% of the total tray flow. However, many sieve tray columns operate, apparently without undue harm to efficiency, with more than half the liquid weeping through the holes.

A modified version of the correlation developed by Fair [*11*] is given in Fig. 11. An operating point above the line representing the desired ratio of open hole area to active tray area (A_h/A_a) is considered a safe design (<25% weeping). A point below the line may be in some doubt but does not necessarily represent a dumping situation. In fact, experience has shown that if the operating point falls anywhere above the curve representing 6–8% open area, the column will most likely operate within acceptable efficiency limits.

For a suitable sieve-tray design, hole velocity (calculated for the vapor) can be estimated from the basic F factor relationship given in Eq. (6). Representative values of F for vacuum columns, atmospheric and moderate-pressure columns, and pressurized columns are 11, 13, and 15, respectively.

Sieve-tray pressure drop combines flow resistance through the perforations with the hydrostatic head of aerated liquid on the tray:

$$h_t = h_h + h_l \qquad (9)$$

where h_t = tray pressure drop, in. liquid; h_h = dry pressure drop across holes, in. liquid; and h_l = aerated liquid head, in. liquid.

Dry pressure drop across the holes can be obtained from the following well-known relationship:

$$h_h = 0.186 \frac{\rho_V}{\rho_L} \left(\frac{U_h}{C_0} \right)^2 \qquad (10)$$

where h_h = dry pressure loss through holes, in. liquid; U_h = hole velocity, ft/s (used in place of v in Eq. 6); and C_0 = dry orifice coefficient (Fig. 12).

The pressure drop across the aerated liquid on the tray is obtained from the following relationship:

$$h_l = \beta(h_w + h_{ow}) \qquad (11)$$

where h_l = aerated liquid head, in. liquid; h_w = head at weir height, in. liquid; h_{ow} = height, in, of liquid over weir = 0.5 $(Q/L_w)^{0.67}$; Q = liquid flow, gpm; L_w = weir length, in; and β = aeration factor (Fig. 13).

In vacuum columns, where performance can be significantly compromised by entrainment, two-phase flow through the holes can raise the dry pressure drop beyond that for pure vapor flow. It is not unusual in low-pressure systems to observe a *total* pressure drop 15 to 25% higher than calculated.

The normal pressure-drop span for sieve trays is from 1.5 to 5 in. H_2O. Outside these limits, trays may weep excessively in vacuum columns, or jet in pressure service. In either case, there could be an appreciable loss in efficiency.

Valve trays

Arranged on a valve tray are liftable caps that act as variable orifices by adjusting themselves to changes in vapor flow. This design (whose installed cost is 15–20% higher than for the equivalent sieve tray) is said to provide turndown over a greater range than is possible for sieve trays.

The valves are actually small metal disks or strips that are lifted above openings in the deck as vapor passes across the trays. The valve caps are restrained by legs or spiders, which limit vertical movement.

Fabricators cite the following advantages for valve trays:

1. A fairly constant pressure drop across a large portion of their operating range.

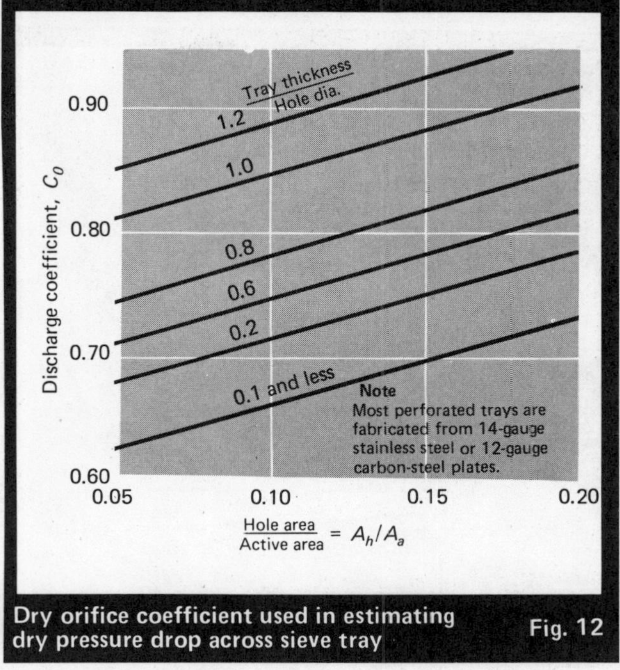

Dry orifice coefficient used in estimating dry pressure drop across sieve tray — Fig. 12

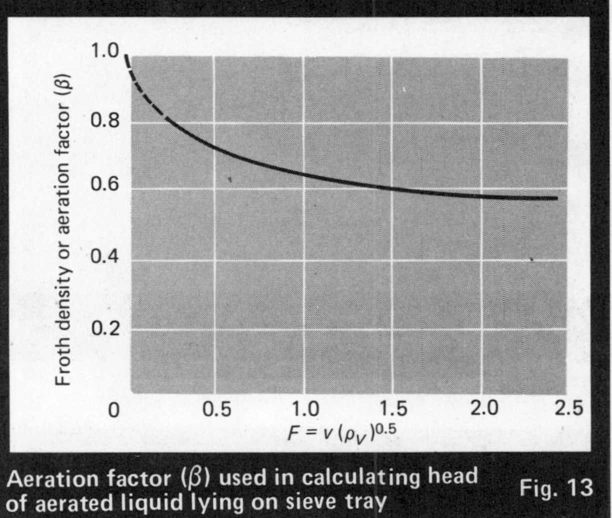

Aeration factor (β) used in calculating head of aerated liquid lying on sieve tray — Fig. 13

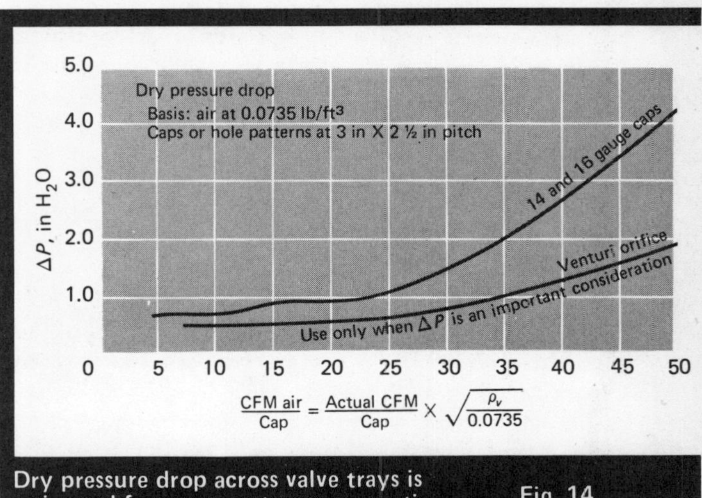

Dry pressure drop across valve trays is estimated from vapor stream properties — Fig. 14

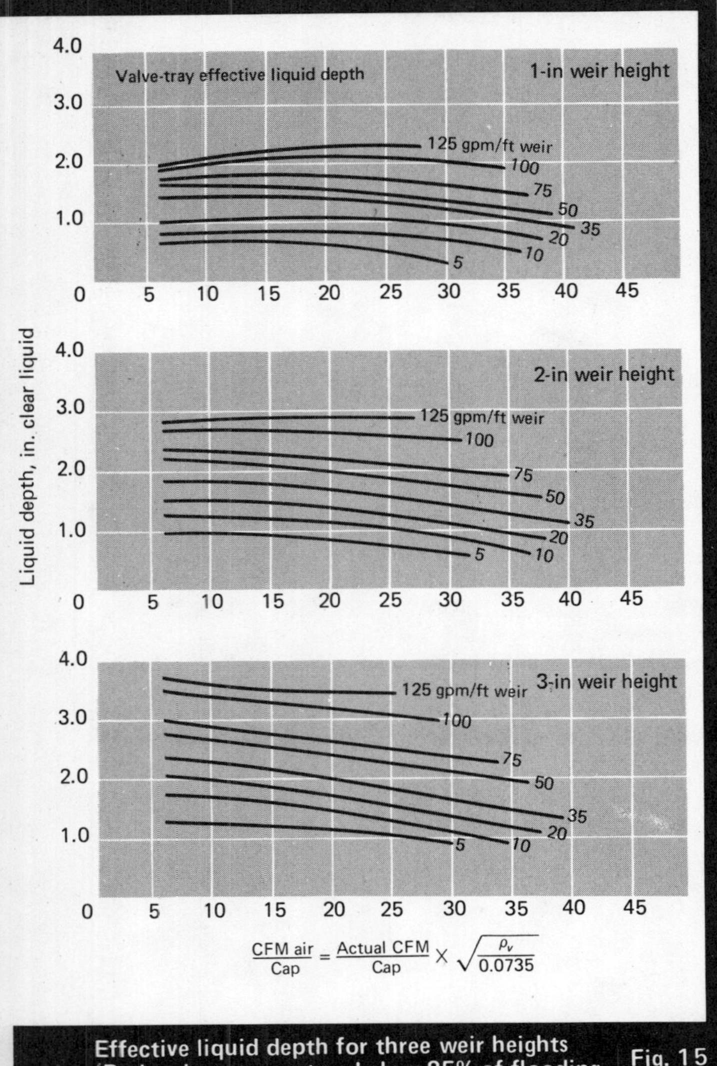

Effective liquid depth for three weir heights (Basis: air-water system below 85% of flooding Fig. 15

2. A high turndown ratio.
3. Operation at about the same capacity and efficiency as sieve trays.

However, it should be noted that good turndown ratio, the most frequently advertised advantage of valve trays, may also be attained with sieve trays by imposing a reasonable pressure drop. Even in vacuum columns, where pressure drop can be critical, the available turndown ratio for sieve trays is adequate for most operations.

A mechanical problem often encountered with valves is wear or corrosion of the retaining lugs or spiders. The constant movement of the valve caps imposes fatigue stresses that are aggravated by operation in a corrosive environment. It is not unusual to find valves missing on the bottom trays where high-boiling and corrosive constituents tend to concentrate.

Valve trays, because of their proprietary nature, are usually designed by the fabricator, although it is possible to estimate some of the design parameters from vendor literature. Tray performance can be predicted from pressure drop charts (Fig. 14 and 15) adapted from the Koch Design Manual [*6*], whereas column and downcomer areas can be calculated by methods outlined above.

The number of caps that can be fitted on a tray is at best an estimate unless a detailed tray layout is prepared. A 3-in × 2½-in cap pattern is the tightest arrangement available and has become the standard for low- and moderate-pressure operations. For such a pattern, about 14 caps/ft^2 fit into the "net" capped area. Active area as defined by Fig. 9b does not take into account stilling sections at the inlet and outlet, edge losses due to column support rings, and unavailable space on top of support beams. It must be assumed that between 6 and 15% of the so-called active areas is not available for valves in large and small columns, respectively.

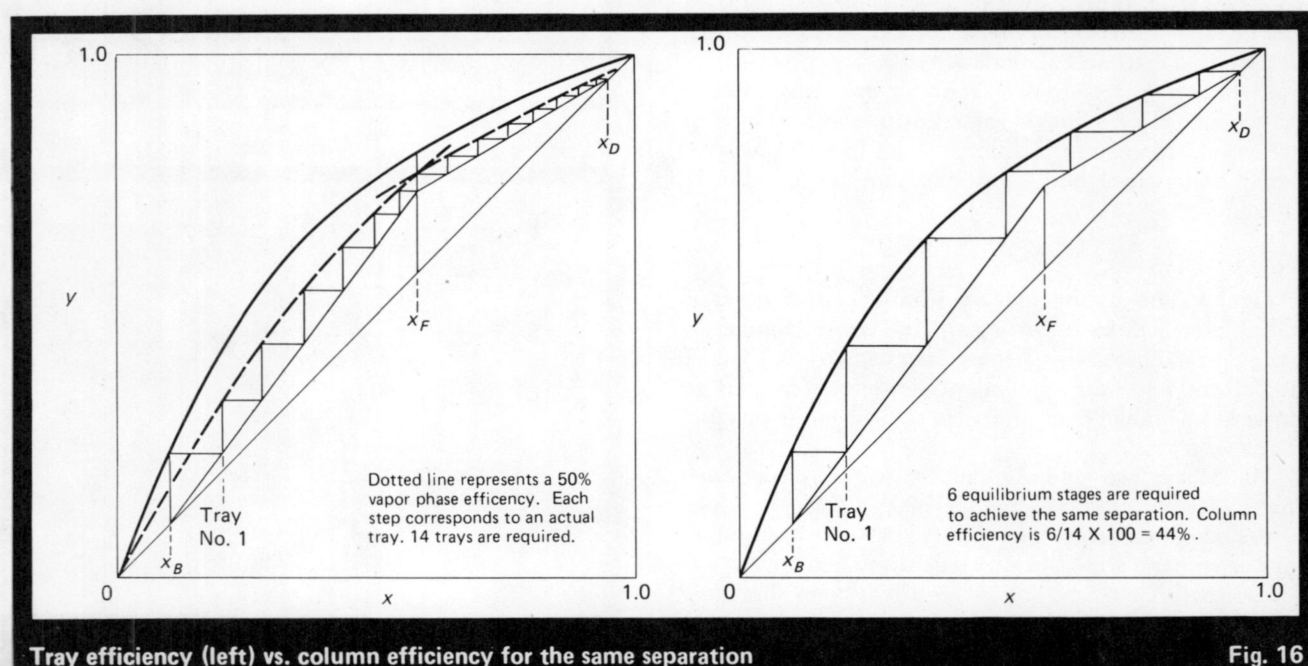

Tray efficiency (left) vs. column efficiency for the same separation Fig. 16

Downcomerless trays

The simplest possible tray design is the sieve tray without downcomers. Its successful operation demands a sufficiently high vapor velocity through the holes to maintain enough liquid on the tray for proper vapor-liquid contact. At low vapor-rates, downflowing liquid is not retained long enough on the tray, and mass transfer becomes quite inefficient since, as is the case in spray columns, the only contact is between the rising vapor and falling liquid droplets.

The major disadvantage of straight, downcomerless trays is that their operating range (turndown ratio) is considerably smaller than that of trays having downcomers. Sufficient vapor flow is required to maintain a liquid level on the tray, but it cannot be so high that it will restrict liquid downflow and therefore flood the column. At their optimum operating point, downcomerless trays have about the same efficiency as sieve trays. However, efficiency will drop drastically as the tray dumps or the floodpoint is approached.

Downcomerless trays may be considered when operating conditions and capacity are not expected to vary; column liquids have a high solids content; and ease of cleaning is important.

Few design procedures for downcomerless trays are published in the open literature, and even the best of these is unlikely to achieve a degree of reliability greater than ±40%. The correlation developed by Sum-Shik et al. [*12*] is cited here as a suitable procedure for estimating an acceptable hole velocity:

$$U_h = 1.15\sqrt{\frac{\rho_L - \rho_V}{\rho_V}}\left(\frac{A_h}{A_c}\right)(1 - \tau) \qquad (12)$$

where: U_h = hole velocity, ft/s
ρ_L, ρ_V = liquid and vapor densities, lb/ft^3
A_h = open hole area, ft^2
A_c = column free cross-sectional area, ft^2
τ = $\dfrac{1}{1 + \left[\dfrac{\Delta P_V}{\Delta P_L} C_o\, k\right]^{0.33}}$
C_o = orifice coefficient (0.62 for holes and slots)
k = pressure loss factor through hole (1.8 for holes $\frac{3}{8}$ to $\frac{7}{8}$ in dia.)
$\dfrac{\Delta P_V}{\Delta P_L}$ = pressure-drop ratio for vapor and liquid flowing through the holes

The column free cross-sectional area can be fixed by applying the same design criteria as were recommended for obtaining the free area in other tray columns. The hole area for a downcomerless tray will be 15–30% of the total area, with a 20–25% open area being considered most effective.

Variations of the basic perforated, downcomerless design include trays that have flat grating instead of perforations (Turbogrid); trays that use sinusoidally shaped, perforated plates (Ripple trays); and trays whose surfaces impart directional flow to the vapor (Kittle trays).

In addition to the trays with and without downcomers described above, a number of other proprietary tray configurations are available for commercial columns (Table II).

Tray and column efficiencies

Tray efficiency compares the separation actually attained to that which is possible in a true equilibrium stage. The Murphree efficiency is defined as:

$$\text{Efficiency} = \frac{\text{Vapor conc. from tray} - \text{Vapor conc. to tray}}{\text{Vapor conc. in equilibrium with liquid} - \text{Vapor conc. to tray}} \qquad (13)$$

This efficiency differs from the overall column effi-

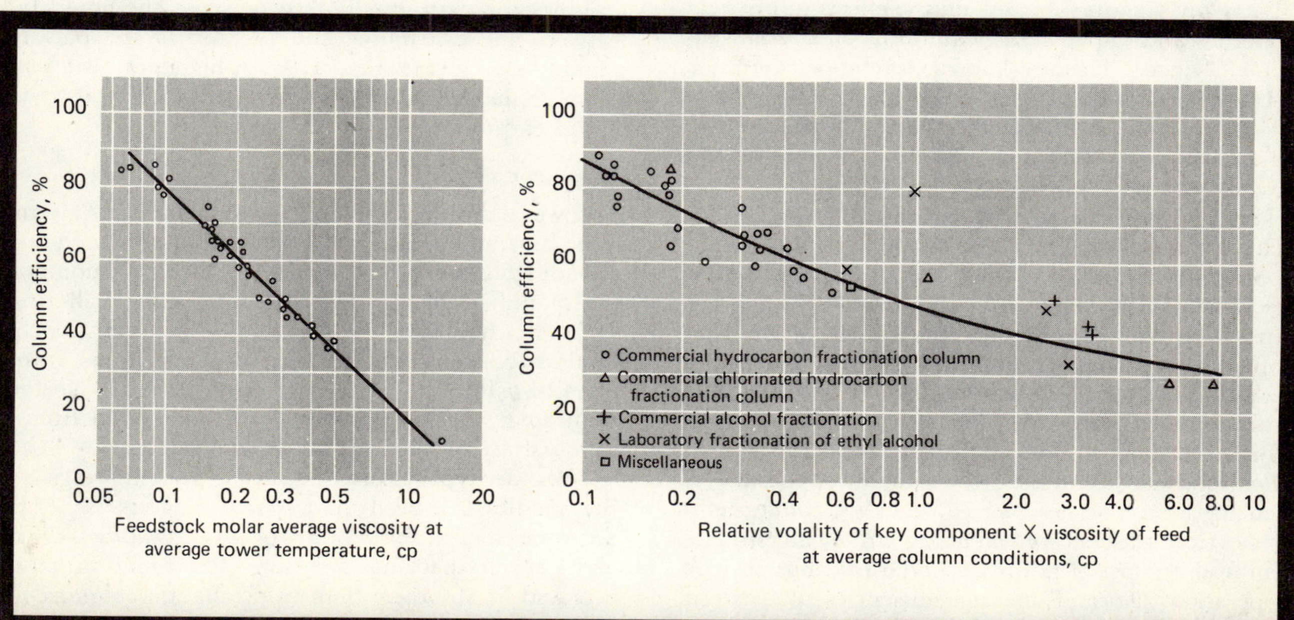

Column efficiencies for various services as a function of feedstock properties

Fig. 17

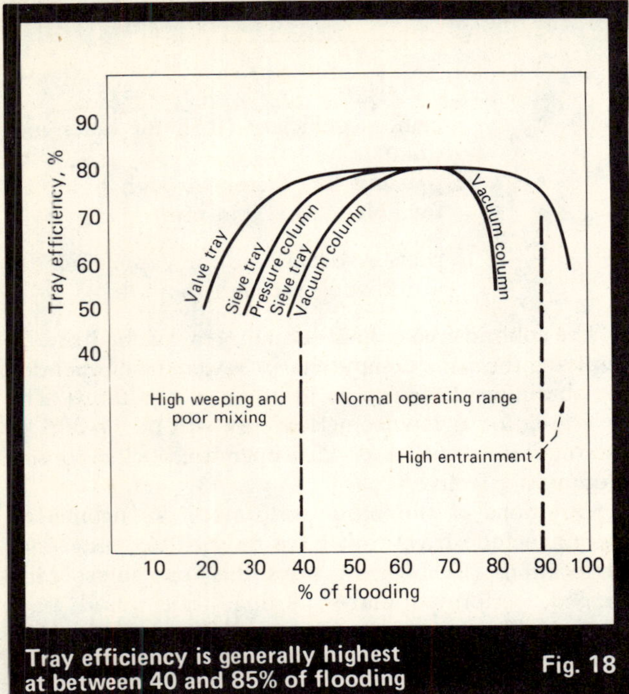

Tray efficiency is generally highest at between 40 and 85% of flooding **Fig. 18**

ciency, which is the number of theoretical equilibrium stages divided by the actual number of column trays.

Overall column efficiencies are usually less than individual tray efficiencies. This is illustrated by the McCabe-Thiele diagram in Fig. 16. The difference becomes most pronounced when a large number of trays are required to achieve a high product purity.

Plate efficiencies are never constant throughout the column, since mass-transfer characteristics change with composition, flowrate and temperature. Especially on the end-trays, where changes in concentration per tray are small, efficiencies drop off drastically as the concentration of the contaminating component falls below 100 ppm.

Early investigators focused on the effects of physical properties and vapor-liquid equilibria on overall column efficiencies. The best-known correlations are those of Drickamer and Bradford [*2*], who, on the basis of refinery column data, related efficiency to average liquid viscosity; and O'Connell [*9*], who also incorporated the effects of relative volatility.

There presently is no truly satisfactory method for accurately calculating tray efficiencies. The only rational procedure available in the open literature is outlined in the classic AIChE monograph on bubble-cap trays [*1*]. Here, investigators combined the effects of physical properties, vapor-liquid equilibria, and tray hydraulics in a diffusion and mass-transfer model that they hoped would represent the conditions of equimolar counter-diffusion through a film.

For efficiency estimates, the simplest approach is often the best. The designer can pick off column efficiencies from the charts given in Fig. 17, or can estimate them from the performance of similar columns already in operation. "Normal" column efficiencies fall between 60 and 85%. They tend toward the lower value when:

- High product purity is specified, and especially when the contaminant concentration must be held substantially below 100 ppm.
- Column performance is controlled by material balance, which is the case if one of the product components is only a small fraction ($<10\%$) of the feed.
- Low contaminant concentrations are specified for both ends of the column.

Tray efficiencies can be drastically reduced by liquid entrainment at one end of the operating range and excessive weeping at the other. Generalized curves in Fig. 18 illustrate that columns usually perform most efficiently between 40 and 85% of flooding.

Packed columns

Dumped or random packings are suited for the following operating conditions:

1. Low-pressure-drop operations (vacuum columns).
2. High liquid/gas ratios (found more frequently in absorbers and scrubbers than in distillation columns).
3. Corrosive environments (plastic or ceramic internals are required).
4. Small-diameter columns.

Liquid distribution in packed towers

Liquid distribution is one of the most critical items in the design of a packed tower. Unless the surface of the packing is adequately wetted, tower efficiency will drop drastically. For good irrigation, there must be at least four liquid-distribution points/ft^2, and the liquid rate should be kept above 2 gpm/ft^2.

It is sometimes difficult to achieve the latter rate in vacuum distillation service; however, many packed columns operate with liquid rates as low as 0.8 gpm, with only a nominal loss in efficiency.

Gravity distribution is the preferred method for introducing liquid onto the packing. Spray nozzles, which in theory assure a more-even distribution, can plug or wear out over prolonged operation, and can be more troublesome than gravity-feed devices. The liquid fall between the distributor and the top of the packing should be no greater than 12 in. A higher fall will tend to break the stream into droplets, and thereby increase liquid entrainment.

Packing types

Tower packings come in many shapes, materials and sizes; in commercial installations, commonly used ones include slotted rings and Intalox saddles. Depending on the manufacturer, slotted rings are called Pall rings (Norton), Ballast rings (Glitsch), Flexirings (Koch), Hy-Pak (Norton), or Cascade Mini-Rings (Mass Transfer, Ltd.). Intalox saddles are only made by Norton. Both rings and saddles can be manufactured from a wide variety of metals, plastics or ceramics.

The older types of packing such as Raschig rings and Berl saddles have been almost completely superseded by the more efficient slotted rings or Intalox saddles. Other high-capacity packings are more often found in scrubbers and in absorbers than in distillation columns.

To ensure proper liquid distribution and to minimize wall effects, it is necessary to place a lower limit on the ratio of column size to packing size. This ratio should equal or exceed 30 for Raschig rings, 15 for saddles, and 10 for slotted rings. The 2-in slotted metal ring is generally found to be best from the standpoint of capacity. In fact, experience shows it to be the most economic packing size and shape for most distillation applications.

Temperature and the corrosive nature of the column environment also influence the choice of packing. Some general guidelines:

1. Metal packing should not be specified if the measured corrosion rate is greater than 10 mils/yr.
2. Plastic packing, while usually resistant to most chemicals, may embrittle with prolonged exposure.
3. If subjected to continuous heat, plastic packing should be fiberglass reinforced.
4. Ceramic packing is usually selected for hot, corrosive service, although it is more fragile than metal or plastic packing. Shifting of the bed can cause the packing to fracture, restricting vapor flow and boosting pressure drop.

Design of packed beds

The generally accepted design procedure for sizing packed columns [*8*] is the modified version of the Sherwood correlation (Fig. 19).

This correlation relates bed diameter to fluid densities, flowrates, and a characterization factor that has been measured for each type and size of packing:

$$d = 1.13\left[\frac{W}{G}\right]^{0.5} \tag{14}$$

where d = bed dia., ft; W = actual vapor flow, lb/s; and G = vapor flow, lb/(s)(ft^2), as in Fig. 19.

The pressure drop across a bed increases with the vapor and liquid rates. If these are raised simultaneously (as is usually the case in distillation), at some point a significant amount of fluid will be held up in the packing voids. The pressure drop across the bed goes up rapidly until downflow of liquid stops entirely (flooding). Unfortunately, this *floodpoint* has not been defined well enough to be incorporated into Fig. 19. Depending on the packing type, the floodpoint falls somewhere within a band of scattered points. For Raschig rings, flooding takes place at a lower pressure drop (1.5 in. H_2O/ft) than it does for some of the more efficient packings such as slotted rings (2 to 2.5 in. H_2O/ft).

In most packed beds, the pressure drop will range between 0.1 and 0.8 in/ft. Actually, for pressure drops of 0.5 to 0.6 in. H_2O/ft, all types of packings seem to be about equally efficient. Below these limits, the efficiency of Raschig rings drops off drastically. Performance of the more efficient packings (e.g., slotted rings and plastic Intalox saddles) does not deteriorate appreciably until the drop falls below 0.1 to 0.2 in. H_2O/ft.

Selecting the right pressure drop for a packed bed depends mostly on its service. In general, it is desirable to apply the highest pressure drop consistent with reliable and economic operation. For moderate- and high-pressure distillation, the drop should be 0.4 to 0.75 in. H_2O/ft; for vacuum distillation, it should be 0.1 to 0.2 in. H_2O/ft; and for absorbers and strippers, 0.2 to 0.4 in. H_2O/ft.

The accuracy of the curves in Fig. 19 becomes somewhat questionable at the extreme ends of the scale. Thus in vacuum stills (equivalent to very low values of X; Fig. 19), actual measured pressure drops can easily be 25 to 30% higher than calculated. This discrepancy is attributed to high entrainment, which (as on trays) appreciably affects the flow behavior of the vapor. At high liquid rates, or at high pressures (when $X > 3.5$), the pressure drop in a functioning column may actually fall above the normally expected floodpoint limit.

Although high-efficiency packings are designed to maintain reasonably good liquid distribution over the cross-section of a bed, there will always be some fraction of the liquid that reaches the column wall. Once there, the liquid will not readily redistribute itself back into the packing, but instead will bypass a large fraction of the vapor stream. Therefore, one should limit the vertical height of a bed to ensure that the major fraction of the liquid will remain on the packing during its passage through the bed. For Raschig rings, maximum bed height should be 2.5 to 3.0 times the bed diameter; for saddles, 5 to 8 times; and for slotted rings, 5 to 10 times. Total bed height should not exceed 20 ft.

Liquid redistributors (wall wipers) are sometimes installed within a bed to bring liquid back into the packing. For slotted rings or saddles, installation of these devices produces little improvement in performance if bed-height criteria have been met. A notable exception occurs when a very high purity must be achieved at the bottom of the column, and even a trace quantity of bypassed liquid can appreciably affect bottoms concentration.

Packing efficiency

Although numerous texts outline theoretical procedures for calculating packing height, there is no reliable method that can be universally applied to distillation systems. Instead, industry today uses the *HETP* concept to convert empirically the number of theoretical stages to packing height.

What makes this concept useful is that *HETP* values are remarkably constant for both organic and inorganic systems. Even with high-surface-tension liquids, good performance is possible so long as the packing wets properly (i.e., liquid rates are kept above 1,000 lb/(h)(ft^2), and difficult-to-wet plastics such as fluorocarbons are avoided.

In commercial columns, values of *HETP* for high-efficiency packings (slotted rings, Intalox saddles) run about 18 in for a 1-in nominal packing size; 26 in for a 1½-in size; and 35 in for a 2-in size. These values are about 6 to 12 in greater than published values derived from the operation of closely controlled pilot-plant columns [*8*].

Because of reduced irrigation efficiency in vacuum columns, it is usually wise to add another 6 in to the listed *HETP*. Absorption systems generally exhibit *HETP*s in the range of 5 to 6 ft. For small columns (dia. <2 ft), an old rule-of-thumb proves surprisingly accurate if the appropriate packing size is used and the packing is properly loaded.

$$HETP = \text{Column diameter} \tag{15}$$

For a particular type of packing, the effectiveness ($HETP/\Delta P$) is fairly constant for all sizes. Little will be

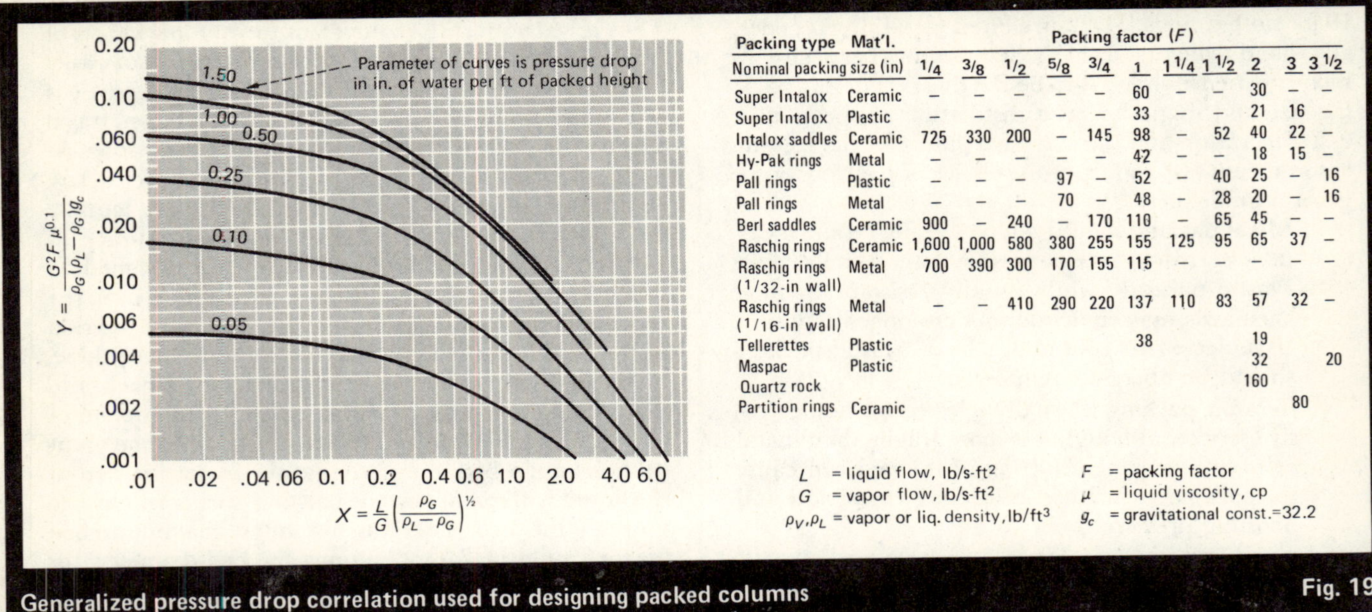

Packing type	Mat'l.	Packing factor (F)										
Nominal packing size (in)		1/4	3/8	1/2	5/8	3/4	1	1 1/4	1 1/2	2	3	3 1/2
Super Intalox	Ceramic	–	–	–	–	–	60	–	–	30	–	–
Super Intalox	Plastic	–	–	–	–	–	33	–	–	21	16	–
Intalox saddles	Ceramic	725	330	200	–	145	98	–	52	40	22	–
Hy-Pak rings	Metal	–	–	–	–	–	42	–	–	18	15	–
Pall rings	Plastic	–	–	–	97	–	52	–	40	25	–	16
Pall rings	Metal	–	–	–	70	–	48	–	28	20	–	16
Berl saddles	Ceramic	900	–	240	–	170	110	–	65	45	–	–
Raschig rings	Ceramic	1,600	1,000	580	380	255	155	125	95	65	37	–
Raschig rings (1/32-in wall)	Metal	700	390	300	170	155	115	–	–	–	–	–
Raschig rings (1/16-in wall)	Metal	–	–	410	290	220	137	110	83	57	32	–
Tellerettes	Plastic						38			19		
Maspac	Plastic									32		20
Quartz rock										160		
Partition rings	Ceramic										80	

L = liquid flow, lb/s-ft²
G = vapor flow, lb/s-ft²
ρ_V, ρ_L = vapor or liq. density, lb/ft³
F = packing factor
μ = liquid viscosity, cp
g_c = gravitational const. = 32.2

Generalized pressure drop correlation used for designing packed columns — Fig. 19

gained by replacing 2-in slotted rings with 1-in slotted rings to improve the *HETP* in a vacuum column, since the higher pressure drop will nearly cancel any reduction in height.

Systematically packed columns

These columns contain preformed sections having a large surface area for a given volume. In order to function as designated, the sections must be assembled within the column according to a prescribed arrangement.

Since systematic packing configurations have not been standardized, no design procedures can be universally applied to them. Engineering correlations are specific for each proprietary packing and are usually supplied by their respective manufacturers.

Koch-Sulzer Packing—This packing consists of parallel layers of corrugated wire gauze arranged in a sloping pattern. Because liquid flow is controlled by capillary action, superficial liquid rates as low as 250 lb/(h)(ft^2) are attained without undue penalty to efficiency.

The packing effectiveness ($HETP/\Delta P$) can be quite high at low pressures, making Koch-Sulzer packing a good candidate for high-vacuum distillation. Another application for which the Koch-Sulzer arrangement should be considered is the replacement of existing trays or packing in vacuum columns in which its lower pressure drop permits raising the column pressure to enhance system capacity. It must be recognized, however, that Koch-Sulzer packing is usually more expensive than any type of dumped packing.

For estimation purposes, it can be assumed that the *HETP* for all sizes of commercial equipment lies between 10 and 12 in, and that the column can be designed for an F factor ($v\rho_V^{0.5}$) between 1.7 and 2.0.

Knit-mesh packing—These packings are offered by a number of fabricators, and tend to have similar construction and substantially the same performance characteristics. Pressure drop is very low, but efficiency is not quite as high as that of Koch-Sulzer packing. *HETP* is somewhat affected by column dia., ranging from 6 in. in small columns to 24 in. in larger columns.

Commercially available mesh packings include Goodloe, Multifil and Hyperfil. Like Koch-Sulzer packing, mesh packing is woven from 6 to 7 mil wire and requires the use of materials of construction that will not corrode under column operating conditions.

Koch Flexipac—Flexipac is similar in construction to Koch-Sulzer packing, but instead of wire mesh uses corrugated metal sheets. Its performance falls somewhere between that of slotted rings and Koch-Sulzer packing.

Kloss and Neo-Kloss packing—These types are suitable for very low pressure operations where separating efficiency is not critical. Since there is no obstruction in the vertical direction (the packings are constructed, respectively, with an assembly of coiled springs, or with concentric cylinders), the pressure drop is extremely low. Their *HETP*s range from 2 to 4 ft.

Glitsch grid—This is an assemblage of stamped panels (having a high open-area) that are stacked on top of each other. The primary advantages of the Glitsch grid are its low pressure drop and high capacity. It is better suited for use in strippers and absorbers where liquid rates are high than in distillation columns where the *HETP* tends toward 6 ft.

Batch Distillation

Batch stills, once quite numerous, are seldom considered today because of the considerable labor and attention that must be lavished on their operation. There are still, however, situations where the choice of batch distillation is justified and favorable economics can be demonstrated. These include: small production runs (usually less than 1 million lb/yr); widely varying feed conditions and product requirements; irregular use of equipment; multiproduct separations; successive production runs with different processes.

Batch stills are mainly used today in pharmaceutical, fine-chemical, dye, cosmetic and liquor processes where the frequent practice is to process a variety of products in relatively small batches.

Single-stage distillation

The simplest example of batch distillation is single-stage, differential distillation, beginning with an initially full stillpot heated at a constant rate. The composition of the vapor leaving the kettle changes continuously but is always in equilibrium with the remaining liquid. Liquid not vaporized is removed as bottoms product at the end of the run.

In 1902, Rayleigh developed an expression that relates distillate and bottoms composition of the single-stage batch still to the fraction of the initial charge that has been vaporized. If the relative volatility is constant, or can be averaged, the following relationship applies:

$$\ln\left(\frac{L_1}{L_2}\right) = \frac{1}{\alpha - 1}\left(\ln\frac{x_1}{x_2} + \alpha\ln\frac{1 - x_2}{1 - x_1}\right) \quad (16)$$

where: L_1 = initial liquid in reboiler, moles; L_2 = final residual liquid, moles; x_1,x_2 = initial, final mole fractions of the more-volatile component in reboiler; y = vapor mole fraction; and α = relative volatility.

Since a simple batch still only provides a single theoretical stage, it is impossible to obtain a complete separation of a pure component unless the relative volatility is infinite. Therefore its application is usually restricted to preliminary recoveries that are later followed by a more rigorous distillation; to production where high purities are not required; and to processing of easy-to-separate mixtures, such as the removal of low boilers or residue from a product.

Distillation with rectification

To achieve a reasonable separation and to obtain product cuts of a desired purity in a batch still, it is necessary to rectify the vapors from the stillpot in a tray or a packed column. Usually such a column does not contain many trays, since a batch still consists solely of a rectifying section. As shown in Fig. 20, the system will pinch, regardless of the number stages, once the low boiler concentration in the bottom approaches the intersection of the operating and equilibrium lines.

As long as the concentration of the more-volatile component in the reboiler stays reasonably high, distillate purity will remain reasonably constant. However, as the component becomes depleted, overhead concentration drops rapidly.

When reflux ratio is held constant during the separation of two key components, the overhead composition will change continuously, and final distillate concentration will be the average of the entire cut collected in the product receiver. Too high a takeoff rate (low reflux ratio) will speed collection of product, but at the expense of product quality. Too low a product withdrawal rate (high reflux ratio) will result in maximum recovery of "on spec" product, but at an unacceptably long batch cycle.

To maintain a constant overhead composition, it is necessary that the reflux ratio be reduced continuously as distillation progresses. This delays onset of a column pinch, but does so at the expense of a longer distillation time. The product run is considered complete when a further increase in reflux ratio will reduce forward flow to such a low level that batch cycle time will be extended beyond acceptable economic limits.

The customary arrangement for operating batch columns is to combine constant and variable reflux ratios. Usually two or three different reflux ratios are applied to each overhead cut. It has been found that for a given number of stages, variations of reflux ratios and collecting times have little influence on the total capacity of a still if they are within reasonable limits established for operation of the still at either constant-reflux or constant-overhead composition.

Considerable effort has been expended by investigators to fix the effect of liquid holdup on separation efficiency in batch stills. For many years, any holdup in a column was considered detrimental. More-recent studies indicate that a limited amount of liquid in the column may actually optimize column capacity.

The existence of a finite holdup creates a "flywheel effect," which causes the composition to change more slowly than "instantaneous" equilibrium conditions tend to indicate. Thus, overhead purity will stay high longer than theoretical calculations might show, especially when the reflux ratio is relatively low (perhaps <8) and the content of the still is turned over rapidly. The flywheel effect is less pronounced when the reflux ratio is high, a condition frequently encountered when high-purity cuts are required.

Literature references seem to indicate that a column holdup of up to 10% may actually improve batch-still performance [3,7]. Above 15%, holdup appears to have a negative effect and column capacity is reduced. Capacity can be further optimized by reducing the external holdup of distillate. The elimination of reflux drums by inline flow-splitters will minimize equilibration time and thereby permit rapid concentration changes between cuts.

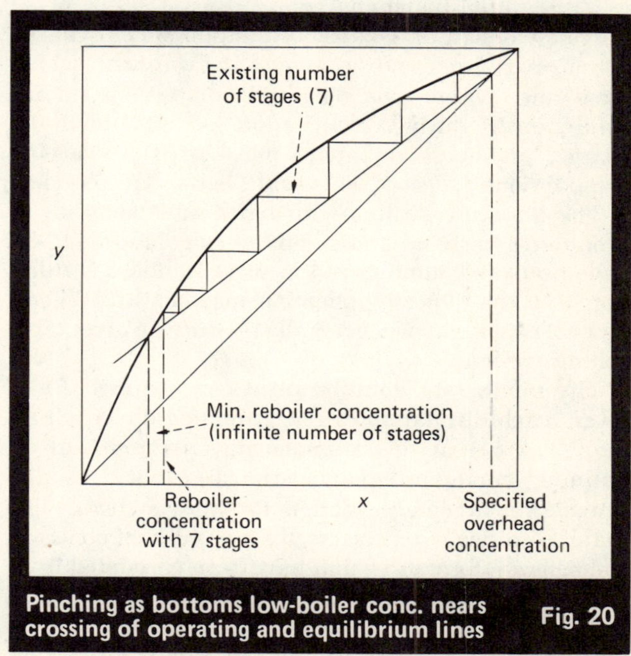

Pinching as bottoms low-boiler conc. nears crossing of operating and equilibrium lines Fig. 20

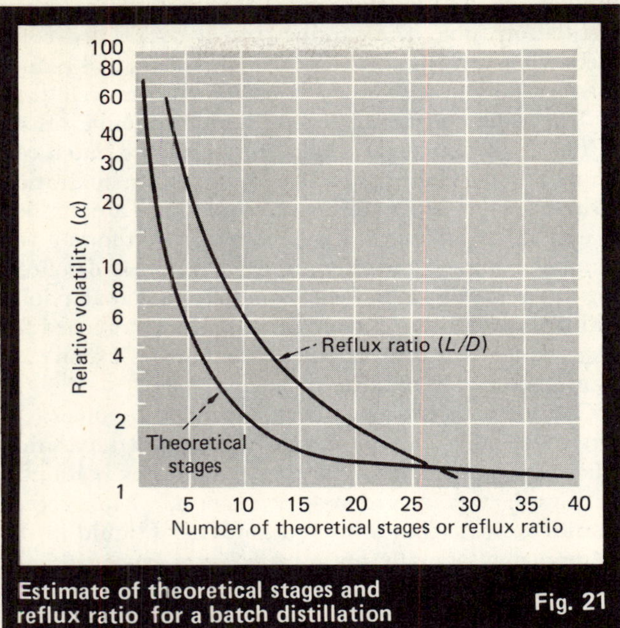

Estimate of theoretical stages and reflux ratio for a batch distillation **Fig. 21**

These rules-of-thumb apply to batch-still operation:

1. Once a column has been installed, capacity for a given product specification is only minimally affected by changes to reflux ratio and to the length of a product cut.
2. As the more-volatile component is removed from the reboiler, separation becomes progressively more difficult.
3. Too low a reflux ratio cannot meet the product specification no matter how many trays are installed.
4. It is impossible to recover in a single operation at high purity a low-boiling component that represents only a small fraction of the initial charge.
5. For optimum separation capacity, minimize or eliminate reflux holdup. Design for a liquid holdup in the column equivalent to 10 to 15% of the initial batch charge.

Precise design of a batch distillation system can be extremely complicated because of the transient behavior of the column. Not only do compositions change continuously during rectification of an individual charge, but successive batches may start from varying compositions as "slop" cuts and "heels" are recycled.

The literature dealing with batch stills abounds in procedures for hand and computer calculations. These have been well summarized in most standard distillation texts, and their application together with liberal doses of experience will generally produce an acceptable column design.

The number of equilibrium stages required for a given batch distillation can be estimated directly from Fig. 21. Whereas this estimate may not result in an optimum configuration in every case, it will usually provide an acceptable design for most systems. The graph empirically represents, for a number of commercial stills, performances that have been correlated from the basic Fenske equation.

This curve is considered suitable for separations requiring a product purity in the range of 300 to 8,000 ppm (0.03 to 0.8 wt%). If more than a single pure overhead product is to be recovered, the number of theoretical stages are determined by the smallest relative volatility between any two of the major system components.

The graph also contains an empirically developed curve for reflux ratios, which permits selection of the highest reflux ratio at which the column will have to operate any time during the batch cycle. Such a value is useful for preliminary selection of a column diameter and estimation of cut time.

Efficiency of batch columns may vary widely during a single cut as the concentration profile in the column shifts over a wide range. For long intervals, operation may take place under pinched conditions where efficiencies as low as 30% have been measured. Without experimental data, a reasonable value to expect for an overall batch column-efficiency is from 50 to 60%.

For packed columns, height will depend on types of internals used and column diameter. The following guidelines are recommended for converting chart values to packing height.

Size and type of packing	Multiply theoretical trays by:
≥2-in dumped (slotted rings or saddles)	3.5 ft
1.5-in dumped (slotted rings or saddles)	3.0 ft
≤1.0-in dumped (slotted rings or saddles)	2.0 ft
Mesh packing (≤12-in dia.)	0.5 ft
Mesh packing (>12-in dia.)	1.0 ft

References

1. AIChE Bubble Tray Design Manual, 1958.
2. Drickamer, H. H., and Bradford, J. R., *Trans AIChE*, Vol. 39, 1943, p. 319.
3. Eichel, F. G., Efficiency Calculations for Binary Batch Rectifications with Holdup, *Chem. Eng.*, July 8, 1966, p. 159.
4. Glitsch, Inc. Bulletin #4900, 3rd ed., 1974.
5. King, C. J., "Separation Processes," McGraw-Hill, New York, 1971.
6. Koch Flexitray Design Manual, Bulletin 960, 1960.
7. Luyben, W. L., Some Practical Aspects of Optimal Batch Distillation Design, *Ind. Eng. Chem. Proc. Des. Develop.*, Vol. 10, No. 1, 1971.
8. Norton Co., "Packed Towers."
9. O'Connel, H. E., *Trans. AIChE*, Vol. 42, 1946, p. 741.
10. Perry and Chilton, C. H., "Chemical Engineers' Handbook," Section 13, McGraw-Hill, New York, 5th ed., 1973.
11. Smith, B. D., "Design of Equilibrium Stage Processes," McGraw-Hill, New York, 1963.
12. Sum-Shik, others, *Chim. Prom.* 13, Vol. 2, 1968, p. 66.
13. Treybal, R. E., "Mass Transfer Operations," 2nd ed., McGraw-Hill, New York, 1968.
14. Van Winkle, M., "Distillation," McGraw-Hill, New York, 1967.
15. Van Winkle, M., and Todd, *Chem. Eng.*, Sept. 20, 1971, p. 136.

The author

Otto Frank is Supervisor of Process Design and Methods at Allied Chemical Corp., where he is responsible for developing and testing design procedures for the Corporate Engineering Dept. With Allied Chemical since 1960, he was formerly employed by Air Products, Inc., and by Du Pont. A registered professional engineer in New York, he holds a B.S. in chemical engineering from Clarkson College of Technology and an M.S. in chemical engineering from Princeton University.

Energy-saving schemes in distillation

Distillation systems can add up to a significant amount of the total energy requirements of a processing scheme, and there are many techniques that can be used to reduce such energy consumption. But before applying any technique, it is necessary to recognize the working relationships between capital, operating costs, and plant operability.

William C. Petterson and **Thomas A. Wells,**
Pullman Kellogg Div., Pullman Inc.

☐ Designing strictly for energy conservation is as ill-advised as ignoring trends in energy prices. Capital is equally critical; it must be effectively used to purchase reliable systems. Gross margins in the chemical process industries (CPI) are such that most potential energy savings cannot offset product loss resulting from an unreliable system.

Here, some basic chemical engineering techniques are discussed, which are being applied to distillation sequences in light-hydrocarbon (i.e., olefins) recovery plants to reduce energy consumption, while maintaining high levels of flexibility and operability.

Utility values

Specification of utility values for a particular evaluation has a controlling effect on the conclusions that are drawn, because one must consider the use of the utility as it relates to the entire pattern of the unit. For the purpose of following our discussion, the stipulations listed below will be used to develop utility values:

■ The plant is highly "work-oriented," with extensive use of high-pressure steam turbines. Low-pressure steam is supplied by turbine exhaust to the low-pressure steam header. The relationship between low-pressure and high-pressure steam costs must account for the work extracted in the turbine in reaching the low-pressure header. It will be assumed that the startup steam requirements for the unit prescribe a high-pressure boiler in excess of any normal steam-combination capacity. The utility-cost factor for steam levels will contain no capital charges for incremental production.

■ The cooling-water system capacity is a direct function of anticipated normal operation. Therefore, incremental water-cooling consumption must be included in capital charges. Table I summarizes utility costs estimated for a plant in the 1979–1980 period.

Plant utility costs in the year 1979-1980 — Table I

Material	Cost
Fuel	$3.00/10^6 Btu
High-pressure steam, 1,500 psig	$4.15/10^3 lb* $3.33/10^6 Btu
Low-pressure steam, 50 psig	$1.75/10^3 lb† $1.60/10^6 Btu
Cooling water, 30°F rise	12 ¢/10^3 gal** 48 ¢/10^6 Btu
Electrical power	3 ¢/kWh

*Based on 90%-efficiency boiler.
†Based on 75%-efficiency turbines.
**Includes capital charges.
Note: Incremental high-pressure steam used for the production of work is for condensing water at a rate of 6.75, which results in $230/[(hp)(yr)].

Originally published September 26, 1977

Interreboiler application — **Table II**

Equipment duty	System with interreboiler	System without interreboiler
Rectifying trays	90	90
No. of stripping trays	45	28
Condenser duty, 10^6 Btu/h	135	135
Interreboiler duty, 10^6 Btu/h	50	0
Reboiler duty, 10^6 Btu/h	50	100
Top section:		
Height, ft	155	155
Diameter, ft	18	18
Thickness, in.	2.2	2.2
Bottom section:		
Height, ft	89	64
Diameter, ft	16	18
Thickness, in.	1.8	2.2
Tower cost, $\$10^6$	5.1	4.8
Exchanger cost, $\$10^6$	3.8	3.7
Refrigerating system cost, $\$10^6$	–0.1	0.0
Total capital cost, $\$10^6$	8.8	8.5
Operating cost, boiler hp,	14,850	17,050
$\$10^6$/yr	4.15	4.77

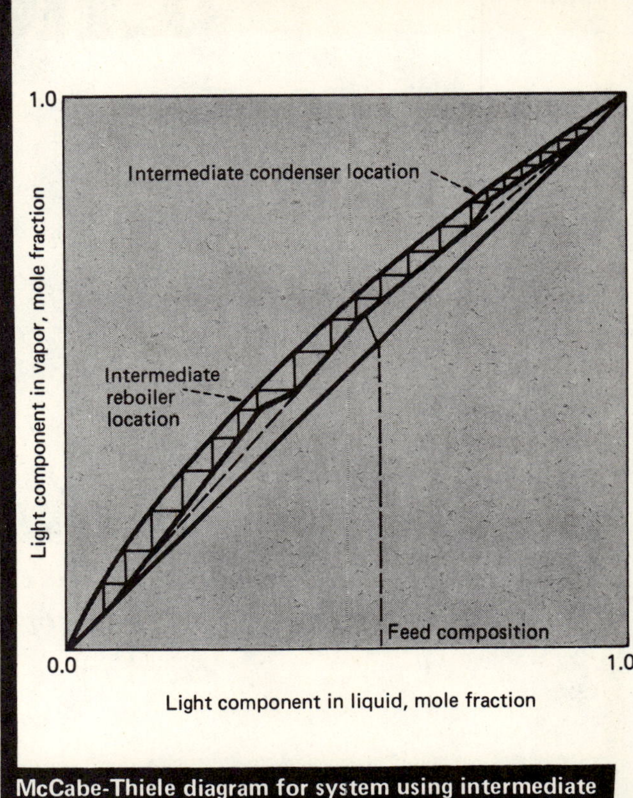

McCabe-Thiele diagram for system using intermediate condenser and intermediate reboiler **Fig. 1**

Interreboilers and intercondensers

Interreboilers and intercondensers—when applied in accordance with typical economic and operating criteria—can produce significant reductions in the operation cost of a distillation system.

The generally accepted approach of applying heat only at the bottom of the tower, and withdrawing heat only at the top, is most often directed by the economic and operability requirements imposed on the design. In situations where energy costs are low, the thermodynamic inefficiencies inherent with this approach are usually not worth reducing.

However, in multistage distillation, it is possible to add and remove heat at numerous locations in the distillation column. It is theoretically possible—but seldom practical—to apply this concept to each equilibrium stage in the column by adding finite quantities of heat to every stripping stage, and removing finite quantities of heat from every rectification stage.

For a specific case where the feed and product rates and purities are constant, and for a particular condenser duty, the total heat applied to the stripping section has a unique value, regardless of the number of places where it is put into the stripping section. The economics of multiple reboiling lie in the ability of the system to utilize multiple levels of heat. In a single reboiler system, the entire heat load must be applied to the base of the column and, therefore, must have a high temperature. Since the cost of heat energy is usually a function of departure from ambient temperature, this single input of high-temperature energy is the most expensive method of reboiling a distillation system.

When the same amount of energy is divided up and added to several intermediate points—between the feed tray and the bottom tray—the temperature levels of the energy can be progressively lower as the feed tray is approached. The temperature of the energy source at a particular location must be higher than the tower

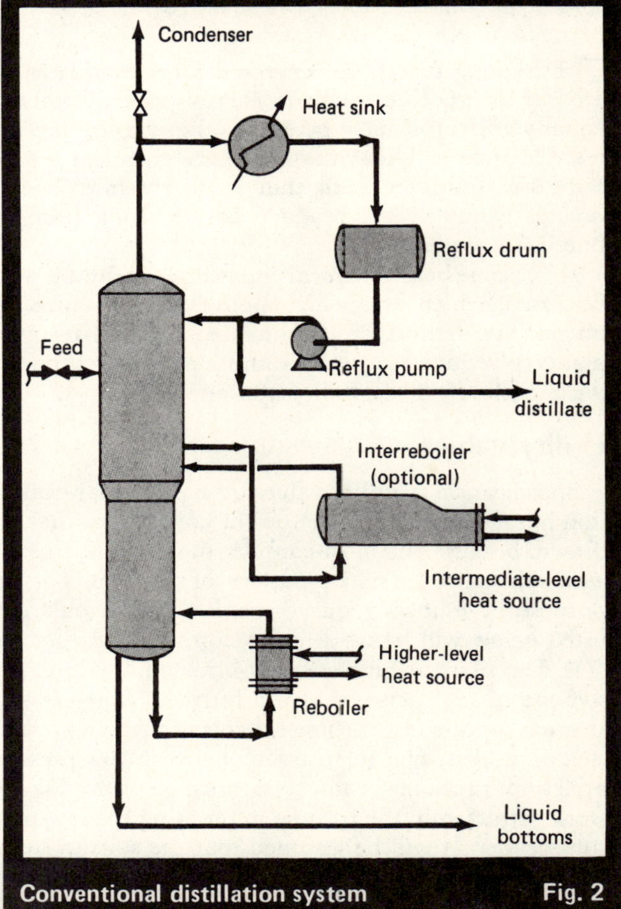

Conventional distillation system **Fig. 2**

Comparison of heat pump and alternative systems — Table III

Equipment costs and dimensions	Cooling-water steam-system	Heat-pump system	Cooling-water/ waste-heat system
Tower overhead temperature, °F	100	40	100
Optimum reflux ratio	6.9	5.6	7.2
Above minimum reflux, %	5	7	13
Tower diameter, ft	15.9	13.9	16.3
Tower height, ft	224	157	190
Shell thickness, in.	1 5/16	9/16	1 3/8
Condenser duty, 10^6 Btu/h	121	–	130
Condenser area, ft^2	61,000	–	65,900
Reboiler duty, 10^6 Btu/h	124	123	134
Reboiler area, ft^2	3,600	123,000	32,000
Compressor requirement, boiler hp	–	5,550	–
Total system capital cost, $\$10^6$	4.2	4.9	4.3
Condensing medium	Cooling water	Heat pump	Cooling water
Medium unit cost, $\$/10^3$ gal	0.12	–	0.12
Total condensing cost, $\$10^6$	1.9	–	2.0
Reboiling medium	Low-pressure steam	–	Circulating hot water
Medium unit cost, $\$10^6$ Btu	1.60	–	(0.07)
Total reboiling medium, $\$10^6$	7.5	–	(0.30)
Total four-year operating cost, $\$10^6$	9.4	5.0	1.7
Total capital + 4-yr operating cost, $\$10^6$	13.6	9.9	6.0
Total annual cost, $\$10^6$	3.4	2.5	1.5

liquid at that point, but only by the amount that results in an economic quantity of heat-transfer area in the intermediate reboiler. The same concepts apply to the rectification section, where instead of using a single low-temperature energy sink, heat can be removed at several locations, which use successively warmer heat sinks as the feed tray is approached.

Using multiple condensers and multiple reboilers can have significant effects on the design of a distillation column itself. The number of distillation stages required to achieve a particular separation is determined by the ratio of liquid-to-vapor flowrates in the column. The column diameter is set by the magnitude of these flows.

Fig. 1 shows a McCabe-Thiele diagram for a binary system, which employs an intermediate condenser and an intermediate reboiler. This diagram shows the significant changes in the liquid-to-vapor ratio that occur above the intercondenser and below the interreboiler. These changes cause an increase in the number of distillation stages required in both the stripping and rectification sections. These increases are a function of the location of the intermediate level systems, and the percentage of the total duty requirements satisfied by these systems.

There is also a potential opportunity to reduce the column diameter above the intercondenser and below the interreboiler, because of the reduced vapor and liquid traffic in these sections. The application of these systems to a distillation column should be studied and optimized considering the following points:

1. The applicable levels of heat or cooling sources determine the point of application of intercondensers and interreboilers. The magnitude of the potential energy savings dictates their applicability to a particular system.

2. The increased loading of the intermediate systems causes changes in the overall tower height and diameter, as well as in the heat-transfer area.

3. The use of less-expensive energy levels reduces the overall operating cost of a particular distillation system. This changes the capital and energy cost relationship, which shifts the optimum operating point to greater percentages above the minimum reflux.

4. The reliability of the heat sources, and the accu-

Economic summary for split-tower operations — Table IV

Equipment cost and dimensions	High-pressure tower	Low-pressure tower
Tower height, ft	225	205
Tower diameter, ft	10.0	9.3
Shell thickness, in.	1 1/16	3/4
Number of trays	123	110
Compressor, boiler hp	5,240	
Capital for total, \$ millions (System including exchangers, pumps compressor, etc.)	5.84	
Operating cost (4 yr), \$ millions	5.13	
Total capital + 4-yr operating cost, \$ millions	10.97	
Total annual cost, \$ millions	2.74	

Note: This system is comparable to the examples treated in Table III.

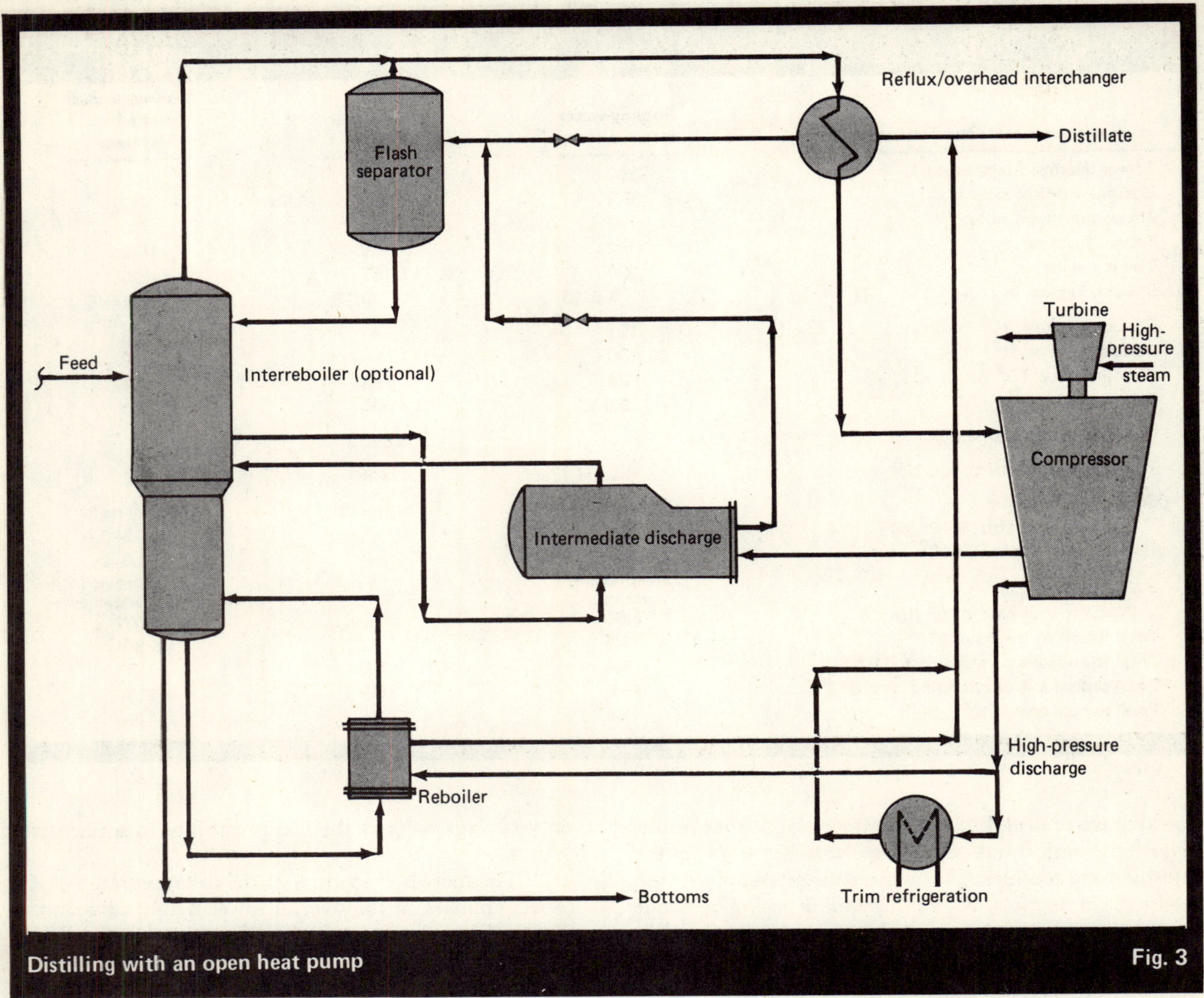

Distilling with an open heat pump
Fig. 3

rate determination to the column internal loading and conditions, must be considered when designing for different levels of flexibility and reliability. The effects of alternate feed composition must also be carefully evaluated, before committing the system to a particular configuration.

As an example of the attractiveness of interreboilers, Table II shows a summary of key factors of two distillation systems. One is equipped with an interreboiler (Fig. 2), while a similar one is not. Each system produces identical products from a given feed. The use of an interreboiler system for this example shows that it takes half a year to recover the capital investment.

It should be emphasized that not every system can economically utilize intermediate reboilers and intermediate condensers. Even less common are systems that can justify more than one intermediate system. The greatest potential exists with systems that use medium- or high-pressure steam, or low levels of refrigeration in the distillation process.

Heat pumps

Distillation systems that use heat pumps have long been used in chemical processing. The recent upsurge in popularity is attributable to potential savings in operating expenses. In a conventional distillation system, energy is used on a once-through basis, entering from a high-temperature heat source and exiting to a low-temperature heat sink. The cost of separation is very high, because the gross energy is totally degraded from a combustion of over 3,500°F in the steam-generation system to ambient temperature in the cooling-water air exhaust. This system owes its existence to simplicity, low investment and cheap energy.

The heat pump, on the other hand, takes the energy from the condenser and uses work to elevate it to a level high enough to be transferred to the column reboiler. This energy—that is totally degraded—is the net work that is required to transfer the energy from the overhead to the reboiler (work of separation), plus the lost work resulting from converting fuel to steam, and then to the work of separation.

The heat pump represents a significant reduction in energy consumption, but at the sacrifice of capital and simplicity. It is important to note that the heat pump configuration is most advantageous when the fractionation system has a low-temperature difference across the column. When the temperature difference expands, the cost of recycling energy increases, and the heat pump loses some of its attractiveness.

Fig. 2 and 3 are a comparison between a conventional distillation system—which utilizes steam in the reboiler and cooling water as a condensing medium—and the same separation carried out in an "open" heat pump. In the open heat-pump system, the gross tower overhead is compressed to a pressure at which its dew-point temperature is high enough to provide a satisfactory temperature approach in the reboiler, where the overhead is condensed while supplying heat to the reboiler.

When there is concern with possible product contamination by compressor oils, the system can be closed by simply providing a conventional overhead condenser to isolate the tower system from the refrigerant system. The closed system imposes the additional thermodynamic inefficiency of a temperature difference between the condensing-tower overhead vapors and the vaporizing refrigerant liquid. The condensed liquid leaving the reboiler—which serves as reflux for the tower—is subcooled against tower overhead vapors. This reduces flash vapors when the reflux is introduced into the column. In most applications, the saving in the reduction of flash vapor recycle is greater than the expense of a superheated compressor suction, but the system should be evaluated for each application.

The heat-pump system requires a certain balance of thermal loads to minimize the use of external heat or refrigeration sources. The thermal balance of the system is particularly sensitive to the thermal condition of the feed and products. For instance, the requirement for a vapor feed and liquid overhead product may impose a case where insufficient reboiler duty is available to condense the necessary reflux plus the product. An external refrigeration source would be required to make up the difference.

Table III shows an analysis of a distillation system that is first developed as a tower—condensed by cooling water and reboiled by steam—and then treated as a heat-pump system. The economic attractiveness of a heat pump is readily apparent when this comparison is made. However, the existence of an alternative free-energy source is not an uncommon occurrence when critically examining an integrated process for energy conservation when a temperature of 175°F is available. This temperature might be too low to provide an attractive heat source in other applications and would, therefore, require cooling by air or cooling water to 110°F before reintroduction to the processing scheme.

As is usually the case, waste-heat recovery represents a very significant improvement in any system when it can be used as a direct replacement for steam. In addition to the utility values given in Table I, the circulating hot water that serves as a heat source is assigned a 7¢/10^6-Btu credit, because its reuse in the process eliminates the operating cost of cooling water and air cooling that must be used to otherwise cool the circulating water to its required temperature. However, these capital items in the hot-water cooling loop must remain in place to provide flexibility.

The results of this brief analysis indicate the advisability of using waste heat to its fullest advantage. However, in situations when this option is not feasible, then the heat pump can be an attractive substitution for steam heat, in spite of the higher capital cost and increased complexity. There are several variations of heat-pump and tower configurations; a brief literature survey will yield several practical techniques and guidelines for their applications.

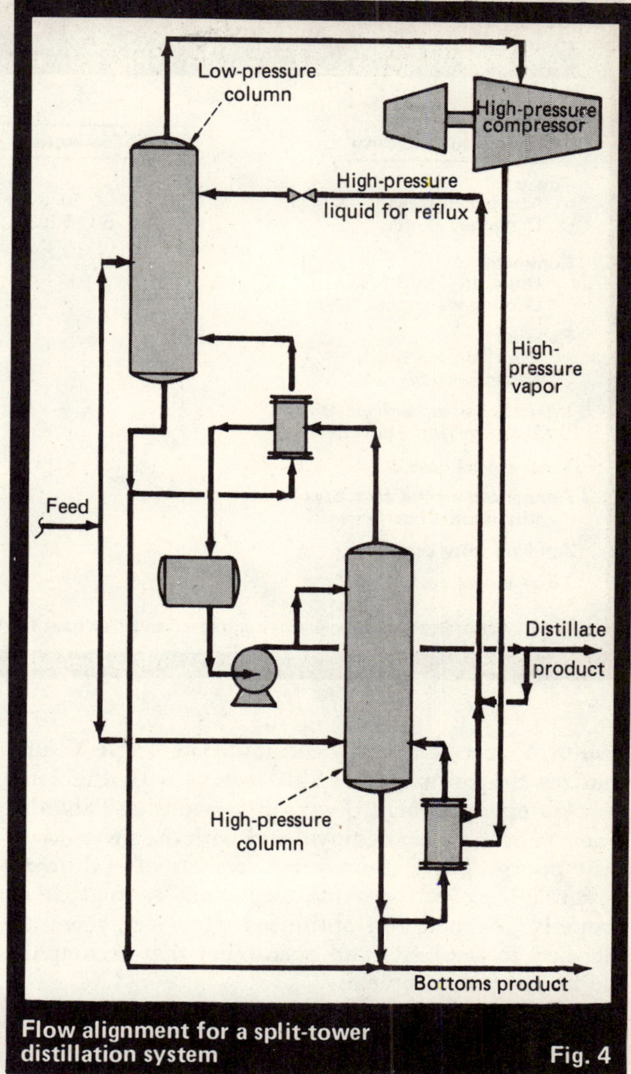

Flow alignment for a split-tower distillation system Fig. 4

Split-tower arrangements

There are several possible arrangements that utilize two separate towers that are "thermally linked." Basically, they consist of a high-pressure tower and a low-pressure tower operating in parallel—such as in Fig. 4—so that the high-pressure tower condenser is used as a source of heat for the low-pressure tower reboiler. For the example considered below, the products from both towers have the same composition, which result from parallel operation on the same feed. The feed-split between the towers depends on the thermal condition of the feed (Table IV).

The linking of the towers allows the same fractionation as the heat-pump case (Table II) with about half the reboiler and condenser duty. However, the much greater temperature difference across the overall system

Interreboiler versus feed heater **Table V**

<table>
<tr><th rowspan="2">Equipment and duty</th><th colspan="2">Feed heater</th><th rowspan="2">Interreboiler</th></tr>
<tr><th>Top section</th><th>Bottom section</th></tr>
<tr><td>Tower</td><td></td><td></td><td></td></tr>
<tr><td>Number of trays</td><td>20</td><td>16</td><td>45</td></tr>
<tr><td>Diameter</td><td>6 ft 6 in.</td><td>5 ft 0 in.</td><td>5 ft 6 in.</td></tr>
<tr><td>Condenser</td><td colspan="2"></td><td></td></tr>
<tr><td>Duty, million Btu/h</td><td colspan="2">26.0</td><td>22.3</td></tr>
<tr><td>Cooling-water cost, $/yr</td><td colspan="2">104,100</td><td>89,500</td></tr>
<tr><td>Reboiler</td><td colspan="2"></td><td></td></tr>
<tr><td>Duty, million Btu/h</td><td colspan="2">10.6</td><td>10.0</td></tr>
<tr><td>Steam cost, $/yr</td><td colspan="2">141,900</td><td>134,300</td></tr>
<tr><td>Interreboiler or feed-heater</td><td colspan="2"></td><td></td></tr>
<tr><td>Duty, million Btu/h</td><td colspan="2">15.3</td><td>12.3</td></tr>
<tr><td>Total capital cost, $</td><td colspan="2">536,500</td><td>524,400</td></tr>
<tr><td>Annualized capital cost, $/yr
(distributed over 4 years)</td><td colspan="2">134,100</td><td>151,400</td></tr>
<tr><td>Annual utility cost, $/yr</td><td colspan="2">246,000</td><td>223,800</td></tr>
<tr><td>Total annual cost, $/yr</td><td colspan="2">380,100</td><td>375,200</td></tr>
</table>

Note: Assuming a cooling-water condenser and free heat for feed heater or interreboiler.

results in increased energy consumption. Table V summarizes the operation of this system with the same heat-pump equipment discussed previously (Table II). The results show near equivalency with the single-tower heat pump. There are several versions of split-tower designs. They can provide significant savings when properly designed and optimized. However, there are sacrifices in flexibility and operability that accompany the increase in complexity.

Feed optimization

A distillation column feed can vary from subcooled liquid to superheated vapor. The thermal condition of the feed is an important parameter in the design of a distillation column because changes in the condition can affect both the capital and operating costs for a given system. The following examples show how operating costs can change significantly with changes in the feed condition indicating that, in the optimization of a distillation system design, the feed condition cannot be ignored.

A distillation column feed may come from one of many types of processing equipment, such as another distillation column, a heat exchanger or a reactor. An example of how the feed condition can be easily modified occurs when the feed stream comes from the overhead condenser of a preceding distillation column. It can be taken either as all vapor (if it can be pressured into the next column) or as all liquid. In other instances, an additional heat exchanger may be needed to either heat or cool the feed stream. Sometimes temperature limitations for exchanger-fouling considerations limit the heat input to a feed stream. These limitations cannot be ignored in the design of a distillation system.

Cooling water, steam, and various other heat sources may be used to modify the feed condition. If steam is used, it must be at a lower pressure, and thus at lower cost than that being used in the column reboiler. The ideal source of heating or cooling is either direct or indirect exchange with another process stream. These sources are usually free.

Consider a typical distillation column that produces 80% of the feed as the overhead product. The changes in condenser and reboiler duties for changes in feed condition are shown in Fig. 5. The condenser duty changes only moderately, while the reboiler duty is more than halved as the feed is shifted from saturated liquid to saturated vapor. For a tower condensed with cooling water and reboiled with 50-psig steam, the utility costs are 37% lower for the all-vapor feed. Capital costs change very little over the range of feed conditions, so the optimum is an all-vapor feed (see Fig. 5). Therefore, if the column feed comes from the overhead of a preceeding tower, it should definitely be taken as all vapor, if possible.

Under other circumstances, such as when the feed is available as a liquid, the appropriate economic analysis must be performed to justify additional capital expenditures. When an essentially free source of heat is readily available—such as from a process stream which must be cooled—it is usually economically justifiable to vaporize, at least partially, an all-liquid feed in systems such as considered in the paragraph above.

A much different situation exists if the fractionation column separates only 20% of the feed into the overhead product. Now, the reboiler duty is almost constant, while the condenser duty almost doubles as the feed changes from all liquid to all vapor (see Fig. 6). For a tower condensed with cooling water, the net annual cost is minimized when the feed is about 40% vapor. If the stream is available as a liquid, the cost of heating it is not justified even if the heat is free. The savings between zero and 40% vaporization are not enough to justify the cost of the feed heater. If the stream is available as a vapor and can be condensed with cooling water, heating should, again, not be considered. The

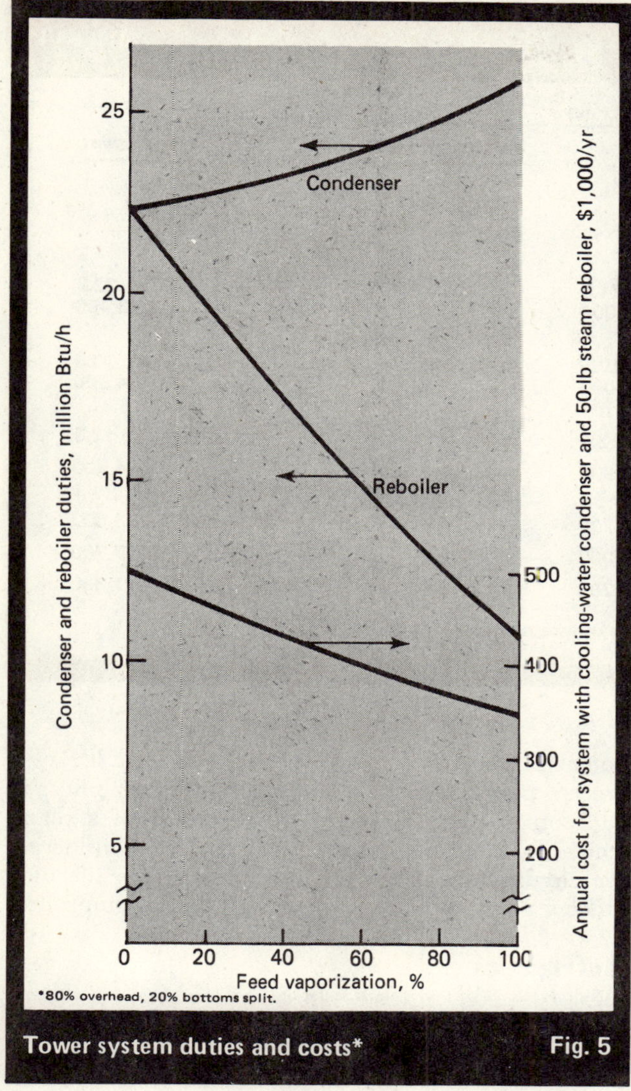

Tower system duties and costs* Fig. 5

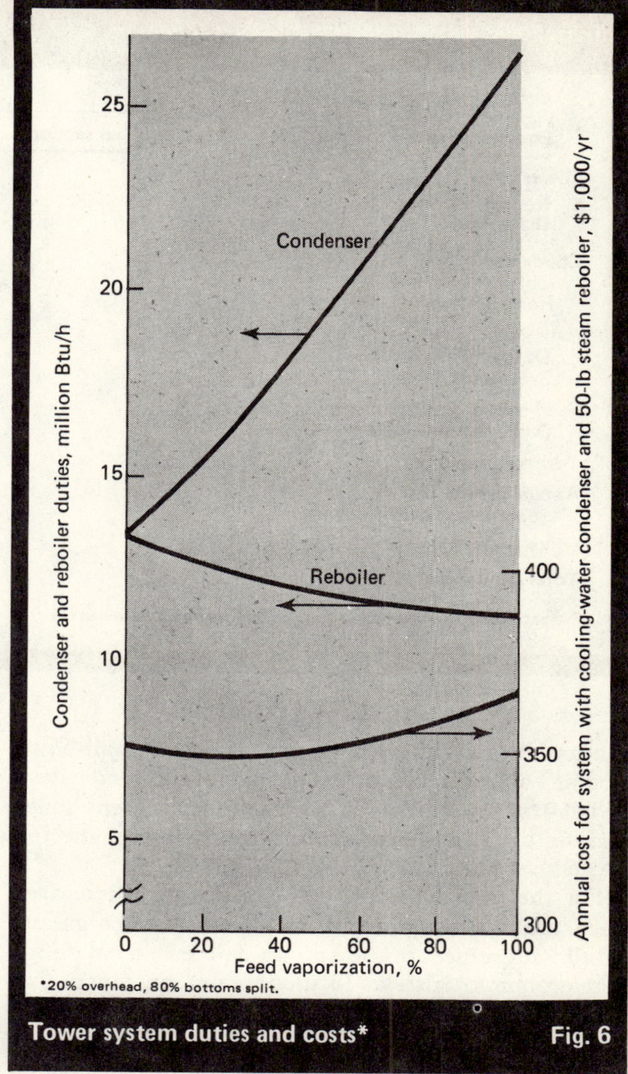

Tower system duties and costs* Fig. 6

cost of exchanger surface, as well as cooling water, exceeds any saving that might be incurred.

However, if the tower has a refrigerated condenser, the high cost of refrigeration, relative to cooling water, greatly effects the economic comparison. Increasing condensing duty significantly affects operating cost to such an extent that it is justifiable to condense the feed in an added cooling-water exchanger, if this is at all possible.

There are many variations that can be considered; it is difficult to make any generalizations as to the optimization of feed condition. The preceding examples provide an indication of the effects of the overhead- and bottoms-product splits, and shows variations in condensing costs on the optimization of the feed condition.

Feed heater versus interreboiler

Both feed heaters and interreboilers provide the use of a lower-level, and thus lower-value, heat source to reduce the total utility cost for a fractionation system. However, both also affect the capital cost of the system. The interreboiler causes an increase in capital costs because of an increased tray requirement. This increase can be partially compensated for by taking advantage of the lower vapor rates below the interreboiler, and reducing the tower diameter in that section. However, reducing the tower diameter does restrict operating flexibility. The feed heater tends to reduce the capital cost of the system.

For a cooling-water, condensed-distillation system producing 80% of the feed as the overhead product, the optimum design for the tower corresponds to a totally vaporized feed, if the heat source is free. This system is compared to a tower with an all-liquid feed and an interreboiler in Table V. The net annual cost for the system, which utilizes a feed heater, is $375,200/yr, compared to $380,000/yr for the system with an interreboiler. In this example, the feed-heater system would probably be selected because of operating advantages over the interreboiler system. The most important of these advantages is the larger temperature approach in the feed heater, as compared to the interreboiler, if the same heat source is used. This results from the lower temperature of the tower feed as compared to any interreboiler feed.

If this tower required a refrigerated condenser, the economics and conclusions would be different. Because of the high cost of refrigeration, the optimum design is

Interreboiler versus feed heater — Table VI

Equipment and duty	Feed heater: Top section	Feed heater: Bottom section	Interreboiler
Tower			
Number of trays	20	16	45
Diameter	6 ft, 6 in.	5 ft, 0 in.	5 ft, 6 in.
Condenser			
Duty, million Btu/h	24.8		22.3
Refrigeratingcost, $/yr	714,000		643,400
Reboiler			
Duty,million Btu/h	12.5		10.0
Steamcost, $/yr	167,000		134,300
Feed heater or interreboiler			
Duty, million Btu/h	12.4		12.3
Total capital, $	576,900		654,400
Annual capital cost, $ (distributed over 4 years)	144,200		163,600
Utility cost, $/yr	881,000		777,700
Total annual cost, $/yr	1,025,200		941,300

Note: Assuming a refrigerated condenser and free heat for feed heater or interreboiler.

compared to an interreboiler system with an all-liquid feed in Table VI. This economic comparison indicates a $38,900/yr (about 8%) advantage for the interreboiler, making it the preferred alternative for use of the free waste-heat source.

For the final selection between design alternatives, the actual systems must be evaluated in much greater detail with respect to the applicable mechanical design and economic criteria. It should also be noted that many of the optimum design points determined in this article exist on very shallow cost curves. Therefore, the penalty for missing the optimum design point is often small.

Because of this, it is desirable to consider both the potential operating costs over the expected life of a process plant and the plant's operating stability, in addition to the criteria for the early life of a plant, which are often the bases for a design. By considering the circumstances that might exist over the life of a process plant, a better design may be accomplished.

References

1. Null, H. R., Heat Pumps in Distillation, *Chem. Eng. Progr.*, July 1976, pp. 58–64.
2. Wolf, W., et al., Energy Costs Prompt Improved Distillation, *Oil and Gas J.*, Sept. 1975 (Reprint).
3. Reus, T. V., and Luyben, W. L., Two Towers Cheaper Than One?, *Hydrocarbon Process.*, July 1975, pp. 93–96.
4. M. W. Kellog Co., U.S. Patent No. 3,000,188, Sept. 9, 1961, Gas Separation (Heat Pump).

The authors

William C. Petterson is a senior process engineer with the Pullman Kellog Div., Pullman Inc., Three Greenway Plaza East, Houston, TX 77046. His responsibilities include process design of petrochemical and specialty chemical processes that have extensive application in fractionation equipment. He holds a B.S.Ch.E from the University of Cincinnati and a M.S.Ch.E from Rice University.

Thomas A. Wells is a senior process engineer with the Pullman Kellogg Div., Pullman Inc., Three Greenway Plaza East, Houston, TX 77046. His activities include process design and development in all areas of the olefins flowscheme. He holds a B.A. in chemistry from Drury College and a B.S. and M.S. in chemical engineering from the University of Missouri-Rolla.

Layout arrangements for distillation columns

Trouble-free operation and maintenance are the principal aims when evaluating design for tower internals, piping connections to tower nozzles, and external piping to other process equipment.

Robert Kern, *Hoffmann - La Roche Inc.* *

☐ To deal with the layout and piping design of specific equipment in chemical process plants, we must know how to plan the overall plant-design activities and meet the requirements of plant design.†

We will start with the layout of distillation columns. In later articles, we will cover the arrangement of heat exchangers, process drums, pumps, heaters, instrument and electrical facilities, and pipe-rack design. Finally, we will take all these components and assemble them into a plot plan for an outdoor process unit or for an enclosed (i.e., housed) chemical plant.

In distillation-column design, we will answer the following:

1. What information is required?
2. What are the design activities when evaluating (a) column internals and pipe connections, and (b) flow diagrams, and process and project data?
3. How do we analyze a distillation-column layout and its piping in plan and elevation?
4. What information is produced?

Required information

For arranging distillation columns, the designer is supplied with the following information:

☐ *Layout and piping specifications* describe minimum access, walkways, platform width and headroom requirements; handling facilities for tower internals, manhole covers, line blinds, relief valves; maximum rise of ladders; pipe-system requirements, such as open or closed relieving systems, minimum line size and required hose stations; access to valves and instruments.

☐ *Design standards* show details of ladder dimensions, ladder and platform position (step-through or side-step

Originally published August 15, 1977

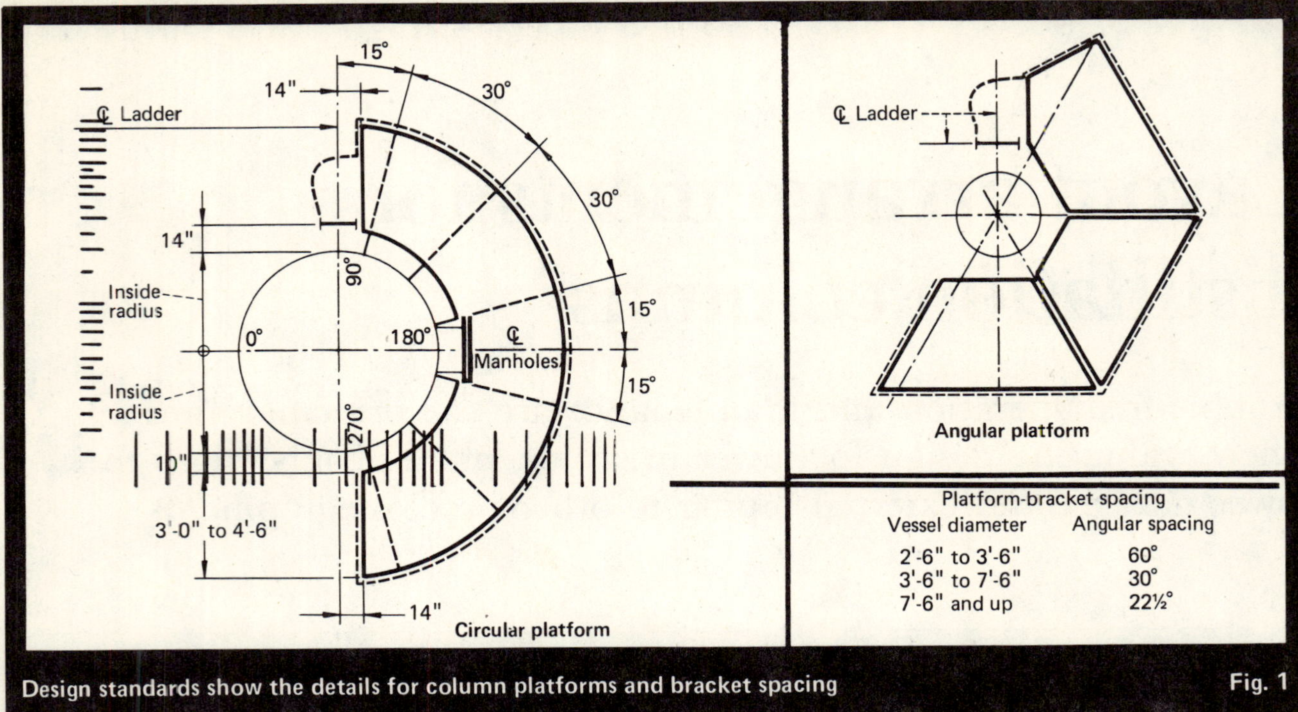

Design standards show the details for column platforms and bracket spacing — Fig. 1

landings); toe-plate, handrail and safety-gate details. For example, Fig. 1 illustrates the details for column platforms and bracket spacing. Design standards usually give the minimum skirt heights related to diameters and operating temperatures.

□ *Process data* come in the form of process and instrumentation (P&I) flow diagrams. These show interconnected equipment and piping; pipe sizes and pipeline components; steam tracing and insulation thickness. Tower elevations, and differences in equipment elevations, are given on the flow diagram or equipment-elevation data sheets.

□ *Plot plan* gives the physical location of a distillation column and its relationship to other equipment, main access, main pipe runs, and pumps.

□ *Instrument sketches* or instrument standards show the location of instrument connections to the tower for gages, level controllers, and level alarms, and location (without orientation) of pressure and temperature connections. The complete instrumentation system is shown on the P&I, or on a separate process-control diagram. (The physical and dimensional details of instruments, equipment, piping and components are available from manufacturers' literature.)

□ *Vessel drawing* for the distillation column gives diameter and length, details and dimensions of internals, and manhole and process-piping connections in elevation (without orientation). Drum, pump and exchanger drawings present details of adjacent process equipment or equipment supported on the column itself.

An integrated layout must be developed from this information. Requirements written in the specifications must be applied. The designed unit should be a functional part of the plant, properly laid out with respect to related equipment and facilities, as shown on the plot plan. Vessel drawings, instrument sketches, design standards, pump or reboiler data, and dimensioned equipment sketches will govern the external arrangement of the distillation-column piping.

Vessel internals and pipe connections

Frequently, interpretation of process requirements inside a tower is more exacting than exterior-piping design. Often, the location of an internal part determines within strict physical limits the location of tower nozzles, instruments, piping and steelwork. The piping-layout designer must concentrate his attention on a large-scale diagram of tower-internal details and the arrangement of process piping.

Let us consider how these factors enter into the layout of a typical distillation column whose dimensions and internals are sketched in Fig. 2. Dimensions and design are developed by the process-vessel specialist. Tower throughput, vapor and liquid flows, working pressures and temperatures determine vessel diameter and wall thickness. The required fractionation determines the number of trays and the tower height.

Depth of liquid in the tower bottom and, consequently, the length of the shell beneath the bottom trapout-boot is largely controlled by the required holding time for the bottom product.

Vapor space beneath the bottom tray is influenced by the reboiler circuit when a reboiler is provided. Reboiler draw-off quantities and vapor-liquid density govern both outlet-nozzle and trapout-boot sizes.

Tower length is further increased by the space above the top tray for accommodating overhead-pipe outlet, reflux-inlet distributor, and manhole. The minimum tray spacing is increased, if necessary, at each manhole location.

The pressure-vessel designer sets the elevation of each nozzle relative to tower internals. The piping designer

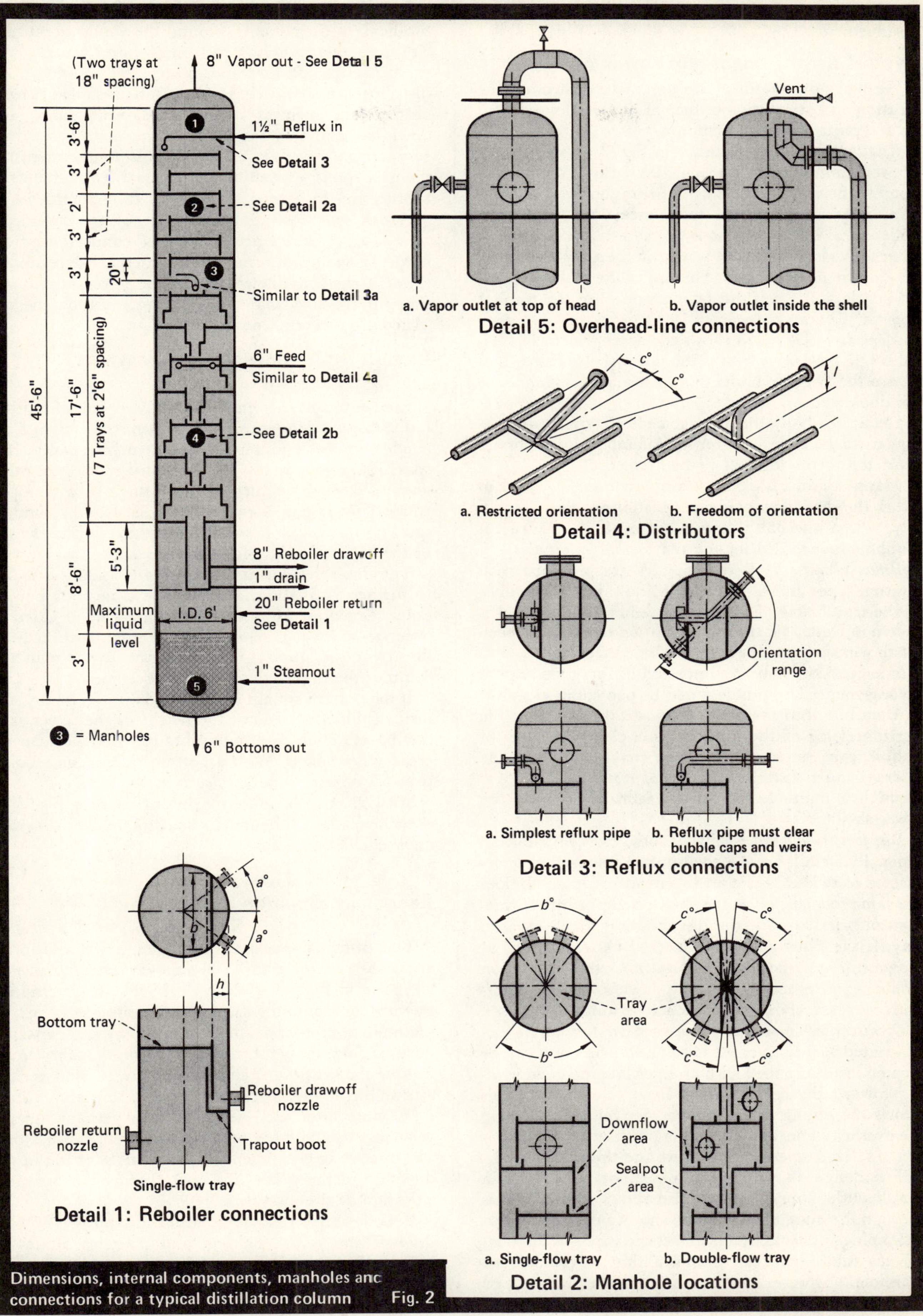

Dimensions, internal components, manholes and connections for a typical distillation column Fig. 2

orients process and utility nozzles to suit the piping arrangements.

Typical arrangements for tower nozzles

We will now examine the parameters affecting the location and orientation of the various nozzles required on a typical distillation column.

Reboiler connections—Detail 1 in Fig. 2 shows reboiler drawoff connections for a single-flow tray. This connection can be very important for arranging tray orientation. The simplest, most economical location for reboiler drawoff connections is shown in Detail 1. Alternative locations may be within the angular limit of $a°$. This angle depends on the size of the reboiler drawoff nozzle and the width (dimension b) of the trapout boot at the tray downflow.

Mannoles— Detail 2a shows the arrangement of typical single-flow trays within the tower shell. The arcs of $b°$ are the angular limits in which a shell manhole or handhole can be placed.

Detail 2b shows the arrangement of typical double-flow trays. Possible areas for shell manhole locations are restricted to the four segments marked $c°$.

Manholes and handholes are not usually placed in either the downflow or sealpot section of trays. Generally, where internal piping is arranged over a tray, a manhole is included for the removal of internals.

Reflux connections—Reflux nozzles are provided with internal pipes that discharge the liquid into the sealpot of the tray below. The simplest reflux-pipe internal is shown in Detail 3a. If tray orientation has already been fixed within prescribed limits by other factors, this nozzle location may be undesirable for good piping arrangement. The problem can be overcome as shown in Detail 3b, but the designer must take care that the horizontal leg of the internal pipe clears the tops of bubble caps or weirs. He must make sure that the internal pipe can be fabricated for easy removal through a manhole, or can be fabricated inside the tower shell.

Distributors—Detail 4 shows double cross-distribution pipes. In Detail 4a, the shell-nozzle orientation is restricted to twice $c°$. The arrangement in Detail 4b does the same job and gives the shell nozzle complete freedom of orientation if the distributor is above the top tray. If the distributor were located between trays, as shown in Fig. 2 (center of the column), the orientation would be governed by the tray downflows. This alternative is possible only where space permits dimension l.

Overhead-line connections—The vapor-outlet nozzle can be located at the top of the head (Detail 5a), or it can be located inside the shell with an internal pipe-bend leading toward the center of the head (Detail 5b). This allows the piping designer more flexibility in locating the overhead line, and arranging platforms for access to vent and instrument connections, and the top manhole.

The design shown in Detail 5a not only shows a simple outlet connection but also allows the best access through the manhole. Detail 5b shows a common internal-piping connection that makes the vapor-outlet nozzle accessible from the platform that serves the top manhole. The vessel's vent connection can also be piped to be accessible from the same platform. This arrangement eliminates the need for a small platform above the top head of the column (around the vapor-outlet nozzle) for access to blinds, instruments and vents.

Internal piping for vapor outlets, reflux inlets, inlet distributors and tray drawoffs must be designed to clear obstructions such as bubble caps, weir dams, and downflow and tray supports. At the same time, these components must fulfill the process requirements at their locations, as well as the physical requirements of fabrication, to permit withdrawal through manholes.

Access, whether internal or external, is very important. This includes accessibility of connections from ladders and platforms, and internal accessibility through shell manholes, handholes or removable sections of trays. A manhole opening must not be obstructed by internal piping.

Layout evaluation of flow diagrams

Feed enters at the center or the lower half of a distillation tower. Light-ends exit as vapor at the top and are condensed. In a gravity-flow system, the condensate splits into two streams: one stream is reflux, the other goes to storage or further processing. In a pumped condensing system, a reflux drum collects the condensate, and the reflux pump delivers the liquid to the top of the column and to further processing.

The liquid residue at the bottom is usually pumped to further distillation. This residue is near its boiling point. Consequently, the column must be elevated in order to meet the required net positive suction head (NPSH) of the pump. The lower the pressure in the column, the higher this elevation becomes.

If the bottom stream is pressure-delivered to, say, the next distillation column, the height of the tower skirt can be minimal. The hydraulics of a reboiler can increase skirt height. With a bottom pump, the reboiler drawoff nozzle usually is sufficiently high.

We will now examine the information in Fig. 3 to show how a layout designer visualizes the flow diagram for a distillation column.

Layout in elevation and in plan

The principal features, manholes, tower platforms and pipe runs of a tower are shown schematically in elevation in Fig. 4a. Nozzle elevations are determined by process requirements, and manhole elevations by maintenance requirements. For economy and easy support, piping should drop or rise immediately upon leaving the nozzle, and should run parallel, and as close as possible, to the tower itself. The horizontal elevation for piping (after the lines leave their vertical runs) is governed by the main pipe-rack elevations. Lines that run directly to equipment at grade, more or less in the direction of the main pipe rack, often have the same elevation as the pipe rack.

Reboiler-line elevations are determined by the drawoff and return nozzles, and their orientation is influenced by thermal-flexibility considerations. Reboiler lines (and overhead lines) should be as simple and as direct as possible. Lines from tower nozzles

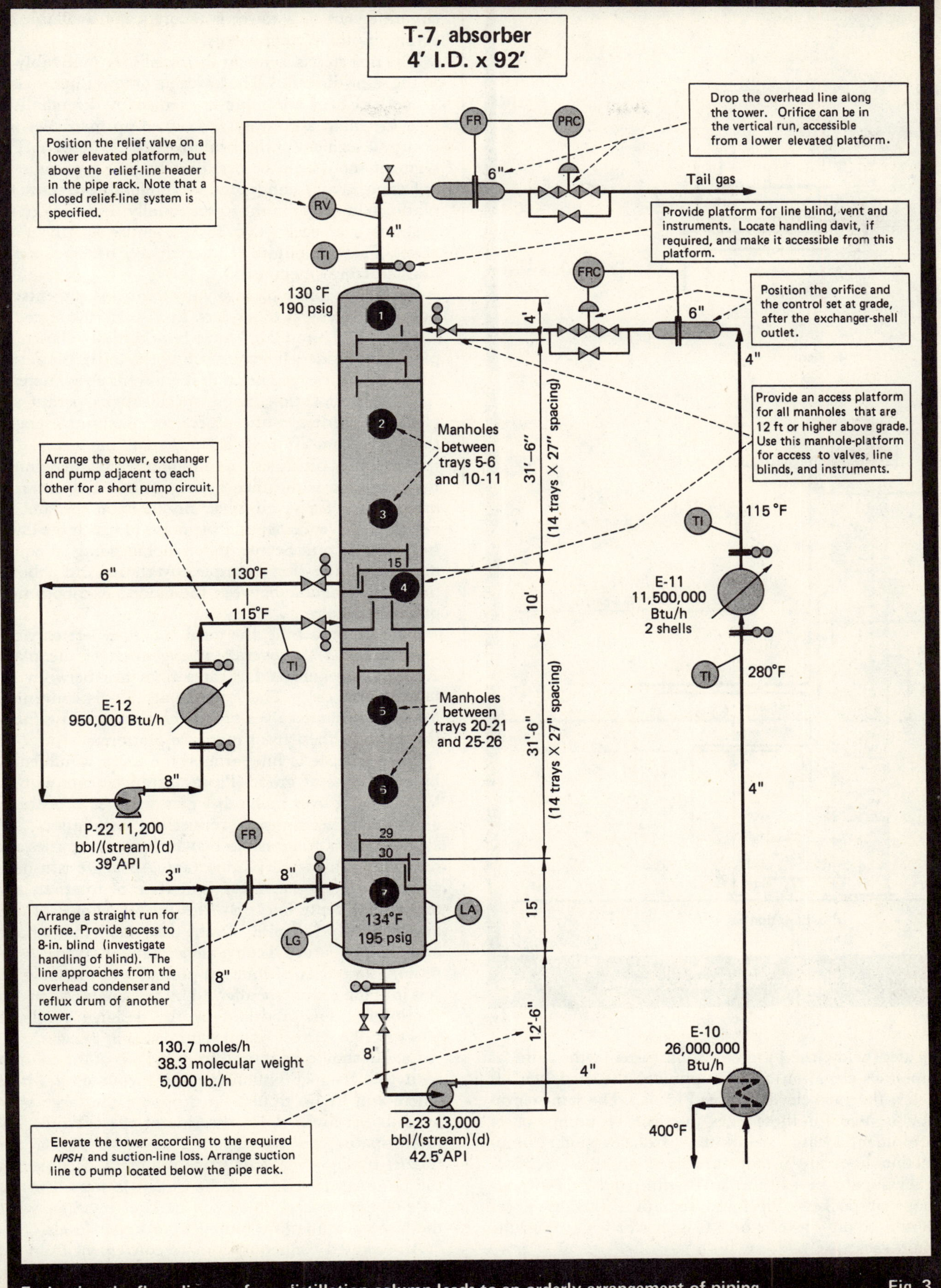

Evaluating the flow diagram for a distillation column leads to an orderly arrangement of piping Fig. 3

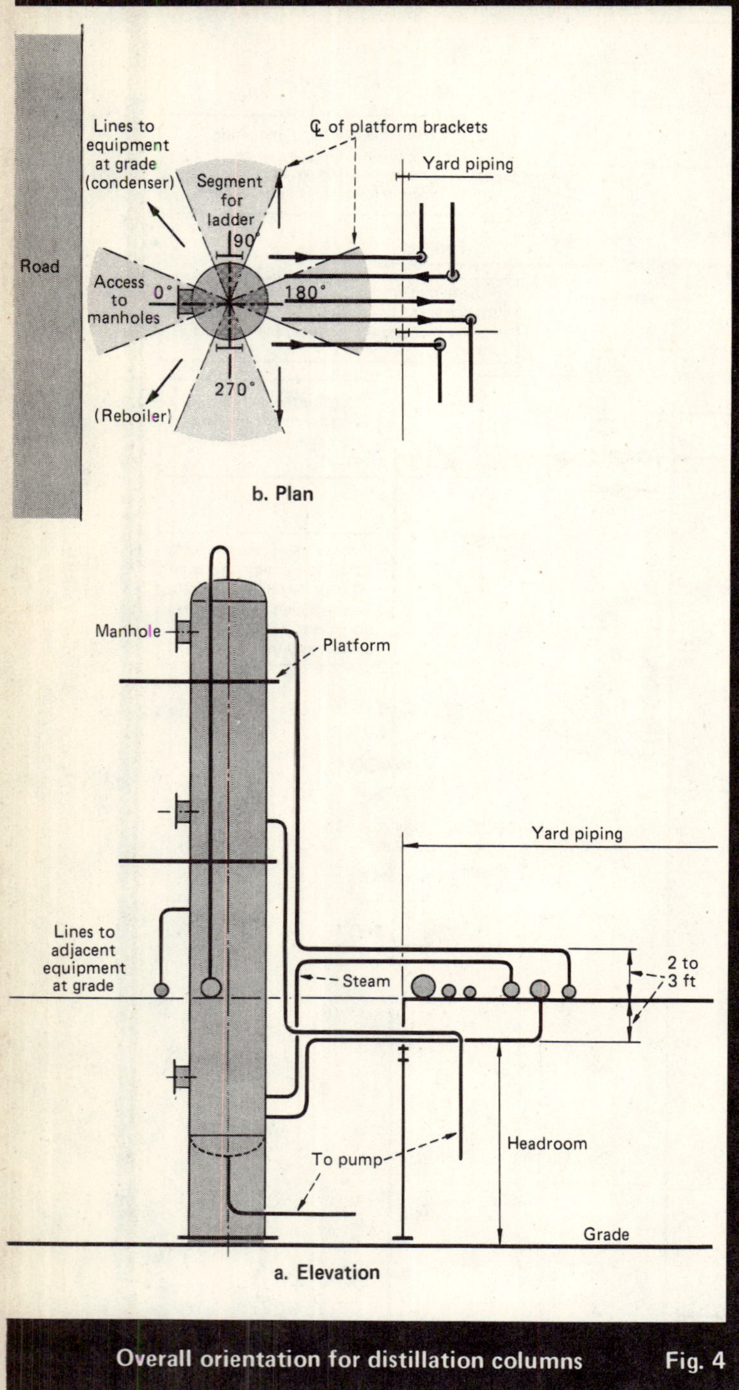

Overall orientation for distillation columns **Fig. 4**

located below the yard elevation (usually the same as pipe-rack elevation) should approach the yard 2 or 3 ft below the yard elevation (see Fig. 4a). The same elevation is used for those lines that run to pumps or to equipment located below the yard elevation. Pump-suction lines are usually arranged on this elevation. They should be as short as possible and run without loops or pockets. Pipelines from pipe rack to tower should be on a level 2 or 3 ft higher or lower than the main rack.

The plan view (Fig. 4b) of a tower shows the segments of its circumference allotted to piping, nozzles, manholes, platform brackets and ladders, in a pattern that usually leads to a well-designed layout. The entire circumference of a tower is theoretically available for arranging these components.

The first step is to orient the manholes, preferably all in the same direction. The lowering of tower internals to grade will be made more convenient if the manholes face the main accessway. The lined-up manholes will occupy a segment of the total tower circumference. This segment should not be occupied by any pipe runs.

From a layout standpoint, it is preferable to space the platform brackets on the tower equally, and to align the brackets over each other for the entire length of the tower. This will minimize interferences between piping and structural members.

Access for tower piping, valves and instruments influences placement of ladders. In Fig. 4b, the segments located at 90° and 270° have been initially chosen for placing of ladders interconnecting the tower platforms. This restricts two segments of the tower's circumference for pipeline location. Some specifications permit single-flight ladders, while others set maximum lengths that vary from 20 to 40 ft.

In routing pipelines, the problem is to interconnect tower nozzles with other remote points. The tentative orientation of a given tower nozzle is on the line between the tower center and the point to which the line is supposed to run. Segments for piping going to equipment at grade (for example, overhead and reboiler lines) are available between the ladders and both sides of the manholes.

Lines approaching the yard can turn left or right, depending on the overall arrangement of the plant. Respective segments for these lines are between the ladders and 180°. The segment at 180° is convenient for lines without valves or instruments, because this is the point farthest from manhole platforms.

The sequence of lines around the tower is influenced by conditions at grade. Piping arrangements without lines crossing over each other give a neater appearance and usually permit more-convenient installation.

The correct relationship between process nozzles and tower internals is very important. An angle is usually chosen between the radial centerline of internals and tower-shell centerlines. By the proper choice of this angle (usually 45° or 90° to the pipe rack), many hours of work and future inconvenience can be saved. Tower piping, simplicity of internal piping, and manhole access into the tower, are affected by this angle. After this, the information produced by the designer results in selecting the correct orientation of tower nozzles.

Layout should be started from the top of the column, with the designer visualizing the layout as a whole. There will be no trouble in dropping the large overhead-line straight down the side of the column. The lower spaces can then be laid out with piping and nozzles by knowing what space is already occupied by the large vertical lines. If the design is started at the base of the column, there will be later revisions when the large pipelines are routed from upper levels.

An elevated condenser is more convenient from the standpoint of tower-piping layout because the large overhead line leaves the tower at a high level and crosses directly to the condenser. This opens a segment

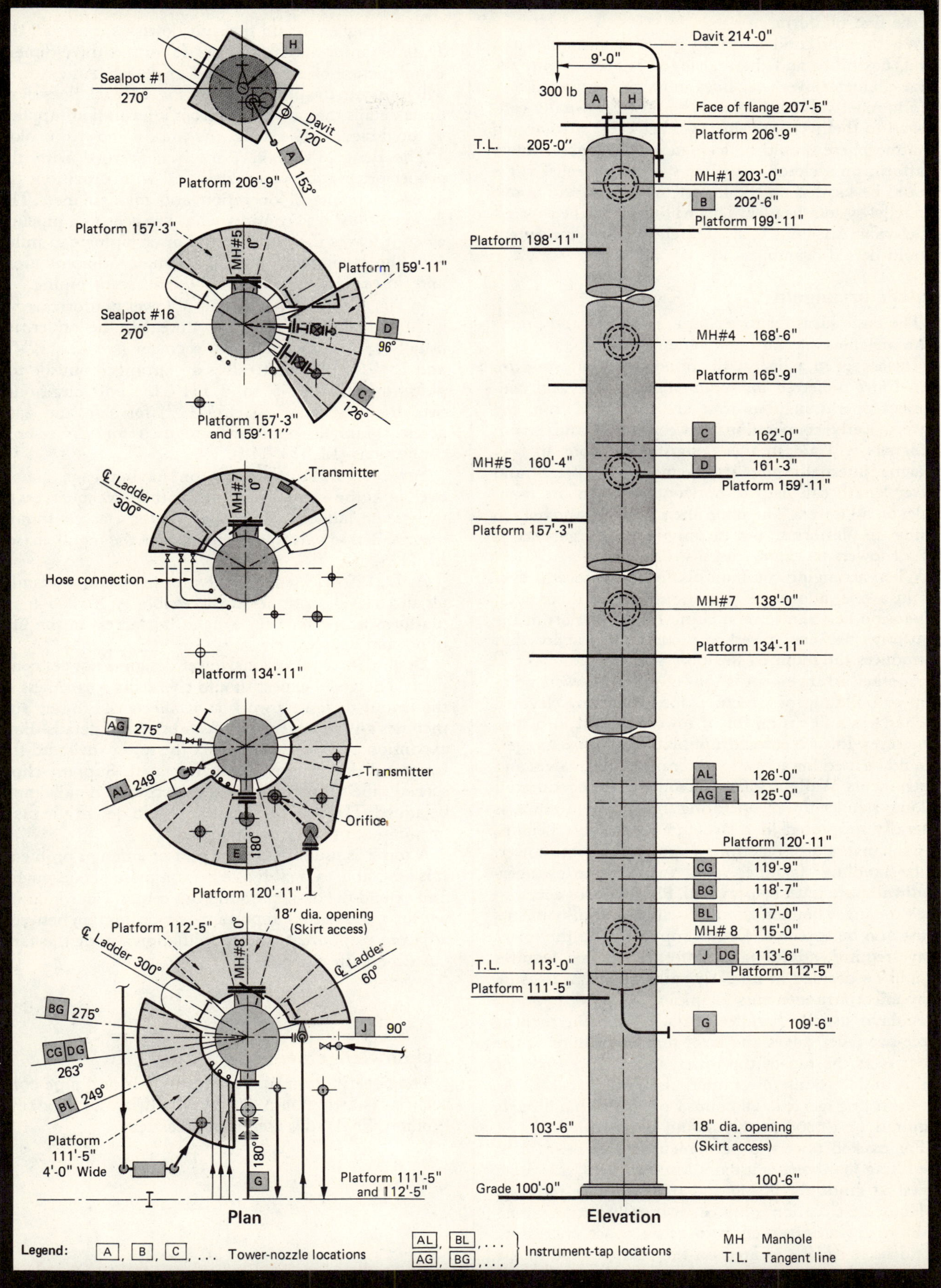

Nozzles and piping must meet process requirements while platforms must satisfy maintenance and operating needs Fig. 5

at lower elevations for piping or for a ladder from grade to the first platform.

Whether the condenser is at grade or at an elevated level, flexibility and thermal load connected with the large-diameter overhead lines must be considered.

The relief valve protecting the tower is usually connected to the overhead line. A relief valve discharging to atmosphere should be located on the highest tower platform. In a closed relief-line system, the relief valve should be located on the lowest tower platform above the relief-system header. This will result in the shortest relief-valve discharge leads. The entire relief-line system should be self-draining.

Tower arrangements

The basic ideas for tower-piping analysis and procedure are the same for one or more towers.

Towers set in a line, with connecting platforms, are sometimes preferred for maintenance access and convenient operation. Platforms are supported from the towers. Early cooperation between vessel and layout designers will aid in this design. Alterations in tray spacing, internal-piping arrangement, skirt height and tower length can help to horizontally align the manholes on all towers. The same alteration will also help in lining up platforms, and in providing common access on all towers to valves and instruments.

When arranging common platforms for several towers in a line, allowance must be made for differential expansion between towers. Using hinges or slots in the platforms between towers is a common practice that introduces the required flexibility.

Another arrangement is that of placing towers adjacent to buildings or structures for supporting elevated exchangers and drums, or of towers framed into steel structures. In such cases, drum platforms or exchangers should be used for access to tower manholes, valves and instruments. Wall thickness can sometimes be reduced if a long, slender tower, operating under low pressure, is laterally supported in a structure at suitable heights.

Exchangers are sometimes supported on the tower. Tube-handling facilities and tube- or exchanger-removal space must be provided. Platforms for access to the exchanger bonnet, channel, valves and instruments must also be specified. Drums supported on the tower may require additional platforms. These features should be decided at an early stage of design because they affect arrangements for piping.

A davit usually handles heavy equipment such as large-size relief valves and large-diameter blinds. If the davit is at the top of the tower, it can also serve for lifting and lowering tower internals to grade. Clearance for the lifting tackle to all points from which handling is required, and good access, should be provided.

For packed towers, a permanent trolley beam over the filling manholes is usually installed, with adequate access at grade for lifting and removing the packing.

Housed distillation columns are usually located along walls where no obstruction to access space is created. Condensers can be located on the roof or close to the ceiling, with gravity-flow reflux arrangement. The reboilers and pumps for housed columns can be located as for an outdoor installation.

Method of design

The designer should first put down on paper all the available information in order to form a three-dimensional picture of the tower and its details. An example will illustrate the procedures. In Fig. 5, black lines show those details that are copied from information supplied to the designer. The designer's work is shown in color.

The designer's main work is concerned with the proper orientation of nozzles, and with provisions for access to points of operation and maintenance. The designer must also consider what happens to a pipeline when it leaves the tower area. Tower piping can influence the layout and piping design in adjacent areas and, in turn, is influenced by the adjacent piping.

In Fig. 5, three 30° segments of tower platforms serve manholes number 5, 7 and 8. A break in the ladder rise usually means an additional segment (see El. 112′-5″ and 134′-11″).* Small valves are arranged outside the platforms (El. 112′-5″ and 134′-11″) and large ones over the platforms (El. 159′-11″) for safe and easy access. Manhole platforms are used for service-hose connections (El. 134′-11″).

Some organizations insist on having gage glasses over, or at the edge of, platforms; others accept access to these from ladders. Additional platform access should be provided to valves located opposite the manhole (see El. 111′-5″).

At El. 120′-11″, a relief valve, orifice plate, transmitter and a level controller are accessible. At El. 206′-9″, a platform is provided for a davit, for access to the line blind and vent valve.

Despite these provisions, tower design is not yet complete. The vessel expert should check the correctness of the layout designer's work as it affects the tower, and then design the tower in all its fabrication details. Pipe expansion and loads that affect the tower shell and the structural design must be investigated. Support clips, special pipe supports, and other necessary details must be added. The structural expert should design the platforming and foundation details.

A tower is usually a major part of a design problem. It is advisable to treat it as a central piece of equipment and extend the design around this center. Coordination within one area is as important as coordination between adjacent plant areas, and coordination among the various experts in plant design.

*El. = elevation.

Acknowledgement

The contribution on vessel internals and pipe connections (starting on p. 70) by Mr. Brian D. Wookey, London, England is greatly appreciated.

Installed cost of a distillation column

If you are weighing the benefits of several distillation-design options, the following cost data and time-saving calculation form will streamline your evaluation.

Photo: Fluor Engineering and Construction, Inc.

J. S. Miller and W. A. Kapella, Union Carbide Corp.

☐ To estimate an installed cost for a new distillation column, the designer must carefully consider the costs of items such as piping and field labor, as well as the cost of the bare column. Given here are data that will let you make a reasonably accurate estimate (within 25%) for a completely installed column if you have fixed the dimensions; number of trays; materials of construction; pressure and temperature (or shell thickness); site location; and fabrication schedule.

The result is the cost of a column that is instrumented, piped to the next item of equipment, insulated, and ready to run. We have not included costs for the ancillary pumps, tanks and heat exchangers that, although part of the distillation system or "module," are usually optimized separately (Fig. 1).

Shell and skirt costs

The column shell is estimated in Fig. 2 as a function of diameter, height and thickness. This figure represents Union Carbide Corp.'s actual purchasing experience, rather than market price surveys. The product of diameter, height and thickness (*DHT*) in in.[3], is very nearly proportional to the volume of metal in the shell, and therefore is proportional to the column weight—the traditional cost-correlation basis for cylindrical vessels.

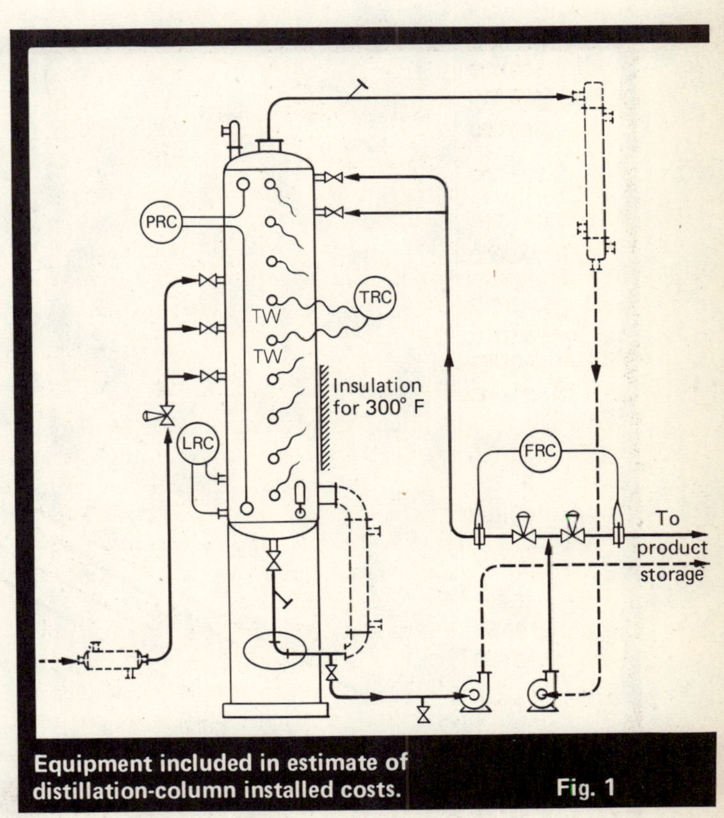

Equipment included in estimate of distillation-column installed costs. Fig. 1

Originally published April 11, 1977

To keep things simple, the height of the column includes the skirt, even though it may be made of a different material. Most skirts are 10–15 ft long and are thicker than the column shell. The thickness of the shell proper can be taken from Fig. 3 or Fig. 4 for either pressure or vacuum service; the latter figure, for vacuum service, should not be extrapolated. Depending on the chemical behavior of the system, an allowance for corrosion may be necessary. Also, if full radiography and stress relief are specified, these items can each add about 5% to the cost.

For the example shown in Box 1, an estimate has been made for the cost of a Type 316L stainless-steel column of the following design: 8 ft dia. x 60 ft overall height x 0.75 in. shell thickness; 30 sieve trays; full radiography required. With a *DHT* of 51,840, a Type 316L shell would cost $64,100, and a carbon steel shell would cost $25,700 at Index 100.

(As we will see later, it was also necessary to estimate the cost of a carbon steel column of similar dimensions (Box 2). We have assumed that both the Type 316L shell and the carbon steel shell are 0.75 in. thick, although there are circum-

Installed cost of Type 316L stainless-steel distillation column — **Box 1**

(1) *DHT* = 8 ft × 60 ft × 0.75 in × 144 = 51,840
(See Fig. 3 and Fig. 4 for values of *T*)

(2) Cost of shell (Index = 100; Fig. 2):
Actual material $61,000 × 1.05 (Radiography or stress relief) = $64,100
Carbon steel $24,500 × 1.05 (Radiography or stress relief) = $25,700

(3) Cost of trays (Index = 100; Fig. 4): $4,000 setup +
Actual material 30 × $870 each × 1.1 Adjustment factor = $32,700
Steel 30 × $440 each × 1.1 Adjustment factor = $18,500

(4) Tray installation (Index = 100; Fig. 5): 30 × $92 each = $2,800

(5) Total cost of column and trays: Actual material = $99,600
Carbon steel = $47,000
Ratio for use in Fig. 7 = 2.12

(6) Bulk-material cost (Index = 100; Fig. 7):
$99,600 Column cost (actual material) × 0.68 = $67,700

(7) Labor manhours, Fig. 7:
$99,600 Column cost (actual material) × 0.066
× 1.30 Multiplier = 8,550 manhours

(8) Indexed material cost:
[$99,600 (5) + $67,700 (6)] × 2.50 Index = $418,000

(9) Labor cost: 8,550 Manhours (7) × $12 Wage rate = $103,000

(10) Construction overhead: $103,000 (9) × 80 % = $82,000

(11) Subtotal = $603,000

(12) Engineering: $420,000 (11) × 20 % = $84,000 *

(13) Subtotal = $687,000

(14) Contingency $687,000 (13) × 25 % = $172,000

(15) Total Installed cost = $859,000

*Same engineering as for a carbon steel column

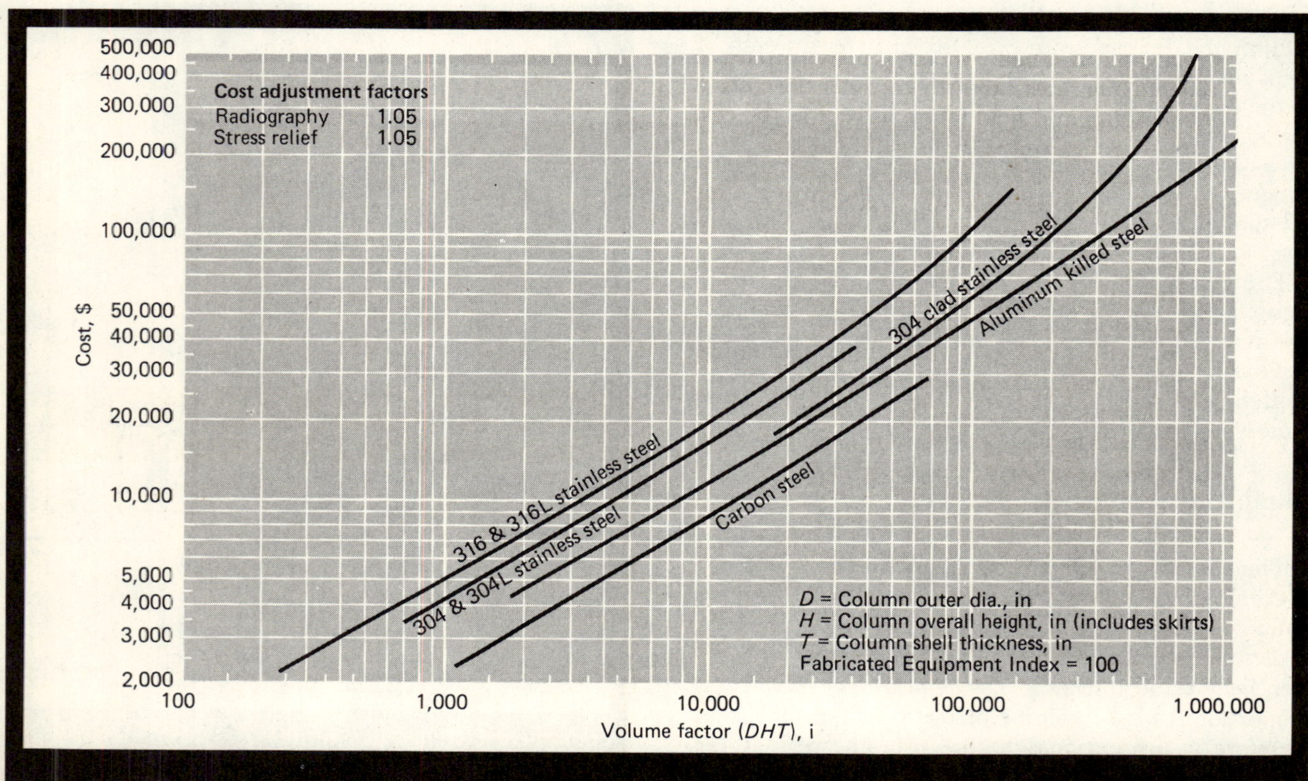

Cost of distillation-column shell as a function of dimensions. **Fig. 2**

Installed cost of carbon-steel distillation column **Box 2**

(1) *DHT* = 8 ft X 60 ft X 0.75 in X 144 = 51,840
(See Fig. 3 and Fig. 4 for values of *T*)

(2) Cost of shell (Index = 100; Fig. 2):
Actual material $___ X ___ (Radiography or stress relief) = $___
Carbon steel $24,500 X 1.05 (Radiography or stress relief) = $25,700

(3) Cost of trays (Index = 100, Fig. 5): $4,000 setup +
Actual material ___ X $___ each X ___ Adjustment factor = $___
Carbon steel 30 X $440 each X 1.1 Adjustment factor = $18,500

(4) Tray installation (Index = 100; Fig. 6): 30 X $92 each = $2,800

(5) Total cost of column and trays: Actual material = $___
Carbon steel = $47,000
Ratio for use in Fig. 6 = 1.00

(6) Bulk-material cost (Index = 100; Fig. 7):
$47,000 Column cost (actual material) X 1.0 = $47,000

(7) Labor manhours, Fig. 7:
$47,000 Column cost (actual material) X 0.14
X 1.30 Multiplier = 8,550 manhours

(8) Indexed material cost:
[$47,000 (5) + $47,000 (6)] X 2.50 Index = $235,000

(9) Labor cost: 8,550 Manhours (7) X $12 Wage rate = $103,000

(10) Construction overhead: $103,000 (9) X 80 % = $82,000

(11) Subtotal = $420,000

(12) Engineering: $420,000 (11) X 20 % = $84,000*

(13) Subtotal = $504,000

(14) Contingency $504,000 (13) X 25 % = $126,000

(15) Total Installed cost = $630,000

*Use this figure for Type 316L column also

stances, especially at elevated temperatures, where they may be different.)

Index 100 refers to the Fabricated Equipment Index published in CHEMICAL ENGINEERING. Its basis is 1957–1959 = 100. You will have to forecast the value for the date that your column is to be delivered. We arbitrarily chose 250 for our example. CHEMICAL ENGINEERING reports a value of 205.9 for Oct., 1976, so 250 would probably correspond to some time in 1978.

Tray installation

The cost of trays (Index 100) comes from Fig. 5, with appropriate allowances for the manufacturer's setup charge and the penalty for having a small order of less than 40 trays. When carbon steel is used, the trays in this example run 70% of the shell cost, but only 50% when the column is constructed of Type 316L. This difference obviously reflects the greater proportion of shop-labor cost in making the carbon steel trays.

Fig. 5 applies well enough to any common type of tray: sieve, valve, or bubble cap. Our experience shows that a majority of new trays are valve trays. Very few bubble-cap trays are specified for new columns these days.

Installing the trays in a new column is nearly always done in the vendor's shop. This cost is estimated in Fig. 6. Field installation requires 25% more labor, and there is a greater chance of damage.

Bulk material and field labor

There is more than one way to estimate the cost of bulk material (Table 1) and field labor. Ideally, estimates should be made from completed civil, mechanical, electrical and control-systems designs. One of the primary objectives of this article, however, is to avoid the need for those designs by providing shortcut estimating factors. These factors for bulk material and field

L = Length between stiffeners
D = Outer diameter of column shell
T = Thickness
(All in consistent units)
800°F
100°F
Most common *L*/*D* = 1.5
L/*D*
D/*T*

Thickness required for vacuum service: carbon and stainless steel cylindrical vessels **Fig. 3**

Breakdown of bulk material (assuming "all steel" construction) **Table I**

Item	% of total
Piping	40.0
Instruments	24.0
Electrical	8.3
Structures	7.0
Insulation	6.7
Concrete	5.7
Fire protection	3.3
Misc. material	3.3
Painting	1.7
	100.0

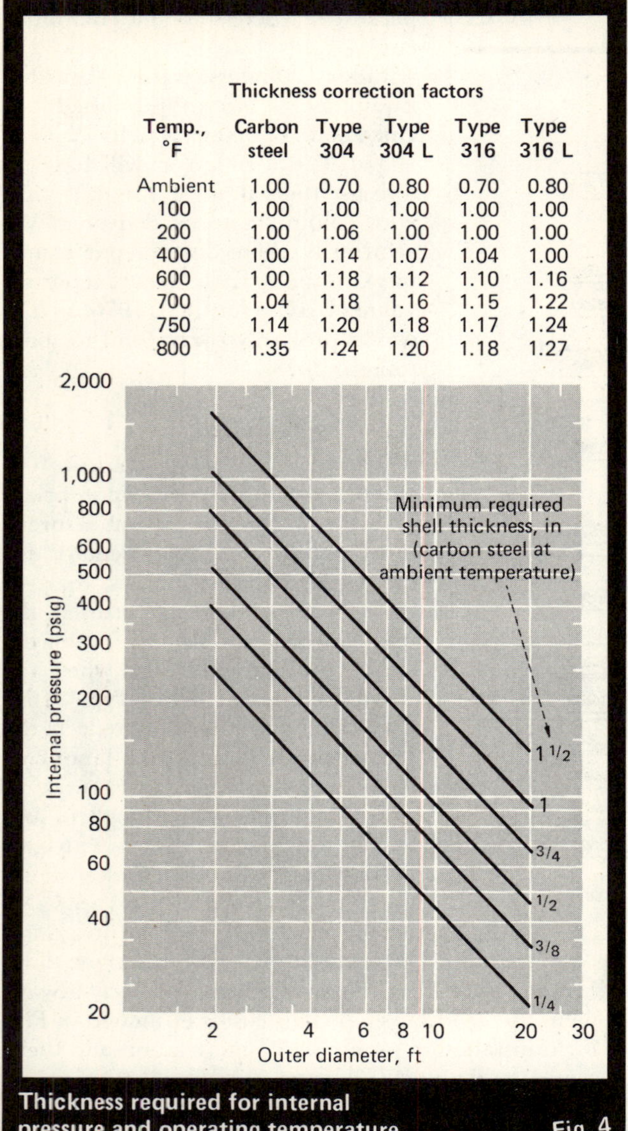

Thickness correction factors

Temp., °F	Carbon steel	Type 304	Type 304 L	Type 316	Type 316 L
Ambient	1.00	0.70	0.80	0.70	0.80
100	1.00	1.00	1.00	1.00	1.00
200	1.00	1.06	1.00	1.00	1.00
400	1.00	1.14	1.07	1.04	1.00
600	1.00	1.18	1.12	1.10	1.16
700	1.04	1.18	1.16	1.15	1.22
750	1.14	1.20	1.18	1.17	1.24
800	1.35	1.24	1.20	1.18	1.27

Thickness required for internal pressure and operating temperature Fig. 4

labor must, of necessity, be related to the cost of the major equipment. The only real choice is whether to estimate the bulk-material components separately or as a whole. We have found that the correlation coefficient is significantly better when bulk materials are estimated as a whole. The same is true of field labor.

In Fig. 7, bulk-material cost and field-labor manhours are plotted against the ratio of total column cost (including internals) for the actual material to that for a carbon steel column. It is necessary to consider this ratio of shell material costs because bulk material and field labor are progressively smaller multiples of column cost as the column becomes increasingly "noble." This is because all bulk materials remain essentially the same, regardless of column material, except for piping, which varies proportionately with column cost. Labor does not change significantly with column material.

Labor is expressed in manhours to simplify the application of wage rates and productivity factors. Our standard productivity factor is based on open-shop construction with less than 300 workers at the construction site. The table of multipliers in Fig. 7 indicates

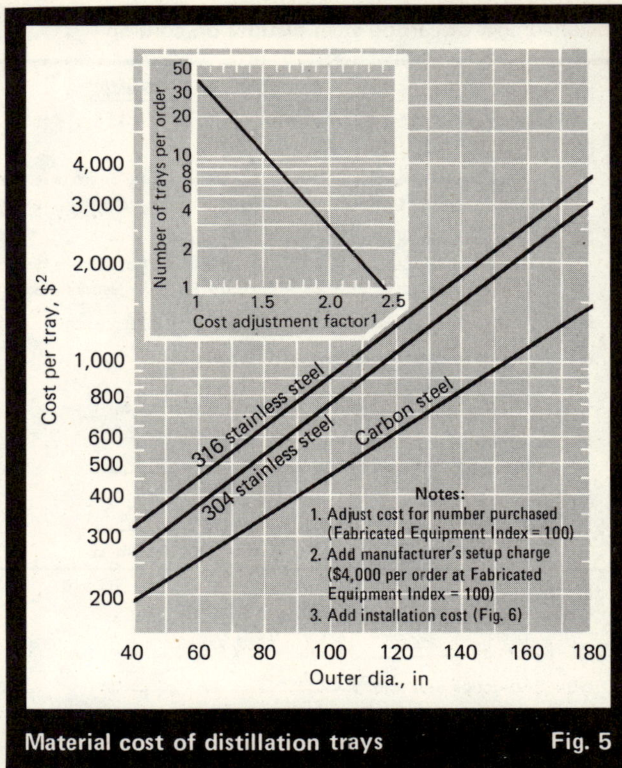

Material cost of distillation trays Fig. 5

generally the decrease in productivity (increase in manhours) that occurs when the size of the field force is increased or when closed-shop labor is involved. These multipliers are applicable along the Texas-Louisiana Gulf Coast. As a rule, productivity is not as high elsewhere. The examples use a productivity multiplier of 1.3 without reference to any particular labor situation.

In the sample calculations, Step 9 is the application of the wage rate to the manhours estimated from Fig. 7. The wage rate should correspond to the time the column is installed. This rate can be either the bare direct wage, or it can include fringes—normally about 20% of the direct wage. The factor for construction overhead, Step 10, must be adjusted accordingly. We roughly derived our $12 per hour (including fringes) for 1978 construction in the New Orleans area from the Dec. 23, 1976 issue of *Engineering News Record* (p. 74).

Overhead; engineering; contingencies

Step 10 estimates construction overhead—also known as field overhead, indirects, or distributables. Including all labor fringes, as well as the cost of construction-equipment leasing and maintenance, overhead will run somewhat more than direct labor. Putting the fringes in the labor as we have done, overhead is about 80% of labor plus fringes.

Engineering should be estimated from a clear, detailed understanding of all the facts and circumstances relative to the project size and execution plan. This, however, is not always feasible with a factored approach. For factored estimating, we find it practical, although not theoretically valid, to express engineering as a function of the sum of all other costs. For a project that includes only a distillation column (as in our example), engineering could be 20 to 25% of the sum of

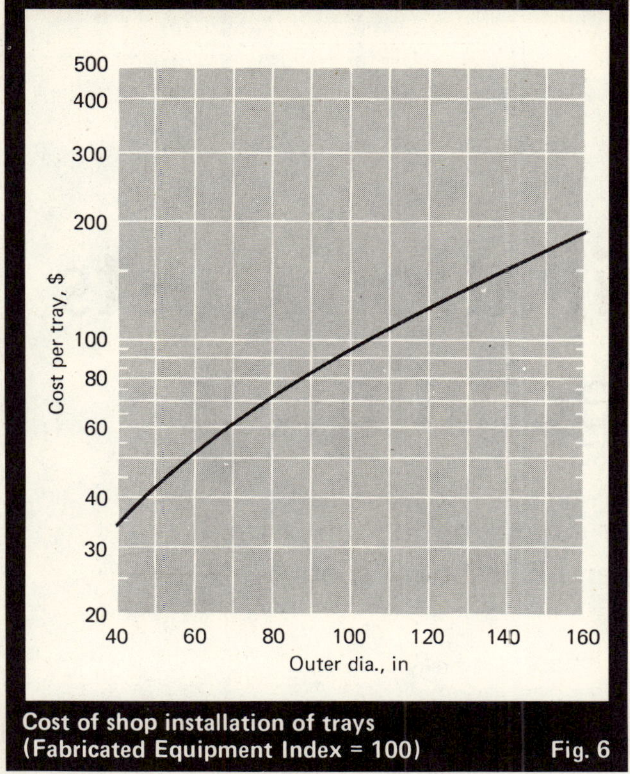

Cost of shop installation of trays (Fabricated Equipment Index = 100) Fig. 6

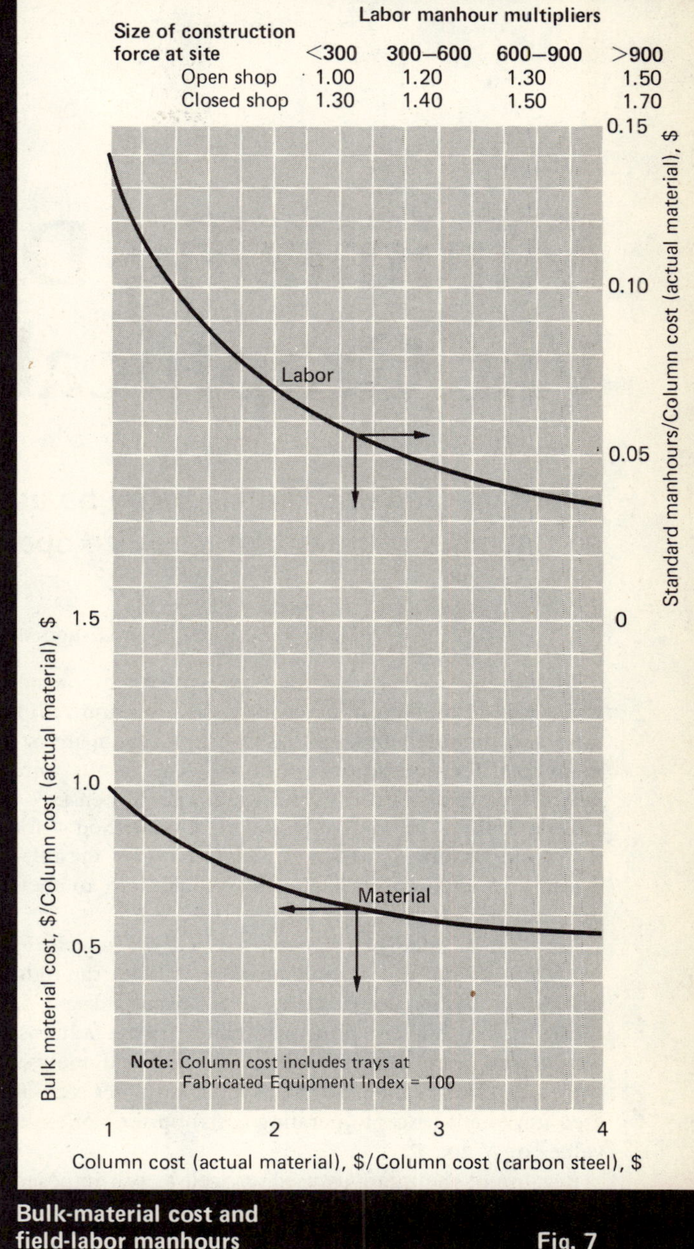

Bulk-material cost and field-labor manhours Fig. 7

other costs, or 16 to 20% of the total cost excluding contingency.

Our definition of engineering includes activities in addition to process design and detailed design and drafting. We include project management, cost engineering, planning and scheduling, purchasing, and vendor expediting. There can also be a duplication of engineering supervision effort when contract engineering firms are employed.

It is probably better to base engineering on a carbon steel system and then assume that the same figure would apply to a system of any other material. For our example, therefore, we calculated the cost of a steel column system on the back of an envelope. We used engineering at 20%, which resulted in an engineering cost of $84,000. This seems like a reasonable figure for the Type 316L system as well.

Every estimate needs a contingency allowance to ensure that there is enough money in the estimate. Our standard factor is 25% for estimates made before engineering is complete. It is easy to use up that amount in design changes alone, not to mention unpredictable outside influences. Management gives no points for underestimating.

The authors

J. S. Miller is a Staff Engineer, Cost Engineering Technology, Engineering Dept., Chemicals and Plastics Div., Union Carbide Corp., P.O. Box 8361, South Charleston, WV 25303, where he is concerned with developing capital-cost estimating data and methods. During his career at Union Carbide, he has been involved in production, design engineering and business analysis. He holds a B.S. in chemical engineering from West Virginia University.

W. A. Kapella is Senior Cost Engineer, Engineering Dept., Chemicals and Plastics Div., Union Carbide Corp., where he is responsible for preparing capital-cost estimates for chemical process plants. He has worked as a design engineer and as a project engineer. A member of AIChE, he holds a B.S. in chemical engineering from Purdue University and an M.S. in business administration from Northern Illinois University. He is registered as a professional engineer in West Virginia.

Minimizing Distillation Costs Via Graphical Techniques

Shortcut, graphical methods can be useful for estimating the reflux ratio and number of trays that minimize operating and investment costs.

MATTHEW VAN WINKLE and WILLIAM G. TODD, University of Texas

In a recent article,[5] the authors presented a graphical, noncomputer method of estimating the optimum reflux ratio and number of stages based on minimum investment cost, for distillation operations. The results were determined by using the shortcut methods of Fenske[1]—for the number of minimum stages; Underwood[4]—for the minimum reflux ratio; and Gilliland[2]—for the relationship of theoretical stages and reflux ratio to their minimum values.

More than 800 cases were studied in which α, $(x_{LK}/x_{HK})_F$, and the percent recovery of both the light key (lower boiling component) in the overhead and of the heavy key (higher boiling component) in the bottoms, were varied over a practical range. A statistical analysis was made to determine the effects of all variables considered important (except operating costs) on the optimum values of N and R.

Because of the methods used, as well as self-imposed restrictions, the correlations presented were limited to: feedrates ranging from 800 to 1,200 lb.-moles/hr.; $\alpha \geqslant 1.5$; and at least 95% recovery of the light and heavy key components.

The article also dealt with the number of trays required for 12 different cases computed by shortcut techniques and the Thiele-Geddes[3] plate-to-plate method. The number of trays predicted by each method were in surprisingly close agreement.

At the conclusion of that study, a number of questions remained unanswered:

- Will operating costs change the evaluation of the optimum† N and R obtained by graphical methods, because of nonlinear costs relationships?
- Will the optimum be changed by variation in feedrate?
- Will the optimum be altered by the inclusion of costs for different materials of construction?

The answers, and the procedures for incorporating them into the graphical correlations, are the basis of this article.

Effects of Operating Cost

To study the questions raised above, it was necessary to develop equations for computing the total operating cost that would reflect current utilities costs (i.e., steam, cooling water, and electricity) and equipment costs (i.e., vessel, heat exchanger, pump, and overhead receiver). To compare the relative effects of these costs on the optimum values of N/N_M and R/R_M, a common time-base of one year (365 days) was selected. Total utility cost for one year was computed and then added to annual equipment-depreciation cost to obtain total yearly cost of operation.

In determining annual depreciation cost, the number of years used for the amortization schedule becomes an important variable. Increasing the depreciation period

Effect of Operating Costs—Table I

(F = 1,000 lb.-moles/hr., 99% separation of both keys)

Depreciation Period, Yr.	$(N/N_M)_{opt.}$ α=1.50	α=1.25	α=1.05	$(R/R_M)_{opt.}$ α=1.50	α=1.25	α=1.05
*	1.73	1.61	1.52	1.42	1.48	1.55
2	2.19	2.09	1.88	1.14	1.15	1.22
4	2.41	2.34	2.06	1.08	1.08	1.14
10	2.67	2.45	2.12	1.04	1.06	1.12

*Based on major-equipment cost only.

†The word optimum is used in this article to indicate minimum operating and investment costs for a given separation.

Originally published March 6, 1972

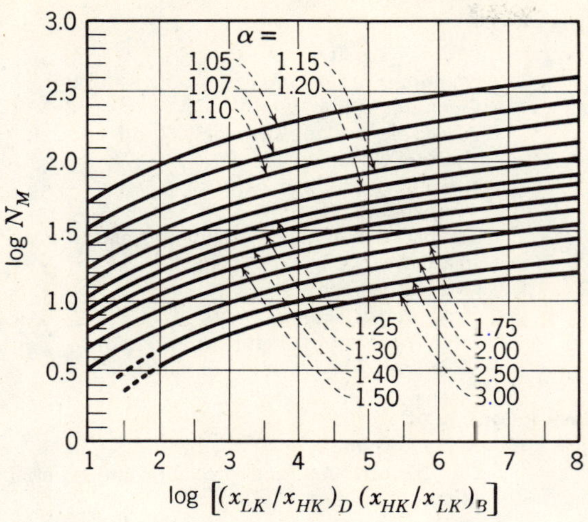

FENSKE-EQUATION for calculating minimum plates, expressed in graphical form—Fig. 1

decreases the effect of investment cost on the total operating cost; decreasing it, increases the effect.

A summary of the utility- and equipment-cost equations used, and the simplifying assumptions made, is given in the box on p. 106, along with the data used. Certain items presented are worthy of comment. In computing the pumping cost (equipment and power), only the cost of pumping the external reflux back to the top of the column has been considered. Since the cost of distillate and bottom-product pumping (assuming thermosiphon reboilers) is a constant, and thus does not vary with the reflux ratio, it has no effect on the minimization of total operating cost. Perforated trays with an assumed efficiency of 80% have been used to compute the actual number of trays. The height of the vessel has been calculated on the basis of a 2-ft. tray spacing.

The use of the graphs presented here as a design technique is limited by the data and approximations employed in calculating the various equipment and operating costs. However, the graphical techniques do serve as

Fractionating-Tower Cost Equations

Major Equipment Cost

All equipment is considered to be plain carbon steel, $M_C = 1.0$.

Vessel material cost

For $2 \leqslant d_i \leqslant 14$,

$$CV = 84.4 - 7.61d_i + 0.269d_i^2$$

For $d_i > 14$, $CV = 30.0$.

Tray cost

$$CT = 12(3.878d_i + 0.0389d_i^2 + 0.0111d_i^3)$$

Tray-cost multiplier

For $N_{act.} \leqslant 50$,

$$m = 1.713 - 0.23(N_{act.}/10) + 0.175(N_{act.}/10)^2.$$

For $N_{act} > 50$, $m = 1.0$.

Vessel weight

From the API-ASME Code,*

$$T_c = Pd_i/(2SE - 1.2P) + I$$

$$T_h = Pd_{avg.}/2SE + I \text{ (Ellipsoidal heads)}$$

I is assumed equal to 0.

$H = 15$ ft. (for bottom skirt height) + 3 ft. (for spacing above top tray) + (number of trays) × (tray spacing in ft.)

$$W_c = (489)T_c \pi\, d_i H/(144)$$

$$W_h = 2(t_h)(\pi\, d^2_{avg.}/2)(489)/(12)^3$$

Total vessel cost

$$CF = m(CT) + CV(W_c + W_h)/(100)$$

*This code has been replaced by Sec. VIII, Unfired Pressure Vessels, ASME Boiler and Pressure Vessel Code. It has been used here as an estimate of the tower and head thicknesses for the cost calculations.—Ed.

Condenser cost

$$CC = k_A(A/1{,}000)e_A$$

Reboiler cost

$$CR = k_A(A/1{,}000)e_A$$

Accumulator cost

$$CA = 1{,}600(V/1{,}000)^{0.56}$$

Pump cost

Only the reflux pump will be considered.

$$CP = k_P(hp/10)e_P$$

Total major-equipment cost

$$CM = M_c(CF + CC + CR + CA + CP)$$

Operating Costs

$$CO = CS + CW + CE$$

Data:

Steam	50¢/1,000 lb.
Cooling water	3¢/1,000 gal.
Electricity	1¢/kwh.
Depreciation	n (a variable)
Δt in water	25 F.
Installation cost	4(*CM*)

Steam cost

$$CS = \left[\frac{(L_o + D)\lambda(mw)}{1{,}000}\right]\left[\frac{(0.50)}{1{,}000}\right](24)(365)$$

Water cost

Assume $Q_C = Q_R$.

$$CW = \frac{[(24)(365)(0.03)][(L_o + D)(\lambda)(mw)]}{(1{,}000)(25)(8.33)}$$

Pumping cost (reflux only)

$$CE = (hp)(0.746)(0.01)(24)(365)$$

Total Yearly Operating Cost

$$CY = 4M_c(CM/n) + CO$$

$$CY = 4M_c(CM/n) + CS + CW + CE$$

Nomenclature

A	Heat-transfer area, sq.ft.
B	Bottoms flowrate, lb.-moles/hr.
CA	Accumulator cost, $
CC	Condenser cost, $
CE	Electricity cost, $
CF	Total column cost, $
CM	Total major equipment cost, $
CO	Yearly operating cost, $
CP	Pump cost, $
CR	Reboiler cost, $
CS	Steam cost, $
CT	Tray cost, $
CV	Vessel cost, $
CW	Cooling-water cost, $
CY	Total yearly cost, $
D	Distillate flowrate, lb.-moles/hr.
d_i	Inside dia., in.
$d_{avg.}$	Average dia., in.
E	Efficiency of longitudinal joint, assumed 80%
e_A	Slope of heat-transfer area vs. cost curve
e_P	Slope of pump horsepower vs. cost curve
F	Feedrate, lb.-moles/hr.
H	Vessel height, ft.
hp	Horsepower
I	Corrosion allowance, in.
k_A	Cost of a 1,000-sq.ft. heat exchanger, $
k_P	Cost of a 10-hp. pump, $
L_o	Reflux flowrate, lb.-moles/hr.
M_c	Material cost factor
m	Tray multiplier
mw	Molecular weight
$N_{act.}$	Actual number of stages
N_M	Minimum number of theoretical stages
$N_{opt.}$	Optimum number of theoretical stages
n	Depreciation period, yr.
P	Internal pressure, psig.
Q_C	Condenser duty, Btu./yr.
Q_R	Reboiler duty, Btu./yr.
q	Heat required to bring 1 lb.-mole of feed to feedplate temperature and to vaporize it, divided by molal latent heat of vaporization of feed
R	Reflux ratio
R_M	Minimum reflux ratio
$R_{opt.}$	Optimum reflux ratio
S	Allowable working stress, 13,000 psi.
T_c	Cylinder thickness, in.
T_h	Head thickness, in.
t	Temperature, °F.
t_1	Temperature of vapor leaving top plate, °F.
V	Internal vapor rate, cu.ft./hr.
W_c	Cylinder weight, lb.
W_h	Head weight, lb.
x	Mole fraction of component in liquid
y	Mole fraction component in vapor

Greek letters

α	Relative volatility
Γ	(N/N_M) correction for depreciation and material of construction
Δ	(R/R_M) correction for depreciation and material of construction
θ	Underwood's parameter
λ	Latent heat of vaporization, Btu./lb.-mole
Σ	Summation
Φ	(N/N_M) correction for feed
Ψ	(R/R_M) correction for feed
μ	$0.55\ \alpha_{LK}$

Subscripts

$avg.$	Average
B	Bottoms
$bot.$	Bottom
D	Distillate
F	Feed
HK	Heavy key component
$HK + 1$	Heavier than heavy key
i	Any component
j	Any component other than i
LK	Light key component
$LK - 1$	Lighter than light key
LK/HK	Relative, LK to HK
M	Minimum
$opt.$	Optimum
r	Reference case

an excellent estimating method. They also should find use as a means of determining starting points for more-rigorous computer calculations.

Inclusion of the utility cost (operating cost) in the minimization of total cost greatly affects the optimum reflux ratio and the number of trays. Tabulated in Table I are the optimum values of N/N_M and R/R_M, with and without operating costs for different depreciation periods and relative volatilities. From these data, it is evident that including operating cost always increases the optimum level of N/N_M and decreases the optimum level of R/R_M.

Data in Table I show that it is more economical to add additional equipment (increase the number of trays) than to increase the utility cost (increase the reflux ratio). In other words, the importance of equipment cost decreases when utility costs are included.

The optimum reflux ratio and number of trays will vary with changes in feedrate. These changes are readily apparent from the data in Table II, which compares the optimum values for various feedrates. In all cases, the optimum value of N/N_M increases with increased feedrate, and the optimum value of R/R_M decreases. It should be pointed out that the diameter of the vessel varies with feedrate and the α of the key components.

Preliminary examination of selected data using these equations led to the following observations:

- Pressure has no appreciable effect on the optimum values of N and R. Pressures of 100, 200 and 300 psia. were investigated for a fixed-diameter column (10 ft.).

Effect of Feedrate—Table II

(6-yr. amortization, 99.9% separation of both keys)

Feedrate, Lb.-moles/hr.	$(N/N_M)_{opt.}$ α=1.50	α=1.25	α=1.05	$(R/R_M)_{opt.}$ α=1.50	α=1.25	α=1.05
100	2.13	2.00	1.64	1.19	1.23	1.35
1,000	2.27	2.22	1.90	1.12	1.12	1.19
5,000	2.48	2.35	2.15	1.05	1.06	1.08

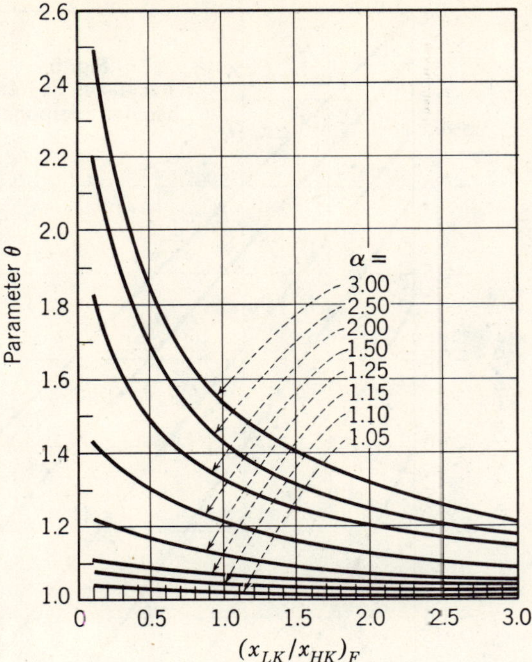

UNDERWOOD'S θ vs. key ratios in feed—Fig. 2

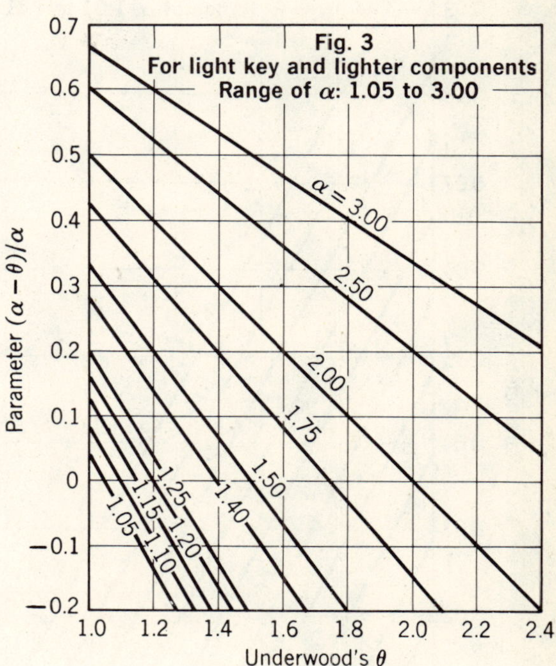

PLOTS for determining value of $\Sigma x_{i,D}/[(\alpha - \theta)/\alpha]$ for calculation of Underwood's $R_M + 1$—Fig. 3, 4, 5.

As would be expected in these cases, feedrate to the column varied with column pressure.

• Material of construction (i.e., plain carbon steel, stainless steel, Monel, etc.) has a direct effect on the optimum values of N/N_M and R/R_M. This is apparent from an inspection of the data in Table III. Increasing the cost of material (M_c) decreases the optimum value of R/R_M. This is the inverse of increasing depreciation years.

Based upon these preliminary observations, a systematic study was undertaken to develop correlations that would reflect the dependence of the optimum values of N/N_M and R/R_M on operating costs, feedrate, equipment life, and material of construction.

Correlations

The same 800 cases defined in the previous study[5] were used for the present work. The basic approach was to correlate $(N/N_M)_{opt.}$ and $(R/R_M)_{opt.}$ for a given reference case and then correlate correction factors that would reflect the effects of feedrates (this is equivalent to varying vessel diameter), material costs, and number of years for amortization. The base case selected was a 10-ft.-dia. column, 100-psia. pressure, 2-yr. equipment amortization, with plain carbon steel as the material of construction.

$(N/N_M)_{opt.}$ for both the reference case and the range of conditions was correlated as a function of the Fenske separation factor, using α as the parameter. This resulted in Fig. 6, which is used to evaluate the $(N/N_M)_r$ values. The correction factors Φ for the effects of various feedrates and α's were calculated and resulted in Fig. 7. Fig. 8 represents the correction factor for the number of years amortized, divided by the material-cost factor, M_c, for various values of α.

The values for Fig. 9, 10 and 11, relating $(R/R_M)_r$ and correction factors for feedrate, Ψ, and amortization, Δ, were calculated in the same manner, using the correction-factor technique.

Effect of Material Cost—Table III

(2-yr. amortization, 10-ft.-dia. column, 99.0% recovery of both keys)

	$(N/N_M)_{opt.}$			$(R/R_M)_{opt.}$		
M_c	α=1.50	α=1.25	α=1.05	α=1.50	α=1.25	α=1.05
1.0	2.28	1.99	1.75	1.11	1.18	1.30
2.0	2.13	2.00	1.72	1.16	1.19	1.33
4.0	2.09	1.90	1.69	1.18	1.24	1.35
6.0	2.04	1.87	1.68	1.20	1.26	1.36

Estimating Procedure

To use Fig. 1 to 11, follow the procedure below to obtain the optimum combination of reflux and number of trays based on minimum yearly operating cost. The system conditions and properties, feedrate, pressure, temperatures, relative volatilities, desired key-component separation, material costs, and amortization years are required input information. The stepwise procedure:

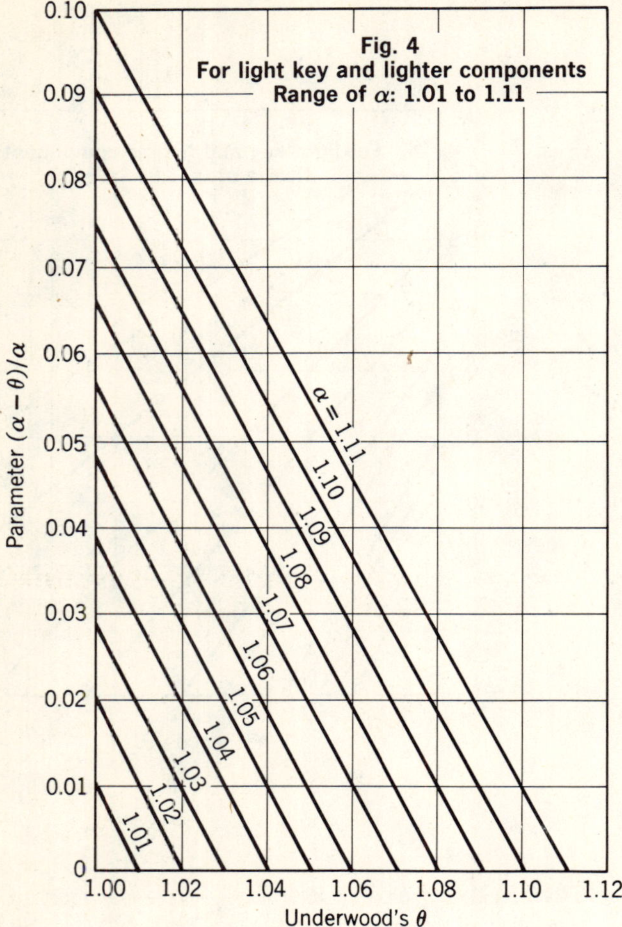

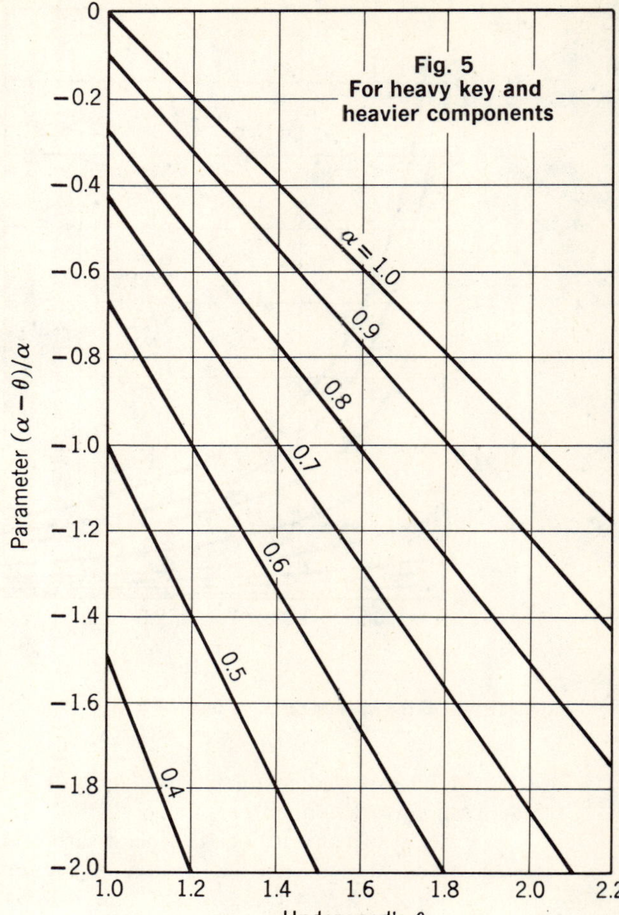

1. Determine the minimum number of plates from Fig. 1, or calculate it by the Fenske Method.[1]

2. Determine the minimum reflux from Fig. 2, 3, 4 and 5 or calculate by the Underwood Method.[4] (A more detailed discussion of the procedure used in steps 1 and 2 can be found in the previous article[5]).

3. Obtain the reference value of $(N/N_M)_r$ from Fig. 6, using the Fenske separation factor for the correct α.

4. For the correct feed and α, obtain the feed correction factor, Φ, Fig. 7.

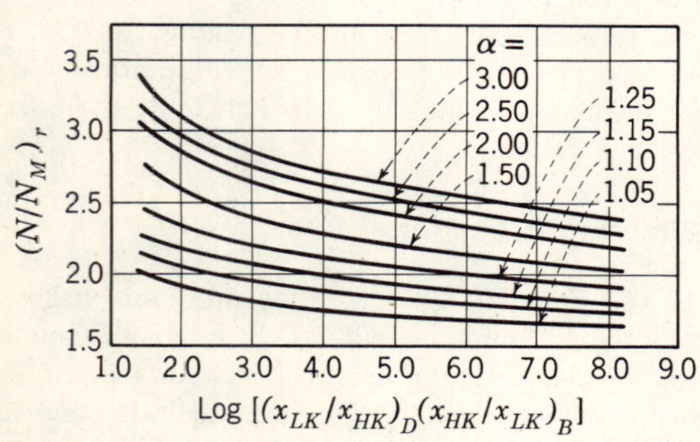

RELATIONSHIP of optimum to minimum number of theoretical trays for the reference case—Fig. 6

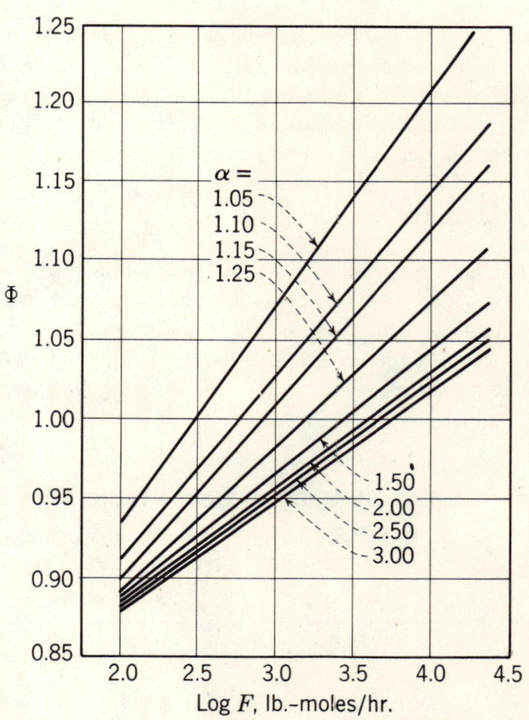

FEEDRATE correction for $(N/N_M)_r$—Fig. 7

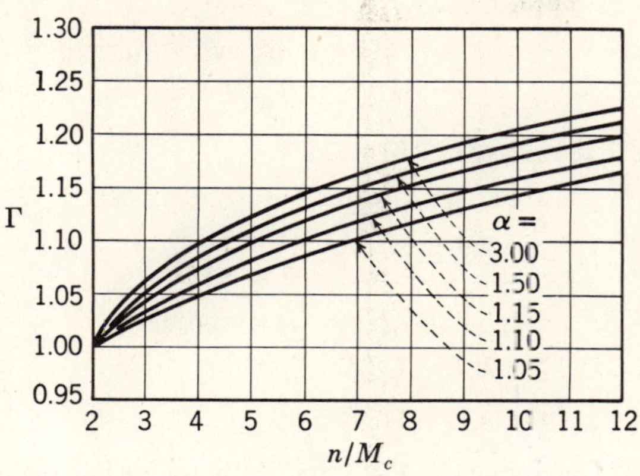

CORRECTION for $(N/N_M)_r$ for the depreciation period and material of construction—Fig. 8

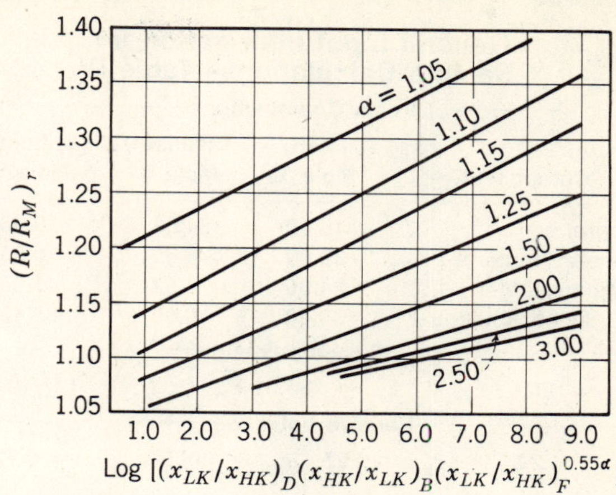

RELATIONSHIP of optimum to minimum reflux ratio for the reference case $(R/R_M)_r$—Fig. 9

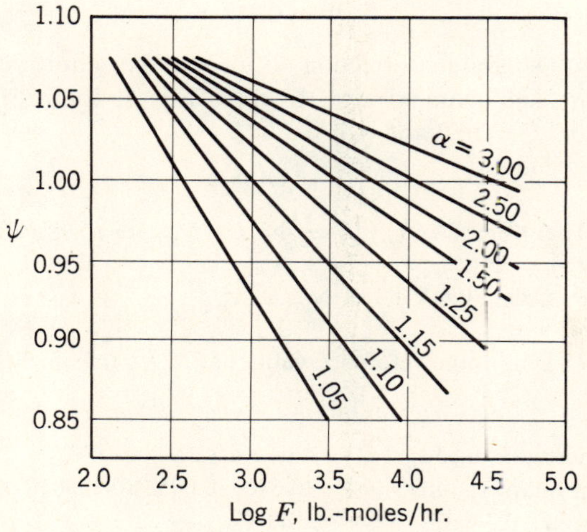

FEEDRATE correction for $(R/R_M)_r$—Fig. 10

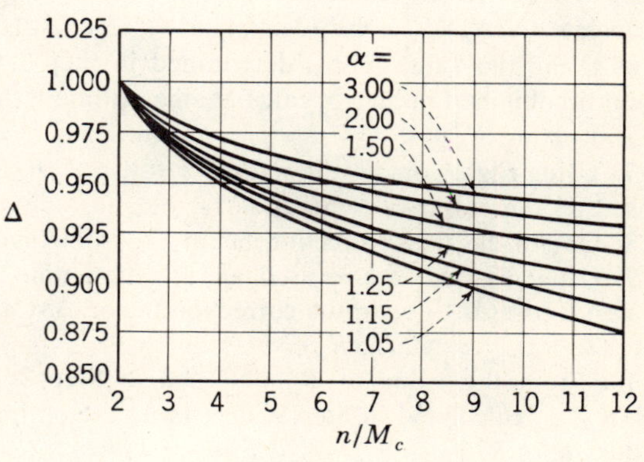

CORRECTION for $(R/R_M)_r$ for the depreciation period and material of construction—Fig. 11

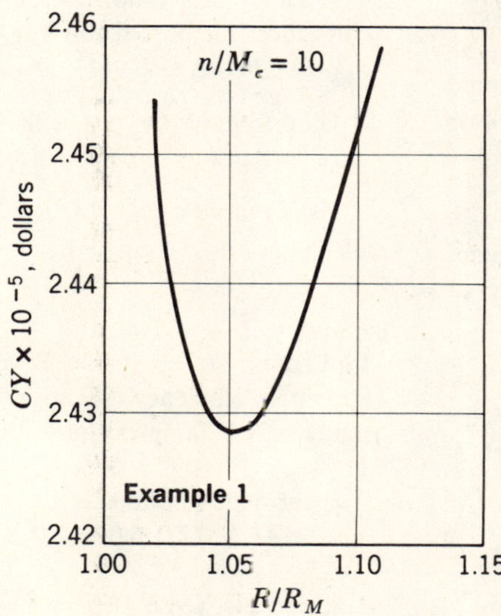

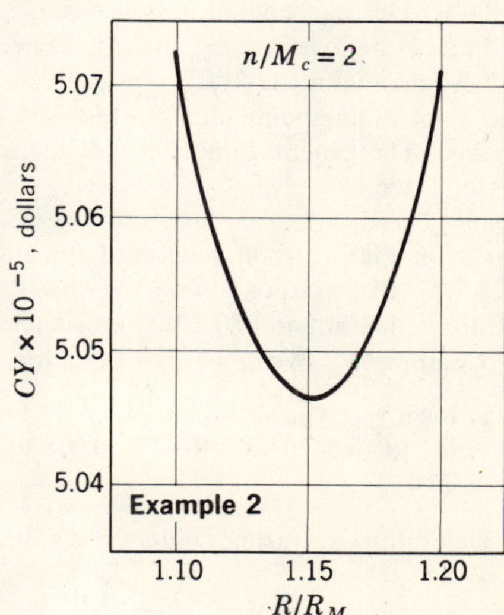

VARIATION of total yearly operating cost with R/R_M for examples 1 and 2 —Fig. 12

General Input Information for Sample Calculations—Table IV

Component	Feed, Mole %	Distillate, Mole %	Bottoms, Mole %
Stream Compositions			
Isoprene	16.00	34.83	0.00
3-methylpentene-1	30.00	64.65	0.55
2-methylpentene-1	24.00	0.52	43.96
2,3-dimethylbutene-2	30.00	0.00	55.49

Temperatures

t_{top} = 120.69 F. $t_{bot.}$ = 153.51 F. $t_{avg.}$ = 137.10 F.

Relative Volatilities

α_{LK-1} = 2.857 α_{LK} = 1.297 α_{HK} = 1.000 α_{HK+1} = 0.728

5. For desired amortization time and material factor cost, M_c, obtain the correction factor, Γ, from Fig. 8. Calculate $(N/N_M)_{opt.} = \Phi\Gamma\ (N/N_M)_r$.
6. Using the value of N_M determined in Step 1 and the ratio obtained in Step 5, calculate the optimum theoretical plates, $N_{opt.}$.
7. Using Fig. 9, obtain the reference value of $(R/R_M)_r$ based on the empirical separation factor.
8. Obtain the feed correction factor, Ψ, from Fig. 10.
9. Using Fig. 11, the desired amortization time and material cost, M_c, obtain the correction factor, Δ. Calculate $(R/R_M)_{opt.} = \Psi\Delta(R/R_M)_r$.
10. From the value of R_M obtained in Step 2 and $(R/R_M)_{opt.}$ calculated in Step 9, determine the optimum reflux ratio, $R_{opt.}$.

Sample Calculations

To illustrate the use of the graphical correlations, we will calculate values of $N_{opt.}$ and $R_{opt.}$ for a given separation for different values of feedrate, amortization years, and material-cost factors. (The third and fourth examples include a method for correcting for tray efficiencies other than 80%.) Desired separation is between 3-methylpentene-1 and 2-methylpentene-1 in the presence of isoprene and 2,3-dimethylbutene-2 at 15 psia. Feed is saturated liquid at its boiling point, and 99% recovery of both keys is desired. The general input data for the examples are given in Table IV.

Example 1

Input data: Plain carbon steel used for all equipment, (M_c = 1.0), amortization on a 2-yr. basis, feedrate of 1,000 lb.-moles/hr., and 80% tray efficiency.

1. Compute N_M by the Fenske Equation.

$$N_M = \log[(x_{LK}/x_{HK})_D(x_{HK}/x_{LK})_B] / \log(\alpha_{lk})_{avg.}$$
$$= \log[(0.6464/0.0052)(0.4396/0.0055)] / \log(1.297)$$
$$= 35.4$$

2. Determine R_M, using Underwood's method.

$$1.0 - q = \sum \frac{\alpha_i (x_i)_D}{\alpha_i - \theta}$$

q = 1.0 for liquid feed at its bubble point.

$$0 = \frac{(2.857)(0.1600)}{2.857 - \theta} + \frac{(1.297)(0.3000)}{1.297 - \theta} + \frac{(1.000)(0.2400)}{1.000 - \theta} + \frac{(0.728)(0.3000)}{0.728 - \theta}$$

By trial and error, θ = 1.123.
Now,

$$R_M + 1 = \sum \frac{\alpha_i (x_i)_D}{\alpha_i - \theta} = \frac{(2.857)(0.3483)}{2.857 - 1.123} + \frac{(1.297)(0.6465)}{1.297 - 1.123} + \frac{(1.000)(0.0052)}{1.000 - 1.1123} + 0 = 5.36$$

R_M = 4.36

3. Determine $(N/N_M)_r$ from Fig. 6, for $\log[(x_{LK}/x_{HK})_D (x_{HK}/x_{LK})_B]$ = 3.997 and α = 1.297.

$$(N/N_M)_r = 2.15$$

4. Determine correction for feedrate, Φ, from Fig. 7, for $\log(F) = \log(1{,}000)$ = 3.0, and α_{LK} = 1.297.

$$\Phi = 0.98$$

5. Determine correction factor for the amortization years and material cost, Γ, from Fig. 8, for n/M_c = 10.0/1.0 = 10.0 and α_{LK} = 1.297.

$$\Gamma = 1.185$$

Therefore, $(N/N_M)_{opt.} = \Phi\Gamma\ (N/N_M)_r$ = (0.98)(1.185)(2.15) = 2.50

6. Calculate $N_{opt.} = N_M (N/N_M)_{opt.}$ = (35.4)(2.50) = 88.55

7. Determine reference value of $(R/R_M)_r$ from Fig. 9,

$$\log(x_{LK}/x_{HK})_D(x_{HK}/x_{LK})_B(x_{LK}/x_{HK})_F^{\mu},$$

where $\mu = 0.55\alpha_{LK}$.
The quantity above = log (0.6465/0.0052)(0.4396/0.0055)(0.3000/0.2400)$^{(0.55)(1.297)}$ = 4.067.

For α = 1.297, $(R/R_M)_r$ = 1.138.

8. Determine feedrate correction factor, Ψ, from Fig. 10, for log (F) = 3.0, and α_{LK} = 1.297. Ψ = 1.023
9. Determine amortization period and material-cost correction factor Δ from Fig. 11 for n/M_c = 10.0 and α_{LK} = 1.297. Δ = 0.9125.

Therefore, $(R/R_M)_{opt.} = \Psi\ \Delta(R/R_M)_r$
= (1.023)(0.9125)(1.138) = 1.06

10. Compute $R_{opt.} = R_M(R/R_M)_{opt.}$ = (4.36)(1.06) = 4.62.

Example 2

Input data: M_c = 6.0, n = 12 yr., F = 1,000 lb.-moles/hr., 80% tray efficiency. (Comments on each step will be made only if the procedure differs from that in Example 1).

1. through 4. Same as Example 1.
5. n/M_c = 12.0/6.0 = 2.0, α_{LK} = 1.297, and thus Γ = 1.0 from Fig. 8.

Therefore, $(N/N_M)_{opt.} = \Phi\Gamma(N/N_M)_r$
= (0.98)(1.0)(2.15) = 2.11

6. $N_{opt.} = N_M(N/N_M)_{opt.}$ = (35.4)(2.11) = 74.69

7. through 8. Same as Example 1: $(R/R_M)_r = 1.138$, $\Psi = 1.023$

9. $n/M_c = 2.0$, $\alpha_{LK} = 1.297$, and $\Delta = 1.0$ from Fig. 11. Thus, $(R/R_M)_{opt.} = \Psi\,\Delta(R/R_M)_r = 1.164$.

10. $R_{opt.} = R_M(R/R_M)_{opt.} = (4.36)(1.164) = 5.08$

Example 3

Input Data: $M_c = 1.0$, $n = 8$ yr., $F = 4{,}000$ lb.-moles/hr., 50% tray efficiency.

1. through 3. Same as Example 1.

4. Log $(F) = \log(4{,}000) = 3.60$, and once again $\alpha_{LK} = 1.297$. Therefore, $\Phi = 1.033$ from Fig. 7.

5. Since the vessel and tray equipment costs are usually much larger than the pump and exchanger costs, the effect of a change in tray efficiency on the optimum value of (N/N_M) can be compensated for by correcting Γ in the following manner:
$(n/M_c)(\text{Efficiency}/80.0) = (8.0/1.0)(50.0/80.0) = 5.0$. A value of Γ corresponding to this corrected (n/M_c) will be selected from Fig. 8.

The effect of the multiplier, (Efficiency/80.0), is to adjust the factor for efficiencies that differ from the base efficiency of 80% used in the development of the correlations.

From Fig. 8 for $(n/M_c) = 5.0$ and $\alpha_{LK} = 1.297$, Γ is determined to be 1.10. Therefore, $(N/N_M)_{opt.} = (N/N_M)_{opt.} = 2.44$.

6. $N_{opt.} = N_M(N/N_M)_{opt.} = 86.38$

7. Same as in Example 1: $(R/R_M)_r = 1.138$

8. Using Fig. 10 for log $(F) = 3.60$ and $\alpha_{LK} = 1.297$, $\Psi = 0.975$.

9. For $(n/M_c)(\text{Efficiency}/80.0) = 5.0$ and $\alpha_{LK} = 1.297$, Δ is found to be 0.9463 from Fig. 11. Therefore, $(R/R_M)_{opt.} = \Psi\,\Delta(R/R_M)_r = 1.05$

10. $R_{opt.} = R_M(R/R_M)_{opt.} = 4.58$.

Example 4

Input data: $M_c = 1.0$, $n = 8$ yr., $F = 200$ lb.-moles/hr., Efficiency = 50%.

1. through 3. Same as Example 1.

4. Log $(F) = 2.3$ and $\alpha_{LK} = 1.297$. From Fig. 7, $\Phi = 0.922$.

5. Same as Example 3, $\Gamma = 1.10$. $(N/N_m)_{opt.} = 2.18$.

6. $N_{opt.} = N_M(N/N_M)_{opt.} = 77.17$

7. Same as in Example 1: $(R/R_M)_r = 1.138$.

8. For log $(F) = 2.3$ and $\alpha_{LK} = 1.297$, $\Psi = 1.08$ from Fig. 10.

9. Same as Example 3, $\Delta = 0.9463$. Thus, $(R/R_M)_{opt.} = 1.16$.

10. $R_{opt.} = R_M(R/R_M)_{opt.} = 5.06$

Discussion of Sample Calculations

Table V contains a summary of the four examples. It illustrates the range of variables studied and the percent deviation between the values of $N_{opt.}$ and $R_{opt.}$ predicted by the graphical methods presented here, and by the shortcut equations of Fenske, Underwood, and Gilliland. The percent deviations reported are well within the limits of accuracy of the graphical correlations. The column diameters are also indicated in Table V, to show that the correlations are applicable for a wide range of column sizes.

Summary of Sample Calculations—Table V

Parameters	Example Number 1	2	3	4
Material-cost factor, M_c	1.0	6.0	1.0	1.0
Amortization years, n	10.0	12.0	8.0	8.0
Feedrate, lb-moles/hr.	1,000.0	1,000.0	4,000.0	200.0
Tray efficiency, %	80.0	80.0	50.0	50.0
Column dia., ft.	10.4	10.8	20.9	4.8
% Deviation in $N_{opt.}$*	−3.7	−1.0	−0.8	+0.4
% Deviation in $R_{opt.}$*	+0.9	+1.4	−1.7	+1.8

*Comparing graphical method presented here with calculations using shortcut equations.

The results of the procedure used to find minimum yearly operating cost by shortcut equations are illustrated in Fig. 12 for Examples 1 and 2. Total yearly operating cost is plotted vs. values of R/R_M. Readily apparent from Fig. 12 are (1) the effect of the number of years used in determining the annual depreciation cost, and (2) the influence of the material of construction ($n/M_c = 10.0$ for Example 1.0, and $n/M_c = 2.0$ for Example 2) on the total yearly operating cost and on the optimum R/R_M. As discussed earlier, the value of n/M_c controls the balance between the importance of equipment and operating costs on the total minimum cost.

The technique for correcting for different tray efficiencies, illustrated in Examples 3 and 4, provides values well within the limits of the basic correlations. It would appear that a similar approach could be used to compensate for different tray types and spacings.

Conclusions

We believe the correlations presented here will give within ±2% the optimum values of the reflux ratio and theoretical stages for the following range of variables: α's of 1.05 to 3.00, feedrates of 100 to 10,000 lb.-moles/hr., saturated-liquid feeds, and tray spacing of 24 in. Changing the values of utility costs has small effect on the correlations. For example, it is doubtful that a change in cost of 5% would cause more than 1% change in any of the correction factors or in the final results.

These correlations should aid design engineers wishing to study a number of cases for economic evaluation. While no claim is made for exact agreement with the iterative procedure, the results should be valuable in locating the approximate optimum from which point more-detailed calculations could be initiated.

References

1. Fenske, M. R., *Ind. Eng. Chem.*, **24,** 482 (1932).
2. Gilliland, E., *Ind. Eng. Chem.*, **32,** 1101 (1940).
3. Thiele, E. W., and Geddes, R. L., *Ind. Eng. Chem.*, **25,** 289 (1933).
4. Underwood A. J. V., *Chem. Eng. Progr.*, **44,** 603 (1948).
5. Van Winkle, M., and Todd, W. G., Optimum Fractionation Design by Simple Graphical Methods, *Chem. Eng.*, Sept. 20, 1971, pp. 136–148.

B. Specific equipment

How to design overhead condensing systems

The state of fluids in pipelines, and the physical arrangement of equipment around the distillation tower, establish the design parameters for meeting hydraulic and piping conditions of the system.

Robert Kern, *Hoffmann - La Roche Inc.*

☐ The state of fluid in the pipelines, and the hydraulic and thermal conditions in condensing systems of distillation columns, are the reverse of those in reboiler circuits.

The inlet line to condensers can carry superheated or saturated vapor, or dispersed vapor-liquid mixtures. Fluid is cooled in the exchanger, and partial or full condensation takes place. The condenser's outlet line can have stratified and dispersed two-phase flow, saturated liquid, or subcooled liquid. In addition, the flowing fluid can be a mixture of two substances. Thus, this type of condensation offers a wide range of classification from a thermodynamic standpoint [*1*].

In contrast, saturated liquid normally flows in the downcomer of reboilers. The liquid is vaporized while passing through the exchanger. The reboiler's outlet line carries a dispersed vapor-liquid mixture having a vapor content of 30 to 90% of total flow.

In this article, we will examine the hydraulics for the following:

1. Condensers with gravity-flow return lines.
2. Condensers with pumped-reflux lines.
3. Two-stage condensation.

Within these groups, hydraulic-design and piping-design conditions vary, depending on the state of fluid in the lines and the physical arrangement of the installation.

Vacuum technology has its own systems, equipment and terminology. Piping design of vacuum-condensing systems are outside the Euler-Bernoulli-Darcy theories and are not included in this article.

Gravity-flow reflux

Horizontal condensers—A condenser in gravity-flow arrangements is located above the level of the terminating point of the condenser's outlet line, as shown in F/1 and F/2. For the horizontal condenser in F/1b, vapor enters the exchanger at the top, and subcooled liquid leaves at the bottom. The looped-outlet pipe ensures a permanent liquid level in the condenser. This liquid level is controlled through the reflux branch and through the takeoff line to storage.

The static-head pressure difference, ΔP_s, between the vertical overhead line and the condenser's outlet line for the arrangements in F/1 can be written as:

$$\Delta P_s = (H/144)(\rho_1 - \rho_2) \quad (1)$$

ΔP_s must be equal to or greater than the sum of (1) the pipe-system resistance, Δp_p, between reference points A and B; (a) exchanger pressure drop, Δp_e; and (3) required pressure difference across the control valve, Δp_{cv}:

$$\Delta P_s \geqq \Delta p_p + \Delta p_e + \Delta p_{cv} \quad (2)$$

The required distance, H, between fractionator inlet and exchanger centerline can be calculated from Eq. (1) as:

$$H = (144\,\Delta P_s)/(\rho_1 - \rho_2) \quad (3)$$

The vapor column can be neglected by assuming $\rho_2 = 0$ in Eq. (1) and (3). All pressures are in psi; densities, ρ, in lb/ft^3; and dimensions, H, in ft.

As Eq. (3) shows, for a minimum of elevation difference between the top of the column and the exchanger centerline, the piping and components resistances must also be minimal.

Originally published September 15, 1975

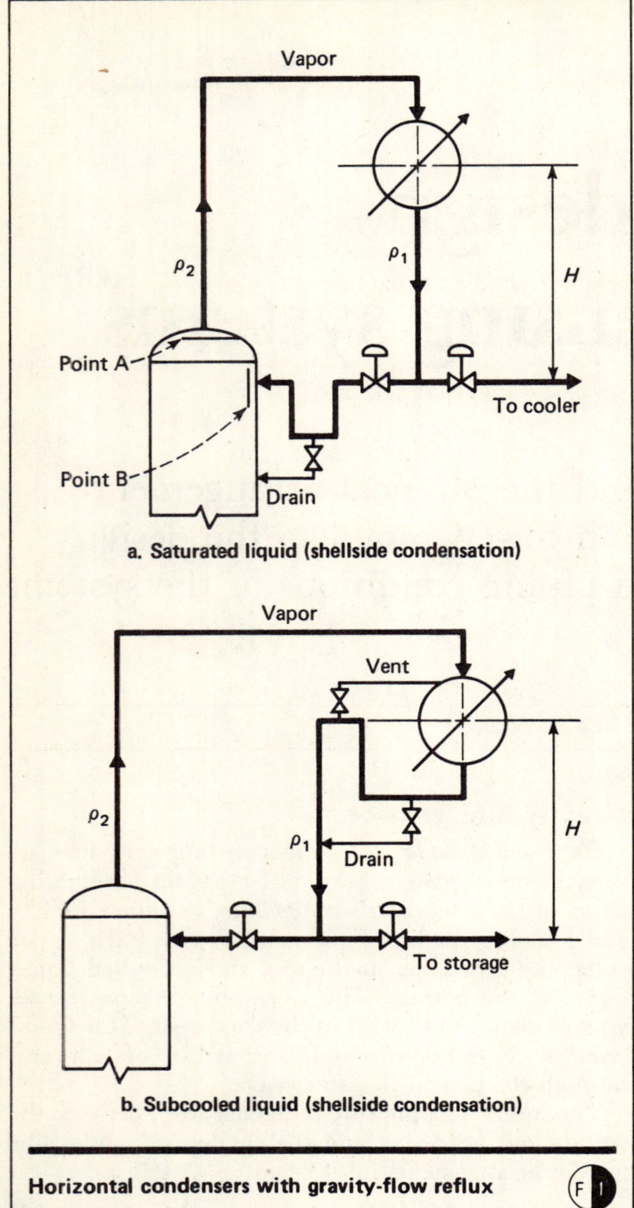

a. Saturated liquid (shellside condensation)

b. Subcooled liquid (shellside condensation)

Horizontal condensers with gravity-flow reflux F 1

Generally, in condensing systems, the unit loss in the piping is low—about a tenth or a hundredth of 1 psi/100 ft. Inlet and outlet resistances to process equipment usually take a considerable portion of the pipeline resistance and should not be ignored in the calculations. (Using three decimal places in the calculations is not unusual.)

In horizontal condensers, condensation takes place in the shell. This gives lower resistance than the tubeside. A baffle (or baffles) in the exchanger is in the horizontal plane through the exchanger's centerline. If necessary, two inlet and two outlet nozzles can halve the total flow, and reduce entrance and exit resistances considerably. In this case, the inlet and outlet piping should be symmetrical.

The subcooled liquid for the arrangement in F/1b can be drained or pumped directly to storage. The product stream for F/1a is usually directed through a cooler before storage.

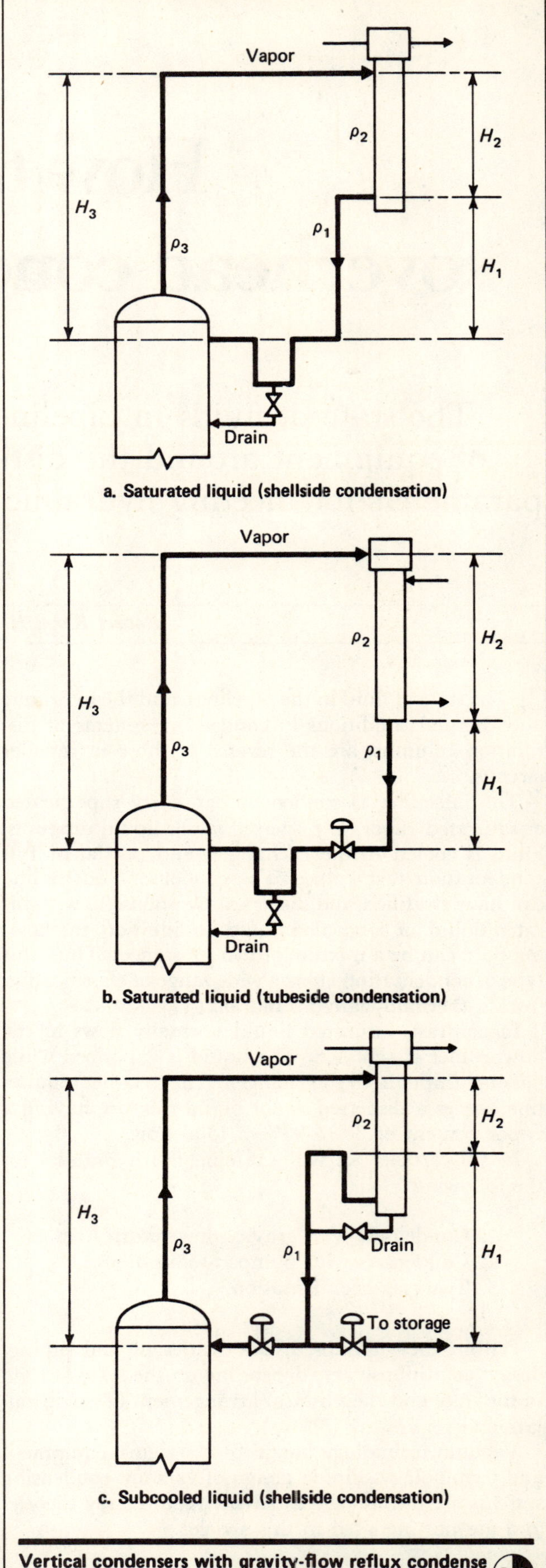

a. Saturated liquid (shellside condensation)

b. Saturated liquid (tubeside condensation)

c. Subcooled liquid (shellside condensation)

Vertical condensers with gravity-flow reflux condense vapors on the shellside or tubeside of exchanger F 2

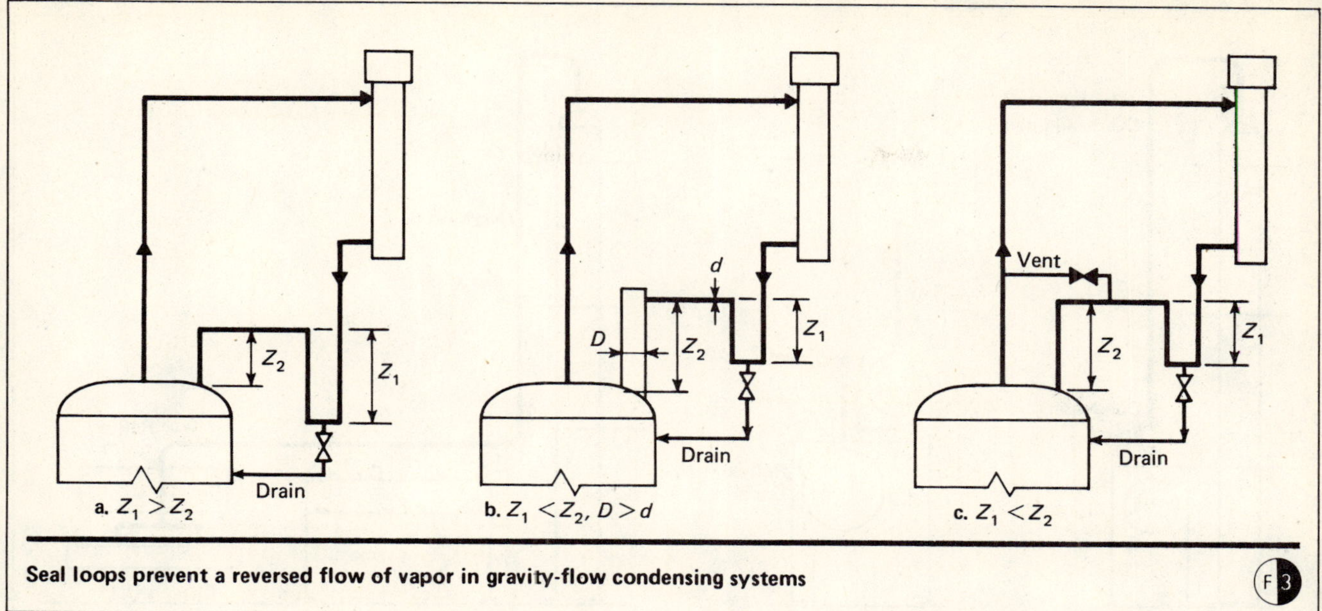

Seal loops prevent a reversed flow of vapor in gravity-flow condensing systems F3

Control valves in these systems should be located at a low point of the return line and product stream. Sufficient static head before the valve inlet will prevent vaporization across the valve. A product cooler should not receive a liquid-vapor mixture.

Vertical condensers—Arrangements for these condensers with gravity-flow outlets are shown in F/2. Condensation can take place in the shell (F/2a), or tubes (F/2b). A single-pass vertical condenser is more suitable for liquid subcooling than a horizontal one. The seal-loop height can be adjusted within a greater range than with horizontal condensers (F/2c). The required liquid level in the exchanger shell is determined by the exchanger's designer.

The hydraulic balance for the arrangements shown in F/2 is:

$$(1/144)(H_1\rho_1 + H_2\rho_2) \geqq (1/144)H_3\rho_3 + \Delta P \quad (4)$$

where ΔP is the sum of piping, Δp_p, exchanger, Δp_e, and control-valve (if any), Δp_{cv}, resistances:

$$\Delta P = \Delta p_p + \Delta p_e + \Delta p_{cv} \quad (5)$$

The elevation difference, as expressed from Eq. (4), between the condenser's outlet and the reflux inlet nozzle will be:

$$H_1 \geqq (1/\rho_1)(H_3\rho_3 - H_2\rho_2 + 144\,\Delta P) \quad (6)$$

where ρ_1 is the density of condensate in the reflux line, ρ_3 is the vapor density in the overhead line, and ρ_2 is the average density in the vertical exchanger:

$$\rho_2 = 0.5(\rho_1 + \rho_2) \quad (7)$$

Seal loop prevents flow reversal

In gravity-flow condensing systems, a seal loop is provided to prevent a reversed flow of vapor in the condenser's outlet line. This loop can be used for holding a liquid level in the condenser, as shown in F/1a and F/2c.

If the gravity-flow reflux line terminates in a vertical leg (Z_2 dimension in F/3), the piping design should be such as to prevent siphoning that can empty the seal. If dimension Z_1 is smaller than Z_2, and the pressure just before the seal loop and at the terminating point after the seal loop is identical (for example, with greatly reduced flow), liquid can be siphoned out of the seal; and intermittently, the condenser will not operate well. This can be prevented if Z_1 is designed to be longer than Z_2 (see F/3a).

For the arrangement in F/3b, the final vertical leg has a larger diameter than the gravity-flow reflux line. Again, this can prevent the siphoning of liquid from the seal loop.

Another arrangement (F/3c) has a closed vent line. This can be opened at reduced condensate flow to keep the seal loop filled with liquid. With this type of venting, the pressure difference across the vent valve should be zero. Therefore, it is essential to connect the end-points of the vent line to locations where pressures are expected to be about equal.

Pumped-reflux arrangements

Typical overhead lines for hydrocarbon distillation columns are shown in F/4. Fluid circulation in the piping is the result of the thermosiphon effect in gravity-reflux condensation. For the systems shown in F/4, there is (and most of the time must be) a pressure difference between the top of the tower and reflux drum. The reflux drum has a pump, which returns the liquid to the top of the tower or sends it to storage.

In these arrangements (besides the sum of the static heads), actual pressure differences, ΔP_p, also enter into the calculations:

$$\Delta P_p = P_1 - P_2 \quad (8)$$

For the dimensions given in F/5, the static-head difference will be:

$$\Delta P_s = (1/144)(\rho_1 H_1 - \rho_2 H_2) \quad (9)$$

where ρ_1 is usually vapor density, and ρ_2 is vapor-liquid

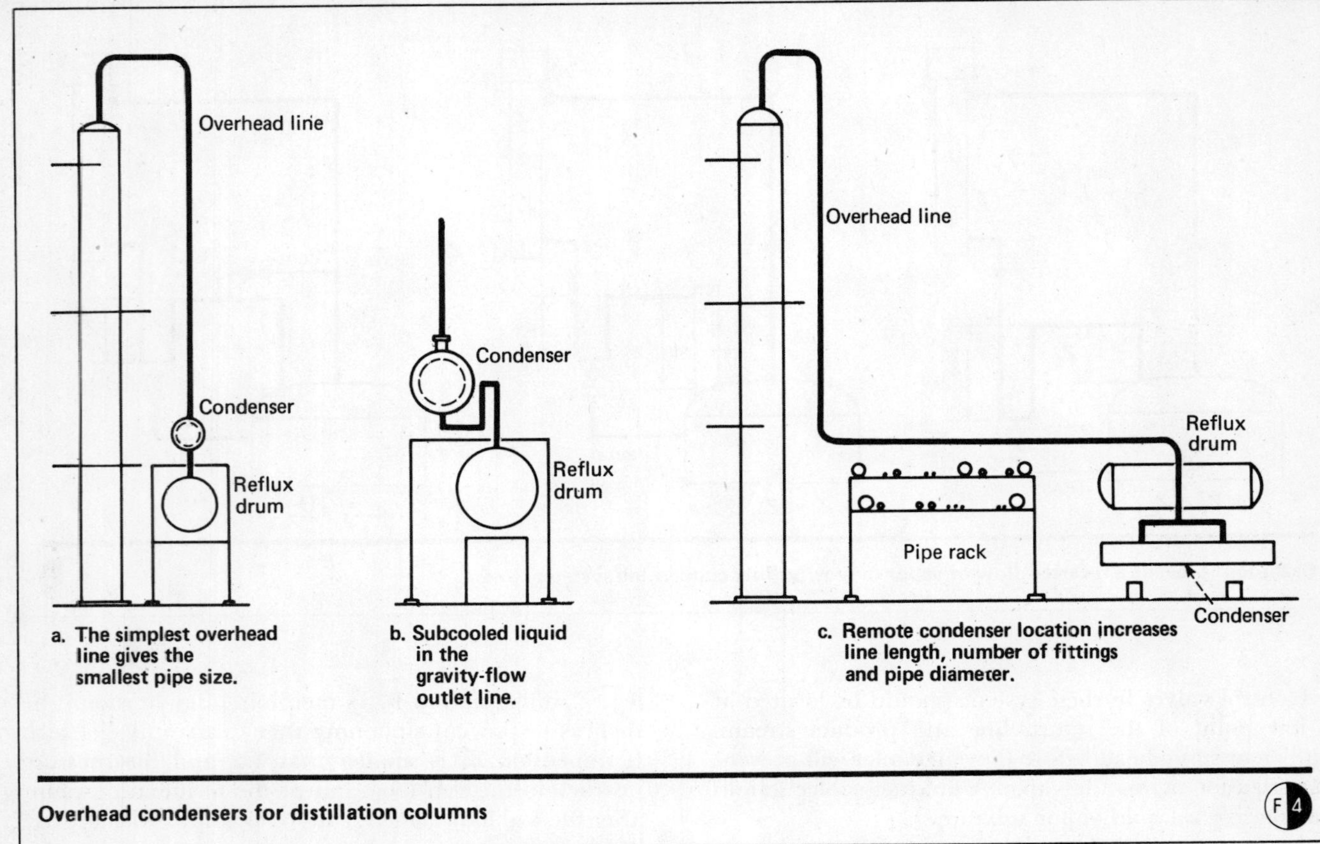

Overhead condensers for distillation columns

F 4

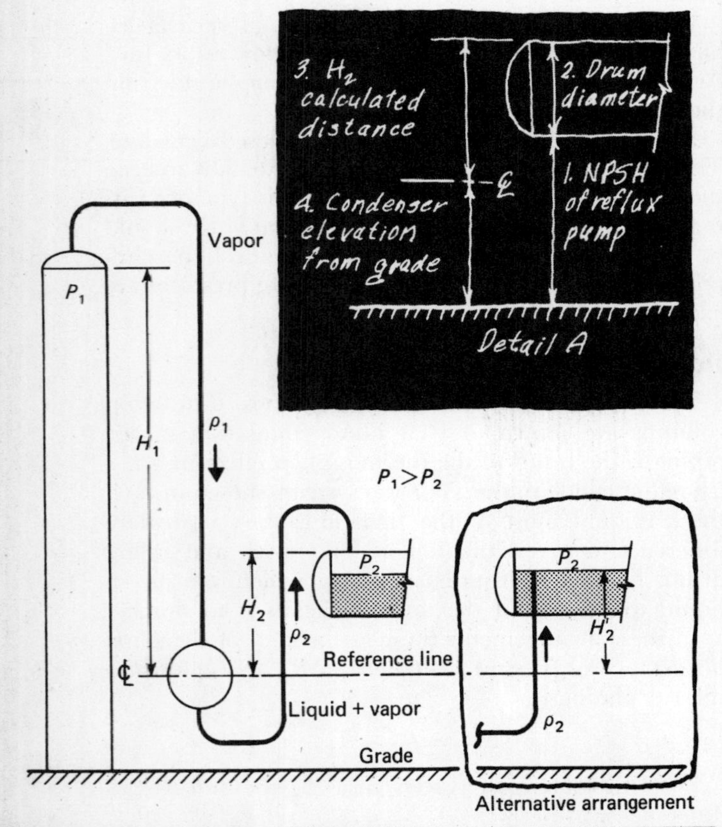

Dimensional relations for a condenser at grade

F 5

mixture (or liquid) density. The overall available ΔP is the sum of Eq. (8) and (9), or:

$$\Delta P = \Delta P_s + \Delta P_p \quad (10)$$

If the righthand side of Eq. (10) becomes negative, the condenser must be placed at an elevation closer to, or above, the reflux drum. A negative value indicates that the static-head backpressure ($\rho_2 H_2$) in Eq. (9) is greater than the sum of (1) pressure difference between the top of the tower and the reflux drum plus (2) the vapor static head ($\rho_1 H_1$) in the overhead line. The greater the condensation, the heavier the mixture becomes in the condenser's outlet line, which results in a greater backpressure. Of course, $\rho_2 H_2$ becomes positive when the line has a gravity-flow arrangement between the condenser outlet and the reflux-drum inlet, as shown in F/4a and F/4b.

Eq. (10) shows the driving force in the overhead system. This must be greater than the sum of piping, inlet and exit losses, Δp_p, and exchanger resistance, Δp_e:

$$\Delta P > \Delta p_p + \Delta p_e \quad (11)$$

Δp_p usually is between 2 to 6 psi, and condenser Δp_e ranges from 0.5 to 5 psi.

The maximum possible condenser-centerline location below the reflux drum (dimension H_2 in F/5) can be calculated from the combination of Eq. (9), (10) and (11), to give:

$$(1/144)(\rho_1 H_1 - \rho_2 H_2) + \Delta P_p - (\Delta p_p + \Delta p_e) = 0 \quad (12)$$

As a safety factor, the positive static-head column pres-

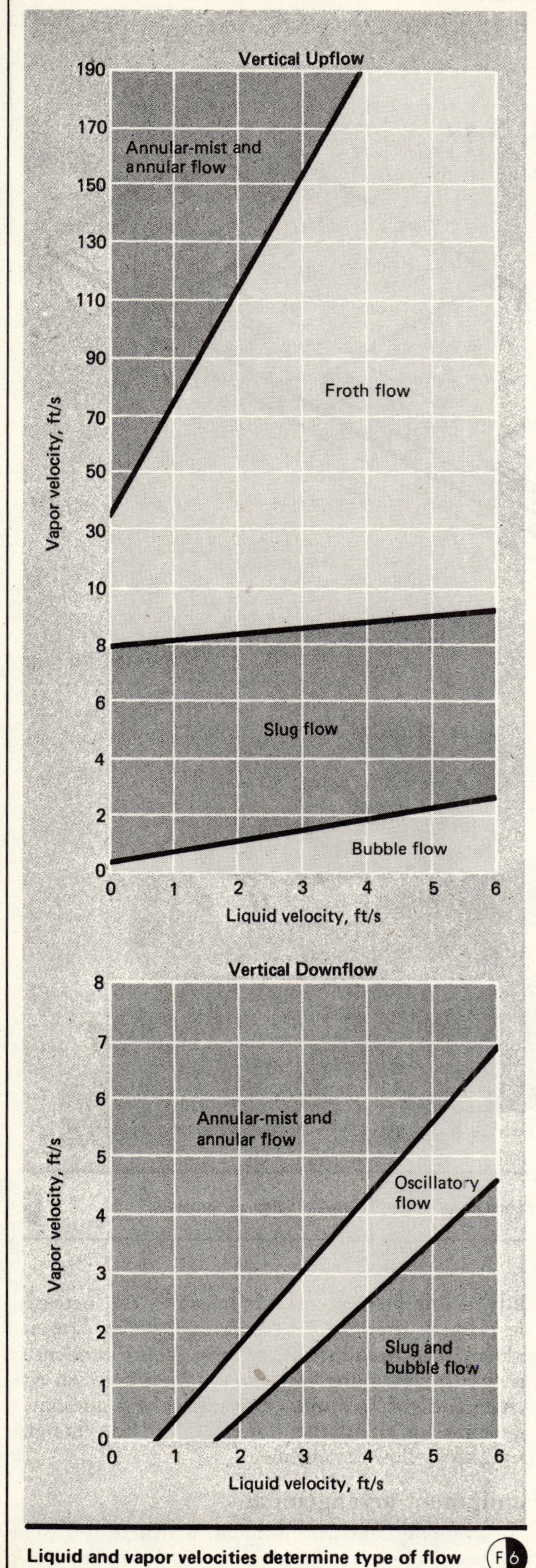

Liquid and vapor velocities determine type of flow F6

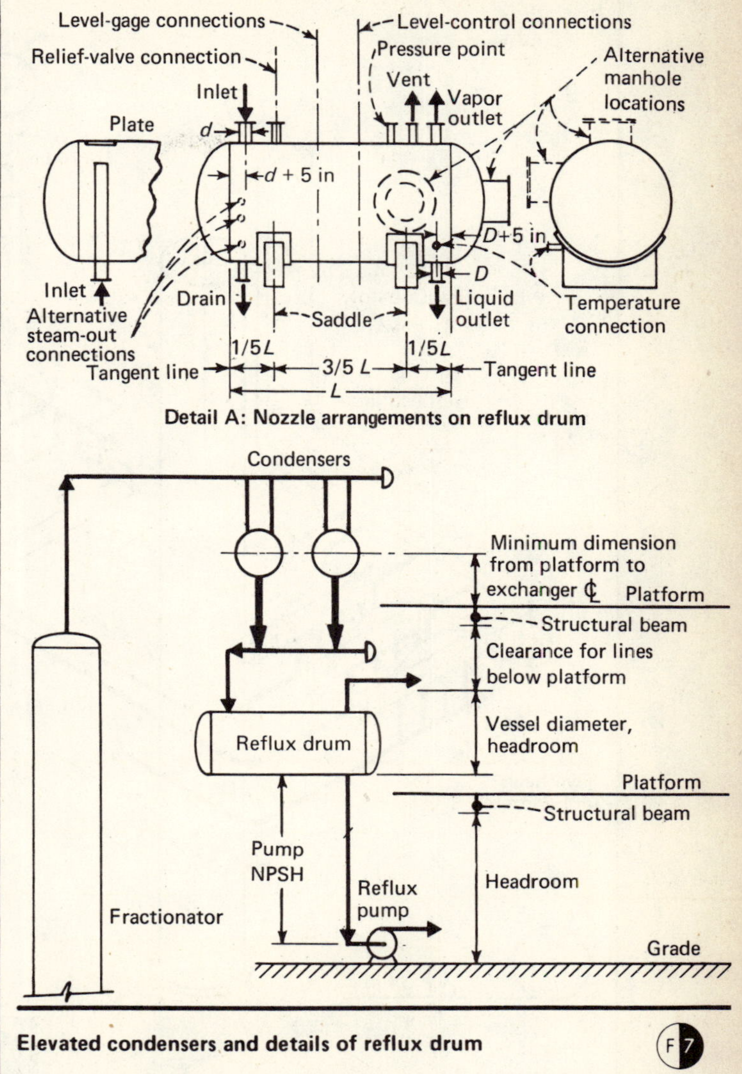

Elevated condensers and details of reflux drum F7

sure of the overhead vapor line can be neglected. Consider:

$$\rho_1 H_1 = 0.$$

Expressing H_2 from Eq. (12) in feet, we get:

$$H_2 = (144/\rho_2)(\Delta P_p - \Delta p_p - \Delta p_e) \qquad (13)$$

In layout design, usually the reflux drum is elevated first in accordance with the required NPSH (net positive suction head) of the reflux pump. The dimensions shown in Detail A of F/5 will establish the condenser elevation from grade.

Slug flow is undesirable

Slug flow can develop in the pocketed condenser-outlet line shown in F/5, depending on vapor-liquid proportion and fluid velocity. Slug flow should be avoided because it can cause undesirable pressure surges.

An empirical relation can be used to estimate the slug-flow region. If the velocity (calculated with two-phase density) in the pipeline is smaller than $(5\rho_1/\rho_v)^{1/2}$, slug flow is possible. The type of flow can also

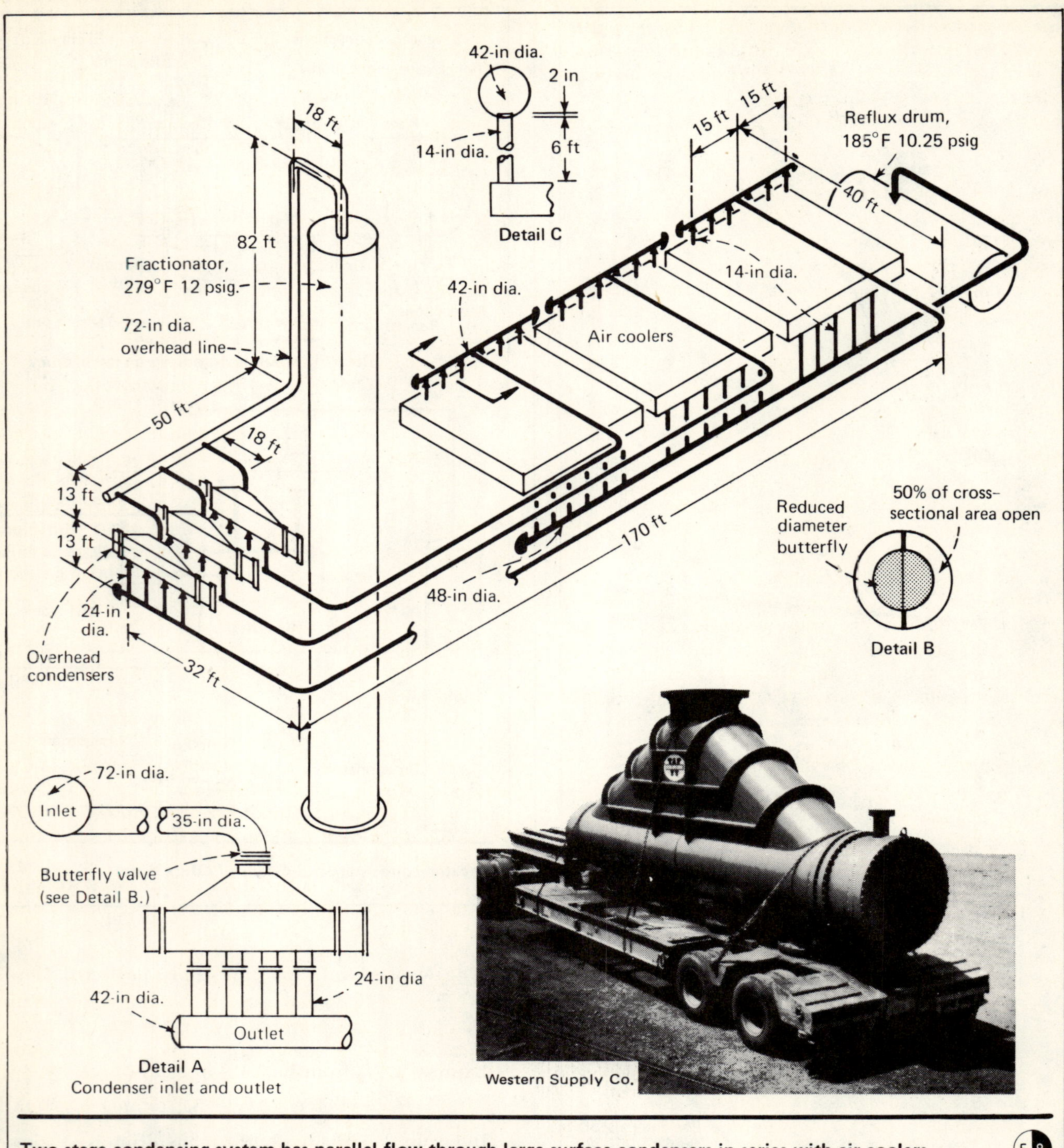

Two-stage condensing system has parallel flow through large surface condensers in series with air coolers F 8

be estimated by calculating the Baker paramaters and using Fig. 1 (*Chemical Engineering*, June 23, 1975, p. 146). Flow-region maps such as are shown in F/6 are also available for vertical flows. To use these maps, the vapor-phase and liquid-phase velocities are calculated separately for the same pipe diameter. The intersection of these in the appropriate diagram of F/6 establishes the flow region.

The general criterion for selecting a suitable line size is that the pipe diameter must be sufficiently small to have the highest possible velocity, but large enough to stay within available pressure differential. The possibility of slug flow can be minimized by (a) increasing the pressure drop in the condenser's outlet line and reducing the resistance of the rest of the condensing system, (b) providing two parallel lines between condenser and reflux drum, (c) using valved alternative pipe runs for alternative flowrates, and (d) changing to a gravity-flow arrangement.

Equipment arrangement

In a gravity-flow arrangement, a structure is often required. Factors that influence layout design are shown in F/7. Below the elevated condensers, a platform is usually provided. Structural members, space for piping below the platform, and reflux-drum diameter or minimum headroom requirement determine the platform level for the reflux drum. Piping, NPSH or headroom will establish the distance between grade and the reflux-drum platform.

The piping arrangement of condensing systems should be as simple as possible. Each elbow in the overhead line represents a sizable percentage of the overall pipe-friction loss. However, lines can become longer because of flexibility considerations. Usually, there is no valve in an overhead line. Pressure loss in fittings and line length have to be compensated for with increased pipe diameter.

Piping design, and access to valves and instruments, also depend on how well nozzles are located and oriented on process vessels. The usual nozzle arrangement of a reflux drum is shown in Detail A of F/7. The condensate inlet is at one end of the top. Alternatively, a bottom inlet with a standpipe in the reflux drum can be designed, eliminating a couple of elbows and reducing pipe resistance. The pump drawoff nozzle is at the bottom and opposite end of the reflux drum. Level-gage and level-controller connections can be in the center. At the center, the liquid level is least agitated by the inlet and outlet streams. Pressure connections are usually in the vapor space, and temperature connections in the liquid space near the pump drawoff nozzle. Alternative manhole locations, and vent, drain and support connections are shown in Detail A of F/7.

Two-stage condensation

We will now describe a large, two-stage, condensing system, and in doing so we will find additional design ideas for the condenser piping.

Where cooling water is expensive, condensing systems can be designed for use with air coolers or with a combination of air coolers and water coolers in series. If close condensate-temperature control is required, the water coolers are located after the air coolers. For optimum heat transfer, a reversed sequence can be chosen.

A large-capacity condensing system handling the overhead from the primary fractionator of an ethylene unit is shown in F/8. The pressure difference between the fractionator and reflux drum is small, $\Delta P = 1.75$ psi. With this limited pressure differential, it is essential to have minimal resistance through exchangers, piping and pipe components.

Resistance of the surface condensers has been minimized by choosing three exchangers, each handling one-third of the total flow; and by providing a flared exchanger inlet, and four large outlets for each shell. Exchanger resistance is given as 0.12 psi, exclusive of inlet and outlet resistances. The photograph in F/8 shows one of the exchangers being delivered.

Air-cooler resistance has been minimized by using numerous single passes and large-diameter finned tubes. Air-cooler pressure loss is estimated as 0.7 psi, excluding nozzles.

Even with all these provisions, close to 50% of the available overall pressure difference has been consumed by the exchangers alone. About 30% of the overall ΔP has been taken up by the inlet and outlet resistances, leaving 20 to 25% of 1.75 psi for pipe resistance. This, and the rather long pipe configuration, mean very large pipe sizes. These are indicated in F/8. Velocities in these lines are less than 50 ft/s. Under these conditions, the piping system from the fractionator to the reflux drum has to be self-draining. Horizontal sections of the piping slope toward the reflux drum.

Saturated vapor flows in the 72-in-dia. overhead line. (Overhead lines are not usually insulated, and it is reasonable to assume that some condensation will take place in them.) Condensate, collected in the overhead line, can drain to the surface condensers through the three tangentially welded branch-connections (see Detail A in F/8). Quite likely, this condensed liquid is not distributed to the three surface condensers equally because of the nonsymmetrical arrangement of the branch lines. Considering the small amount of liquid, and that this liquid does not affect the heat-transfer duty of the exchangers, the nonsymmetrical arrangement is accepted.

For equal vapor distribution to the three surface condensers, butterfly valves are installed at the inlet flanges of the flared exchanger sections (see Detail B in F/8). The butterfly flap in each valve has a reduced diameter—leaving 50% of the cross-sectional area of the inlet nozzle permanently open. In this way, it is assumed that, with less resistance, a more-sensitive regulation can be accomplished than with a valve having a full-diameter flap.

Pipelines between the surface condensers and air coolers have two-phase dispersed flow. Because of the low velocity, elbows and tee connections, some liquid separation (with stratified flow) at the bottom of the lines is also assumed. (Under such conditions, and with these line sizes, it is difficult to assess how valid two-phase-flow theories are.)

The outlet lines of the surface condensers connect centrally to the inlet headers of the air coolers. Equal fluid distribution is assured by the relatively great difference in resistance between the very-low header loss and the high entrance resistance to the air cooler branch-connections. For equal distribution of the stratified liquid at the bottom of the pipe, each branch-connection extends about 2 in up into the horizontal air-cooler header (see Detail C in F/8). These extensions dam up the liquid in the bottom of the pipe, and the overflow facilitates equal liquid distribution across each air-cooler pass. Symmetrical piping at the outlet side of air coolers is not considered as essential as at the inlet side.

References

1. Kern, D. Q., "Process Heat Transfer," McGraw-Hill, New York, 1950.
2. Golan, L. P. and Stenning, A. H., Two-Phase Vertical Flow Maps, Joint Symposium on Fluid Mechanics and Measurement in Two-Phase Flow Systems, at University of Leeds, Leeds, England, Sept. 24–25, 1969; *Proc. Inst. Mech. Engrs. (London)*, Vol. 184, Part 3C (1970).
3. Cady, P. D., How To Stop Slug Flow in Condenser Outlet Piping, *Hydrocarbon Process.*, Sept. 1963.
4. Greene, J. L., Symmetrical Piping Arrangements Solve Two-Phase Flow Distribution Problems, *Hydrocarbon Process.*, Feb. 1967.

How to Design Reflux Drums

Here is a method for designing distillation reflux drums to an optimum without going through the usual trial-and-error calculations.

B. SIGALES, Instituto Petrolquímica Aplicada

Ordinarily, the dimensions of a distillate-reflux drum must be fixed to allow: (1) droplet disengagement in the vapor phase; (2) enough surge time for the liquid phase; and, often, (3) settling of small amounts of a second heavy-liquid phase (usually water).

In the vapor, droplet separation is attained by limiting vapor velocity, according to Newton's law. Therefore, vapor flow sets the minimum vapor-passage area, i.e. the section free of liquid. To some extent, it can be said that the diameter of drums is governed by the vapor flow. (This is always true for vertical drums.)

Maximum vapor velocity is expressed by:

$$v = K\sqrt{\frac{\rho_L - \rho_V}{\rho_V}} \cong K\sqrt{\frac{\rho_L}{\rho_V}} \qquad (1)$$

where v = vapor velocity; K = velocity constant; ρ_V = vapor density at design conditions; ρ_L = liquid density at design conditions (normally from 20–40 times ρ_V). Ordinary values for K are 0.13 ft/s for horizontal vessels, and 0.26 ft/s for horizontal vessels having a wire-mesh demister.

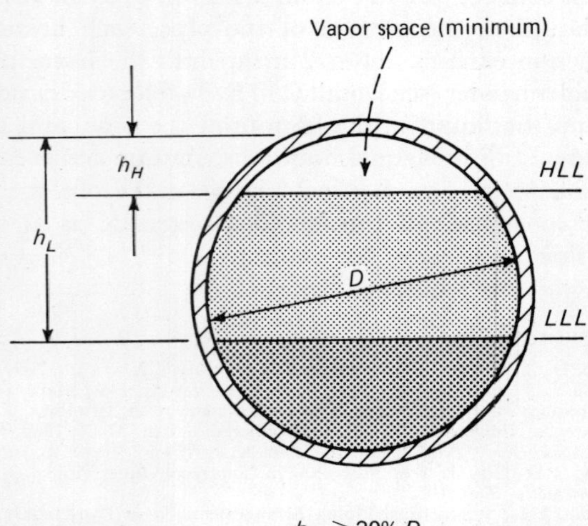

CROSS-SECTION of reflux drum shows L_{HL} and L_{LL}—Fig. 1

Knowing the value of v, one can determine the minimum free area for vapor passage. This area should also include a minimum clearance, h_H, above the high liquid level, L_{HL}, (Fig. 1), equal to 20% of the drum diameter. If possible, this clearance should be greater than 10 in.

How To Fix Surge (Residence) Time

Liquid streams require a minimum surge, or residence, time in the drum to facilitate control, absorb process upsets and fluctuations, maintain safe operations, etc. Sometimes, a maximum residence-time limit is fixed for fouling liquids. Surge time can be defined as the time required to empty the vessel if inflow is stopped. In some instances, drum volume is governed by liquid flow.

Fixing of surge time requires consideration of instrumentation characteristics, because vessel dimensions affect the time constant of the system. Ordinarily, "cookbook" figures are used [*1,2,3*], but good judgment must be applied in each situation because of the interrelationship between the drum and other process steps. For rigorous instrumentation-based methods, see Ref. 4.

Surge time is dependent on: (a) holdup time with the drum half-full; and (b) holdup time between L_{HL} and L_{LL}. This approach excludes the liquid below the L_{LL}, which makes for a conservative design.

Usually, half-full holdup time will set the normal liquid level (L_{NL}); therefore, surge time is about half the time needed for holdup between high and low liquid levels. This, however, can be misleading, as will be shown in the worked-out example presented further on.

On the basis of holdup between high and low liquid levels, the following typical values will be assumed:*

Stream	Hold from L_{HL} to L_{LL}, Min
Tower reflux	5
Product to storage	2
Product to heat exchanger along with other process streams	5
Product to heater	10

Note that only the larger volume must be used; do not add two volumes (e.g., reflux plus product).

* Level-recording-controller and level-alarm instrumentation are also assumed.

Originally published March 3, 1975

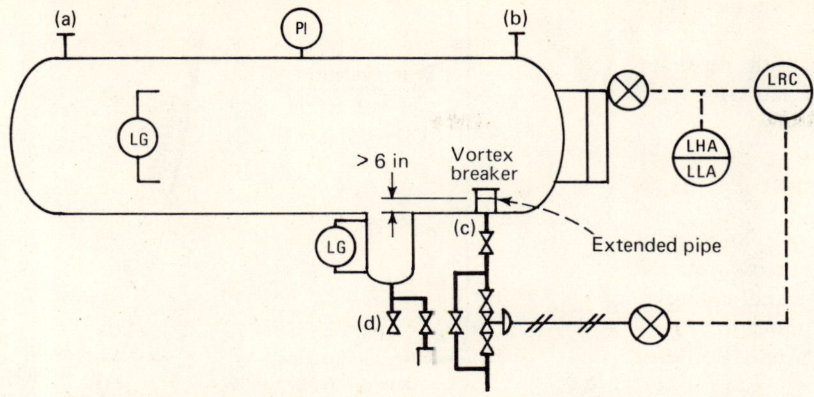

Pipe Size, In.

Drum Diameter Ft	Feed (a)	Ovhd. (b)	Bottoms (c)	Drain (d)	Accesways Quant.-in.
3	4	2	3	1,5	2-6 Handhole
4	6	3	4	2	1-18 Mandhole
6	10	4	8	2	1-18 Mandhole
8	12	6	10	3	1-18 Mandhole
10	16	8	12	3	2-18 Mandhole
13	20	10	16	4	2-18 Mandhole
16	24	12	20	4	2-18 Mandhole

REFLUX DRUM with pot, piping and instrumentation in schematic arrangement corresponding to actual design—Fig. 2

Watkins [*5*] provides a method to fix the surge time (drum half-full) that takes into account instrument types and a labor factor. Even so, there is still a lot of guessing in determining the final figure. Nevertheless, the Watkins approach is a good exercise in applying good judgment to specific situations.

To separate a second liquid phase, a maximum liquid velocity (based on Stokes law) could be set. Normally, for small amounts of heavy liquids, a residence time is fixed in a way that a total holdup time is obtained from the controlled surge time, t_s (time needed for liquid to descend from L_{HL} to L_{LL}, plus the settling time, t_d, when the liquid is below L_{LL}). For water in hydrocarbons, a reasonable settling-time figure is 5 min, based on total hydrocarbon volume. For other products, the time in minutes required for them to separate can be estimated from:

$$t_d = 3\mu/\Delta \text{sp gr} \quad (2)$$

where μ = viscosity of the predominant or continuous liquid phase, cp; and Δsp gr = difference of specific gravities (or densities in g/cm^3) between the two liquid phases.

Summarizing, the total holdup time, t, for the drum is equal to the surge time (for a single liquid phase) plus the settling time:

$$t = t_s + t_d \quad (3)$$

If no second liquid phase is to be settled, L_{LL} is situated above vessel bottom at a distance, $D - h_L$ (Fig. 1), equal to 10% of the drum diameter. If possible, this distance should be greater than 5 in. This circular segment of height 10% of dia. has an area equal to 5% of the total vessel cross-sectional area (Fig. 1).

The dimensions of the vessel (L and D) are set by economic criteria, although plant or piping-layout considerations may influence the final dimensions. An L/D ratio determination is complex, because it depends on standard plate thickness and dimensions, welding techniques, heat treatment, etc. [*6*]. A good approximation might be the minimum vessel weight, with a correction factor for the heads. In this way, a typical formula for vessels with 2 : 1 elliptical heads is [*1*]:

$$[(P/(2SE - P)]D^4 + 0.083D^3 - 0.0905V = 0 \quad (4)$$

where D = vessel diameter, ft; V = vessel volume, gal; P = design pressure, psig; S = allowable stress of the shell material, psi; and E = joint efficiency, $\leqslant 1$.

Normally, distillate-reflux drums are installed horizontally because the liquid load is important. Because of this, the design method that follows applies only to horizontal drums. For economical evaluations, typical drum-nozzle sizes and piping and instrument (P&I) data are given in Fig. 2. Main nozzles (a, b and c in Fig. 2) must be placed at minimum distances from the tangent lines of the dished heads to the body of the vessel. Ref. 7 provides information on vortex breakers, whereas drum arrangement and elevation (head room must be provided) are discussed in Ref. 8. Vessels should be sloped (0.5% is usual) to facilitate draining.

When a small quantity of a second liquid phase is settled, a drawoff pot is normally provided to make separation of the heavy liquid easier. The pot diameter is ordinarily determined for heavy-phase velocities of 0.5 ft/min. Minimum length is 3 ft, to allow space for interface-level controller connections. A minimum pot diameter for reflux drums having 4–8 ft diameters is 16 in; for drum diameters above 8 ft, diameters of at least

24 in are used. The pot must also be placed at a minimum distance from the tangent line that joins the head with the body of the vessel.

Although it is possible to size a drum with the above criteria, the conventional way is by trial and error [2,5]. A method that avoids trial and error follows.

Recommended Calculation Method

First, determine v from Eq. (1), then t_s, t_d and t from Eq. (3). Calling x_L the percent of the drum cross-sectional area occupied by the liquid at its L_{HL}, and calling x_v the minimum, free, vapor-passage area above the L_{HL}—expressed as percentage of the drum's cross-sectional area—we can write:

$$x_L + x_v = 100 \qquad (5)$$

Neglecting the head volumes, vapor velocity = (flow of vapor)/(vapor-passage area), or:

$$v = \frac{M_v/(3{,}600\rho_v)}{(x_v/100)(\pi D^2/4)}, \text{ ft/s} \qquad (6)$$

and holding time = (liquid volume)/(liquid flow), or:

$$t = \frac{(x_L/100)L(\pi D^2/4)}{M_L/(60\rho_L)}, \text{ min} \qquad (7)$$

where M_v = vapor flowrate, lb/h; ρ_v = vapor density at design conditions, lb/ft^3; M_L = liquid flowrate, lb/h; ρ_L = liquid density at design conditions, lb/ft^3; and $t = t_s + t_d$, min.

Combining Eq. (6) and (7) and rearranging:

$$x_v/x_L = (LM_L\rho_L)/(vtM_v\rho_v)$$

Substituting Eq. (1) in the above expression:

$$x_v/x_L = [(LM_v)/(KtM_L)](\rho_L/\rho_v)^{1/2} = (LM_v a)/(KtM_L)$$

where $a = (\rho_L/\rho_v)^{1/2}$, dimensionless.

Rearranging and taking into account Eq. (5):

$$L = \left(\frac{x_v}{100 - x_v}\right)\left(\frac{60tKM_L}{aM_v}\right), \text{ ft} \qquad (8)$$

If no second liquid phase is to be settled, $t = t_s$; therefore use the expression:

$$L = \left(\frac{x_v}{95 - x_v}\right)\left(\frac{60tKM_L}{aM_v}\right) \qquad (8a)$$

Solving Eq. (7) for L, and combining with Eq. (8):

$$\frac{x_v K}{M_v a} = \frac{1}{9\pi D^2 \rho_L}, \text{ or:}$$

$$D = (1/3\pi^{1/2})(M_v a/x_v K\rho_L)^{1/2} = 0.188(M_v a/K\rho_L)^{1/2}(1/\sqrt{x_v}), \text{ ft} \qquad (9)$$

Giving values to x_v in Eq. (8) and (9), we obtain sets of drum lengths and diameters that fulfill exactly the requirements of maximum allowable vapor velocity and liquid-holding time (surge plus settling). Plotting these values (three points are sufficient) in a length-versus-diameter graph, the curve divides the graph into two regions: (1) where values of L and D will not meet the requirements, and (2) where the requirements will be exceeded. Therefore, the minimum dimensions of the drum lie on the curve. Let us now consider the optimum vessel dimensions as expressed in Eq. (4). Substituting V by its value (2:1 ellipsoidal heads):

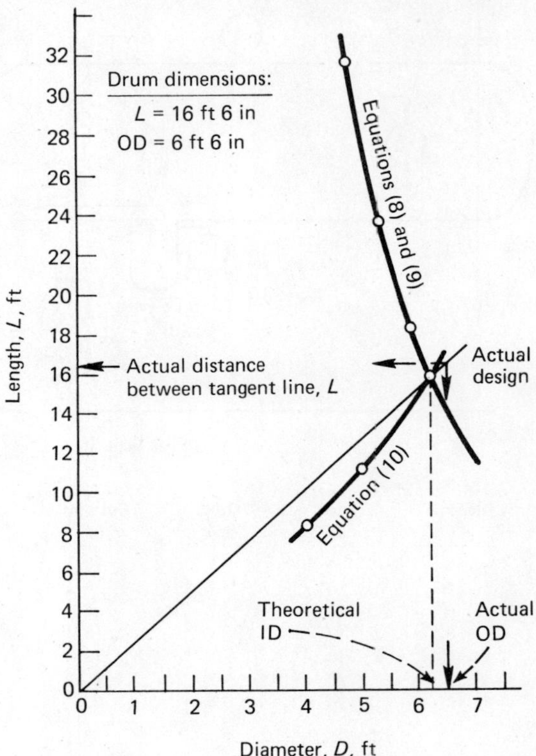

L-D PLOT shows optimum drum dimensions—Fig. 3

$$V = 7.48(\pi D^2/4)[L + (D/3)], \text{ gal}$$

Solving for L, Eq. (4) becomes:

$$L = \frac{PD^2}{0.053(2SE - P)} + 1.227D, \text{ ft} \qquad (10)$$

Plotting Eq. (10) in the L-D chart (three points suffice), the intersection of the L and D curves will give us the optimum theoretical L and D. Joining the intersection point with the origin, the straight line obtained will cut the vertical on the next O.D. commercial head. This is the final value of L (Fig. 3). Note that D is really the I.D. Taking into account the vessel thickness, the larger O.D. will give an I.D. approximately equal to the theoretically optimum D.

If another method (other than Eq. 4) is desired to obtain the optimum dimensions, use L as a function of D in a similar way, and plot in the L-D graph. For example, the rule of thumb where $L/D = 3$ provides a straight line from the origin, with a slope of 3. In any event, the curve obtained with Eq. (8) and (9) allows the selection of dimensions according to standard or available plates and heads.

Sample Problem

Determine the reflux-drum dimensions for the following conditions: operating pressure = 250 psig; reflux =

87,400 lb/h; product to heater = 28,300 lb/h; liquid density at operating conditions = 39.90 lb/ft³; vapor load = 14,100 lb/h; vapor density at operating conditions = 1.15 lb/ft³; liquid water = 3,500 lb/h; S = 13,750 psi; E = 0.85; corrosion allowance (C) = 0.125 in. The vessel is to be constructed of a A-285 Gr C steel, X-ray spot checked.

To determine the surge time, the controlling liquid will be the reflux. Therefore, $t_s = 5$ min, and M_L = 87,400 lb/h. Settling time shall be 5 min on total hydrocarbon, or 87,400 + 28,300 = 115,700 lb/h. Therefore, t_d (referred to M_L) = 5(115,700/87,400) = 6.5 min, and $t = 5 + 6.6 = 11.6$ min.

If no demister is used, $K = 0.13$ and $a = (39.9/1.15)^{1/2} = 5.89$. Substituting values in Eq. (8) and (9):

$$L = \left(\frac{60 \times 11.6 \times 0.13 \times 87{,}400}{5.89 \times 14{,}100}\right)\left(\frac{x_v}{100 - x_v}\right) = (95.22)\left(\frac{x_v}{100 - x_v}\right), \text{ ft}$$

$$D = 0.188\sqrt{\frac{14{,}100 \times 5.89}{0.13 \times 39.9}}\left(\frac{1}{\sqrt{x_v}}\right) = (23.79)\frac{1}{\sqrt{x_v}}, \text{ ft}$$

If x_v = 16%: L = 18.14 ft, and D = 5.95 ft
x_v = 20%: L = 23.81 ft, and D = 5.32 ft
x_v = 25%: L = 31.74 ft, and D = 4.76 ft

Making the Plot

The plotted L-D curve is shown in Fig. 3. The minimum-cost L-D relationship will be according to Eq. (10):

$$L = \frac{250D^2}{0.053(2 \times 0.85 \times 13{,}750) - 250} - 1.227D = 0.204D^2 - 1.227D$$

If D = 4 ft, L = 8.17 ft
D = 5 ft, L = 11.24 ft
D = 6 ft, L = 14.70 ft

Plotting these values in Fig. 3 gives a theoretical I.D. of 6 ft 3 in. Connecting the curve intersection point with the origin, the straight line cuts the O.D. line at 6 ft 6 in, which is the next commercial O.D., and at L = 16 ft 6 in, which is the distance between the tangent lines. This means two standard 8-ft-wide plates, plus 3 inches from tangent line to weld line on each head. Note that for the indicated pressures and material of construction, Eq. (4) yields an L/D ratio lower than 3.

To determine vessel thickness using the standard formula (outside diameter, D_o):

$$t = \left(\frac{PD_o}{2SE + 0.8P}\right) + C = \frac{(250 \times 78)}{(2 \times 0.85 \times 13{,}750) + (0.8 \times 250)} + 0.125 = 0.827 + 0.125 = 0.952 \text{ in}$$

Therefore, select a t value of 1 in. The vessel's I.D. = 78 − 2 = 76 in, or 6.33 ft.

To determine liquid levels (Fig. 1), use Eq. (9):

$$x_v = (23.79/6.33)^2 = 14\%$$

The liquid volume below L_{LL} (5 min on total hydrocarbon) = (5 × 115,700)/(60 × 39.9) = 241.6 ft³, and percent of segment area (A_s):

$$A_s = \frac{(241.6/16.5)}{(\pi 6.33^2/4)}(100) = 46.5\%$$

The L_{HL} distance to the top of the vessel could be determined by means of Fig. 2 of Ref. 2:

h_H = (20/100)(76) = 15 in, which provides a clearance of 20% of the value of D, and is greater than 10 in.

The L_{LL} distance to the top of the vessel (above L_{LL} percentage area = 100 − 46.5 = 53.5%) is:

h_L = (51/100)(76) = 39 in. Thus, 76 − 39 = 37 in.

The head volumes have been neglected. We can now determine how much additional safety they represent by means of the equation provided in Ref. 9. The liquid in 2:1 ellipsoidal heads at L_{HL} is:

$$0.000304[(2)(76^3/4) - 15^2][(1.5 \times 76) - 15] = 60 \text{ ft}^3$$

The liquid in the heads at L_{LL} is:

$$(0.000304 \times 37^2)[(1.5 \times 76) - 37] = 32 \text{ ft}^3$$

Extra settling time = (60 × 32 × 39.9)/115,700 = 0.66 min, and extra surge time:

$$(60 - 32)(60 \times 39.9)/87{,}400 = 0.77 \text{ min}$$

Finally, to determine the pot diameter, the water velocity will be 0.5 ft/min, which needs a section of:

$$[35{,}000/(60 \times 62.4)]/0.5 = 1.9 \text{ ft}^2$$

Since 3 ft of 20-in Schedule 20 pipe has a passage area of 2.02 ft², (Perry's Chemical Engineers' Handbook, 4th ed., Table 6-2, McGraw-Hill, New York, 1963), the 1.9 ft² measurement fits. With the aid of Fig. 2, a final sketch can now be prepared.

References

1. Happel, J., "Chemical Process Economics," pp. 250–254, Wiley, New York (1958).
2. Scheiman, A. D., Use Nomographs to Size Horizontal Vapor-Liquid Separators, *Hydrocarbon Process.* and *Petrol. Refiner*, **43,** No. 5, p. 155, May 1964.
3. Oliver, E. D., "Diffusional Separation Processes," pp. 400–404, Wiley, New York (1966).
4. Harriot, P., "Process Control," McGraw-Hill, New York (1964).
5. Watkins, R. N., Sizing Separators and Accumulators, *Hydrocarbon Process.*, **46,** No. 11, p. 253, Nov. 1967.
6. Furman, T. T., and Cheers, R. F., Optimizing Pressure Vessels, *British Chem. Eng.*, **16,** No. 6, p. 478, June 1971.
7. Patterson, F. M., Vortexing can be prevented in process vessels and tanks, *Oil Gas J.*, **67,** No. 31, p. 118, Aug. 4, 1969.
8. Kern, R., How to Arrange Process Drums, *Petrol. Refiner*, **40,** No. 5, p. 195, May 1961.
9. Kaplan, F., Capacity of Part-Full Tank Heads, *Hydrocarbon Process.*, **47,** No. 7, p. 176, July 1968.

Meet the Author

B. Sigalés is presently chemical-participations coordinator and chemical-companies board advisor for the Banco Industrial de Catalana (Barcelona, Spain), as well as professor of process-vessel design at both the Polytechnic University of Barcelona and the National Assn. of Spanish Engineers for Industry, and professor of crude-oil evaluation and workup at Polytechnic University of Barcelona. He has done process engineering, mechanical engineering, refinery procurement, and corporate planning for three of Europe's large process companies. A graduate with a doctorate in engineering, he has published papers in Spain and Great Britain.

More on design of reflux drums

Here is a refinement to the article on designing reflux drums that appears on page 180. It presents a simple way of handling the settling time of a second liquid phase.

B. Sigalés, *Instituto Petrolquimica Aplicada, Barcelona, Spain*

The previous article [1] presented a method of avoiding trial-and-error solutions for sizing reflux accumulators on distillation towers. The method is rigorous only if there is no second liquid phase to settle. However, if a second phase (usually water from stripping steam) exists, the method uses a rather conservative empirical residence time for settling.

If we neglect the volume of the drum's dished heads, we can (by a relatively simple mathematical procedure) obtain a rigorous solution to the problem of determining the liquid settling space required, by using the settling velocity of the heavy-phase drops. This method is suitable for minicomputer programming.

The rising of the light-phase drops in the heavy phase is not considered. Usually the heavy phase exists in comparatively small amounts, and if a light-phase drop exists in it, the settling takes place in a pot [1].

The basic parameters in setting the drum's dimensions, and the H_{LL} and L_{LL} positions, are shown in Fig. 1 (F/1).

The drum must provide space to allow the disengagement of the drops of liquid being carried out by the vapor [1]. Noting that x_v represents the percentage of cross-flow area of the drum above the H_{LL}, the maximum allowable vapor velocity can be expressed by:

$$V = K\left(\frac{\rho_1 - \rho_v}{\rho_v}\right)^{1/2} \simeq K\left(\frac{\rho_1}{\rho_v}\right)^{1/2} = \frac{\text{Vapor load}}{\text{Min. vapor cross-section area}}$$

$$= \frac{M_v/3{,}600\rho_v}{\dfrac{x_v}{100}\dfrac{\pi D^2}{4}} \text{ ft/s} \qquad (1)$$

Also, the drum must provide surge time, t, for the liquid, which must range between the maximum (H_{LL}) and minimum (L_{LL}) operational liquid levels [1]. If x_1

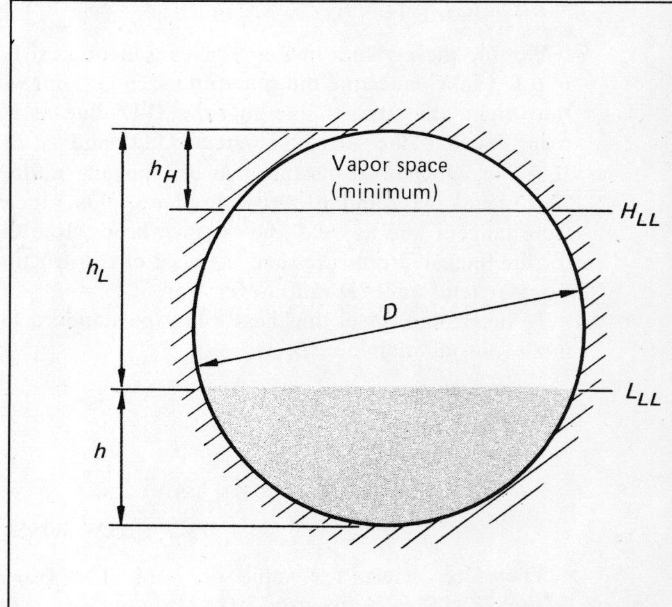

Reflux drum cross-section, showing levels

Originally published September 29, 1975

is the percentage of the drum cross-sectional area situated between H_{LL} and L_{LL}:

$$t = \frac{\text{Volume between } H_{LL} \text{ and } L_{LL}}{\text{Flow of liquid}}$$

$$= \frac{(x_1)(\pi D^2/4)(L)}{M_1/60\rho_1} \text{ min} \quad (2)$$

Finally, below the L_{LL}, there must be enough holdup time to allow the drops of heavy-phase liquid to reach the bottom of the drum (or the pot) before the dispersing phase that contains such drops leaves the drum [2]. With x being the cross-sectional area of the drum below the L_{LL}, expressed as percent of the total cross-section, then:

$$\frac{\text{Distance from } L_{LL} \text{ to drum bottom}}{\text{Settling velocity}} = \frac{h}{s}$$

$$= \frac{(x/100)(\pi D^2/4)(L)}{M/60\rho} \text{ min} \quad (3)$$

where, from Stokes' Law [2]:

$$s = 1.733\frac{\rho_h - \rho_1}{\mu} \text{ ft/min}$$

M is the total liquid entering the drum (including both liquid phases) but, normally, $\rho \simeq \rho_1$.

Eq. (1), (2) and (3) can be put into the form:

$$\frac{x_v}{100} = \frac{M_v}{900\pi K(\rho_1\rho_v)^{1/2}}\frac{1}{D^2} \quad (1')$$

$$\frac{x_1}{100} = \frac{tM_1}{15\pi\rho_1}\frac{1}{L(D^2)} \quad (2')$$

$$x = \frac{M}{15\pi s\rho_1}\frac{y}{LD} \quad (3')$$

where y is height, h, expressed as a percentage of D.

Taking $L/D = r$, and dividing Eq. (1') and (2') by (3'):

$$\frac{x_v}{x} = \frac{5M_v sr(\rho_1/\rho_v)^{1/2}}{3KM}\frac{1}{y} = \frac{C}{y} \quad (4)$$

$$\frac{x_1}{x} = \frac{100M_1 ts}{M}\frac{1}{Dy} = \frac{E}{Dy} \quad (5)$$

where C and E group the data and the parameter, r.

Adding Eq. (4) and (5):

$$\frac{x_v + x_1}{x} = \frac{1}{y}\left(C + \frac{E}{D}\right)$$

but, because $x_v + x_1 + x = 100$:

$$\frac{100}{x} = \frac{1}{y}(C + \frac{E}{D}) + 1 = \frac{C + (E/D) + y}{y}$$

or:

$$D = \frac{E}{100(y/x) - y - C} \quad (6)$$

and, from Eq. (3'):

$$D = \left(\frac{M}{15\pi s\rho_1 r}\frac{y}{x}\right)^{1/2} = \left(F\frac{y}{x}\right)^{1/2} \quad (7)$$

F being a function only of r and the data.

Eq. (6) and (7), together with the relationship between x and y shown in F/2, allow us to determine D and to determine the levels H_{LL} and L_{LL}.

The trial-and-error procedure is started by assuming a value of y, which gives the corresponding value of x on Fig. 2, and then using these values to calculate D from Eq. (6) and (7). To facilitate these calculations, F/3 shows y/x versus y.

Supplementary design rules are that h shall not be smaller than 10% of the drum diameter, D, or 5 in., whichever is greater. Also, h_H must not be smaller than 20% of D. To allow for level-instrument mounting, the distance between H_{LL} and L_{LL} must be 14 in. or more.

Solving Eq. (6) and (7) for a series of values for r permits construction of the minimum-dimensions L–D curve [1]. This curve permits one to set the optimum drum dimensions according to any desired criteria: economics, space, dished-head availability, etc.

It must be emphasized that, except for the case of pure holding time (or storage), the function of the drum—either in vapor-liquid or liquid-liquid service—depends on its cross-sectional area, D, and its length, not on the vessel volume. Therefore, the proper way to optimize drums is to determine sets of values of D and L that exactly fulfill the conditions of fluid velocities and residence times. These values will be the minimum dimensions of the drums that meet all the operational requirements [1]; in the conventional design procedure that fixes the drum volume and then "optimizes" the L/D ratio [3,4,5], all or some of the operational requirements would be exceeded.

Nomenclature

- C Parameter defined by Eq. (4)
- D Drum inside dia., ft
- E Parameter defined by Eq. (5)
- F Parameter defined by Eq. (7)
- h Drum cross-section-segment height (defined in F/1 and by subscripts)
- L Drum length between tangent lines, ft
- M Mass flowrate, lb/h
- r L/D ratio
- s Settling velocity, ft/min
- t Surge time, min
- V Maximum allowable vapor velocity, ft/s
- x Drum cross-sectional-area segment, percent of total drum cross-section, $\pi D^2/4$
- y Drum cross-sectional-segment height, percent of drum dia., D
- μ Viscosity of the continuous phase, cp
- π Ratio of circle circumference to dia., 3.14159 . . .
- ρ Density, lb/ft^3 (defined by subscripts), except in the formula to determine the settling velocity, s, where the units are g/cm^3

Subscripts

- H Refers to high liquid level, H_{LL}
- h Heavy liquid phase (drops)
- L Refers to low liquid level, L_{LL}
- l Light liquid phase (continuous phase)
- v Vapor

We will now demonstrate use of this method by solving the same problem that was handled in Ref. *1*, which used the empirical settling-time method.

Sample problem [*1*]

Determine the reflux-drum dimensions for the following conditions:

Operating pressure = 250 psig; reflux = 87,400 lb/h; product to heater = 28,300 lb/h; liquid density at operating conditions = 39.90 lb/ft^3; vapor load = 14,100 lb/h; vapor density at operating conditions = 1.15 lb/ft^3; liquid water = 3,500 lb/h; continuous-hydrocarbon-phase viscosity at operating conditions = 0.25 csk.

From Ref. *1*, the residence time, based on reflux of 87,500 lb/h, will be $t = 5$ min, and $K = 0.13$.

The water settling velocity will be:

$$s = 1.733\frac{\left(1 - \frac{39.9}{62.4}\right)}{0.25\frac{39.9}{62.4}} = 4 \text{ ft/min}$$

This value of 4 ft/min is greater than 0.833 ft/min, so we will use the latter value.

In Ref. *1*, the theoretical solution to this problem was $D = 6.25$ ft and $L = 15.865$ ft, or $r = 2.5385$. Using this value of r and the data, we have:

$$C = \frac{(5)(14{,}100)(0.833)(39.90/1.15)^{1/2}(2.5385)}{(3)(87{,}400 + 28{,}300 + 3{,}500)(0.13)} = 18.888$$

$$E = \frac{(100)(87{,}400)(5)(0.833)}{119{,}200} = 305.3867$$

$$F = \frac{119{,}200}{(15)\pi(0.833)(39.90)(2.5385)} = 29.9805$$

and in Eq. (6) and (7):

$$D = \frac{305.3867}{100\frac{y}{x} - y - 18.888} \text{ ft} \tag{6}$$

$$D = [29.9805(y/x)]^{1/2} \tag{7}$$

Using F/3, the following table can be prepared:

Function	y, % 20	40	35	27.2*
y/x (F/3)	1.4047	1.0709	1.125	1.097
D, from Eq. (6), ft	3.00	6.33	5.21	5.72
D, from Eq. (7), ft	6.49	5.67	5.81	5.72
x, %				33.9

*Graphically, plotting D vs. y.

Therefore, $D = 5.72$ ft and $L = 5.72(2.5385) = 14.52$ ft or, say, 14.5 ft.

The theoretical savings in volume and weight of the drum (weight, because the thickness for equal pressures is proportional to the diameter) compared with the solution obtained in Ref. *1* ($D = 6.25$ ft) would be:

$$\left[\left(\frac{6.25}{5.72}\right)^3 - 1\right](100) = 30\%$$

To determine H_{LL} and L_{LL} from Eq. (1′) and (2′):

$x_v = 18.88/1.097 = 17.2\%$; from F/2, $y_H = 22.9\% > 20$, O.K.

$$x_1 = \frac{305.3867}{5.72(1.097)} = 48.7\%$$

$x_v + x_1 = 17.2\% + 48.7\% = 65.9\%$; from F/2, $y_L = 62.2\%$

Check (F/1):

$x + x_1 + x_v = 33.9 + 48.7 + 17.2 = 99.8 \simeq 100\%$, O.K.

$y + y_L = 37.2 + 62.6 = 99.8 \simeq 100\%$, O.K.

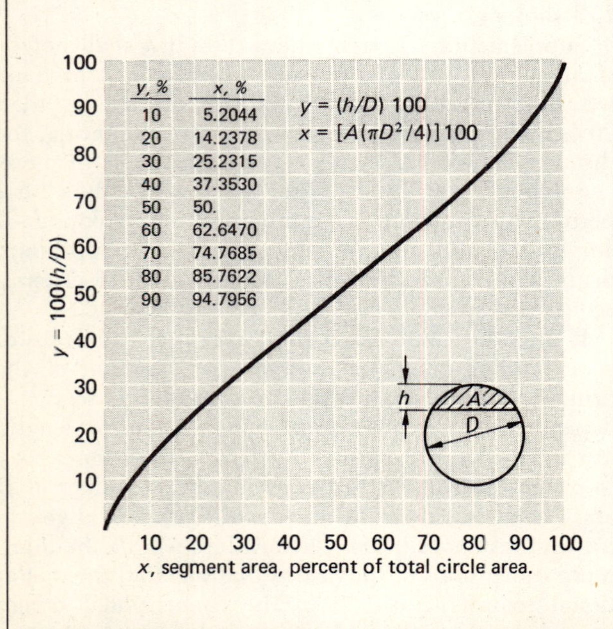

Segment height as a function of segment area (F2)

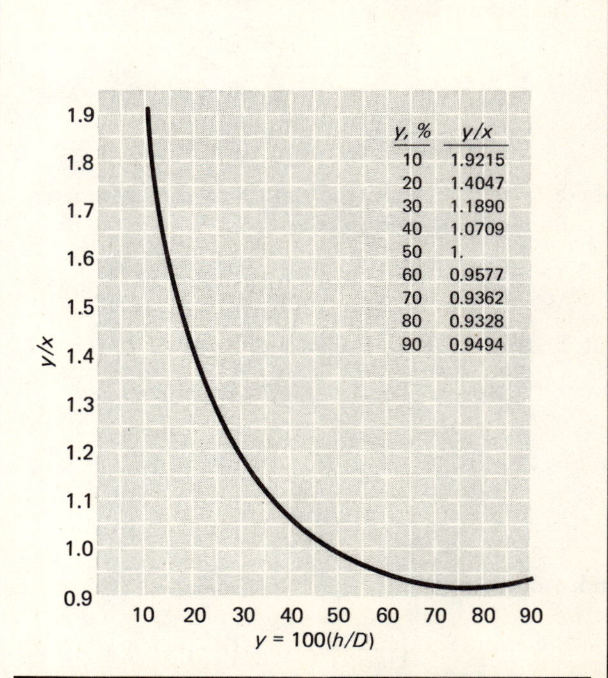

Ratio of segment height/area vs. segment height (F3)

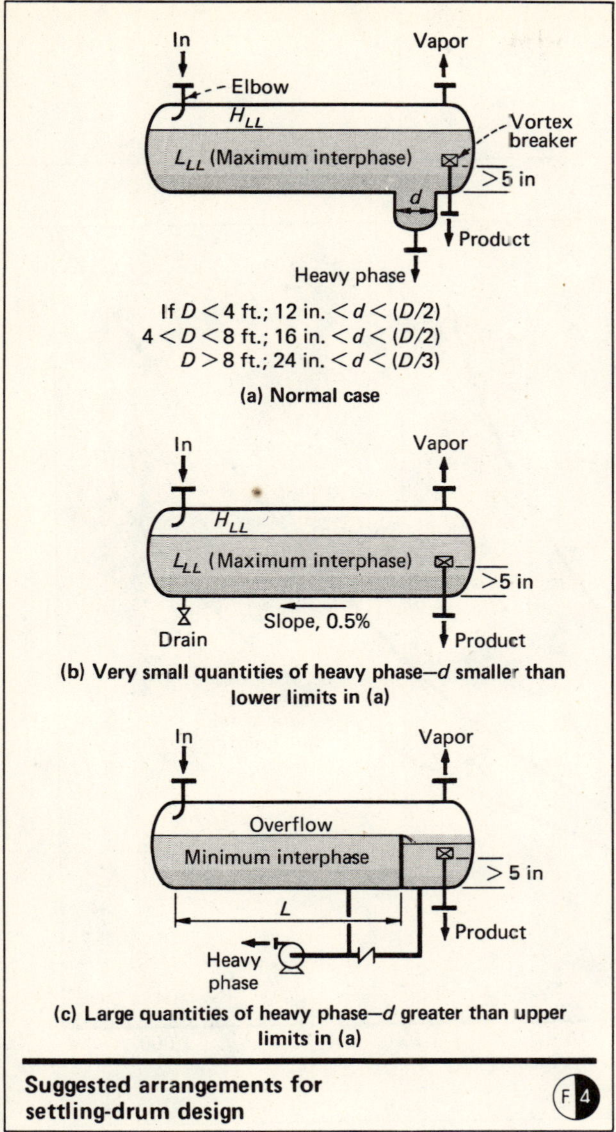

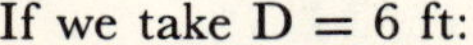

Suggested arrangements for settling-drum design F4

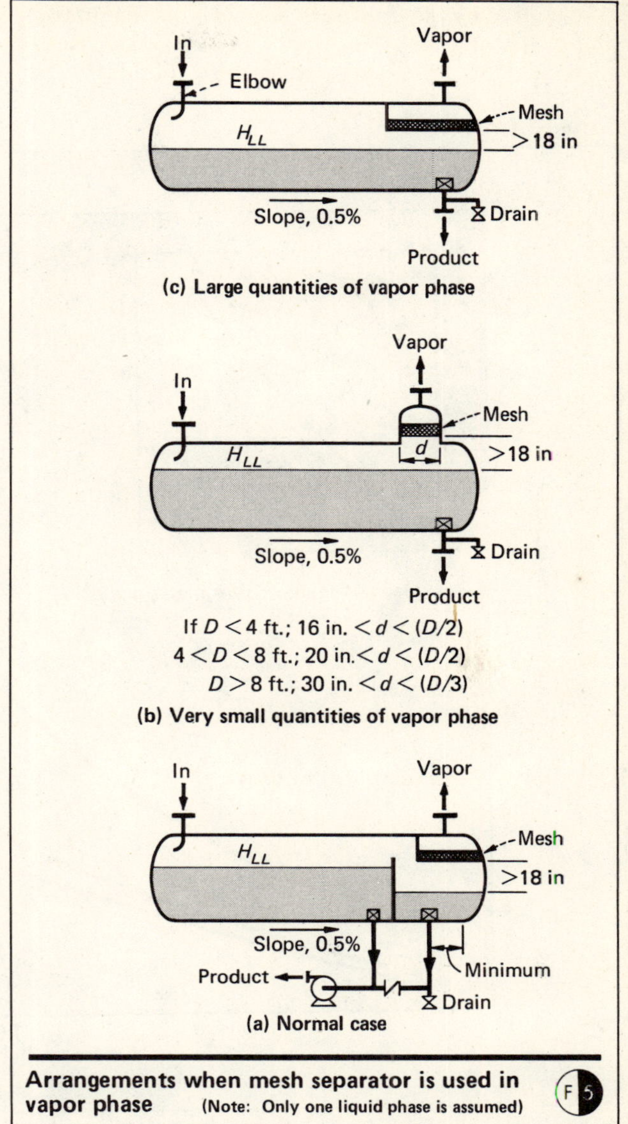

Arrangements when mesh separator is used in vapor phase (Note: Only one liquid phase is assumed) F5

If we take D = 6 ft:

$$h_H = (22.9/100)(6) = 1.37 \text{ ft} \simeq 16.5 \text{ in} > 10 \text{ in, O.K.}$$

$$h_L = (62.6/100)(6) = 3.76 \text{ ft} \simeq 45 \text{ in}$$

$$h_L - h_H = 45 - 16.5 = 28.5 \text{ in} > 14 \text{ in, O.K.}$$

A 32-in-maximum-range displacement level controller can be installed.

The actual water-settling time will be:

$$\frac{(33.9/100)(36\pi/4)(14.5)}{\dfrac{115{,}700}{(60)(39.90)}} = 2.9 \text{ min}$$

which is much lower than the empirical rule of 5 min.

Typical arrangements

F/4 shows some suggested arrangements according to the relative amounts of the two liquid phases. For additional information, see Ref. *1*. F/5 shows arrangements if a wire-mesh separator is used for the vapor phase (normal when a compressor is installed downstream).

References

1. Sigalés, B., How to Design Reflux Drums, *Chem. Eng.*, Mar. 3, 1975, p. 157. (Note: In above reference, the extended pipe length in Fig. 2 should be 5 in, not 6 in as shown.)
2. Sigalés, B., How To Design Settling Drums, *Chem. Eng.*, June 23, 1975, p. 141.
3. Happel, John, "Chemical Process Economics," Wiley, New York, 1958, pp. 179–182.
4. Brownell, L. E. and Young, E. H., "Process Equipment Design," Wiley, New York, 1959, pp. 80–81.
5. Abakians, K., Nomograph Gives Optimum Vessel Size, "Nomograph Handbook," *Hydrocarbon Process,* Gulf Pub. Co., Houston, Tex.

The author

B. Sigalés is a professor at Instituto Petrolquimica Aplicada, Polytecnic University of Barcelona, Av. Generalisimo 647, Barcelona-14, Spain, where he specializes in crude-oil evaluations and workup and process-vessel design. He is also a professor at Centro Perfeccionamiento del Ingeniero, the National Assn. of Spanish Engineers for Industry (Barcelona and Madrid), and he is Chemical Participations Coordinator, Chemical Companies Board Advisor, Banco Industrial de Catalunya, Barcelona. He has a doctorate in engineering from Escuela Tecnica Superior Ingenieros de Industriales de Barcelona.

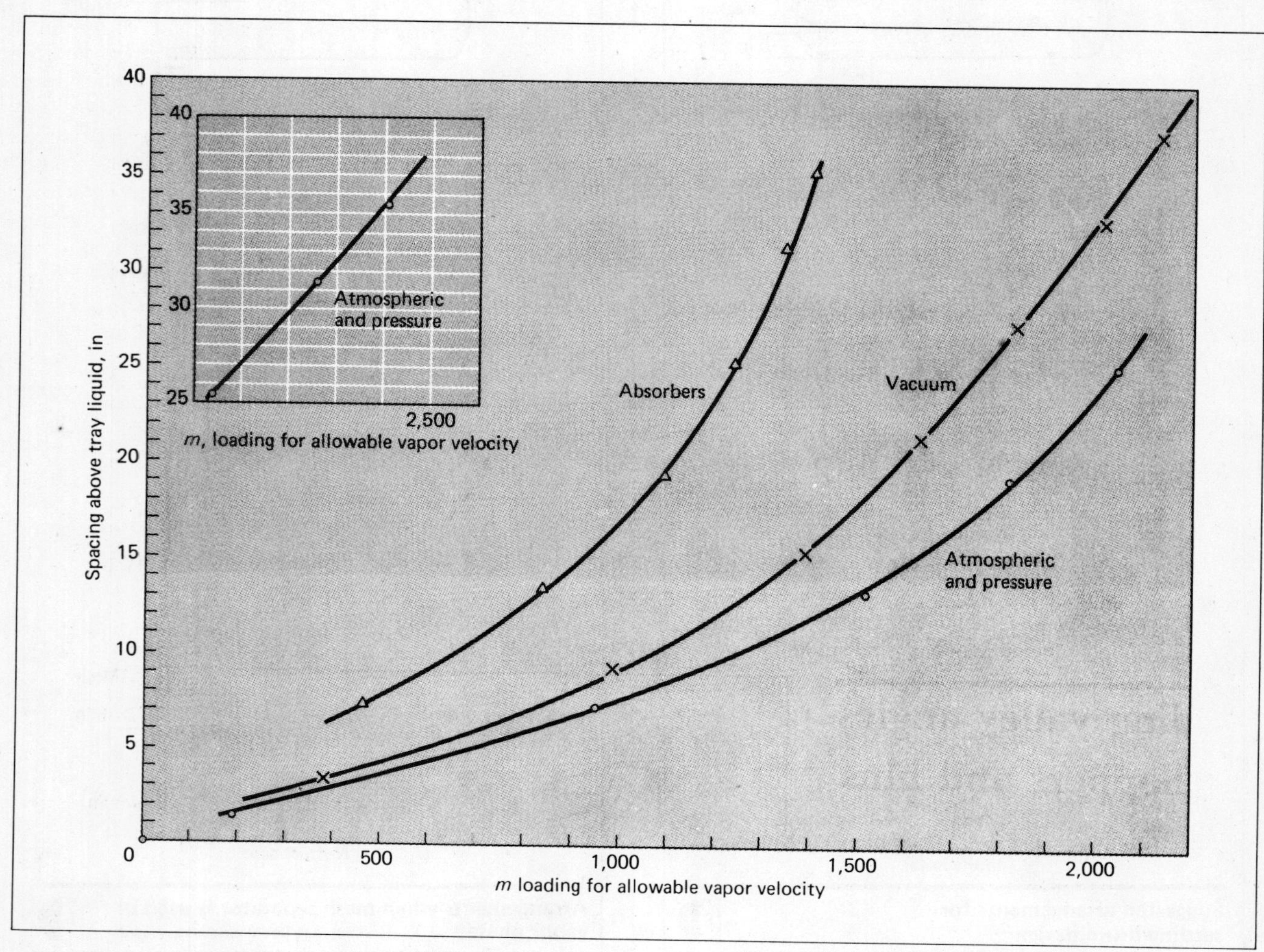

Demonstrated data for tower design

A. R. Chimes, New York City

☐ Process engineers faced with the problem of designing distillation and absorption towers may be presented with a variety of proprietary tray designs, each with its own claims for allowable loadings. While the calculations to justify these designs are often impressive, they may be no more than elaborate solutions to unimportant problems. The critical calculation in fractionating tower design consists of finding the tower diameter required for the allowable vapor velocity between the trays.

There are a number of well-established formulas for calculating the allowable mass velocity of vapors rising above a fractionating tray. The following formula, which resembles many of the others, neglects the conventional vapor-density correction for compressibility and thus is not applicable at very high pressures. However, it has the virtue of proof through application in many hundreds of distillation towers that were bid competitively and are operating today.

This formula solves for the area required between downcomers for disentrainment, by first calculating the allowable mass velocity, G, as:

$$G = m\left(\frac{MPS}{T}\right)^{1/2}$$

where: G = allowable mass velocity of vapor, lb/(h)(ft^2)

M = molecular weight of vapor

P = absolute pressure, psia

S = specific gravity of liquid on the tray

T = temperature on the tray, °R

Values for m are obtained from the chart, according

Originally published October 27, 1975

to the fractionation service, whether vacuum, atmospheric and pressure, or absorption.

The required vapor area then equals W/G, where W is the lb/h of vapors.

To this required area must be added the area of the downcomers, $A_l = L/10{,}770$, where A_l is the required area, ft^2, at the top of the downcomer and L is the volume-flow of liquid reflux at the tray temperature, gal/h. Since there must be an inlet as well as outlet downcomer, A_l should be doubled to calculate the total required area.

In many instances where the tower diameter ranges 5–12 ft, the downcomer may be constructed as a cord across the tower's projected circle. In such instances, the downcomer may provide the required area (necessary for disengaging vapors from the liquid) but may be so narrow as to have the overflowing liquid project against the inside wall and thus prevent disengaged vapors from escaping. An empirical rule for determining the minimum width to avoid this is:

$$C = (ht)^{1/2}$$

where C is the minimum width of the downcomer, in, h is the height of liquid overflowing the edge of the downcomer, in, and t is the deck spacing, in.

Note: The ordinate on the chart of m values is the spacing above the tray liquid; actual tray spacing will be 2–5 in more than this, depending on the tray. The curves in the chart of m values imply that there is an optimum relation between diameter and tray spacing for minimum overall cost; but tower sizing is more complicated than that. The vapors rising above fractionating trays vary from a maximum near the top of a section to a minimum near the bottom, as a consequence of temperature gradient and the corresponding added reflux necessary to remove sensible heat. Thus the practical problem is often that of finding a tray with a enough flexibility to suit both top and bottom loadings. Any optimization should allow for maximum and minimum vapor velocities through the tray openings—with the suitable tray, carrying an acceptable number of openings, fitted into the chosen tower diameter.

Moreover, there must be access for workmen to climb up and down through the tower during construction and turnaround, so the layout of internal manways through the trays usually influences tray design as much as the hydraulics. The selected number of tray openings should allow space between them for bolting the flanges of an internal manway.

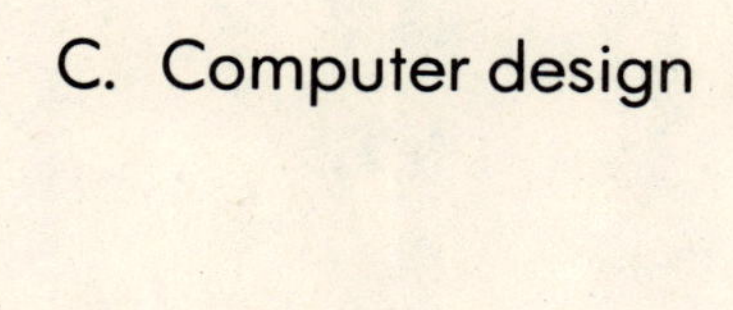

C. Computer design

A fast computer method for distillation calculations

A new method for steady-state and transient behavior of complex distillation columns handles any number of feed, side-product and heat streams. Stage efficiencies and any type of phase-equilibrium relationship can also be accommodated in the solution algorithm.

Alexander P. Economopoulos, Ministry of Social Services, Greece

☐ The distillation program is the focal point for process design by computer-simulation packages.

Factors to be considered for evaluating a distillation algorithm are flexibility when solving complex problems, stability when dealing with nonideal component properties, and efficiency in computer-time and storage requirements. Available solution methods combine, to a limited degree, some of these desirable characteristics.

For example, the method of Lewis and Matheson [*1*], as modified by Bonner [*2*]; and that of Thiele and Geddes [*3*], as modified by Lyster, et al. [*4*] are well known for the stability and efficiency of their solution, but lack flexibility in handling complex problems having multiple feeds, side-products and heat leaks. On the other hand, the method of Amundson and Pontinen [*5*] provides an elegant and flexible approach, well suited for complex problems but difficult to converge for components with nonideal properties.

Our objective here is to provide a method that combines stability and efficiency with inherent flexibility. Formulation of the mathematical model is based on a generalized tower, each stage of which accommodates material and heat leaks. An iterative scheme solves the model, with vapor-liquid equilibria linearized at certain points along the distillation path. At each point, the component fractions in the vapor are expressed as linear functions of the plate efficiency and all mutually independent liquid fractions.

Solving equations for a transient model adds little to program complexity and running costs over the steady-state one. This allows the use of a dynamic solution algorithm for all normal design tasks, and also yields these advantages:

1. Increased stability of the solution in problems that are difficult to converge. This can be done by approaching the desirable steady-state dynamically from any arbitrary initial condition. In this way, the physical stability of the column is reflected in the transient solution of the model.

2. Simulation of transient operations such as batch distillation, or automatic-control and startup studies.

Originally published April 24, 1978

Generalized stage for distillation

Depending on the value of the external material and heat leaks, Fig. 1 can represent any type of stage encountered in distillation operations, such as a simple stage, a feed stage, a side-product takeoff stage, a reboiler, or a partial condenser.

The feed and side-product streams in Fig. 1 account for all external factors affecting the material balances around the stage. For chemically reactive systems, the effect of the reaction on the material balances can be represented by a hypothetical feed stream—adding the reaction products to, and removing the reactants from, the stage.

The heat leak accounts for all external factors affecting the enthalpy balance around the stage, such as heat losses, or reboiler and partial-condenser duties. Enthalpy content of the feed and side-product streams are included as separate terms in the enthalpy-balance equations of the model. In chemically reactive systems, the heat of reaction can be represented by a hypothetical heat leak.

The material leaks, shown in Fig. 1, have been separated into a feed stream and vapor and liquid side-

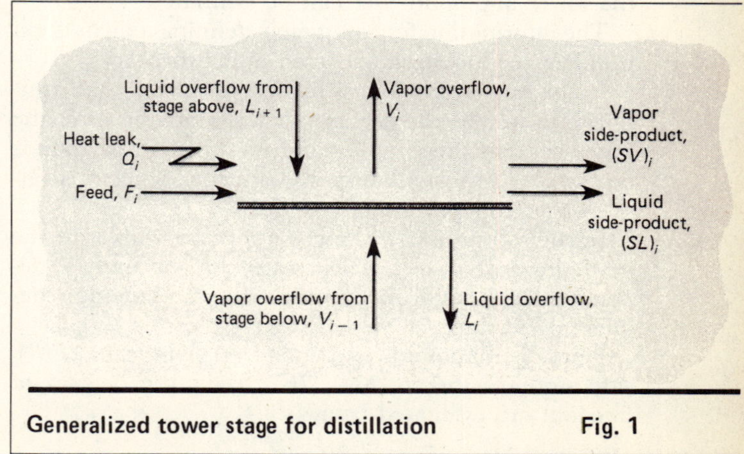

Generalized tower stage for distillation **Fig. 1**

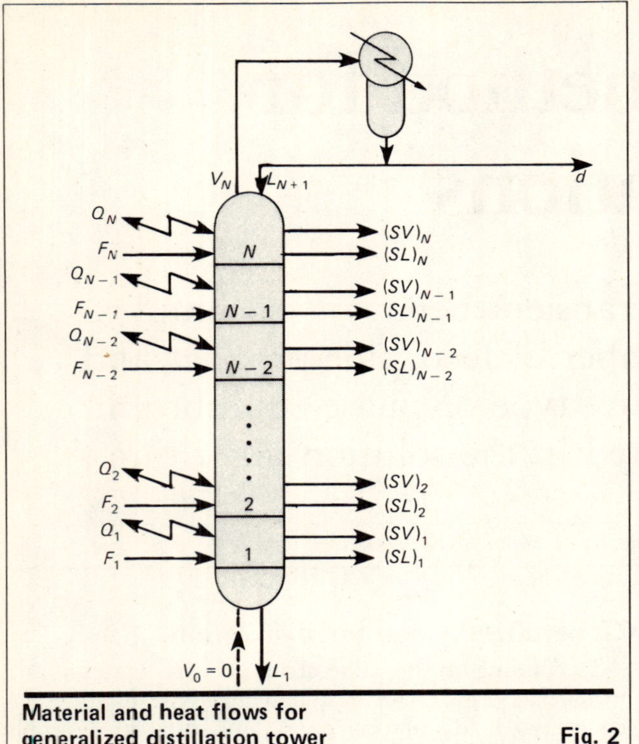

Material and heat flows for generalized distillation tower **Fig. 2**

Required phase properties and equilibria **Table I**

Property or equilibrium	Function (explicit or implicit)
Distillation equilibria for phase	$Y_j = f_1\ (\bar{x}, P)$
Enthalpy for phase	$H = f_2\ (\bar{x} \text{ or } \bar{y}, T, P)$
Chemical-reaction kinetics (for chemically reactive systems only)	---
Phase density, viscosity, diffusivity, surface tension, etc., as required for stage design and performance estimation	---

Performance characteristics for stages **Table II**

Characteristic	Function
Plate efficiency	$(EMV)_{j,i} = f_3\ (L, \bar{x}, V, \bar{y}, \text{phase properties})$
Stage pressure drop	$\Delta P_j = f_4\ (L, \bar{x}, V, \bar{y}, \text{phase properties})$
Condenser at overhead-vapor-line pressure drop	$U_j = f_5\ (L, \bar{x}, V, \bar{y}, \text{phase properties})$
Stage and condenser holdups	$U_D = f_6\ (L, \bar{x}, V, \bar{y}, \text{phase properties})$

product streams. The compositions of the last two streams are identical with the respective vapor and liquid stage-overflows. These identities are taken into consideration in the material-balance equations to facilitate their solution and reduce the number of external specifications [*6*].

The diagram for the distillation tower of Fig. 2 comprises a total condenser and N generalized stages (such as that in Fig. 1). The reboiler and partial condenser, if any, are counted as normal stages. Depending upon the values of the feed, side-product and heat streams, Fig. 2 can represent any simple or complex distillation tower.

Defining the problem

A problem is completely defined when a unique solution for its mathematical model exists and all of the heat or material leaks entering or leaving each stage of the tower and condenser can be estimated.

The information for completely defining a distillation problem [*6*] has been classified into two parts:

In the first part belong all phase properties and equilibria, as well as the performance characteristics of the tower components (Table I and II). This information is necessary for formulating the independent internal-stage and condenser relationships.

In the second part belong the input variables of the mathematical model. These must be mutually independent and equal in number with the column's degrees of freedom.

For a distillation tower having $(IQ)_t$ heat leaks, $(ISP)_t$ side-product streams, and $(IF)_t$ feeds, the degrees of freedom are estimated from [*6*]:

$$(Nf)_{t/c} = 3 + (IQ)_t + (ISP)_t + (M+1)(IF)_t + (Nr)_t \quad (1)$$

We can apply Eq. (1) directly to the tower shown in Fig. 2, for which $(IQ)_t = N$, N heat leaks exist, one in each stage; $(ISP)_t = 2N$, $2N$ side-product streams exist, one liquid and one vapor for each stage; $(IF)_t = N$, N feed streams exist, one for each stage; and $(Nr)_t = 1$, one decision having to be made regarding the number of tower stages. Thus, for the distillation diagram of Fig. 2:

$$(Nf)_{t/c} = 4 + 4N + NM \quad (2)$$

The input-variable set (defining each individual problem for the mathematical model of this study) is shown in Table III.

Heat and material leaks, shown in Fig. 2, that do not exist in the corresponding stages of a given problem should be assigned a value of zero.

For solving problems involving columns without total condensers, the reflux ratio, R, in the model should be set equal to zero.

Computer program

The general organization of the computer program requires interfacing the design variables, phase properties, stage-design and stage-performance correlations with the solution algorithm. The overall scheme is independent of the distillation method, and is outlined below for interfacing:

1. *Design variables*—In the solution of the mathematical model, the input variables listed in Table III are assumed known. In practice, these variables have to be estimated from a set of external tower specifications, called design variables, that differ from problem to problem. The calculations, which also involve dependent variables, have to be performed iteratively along the iterations of the solution method. Subroutine INPUT (to

Nomenclature

$A_{j,i}$	Coefficient in Eq. (51), defined by Eq. (52) and (53)
$B_{j,i}$	Coefficient in Eq. (51), defined by Eq. (54) and (55)
C_i	Coefficient in Eq. (51), defined by Eq. (56) and (57)
$c_{j,i}$	Molar fraction of component j in the feed of stage i
$D_{j,i}$	Coefficient in Eq. (51), defined by Eq. (58) and (59)
d	Distillate molar rate
$E_{j,i}$	Vaporization efficiency of component j in stage i
$(EMV)_{j,i}$	Modified Murphree stage efficiency of component j in stage i
F_i	Molar rate of feed to stage i
$(HF)_i$	Enthalpy of feed to stage i
$(HL)_i$	Enthalpy of liquid overflow from stage i
$(HV)_i$	Enthalpy of vapor overflow from stage i
$(IF)_t$	Total number of feed streams in the tower
I_m	Total number of points along the distillation path where vapor-liquid equilibria are linearized
$(IQ)_t$	Total number of heat leaks in the tower
$(ISP)_t$	Total number of side-product streams in the tower
l	Variable related to the frequency that the matrix [Φ] is estimated along the distillation path ([Φ] is estimated on every other lth stage)
L_i	Molar rate of the liquid overflow from stage i
l_s	The number of the first tower-stage from the bottom where the vapor-liquid equilibria are linearized
M	Total number of components
N	Total number of stages in the tower, counting the reboiler and partial condenser
$(Nf)_{t/c}$	Degrees of freedom of the tower/condenser system
$(Nr)_t$	Number of decisions regarding the repetition of elements and the location of heat and material leaks
$(Ol)_i$	Variable defined from Eq. (24) and (25)
$(Ov)_i$	Variable defined from Eq. (23)
$(Ox)_{j,i}$	Variable defined from Eq. (26) and (27)
P_i	Pressure on stage i
$(Pr)_{j,i}$	Variable defined from Eq. (60) and (61)
Q_i	Heat leak into stage i
R	Reflux ratio
$(SL)_i$	Liquid side-product stream from stage i
$(SV)_i$	Vapor side-product stream from stage i
t	Time
T_i	Temperature on stage i
U_D	Liquid holdup of the condenser system
U_i	Liquid holdup of stage i
V_i	Vapor overflow from stage i
$x_{j,i}$	Molar fraction of component j in the liquid overflow from stage i
$y_{j,i}$	Molar fraction of component j in the vapor overflow from stage i
$Y_{j,i}$	Molar fraction of component j in the vapor that is in equilibrium with the liquid leaving stage i
ΔP_i	Pressure drop of stage i
Δt	Time increment
ε	Coefficient defined from Eq. (40)
μ	Factor assigning different weights in each time level
ξ	Variable used in Eq. (16)
$[\Phi]_{j,k}$	Term of the matrix [Φ], defined by Eq. (36)
ω	Variable used in Eq. (16)

Subscripts

i	Stage number. Stage counting starts from the reboiler ($i = 1$)
j	Component number
I	Integer variable defined for a given stage from Eq. (46)
k	Component number
n	Number of a liquid solution
t	Tower
t/c	Tower/condenser system

Superscripts

o	Value of the variable in previous time level
$'$	Value of the variable in previous iteration

be supplied by the user for each problem) is assumed here to perform the design-variable interfacing calculations [7] and provide the input variables in each iteration of the solution method.

2. *Phase properties*—Sophisticated property packages find increasing use in distillation calculations but also put a strain on computer time and storage requirements. A study of these problems has shown that the most appropriate interfacing method, which also allows maximum flexibility, is the development of a series of property subroutines, each performing the required property calculations [7]. For this study, it will be assumed that a subroutine is available for estimating each property listed in Table I.

3. *Stage-design and stage-performance correlations*—Inter-

Input variables for mathematical model **Table III**

Input variable	Specifications, No.
Number of stages	1
Condenser pressure	1
Reflux ratio	1
Reflux temperature	1
Rate of each heat leak	N
Rate of each liquid side-product stream	N
Rate of each vapor side-product stream	N
Rate composition and temperature of feed streams	$(M+1)N$
Total	$(Nf)_{t/c} = 4 + 4N + MN$

action between the type and design of each stage with the solution algorithm is handled by having each stage designed, and its performance estimated periodically, along the iterations of the solution method [*7*]. This yields more-accurate results and makes detailed tower design easier.

Simplifying assumptions

In order to simplify the complexity of the equations, the following assumptions are made:

1. Vapor holdups are negligible compared to the liquid ones. This assumption, which affects only the transient calculations, is valid when the tower operating pressure is not close to the critical component pressures.

2. Liquid on each stage is perfectly mixed, i.e., the mean fraction of all components in the liquid holdup is equal to those in the liquid overflow stream. This assumption does not normally introduce appreciable error because the effects of imperfect mixing are taken into consideration by stage-efficiency coefficients.

The mathematical model

Let us now establish the several equations to represent the typical stage, total condenser and reboiler:

Typical stage—Subject to the preceding assumptions, the mass balance for component j around stage i (Fig. 1) yields:

$$V_{i-1}y_{j,i-1} - V_i y_{j,i} + L_{i+1}x_{j,i+1} - L_i x_{j,i} + F_i c_{j,i} - (SV)_i y_{j,i} - (SL)_i x_{j,i} = \frac{d(U_i x_{j,i})}{dt} \quad (4)$$

Similarly, the overall material and enthalpy balances for this stage are:

$$V_{i-1} - V_i + L_{i+1} - L_i + F_i - (SV)_i - (SL)_i = \frac{dU_i}{dt} \quad (5)$$

$$V_{i-1}(HV)_{i-1} - V_i(HV)_i + L_{i+1}(HL)_{i+1} - L_i(HL)_i + F_i(HF)_i - (SV)_i(HV)_i - (SL)_i(HL)_i = \frac{d[U_i(HL)_i]}{dt} \quad (6)$$

Total condenser—The overall material balance and the component j mass balance, around the condenser, yield:

$$V_N - L_{N+1} - d = \frac{dU_D}{dt} \quad (7)$$

$$V_N y_{j,N} - (L_{N+1} + d)x_{j,N+1} = \frac{d(U_D x_{j,N+1})}{dt} \quad (8)$$

Column reflux ratio is given by:

$$R = L_{N+1}/d \quad (9)$$

In order to eliminate d, we introduce Eq. (9) into (7) and Eq. (7) into (8), and rearrange to get:

$$V_N - \left(\frac{R+1}{R}\right)L_{N+1} = \frac{dU_D}{dt} \quad (10)$$

$$V_N y_{j,N} - \left(\frac{R+1}{R}\right)L_{N+1}x_{j,N+1} = \frac{d(U_D x_{j,N+1})}{dt} \quad (11)$$

Reboiler—From Fig. 2, we find that no vapor stream enters the reboiler from a lower stage. Hence:

$$V_0 = 0 \quad (12)$$

Eq. (4) to (6) and (10) to (12) for $j = 1, 2, \ldots, M-1$; and $i = 1, 2, \ldots, N$—together with Eq. (13) and (14)—constitute the mathematical model of the generalized distillation process of Fig. 2.

$$\sum_j x_{j,i} = 1 \quad (13)$$

$$\sum_j y_{j,i} = 1 \quad (14)$$

This model is to be solved in conjunction with the phase-property and stage-performance correlations of Table I and II, assuming that the input variables of Table III are defined.

Numerical form of mathematical model

Using a two-point implicit formula, a differential equation of the form

$$f(\omega_1, \omega_2, \ldots, \xi) = \frac{d(U\xi)}{dt} \quad (15)$$

can be approximated numerically by:

$$\mu f(\omega_1, \omega_2, \ldots, \xi) + (1-\mu)f(\omega_1^o, \omega_2^o, \ldots, \xi^o) = \frac{U\xi - U^o\xi^o}{\Delta t} \quad (16)$$

where the superscript o denotes the values of the variables $\omega_1, \omega_2, \ldots, \xi$ and U in the previous time level. The factor μ assigns different weights in each time level, and its value falls in the interval $0 \le \mu \le 1$.

With $\mu = 0$ or with $\mu = 1$, Eq. (16) provides a first-order explicit or implicit approximation, respectively, of Eq. (15). With $\mu = 0.5$, a second-order approximation is obtained.

As the time steps, Δt, are allowed to increase, the solution remains stable, i.e., the inherited error always remains bounded, provided that:

$$0.5 < \mu = 1 \quad (17)$$

Under this condition, a steady-state solution at the end of the first time step may be obtained by choosing a value of Δt that is sufficiently large.

Using the two-point implicit formula to approximate the differential equations of the mathematical model, and rearranging terms, we obtain:

From Eq. (6):

$$V_i = \frac{1}{(HV)_i}\left\{(HV)_{i-1}V_{i-1} - (HV)_i(SV)_i - (HL)_i[L_i + (SL)_i] + (HL)_{i+1}L_{i+1} + (HF)_iF_i + (Ov)_i - \frac{U_i(HL)_i}{\mu\Delta t}\right\} \quad (18)$$

From Eq. (10):

$$L_{N+1} = \frac{R}{R+1}\left[V_N + (Ol)_{N+1} - \frac{U_D}{\mu\Delta t}\right] \quad (19)$$

From Eq. (5):

$$L_i = V_{i-1} - [V_i + (SV)_i] - (SL)_i + L_{i+1} + F_i + (Ol)_i - \frac{U_i}{\mu \Delta t} \quad (20)$$

From Eq. (11):

$$x_{j,N+1} = \left[\frac{1}{\frac{R+1}{R} + \frac{U_D}{\mu L_{N+1} \Delta t}} \right] \times [V_N y_{j,N} + (Ox)_{j,N+1}] \quad (21)$$

and, from Eq. (4):

$$V_{i-1} y_{j,i-1} - [V_i + (SV)_i] y_{j,i} - \left[L_i + (SL)_i + \frac{U_i}{\mu \Delta t} \right] x_{j,i} + L_{i+1} x_{j,i+1} = -F_i c_{j,i} - (Ox)_{j,i} \quad (22)$$

where:

$$(Ov)_i = \frac{1-\mu}{\mu} \{ V_{i-1}{}^o (HV)_{i-1}{}^o - [V_i{}^o + (SV)_i{}^o](HV)_i{}^o - [L_i{}^o + (SL)_i{}^o](HL)_i{}^o + L_{i+1}{}^o (HL)_{i+1}{}^o + F_i{}^o (HF)_i{}^o \} + \frac{U_i{}^o (HL)_i{}^o}{\mu \Delta t} \quad (23)$$

$$(Ol)_{N+1} = \frac{1-\mu}{\mu} \left[V_N{}^o - \left(\frac{R^o + 1}{R^o} \right) L_{N+1}{}^o \right] + \frac{U_D{}^o}{\mu \Delta t} \quad (24)$$

$$(Ol)_i = \frac{1-\mu}{\mu} \{ V_{i-1}{}^o - [V_i{}^o + (SV)_i{}^o] - [L_i{}^o + (SL)_i{}^o] + L_{i+1}{}^o + F_i{}^o \} + \frac{U_i{}^o}{\mu \Delta t} \quad (25)$$

$$(Ox)_{j,N+1} = \frac{1-\mu}{\mu} \left\{ V_N{}^o y_{j,N}{}^o - \left(\frac{R^o - 1}{R^o} \right) L_{N+1}{}^o x_{j,N+1}{}^o \right\} + \left(\frac{U_D{}^o}{\mu \Delta t} \right) x_{j,N+1}{}^o \quad (26)$$

$$(Ox)_{j,i} = \frac{1-\mu}{\mu} \{ V_{i-1}{}^o y_{i,i-1}{}^o - [V_i{}^o + (SV)_i{}^o] y_{j,i}{}^o - [L_i{}^o + (SL)_i{}^o] x_{j,i}{}^o + L_{i+1}{}^o x_{j,i}{}^o + F_i{}^o c_{j,i}{}^o \} + \frac{U_i{}^o x_{j,i}{}^o}{\mu \Delta t} \quad (27)$$

The terms (Ov), (Ol) and (Ox) are the only variables that have to be stored in memory from the previous time level. For the steady-state solution, it is sufficient to set (Ov), (Ol) and (Ox) equal to zero and make Δt large, or the U_i's zero.

Numerical solution of the model

The computer flow diagram (Fig. 3) interfaces the design variables, phase properties, and stage-design and stage-performance correlations with the solution method according to the program scheme [7] previously outlined.

The algorithm for the numerical solution of the mathematical model comprises the following consecutive steps:

1. *Initialization*—Liquid concentrations, $x_{j,i}$, and condenser pressure, P_D, have to be initialized. Following this, approximate values for the pressure drop, ΔP_i, and efficiency, $(EMV)_{j,i}$, are obtained for each stage from subroutine DESPER, assuming normal operating conditions [7].

Where multiple design alternatives are examined, the previous solution is usually a good starting point for the new ones, and the initialization step is skipped.

2. *Vapor compositions*—From the phase-equilibrium relations, the vapor in equilibrium with the liquid on each stage is calculated. Based on this, the composition of the vapor leaving each stage is computed from the stage-efficiency equation.

3. *Vapor and liquid flowrates*—The rate of vapor leaving each stage is computed from Eq. (18) for $i = 1, 2, \ldots, N$, while liquid rates are estimated from Eq. (19) and (20) for $i = N, N-1, \ldots, 1$. These calculations are repeated until convergence is reached due to the interaction between the calculated vapor and liquid rates and stage holdups. In the first iteration of the steady-state solution (where liquid rates are not known), the calculations start by assuming negligible liquid enthalpies, and estimating the vapor overflows from:

$$V_i = \left[\frac{(HV)_{i-1}}{(HV)_i} \right] V_{i-1} - (SV)_i + \left[\frac{(HF)_i}{(HV)_i} \right] F_i \quad (28)$$

The vapor and liquid rates estimated by this scheme are functions of the liquid and vapor compositions used in the enthalpy estimations.

4. *Liquid compositions*—Liquid compositions are estimated by solving Eq. (21) and (22) in conjunction with the vapor-liquid equilibria and the stage-efficiency relations. This is accomplished by a procedure whereby any type of vapor-liquid equilibrium relations is linearized at certain points along the distillation path. These linear equations, together with stage-efficiency relations, are used to eliminate the vapor fractions from Eq. (21) and (22). The resulting equations can then be easily solved for the liquid compositions. The procedure outlined here is the most important part of the method and will be described in detail.

5. *Liquid compositions*—The estimated liquid compositions are normalized by the following relation, after setting all negative concentrations equal to zero:

$$x_{j,i} \Big/ \sum_{k=1}^{M} x_{k,i} \Rightarrow x_{j,i} \quad (29)$$

6. *Convergence*—If convergence of the solution algorithm has not been reached, repeat the calculations from step 2.

7. *New time level*—If a solution on a new time level is required, calculate the variables (Ov), (Ol) and (Ox) from Eq. (23) to (27), and start the calculations from step 2.

Linearization of vapor-liquid equilibria

A variety of methods are used for predicting vapor-liquid equilibria in distillation problems. Most of these

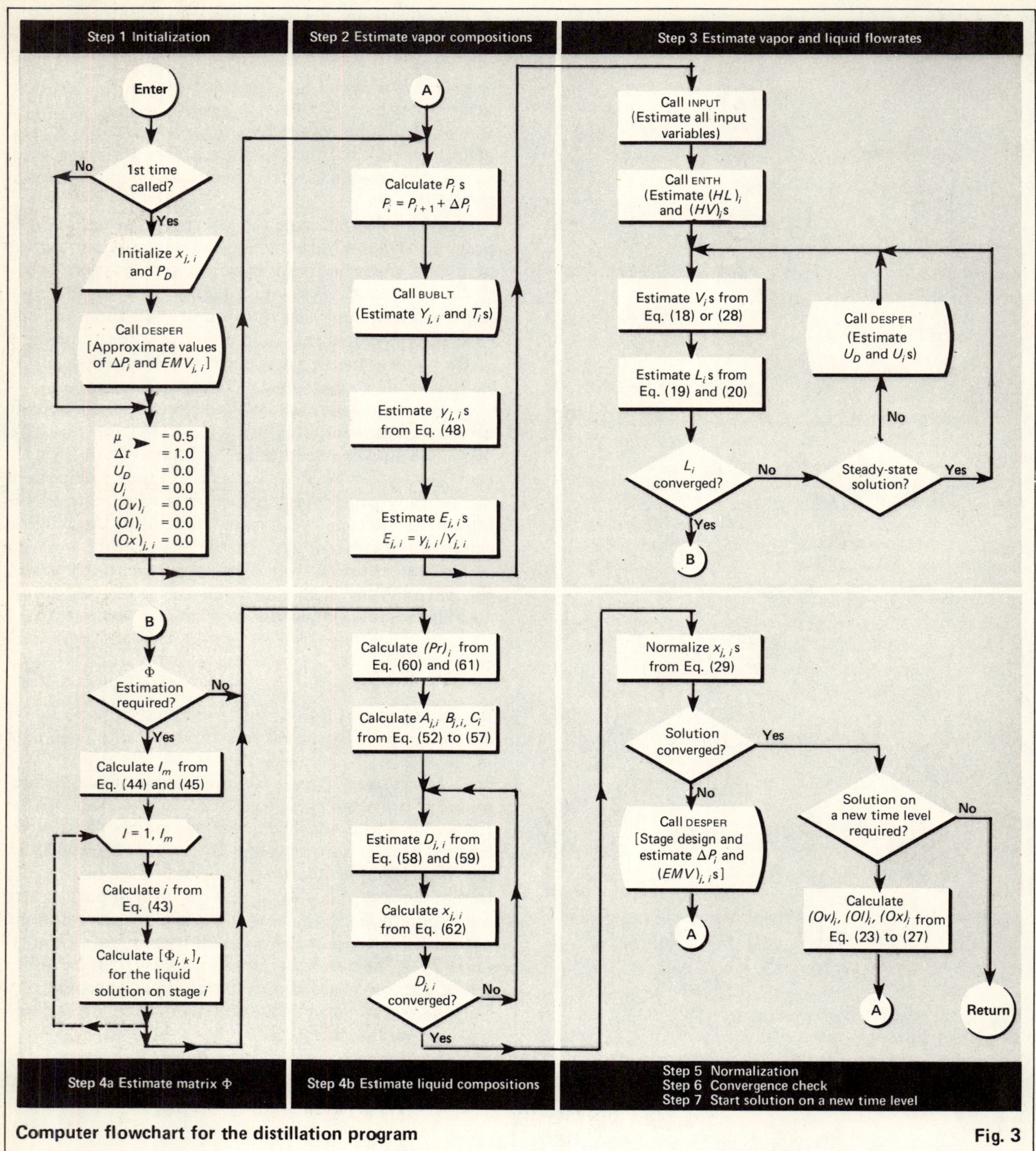

Computer flowchart for the distillation program **Fig. 3**

use complex relations and iterative computational schemes. In terms of flexibility, it is essential not to impose any restrictions on the method used for the vapor-liquid equilibrium because this may pose considerable inconvenience for the designer.

For this study, it will be assumed that given the pressure P and the composition $x_1, x_2, \ldots, x_M$ of a liquid solution, x, a subprogram can be called that performs the bubble-point temperature calculations to provide the boiling point T and the composition $Y_1, Y_2, \ldots, Y_M$ of the vapor, Y, in equilibrium with the liquid.

In the general case, the fraction, Y_j, of component j in the vapor is a function of the pressure and the $(M - 1)$ mutually independent component fractions of the liquid solution x. Thus:

$$Y_j = f(x_1, x_2, \ldots, x_{M-1}, P)$$

Assuming that the pressure remains constant, and differentiating the above equation, we get:

$$dY_j = \sum_{k=1}^{M-1} \frac{\partial Y_i}{\partial x_k} dx_k \tag{30}$$

In numerical form, we can represent Eq. (30) by the following relations:

$$(\Delta Y_j)_n = \sum_{k=1}^{M-1} \Phi_{j,k}(\Delta x_k)_n \tag{31}$$

$$(\Delta Y_j)_n = (Y_j)_n - (Y_j)_o \tag{32}$$

$$(\Delta x_j)_n = (x_j)_n - (x_j)_o \tag{33}$$

$$\Phi_{j,k} = \partial Y_j / \partial x_k \tag{34}$$

where $(x)_o$ is the base solution, around which the phase equilibria are to be linearized; $(x)_n$ is any given liquid solution identified with the number n in the neighborhood of $(x)_o$; $(Y)_o$ is the vapor in equilibrium with $(x)_o$; and $(Y)_n$ is the vapor in equilibrium with $(x)_n$.

Knowing $(x)_o$, $(Y)_o$ and the matrix $[\Phi]$, we can use the linear relation, Eq. (31), to estimate $(Y)_n$ for any given $(x)_n$ in the neighborhood of $(x)_o$.

Calculation of the matrix

Eq. (32) for $j = 1, 2, \ldots, M-1$ can be written in matrix form as:

$$\begin{bmatrix} \Phi_{1,1} & \Phi_{1,2} & \cdots & \Phi_{1,M-1} \\ \Phi_{2,1} & \Phi_{2,2} & \cdots & \Phi_{2,M-1} \\ \cdot & \cdot & & \cdot \\ \cdot & \cdot & & \cdot \\ \cdot & \cdot & & \cdot \\ \Phi_{M-1,1} & \Phi_{M-2,2} & \cdots & \Phi_{M-1,M-1} \end{bmatrix} \bullet \begin{bmatrix} (\Delta x_1)_n \\ (\Delta x_2)_n \\ \cdot \\ \cdot \\ \cdot \\ (\Delta x_{M-1})_n \end{bmatrix} = \begin{bmatrix} (\Delta Y_1)_n \\ (\Delta Y_2)_n \\ \cdot \\ \cdot \\ \cdot \\ (\Delta Y_{M-1})_n \end{bmatrix} \tag{35}$$

Eq. (32) for $n = 1, 2, \ldots, M-1$ can be written in matrix form as:

$$\begin{bmatrix} \Phi_{1,1} & \Phi_{1,2} & \cdots & \Phi_{1,M-1} \\ \Phi_{2,1} & \Phi_{2,2} & \cdots & \Phi_{2,M-1} \\ \cdot & & & \\ \cdot & & & \\ \cdot & & & \\ \Phi_{M-1,1} & \Phi_{M-1,2} & \cdots & \Phi_{M-1,M-1} \end{bmatrix} \bullet \begin{bmatrix} (\Delta x_1)_1 & (\Delta x_1)_2 & \cdots & (\Delta x_1)_{M-1} \\ (\Delta x_2)_1 & (\Delta x_2)_2 & \cdots & (\Delta x_2)_{M-1} \\ \cdot & \cdot & \cdots & \\ \cdot & \cdot & \cdots & \\ \cdot & \cdot & \cdots & \\ (\Delta x_{M-1})_1 & (\Delta x_{M-1})_2 & \cdots & (\Delta x_{M-1})_{M-1} \end{bmatrix} =$$

$$\begin{bmatrix} (\Delta Y_1)_1 & (\Delta Y_1)_2 & \cdots & (\Delta Y_1)_{M-1} \\ (\Delta Y_2)_1 & (\Delta Y_2)_2 & \cdots & (\Delta Y_2)_{M-1} \\ \cdot & \cdot & & \\ \cdot & \cdot & & \\ \cdot & \cdot & & \\ (\Delta Y_{M-1})_1 & (\Delta Y_{M-1})_2 & \cdots & (\Delta Y_{M-1})_{M-1} \end{bmatrix} \tag{36}$$

In compact notation, Eq. (36) is represented by:

$$[\Phi][\Delta x] = [\Delta Y] \tag{37}$$

Assuming that the composition of the solutions $(x)_o$, $(x)_1$, $(x)_2$, . . . , $(x)_{M-1}$ is known, and estimating the vapor $(Y)_o$, $(Y)_1$, . . . , $(Y)_{M-1}$ by calling up the vapor-liquid equilibrium routine, we can use Eq. (32) and (33) to estimate the terms of the matrices $[\Delta x]$ and $[\Delta Y]$. Thus, we can estimate the matrix $[\Phi]$ from Eq. (37) by writing it in the form:

$$[\Phi] = [\Delta Y][\Delta x]^{-1} \tag{38}$$

In order to facilitate the inversion of the $(M-1) \times (M-1)$ matrix—i.e., $[\Delta x]$ in Eq. (38)—we choose the composition of the liquid solution n, with $n = 1, 2, \ldots, M-1$, as follows:

$$(x_j)_n = \begin{cases} (x_j)_o & \text{for } j = 1, 2, \ldots, M-1 \quad j \neq n \\ (x_n)_o + \varepsilon_n & \text{for } j = n \end{cases} \tag{39}$$

$$\varepsilon_n = \begin{cases} 0.01 & \text{if } (x_n)_o + 0.01 \leq 1.0 \\ -0.01 & \text{otherwise} \end{cases} \tag{40}$$

Thus:

$$\Delta x = \begin{bmatrix} \varepsilon_1 & & & & \\ & \varepsilon_2 & & & \\ & & \cdot & & \\ & & & \cdot & \\ & & & & \varepsilon_{M-1} \end{bmatrix} \tag{41}$$

And, from Eq. (38):

$$\Phi_{j,k} = \frac{1}{\varepsilon_k}[(Y_j)_k - (Y_j)_o] \tag{42}$$

Summarizing, to linearize the vapor-liquid equilibria around the liquid solution $(x)_o$, we:

■ Define $(M-1)$ liquid solutions from Eq. (39) and (40).

■ Call the subprogram that performs the vapor-liquid equilibrium calculations, to estimate the composition of the vapor in equilibrium with each of the liquid solutions.

■ Use Eq. (32) and (42) to estimate the terms of matrix $[\Phi]$.

Vapor-liquid equilibria

Linearization of the vapor-liquid equilibria along the distillation path requires that a linear relation between Y and x [such as that of Eq. (32)] be used for every stage of the column, to simplify the solution of the component material-balance equations. The terms of the matrix $[\Phi]$, however, do not necessarily change appreciably from stage to stage. Hence, they may not have to be computed and kept stored in the computer memory for all stages of the column.

For most multicomponent systems, estimation of the matrix $[\Phi]$ in four to eight stages along the distillation path is sufficient. When the total number of stages is not more than eight, the matrix $[\Phi]$ may be estimated for the liquid solutions on all tower stages. For columns with more stages, consideration should be given for estimating the matrix $[\Phi]$ on every other second, third, etc., stage to allow more than one neighboring stage to share the same values of $[\Phi]$. This saves both computer time and core-memory requirements without appreciably reducing the speed of convergence.

Computer implementation of the above linearization when the matrix [Φ] is to be estimated on every other lth stage of the tower can be accomplished by the following scheme:

1. Estimate the matrix $[\Phi]_I$ for tower stages:

$$i = l_s + (I - 1)l \quad I = 1, 2, \ldots, I_m \quad (43)$$

where l_s, the number of the first tower stage for which [Φ] is estimated, is given by:

$$l_s = Integer[1/2] + 1 \quad (44)$$

and I_m, the total number of stages along the distillation path for which [Φ] is estimated, is given by:

$$I_m = Integer\left[\frac{N - l_s}{l}\right] + 1 \quad (45)$$

Eq. (45) is derived from Eq. (43), with $i = N$, and solved for I.

2. During the solution of the mathematical model for tower stage i, use $[\Phi]_I$ with

$$I = Integer[i/l] \quad (46)$$

3. Update from time to time the matrices [Φ] as the solution progresses and as the composition profiles change from their starting values.

Plate efficiencies

The plate-efficiency coefficients take into account the nonideal behavior of real plates, and the departure from equilibrium of the vapor and liquid stage-overflow streams.

The modified Murphree efficiency coefficient, Eq. (47), written in the form of Eq. (48), permits the estimation of $y_{j,i}$ for $i = 1, 2, \ldots, N$ as a function of $(EMV)_{j,i}$, $Y_{j,i}$, and $y_{j,i-1}$.

$$(EMV)_{j,i} = \frac{y_{j,i} - y_{i,i-1}}{Y_{j,i} - y_{j,i-1}} \quad (47)$$

$$y_{j,i} = (EMV)_{j,i}Y_{j,i} + [1 - (EMV)_{j,i}]y_{j,i-1} \quad (48)$$

Eq. (48) will be used to establish the vapor-composition profile after each iteration of the solution algorithm (Fig. 3). For the reboiler, it is assumed that $(EMV)_{j,1} = 1$.

Following this, the vaporization efficiencies [8] can be estimated from:

$$E_{j,i} = y_{j,i}/Y_{j,i} \quad (49)$$

Introducing Eq. (49), (32) and (33) into Eq. (31), and rearranging, we obtain:

$$y_{j,i} = y_{j,i}' + E_{j,i} \sum_{k=1}^{M-1} (\Phi_{j,k})_I(x_{k,i} - x_{k,i}') \quad (50)$$

Eq. (50) gives the composition of the vapor leaving each tower stage, taking into account stage-vaporization efficiency and linearized vapor-liquid equilibria.

Liquid concentrations

To calculate the liquid concentrations, we use Eq. (50) to eliminate $y_{j,i-1}$ and $y_{j,i}$ from Eq. (21) and (22), and we use Eq. (21) to eliminate $x_{j,N+1}$ from Eq. (22). After some rearranging, we obtain:

$$A_{j,i}x_{j,i-1} + B_{j,i}x_{j,i} + C_i x_{j,i+1} = D_{j,i} \quad (51)$$

where:

$$A_{j,1} = 0 \quad (52)$$

$$A_{j,i} = V_{i-1}E_{j,i-1}(\Phi_{j,j})_I \qquad i = 2, 3, \ldots, N \quad (53)$$

$$B_{j,i} = -[V_i + (SV)_i]E_{j,i}(\Phi_{j,j})_I - [L_i + (SL)_i] - \frac{U_i}{\mu\Delta t} \qquad i = 1, 2, \ldots, N-1 \quad (54)$$

$$B_{j,N} = -[V_N + (SV)_N]E_{j,N}(\Phi_{j,j})_I - [L_N + (SL)_N] - \frac{U_N}{\mu\Delta t} + \frac{V_N E_{j,N}(\Phi_{j,j})_I}{\frac{R+1}{R} + \frac{U_D}{\mu L_{N+1}\Delta t}} \quad (55)$$

$$C_i = L_{i+1} \quad (56)$$

$$C_N = 0 \quad (57)$$

$$D_{j,i} = -F_i c_{j,i} - (Ox)_{j,i} - (Pr)_{j,i} - V_{i-1}E_{j,i-1} \sum_{\substack{k=1\\k\neq j}}^{M-1} (\Phi_{j,k})_I x_{k,i-1} + [V_i + (SV)_i]E_{j,i} \sum_{\substack{k=1\\k\neq j}}^{M-1} (\Phi_{j,k})_I x_{k,i} \qquad i = 1, 2, \ldots, N-1 \quad (58)$$

$$D_{j,N} = -F_N c_{j,N} - (Ox)_{j,N} - (Pr)_{j,N} - V_{N-1}E_{j,N-1} \sum_{\substack{k=1\\k\neq j}}^{M-1} (\Phi_{j,k})_I x_{k,N-1} + [V_N + (SV)_N]E_{j,N} \sum_{\substack{k=1\\k\neq j}}^{M-1} (\Phi_{j,k})_I x_{k,N} - \left[\frac{1}{\frac{R+1}{R} + \frac{U_D}{\mu L_{N+1}\Delta t}}\right]\left[(Ox)_{j,N+1} + (Pr)_{j,N+1} + V_N E_{j,N} \sum_{\substack{k=1\\k\neq j}}^{M-1} (\Phi_{j,j})_I x_{k,N}\right] \quad (59)$$

$$(Pr)_{j,i} = V_{i-1}[y_{j,i-1}' - E_{j,i-1} \sum_{k=1}^{M-1} (\Phi_{j,k})_I x_{k,i-1}'] - [V_i + (SV)_i](y_{j,i}' - E_{j,i} \sum_{j=1}^{M-1} (\Phi_{j,k})_I x_{k,i}') \qquad i = 1, 2, \ldots, N \quad (60)$$

$$(Pr)_{j,N+1} = V_N\left[y_{j,N}' - E_{j,N} \sum_{j=1}^{M-1} (\Phi_{j,k})_I x_{k,N}'\right] \quad (61)$$

Eq. (51) for $i = 1, 2, \ldots, N$ can be written in the following tridiagonal matrix form:

$$\begin{bmatrix} B_{j,1} & C_1 & & & & \\ A_{j,2} & B_{j,2} & C_2 & & & \\ & A_{j,3} & B_{j,3} & C_3 & & \\ & & \cdot & \cdot & \cdot & \\ & & & A_{j,N-1} & B_{j,N-1} & C_{N-1} \\ & & & & A_{j,N} & B_{j,N} \end{bmatrix} \bullet \begin{bmatrix} x_{j,1} \\ x_{j,2} \\ \cdot \\ \cdot \\ \cdot \\ x_{j,N} \end{bmatrix} = \begin{bmatrix} D_{j,1} \\ D_{j,2} \\ \cdot \\ \cdot \\ \cdot \\ D_{j,N} \end{bmatrix} \quad (62)$$

The parameters $A_{j,i}$, $B_{j,i}$ and C_i of Eq. (62) are independent of the liquid compositions sought, and their estimation is straightforward. The parameters, $D_{j,i}$ and $D_{j,N}$, given by Eq. (58) and (59) are functions of the liquid compositions sought, and their calculation is iterative, as shown in the diagram of Fig. 3.

The tridiagonal matrix, Eq. (62), can easily be solved separately for each component to yield the liquid-composition profile in the column [9].

Analysis of the algorithm

From our preceding discussion and the diagram of Fig. 3, we can see that the solution algorithm contains three loops that involve iterative calculations:

1. The inner loop for calculating the vapor and liquid flowrates (Step 3). The number of iterations depends on the stage holdup functions and on the relative values of the vapor and liquid enthalpies. Generally, few iterations are required to reach convergence.

2. The inner loop for calculating the tridiagonal matrix terms, $D_{j,i}$ (Step 4b). The number of iterations depends on the relative values of $\Phi_{j,j}$ and $\Phi_{j,k}$, $j \neq k$. For example, if $\Phi_{j,k} = 0$ for $j \neq k$, the estimation of $D_{j,i}$ is straightforward. Generally, few iterations are required to reach convergence.

3. The overall loop of the solution algorithm. The number of iterations here depends mainly on:

a. Inaccuracies introduced by the linearized vapor-liquid equilibria. These inaccuracies are not significant provided that the matrix $[\Phi]$ is estimated at frequent intervals along the distillation path and updated periodically during the calculations.

b. Composition dependence of the vapor and liquid flowrate profiles. This exerts some influence on the rate of convergence for systems with dissimilar latent heats of vaporization. In problems where repeated use of the solution method is expected, it might be advantageous to use modified molecular weights, selected to yield equal modified-molar heats of vaporization for all components. This reduces the composition dependence of the vapor and liquid flowrates and speeds up the overall rate of convergence.

c. Interfacing of the design variables. For the general case, estimation of the model's input variables from the given set of design variables in each problem involves dependent variables as well. Thus, estimating the input variables is iterative in nature and is performed in parallel with the iterations of the solution method [7]. However, this tends to slow down the overall rate of convergence, and depends on the degree of compatibility between the design and input variable sets.

The proposed method compares favorably with existing distillation methods in relation to flexibility for solving complex problems, and stability for systems having nonideal component properties.

Computer storage

An assessment of the increased number of variables that have to be kept in the computer memory by this method follows:

1. Solving the transient rather than the steady-state model requires storage of the variables $(Ov)_i$, $(Ol)_i$ and $(Ox)_{j,i}$ from the previous time level. Thus, a total of $NM + N + M$ variables must be stored (M, total components, and N, total stages).

2. Formulating the mathematical model on the basis of Fig. 2 provides the possibility of having a feed stream, vapor and liquid side-product streams, and a heat leak, in all tower stages. This increases the storage requirements by a maximum of $NM + 4N$ variables.

3. Linearizing the vapor-liquid equilibria at certain points (maximum of 8) along the distillation path increases the storage requirements by a maximum of $8(M-1)^2$ variables.

4. Solving the tridiagonal matrix equation for the liquid compositions requires calculating the terms of $A_{j,i}$, $B_{j,i}$, C_i and $D_{j,i}$. This increases the storage requirements by $3NM - 2N$ variables.

Hence, the total increase in the computer-storage requirements is less than $5NM + 3N + M + 8(M-1)^2$. For example, for a tower with 30 stages ($N = 30$) and 5 components ($M = 5$), the increase in the varia-

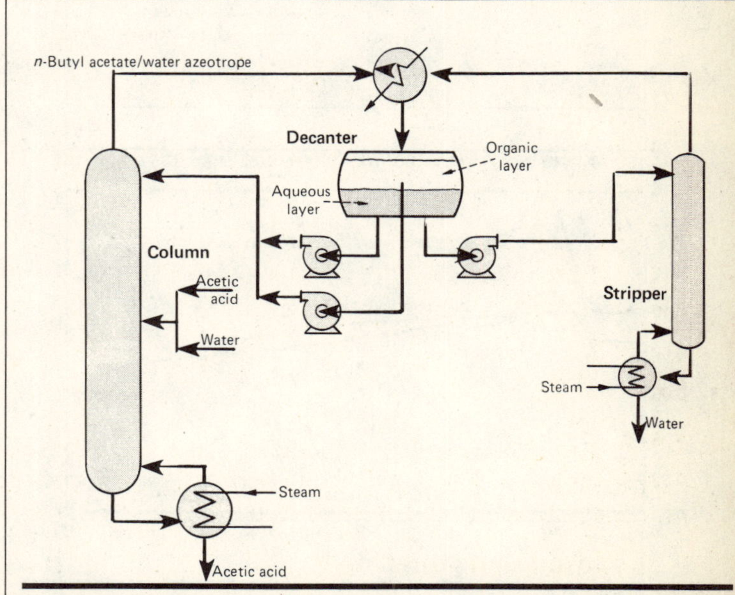

Dehydration plant for acetic acid **Fig. 4**

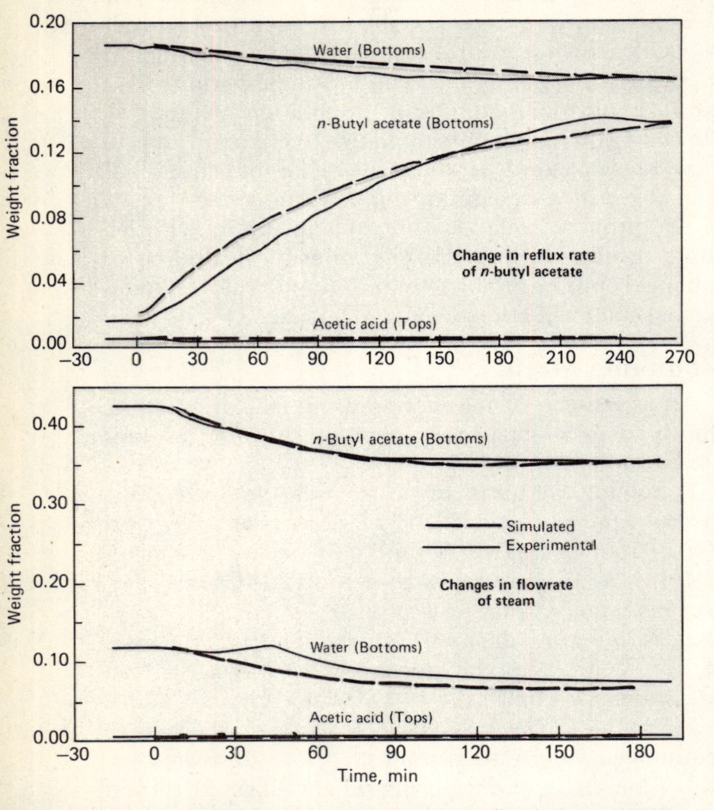

Open-loop plant responses to step changes **Fig. 5**

Plant response to step change in set point of fraction controller for *n*-butyl acetate **Fig. 6**

bles to be stored is less than 973, and for most computer systems is rather insignificant.

Simulation for a complex distillation

The pilot plant, illustrated in Fig. 4, separates acetic acid from water azeotropically by using *n*-butyl acetate as the entrainer. The simulation of this plant provides a test of the stability and rate of convergence of the distillation program for computer simulation. This example combines the following adverse factors:

1. Vapor-liquid equilibrium relations are nonideal.
2. Substantial differences exist between the molar heats of vaporization of the components.
3. Reflux to the separation column is not a simple function of the overhead-product composition because the decanter and stripper further process the overhead.

In order to simulate this plant, the reflux ratio, R, must be set equal to zero. Furthermore, the reflux from the two layers in the decanter is treated as a feed to the top stage of the separation column, and calculated in each iteration of the solution method by solving the condenser-decanter-stripper model [*10*].

The solution method is shown to yield a fast rate of convergence and to be capable of satisfactorily simulating the open response and the control response of the plant. Fig. 5 presents the experimental and simulated plant-responses to a step change in the flowrates for *n*-butyl acetate reflux and steam. Fig. 6 compares the measured and simulated plant-responses [*10*] to a step change in the controller for the *n*-butyl acetate fraction.

Acknowledgements

The author wishes to acknowledge the financial assistance received from the National Research Council of Canada and the University of Calgary for completing part of this study.

References

1. Lewis, W. K. and Matheson, G. L., *Ind. Eng. Chem.*, Vol. 24, p. 494 (1932).
2. Bonner, J. S., *Chem. Eng. Prog. Symp. Ser.*, Vol. 55, No. 21, p. 87 (1959).
3. Thiele, E. W. and Geddes, R. L., *Ind. Eng. Chem.*, Vol. 25, p. 289 (1933).
4. Lyster, W. N., Sullivan, Jr., S. L., Billingsley, D. S. and Holland, C. D., *Pet. Refiner*, June 1959, p. 221; July 1959, p. 151; Oct. 1959, p. 139.
5. Amundson, N. R. and Pontinen, A. J., *Ind. Eng. Chem.*, Vol. 50, p. 370 (1958).
6. Economopoulos, A. P., A Generalized Method for Calculating the Degrees of Freedom of Staged Operations Systems. Unpublished paper.
7. Economopoulos, A. P., General Design of Stage Operations Solution Packages. Unpublished paper.
8. Holland, C. D., "Multicomponent Distillation," Prentice-Hall, Englewood Cliffs, N.J., 1963.
9. Carnahan, B., Luther, H. A. and Wilkes, J. O., "Applied Numerical Methods," Wiley, New York, 1969.
10. Economopoulos, A. P., Distillation Modeling and Control, Ph.D. Thesis, University of Calgary, Canada, 1973.

The author

Alexander P. Economopoulos is chief of the source inventory section, Ministry of Social Services, Environmental Pollution-Control Project, 147 Patission St., Athens (814), Greece. Previously, he was a lead process-design engineer with Canatom Mon/Max Montreal (a joint venture of two private engineering companies). His research and industrial experience includes process design and development, computer simulation, optimization and automatic control. He has a Dipl. Chem. Eng. from the National Technical University of Athens, and a Ph.D. in chemical engineering from the University of Calgary, Canada.

Computer design of sieve trays and tray columns

An evaluation of methods for designing and estimating the performance of sieve trays and tray columns leads to a computer program that can quickly establish the necessary parameters.

Alexander P. Economopoulos, Ministry of Social Services, Greece

☐ A generalized computer procedure for individual trays and complex tray columns yields the optimum design for the tray parameters, has a desirable degree of operating flexibility, and allows accurate estimation of performance.

The existing literature correlations are supplemented by equations formulated from graphs and tables of well-established manual methods for estimating tray design and performance. This set of equations forms the basis for the analysis and development of a coherent package for tray-column calculations.

The computer procedure also offers a convenient tool for quickly sizing various design alternatives or assessing the capability of an existing column to function efficiently with the imposed operating conditions. Additional design methods, suitable for sieve trays or other types of trays, can be easily accommodated in the program. Refinements to the basic approach presented here, reflecting the user's experience, should not be difficult to introduce.

Technique for tray design and performance

The tray-performance parameters required [6] for the solution of staged-operations problems are: stage-pressure drop, ΔP_i; liquid holdup, U_i; and tray efficiency, $E_{(MV)j,i}$. As the flowchart of Fig. 1 shows, these parameters can be estimated once the required tray-design data are fixed and the physical properties given.

Intermediate tray-performance parameters such as dry-tray pressure drop, h_d; surface-tension head, h_σ; liquid crest over weir, h_{ow}; clear liquid height in the bubbling area, h_l; downcomer backup, h_{ud}; and fractional vapor entrainment, E, are used for estimating the tray-performance parameters, as well as for designing the trays. The input/output variables for calculating these intermediate parameters are listed in Table I.

For designing new columns, the algorithm calculates all parameters involved in the tray-performance estimations. These parameters (Fig. 1) are the number of flow passes, N_P; active and downcomer areas, A_a and A_d; perforated area, A_h; weir and flowpath lengths, W_L and F_{PL}; flowpath width, W_{FP}; and tray spacing, t_s.

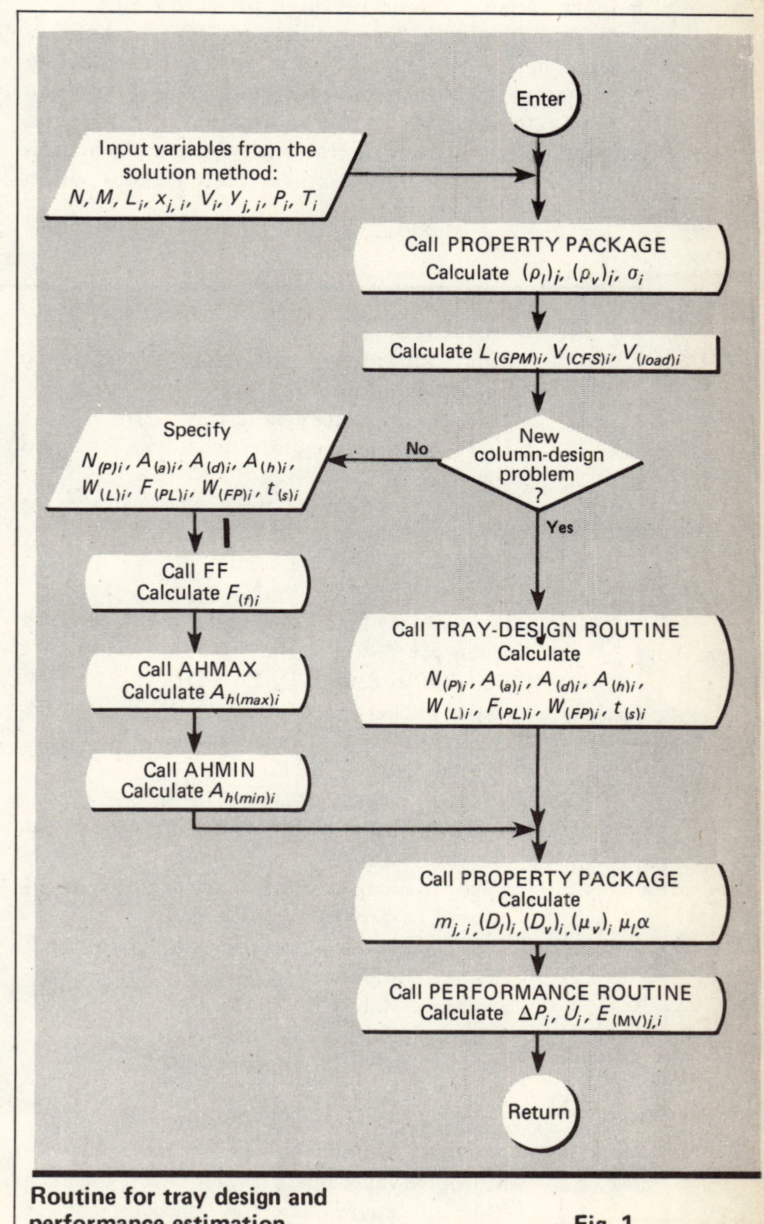

Routine for tray design and performance estimation — Fig. 1

Originally published December 4, 1978

The input/output variables for these tray-design correlations are also listed in Table I.

For simulating existing columns, the tray-design routines are used to calculate the flood factor, F_f; maximum and minimum allowable hole areas, $A_{h(max)}$ and $A_{h(min)}$; and the stage performance. This enables the user to quickly determine whether a feasible solution has been obtained within the actual operating limitations of the sieve trays.

Design parameters for new columns

In new-column design (Table I), the user has to make some decisions in order to fix a number of tray-design parameters:

- Desirable operating flexibility of single trays or tray columns is specified by fixing the flood and weep factors, F_f and F_w. Low flood factors are known to yield large tray sizes and increased tower-vessel volumes. Low weep factors are shown to cause increased tower-vessel heights and/or undesirably high pressure drops.
- Weir height, h_w; deck thickness, t_d; and hole diameter, d_h, are tray construction parameters affecting the dry-tray pressure drop and the clear liquid height.
- Maximum tray pressure drop, ΔP_{max}, becomes important in vacuum operations where the accumulated pressure in the bottom stages has to be kept within limits.
- System derating factor, S_F, depends mainly on the foaming tendency of the system. It fixes the degree of expected derating of the tray capacity.
- Maximum tray-spacing value, $t_{s(max)}$, is used for the initial sizing of each tray in the column. As smaller trays are subsequently increased in size to match the larger ones, their tray spacing can be decreased in order to maintain a uniform flood factor across the tower while reducing the tower-vessel volume to a minimum. The resulting tray spacing cannot be less than the specified $t_{s(min)}$ that is required for tray maintenance and repair. The externally specified parameters, $t_{s(max)}$ and $t_{s(min)}$, are important for they affect the shape of the column and bear an impact on the capital investment. Higher values for $t_{s(max)}$ result in slender and taller towers with smaller tray area but increased overall vessel volume. The difference between $t_{s(max)}$ and $t_{s(min)}$ must be such as to allow taking up the slack of the reduced flood factors from the increase in the size of small trays in the column. Studies by the author with vacuum towers has shown that for $t_{s(min)} = 12$ in., the optimum value of $t_{s(max)} = 18$ in. This may vary from problem to problem and is the only tray-design parameter that must be externally optimized.

For normal problems, the designer may refrain from specifying the above parameters—in which case, the program can use the default values listed in Table II.

Nomenclature

a	Interfacial area per unit volume of vapor and liquid holdup, ft²/ft³
A_a	Active or bubbling area of tray, ft²
A_d	Downcomer area, cross-sectional area for liquid downflow, ft²
$A_{d'}$	Intermediate value of the downcomer area, as defined by Eq. (56), ft²
A_h	Hole or perforation area, ft²
A_t	Total tray area, ft²
A_{ud}	Free area for liquid flow under the downcomer apron, ft²
b	Variable, as defined by Eq. (43)
C_0	Vapor discharge coefficient for dry tray, as calculated by Eq. (2), dimensionless.
c_1	Variable, defined by Eq. (66)
c_2	Variable, defined by Eq. (67)
C_1	Vapor discharge coefficient for dry tray, as calculated by Eq. (4), dimensionless
C_2	Vapor discharge coefficient for dry tray, as calculated by Eq. (6), dimensionless.
C_{AF_o}	Tray capacity factor, as calculated by Eq. (62), (65) or (68), dimensionless
D	Molecular diffusivity, ft²/h
D_E	Eddy diffusivity of backmixing, ft²/h
d_h	Hole or perforation diameter, in.
D_t	Tray diameter, ft
e	Liquid entrainment, lb-moles/h
E	Fractional entrainment, as defined by Eq. (19)
E_{MV}	Murphree vapor-phase tray efficiency, fractional
E^*_{MV}	Murphree vapor-phase tray efficiency not corrected for entrainment, fractional
E_O	Overall tray efficiency, fractional
E_{OV}	Point vapor-phase tray efficiency, fractional
f	Fanning friction factor, dimensionless
F	Weir-construction correction factor, fractional
F_f	Tray flood factor, fractional
F_{fd}	Downcomer flood factor, fractional
F_{lv}	Liquid-vapor flow parameter, as defined by Eq. (20)
F_{PL}	Length of liquid flowpath on the tray, ft
F_s	Vapor-flow parameter based on active area defined by Eq. (11)
F_w	Tray weep factor, fractional
H	Chord height or downcomer width, ft
h_d	Dry-tray vapor pressure drop, as head of clear liquid, in.
h_f	Froth height in the active area, in.
h_{fd}	Froth height in the downcomer, in.
h_l	Clear or effective liquid height in the tray, in.
h_{ld}	Clear or effective liquid height in the downcomer, in.
h_{ow}	Height of liquid crest over weir, in.
h_t	Total head loss of vapor through the tray, as head of clear liquid, in.
h_{ud}	Head loss due to liquid flow under downspout apron, in.
h_w	Weir height, in.
h_σ	Head loss due to bubble formation, as head of clear liquid, in.
L	Liquid overflow rate, lb-moles/h
L_{GPM}	Liquid overflow rate, gpm
m	Slope of the vapor-liquid equilibrium chord = dy/dx

Intermediate-tray performance

Dry-tray pressure drop—Hughmark and O'Connell [13,19] used the orifice equation to estimate the dry-tray pressure drop:

$$h_d = \frac{0.186}{(C_0)^2}\left(\frac{V_{CFS}}{A_h}\right)^2\left(\frac{\rho_v}{\rho_l}\right)\left[1-\left(\frac{A_h}{A_a}\right)^2\right] \quad (1)$$

They also presented a plot of the orifice coefficient, C_0, which is fitted by the equation:

$$C_0 = \frac{\left[880.6 - 67.7\left(\frac{d_h}{t_d}\right) + 7.32\left(\frac{d_h}{t_d}\right)^2 - 0.338\left(\frac{d_h}{t_d}\right)^3\right]}{1{,}000} \quad (2)$$

Hunt et al. [14,23] presented the following formula:

$$h_d = 0.186C_1\left(\frac{V_{CFS}}{A_h}\right)^2\frac{\rho_v}{\rho_l} \times \left[0.5 - 0.4\frac{A_h}{A_a} + 4f\left(\frac{t_d}{d_h}\right) + \left(1-\frac{A_h}{A_a}\right)^2\right] \quad (3)$$

where the Fanning friction factor, f, can be assumed for our purposes equal to 0.01, and the orifice coefficient, C_1, for $0.5 \leq d_h/t_d \leq 5$ can be calculated from:

$$C_1 = 1.09(d_h/t_d)^{0.25} \quad (4)$$

Leibson et al. [9,17] used the orifice equation in the form:

$$h_d = \frac{0.186}{C_2^2}\left(\frac{V_{CFS}}{A_h}\right)^2\frac{\rho_v}{\rho_l} \quad (5)$$

From experimental dry-tray pressure-drop data in the literature, they produced a graph for the orifice coefficient C_2. Based on this graph, the following correlation was derived:

$$C_2 = \left(0.836 + 0.273\frac{t_d}{d_h}\right)\left(0.674 + 0.717\frac{A_h}{A_a}\right) \quad (6)$$

The correlations of Hughmark and O'Connell and Hunt et al. give comparable results, while that of Leibson et al. yields somewhat higher pressure-drop values.

Surface-tension head—The surface-tension head, defined as the necessary pressure to form a vapor bubble through the tray hole, is given [9] by:

$$h_\sigma = 0.04\sigma/\rho_l d_h \quad (7)$$

Liquid crest over weir—The liquid crest over the weir can be estimated by the Francis weir formula [1]:

$$h_{ow} = 0.092F(L_{GPM}/W_L)^{2/3} \quad (8)$$

where the correction factor, F, for multipass trays can be

M	Total number of components in the system
N	Total number of trays in the column
N_L	Number of liquid-phase transfer units
N_P	Number of tray flow-passes
N_{Pe}	Peclet number for liquid mixing, as defined by Eq. (38), dimensionless
N_{Sc}	Schmidt number for the vapor phase, as defined by Eq. (28), dimensionless
N_V	Number of vapor-phase transfer units
P	Tray pressure, psia
S_F	System derating factor
t	Residence time, h
t'	Contact time, h
T	Tray temperature, deg. F.
t_d	Tray deck thickness, in.
t_s	Tray spacing, in.
$t_{s'}$	Intermediate tray-spacing parameter, as defined by Eq. (60), in.
$t_{s(max)}$	Tray spacing on which the trays are initially sized, in.
$t_{s(min)}$	Minimum tray spacing required for tray maintenance, in.
T_{td}	Tray turndown ratio, as defined by Eq. (47)
U	Mass holdup of liquid on the tray, lb
V	Vapor overflow rate, lb-moles/h
V_{CFS}	Vapor flowrate, ft^3/s
V_d	Velocity of liquid in the downcomer, gpm/ft^2 of downcomer area
V_{load}	Vapor load factor, as defined by Eq. (53)
W_{FP}	Average width of liquid flowpath on the tray, calculated from Eq. (55)
W_L	Weir length, ft
x	Mole fraction of a component in the liquid phase
y	Mole fraction of a component in the vapor phase
α	Relative volatility
β	Aeration factor of the liquid in the active tray area, dimensionless
ΔP	Total pressure loss of vapor through the tray, psi
η	Variable defined by Eq. (37)
λ	Variable defined by Eq. (45)
μ	Viscosity, cP
ρ	Density, lb/ft^3
σ	Liquid surface tension, dyne/cm
φ	Aeration factor of the liquid in the downcomer, dimensionless

Subscripts

i	Stage number (stage counting starts from bottom)
j	Component number
k	Stage number used instead of i
l	Liquid phase
mc	Minimum value of a variable arising from the tray pressure-drop limitation ΔP_{max}
Mc	Maximum value of a variable arising from the tray pressure-drop limitation ΔP_{max}
md	Minimum value of a variable originating from tray flooding due to downcomer backup
Md	Maximum value of a variable originating from tray flooding due to downcomer backup
sd	Side downcomer
v	Vapor phase
z	Tray zone

Input/output variables of the system routines **Table I**

System routines	Output variables	Input variables: Flowrates	Physical properties	Calculated performance parameters*	Estimated design parameters†	Specified design parameters‡
Tray design						
1. Active area (AA)	A_a	L_{GPM}, V_{CFS}	ρ_l, ρ_v, σ		W_L, F_{PL}	S_F, F_f, $t_{s(max)}$
2. Tray spacing (TS)	t_s	L_{GPM}, V_{CFS}	ρ_l, ρ_v, σ		A_a, W_L, F_{PL}	S_F, F_f, $t_{s(min)}$
3. Flood factor (FF)	F_f	L_{GPM}, V_{CFS}	ρ_l, ρ_v, σ		A_a, W_L, F_{PL}, t_s	S_F
4. Downcomer area (AD)	A_d	L_{GPM}	ρ_l, ρ_v		A_a	S_F, F_f, $t_{s(max)}$
5. Tray geometry (TG)	W_L, F_{PL}, W_{FP}				A_a, A_d, N_p	
6. Maximum hole area (AHMAX)	$A_{h(max)}$	V_{CFS}	ρ_l, ρ_v	h_{ow}, h_l, h_σ	A_a	h_w, F_w, t_d, d_h
7. Minimum hole area (AHMIN)	$A_{h(min)}$	V_{CFS}	ρ_l, ρ_v	h_{ow}, h_l, h_σ, h_{ud}	A_a	h_w, t_s, F_f, ΔP_{max}, t_d, d_h
8. Entrainment (FRE)	E	L_{GPM}, V_{CFS}	ρ_l, ρ_v			F_f
Tray performance						
1. Crest over weir	h_{ow}				A_t, W_L	
2. Clear-liquid height	h_l	L_{GPM}, V_{CFS}	ρ_l, ρ_v	h_{ow}, h_σ	A_t, W_{FP}	h_w
3. Dry pressure drop	h_d	V_{CFS}	ρ_l, ρ_v		A_a, A_h	t_d, d_h
4. Surface-tension head	h_σ		ρ_l, σ			d_h
5. Head loss under downcomer	h_{ud}	L_{GPM}			A_d	
6. Stage efficiency	$E_{(MV)j,i}$	L_{GPM}, V_{CFS}	ρ_l, ρ_v, σ, m, D_l, D_v, μ_v, μ_l, α	h_{ow}, h_σ	A_t, A_a, A_h, F_{PL}, W_{FP}, t_s	h_w

*Tray performance parameters calculated by other system routines.

†Tray design parameters estimated by system routines in new column-design problems, or specified by the designer in existing column-simulation problems.

‡Tray design parameters to be specified by the user. Table II lists typical default values.

assumed equal to 1.0, while for one-pass trays can be computed by the Bolles [1] relation solved numerically for F:

$$\frac{(L_{GPM})^{2/3}}{(W_L)^{5/3}} = 61.73\,\frac{\sqrt{1-\left(\frac{W_L}{D_t}\right)^2}/F^3-\sqrt{1-\left(\frac{W_L}{D_t}\right)^2}}{\left(\frac{W_L}{D_t}\right)F} \quad (9)$$

Clear liquid height—Fair [6] correlated the effective liquid head to the operating liquid seal at the tray outlet weir ($h_w + h_{ow}$) by means of an aeration factor, β, as follows:

$$h_l = \beta(h_w + h_{ow}) \quad (10)$$

The β-factor is estimated as a function of a vapor kinetic-energy parameter, F_s:

$$F_s = (V_{CFS}/A_a)(\rho_v)^{0.5} \quad (11)$$

through a graph, based on which the following correlation was derived:

$$\beta = 0.977 - 0.619F_s + 0.341(F_s)^2 - 0.0636(F_s)^3 \quad (12)$$

Foss and Gerster [10,23] recommended the following equation for estimating the clear liquid height:

$$h_l = 0.24 + 0.725h_w - 0.29h_w\frac{V_{CFS}}{A_a}\sqrt{\rho_v} + 0.01\frac{L_{GPM}}{W_{FP}} \quad (13)$$

Hughmark and O'Connell [13,19] presented two curves relating the effective liquid head to the operating seal at the outlet weir—one for high and the other for low vapor loadings.

From their graph, the following relations were derived. For $F_s \geq 1.4$, where F_s is given by Eq. (11):

$$h_l = 0.377 + 0.955(h_w + h_{ow}) - 0.221(h_w + h_{ow})^2 + 0.024(h_w + h_{ow})^3 - h_\sigma \quad (14)$$

For $F_s < 1.4$:

$$h_l = 0.374 + 1.12(h_w + h_{ow}) - 0.266(h_w + h_{ow})^2 + 0.027(h_w + h_{ow})^3 - h_\sigma \quad (15)$$

The correlations of Fair [6] and Foss and Gerster [10,23] provide comparable predictions, while that of Hughmark and O'Connell [13,19] yields higher values, especially in vacuum operations.

Head loss under downcomer—The liquid head loss for flow under the downcomer apron is calculated from [9]:

$$h_{ud} = 0.558\left(\frac{L_{GPM}}{448.8A_{ud}}\right)^2 \quad (16)$$

Normal tray-construction practice yields approximately:

$$A_{ud} = 0.42A_d \quad (17)$$

Downcomer backup—The clear liquid height in the downcomer is equal to the height of liquid immediately

Default values for design specifications **Table II**

F_f	= 0.82	(nonvacuum service, tray diameter ≥ 36 in.)
F_W	= 0.60	(for a minimum turndown ratio of 2)
h_W	= 2.0in.	(nonvacuum service)
t_d	= 0.078in.	(14-gage deck)
d_h	= 3/8 in.	
ΔP_{max}	= 0.15psi	(nonvacuum service)
S_F	= 1.0	(nonfoaming, normal system)
$t_{s(min)}$	= 12 in.	
$t_{s(max)}$	= 18 in.	

outside the downcomer ($h_w + h_{ow}$) increased by the head loss of the liquid flowing under the downcomer and the total tray pressure drop:

$$h_{ld} = h_w + h_{ow} + (h_t + h_{ud})[\rho_l/(\rho_l - \rho_v)] \quad (18)$$

Entrainment—The fractional entrainment, E, is defined as:

$$E = e/(L + e) \quad (19)$$

It can be calculated from a graph presented by Fair [8,9,23] as a function of the tray flood factor, F_f, and a parameter, F_{lv}, where:

$$F_{lv} = \frac{L_{GPM}}{448.8V_{CFS}}\sqrt{\frac{\rho_l}{\rho_v}} \quad (20)$$

This graph was correlated by:

$$E = \exp[-(6.692 + 1.956F_f)(F_{lv})^{(-0.132+0.654F_f)}] \quad (21)$$

Tray performance

The following tray-performance parameters are required for the solution method:

Total tray pressure drop—The simple assumption that the total tray pressure-drop is equal to the sum of the dry, effective liquid and surface-tension heads yields surprising accuracy, and has proved reliable for design purposes. Thus:

$$h_t = h_d + h_l + h_\sigma \quad (22)$$

$$\Delta P = h_t(\rho_l/1{,}728) \quad (23)$$

Liquid holdup—The liquid holdup on the tray is related to the effective liquid head in the active and downcomer areas by:

$$U = (h_l A_a + h_{ld} A_d)(\rho_l/12) \quad (24)$$

A vertical downcomer apron without seal or drawoff pan has been assumed.

Tray efficiency: two-film theory—The Murphree point efficiency has been related to the vapor and liquid mass-transfer units by:

$$E_{OV} = 1 - \exp\left\{-\left[\frac{1}{N_V} + \frac{mV}{L}\left(\frac{1}{N_L}\right)\right]\right\} \quad (25)$$

The AIChE Manual [2] presents the following empirical expressions for estimating N_V and N_L:

$$N_V = \frac{\left(0.776 + 0.116h_w - 0.29F_s + 0.0217\frac{L_{GPM}}{F_{PL}}\right)}{(N_{Sc_v})^{0.5}} \quad (26)$$

$$N_L = 7.31(10^5)D_l^{0.5}(0.26F_s + 0.15)t_l \quad (27)$$

where:

$$N_{Sc_v} = 2.42\mu_l/\rho_v D_v \quad (28)$$

$$t_l = \frac{(A_h/12)A_a}{3{,}600\left(\frac{L_{GPM}}{448.8}\right)} = \frac{h_l A_a}{96.3L_{GPM}} \quad (29)$$

Hughmark [12] derived different correlations for estimating N_V and N_L that are claimed to have better accuracy especially for systems controlled by the liquid phase:

$$N_V = 2a\left(\frac{D_v t_v}{\pi}\right)^{0.5}\left(\frac{h_f}{h_f - h_l}\right) \quad (30)$$

$$N_L = 2a\left(\frac{D_l}{\pi}\right)^{0.5}\frac{h_f}{h_l}\left(\frac{t_l}{(t_l')^{0.5}}\right) \quad (31)$$

where:

$$t_v = (h_f - h_l)/(12)(3{,}600)\left(\frac{V_{CFS}}{A_a}\right) \quad (32)$$

$$\frac{t_l}{(t_l')^{0.5}} = \left(\frac{2}{h_w}\right)^{1/8}\left(\frac{18}{100A_h/A_a}\right)^{1/8} \times \exp(1.065 - 1.8h_{ow} - 1.517F_s + 0.67h_{ow}F_s) \quad (33)$$

$$a = \left(91 + 266F_s - 92(F_s)^2 - \frac{75}{h_w} + 40h_{ow}\right) \times \left(900\frac{\rho_v}{\rho_l}\right)^{1/6}\left(\frac{0.67}{\mu_l}\right)^{1/12}\left(\frac{70}{\sigma}\right)^{1/10} \quad (34)$$

$$h_f = -0.61 + 115\left(\frac{V_{CFS}}{A_a}\right)^2\left(\frac{\rho_v}{\rho_l - \rho_v}\right) + 1.64h_w + 1.49h_{ow} \quad (35)$$

Based on a tray liquid-mixing model, the AIChE Manual correlates the Murphree point efficiency to the Murphree overall tray efficiency, as follows:

$$E_{MV}^* = E_{OV}\left[\frac{1 - \exp[-(\eta + N_{Pe})]}{(\eta + N_{Pe})[1 + (\eta + N_{Pe})/\eta]} + \frac{\exp(\eta) - 1}{\eta[1 + \eta/(\eta + N_{Pe})]}\right] \quad (36)$$

where:

$$\eta = \frac{N_{Pe}}{2}\left[\left(1 + \frac{4mVE_{OV}}{LN_{Pe}}\right)^{0.5} - 1\right] \quad (37)$$

$$N_{Pe} = (F_{PL})^2/D_E t_l \quad (38)$$

$$D_E = \left[0.774 + 1.026\frac{V_{CFS}}{A_a} + 0.15\frac{L_{GPM}}{F_{PL}} + 0.9h_w\right]^2 \quad (39)$$

Tray efficiency, as calculated from Eq. (36), has to be corrected for liquid entrainment. Colburn [3] has shown that the effect of vapor entrainment is given by:

$$E_{MV} = E_{MV}^*/[1 + E_{MV}^* E/(1 - E)] \quad (40)$$

where E, the fractional entrainment, is estimated from Eq. (21).

Tray efficiency from overall tray efficiency—Drickamer and Bradford [4,22] correlated graphically the overall efficiency data from 54 refinery columns as a function of the feed viscosity. Their graph for the Murphree efficiency is fitted by the following equation:

$$E_O = 0.17 - 0.265 \ln (\mu_{feed}) \qquad (41)$$

O'Connell [21,22,23] presented graphs for distillation towers and absorbers, which are better suited for systems having high-relative-volatility components. The graph for distillation towers is correlated by:

$$E_O = 0.485 - 0.129b + 0.018b^2 + 0.001b^3 \qquad (42)$$

where:
$$b = \ln (\alpha\mu_{feed}) \qquad (43)$$

The overall tray efficiency estimated from either the Drickamer and Bradford or O'Connell methods can be used to calculate the Murphree stage efficiency [18,22] from:

$$E_{MV} = (\lambda^{E_O} - 1)/(\lambda - 1) \qquad (44)$$

where:
$$\lambda = mL/V \qquad (45)$$

The two-film theory provides the possibility of investigating the effect of changes in the tray design, e.g., an increase in the number of flow passes, or the effect of operating conditions on tray efficiency. It can also be used for obtaining efficiency values for systems where no actual tray-efficiency data exist. However, the calculated efficiencies should be checked against experimental efficiencies of similar systems whenever possible.

The confidence one can place on efficiency predictions by either correlation depends upon the type of system involved. Hence, the original references should be checked.

System derating factor

Correlations for estimating the active and downcomer areas involve only the vapor and liquid densities. The effect of all other properties on tray-sizing calculations is introduced through the system derating factor, S_F, that derates the tray's load-handling capacity. Fair [9] related this factor to a phase property, liquid surface tension, through:

$$S_F = (\sigma/20)^{0.2} \qquad (46)$$

In the Glitsch [11] and Koch [16] manuals, S_F is related to the foaming tendency of the system. Derating factors for various systems are:

Derating factor, S_F	System condition	Examples
1.0	Nonfoaming	Regular
0.85	Moderate foaming	Oil absorbers, amine and glycol regenerators, etc.
0.73	Heavy foaming	Amine and glycol contactors, etc.
0.60	Severe foaming	Methyl ethyl ketone units, etc.
0.30	Foam-stable	Caustic regenerators, etc.

Flood and weep factors and turndown ratio

The flood factor relates the design or operating loads to loads at the flood point. Recommended design values for the flood factor from the Glitsch manual [11] are 0.82 for normal problems and 0.77 for vacuum towers. Towers having a diameter less than 36 in. should have the flood factor further decreased by 0.1.

The weep factor relates the design or operating loads to those at the weep point. This factor is introduced to facilitate the design of tray columns by ensuring uniform tray-operating flexibility.

The tray turndown ratio (a measure of tray-operating flexibility) is related to the flood and weep factors by:

$$T_{td} = 1/F_f F_w \qquad (47)$$

Correlations for tray sizing

To obtain the necessary tray size, various factors must be considered. These include: number of flow passes; weir length, flowpath length and width of flowpath; downcomer area; active area; tray spacing; tray flood factor; and total tray area. Let us discuss the effects of each of these factors on tray size and provide the appropriate correlations.

Number of flow passes—The number of tray passes affects the weir length of the tray and, hence, the weir liquid loading. For a given tray, the number of flow passes is increased, as necessary, to bring the weir loading within proper design limits:

$$L_{GPM}/W_L \leq 96 \qquad (48)$$

Depending on the size of tray, there is a limit to the number of flow passes. This limit is necessary in order to be able to construct internal manways and obtain a satisfactory tray efficiency. Both conditions require flowpath lengths of adequate size. The recommended maximum number of flow passes is given by:

$$N_{P(max)} = 0.377\sqrt{A_t} \qquad (49)$$

where $N_{P(max)}$ is rounded to the next higher integer but is never less than 1.

In some problems, the above limit on the maximum number of flow passes may prevent the reduction of the liquid loading of the weir within the normal design limits, as defined by Eq. (48). In such cases, loadings as high as 240 gpm/ft of weir length can be tolerated.

Weir and flowpath lengths, and width of flowpath—Interior downcomers can be assumed to have rectangular shapes and are placed in positions to divide the tray into equal flowpath lengths. Based on these considerations, the following correlations can be used to obtain a good approximation of the tray's weir and flowpath lengths:

$$W_L = W_{L(sd)} + D_t(N_P - 1)^{0.946} \qquad (50)$$

$$F_{PL} = \frac{\left[D_t - 2H_{sd} - 2\left(\frac{A_d}{D_t}\right)(N_P - 1)^{0.054}\left(1 - \frac{A_{d(sd)}}{A_d}\right)\right]}{N_P} \qquad (51)$$

We must now calculate the weir length, $W_{L(sd)}$, and chord height, H_{sd}, of the side downcomers. Let us assume that the area of the side downcomer, $A_{d(sd')}$, is known. We can now calculate H_{sd} by numerically solving the following nonlinear equation:

$$\frac{A_{d(sd)}}{A_t} = \frac{1}{\pi}\left[\cos^{-1}\left[1 - 2\left(\frac{H_{sd}}{D_t}\right)\right] - 2\left[1 - 2\left(\frac{H_{sd}}{D_t}\right)\right]\sqrt{\frac{H_{sd}}{D_t}\left(1 - \frac{H_{sd}}{D_t}\right)}\right] \quad (52)$$

while $W_{L(sd)}$ is estimated from:

$$W_{L(sd)} = 2\sqrt{H_{sd}(D_t - H_{sd})} \quad (53)$$

The total downcomer area is distributed among the side, center, off-center and off-side downcomers in multipass trays in proportion to the fraction of the active area served by each. Based on this principle, the following correlation has been derived for estimating the side downcomer area as a function of the overall downcomer area and the number of flow passes:

$$A_{d(sd)} = A_d(N_P)^{-(0.916+0.0476N_P)} \quad (54)$$

The average width of flow can be calculated as a function of the tray active area and the flowpath length from:

$$W_{FP} = A_a/F_{PL} \quad (55)$$

The design correlations recommended by the Glitsch manual [*11*] for sizing downcomers is summarized by:

$$A_d = \text{the larger of}\begin{cases} A_{d'} \\ \text{the smaller of} \quad \begin{matrix} 2A_{d'} \\ 0.11A_a \end{matrix} \end{cases} \quad (56)$$

where:

$$A_{d'} = L_{GPM}/V_d F_f \quad (57)$$

$$V_d = \text{the smaller of}\begin{cases} 250\, S_F \\ 41\sqrt{\rho_l - \rho_v}S_F \\ 7.5\sqrt{t_s(\rho_l - \rho_v)}S_F \end{cases} \quad (58)$$

From the downcomer sizing method and the velocity graph given in the Koch manual [*16*], the following equations were derived:

$$A_d = L_{GPM}/V_d F_f$$

where:

$$V_d = \text{the smaller of}\begin{cases} 8.578 t_{s'} S_f \\ 0.533 t_{s'}(\rho_l - \rho_V)^{0.82} S_F \end{cases} \quad (59)$$

$$t_{s'} = \text{the smaller of } t_s \text{ and } 30 \quad (60)$$

Active area—Based on the method for ballast trays shown in the Glitsch manual [*11*], the tray active area is calculated from:

$$A_a = \frac{V_{load} + L_{GPM}(F_{PL}/1{,}083)}{C_{AF_o} S_F F_f} \quad (61)$$

The graph for the capacity factor, C_{AF_o}, has been correlated by the following:

$$C_{AF_o} = \text{the smaller of}\begin{cases} (t_s)^{0.65}(\rho_v)^{0.167}/12 \\ 0.3174 + 0.04122(t_s - 12)^{0.483} - 10^{-6}\rho_v(245 + 661t_s) \\ 0.595 - 0.0596\rho_v \end{cases} \quad (62)$$

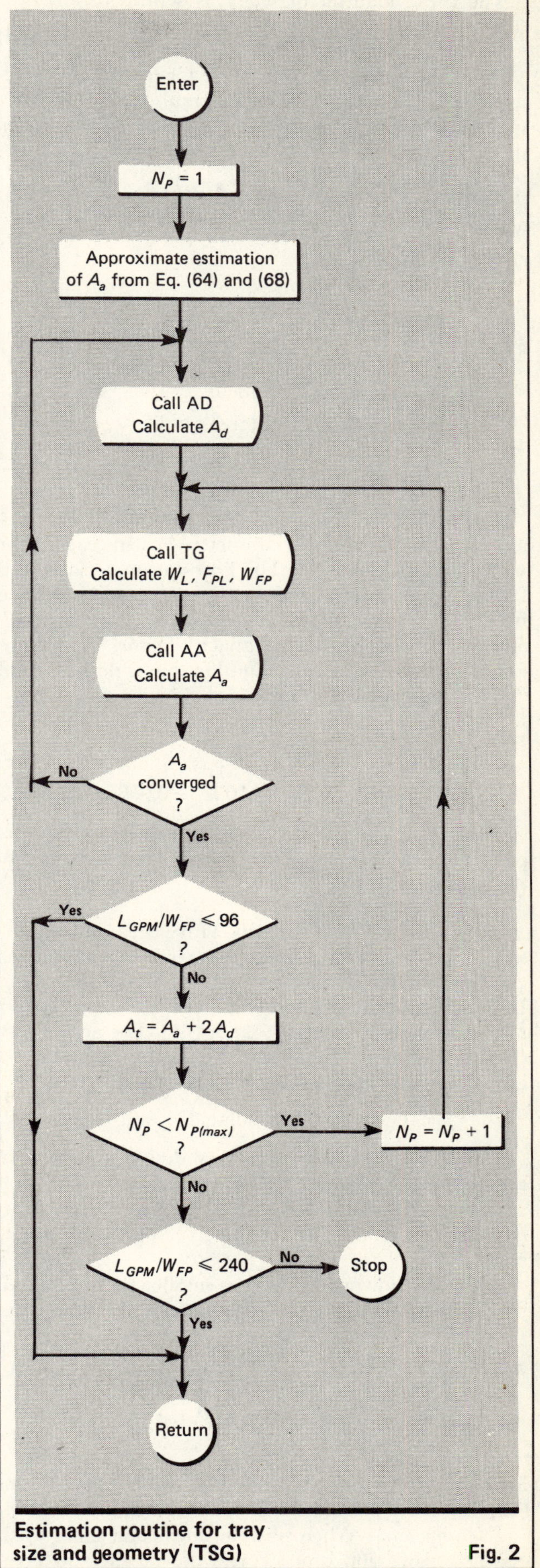

Estimation routine for tray size and geometry (TSG) **Fig. 2**

The vapor load factor, V_{load}, is defined as:

$$V_{load} = V_{CFS}\sqrt{\rho_v/(\rho_l - \rho_v)} \quad (63)$$

From the estimation procedure for (a) the tray's active area, (b) the capacity factor graph and (c) the tray spacing tables (as described in the Koch manual [16]), the following correlations have been derived:

$$A_a = V_{load}/C_{AF_o}S_F F_f \quad (64)$$

where:

C_{AF_o} = the smaller of

$$\begin{cases} 0.1667 + t_s/82.3 \\ c_1(t_s)^{c_2}\left(0.553 - \dfrac{L_{GPM}}{479W_L F_f c_1(t_s)^{c_2}}\right) \end{cases} \quad (65)$$

$$c_1 = \begin{cases} 0.153 \text{ for } \rho_v \leq 1.5 \\ 0.172 \text{ for } \rho_v > 1.5 \end{cases} \quad (66)$$

$$c_2 = \begin{cases} 0.587 \text{ for } \rho_v \leq 1.5 \\ 0.546 \text{ for } \rho_v > 1.5 \end{cases} \quad (67)$$

The Fair correlation [8,9] for estimating the tray active area used a flow parameter, F_{lv}, instead of the vapor load factor, V_{load}. The flow parameter accounts for the liquid/vapor kinetic energy effects and is defined by Eq. (20). The active area is estimated from Eq. (64), and S_F from Eq. (46). The capacity factor, C_{AF_o}, is obtained from the following equation that is derived from Fair's flood-capacity graph:

C_{AF_o} = the smaller of

$$\begin{cases} 0.118 \exp(0.0479t_s) \\ 0.425 \exp(0.0479t_s)[0.1092 - 0.058 \ln(F_{lv})] \end{cases} \quad (68)$$

The active area calculated by Fair's method is independent of the number of tray passes. The results are on the conservative side.

Tray spacing—When the tray active area and the desirable flood factor are given, the required tray spacing can be estimated by solving Eq. (62), (65) or (68) for t_s. Depending on the method employed, the capacity factor is computed from Eq. (61) or (64). The actual tray spacing cannot be smaller than the minimum t_s required for tray-accessibility reasons. Thus:

$$t_s = \text{the larger of calculated } t_s \text{ and } t_{s(min)} \quad (69)$$

Tray flood factor—When the tray design and spacing are specified, we can use Eq. (61) or (64) in conjunction with the corresponding Eq. (62), (65) or (68) to calculate the tray flood factor for the given operating conditions.

When the downcomer area is smaller than required, the Glitsch manual [11] recommends the following:

$$F_f = \left[\frac{V_{load}}{\left(A_a C_{AF_o} S_F - \dfrac{A_d V_d F_{PL}}{1{,}083}\right)\left(\dfrac{A_d V_d}{L_{GPM}}\right)^{0.6}}\right]^{0.625} \quad (70)$$

Eq. (70) is to be used in conjunction with Eq. (57) and (62).

Total tray area—When straight downcomers are used, or when using sloped or stepped downcomers with recessed inlet areas or drawoff sumps, the total tray area is given by:

$$A_t = A_a + 2A_d \quad (71)$$

Tray hole area

Determination of the tray hole area affects the tray operating flexibility through the dry-tray pressure drop. Reduction in the hole area reduces the weep point and, up to a certain point, increases the tray turndown ratio. Beyond this point, however, it can cause premature flooding due to downcomer backup and/or undesirably high tray pressure drop. The objective is to find the hole-area range that will ensure both satisfactory tray-performance characteristics and the required operation flexibility.

Calculation of a range rather than a single value of the hole area is important in the design of tray columns. For construction reasons, the objective is to calculate zones so that as many trays as possible will share identical design parameters.

To calculate the hole area that corresponds to a given overall pressure drop, we can solve Eq. (1), (3) or (5) for the dry pressure drop. For example, Eq. (1) for the hole area yields:

$$A_h = \frac{V_{CFS}}{\sqrt{\left(\dfrac{V_{CFS}}{A_a}\right)^2 + 5.38C_0^2\left(\dfrac{\rho_l}{\rho_v}\right)(h_t - h_l - h\sigma)}} \quad (72)$$

where h_d in Eq. (1) was substituted from Eq. (22), and C_0 is calculated from Eq. (2).

Maximum tray-hole area

Operation with vapor loads below those at the weep point is not predictable. Hence, the weep point is considered the lower design limit of a given tray. The flood point is the upper limit.

Existing literature correlations for this parameter are not sufficiently reliable since experimental observations are scarce and not in good agreement.

Fair [9], using the weep-point measurements of Mayfield et al. [20] and Hutchinson et al. [15], presented a graphical correlation between the operating liquid seal at the tray outlet weir $(h_w + h_{ow})$ (which can be viewed as the driving force for liquid weepage) versus the surface-tension and dry pressure-drop heads (which resist weepage). This graph has been fitted by:

$$(h_d + h_\sigma) = 0.35(h_w + h_{ow})^{0.573} \quad (73)$$

Based on another tray weep-point correlation by Hughmark and O'Connell [13], the following equation was derived:

$$(V_{CFS}/A_h)(\rho_v)^{0.5} = 3.26 + 3.37(h_d + h_l + h_\sigma) \quad (74)$$

Eq. (74) does not always have a solution because it is not always possible to find a pair of values for h_d and A_h that are related through Eq. (72) and satisfy Eq. (74). In these cases, the square-root term of Eq. (77) is imaginary. Eq. (73) yields reasonable results, and in the absence of a more appropriate relation, its use is recommended.

Substituting Eq. (22) into Eq. (73) and solving for h_t, we obtain the minimum pressure drop required to maintain the operating conditions at the weep point:

$$h_{t(min)} = h_l + 0.35(h_w + h_{ow})^{0.573} \quad (75)$$

Using Eq. (1) to eliminate A_h from Eq. (74) and making the simplifying assumption that:

$$1 - (A_h/A_a)^2 = 0.99 \quad (76)$$

we can solve for h_t to obtain an expression for the minimum pressure drop required to maintain the tray at the weep point:

$$h_{t(min)} = -0.967 + 0.238(C_0)^2 \times \rho_l\left[1 - \sqrt{1 - \frac{2.48}{(C_0)^2\rho_l}[3.26 + 3.37(h_l + h_c)]}\right] \quad (77)$$

An increase in the vapor loads tends to reduce the liquid aeration factor, β, in Eq. (12), while the liquid crest over the weir increases with the liquid overflow rates, as given by Eq. (8). These tend to cancel each other out for constant L/V ratios. Hence, within the accuracy of Eq. (75) or (77), it can be assumed that h_l (and also $h_{t(min)}$) is rather independent of the operating conditions. The significance of this is that the tray has to be designed to deliver the minimum pressure drop even at the lowest design vapor load.

From Eq. (72), we can obtain the maximum permissible hole area that will keep the tray above the weep point at the minimum design vapor rates:

$$A_{h'(min)} = \frac{F_w V_{CFS}}{\sqrt{\left(\frac{F_w V_{CFS}}{A_a}\right)^2 + 5.38(C_0)^2\left(\frac{\rho_l}{\rho_v}\right)(h_{t(min)} - h_l - h_\sigma)}} \quad (78)$$

For tray hydraulic stability, the tray hole area is not allowed to exceed 15% of the tray active area, even if this is permissible from Eq. (78). Thus:

$$A_{h(max)} = \text{the smaller of } A_{h'(max)} \text{ and } 0.15A_a \quad (79)$$

Minimum tray-hole area

The downcomer backup increases with the tray pressure drop until all of the downcomer is filled with froth. Further increase in the tray pressure drop may cause tray flooding because the downcomer may not be able to handle the liquid overflow.

Assuming that the downcomer aeration factor is φ, the downcomer froth height is given by:

$$h_{fd} = h_{ld}/\varphi \quad (80)$$

This height should not be allowed to exceed the downcomer height, and at limiting conditions must satisfy the relation:

$$h_{ld}/\varphi F_{fd} = t_s + h_w \quad (81)$$

where F_{fd} is the downcomer flood factor.

There are no data in the literature for estimating the downcomer aeration factor. However, it is generally accepted [9,19,23] that a value given by the following relation provides satisfactory tray design:

$$\varphi F_{fd} = 0.5 \quad (82)$$

Based on Eq. (81) and (82), Eq. (18) solved for h_t yields the maximum permissible pressure drop that does not cause flooding by downcomer backup:

$$h_t = (0.5t_s - 0.5h_w - h_{ow})\left(\frac{\rho_l - \rho_v}{\rho_l}\right) - h_{ud} \quad (83)$$

An increase in liquid loads as the flood point is approached increases both h_{ow} and h_{ud}, with a subsequent reduction of the maximum acceptable tray pressure drop. To avoid premature tray flooding, the maximum permissible design pressure drop has to be calculated at the maximum expected liquid rate (i.e., L_{GPM}/F_f). Taking into consideration Eq. (8) and (16), Eq. (83) yields:

$$h_{t(Md)} = \left(0.5t_s - 0.5h_w - \frac{h_{ow}}{(F_f)^{2/3}}\right)\left(\frac{\rho_l - \rho_v}{\rho_l}\right) - \frac{h_{ud}}{(F_f)^2} \quad (84)$$

The value of the tray pressure drop from Eq. (84) should not be exceeded even at the incipient jet-

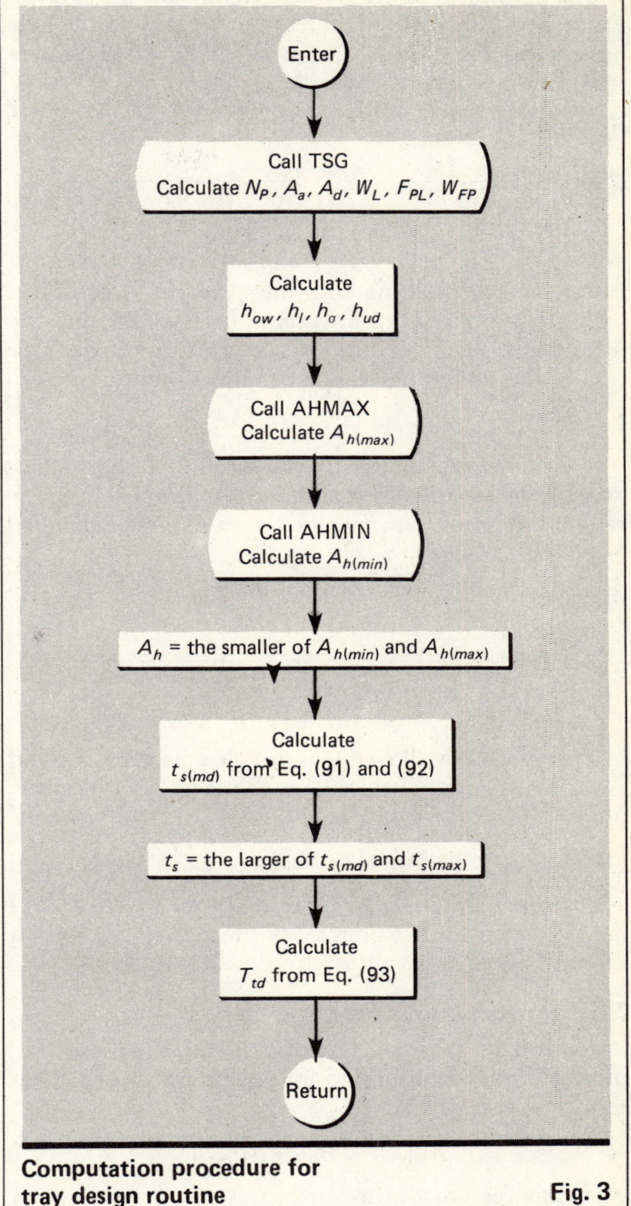

Computation procedure for tray design routine **Fig. 3**

flooding conditions where the vapor rate is V_{CFS}/F_f. From Eq. (72), we can calculate the minimum hole area, which for the maximum vapor flow will cause a tray pressure drop equal to $h_{t(Md)}$:

$$A_{h(md)} = \frac{V_{CFS}/F_f}{\sqrt{\left(\frac{V_{CFS}}{F_f A_a}\right)^2 + 5.38(C_0)^2\left(\frac{\rho_l}{\rho_v}\right)(h_{t(Md)} - h_l - h_\sigma)}} \quad (85)$$

For towers operating with low pressure or equipped with many stages, a maximum permissible pressure drop across each tray, ΔP_{max}, is often specified. This creates the additional vapor head-loss limitation:

$$h_{t(Mc)} = (1{,}728/\rho_l)\Delta P_{max} \quad (86)$$

The value calculated from Eq. (86) is the highest acceptable pressure drop at design conditions. The corresponding minimum tray-hole area can be calculated from Eq. (72) as:

$$A_{h(mc)} = \frac{V_{CFS}}{\sqrt{\left(\frac{V_{CFS}}{A_a}\right)^2 + 5.38(C_0)^2\left(\frac{\rho_l}{\rho_v}\right)(h_{t(MC)} - h_l - h_\sigma)}} \quad (87)$$

The minimum allowable tray-hole area that does not cause downcomer backup or excessive tray pressure drop is then:

$$A_{h(min)} = \text{the larger of } A_{h(md)}, A_{h(mc)}, 0.05A_a \quad (88)$$

For operating stability, the hole area in Eq. (88) is not allowed to be less than 5% of the active area, even if a larger value is permitted by Eq. (85) and (87).

Minimum tray spacing

Assuming that $A_{h(max)} \geq A_{h(min)}$, Eq. (79) and (88) provide the minimum and maximum limits of the design hole area. Thus:

$$A_{h(min)} \leq A_h \leq A_{h(max)} \quad (89)$$

In this region, the tray weep factor is equal to or better than the design value, the downcomer does not flood prematurely, and the total tray pressure drop does not exceed the maximum specified.

In the event that $A_{h(max)} < A_{h(min)}$, we set:

$$A_h = A_{h(max)} \quad (90)$$

so that the tray does not weep prematurely.

From Eq. (88), we can see that either $A_{h(md)}$ or $A_{h(mc)}$, or both, are larger than $A_{h(max)}$. The significance of the former is that the pressure drop generated by the hole area, $A_{h(max)}$, is higher than the downcomer can withstand. Unless the tray spacing is increased, the tray will flood prematurely by downcomer backup. The significance of the latter is that the pressure drop generated by the hole area, $A_{h(max)}$, at design operating conditions exceeds the maximum specified ΔP_{max}.

To remedy this situation, we could reduce the specified weir height and/or increase the weep factor. These are outside decisions. For a given problem, the best that can be done is to increase the tray spacing so that the tray does not flood prematurely by downcomer backup, while designing the tray to deliver the minimum pressure drop.

The approach of not stopping the calculations when the maximum specified pressure drop is exceeded, but settling instead for the minimum pressure drop obtainable, allows us to design a tower by simply specifying a small ΔP_{max}, and have the program through Eq. (90) select the maximum allowable hole area for each tray.

To calculate the minimum tray spacing, $t_{s(md)}$, required to avoid premature tray flooding by downcomer backup, when the tray hole area is A_h, we solve Eq. (85) for $h_{t(Md)}$:

$$h_{t(Md)} = h_l + h_\sigma + \frac{0.186}{(CO)^2}\left(\frac{V_{CFS}/F_f}{A_h}\right)^2\left(\frac{\rho_v}{\rho_l}\right)\left[1 - \left(\frac{A_h}{A_a}\right)^2\right] \quad (91)$$

Introducing $h_{t(Md)}$ from Eq. (91) into Eq. (84) and solving for t_s, we obtain the required minimum tray spacing:

$$t_{s(md)} = 2\left(\frac{\rho_l}{\rho_l - \rho_v}\right)\left(h_{t(Md)} + \frac{h_{ud}}{(F_f)^2}\right) + h_w + 2\left(\frac{h_{ow}}{(F_f)^{2/3}}\right) \quad (92)$$

Depending upon the chosen value of A_h, the tray turndown ratio is given by:

$$T_{td} = \frac{1}{F_f F_w}\left(\frac{A_{h(max)}}{A_h}\right) \quad (93)$$

Thus, when A_h varies between $A_{h(max)}$ and $A_{h(min)}$, the tray turndown ratio falls in the interval:

$$\frac{1}{F_f F_w} \leq T_{td} \leq \frac{1}{F_f F_w}\left(\frac{A_{h(max)}}{A_{h(min)}}\right) \quad (94)$$

Design of individual trays

From the tray-sizing correlations [Eq. (48) through (71)], it would appear that there is an interaction between the number of flow passes, the weir and flowpath lengths, the downcomer area and the active area. We can also see this interaction from Table I, in which the required inputs for each correlation are listed.

As a result of this interaction, an iterative tray design procedure has to be followed. The flowchart in Fig. 2 shows an algorithm performing this task.

The calculations start by assuming a one-pass tray, and estimating the tray active area by Fair's correlation [Eq. (64) and (68)], using as inputs only the tray overflow rates, phase densities, and liquid surface tension.

Following this, we estimate the tray active area and the downcomer area by the chosen method, and calculate the associated weir and flowpath lengths. Due to interaction of these variables, the calculation procedure is iterative and proceeds until convergence.

Liquid loading of the weir is calculated in the final step to assess whether it falls within the recommended design limits [Eq. (48)]. If not, the number of tray passes will have to be increased by one, provided this is permitted by the size of the tray [Eq. (49)]. With new N_P, calculations for the tray active and downcomer areas and for weir and flowpath lengths are repeated.

Having fixed the tray size and geometry, the maximum and minimum hole areas are estimated, as the

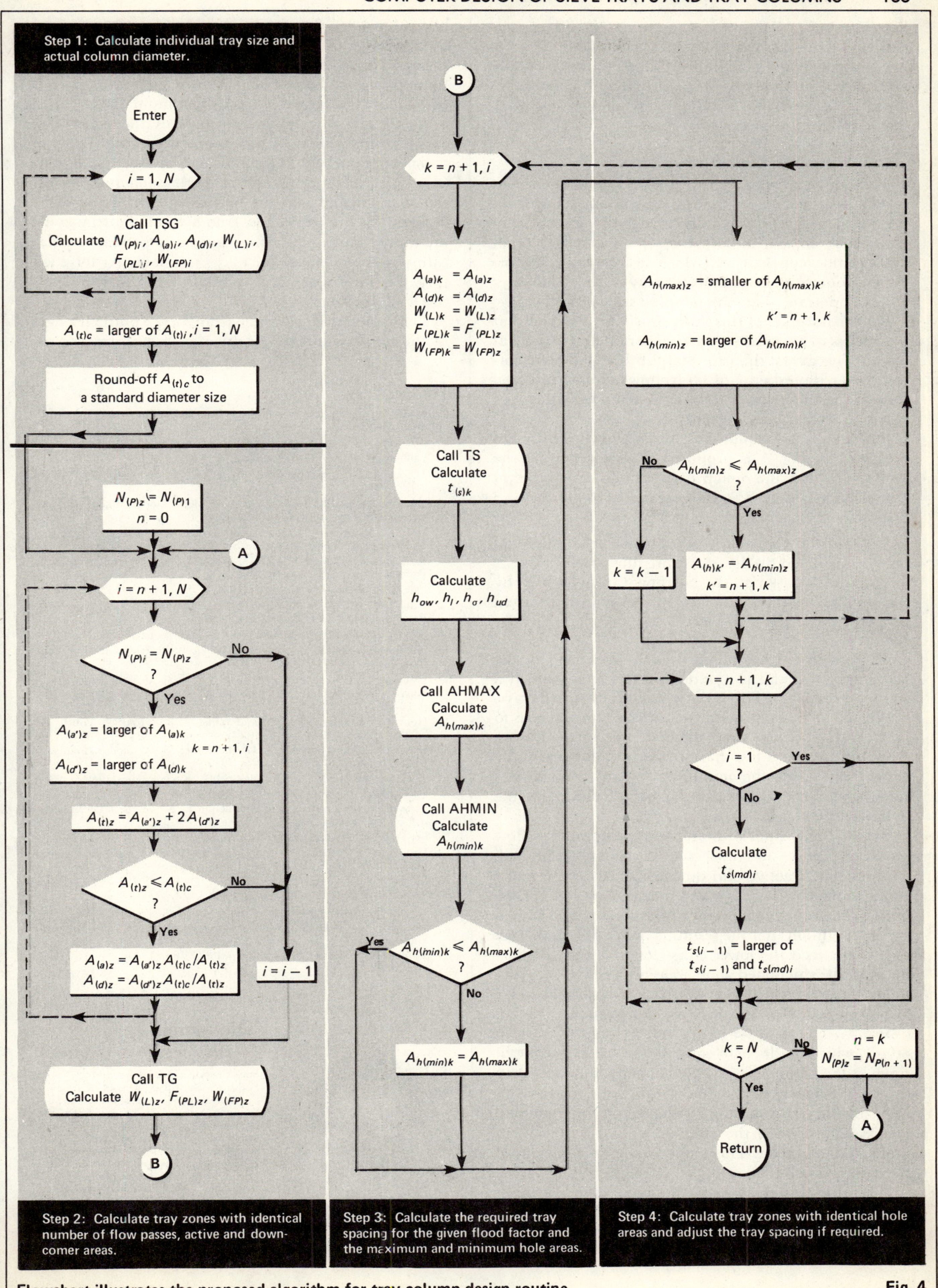

Flowchart illustrates the proposed algorithm for tray-column design routine **Fig. 4**

flowchart in Fig. 3 shows. If $A_{h(min)} < A_{h(max)}$, the tray is designed with the minimum permissible hole area for maximum operating flexibility. Otherwise, the tray hole area is made equal to $A_{h(max)}$, so that the desirable operating flexibility is maintained while the tray spacing is adjusted if necessary.

Design of tray columns

The design of a tray in a tray column is subject to certain restrictions, originating on the one hand from the tower vessel [which imposes equality in the tray diameters], and on the other by the tray construction and erection convenience [which is best served by the design uniformity of trays in the tray zones]. The latter is less important than the required tray flexibility.

For the purposes of this study, it will be assumed that it costs less to have the tray design parameters change as often as necessary than to establish better design uniformity at the expense of larger tower-vessel and tray sizes. Nonuniformity in the tray spacing is assumed to create no construction problems.

Based on these conditions, the tray-column design flowchart of Fig. 4 is proposed. This scheme can be modified to suit the design practice and experience of the individual user. The calculations in Fig. 4 proceed through the following steps:

Step 1—The size and geometry of each individual tray is calculated by the procedure illustrated in Fig. 2. The design is based on the overflow rates provided by the solution method (see Fig. 1), the required system properties, and the tray spacing, $t_{s(max)}$, as specified by the user.

The tower cross-sectional area is set equal to the larger tray size and is rounded off to correspond to a standard diameter. This is the common size for all trays.

When a tower vessel with two different diameters is to be used, the tray and vessel diameter in each section are estimated as above, but the optimum cutoff point of the vessel that yields the minimum overall tray area is calculated. This option has not been included in the flowchart of Fig. 4.

Step 2—Tray zones with an identical number of flow passes, and active and downcomer areas are established. The zone, and active and downcomer areas are made equal to the largest active and downcomer areas of the trays in the zone. The extent of the zone is determined by the condition that the resulting area of the zone tray does not exceed the tower cross-sectional area. Following this, the calculated zone areas are normalized, and the associated weir length, flowpath length, and width of flowpath of the zone tray are determined.

Step 3—The increase in the size of smaller trays in Step 1, which made them equal to the size of the larger trays in the column, has proportionally enlarged their capacity. Advantage is taken of this extra capacity by reducing the tray spacing to the minimum required. Hence, all trays in the column will operate with a uniform flood factor. This reduces the overall height of the tower vessel.

Following this evaluation, the maximum and minimum hole areas of each tray are estimated.

Step 4—Tray zones with identical design parameters, including the hole area, are established. The maximum allowable hole area of the tray zone, $(A_{h(max)})_z$, is equal to the smaller of the maximum hole areas, $(A_{h(max)})_k$, of the individual trays in the zone. Similarly, the minimum permitted hole area of the tray zone, $(A_{h(min)})_z$, is equal to the larger of the minimum hole areas, $(A_{h(min)})_k$, of the individual trays in the zone. The extent of the zone, which contains trays with identical design parameters, is established by the condition: $(A_{h(max)})_z \geq (A_{h(min)})_z$, but in any case cannot exceed the extent of the zone established in Step 2.

Based on the calculated hole area of the tray zone, the minimum tray spacing required to prevent premature downcomer backup, $t_{s(md)}$, is calculated for each tray. The tray spacing, t_s, established in Step 3 is adjusted if necessary.

The calculations continue from Step 2 until all stages of the tray-column have been classified in tray zones.

References

1. Bolles, W. L., Optimum Bubble-Cap Tray Design, *Pet. Process.*, Feb., Mar., Apr., May 1956.
2. "Bubble-Tray Design Manual," AIChE, New York, 1958.
3. Colburn, A. P., *Ind. Eng. Chem.*, Vol. 28, 526 (1936).
4. Drickamer, H. G. and Bradford, J. R., *Trans. Am. Inst. Chem. Eng.*, Vol. 39, 319 (1943).
5. Economopoulos, A. P., A Fast Computer Method for Distillation Calculations, *Chem. Eng.*, Apr. 24, 1978, p. 91.
6. Economopoulos, A. P., A Generalized Method for Calculating the Degrees of Freedom of Staged Operations Systems, unpublished paper.
7. Economopoulos, A. P., General Design of Stage Operations Solution Packages, *Chem. Eng. (London)*, Nov. 1978.
8. Fair, J. R., *Petro/Chem. Eng.*, Oct. 1961, p. 45.
9. Fair, J. R., Chap. 15 in Smith, B. D., "Design of Equilibrium Stage Processes," McGraw-Hill, New York, 1963.
10. Foss, A. S. and Gerster, J. A., *Chem. Eng. Prog.*, Jan. 1956, p. 28.
11. "Ballast Tray Design Manual," Bulletin No. 4900, F. W. Glitsch & Sons, Inc., Dallas, Tex., 1970.
12. Hughmark, G. A., *Chem. Eng. Prog.*, July 1965, p. 97.
13. Hughmark, G. A. and O'Connell, H. E., *Chem. Eng. Prog.*, Mar. 1957, p. 127-M.
14. Hunt, C. d'A., Hanson, D. N. and Wilke, C. R., *AIChE J.*, Vol. 1, 441 (1955).
15. Hutchinson, M. H., Buron, A. G. and Miller, B. P., paper presented at AIChE meeting, Los Angeles, May 1949.
16. "Flexitray Design Manual," Bulletin 960, Koch Engineering Co., New York, 1960.
17. Leibson, I., Kelley, R. E. and Bullington, L. A., *Pet. Refiner*, Feb. 1957, p. 127; March 1957, p. 288.
18. Lewis, W. K., *Ind. Eng. Chem.*, Vol. 28, 399 (1936).
19. Ludwig, E. E., "Applied Process Design for Chemical and Petrochemical Plants," Vol. 2, Gulf Publishing, Houston, 1964.
20. Mayfield, F. D., Church, W. L., Green, A. C., Lee, D. C. and Rasmussen, R. W., *Ind. Eng. Chem.*, Vol. 44, 2238 (1952).
21. O'Connell, H. E., *Trans. Am. Inst. Chem. Eng.*, Vol. 42, 741 (1946).
22. Smith, B. D., "Design of Equilibrium Stage Processes," McGraw-Hill, New York, 1963.
23. Treybal, R. E., "Mass Transfer Operations," 2nd ed., McGraw-Hill, New York, 1968.

The author

Alexander P. Economopoulos is chief of the source inventory section, Ministry of Social Services, Environmental Pollution-Control Project, 147 Patission St., Athens (814), Greece. Previously, he was a lead process-design engineer with Canatom Mon/Max Montreal (a joint venture of two private engineering companies). His research and industrial experience includes process design and development, computer simulation, optimization and automatic control. He has a Dipl. Chem. Eng. from the National Technical University of Athens, and a Ph.D. in chemical engineering from the University of Calgary, Canada.

Heavy-oil distillation via computer simulation

New computer-simulation technology enables engineers having little experience to quickly design heavy-oil distillation equipment that will not be oversized but will still provide needed flexibility.

Keith J. Ritchey, Gulf Oil Corp.; Frank B. Canfield, ChemShare Corp.; and Timothy B. Challand, ChemDesign Inc.

☐ This article outlines the basic steps necessary for the confident application of third generation programming to the design, revamping or performance evaluation of heavy-oil distillation equipment. Atmospheric and vacuum distillation of crude oil are presented as examples of how to use the general approach.

Its purpose is to fill the void between the theoretical approach to crude-oil and heavy-oil separation and the totally pragmatic approach that relies on rules-of-thumb from previous designs. Much has been written on the theoretical aspects of distillation, and on how to practically apply computer simulation.

With computer programs available for crude- and heavy-oil separation, it is almost mandatory that the engineer use computer simulation to arrive at a process design that is not oversized but that still provides the degree of flexibility required.

Characterization of heavy oils

Traditionally, heavy oils, such as crude oil, have been characterized by their true boiling point (TBP) distillations, and a component-by-component analysis of some of the lighter components. Methods for performing the heavy-oil TBP distillation have been described in detail elsewhere [*1*].

The objective of the TBP distillation is to separate the oil into several narrow boiling-point fractions, from which the properties of each may be thoroughly examined. Generally, the properties of interest for each fraction include specific gravity, sulfur content, pour point. In some cases, further distillation tests are performed. The results of these various tests constitute the oil assay.

Recently, other techniques have been applied to the analysis of the products of crude oils [*2*, *3*], and it is expected that this will be increasingly the case. At this time, however, the TBP distillation analysis is normally the only reasonably accurate source of data available to the crude- and vacuum-unit designer.

A typical crude-oil assay, that for West Texas sour crude oil, is shown in Table I. This assay contains sufficient information to allow the design of an atmospheric distillation unit to process this crude. If, however, a vacuum tower is also to be designed, additional vacuum equilibrium data are required.

Normally, it is not necessary to put into a program all of the information in the assay, but only properties that can be blended by weight or volume.

Sulfur content is a good example of such a property. Thus, for the crude oil described in Table I, a naphtha product with TBP cut-points between 170°F and 375°F would have a sulfur content of 0.25%. Other properties, such as viscosity and pour point, require special blending methods built into the program.

Fig. 1 shows a section of the input to a computer program that can handle a wide range of crude-oil properties. As can be seen, the input is both free-format and conversational, so that it is clear which crude-oil properties are being supplied.

The TBP points shown were produced by plotting the eight distillate fractions given in the crude assay, and drawing a smooth curve through them. It is recommended that data be graphically smoothed so that the points put into the program are evenly spaced and cover the whole range of the crude distillation. Because extrapolations and interpolations are usually carried out by quadratic or linear means, unevenly spaced or unsmoothed data can lead to poor results. Although this approach may seem laborious, it does ensure that the designer has complete control over the computer interpretation of the crude assay.

The TBP 100% point was set at 1,400°F by extrapolation. West Texas sour crude oil is relatively light and this end-point appears reasonable [*4*, *5*].

When the vacuum equilibrium flash-vaporization data are available, they may be used to help predict a consistent end-point. It is not unreasonable to expect

Originally published August 2, 1976

Typical crude oil assay—35.0 °API gravity West Texas sour crude oil — Table I

General properties	
Gravity, °API	35.0
Viscosity, Suv, sec. at 77°F	45.9
100°F	40.8
Pour point, °F	–5
Vapor pressure, Reid, lb	6.4
Water and sediment, % vol.	0.7
Sulfur, GRM1156, % wt.	1.42
Hydrogen sulfide, GRM682, % wt.	0.002
Total halides as NaCl, lb/1,000 bbl crude	5
Carbon residue, Con, % wt.	2.66
Ash content, % wt.	0.002
Nickel, ppm wt.	4
Vanadium, ppm wt.	5
Lead, ppm wt.	1.2
Sodium, ppm wt.	0.6

Light-ends analysis	Vol. %	Wt. %
Methane	0.005	0.0014
Ethane	0.13	0.06
Propane	0.95	0.57
Isobutane	0.45	0.30
Normal butane	1.68	1.16
Isopentane	1.33	0.98
Normal pentane	1.58	1.17

Distillation, D86 (modified)			
Overpoint,	116°F		
At °F	122	% Condensed	<1
	176		5
	212		9
	284		20
	320		25
	392		33
	428		37
	464		41
	500		45
	518		47
	554		51
	590		55.5
After	590	% Residue	44.0
		% Loss	0.5

TBP assay	C_4 and lighter	Distillates								Residues	
TBP cut point, °F	57	170	310	375	420	520	680	740	800	680+	800+
% wt.	2.09	5.28	13.58	7.20	4.26	10.69	15.32	5.44	5.88	41.41	30.09
Yield on crude, % vol.	3.22	6.74	15.46	7.77	4.48	10.93	14.96	5.16	5.53	37.20	26.51
Gravity, °API	–	80.6	58.1	48.3	43.8	38.8	31.2	26.8	25.2	18.1	15.2
Sulfur, % wt.	–	0.135	0.211	0.323	0.395	0.575	0.911	1.20	1.36	1.94	2.07
Viscosity, Suv, sec. at 32°F	–	–	–	1.56 (Cst)	2.52 (Cst)	–	–	–	–	–	–
77	–	–	–	1.07 (Cst)	1.58 (Cst)	35.0	–	–	–	–	–
100	–	–	–	–	–	33.1	44.3	81.6	160.0	–	–
210	–	–	–	–	–	–	31.9	37.1	43.3	122.8	326
250	–	–	–	–	–	–	–	–	–	72.2	151.8
Viscosity, Furol, sec. at 122°F	–	–	–	–	–	–	–	–	–	110	540
UOP characterization factor	–	12.62	11.88	11.70	11.68	11.70	11.65	11.75	11.84	–	–
Pour point, Fed 1411.3,°F	–	–	–	–	–65	–40	+15	+55	+75	+70	+80
Freezing point, Fed 1411.3,°F	–	–	–	<–76	–76	–40	–	–	–	–	–
Nitrogen total, ppm	–	–	20.2	0.3	1.0	2.0	–	–	–	–	–
Nitrogen total, % wt.	–	–	–	–	–	–	0.009	0.061	0.063	–	–
Mercaptan, sulfur, % wt.	–	0.070	0.069	0.167	0.135	0.100	0.046	0.011	0.009	–	–
Hydrogen sulfide, % wt.	–	0.008	0.008	20.001	–	–	–	–	–	–	–
Carbon residue, Rams, % wt.	–	–	–	–	–	–	0.09	0.10	0.11	6.42 Con	8.84 Con
Aniline point, °F	–	–	123.9	123.6	132.7	141.5	154.2	167.1	176.5	–	–
Knock rating, octane no. motor, clear	–	68.4	53.5	43.2	35.4	–	–	–	–	–	–
Ash, % wt.	–	–	–	–	–	–	–	–	–	0.005	0.007
Nickel, ppm	–	–	–	–	–	–	–	–	–	9.9	13.6
Vanadium, ppm	–	–	–	–	–	–	–	–	–	12.0	16.6
Vapor pressure, Reid, lb	–	11.3	4.2	2.4	–	–	–	–	–	–	–

that heavy crude oils could have end-points as high as 2,000°F. Whatever the end-point input, it is important to bear in mind that the physical-property generation routines of most computer programs have been based on an upper temperature limit of approximately 1,200°F. Although extrapolations to higher temperatures do not normally produce significant errors, there are occasions when predicted values for properties, such as critical pressure and molecular weight, are suspect [*6*]. If this should occur, the computer output may be corrected by manually smoothing the data points and again putting them into the program.

ASTM vs. TBP data

An alternative method of representing crude oil distillation data is by means of an ASTM D86 distillation. Although this method is generally applied to petroleum products rather than to crude oil, it is fairly common to find ASTM distillation data in crude assays. Whenever possible, it is preferable to use TBP data, with the representation of the crude oil for design calculations being based on the TBP distillation.

Methods are available for converting ASTM data into TBP data, although this procedure can immediately introduce an error of ±8°F [*7*]. Because only seven temperature points are used in converting from ASTM to TBP data, more data points do not increase the accuracy of the representation.

The crude-oil API gravity data were similarly plotted and a smooth curve drawn through them. It is important to note that gravities are put into the program as a function of mid-volume percent, rather than

```
*35.0 API GRAVITY WEST TEXAS SOUR CRUDE ASSAY*
FEED(BBL/DAY) 1=50000.,
COMPONENTS=2,3,4,5,6,7,8,
FEED REAL(VOL) 1=0.005,0.13,0.95,0.45,1.68,1.33,1.58,
FEED TBP(F)1=170.,267.,345.,437.,535.,644.,753.,873.,1057.,1400.,
FEED VOLUME 1=10.,20.,30.,40.,50.,60.,70.,80.,90.,100.,
FEED GRAVITY(API)1=69.7,56.,47.7,40.9,35.,29.4,25.6,19.7,13.8,5.,
FEED GRAVITY VOL 1=10.,20.,30.,40.,50.,60.,70.,80.,90.,100.,
TEMPERATURE INCREMENTS(F)=25.,50.,100.,
TEMPERATURE BREAKS(F)=700.,1000.,
FEED VISCOSITY(F,CST)1=1,100.,2.,5.6,16.,34.5,
FEED VISCOSITY VOL 1=1,43.64,56.58,66.14,71.49,
FEED VISCOSITY(F,CST) 1=2,210.,1.5,3.4,5.25,
FEED VISCOSITY VOL 1=2,56.58,66.14,71.49,
FEED PROPERTY(WGT)1=1,SULPHUR WT _,0.13,0.21,0.32,0.4,0.57,0.91,1.2,1.36,1.94,
    2.07,
FEED PROPERTY VOL 1=1,6.59,17.69,29.31,35.43,43.64,56.58,66.14,71.49,81.78,
    87.13,
FEED PROPERTY(PPI)1=2,POUR POINT DEG F,-65.,-40.,15.,55.,75.,
FEED PROPERTY VOL 1=2,35.43,43.64,56.58,66.14,71.49,
''
```

Portion of computer input program describes crude oil feed **Fig. 1**

as volume percent over, as was the case for TBP data. Problems can occur when both distillate and residue gravities are used to describe the whole crude.

There is a characteristically distinct break on a plot of API gravity versus mid-volume percent at the point where distillate and residue properties overlap. Generally, it is reasonable to put this discontinuity into the program to maintain the overall mass balance. However, this procedure can result in small gravity errors for those fractions around the break point.

Gravity data have not been given above the 90% volume point, to allow the program to adjust the gravities of fractions in this area. This adjustment is made so that the API gravity produced by summing the component fractions is equal to the bulk API gravity of the feed. It is not unusual for crude oil API gravity plots to exhibit localized maxima and minima along the curve. Should this be the case, sufficient data points should be put into the program so that it may account for this effect.

If API gravity versus mid-volume percent data are not available, bulk API gravity may be used. From the bulk gravity and the crude oil's mean-average boiling point, the UOP characterization constant may be calculated. This constant can then be applied for all fractions of the crude oil. Obviously, this is a gross assumption that can lead to errors, particularly at the front end of the TBP.

Most full-range crude oils contain identifiable low-boiling pure components, and the amounts of these components are usually given in the assay. As can be seen in Fig. 1, these pure components may be put in by reference to code numbers, and their amounts given as volume percent of crude.

Other physical properties have been put in directly from the crude assay as a function of mid-volume percent. In the case of viscosities, the reference temperature at which the viscosities have been measured must also be given.

As mentioned previously, viscosity and pour-point data are handled by means of special blending techniques. For other properties, the blending basis must be given—e.g., sulfur is blended on a weight basis.

It is important to note that most physical properties of the crude are only relevant over specific areas. Thus, pour points have no significance for gasoline fractions. However, most computer programs will extrapolate pour-point data over the entire range of the crude. Consequently, physical-data output by the program for areas not given in the input is frequently meaningless.

Fig. 2 represents a section of computer output that resulted from the input of Fig. 1. The program has divided the TBP distillation section of the crude into average boiling point (ABP) fractions. Each such fraction is treated as a pure component for all further design calculations, and the figure given for each fraction is its average boiling point. This method of representing crude-oil streams has been described previously [*8*].

Once the ABP components have been assigned average boiling points, and the API gravities have been derived from linear interpolation of the input data, molecular weights and critical properties may be predicted by standard correlations [*8*, *9*]. The API gravities of those ABP components comprising the last 10% by volume of the crude oil have been adjusted so that an overall API gravity of 35.0 is achieved.

The distribution of sulfur content over the crude has been likewise adjusted to give an overall value of 1.42 wt %. Because pour-point data have a relatively narrow range of applicability, it is not advisable to attempt to adjust properties of the individual components to arrive at an overall value of –5°F (Table I). This also applies to viscosity data.

The end-result of the crude-assay processing step is a

*** 35.0 API WEST TEXAS SOUR CRUDE ANALYSIS**

CRUDE ASSAY 1(CONTINUED)

DETAILED FEED COMPONENT ANALYSIS OF FEED NO. 1

NO	COMP. NAME	CUM VOL	MOL WT.	API	VISCOSITY IN CST 100 F	210 F	SULPHUR WT BASIS	POUR POINT
1	WATER	.00	18.02	10.0	.0000	.0000	.0000	.0000
2	METHANE	.01	16.04	339.5	.0000	.0000	.0000	.0000
3	ETHANE	.14	30.07	265.8	.0000	.0000	.0000	.0000
4	PROPANE	1.09	44.09	147.5	.0000	.0000	.0000	.0000
5	I-BUTANE	1.54	58.12	120.0	.0000	.0000	.0000	.0000
6	N-BUTANE	3.22	58.12	110.8	.0000	.0000	.0000	.0000
7	I-PENTANE	4.55	72.15	95.1	.0000	.0000	.0000	.0000
8	N-PENTANE	6.13	72.15	92.8	.0000	.0000	.0000	.0000
9	131 ABP	9.79	83.53	72.9	.1169	.0102	.1945	-148.7
10	180 ABP	12.58	95.60	67.7	.1512	.0148	.2221	-138.8
11	210 ABP	15.67	102.76	63.2	.1910	.0207	.2499	-129.9
12	240 ABP	18.76	110.55	59.2	.2443	.0293	.2820	-120.5
13	270 ABP	22.31	118.83	55.4	.3181	.0423	.3200	-110.4
14	300 ABP	26.15	127.58	52.0	.4269	.0631	.3669	-99.10
15	330 ABP	30.00	137.03	49.0	.5797	.0950	.4211	-87.39
16	360 ABP	33.26	147.26	46.5	.7692	.1376	.4754	-76.57
17	390 ABP	36.52	158.12	44.2	.9970	.1919	.5285	-66.64
18	420 ABP	39.78	169.60	42.1	1.292	.2661	.6008	-57.56
19	450 ABP	42.86	181.74	40.1	1.663	.3634	.6953	-47.84
20	480 ABP	45.92	194.56	38.2	2.097	.4886	.8072	-37.34
21	510 ABP	48.98	208.01	36.4	2.584	.6927	.9113	-25.79
22	540 ABP	51.83	222.19	34.8	3.240	.8728	1.016	-13.62
23	570 ABP	54.59	237.02	33.2	4.102	1.110	1.117	-1.157
24	600 ABP	57.34	252.51	31.6	5.278	1.420	1.217	11.95
25	630 ABP	60.09	268.62	30.1	6.931	1.809	1.325	23.98
26	660 ABP	62.84	285.63	28.7	9.289	2.295	1.437	35.52
27	690 ABP	65.60	303.76	27.6	12.71	2.897	1.552	47.01
28	735 ABP	71.00	332.85	26.1	21.48	4.060	1.722	63.27
29	795 ABP	76.00	372.42	23.7	47.62	6.154	1.961	82.07
30	855 ABP	80.65	412.15	20.8	111.2	8.925	2.305	98.06
31	915 ABP	83.91	455.08	18.3	214.3	11.98	2.666	110.1
32	975 ABP	87.17	501.81	16.4	481.8	15.17	2.788	119.4
33	1080 ABP	92.86	572.90	11.0	1345.	20.75	2.944	131.0
34	1230 ABP	97.23	616.06	-2.0	4757.	29.09	3.143	142.7
35	1352 ABP	100.00	611.70	-12.7	.1251+05	36.65	3.222	150.1
			193.48	35.0	3.722	.9678	1.420	97.56

Section of computer output represents results from Fig. 1 input

Fig. 2

detailed characterization of the crude oil that can be applied in design studies for either a new unit or a revamping of an old one.

Simulation of new units

The design of a new crude atmospheric or vacuum unit, or both, must be carried out with regard to the requirements of the particular refinery and its location. Before the design of an individual unit can begin, several overall project decisions have to be made. The availability and selection of crude-oil feedstocks, product demand and specifications, and the source and availability of utilities are just some of the factors that have to be resolved.

These decisions lead to a determination of the refinery processing flowscheme and the finished-products slate. By the time the designer begins work, the target yields and product specifications for the unit are fairly well defined. The following discussion assumes that this stage has been reached.

Initially, the feasibility of the unit-design mass balance should be checked against the characteristics of the crude feedstocks. This is most simply done with a short-cut distillation program. Generally, such a program uses the Geddes fractionation index method to calculate product properties [*10*, *11*].

For new-unit designs, it is advisable to specify product yields and major specifications and allow the pro-

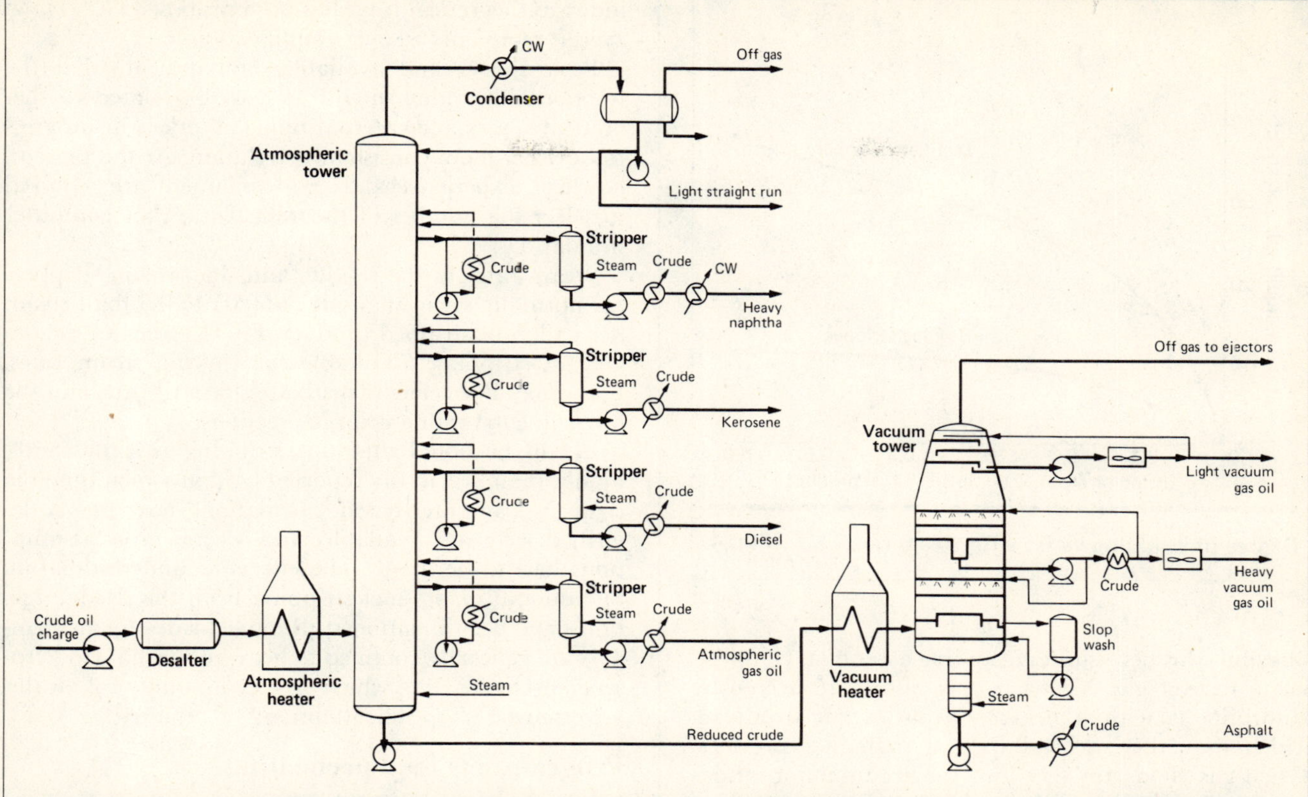

Flowscheme represents typical atmospheric and vacuum units for distillation of crude oil Fig. 3

gram to calculate the fractionation requirements. Table II indicates some typical product-property specifications. The fractionation index is a measure of the degree of separation between products and a useful indicator of the feasibility of the operation.

For example, the maximum fractionation index normally attainable in a conventional atmospheric crude-unit for the separation of residue and heavy gas oil is around 3.0. If the computer program calculates, for the yields and product specifications given, that the fractionation index for this separation is in the region of 4.0 or higher, then the separation is not likely to be feasible, and product yields or specifications should be amended.

It is important to note that the maximum practical value of 3.0 for fractionation index applies only to the separation of residue and heavy gas oil. Higher values are attainable for the separation of, for example, kerosene and light gas oil.

The results of the short-cut crude analysis may be used for the unit-design mass balance. The definitive design of the atmospheric or vacuum unit can then be made within these limits.

It is normal for a unit to be designed to produce a series of different products from the same feedstock, or to be capable of processing several different crude oils. Because of the low cost of using a short-cut distillation program, all possible combinations of product and crude feeds can be investigated. The results of these studies will quickly highlight those areas that should be checked by rigorous design methods.

Preliminary design steps

At this stage, the designer must decide on the unit's configuration and the major operating parameters. Fig. 3 shows a typical flowscheme for an atmospheric and vacuum unit.

Because of the number of degrees of freedom involved, rigorous-design computer programs require that certain parameters be fixed. These include tower pressure, number of theoretical stages, location of side-draws and vapor returns, the use of steam or reboilers for side-stream strippers, pumparound type and location, etc.

Second-generation design programs also required that initial estimates be made for column temperature and vapor profiles. However, third-generation programs now include computational techniques that obviate the need for these estimates.

The design pressure for an atmospheric crude unit is normally set as near to atmospheric pressure as possible. This lowers the flash zone pressure, maximizing distillate production for a given flash-zone temperature. Any gain in distillate production must be balanced against a larger quantity of gas in the overheads accumulator.

The accumulator pressure is usually set at 5 to 10 psig, with 5 psi allowed for pressure drop through the overhead condensation system. Tray pressure drop depends on the tray design, although it is usually 0.2 psi per tray.

An accumulator pressure as low as 0.5 psig is pos-

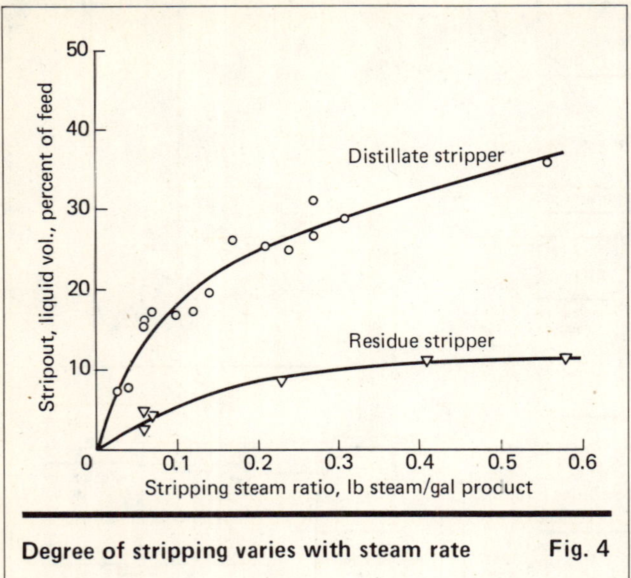

Degree of stripping varies with steam rate **Fig. 4**

sible, but this necessitates installing a vent gas compressor to recover gas. Vent gas may also be recovered by absorbing it with a heavier naphtha cut overhead, which may then be stabilized at a higher pressure. When gas yields are very low, the accumulator off-gas may be flared or routed to a low-pressure burner in a fired heater.

The operating pressure for a vacuum unit depends to a large extent on the atmospheric residue properties and the desired products. Generally, however, vacuum towers operate with a top pressure of between 50 and 60 mm Hg, with a practical minimum of 10 mm Hg.

Designing for feed flashing

As has been mentioned, tower pressures directly affect the amount of feed vaporization in the column flash zone. The amount of vaporization at the inlet to the column in excess of distillate products is termed "overflash." The overflash is generally given on the basis of volume percent of crude and is traditionally set at around 5%.

Computer design techniques have shown that the overflashing of crude feeds can lead to conservative designs. The vapor stripped out in the residue stripping section of an atmospheric unit can, in most cases, provide more than enough reflux to wash the trays above the flashing zone and prevent discoloration of the heaviest distillate draw. In some recent designs, crude feed has, in fact, been underflashed with regard to distillate products.

A useful technique in computer modeling of these units is to place a heater on the feed tray and allow the duty to vary with the quantity of liquid that runs back from the tray above the flash zone. This enables the overflash to be set and allows the computer to calculate the required fired-heater duty, exit temperature and vaporization.

The number of theoretical trays for an atmospheric or vacuum column can be indirectly determined from the fractionation index calculated in the short-cut program. A correlation that directly relates fractionation index to theoretical trays has been proposed [*12*]. However, it represents an oversimplification .

Basic theory and available data indicate that the fractionation index should at least be related to theoretical stages and internal reflux. Work is in progress to devise a more consistent correlation. At the present, however, experience and good judgment are required to select the number of theoretical trays for computer simulations.

Steam rates to the residue and side-stream strippers are normally set in the range of 0.05 to 0.3 lb of steam per gallon of stripped product. Fig. 4 indicates the degree of stripping attainable for varying steam rates. The upper line refers to distillate side-strippers, and the lower line to residue stripping sections.

It will be noted that this stripping is significantly higher than previously reported [*13*], at which time the rigorous computer-based calculations now being described were not available, and several gross assumptions had to be made. The effects of underestimating the amount of product stripping from the residue section have been mentioned. Reboiled side-stream strippers are generally confined to heavy naphtha and kerosene products, in which trace amounts of water adversely affect specifications.

Program sets feed preheating

In order to minimize the diameter of the main fractionating tower and to improve heat recovery, it is desirable that the tower operate with several different internal reflux ratios. This is different from the conventional distillation column, in which reflux is normally only provided by returning a fraction of the condensed overhead vapors.

Fig. 5 shows pumparound and pumpback systems, which are the two most common methods of adjusting reflux ratios within the different sections of the tower. It can be seen that by adjusting the amount of heat removed from the circulating liquid streams, the reflux ratio in the section of the column below the liquid return tray can be controlled. This enables the degree of fractionation between adjacent products to be varied between certain limits.

The location and quantity of heat removal from pumparound or pumpback systems can present problems because they must be defined before the computer simulation can begin. Pumparound systems are generally favored to pumpback ones for most atmospheric and vacuum towers, because of the reduction in size of the side-stream strippers. In order to determine an initial estimate for pumparound heat-removal rates, hand calculations are performed to determine the heat available for reflux, given a feed vaporization. The short-cut distillation computer program can calculate approximate product temperatures and feed-vaporization curves to assist in this work.

It should be recognized that pumparound heat-removal duties and temperature levels are invariably related to the feed-preheat-train requirements. This means that the column design should proceed in parallel with feed preheat design in order to optimize heat recovery. To determine initial estimates for pumparound rates, the temperature drop over the circuit should range between 100°F and 150°F.

There are heat-exchanger network optimization programs that will determine an optimum feed-preheat system, given all the available hot streams from the column. The hot streams that would be uneconomical to use are rejected, and the optimum order of exchange is determined for the remainder.

These programs also generate capital cost data, operating costs, etc. However, it is possible to take some initial steps toward optimizing the heat-recovery system by manual methods. One approach relies on a graphical trial and error technique in which the feed heating curve is plotted, and the cooling curves of the various hot streams are backplotted against it. To facilitate the calculation, a desalter temperature of between 250°F and 260°F, and a minimum temperature approach of between 5°F and 10°F, are assumed. Several trial calculations are required before a satisfactory arrangement is arrived at.

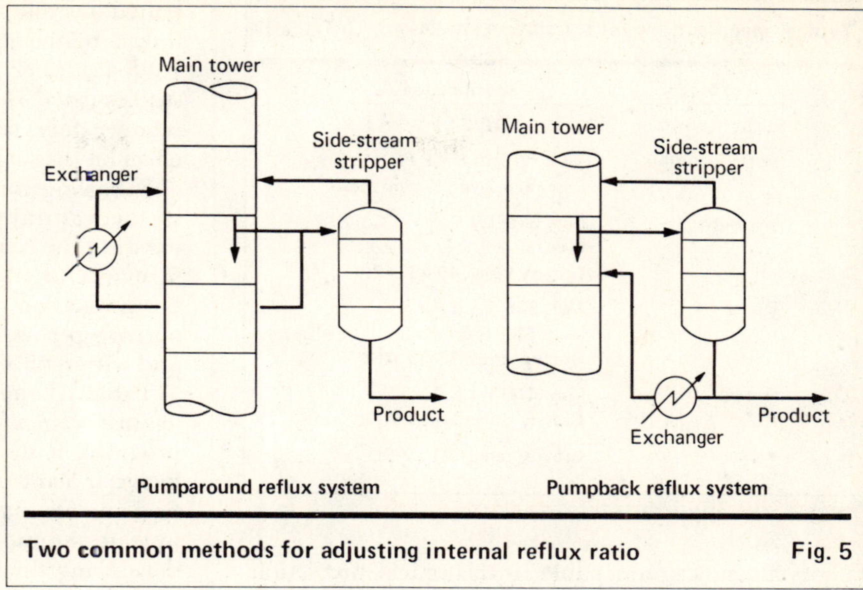

Two common methods for adjusting internal reflux ratio **Fig. 5**

An atmospheric unit similar to that shown in Fig. 3 was simulated on a rigorous computer-design program. From the results, it was possible to determine the changes necessary to optimize the design with respect to column diameter and heat recovery. The results indicated that the temperature drop over the lower pumparound was excessive. To amend this, the pumparound rate could have been increased. However, because the lower pumparound section was the most heavily loaded section of the column, raising the pumparound rate would have aggravated the problem.

One solution would have been to increase the upper pumparound cooling duty at the expense of the lower system. This had several effects other than evening out the internal loadings in the column. The separation between light and heavy gas oil was improved and the lower pumparound temperature raised, but the heat available at this temperature was reduced. In this case, the increase in upper-pumparound reflux duty was useful because it matched that in the crude preheat train, where the crude was to be desalted at 250°F.

A further improvement that might be made would be to provide an external pumparound circuit at the top of the column and eliminate cold reflux from the overhead accumulator. This allows heat recovery from the overheads without the problems of crude-oil leakage into finished products and the possibility of water condensation on the top trays.

As is apparent from the foregoing discussion, the initial computer calculations are only the first step, and three or four runs are usually required before the final design can be settled upon.

When the unit is to be designed to handle a number of different crudes or products, rigorous calculations are generally performed for only the lightest and heaviest crude cases, to size the top and bottom sections.

For the intermediate cases, the short-cut distillation computer program has proven effective. The designer can use these runs to determine whether a wide mid-distillate product or an intermediate crude feed could create problems. Changes in product yields could also affect the feed preheat train, and each case should be checked against the proposed network.

Revamping atmospheric and vacuum units

Unlike when he designs a new unit, the engineer does not always have the option of optimizing product yields or of selecting the best equipment configuration when he revamps a unit. Generally, the design margins for revamping are also much narrower.

The initial step when revamping is to evaluate product yields and properties for the crude feedstock. Usually, the unit must be capable of processing a range of feedstocks, either as individual crudes or as mixtures of several different crude oils. This evaluation is most easily performed with a short-cut distillation computer program, as has been described.

Because the target capacity for the unit is undefined at this stage, the calculations can be carried out on a unit basis, say 1,000 bbl/stream-day. For an existing refinery, product specifications are very clearly defined in order to meet local market requirements.

The short-cut computer program will generate product properties expressed in terms of their distillation curves, gravities, etc. Table II shows a typical set of such properties that have been used in a revamping. When other product properties, such as pour point or penetration, must be met, they must be correlated back to computer-derived properties.

The collection of plant operating data is of the utmost importance in determining the target increase in unit capacity and in checking the validity of computer simulation models. It is essential that data for the unit when operating at maximum capacity be included in the design basis. Although it is possible to predict tower flooding as a capacity limitation from computer-derived vapor and liquid tray loadings, other limiting

Typical specifications for simulation studies — **Table II**

Product	Specification
Light naphtha	Max. 300°F, E.P.
Heavy naphtha	Max. 385-400°F, E.P. Less than 2 vol. % pentanes
Kerosene	Min. 310°F, I.B.P. Max. 530°F, E.P. Gravity range 40-45°API.
Diesel	Min. 375°F, I.B.P. Max. 650°F, E.P. Gravity range 31-35°API
Gas oil	Min. 475°F, I.B.P. Gravity range 18-28°API.
Asphalt	Gravity range 5-10°API.

factors are not so amenable to theoretical prediction. An example would be the limitation of a fired heater's duty due to flame impingement on the heater tubes.

After the analysis of plant bottlenecks and product yields and properties has been completed, the target capacity-increase can be set. In some revamping studies, additional external factors may have an influence on the final plant capacity. These include a limited market for certain products, and capacity limitations in downstream units.

In addition to plant yield and operating data, it is necessary that design specifications and actual performance data of existing equipment be tabulated. These data would include column diameters and internals details, heat-exchanger specifications, pump capacity curves and fired-heater details. In order to produce an economic revamping design, it is necessary to reuse as much of the existing equipment as possible.

Finally, a survey of the existing plant's utility systems must be made. It is most important that there be adequate provision for the higher loads—on effluent treatment plants, cooling towers, electrical systems, storage facilities and boiler plants—occasioned by the increase in plant capacity. Although these factors often directly influence certain design decisions, such as using electric or steam drives, it is unusual for them to directly limit an increase in plant capacity.

Ready to develop the model

Once the foregoing design information has been obtained, the atmospheric- or vacuum-column simulation model can be developed. Initially, the computer model is set up to reproduce the observed plant operating conditions. A comparison between actual and computer-generated operating parameters serves to validate the simulation model and to check the accuracy of the feedstock analysis.

The number of theoretical trays to be used in the model is based on the number of actual trays in the tower. As a starting point, tray efficiencies in fractionating sections are taken as about 75%, and stripping trays as about 50%.

Pumparound section trays are not normally counted as contributing toward fractionation and are represented by one theoretical tray. Side-stream strippers linked to the main fractionating tower can be simulated accurately, using only two theoretical stages. Case studies have shown that for a given stream rate or reboiler duty, the use of more than two stages has little effect on the stripping efficiency.

Following the initial computer run, the number of theoretical trays and internal reflux ratios can be varied until the results are compatible with the actual performance of the tower. Experience has indicated that the simulation results can match actual product yields and properties within the accuracy of measurement and within ±5°F for tray temperatures.

Table III shows a comparison of actual plant data against values generated by a computer simulation program. It should be emphasized that a close match between plant and computer data is only possible if the feed analysis is accurate. The following case histories illustrate the use of a computer simulation program in developing flowschemes. Although the studies are by no means exhaustive, they do detail the more commonly adopted plant modifications.

Two ways of solving column top bottleneck

Sometimes, the top section of an atmospheric column is limiting. This often occurs when a feedstock lighter than the original crude oil must be processed. The problem may be solved in one of two ways:

First—for relatively small capacity increases, or for crude oils that have an abnormally high light-ends content—a crude flash-drum may be installed. The vapor product removed from the flash-drum may then be recombined with the overheads from the main tower before being condensed in the overheads system. This approach may also be adopted when the capacity of the atmospheric heater is limiting. In this case, the flashed vapor can be fed to the main tower above the flash zone. The computer program determines the amount of flash vapor produced and checks the revised main-tower tray loadings for flooding.

Second, for larger increases in crude capacity, it may be necessary to install a preflash tower. In this example, it was decided to take 15% of the crude charge overhead as a light straight-run gasoline. The column was set to operate with 15 theoretical trays and an overhead accumulator pressure of 35 psig, so that essentially all of the light ends in the feed are recovered in the overhead liquid product. The preheated crude feed entered the preflash tower at 425°F. If additional feed vaporization is required, this may be supplied by either adding a separate fired heater reboiler or a fired heater on the feed.

In this case, a reboiler was added, with the circulation rate set so that the outlet vaporization did not exceed 20%. For the 35° API crude feed, the reboiler outlet temperature is around 575°F. The amount of light product removed in this way is limited by the main-tower flash-zone temperature. For most crudes, the flash-zone temperature should not exceed 720°F, to prevent product contamination due to thermal cracking. As before, the computer model is used to design the preflash tower and to check the operation of the atmospheric column.

When a preflash unit is added before an atmospheric tower, the products from the tower will change. The main tower will produce a heavy naphtha overhead, and the separation characteristics for the other products will be altered. Most of the light ends in the crude feedstock are recovered in the light straight-run gasoline from the preflash tower. It is often necessary to alter the draw locations on the main tower, and retraying is usually required. The computer simulation program is used to predict the new tray loadings and to determine the location of side-draws. The circulating reflux systems will also have to be amended with the object of reusing as many of the existing exchangers as possible.

Computer vs. test run data, West Texas sour crude oil — Table III

Products	Gas and L.S.R. Gasoline	Naphtha	Kerosene	Diesel	Heavy gas oil	Residuum
Yield, vol % (actual)	18.1	7.0	15.5	15.6	8.4	35.4
(simulation)	18.1	7.0	15.5	15.7	8.4	35.3
Draw tray temperature, °F (actual)	221*	296	375	528	598	633
(simulation)	238*	295	368	531	602	628
5-95% Gap, °F (actual)	−11	−4	+8	−51		
(simulation)	−12	−7	−8	−59		
ASTM 50% point, °F (actual)	200	294	408	550	732	965
(simulation)	201	295	405	553	731	967
Gravity, °API (actual)	65.9	51.3	43.5	33.6	27.5	15.7
(simulation)	69.7	52.3	43.7	34.0	27.5	14.2

*Actual charge to crude unit contains more light ends than crude assay data used in simulation. Gap and gravity data apply to L.S.R. gasoline only.

When the middle-distillates section limits

Occasionally the middle-distillates section of the atmospheric tower may be limiting. In this case, it is sometimes possible to use excess capacity in the vacuum unit. In particular, a heavy gas-oil product that is being taken from the column may be left in the atmospheric residue and routed to the vacuum unit. This effectively reduces the vaporization necessary in the flash zone and so unloads the fractionating trays. It may be necessary to make some adjustments to the residue stripping trays. This technique also reduces the duty required from the atmospheric heater. However, this does assume sufficient excess capacity in the existing vacuum unit from which heavy gas oil may be taken as a light gas oil. The computer simulation model is used to determine the amount of heavy gas oil to be left in the atmospheric residue.

Limitations in vacuum units

Capacity limitations within the vacuum unit can be overcome with the installation of a vacuum flasher. In this case, a vacuum flasher is located after the atmospheric tower to remove light gas oils. Approximately 20% by volume of the atmospheric residue is taken overhead as a light vacuum gas oil, which may then be taken as a separate product after cooling to around 175°F. The off-gas from the vacuum flasher is fed to the top section of the vacuum column. This configuration has the advantage of using the existing ejector system on the vacuum tower to obtain subatmospheric pressures in the vacuum flasher.

In the computer simulation model for the vacuum flasher, the unit is usually set up with two or three theoretical trays, with either a single overhead product or two side-stream products. As with the preflash tower, the amount of lighter material that can be removed upstream of the main column is limited by the flash-zone temperature, which generally should not exceed 785°F.

Sometimes, capacity can be increased significantly when revamping vacuum units by switching from trayed to packed sections. Heat-transfer rates within a vacuum unit are as significant a factor as fractionation power. Packing or proprietary meshes can provide considerable capacity increases for a relatively small capital expenditure. The computer model for the vacuum tower is used to determine the internal vapor and liquid loadings. In general, tray efficiencies are quite low, and the model will contain no more than five or six theoretical trays.

The computer simulation of vacuum units can be accurate, but significant errors can result from inadequate feed data. Frequently, the crude assay TBP distillation and API gravity data will not extend beyond the 50 or 60 volume percent points. Consequently, the vacuum unit designer must extrapolate the available data into the higher volume percent ranges. Operating-plant data are helpful in validating the extrapolation.

Another area that can present some uncertainty is the amounts of inerts and light ends, from thermal cracking, present in the unit. These factors have been found to be a function of charge rate and feed temperature. As a first approximation, the light components can be taken as 0.33 volume percent of the reduced crude charge. Gases from cracking can be based on the generation of 0.8 lb gas/bbl at 900°F and a 50% reduction for each 25°F lowering of the flash-zone temperature. Air leakage can be estimated as 6 lb/h for a 1,000-bbl/stream-d unit, with leakages for larger units being based on the square root of the proportional increase over 1,000 bbl/stream-d.

Heat-transfer problems

Apart from bottlenecks in the fractionating equipment, it is frequently the heat-transfer equipment that is limiting. Fired heaters are often a problem, and a detailed analysis is required for an optimum solution. Some of the possible techniques for increasing fired-heater capacities include adding a new convection section, increasing the number of tube passes, replacing existing burners with new high-intensity burners, and converting from fuel-oil to fuel-gas firing. Bottlenecks in the heat-exchanger train can be overcome by the addition of new exchangers in parallel with existing units, or by the installation of a completely new exchanger train to operate in parallel with the old train.

Because fuel costs have substantially increased, there is now more justification for additional surface area in the preheat train.

Approximate vs. rigorous methods

Despite the elimination of hand calculation in basic design, such calculation is essential for checking the validity of computer results, and for determining the direction of further case studies. Whatever the computer method, it is good practice to manually check some of the output results.

At the very least, an overall heat-and-mass balance should be performed. When determining which column parameters to vary for additional case studies, manual methods can be useful. An example would be when the degree of fractionation predicted by the computer is inadequate and it is necessary to increase the internal reflux. At this time, Packie's F-factor charts may be used [14], although a more useful correlation is being developed. The value of the F-factor for theoretical trays will be different from that predicted for real trays. However, the shapes of the curves are similar and may be used to extrapolate to a new value corresponding to the desired degree of fractionation. From the F-factor determined in this manner, the new value for the internal reflux may be calculated.

Computerized short-cut distillation programs are extremely valuable in conjunction with either rigorous design methods or operating-plant data. For the design of a new crude or vacuum unit, shortcut methods may be used to predict product properties and fractionation requirements. Although most shortcut methods do not provide enthalpy balances around a unit, they do produce sufficient data to determine the basic column configuration. The accuracy of these shortcut methods for predicting product properties is generally good. However, discrepancies do occur, particularly in the prediction of front- and back-end distillation properties.

Estimates of product yields and properties from a given crude oil in conjunction with unit capacity data are essential elements in the preparation of refinery economic evaluations and planning studies. Rigorous and short-cut computer simulations to predict the operation of atmospheric and vacuum towers can greatly improve the accuracy of the product estimates, thus improving the quality of the results of these studies.

The processing capacity of a given atmospheric or vacuum unit will vary slightly from one crude-oil feedstock to another. Rigorous computer simulation-programs and plant data permit the accurate analysis of capacity for a given combination of unit and feedstock. The results of this analysis are separation characteristics that may be used in short-cut computer distillation programs. This means that the short-cut program results will reflect actual yields from that unit, rather than a hypothetical breakdown from the crude assay.

After validation of the yield-results against actual plant data, the model may be used in the refinery linear-programming models as a realistic capacity restraint. This improvement in yield information is extremely valuable in the evaluation of crude-oil purchases and scheduling considerations.

References

1. Nelson, W.L., "Petroleum Refinery Engineering," McGraw-Hill, 1968, p. 95.
2. Dooley, J.E., Hirsch, D.E., Thompson, C.J. and Ward, C.C., Analyzing Heavy Ends of Crude, *Hydrocarbon Proc.,* Vol. 52, No. 9 (Sept. 1973), pp. 123-130.
3. Walsh, R.P. and Mortimer, J.V., New Way to Test Product Quality, *Hydrocarbon Proc.,* Vol. 51, No. 9 (Sept. 1971), pp. 153-158.
4. Nelson, W.L., What are Average Distillation Curves of Crude Oils?, *Oil and Gas J.,* Oct. 14, 1968, pp. 126-127.
5. Nelson, W.L., Does Crude Boil at 1400°F?, *Oil and Gas J.,* Mar. 25, 1968, pp. 125-126.
6. Bardone, E., Mori, P., and Ferroni, E., Vacuum Design vs. Distillation Tests, *Hydrocarbon Proc.,* Vol. 59, No. 12 (Dec. 1973), pp. 71-72.
7. Documentation of the Basis for Selection of the Contents of Chapter 3 in the Technical Data Book—Petroleum Refining, Documentation Report No. 3-66, American Petroleum Institute.
8. Hariu, O.H. and Sage, R.C., Crude Split Figured by Computer, *Hydrocarbon Proc.,* Vol. 49, No. 4 (Apr. 1969), pp. 143-148.
9. Cavett, R.H., Physical Data for Distillation Calculations—Vapor-Liquid Equilibria, 27th Mid-year meeting of the American Petroleum Institute, Div. of Refining, Vol. 42, No. 3, 1962, pp. 351-366.
10. Geddes, R.L., A General Index of Fractional Distillation Power for Hydrocarbon Mixtures, AIChE J., (Dec., 1958), pp. 389-392.
11. Gilbert, R.J.H., Leather, J., and Ellis, J.F.G., The Application of the Geddes Fractionation Index to Crude Distillation Units, *AIChE J.,* Vol. 12, No. 3, May 1966, pp. 432-437.
12. Jakob, R.R., Estimate Number of Crude Trays, *Hydrocarbon Proc.,* Vol. 41, No. 9, May 1971, pp. 149-152.
13. Nelson, W.L., Amount of Steam Required for Stripping, *Oil and Gas J.,* Mar. 2, 1944, pp. 72-75.
14. Packie, J.W., Distillation Equipment in the Oil Refining Industry, *AIChE Transactions,* Vol. 37, 1941, pp. 51-78.

The authors

Keith J. Ritchey is a Senior Process/Project Engineer in the Engineering Div. of Gulf Oil Corp. (P. O. Box 1357, Houston, TX 77001). His responsibilities include process design and project engineering of chemical plants and refineries. His previous experience has been in economics and planning, plant operation and technical computer applications. He holds a B.S. in chemical engineering from the University of Texas, and an M.S. in industrial engineering from the University of Houston.

Frank B. Canfield is Chairman of the Board of ChemShare Corp. (2500 Transco Tower, Houston, TX 77027). Previously, he was Chairman of the Dept. of Chemical Engineering at the University of Oklahoma, where he was a professor for 12 years. He holds a B.S. in chemical engineering from the University of Arkansas and a Ph.D. from Rice University, is a member of AIChE, ACS, and is listed in "Who's Who in America."

Timothy B. Challand is a Senior Process Engineer responsible for all process engineering consulting for ChemDesign Inc. (6300 W. Loop South, Suite 230, Bellaire, TX 77401). Until recently, he held a similar position with ChemShare Corp. Previously, he was a process engineer in Humphreys and Glasgow Ltd.'s oil and petrochemical group, and subsequently a lead process engineer for a large polyester project for MonMax—H & G Services Ltd., Humphreys and Glasgow's Canadian affiliate. He graduated with a B. Sc. in chemical engineering from the University of Birmingham, England, and is a Chartered Engineer, U.K., a registered engineer in the Province of Quebec, and a member of IChE, U.K., and AIChE.

D. Calculator programs

Streamline flash computations with calculator program

Flash vaporization of a continuous feed is the starting point for countless distillation and fractionation operations. This convenient program performs flash calculations on mixtures containing up to nine components.

Sohrab Mansouri, *University of Michigan*

☐ The simplest continuous-distillation technique is the single-stage flashing of a feed liquid. Once the temperature and pressure within the flash tank have been set, the equilibrium phase concentrations in the liquid and gaseous product streams can then be determined.

This program, written for the Hewlett-Packard HP-67 or 97 calculators, performs flash calculations for multicomponent mixtures containing up to nine components. In order to define the equilibrium properties of the mixture, the user must know the K-values of the components at the flash conditions.

Calculation method

Assume that F moles/h of an n-component stream are introduced as feed into the flash vaporization tank shown in the figure. The resulting vapor and liquid streams are withdrawn at the rates of V and L moles/h. The mole fractions of the components in the feed, vapor and liquid streams are designated as z_i, y_i and x_i, respectively (where $i = 1, 2, \ldots n$).

Assuming a vapor-liquid equilibrium and steady-state operation, we have the following series of algebraic relationships:

Overall balance: $F = V + L$
Component balance: $z_i F = x_i L + y_i V$
Equilibrium: $K_i = y_i / x_i$

K_i is the equilibrium constant for the ith component, at the temperature and pressure in the tank. (The K-values can be found in Ref. [*2*].) From the above equations, and the fact that

$$\sum_{i=1}^{n} x_i = \sum_{i=1}^{n} y_i = 1$$

it can be shown that:

$$\sum_{i=1}^{n} \frac{z_i(K_i - 1)}{V(K_i - 1) + F} = 0 \qquad (1)$$

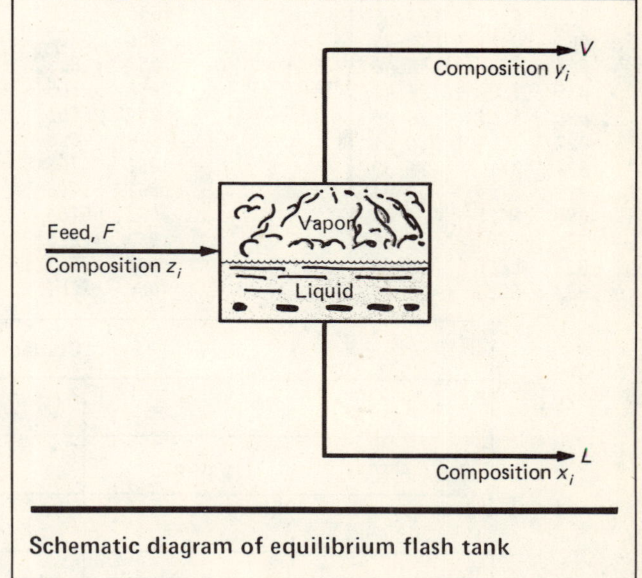

Schematic diagram of equilibrium flash tank

Newton's method will be used to solve this equation for V. The algorithm for this iterative method can be written as:

$$V_{k+1} = V_k - \frac{f(V_k)}{f'(V_k)} \qquad (2)$$

where
$$f(V_k) = \sum_{i=1}^{n} \frac{z_i(K_i - 1)}{V_k(K_i - 1) + F} \qquad (3)$$

$$f'(V_k) = \sum_{i=1}^{n} \frac{z_i(K_i - 1)^2}{[V_k(K_i - 1) + F]^2} \qquad (4)$$

A reasonable first estimate for V is needed to start the iteration.

We will use the following test criterion for terminating the iteration procedure:

Originally published August 27, 1979

Program listing and summary of calculation logic **Table I**

Step	Key entry	Key code	Comments
001	*LBL1	21 01	Initalize and store data
002	CLRG	16-53	
003	2	02	
004	5	05	
005	STOI	35 46	
006	STO4	35 04	
007	*LBL0	21 00	
008	R/S	51	
009	DSZI	16 25 46	
010	1	01	
011	-	-45	
012	STOi	35 45	
013	DSZI	16 25 46	
014	R/S	51	
015	STOi	35 45	
016	GTO0	22 00	Start calculation
017	*LBL2	21 02	
018	RCLI	36 46	
019	STO5	35 05	
020	*LBL4	21 04	Calculate $f(V_k)$ from Eq. 3
021	RCLi	36 45	
022	GSB5	23 05	
023	×	-35	
024	ST+6	35-55 06	
025	ISZI	16 26 46	
026	RCLI	36 46	
027	RCL4	36 04	
028	X≠Y?	16-32	
029	GTO4	22 04	
030	RCL5	36 05	(See next column)
031	STOI	35 46	
032	*LBL6	21 06	
033	RCLi	36 45	

Step	Key entry	Key code	Comments
034	GSB5	23 05	Calculate $f'(V_k)$ from Eq. (4)
035	X^2	53	
036	×	-35	
037	ST+0	35-55 00	
038	ISZI	16 26 46	
039	RCLI	36 46	
040	RCL4	36 04	
041	X≠Y?	16-32	
042	GTO6	22 06	
043	RCL6	36 06	Calculate V_{k+1} from Eq. 2
044	RCL0	36 00	
045	÷	-24	
046	STO0	35 00	
047	RCL2	36 02	
048	+	-55	
049	STO2	35 02	
050	RCL0	36 00	Check for convergence using Eq. 5
051	RCL2	36 02	
052	÷	-24	
053	ABS	16 31	
054	RCL3	36 03	
055	X>Y?	16-34	
056	GTO7	22 07	
057	*LBL3	21 03	Set initials for another iteration
058	RCL5	36 05	
059	STOI	35 46	
060	CLX	-51	
061	STO6	35 06	
062	STO0	35 00	
063	GTO4	22 04	
064	*LBL5	21 05	(See next column)
065	ISZI	16 26 46	
066	RCLi	36 45	

Step	Key entry	Key code	Comments
067	RCLi	36 45	Subroutine for $f(V_k)$ and $f'(V_k)$
068	RCL2	36 02	
069	×	-35	
070	RCL1	36 01	
071	+	-55	
072	÷	-24	
073	RTN	24	
074	*LBL7	21 07	Display V
075	RCL2	36 02	
076	R/S	51	
077	*LBL8	21 08	Calculate and display x_i
078	DSZI	16 25 46	
079	RCLi	36 45	
080	1	01	
081	+	-55	
082	RCLi	36 45	
083	RCL2	36 02	
084	×	-35	
085	RCL1	36 01	
086	+	-55	
087	1/X	52	
088	RCL1	36 01	
089	×	-35	
090	DSZI	16 25 46	
091	RCLi	36 45	
092	×	-35	
093	R/S	51	
094	×	-35	Display y_i
095	R/S	51	
096	GTO8	22 08	
097	R/S	51	

Content of registers

0 Used	1 F	2 V	3 ϵ	4 Used
5 Used	6 Used	7 z_9	8 K_9-1	9 z_8
S0 K_8-1	S1 z_7	S2 K_7-1	S3 z_6	S4 K_6-1
S5 z_5	S6 K_5-1	S7 z_4	S8 K_4-1	S9 z_3
A K_3-1	B z_2	C K_2-1	D z_1	E K_1-1
I Used	—	—	—	—

$$\left|\frac{V_{k+1}-V_k}{V_{k+1}}\right| \leqslant \epsilon \qquad (5)$$

Here ϵ is a small positive number. For $\epsilon = 10^{-N}$, the final value of V should be accurate to approximately N significant figures.

Note that Eq. 1 has n roots (n is number of components). Only one root lying within the interval $0 < V < F$ is of interest to us.

If a bad guess value is chosen, the Newton iteration procedure may converge to the other, unwanted roots instead of the desired solution. In such a case, make another estimate of V and start the iteration again (see the example below).

After computing the vapor flowrate V, we can find the values of x_i and y_i, the mole fractions of a component in the equilibrium phases, from the following equations:

$$x_i = \frac{z_i F}{V(K_i - 1) + F}$$

$$y_i = x_i K_i$$

Users' instructions for flash-calculation program **Table II**

Input procedure

Step	Instruction	Input data	Keys
1	Key in program		
2	Initialize		GSB 1
3	Store variables		
	Feedrate	F	STO 1
	Initial guess	V	STO 2
	Tolerance	ϵ	STO 3
		K_1	R/S
		z_1	R/S
		K_2	R/S
		z_2	R/S
		•	•
		•	•
		K_n	R/S
		z_n	R/S
4	Run		GSB 2
If the initial guess value for V is unsuccessful, choose another value, then proceed as follows:			
5	Enter another initial guess for V		STO 2
6	Run		GSB 3

Output procedure

Step	Instruction	Keys	Output data
1	Computation stops		V
2	Mole fraction in liquid	R/S	x_1
3	Mole fraction in vapor	R/S	y_1
4		R/S	x_2
5		R/S	y_2
•		•	•
•		•	•
•		R/S	x_n
		R/S	y_n

Using the program

The program (Tables I and II) reads values for F, the K_i, and the z_i as data.

A small positive number—the tolerance that is given by ϵ—and the first estimate of V should also be read in.

When computation ceases, the final value of V is displayed. Pressing the R/S button causes the calculator to display the mole fraction of component 1 in the liquid product, x_1. Pressing the R/S button again causes the calculator to display the mole fraction of this component in the vapor, y_1. Pressing the button repeatedly displays x_2, y_2, and so on, until the last component mole fraction is displayed.

Computation time is about 1 to 5 min, depending on the number of components and the magnitude of the test criterion ϵ.

Options. If the first estimate of V fails, or if different values of F and ϵ are needed, just key in the new values in the proper registers (register 1 for F, register 2 for V, and register 3 for ϵ) and press GSB 3.

An example

The composition of the feed to a natural-gas liquefaction plant [3] is given below. The feed will be flashed at 600 psia and 20°F. If the flowrate is 1,000 moles/h, determine the flowrates of the liquid and vapor streams and their compositions.

Component	i	K_i	z_i
Carbon dioxide	1	0.90	0.0112
Methane	2	2.70	0.8957
Ethane	3	0.38	0.0526
Propane	4	0.098	0.0197
Isobutane	5	0.038	0.0068
n-Butane	6	0.024	0.0047
Pentanes	7	0.0075	0.0038
Hexanes	8	0.0019	0.0031
Heptanes and heavier	9	0.0007	0.0024
			1.0000

Assume an initial guess value of $V = 800$ moles/h. To terminate the calculation, use a convergence test such that $\epsilon = 10^{-4}$.

The results of this attempt give $V = 1{,}511.1$ moles/h. Since V cannot be larger than 1,000, we will try a second guess value of 950. When this is used in the calculation, we obtain an acceptable result of $V = 959.17$ moles/h. The compositions of the liquid and vapor streams are as follows:

Component i	x_i	y_i
1	0.0124	0.0111
2	0.3405	0.9193
3	0.1298	0.0493
4	0.1461	0.0143
5	0.0880	0.0033
6	0.0736	0.0018
7	0.0791	0.0006
8	0.0727	0.0001
9	0.0578	4.05×10^{-5}

References

1. Carnahan, B., Luther, H. A., and Wilkes, J. O., "Applied Numerical Methods," John Wiley and Sons, 1969.
2. Katz, D. L., et al., "Handbook of Natural Gas Engineering," McGraw-Hill, New York, 1959.
3. Woicik, J. F., Equilibrium-flash calculation, *Chem. Eng.*, Aug. 16, 1976, p. 89.

The author

Sohrab Mansouri, 1230 Hubbard St., Stanley 2203, Ann Arbor, MI 48109, is completing work toward a Ph.D. in bioengineering at the University of Michigan. He took his M.S. in chemical engineering at the same university and a B.S. in chemical engineering at Tehran University in Iran. He is a member of AIChE.

Calculator program for sour-water-stripper design

Hand calculators take the drudgery out of lengthy design mathematics. This program for the Hewlett-Packard HP97 determines the number of trays required to reduce the ammonia and hydrogen sulfide content of a sour-water stream within given specifications.

Norman H. Wild, Davy International Ltd.

☐ Designing a sour-water stripper can be a laborious procedure, involving lengthy tray-to-tray calculations and trial-and-error methods. This article outlines the functions of a sour-water stripper and the theory of the design calculations, and presents the HP97 program that has been devised. With this program, it is possible to punch in the basic design data in a few seconds and then have the calculator print out the number of trays required a few minutes later. A worked example is given and explained.

Sour-water strippers

Sour-water strippers are widely used in oil refineries and petrochemical plants to clean up fouled water streams prior to sending these to rivers or reusing them in the plant. Such waters commonly arise from the washing of reactor products that have been hydrogen treated in hydrodesulfurization or hydrocracking operations. The waters usually contain ammonia (NH_3) and hydrogen sulfide (H_2S). Other components present may include phenols and cyanides, but removal of these is beyond the scope of this article.

A stripper may consist merely of a packed or trayed column down which the sour water is passed countercurrently to "open" steam. The excess steam and foul vapors from the top of such a column may then be burned. Legislation restricting atmospheric pollution, and especially the high cost of energy, has rendered such a crude arrangement unsuitable for new designs.

To conserve the steam condensate in pure form, it is preferable to energize the stripper, using a reboiler instead of "open" steam. Also, passage of the hot overhead vapors directly to a furnace is likely to have an adverse effect on combustion there, due to the diluent steam present. Thus an overhead condenser will usually be provided to remove the excess steam, even though the resultant reflux is rich in NH_3 and H_2S and must be returned to the stripper, thereby increasing the stripping load.

The reflux may be returned to a separate rectifying section above the main stripping column. This is theoretically more efficient, but it complicates the design somewhat since the column diameter for the rectifying zone should be reduced in accordance with the reduced liquid loading. Also, severe corrosion and foaming problems have sometimes been reported in rectifying sections where NH_3 and H_2S concentrations are necessarily high.

In general, it is preferable to return the reflux directly on to the feed tray, which is then designated Tray 1, the top tray.

The feed is conveniently preheated by external heat exchange with the bottoms before entering the stripper.

A stripper designed along these lines is shown in Fig. 1, with sample data written in. This forms the basis for the calculation procedure of this article. Naturally, the procedure may be modified to agree with other basic types of stripper design.

Background to calculations

The Beychok/Van Krevlin correlations [*1* (p. 163)], covering ammonia / hydrogen-sulfide / steam / water equilibria appear to provide the most practical basis for designing sour-water strippers. Researchers for the American Petroleum Institute [*2*] have more recently produced other correlations but these are difficult to use for an initial design. However they could possibly be applied at a later stage for checking equilibria at critical points.

Commercially-available computer programs use the Beychok method. However, those examined by the author did not calculate the number of trays required for stripping to a specification (which is the designer's problem) but merely tested the effect of a given number of trays.

By using Beychok's suggestions (p. 173) for simplifying the equations to avoid using trial-and-error charts, a calculator program was devised for determining the

Originally published February 12, 1979

Nomenclature

Data Constants (given and directly-derived data)

- a NH_3 in feed stream, lb/h
- b H_2S in feed stream, lb/h
- c H_2O in feed stream, lb/h
- d NH_3 in bottoms product, ppm (specified)
- e H_2S in bottoms product, ppm (specified)
- f H_2O in bottoms product, ppm (derived)
- h NH_3 in tail gas, lb/h (derived)
- j H_2S in tail gas, lb/h (derived)
- k H_2O in tail gas, lb/h (derived)
- p NH_3 in bottoms product, lb/h
- q H_2S in bottoms product, lb/h
- r H_2O in bottoms product, lb/h
- *stm* H_2O in stripper overhead vapor, lb/h (i.e., stripping-stream rate)
- P_n Pressure above Tray "n", psia
- P_{tg} Pressure of tail gas above reflux, psia
- T_{tg} Temperature of tail gas/reflux in reflux drum, °F
- T_{fd} Temperature of feed entering stripper, °F (after external preheat)
- T_{bm} Temperature of bottoms leaving stripper reboiler, °F
- ΔP_{tt} Pressure drop across one theoretical tray, psi

Data Variables (indirectly derived data)

- L_r H_2O liquid content of reflux, lb/h
- L_n H_2O content of liquid leaving Tray "n", lb/h
- V_r H_2O content of vapor leaving reflux drum, lb/h (note $V_r = k$)
- V_n H_2O content of vapor leaving Tray "n", lb/h (note $V_1 = stm$)
- x_r NH_3 content of liquid leaving reflux drum, lb/h
- x_n NH_3 content of vapor leaving Tray "n", lb/h
- y_r H_2S content of liquid leaving reflux drum, lb/h
- y_n H_2S content of vapor leaving Tray "n", lb/h
- A_r NH_3 content of liquid leaving reflux drum, ppm
- A_n NH_3 content of liquid leaving Tray "n", ppm
- S_r H_2S content of liquid leaving reflux drum, ppm
- S_n H_2S content of liquid leaving Tray "n", ppm
- AP_r NH_3 partial pressure above reflux, psia
- AP_n NH_3 partial pressure above Tray "n", psia
- SP_r H_2S partial pressure above reflux, psia
- SP_n H_2S partial pressure above Tray "n", psia
- WP_r H_2O partial pressure above reflux, psia
- WP_n H_2O partial pressure above Tray "n", psia
- T_n Temperature of liquid leaving Tray "n", °F

Miscellaneous

- K, C, H_o Temperature-dependent variables used in Beychok/Van Krevlin equations

number of theoretical stripping trays required to meet given specifications under given conditions.

The program was originally stored on one magnetic card (program C) to cover approximately 200 steps (the HP97 has a program memory consisting of 224 steps). Two other cards, A & B, were added later so that the preparatory mass-and-heat-balance calculations could be done automatically, ahead of the main tray-to-tray calculations.

Although the programs are designed for the Hewlett-Packard printing calculator HP97, this machine is interchangeable in its use of program cards with the HP67 model. Therefore, the HP67 could also be used, though it is somewhat less convenient. For example, the number of trays calculated would be held in the I register, R_I, instead of being printed out.

The program calculates the number of trays required directly. It also gives the reboiler duty. It is possible to optimize the design by repeated calculator runs, which, of course, cost nothing if one has ready access to a hand calculator. The final design may be checked with a commercial computer program in a single run. This will also provide other data, such as column size and detailed tray loadings, as required.

Design correlations

The method of tray-to-tray design calculation to determine the number of stripping trays required is a development of Beychok's manual procedure (p. 182). Since the known conditions are in the reflux drum and at the top of the stripper, the partial pressures for NH_3 and H_2S above liquid at a known temperature are easily found. The Beychok/Van Krevlin equations then give the NH_3 and H_2S concentrations in the equilibrium liquid phase as parts per million. By a mass balance, the partial pressures for NH_3 and H_2S in the vapor from the next tray are found as, in turn, are the concentrations in the liquid. Thus it is possible to work

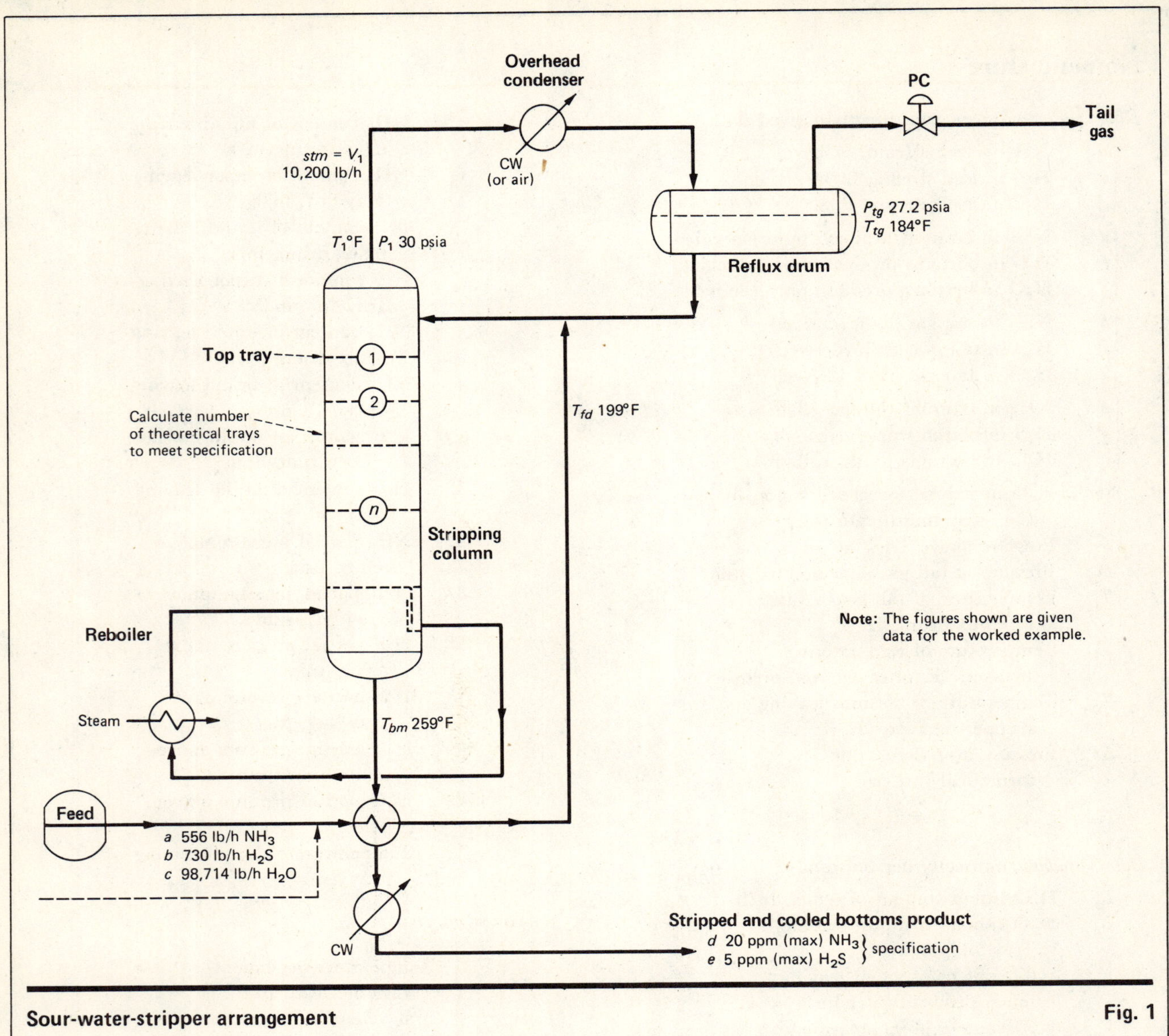

Sour-water-stripper arrangement **Fig. 1**

down the stripper until sufficient tray equilibria have been calculated to achieve the specified concentrations of NH_3 and H_2S in the bottoms liquid.

Two points concerning the correlations should be noted. First, Beychok proposes (p. 173) that his equations may be simplified, provided that the H_2S concentration is less than about 5,000 ppm. (This simplification is very suitable for application in a calculator program and avoids the use of Beychok's charts in the normal trial-and-error manner.) Second, the Van Krevlin work is only held to be valid for molar ratios of NH_3 to H_2S exceeding 1.5.

For practical use, both of these restrictions are relatively unimportant, since by the time the feed and reflux have passed through about two theoretical trays such limitations will have been removed anyway. The H_2S is relatively so volatile that the molar ratio of 1.5 for NH_3 to H_2S will be rapidly exceeded; also, the concentration of H_2S will be quickly brought below 5,000 ppm.

In cases of doubt—where few trays are being employed in a design; where the feed is very rich; or where an inadequate safety factor is being used to cover for variations in feed composition—the equilibria occurring at the top of the stripper should be checked out by independent correlations, such as those derived by Miles and Wilson [*2*]. These correlations, although derived more recently than Beychok's, are, unfortunately, very difficult to use when attempting to obtain the liquid composition from that of the vapor—instead of the reverse.

(Text continues on p. 157)

Program

The sour-water-stripper program for the HP97 consists of three magnetic cards, A, B and C. With these three cards, plus the basic design specifications and the procedure outlined in this article, it is possible to obtain the number of trays required to meet the design specifications within minutes.

Magnetic program card A

Locations	Keys	Codes	Comments
000–006	*LBLE PREG P⇌S PREG P⇌S RTN	21 15 16-13 16-51 16-13 16-51 24	Label "E": This section prints out the stored data so that the operator can check that they have been correctly entered—see worked example.
007–014	*LBLA EEX 6 RCL3 – RCL4 – STO5	21 11 –23 06 36 03 –45 36 04 –45 35 05	Label "A": This heads the calculation run. The water content of the bottoms is calculated as ppm H_2O, and the figure is stored in PR_5 (Primary Register 5).
015–036	*LBL0 RCLD 3 8 3 + 7 0 0 7 X⇌Y ÷ 1 4 . 4 6 5 X⇌Y – ϵ^x STOC	21 00 36 14 03 08 03 –55 07 00 00 07 –41 –24 01 04 –62 04 06 05 –41 –45 33 35 13	Label "0": Eq. (5) is applied to calculate WP_r, the water vapor pressure (psia) above the reflux, corresponding to T_{tg}°F, the reflux temperature stored in register R_D. For this purpose, the reflux is assumed to have the same vapor pressure as water.
037–073	*LBL1 1 8 RCLC x 3 4 ÷ RCL8 RCLC – STOA ÷ 2 RCL0 x RCL1 + 2 RCL3 x RCL4 + RCL5 ÷ STO9 RCL2 x – x 1 RCL9 R↑ x – ÷ STO9	21 01 01 08 36 13 –35 03 04 –24 36 08 36 13 –45 35 11 –24 02 36 00 –35 36 01 –55 02 36 03 –35 36 04 –55 36 05 –24 35 09 36 02 –35 –45 –35 01 36 09 16–31 –35 –45 –24 35 09	Label "1": Eq. (13) is used to calculate k, the quantity of water in the tail gas (lb/h).
074–101	*LBL2 RCL0 PRTX RCL1 PRTX RCL2 PRTX SPC RCL9 – RCL5 ÷ STO6 RCL3 x STOI PRTX RCL6 RCL4 x STO4 PRTX RCL2 RCL9 – STO5 PRTX SPC	21 02 36 00 –14 36 01 –14 36 02 –14 16-11 36 09 –45 36 05 –24 35 06 36 03 –35 35 46 –14 36 06 36 04 –35 35 04 –14 36 02 36 09 –45 35 05 –14 16-11	Label "2": Eq. (15) and (16) are applied, and the feed and bottoms component quantities are printed out (lb/h).
102–116	*LBL3 RCL4 RCL1 X⇌Y – RCL0 RCL3 RCL6 x – STO6 PRTX X⇌Y STO7 PRTX	21 03 36 04 36 01 –41 –45 36 00 36 03 36 06 –35 –45 35 06 –14 –41 35 07 –14	Label "3": The NH_3 and H_2S quantities, (h and j, lb/h), in the tail gas, are found.
117–113	*LBL4 ÷ 2 x 1 + 1/X RCLA x STOB RCLA X⇌Y – STOA RCL9 PRTX SPC	21 04 –24 02 –35 01 –55 52 36 11 –35 35 12 36 11 –41 –45 35 11 36 09 –14 16-11	Labels "4" and "5": The partial pressures of NH_3, H_2S and H_2O above the reflux are found (AP_r, SP_r, WP_r psia).
134–141	*LBL5 RCLA PRTX RCLB PRTX RCLC PRTX SPC	21 05 36 11 –14 36 12 –14 36 13 –14 16-11	Same as above
142–144	*LBL6 RCLI STO3	21 06 36 46 35 03	Label "6": The ammonia content of the bottoms (p lb/h) is transferred from register R_I to PR_3.
145–160	*LBL7 1 2 1 EEX 3 ENT↑ 1 . 0 1 6 3 RCLD Y^x ÷	21 07 01 02 01 –23 03 –21 01 –62 00 01 06 03 36 14 31 –24	Label "7": Eq. (1) is applied to calculate K.
161–185	*LBL8 1 . 2 7 RCLD 3 2 – . 6 4 8 Y^x Y^x 2 . 8 4 EEX CHS 7 x X⇌Y ÷	21 08 01 –62 02 07 36 14 03 02 –45 –62 06 04 08 31 31 02 –62 08 04 –23 –22 07 –35 –41 –24	Label "8": Eq. (2) is applied to calculate C.
186–193	*LBL9 RCLA RCLB x X⇌Y ÷ √X STOE	21 09 36 11 36 12 –35 –41 –24 54 35 15	Label "9": Eq. (3) is applied to calculate S_r (ppm H_2S).
194–206	*LBLa 2 ÷ RCLA R↑ x + STOI PRTX RCLE PRTX SPC SPC	21 16 11 02 –24 36 11 16-31 –35 –55 35 46 –14 36 15 –14 16-11 16-11	Label "a": Eq. (4) is applied to calculate A_r (ppm NH_3).
207–217	*LBLb P⇌S RCL0 P⇌S RCL9 – P⇌S STO4 P⇌S RTN R/S	21 16 12 16-51 36 00 16-51 36 09 –45 16-51 35 04 16-51 24 51	Label "b": The amount of water in the reflux (L_r lb/h) is calculated from (stm – k) and stored in SR_4 (Secondary Register 4).

Magnetic program card B

Locations	Keys	Codes	Comments
001–026	*LBLA P⇌S RCL4 P⇌S ENT↑ ENT↑ EEX 6 RCLI – RCLE – ÷ STO8 RCLI x STOA PRTX RCL8 RCLE x STOB PRTX R↑ PRTX SPC	21 11 16-51 36 04 16-51 –21 –21 –23 06 36 46 –45 36 15 –45 –24 35 08 36 46 –35 35 11 –14 36 08 36 15 –35 35 12 –14 16-31 –14 16-11	Label "A": This heads the calculation run. The values for NH_3 and H_2S quantities (lb/h) in the reflux are calculated and stored in registers R_A and R_B and printed out.
027–044	*LBL0 RCL7 RCLB + RCL6 RCLA + P⇌S STO5 PRTX 2 x X⇌Y STO6 PRTX RCL0 PRTX SPC	21 00 36 07 36 12 –55 36 06 36 11 –55 16-51 35 05 –14 02 –35 –41 35 06 –14 36 00 –14 16-11	Label "0": The overhead components' quantities (lb/h) are calculated by addition of tail gas to reflux. These are stored and printed out.
045–060	*LBL1 1 . 8 9 x ENT↑ R↓ + + ÷ RCL1 STO8 x STOC PRTX	21 01 01 –62 08 09 –35 –21 –31 –55 –55 –24 36 01 35 08 –35 35 13 –14	Label "1": They are used in Eq. (10) to determine WP_1 (psia), the water-vapor partial pressure in the overheads.
061–083	*LBL2 7 0 0 7 ENT↑ 1 4 . 4 6 5 RCLC LN – ÷ 3 8 3 – STOI PRTX SPC	21 02 07 00 00 07 –21 01 04 –62 04 06 05 36 13 32 –45 –24 03 08 03 –45 35 46 –14 16-11	Label "2": Eq. (5) is used to calculate T_1°F, the temperature of the top tray (Tray 1) liquid, corresponding to WP_1. This is stored in register R_I.
084–101	*LBL3 P⇌S RCL2 RCL1 . 2 x + RCL0 . 5 x + RCLI P⇌S RCL2 – x	21 03 16-51 36 02 36 01 –62 02 –35 –55 36 00 –62 05 –35 –55 36 46 16-51 36 02 –45 –35	Label "3": The sensible heat (Btu/h) absorbed by the feed in heating from its entry temperature, T_{fd}°F, to the Tray 1 temperature, T_1°F, is calculated.
102–118	*LBL4 RCL4 RCLA . 5 x + RCLB . 2 x + RCLI RCLD – x +	21 04 36 04 36 11 –62 05 –35 –55 36 12 –62 02 –35 –55 36 46 36 14 –45 –35 –55	Label "4": The sensible heat (Btu/h) absorbed in heating the entering reflux from T_{tg} to T_1°F is similarly calculated and added on to the previous value for the feed. For sensible-heat calculations the specific heat for NH_3 in solution is taken as 0.5, and for H_2S as 0.2, while for water it is 1.0.
119–134	*LBL5 RCL5 3 5 0 x STOE . 7 x RCL6 2 0 0 x +	21 05 36 05 03 05 00 –35 35 15 –62 07 –35 36 06 02 00 00 –35 –55	Label "5": The heat required to vaporize all the H_2S (at a latent-heat value of 200) and 70% of the NH_3 (at a latent-heat value of 350 Btu/lb) on the top tray (Tray 1) is calculated.
135–137	*LBL6 + PRTX	21 06 –55 –14	Label "6": The previous values are added to give the total heat absorbed on Tray 1 (Btu/h). This is printed out.
138–145	*LBL7 9 5 0 ÷ STO9 PRTX SPC	21 07 09 05 00 –24 35 09 –14 16-11	Label "7": The total heat is divided by 950, the mean latent heat for steam (Btu/lb), to give the steam condensed (lb/h) on Tray 1. This is stored in SR_9 and printed.
146–164	*LBL8 RCL4 P⇌S RCL2 + P⇌S RCL9 + STO4 RCL3 RCLI – x RCLE . 3 x + PRTX	21 08 36 04 16-51 36 02 –55 16-51 36 09 –55 35 04 36 03 36 46 –45 –35 36 15 –62 03 –35 –55 –14	Label "8": The total liquid H_2O quantity (lb/h) on Tray 2 is found and stored in register SR_4. This is multiplied by the temperature rise to the bottom of the stripper, and this heat duty is added to that needed to vaporize the assumed remaining 30% of the NH_3. The total is the heat absorbed below Tray 1 (Btu/h). This is printed.

Magnetic program card B (cont.)

Locations	Keys	Codes	Comments
165–173	*LBL9 9 5 0 ÷ STOE PRTX SPC SPC	21 09 09 05 00 –24 35 15 –14 16-11 16-11	Label "9": The previous value (Btu/h) is divided by 950 to give the equivalent steam condensed below Tray 1. This value is stored in R_E for use under label "A" of card C.
174–181	*LBLa RCL7 RCL9 RCL0 + STO7 X⇌Y STO9	21 16 11 36 07 36 09 36 00 –55 35 07 –41 35 09	Label "a": The total water content of the vapor under Tray 1 (i.e., leaving Tray 2) is found (V_2 lb/h). This is stored in register SR_7 after transferring the value previously in SR_7 to SR_9.
182–192	*LBLb P⇌S RCL4 RCL3 P⇌S STO1 X⇌Y STO2 P⇌S RTN R/S	21 16 12 16-51 36 04 36 03 16-51 35 01 –41 35 02 16-51 24 51	Label "b": The data for NH_3 and H_2S (lb/h) in the bottoms liquid (p and q) are transferred from registers PR_3 and PR_4 to SR_1 and SR_2, respectively, for convenient use in the tray-to-tray calculations of card C.

Magnetic program card C

Locations	Keys	Codes	Comments
001–008	*LBLA P⇌S RCLE RCL3 RCLI – ÷ STO3	21 11 16-51 36 15 36 03 36 46 –45 –24 35 03	Label "A": This heads the calculation run that goes down to, but stops before, label "C." From the value for steam condensed below Tray 1 is found the steam condensed per °F rise in temperature of the descending liquid. This value is stored in register SR_3 for use in calculating L and V values on lower trays.
009–030	*LBL0 RCL8 RCLC – RCL5 2 × STO5 RCL6 + ÷ STOB RCL5 × STOA PRTX RCLB RCL6 × STOB PRTX SPC	21 00 36 08 36 13 –45 36 05 02 –35 35 05 36 06 –55 –24 35 12 36 05 –35 35 11 –14 36 12 36 06 –35 35 12 –14 16-11	Label "0": The partial pressures for NH_3 and H_2S above Tray 1 are calculated (using Eq. 11 and 12), and stored in R_A and R_B. The values are also printed.
031–033	*LBL1 RCLI STOD	21 01 36 46 35 14	Label "1": The temperature, T_1°F, for Tray 1 is stored in R_D.
034–096	*LBLB 1 2 1 EEX 3 ENT↑ 1 . 0 1 6 3 RCLD Yˣ ÷ *LBL2 1 . 2 7 RCLD 3 2 – . 6 4 8 Yˣ Yˣ 2 . 8 4 EEX CHS 7 × X⇌Y ÷ *LBL3 RCLA RCLB × X⇌Y ÷ √X STOE *LBL4 2 ÷ RCLA R↑ × + STO0 PRTX RCLE PRTX SPC SPC RTN	21 12 01 02 01 –23 03 –21 01 –62 00 01 06 03 36 14 31 –24 21 02 01 –62 02 07 36 14 03 02 –45 –62 06 04 08 31 31 02 –62 08 04 –23 –22 07 –35 –41 –24 21 03 36 11 36 12 –35 –41 –24 54 35 15 21 04 02 –24 36 11 16-31 –35 –55 35 00 –14 36 15 –14 16-11 16-11 24	Label "B": This heads a subroutine down to label "C." This is the same sequence through labels "2," "3" and "4" as described for program card A through labels "7," "8," "9" and "a." But here, A_1 and S_1 for Tray 1 liquid (ppm NH_3 and H_2S contents) are calculated.
097–099	*LBLC 1 STOI	21 13 01 35 46	Label "C": This heads the main iterative calculation, which, however, loops back through label "D" (not label "C"). To begin, 1 is entered to register R_I, to signify that Tray 1 has been calculated. (The register R_I is used as a tray counter)

Magnetic program card C (cont.)

Locations	Keys	Codes	Comments
100–131	*LBLD ISZI RCLI PRTX RCL4 PRTX SPC EEX 6 RCL0 – RCLE – ÷ STO6 *LBL5 RCL0 × RCL1 – PRTX 2 × STO5 *LBL6 RCL6 RCLE × RCL2 – STO6 PRTX	21 14 16 26 46 36 46 –14 36 04 –14 16-11 –23 06 36 00 –45 36 15 –45 –24 35 06 21 05 36 00 –35 36 01 –45 –14 02 –35 35 05 21 06 36 06 36 15 –35 36 02 –45 35 06 –14	Label "D": This begins the calculation for the next tray by adding 1 to R_I and printing out the tray number. Eq. (6) and (7) are applied through labels "D," "5" and "6" to calculate and print out values x_2 and y_2, the lb/h of NH_3 and H_2S rising from Tray 2. Values $2x_2$ and y_2 are stored in registers SR_5 and SR_6.
132–134	*LBL7 RCL9 ST+8	21 07 36 09 35-55 08	Label "7": The pressure above the tray is updated from P_1 to P_2 psia for Tray 2 in register SR_8.
135–170	*LBLa RCL6 RCL5 + RCL7 PRTX SPC *LBL8 1 . 8 9 × STOE + RCL8 PRTX X⇌Y ÷ STO0 RCL5 × STOA PRTX RCL0 RCL6 × STOB PRTX *LBL9 RCL0 RCLE × STOC PRTX SPC	21 16 11 36 06 36 05 –55 36 07 –14 16-11 21 08 01 –62 08 09 –35 35 15 –55 36 08 –14 –41 –24 35 00 36 05 –35 35 11 –14 36 00 36 06 –35 35 12 –14 21 09 36 00 36 15 –35 35 13 –14 16-11	Label "a": From labels "a" through "8" and "9," Eq. (8), (9) and (10) are used to calculate AP_2, SP_2 and WP_2 psia, the component vapor pressures above Tray 2.
171–194	*LBLb 7 0 0 7 ENT↑ 1 4 . 4 6 5 RCLC LN – ÷ 3 8 3 – ENT↑ ENT↑ PRTX SPC	21 16 12 07 00 00 07 –21 01 04 –62 04 06 05 36 13 32 –45 –24 03 08 03 –45 –21 –21 –14 16-11	Label "b": From label "b," Eq. (5) is used to calculate T_2°F, the temperature on Tray 2.
195–203	*LBLc RCLD – RCL3 × ST+4 ST+7 X⇌Y STOD	21 16 13 36 14 –45 36 03 –35 35-55 04 35-55 07 –41 35 14	Label "c": From label "c," the change in tray temperature from Tray 1 to Tray 2 is used to update the liquid and vapor quantities in SR_4 and SR_7, which will affect the next tray. The new temperature is finally placed in R_D.
204–205	*LBLd GSBB	21 16 14 23 12	Label "d": The subroutine under label "B" is performed to calculate the liquid phase composition on Tray 2, i.e. A_2 and S_2.
206–212	*LBLe RCL5 X<0? RCL6 X<0? GTOE GTOD	21 16 15 36 05 16-45 36 06 16-45 22 15 22 14	Label "e": Having completed the calculations for the tray in question (Tray 2), the NH_3 and H_2S quantities leaving this tray, x_2 (as $2x_2$) and y_2 lb/h are checked to find if they are both less than zero—i.e., if the specified degree of stripping has been achieved. If so, the loop is broken by passing out to label "E." If not, the calculation continues through the loop entry at label "D."
213–222	*LBLE RCLI PRTX SPC SPC PREG P⇌S PREG RTN R/S	21 15 36 46 –14 16-11 16-11 16-13 16-51 16-13 24 51	Label "E": From label "E," the calculation is completed. The number of trays is printed out from register R_I. The total register contents are then printed out, the secondary register first, to reveal valuable data concerning the possible reboiler. For example, the vapor rate as stored in SR_7, the pressure in SR_8 and the temperature in R_D.

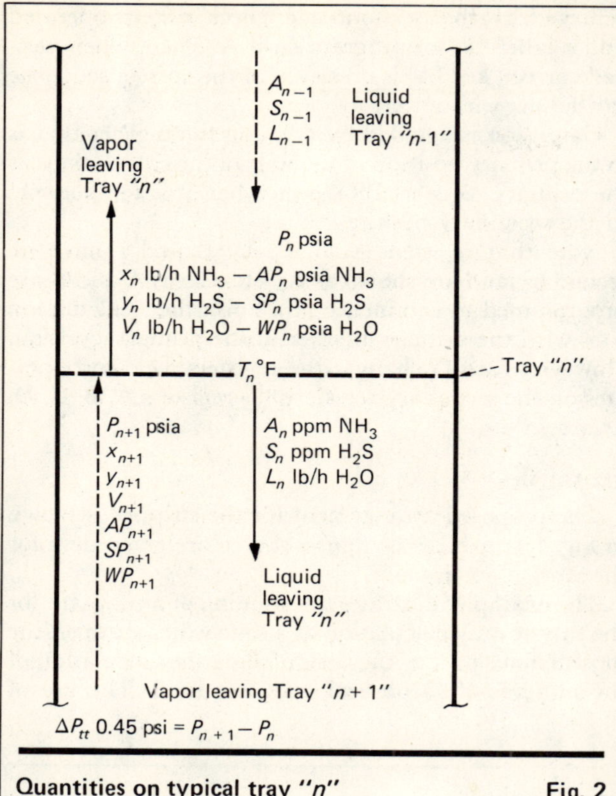

Quantities on typical tray "n" **Fig. 2**

Data entry **Table I**

Primary registers

Symbol	Item	Quantity	Register	Units
a	NH_3 flowrate	556.00	0	lb/h
b	H_2S flowrate	730.00	1	lb/h
c	H_2O flowrate	98,714.00	2	lb/h
d	NH_3 content	20.00	3	ppm
e	H_2S content	5.00	4	ppm
		0.00	5	
		0.00	6	
		0.00	7	
P_{tg}	Reflux drum pressure	27.20	8	psia
		0.00	9	
		0.00	A	
		0.00	B	
		0.00	C	
T_{tg}	Reflux drum temperature	184.00	D	°F
		0.00	E	
		0.00	I	

Secondary registers

Symbol	Item	Quantity	Register	Units
stm	Stripping steam	10,200.00	0	lb/h
P_1	Tower top pressure	30.00	1	psia
T_{fd}	Feed temperature	199.00	2	°F
T_{bm}	Bottoms temperature	259.00	3	°F
		0.00	4	
		0.00	5	
		0.00	6	
ΔP_{tt}	Pressure drop per tray	0.45	7	psia/tray
		0.00	8	
		0.00	9	

For convenience, the correlations given in Beychok's book have been modified to use temperature in °F throughout, instead of a mixture of °F and °C. Thus, from Fig. 25 and 26, page 163 and 164, and also from Eq. (14) and (15) may be derived:

$$K = 121{,}000/(1.0163)^T \tag{1}$$

$$C = \left(\frac{2.84 \times 10^{-7}}{K}\right)(1.27)^{[(T-32)^{0.648}]} \tag{2}$$

where T is the temperature of the liquid/vapor equilibrium mixture in °F, and K, C are temperature-dependent variables. Also, Beychok's Eq. 16 applies:

$$S = \sqrt{\frac{AP \times SP}{C}} \tag{3}$$

where S = ppm H_2S in the liquid, and AP and SP are the partial pressures (psia) of NH_3 and H_2S above the liquid. By substituting $K = (8.8 \times 10^5)H_o$ into Beychok's equations, one obtains:

$$A = K \cdot AP + S/2 \tag{4}$$

where A = ppm NH_3 in the liquid and H_o is a temperature-dependent variable. In order to correlate water vapor pressure (which frequently occurs under the description, H_2O partial pressure or WP psia) with temperature (T°F), the equation quoted by Miles and Wilson, page 24, relating mm Hg to °C, has been modified and also adjusted to conform better with steam-table data, to give:

$$T = 7{,}007/(14.465 - \log_e WP) - 383 \tag{5}$$

Other equations used include, for tray-to-tray mass balance:

$$x = A\left(\frac{L}{10^6 - A - S}\right) - p \tag{6}$$

$$y = S\left(\frac{L}{10^6 - A - S}\right) - q \tag{7}$$

where x = NH_3 content of vapor (lb/h)
y = H_2S content of vapor (lb/h)
p = NH_3 in bottoms product (lb/h)
q = H_2S in bottoms product (lb/h)

From the x, y and V values, partial pressures are calculated by, for example:

$$AP = \frac{x}{17}\left(\frac{P}{x/17 + y/34 + V/18}\right)$$

where P = pressure, psia and V = H_2O content of vapor, lb/h. Molecular weight of NH_3 = 17, H_2S = 34 and H_2O = 18.

This reduces to:

$$AP = 2x\left(\frac{P}{2x + y + 1.89V}\right) \tag{8}$$

Similarly:

$$SP = y\left(\frac{P}{2x + y + 1.89V}\right) \tag{9}$$

$$WP = 1.89V\left(\frac{P}{2x + y + 1.89V}\right) \tag{10}$$

and:

$$AP_1 = 2x_1\left(\frac{P_1 - WP_1}{2x_1 + y_1}\right) \tag{11}$$

$$SP_1 = y_1\left(\frac{P_1 - WP_1}{2x_1 + y_1}\right) \tag{12}$$

Equations used for the overall mass balance include:

$$k = \frac{(2a + b - Mc)N}{1 - MN} \tag{13}$$

where $M = \dfrac{2d + e}{f}$ and $N = \dfrac{18}{34}\left(\dfrac{WP_r}{P_{tg} - WP_r}\right)$

$$p = d\left(\frac{c - k}{f}\right) \tag{14}$$

$$q = e\left(\frac{c - k}{f}\right) \tag{15}$$

$$r = c - k \tag{16}$$

Description of program

The full program is stored on three magnetic program cards—A, B and C. This leaves the main storage registers free for the entry of data. The cards are entered successively as required by the calculation run. If desired, program card C may be used on its own. However in this case, the mass and internal heat balances corresponding to program cards A and B must first be worked up manually.

Card A contains two operational labels, "E" and "A", and labels "0" through "9", together with "a" and "b" for identification of calculation sections. When key [E] is pressed in the program run, the data in the storage registers are printed out for checking, but no calculation is made. When [A] is pressed, all the calculations are done.

Card B contains only one operational label, "A", for running the program. Labels "0" through "9", with "a" and "b", are used for identification.

Card C contains five labels, "A" through "E", for use operationally. There are also fifteen labels "0" through "9" and "a" through "e", which are used for identification. Label "a" (activated by keys [f] and [A]), is used only in the special case of manual mass-and-heat-balance calculation, where cards A and B have not been used and card C must be used for any programmed calculations. The sequence between labels "B" and "C" is used also as a subroutine when the loop between labels "D" and "E" is functioning for tray-to-tray calculation.

In the tray-to-tray calculation, register R_I acts as a tray counter. At the end of each calculation cycle, the values stored in secondary registers SR_5 and SR_6 are tested. If both are negative, the loop is broken and the calculation is completed by printing out the number of trays and the contents of the storage registers.

Inasmuch as the HP97 (and HP67) calculator uses a system of ten primary and ten secondary storage registers, numbered 0 to 9, which are interchangeable, it is very important to have these in the correct (not reversed) position when running calculations. This is achieved by the printout and check system operated under label "E" of program card A. Then, when verified correct, key [A] is pressed and the correct sequence can follow without error.

Confusion is avoided by defining the register that is in the primary position at the start of the calculations as the primary register (PR), even when it is temporarily in the secondary position.

Note that registers A, B, C, D, E and I have no secondary and are shown as R_A etc. Cards A and B are programmed to commence and finish their calculation runs with the primary register in the primary position. However, card C changes the registers over and operates on the secondary register plus registers A, B, C, D, E and I.

Example

The proposed arrangement for the stripper is shown in Fig. 1, in which the figures shown are given data for the worked example.

The example illustrates the running of a program for the tray-to-tray calculation of a sour-water stripper, the most tedious part of the design being the calculation of the number of theoretical trays required. The use of

Data printout **Table II**

	Feed, lb/h	Bottoms, lb/h	Tail gas, lb/h
NH_3	556.00 *a*	1.97 *p*	554.03 *h*
H_2S	730.00 *b*	0.49 *q*	729.51 *j*
H_2O	98,714.00 *c*	98,292.43 *r*	421.57 *k*

	Reflux drum partial pressure, psia	Reflux composition, ppm
NH_3	11.44 AP_r	101,549.83 A_r
H_2S	7.53 SP_r	61,741.12 S_r
H_2O	8.22 WP_r	

(End of card A)

	Reflux components, lb/h	Overheads components, lb/h	Overheads conditions
NH_3	1,186.79 x_r	1,740.82 x_1	23.89 WP_1, psia
H_2S	721.55 y_r	1,451.06 y_1	237.55 T_1, °F
H_2O	9,778.43 V_r	10,200.00 stm = V_1	

Thermal loading

	On Tray 1	Below Tray 1
Heat absorbed, Btu/h	5,101,400.29	2,625,365.56
Equivalent steam condensed, lb/h	5,369.90	2,763.54

(End of card B)

	Partial pressures, psia	Liquid composition, ppm
NH_3	4.31 AP_1	14,286.28 A_1
H_2S	1.80 SP_1	6,150.78 S_1

(End of card C A/B)

Data stored in secondary registers — **Table III**

Symbol	Units	Item	Quantity	Register
$^*V_1$	lb/h	H_2O in vapor from Tray 1	10,200.00	0
p	lb/h	NH_3 in bottoms	1.97	1
q	lb/h	H_2S in bottoms	0.49	2
T_{bm}	°F	Bottoms temperature	259.00	3
L_1	lb/h	H_2O in liquid from Tray 1	113,862.33	4
x_1	lb/h	NH_3 in vapor from Tray 1	1,740.82	5
y_1	lb/h	H_2S in vapor from Tray 1	1,451.06	6
V_2	lb/h	H_2O in vapor from Tray 2	15,569.90	7
P_1	psia	Pressure above Tray 1	30.00	8
ΔP_{tt}	psi	Pressure drop per tray	0.45	9
$^*x_r$	lb/h	NH_3 in reflux	1,136.79	A
$^*y_r$	lb/h	H_2S in reflux	721.55	B
WP_1	psia	H_2O partial pressure above Tray 1	23.89	C
$^*T_{tg}$	°F	Reflux temperature	184.00	D
	lb/h	Steam condensed below Tray 1	2,763.54	E
T_1	°F	Temperature of liquid leaving Tray 1	237.55	I

*The value shown as stored is not used in any subsequent calculation.

preceding programs for calculating the mass and heat balances is also included.

Problem

A sour-water stripper is to be designed to remove H_2S and NH_3 down to levels of 5 ppm and 20 ppm, respectively, from a feed stream of 100,000 lb/h containing 0.73 wt % H_2S and 0.556 wt % NH_3.

Stripped overheads are to be condensed, with the liquid portion to be returned as reflux to the top tray (feed tray) of the tower, while the reflux and tail gas will separate in the reflux drum at a pressure of 27.2 psia and temperature of 184°F. The tail gas will be sent to a sulfur recovery plant.

It is estimated that a tower-top operating pressure of 30 psia will be suitable, allowing for reboiler operation at 35 psia and a bottoms temperature (ex-reboiler) of 259°F. The pressure drop per theoretical tray is assumed to be 0.45 psi.

The stripping steam rate is to be taken as 10,200 lb/h (i.e., 10.2% of feed).

In consideration of heat exchange with the bottoms stream, the feed is assumed to be preheated, to 199°F, with no vaporization, before entering the feed tray (Tray 1).

Data entry

The calculator is switched on and put in the program "Run" mode. This mode is maintained throughout.

The magnetic program card A is passed through the card reader twice to read both edges.

The data for the secondary storage register is now entered, in accordance with Table I and as follows:

The stripping steam rate, *stm* (or V_1), which is 10,200 lb/h, is put in storage register SR_0.

The tower-top operating pressure, P_1, which is 30 psia, is put in SR_1.

The feed temperature, T_{fd}, which is 199°F, is put in SR_2.

The bottoms temperature, T_{bm}, which is 259°F, is put in SR_3.

The pressure drop per theoretical tray, ΔP_{tt}, which is 0.45 psi, is put in SR_7.

The temperature in the reflux drum, T_{tg}, which is 184°F, is put in R_D.

This completes entry to the secondary register and the alphabetic registers (A, B, C, D, E, I).

The primary and secondary registers are interchanged by pressing the keys for P ⇆ S.

Entry is now made to the primary register.

The feed components are entered:

NH_3 in feed, *a*, which is 556 lb/h, is put in PR_0.

H_2S in feed, *b*, which is 730 lb/h, is put in PR_1.

H_2O in feed, *c*, which is 98,714 lb/h, is put in PR_2.

The specified levels of components in the bottoms are entered:

NH_3 in bottoms, *d*, which is 20 ppm, is put in PR_3.

H_2S in bottoms, *e*, which is 5 ppm, is put in PR_4.

Finally, the pressure in the reflux drum, P_{tg}, which is

Tray-to-tray calculation printouts (first two and last two full printouts) — **Table IV**

Symbol	Units	Item	Printouts			
		Tray number "n"	2.00	3.00	11.00	12.00
L_{n-1}	lb/h	H_2O in liquid from Tray "(n-1)"	113,862.33	114,679.95	116,450.36	116,551.94
x_n	lb/h	NH_3 in vapor leaving Tray "n"	1,658.64	977.27	0.84	−1.55
y_n	lb/h	H_2S in vapor leaving Tray "n"	714.46	315.77	0.19	−0.40
V_n	lb/h	H_2O in vapor leaving Tray "n"	15,569.90	16,387.52	18,157.93	18,259.51
P_n	psia	Pressure above Tray "n"	30.45	30.90	34.50	34.95
AP_n	psia	NH_3 partial pressure above Tray "n"	3.02	1.82	1.69	−3.15
SP_n	psia	H_2S partial pressure above Tray "n"	0.65	0.29	1.87	−4.01
WP_n	psia	H_2O partial pressure above Tray "n"	26.78	28.79	34.50	34.95
T_n	°F	Temperature on Tray "n"	243.89	247.98	258.43	259.20
A_n	ppm	NH_3 in liquid leaving Tray "n"	8,443.49	4,643.98	3.54	−4.95
S_n	ppm	H_2S in liquid leaving Tray "n"	2,726.99	1,310.96	0.82	1.62
					Trays required	12.00

27.2 psia, is put in PR_8. (T_{tg} has already been entered into R_D).

Printouts

As this completes data entry, key [E] is pressed.

The data are printed out and checked off to ensure that they are exactly as listed in Table I.

Key [A] is pressed to run the program.

Data are printed out in four groups of three items and one group of two, making fourteen lines altogether. In each group, NH_3 is first, H_2S second and H_2O third (when there is a third item).

In the first group are the feed quantities (lb/h) as entered.

In the second group are the bottoms quantities (lb/h).

In the third group are the tail-gas quantities (lb/h).

In the fourth group are the reflux-drum partial pressures (psia).

In the fifth group are A_r and S_r, which are the NH_3 and H_2S ppm in the reflux. These printouts are shown in Table II.

Magnetic program card B is now passed through the card reader twice to read both edges. This replaces the previous program completely.

Key [A] is pressed to run the program.

Data are printed out in two groups of three items (NH_3, H_2S, H_2O).

In the first group are the quantities in the reflux (lb/h).

In the second group are the quantities in the overheads (lb/h).

Further data are printed out as three groups of two items.

These are:

WP_1, the partial pressure of H_2O (psia) above Tray 1 (i.e., the partial pressure of H_2O in the overheads).

T_1, the corresponding liquid temperature (°F) on Tray 1.

There follows the heat absorbed (Btu/h) on Tray 1 and the equivalent steam condensed there (lb/h).

Then follows the heat absorbed (Btu/h) below Tray 1 and the equivalent steam condensed there (lb/h).

NB: At this point, if the contents of the secondary storage registers and the alphabetic registers R_A to R_I are printed out, they appear as in Table III.

Magnetic program card C is now passed through the card reader twice to read both edges. This replaces the previous program completely.

Key [A] is pressed to run the first section of the program down to label "C".

Note that the primary storage registers, PR_0 to PR_9, are not used in running card C.

Data are printed out in two groups of two items.

These are:

AP_1, the partial pressure of NH_3 (psia) above Tray 1

SP_1, the partial pressure of H_2S (psia) above Tray 1

Then follow:

A_1, the NH_3 ppm in the liquid on Tray 1

S_1, the H_2S ppm in the liquid on Tray 1

All these quantities are listed in Table II.

The main program is now put into operation from label "C" by pressing key [C].

The printouts for Trays 2, 3, 11 and 12 then appear as shown in Table IV. These are typical for the tray-to-tray calculation. The printouts for trays 4 through 10 have been excluded merely to save space.

The program execution stops only after values for NH_3 and H_2S are found to be negative, indicating that specifications have been met. In the example, this occurs after twelve trays have been calculated, and means that no more than twelve trays are necessary. The calculator prints "12" and the register contents.

Semiautomatic procedure

If it is desired to do the mass and internal heat balance manually, it is possible to calculate the number of trays by using only program card C. From the given temperature, T_{tg}, calculate the H_2O partial pressure in the reflux drum. From the overall mass balance, calculate the quantities of the components in the bottoms and the tail gas. Then the internal heat balance is calculated, and the quantity of steam condensed on Tray 1 is found. Various intermediate values are stored in the registers, and program card C is run through the calculator in much the same way as described for the fully automatic procedure.

Comparison with commercial programs

Examination of the printout (Table IV) shows that the specified stripping was really achieved on Tray 11. The steam rate rising from Tray 11 is 18,158 lb/h, which may be regarded as the steam rate rising from the reboiler if Tray 11 is the reboiler.

This may be compared with the results from a commercial computer program that was run on this problem. With the commercial program and 11 chosen trays (including the reboiler), the bottoms specification was met.

Also, the steam rate from the reboiler was given as 1,004.2 lb mole/h of 100% H_2O—i.e., 18,076 lb/h steam. The compatibility of these two figures (18,158 and 18,076 lb/h) confirms that the use of the simplified thermal data in the HP97 calculation is sufficiently accurate in this instance.

References

1. Beychok, M. R., "Aqueous Wastes from Petroleum and Petrochemical Plants," John Wiley & Sons, New York. 1967.
2. Miles, D. H., and Wilson, G. M., Vapor-Liquid Equilibrium Data for Design of Sour Water Strippers, Annual Report for 1974 (sponsored by the American Petroleum Institute), Center for Thermochemical Studies and Dept. of Chemical Engineering, Brigham Young University, October 1975.

The author

Norman H. Wild is Process Group Head—Offsites, in the Technical Operations department of Davy International Ltd., Baker Street, London W.1, England. Tel: 01-486-6677. He has covered a wide range of refinery and petrochemical plant operations and design and also has had experience in designing and operating pilot plants. He obtained a London University external B.Sc. in Chemistry and later chose to take up chemical engineering as a process engineer in the south of England with Esso Petroleum Co. at the Fawley Oil Refinery. He is a member of the Royal Institute of Chemistry and the Institution of Chemical Engineers in England.

Determining ideal stages on a pocket calculator

*H. Tan, Toronto, Ontario**

☐ Design calculations of the stages of binary distillation may be performed either graphically or analytically. Graphical methods lack precision, while algebraic calculations are time-consuming. A programmable pocket calculator may be an attractive alternative for simpler problems.

The following program calculates the number of ideal stages of distillation with a HP-25 hand-held calculator. Because of the limiting capacity of the HP-25 (49 merged program steps and 8 memory registers), the program is applicable only to binary systems with constant molar flow rates and constant relative volatilities. For such a problem, the data are usually given as:

X = mol-fraction of lighter component in the feed material
q = total heat required to convert 1 mol of feed into vapor, divided by its molal latent heat
q' = $(1 - q)$
X_D = mol-fraction of lighter component in overhead product
X_B = mol-fraction of lighter component in bottoms
R = reflux ratio
α = relative volatility of light to heavy component

Before running the program, the following have to be calculated and stored in the memory registers: $1/\alpha$, $X_D/(R+1)$, $R/(R+1)$, X_i, and $(Y_i - X_B)/(X_i - X_B)$, where X_i and Y_i a e the coordinates of the point of intersection of the two operating lines and the q-line, and are calculated as:

$$X_i = \frac{X_f - q'X_D/(R+1)}{q + q'R/(R+1)}$$

$$Y_i = X_iR/(R+1) + X_D/(R+1)$$

The output of the program gives both the total number of ideal stages and the position of the feed stage. During execution, the program also pauses to display the $X - Y$ composition of each stage.

*65 Sussex Ave., Toronto, Ont., Canada M5S 1J8.

HP-25 program to calculate number of ideal stages in distillation

Step	Key entry	Step	Key entry	Registers
01	f FIX 4	26	RCL 5	R_0 : n
02	RCL 0	27	+	R_1: X_D, Y_n, X_n
03	f PAUSE*	28	STO 1	R_2 : X_B
04	RCL 1	29	GTO 01	R_3 : $1/\alpha$
05	f PAUSE*	30	RCL 1	R_4 : $R/(R+1)$
06	RCL 3	31	RCL 7	R_5 : $X_D/(R+1)$
07	X	32	X	R_6 : X_i
08	f LAST x	33	1	R_7 : $\frac{(Y_i - X_B)}{(X_i - X_B)}$
09	1	34	RCL 7	
10	−	35	−	
11	RCL 1	36	RCL 2	
12	X	37	X	*Replace steps 03, 05 and 16 with g NOP to suppress intermediate output (n, Y_n, X_n).
13	1	38	+	
14	+	39	STO 1	
15	÷	40	RCL 2	
16	f PAUSE*	41	f x ≥ y	
17	STO 1	42	GTO 48	
18	RCL 6	43	EEX	
19	f x ≥ y	44	4	
20	GTO 30	45	CHS	
21	1	46	STO + 0	
22	STO + 0	47	GTO 01	
23	RCL 1	48	RCL 0	
24	RCL 4	49	GTO 00	
25	X			

User instructions

Step	Instructions	Input data units	Keys					Output data/units
1	Key in program							
2	Store data in registers	n*	STO	0				n
		X_D	STO	1				Y_1
		X_B	STO	2				X_B
		α	g	1/x	STO	3		$1/\alpha$
		R	↑	↑	1	+	÷	
			STO	4				R/(R+1)
		X_D	↑					
		R	↑	1	+	÷	STO	
			5					$X_D/(R+1)$
		X_f	RCL	5				
		q'	X	−				
		q'	RCL	4	X			
		q	+	÷	STO	6		X_i
			RCL	4	X	RCL	5	
			+	RCL	2	−	RCL	
			6	RCL	2	−	÷	$\frac{Y_i - X_B}{X_i - X_B}$
			STO	7				
3	Run program		f	PRGM	R/S			
	stage number							n
	vapor composition							Y_n
	liquid composition							X_n
	total number of stages in enriching section, N_e total number of stages in stripping section, $O.N_s(10)^4$							N_e . N_s

*Set initial value of number of stages, n = 1.0000

Originally published March 14, 1977

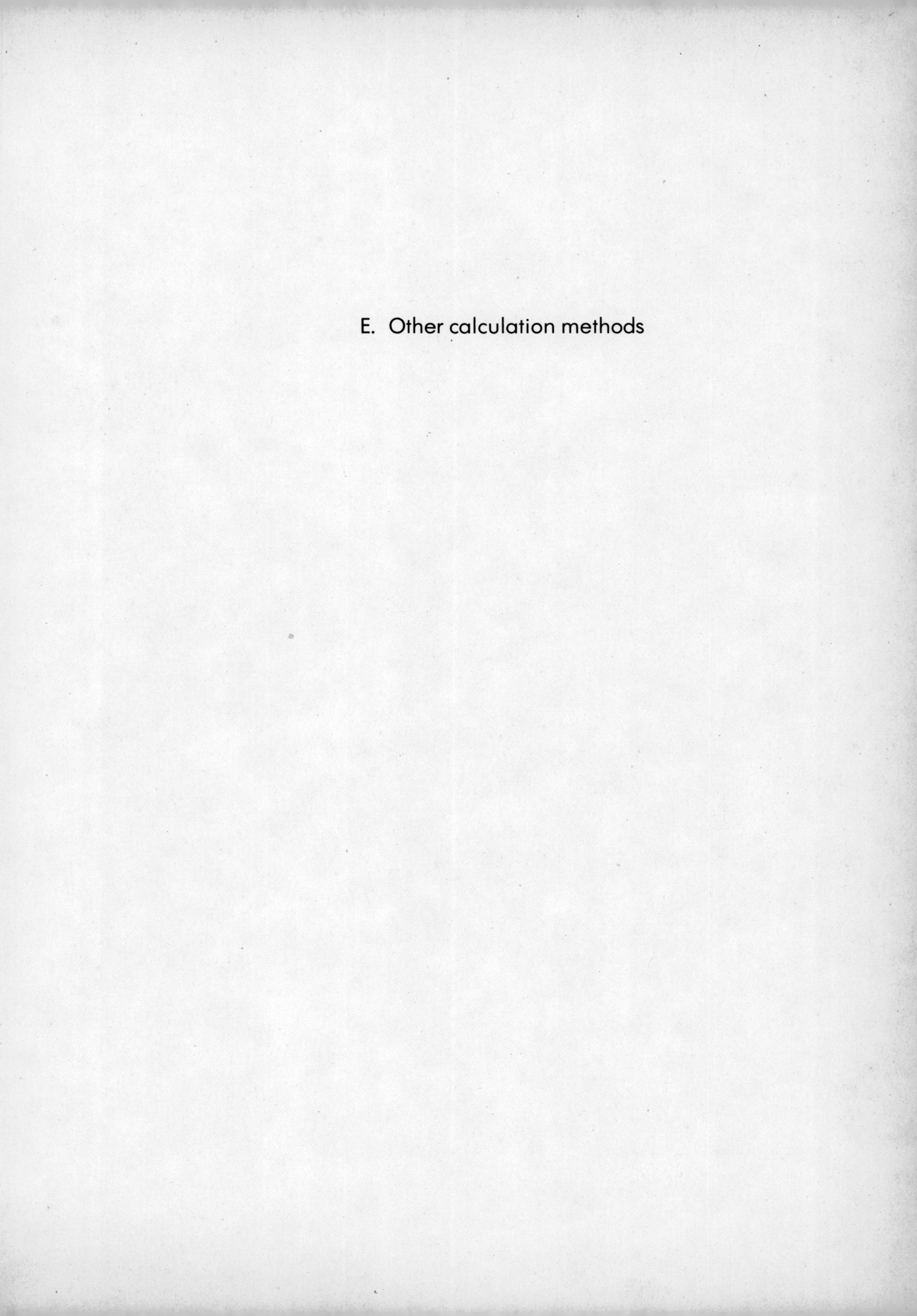

E. Other calculation methods

Estimating recoveries in multicomponent distillation

To find the amounts of non-key components in the distillate and bottoms usually requires a plate-to-plate calculation. Here is a short-cut graphical method of estimation that needs figures only for the desired recovery of the two key components and the relative volatility of all components.

Carl L. Yaws, Cheng-Shen Fang and Prabodh M. Patel, Lamar University.

☐ Multicomponent distillation separates the components of a mixture into products of higher purity. The primary components to be separated are often called the light key (LK) and heavy key (HK). The LK component is recovered at higher purity in the distillate, and the HK component is recovered at higher purity in the bottoms.

The other components in the mixture are also recovered in the distillate and bottoms, and it is important to know the extent of this recovery. Fig. 1 is a simplified diagram of a typical distillation, which shows how the distillate and bottoms compositions compare with the feed.

This article presents a graphical short-cut method for estimating non-key-component recoveries. The short-cut results compare favorably with plate-to-plate calculations and provide an excellent first-order estimate for the engineer who either does not have the time to do a plate-to-plate calculation or does not need the extra precision. A nomograph is included to help in rapidly estimating non-key-component recoveries.

Component distribution equations

The distribution of components between the distillate and bottoms is given by the Hengstebeck-Geddes equation [*1,2,3*]:

$$\log (d_i/b_i) = a + b \log \alpha_i \tag{1}$$

where d_i = moles of component i in distillate
b_i = moles of component i in bottoms
α_i = relative volatility of component i
a, b = correlation constants

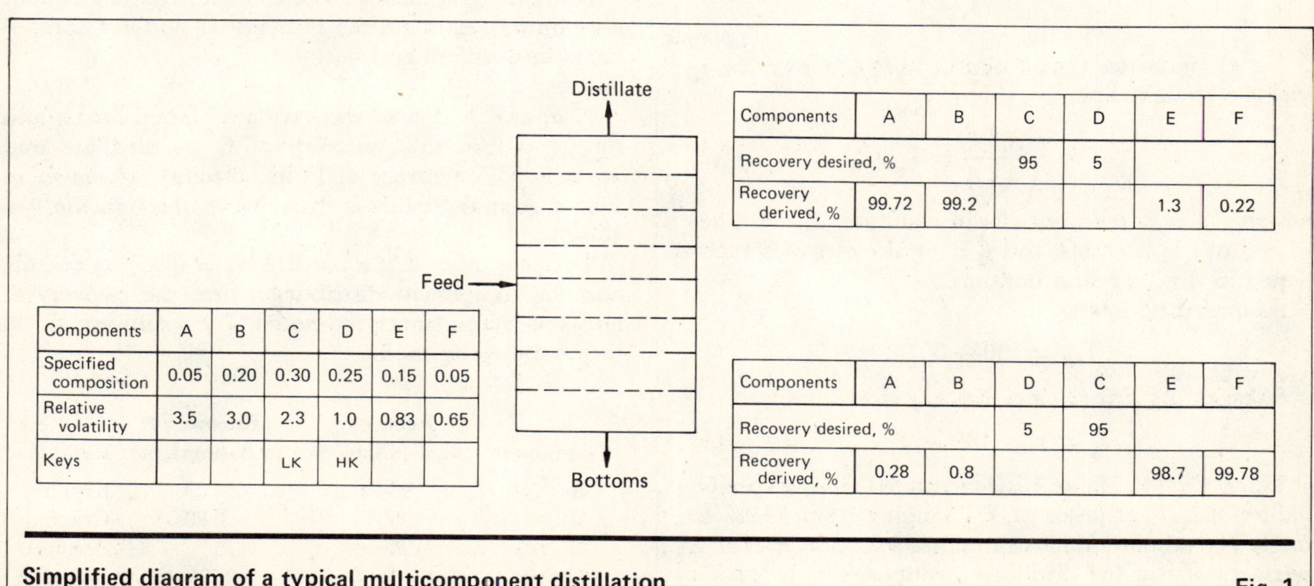

Components	A	B	C	D	E	F
Specified composition	0.05	0.20	0.30	0.25	0.15	0.05
Relative volatility	3.5	3.0	2.3	1.0	0.83	0.65
Keys			LK	HK		

Components	A	B	C	D	E	F
Recovery desired, %			95	5		
Recovery derived, %	99.72	99.2			1.3	0.22

Components	A	B	D	C	E	F
Recovery desired, %			5	95		
Recovery derived, %	0.28	0.8			98.7	99.78

Simplified diagram of a typical multicomponent distillation **Fig. 1**

Originally published January 29, 1979

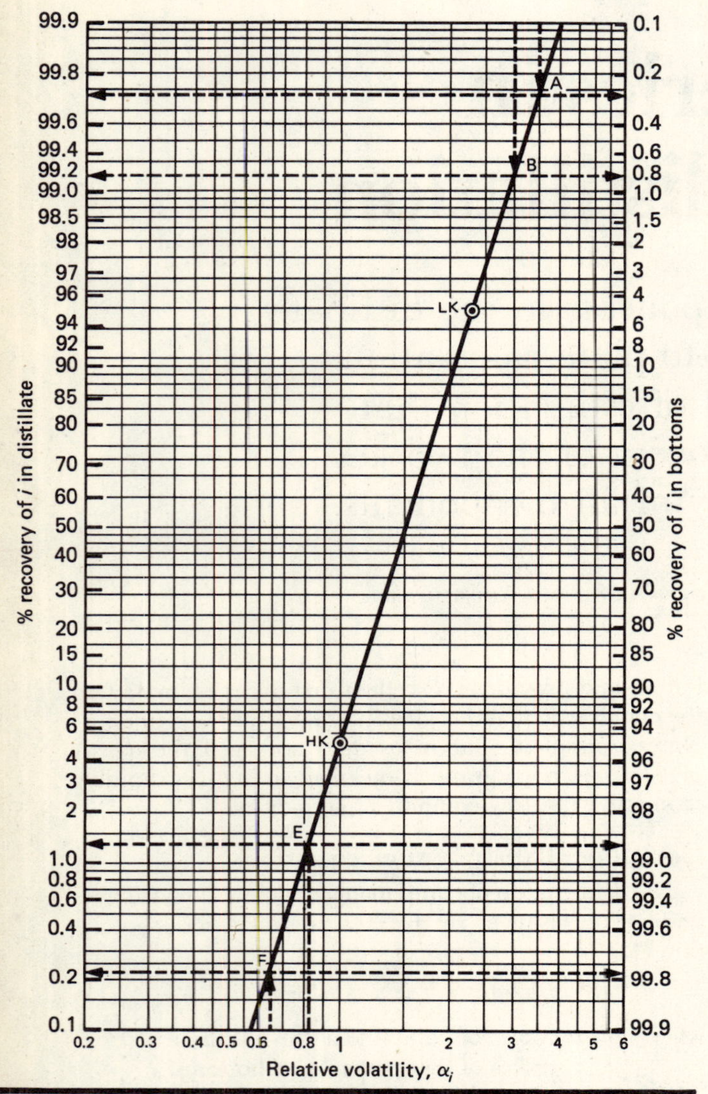

Estimation of recovery of non-key components (A, B, E, F) **Fig. 2**

Eq. (1) indicates that a plot of log (d_i/b_i) vs. log α_i yields a straight line.

Now,

$$d_i/b_i = \frac{d_i/f_i}{b_i/f_i} = Y_{iD}/Y_{iB} \tag{2}$$

where Y_{iD} = % recovery of i in distillate, Y_{iB} = % recovery of i in bottoms, and f_i = total moles of component i in distillate and bottoms.

Assuming no losses,

$$Y_{iB} = 100 - Y_{iD} \tag{3}$$

Substituting Eq. (3) into Eq. (2) yields:

$$d_i/b_i = Y_{iD}/(100 - Y_{iD}) \tag{4}$$

Using Eq. (4), Table I is constructed, giving specific values of d_i/b_i at levels of Y_{iD} ranging from 99.9% to 0.1%. The tabulation illustrates that a specific level of % recovery of i in the distillate corresponds to a specific d_i/b_i value for any component in a multicomponent distillation.

Values of d_i/b_i at various levels of % recovery of i in distillate **Table I**

Y_{iD}	d_i/b_i	Y_{iD}	d_i/b_i	Y_{iD}	d_i/b_i	Y_{iD}	d_i/b_i
99.9	999	96	24.0	40	0.6670	2	0.02040
99.8	499	94	15.7	30	0.4290	1.0	0.01010
99.6	249	92	11.5	20	0.2500	0.8	0.00806
99.4	166	90	9.00	15	0.1760	0.6	0.00604
99.2	124	85	5.67	10	0.1110	0.4	0.00402
99.0	99.0	80	4.00	8	0.0870	0.2	0.00200
98.5	65.7	70	2.33	6	0.0638	0.1	0.00100
98.0	49.0	60	1.50	4	0.0417		
97.0	32.3	50	1.00	3	0.0309		

Component recovery nomograph

Recalling that log (d_i/b_i) vs. log α_i yields a straight line in accordance with Eq. (1), a nomograph is constructed by plotting d_i/b_i vs. α_i on log-log graph paper and then superimposing a Y_{iD} scale over the d_i/b_i scale, according to the values given in Table I. The resulting nomograph, relating component recovery and component relative volatility, is given in Fig. 3. This may be used to estimate component recovery in distillate and bottoms, as follows:

1. Plot relative volatility (α_i) and % desired recovery for LK and HK. Draw a straight line through these two points. The non-key component points will also be on this straight line.
2. Using α_i and the component distribution line, estimate % recovery of non-key components in distillate and bottoms.

The procedure is illustrated in the following example.

Problem:

Component C is to be separated from Component D by distillation. A 95% recovery of both key components (LK, HK) is desired. Saturated-liquid feed composition and relative volatilities (at average column conditions) are given in Fig. 1.

Using the graphical short-cut method for component distribution, estimate the recovery of non-key components in distillate and bottoms.

Solution:

1. α_i and % desired recovery are plotted for LK and HK ($\alpha_c = 2.3$, 95% recovery of C in distillate and $\alpha_D = 1$, 95% recovery of D in bottoms), as shown in Fig. 2. A straight line is then drawn through the two points.
2. Using $\alpha_A = 3.5$, $\alpha_B = 3.0$, $\alpha_E = 0.83$, $\alpha_F = 0.65$, and the component distribution line, the recovery of non-key components is estimated. The results are shown in the following table:

Component	Recovery in distillate, %	Recovery in bottoms, %	Remarks
A	99.72	0.28	Graph
B	99.20	0.80	Graph
C (LK)	95	5	Specified
D (HK)	5	95	Specified
E	1.3	98.7	Graph
F	0.22	99.78	Graph

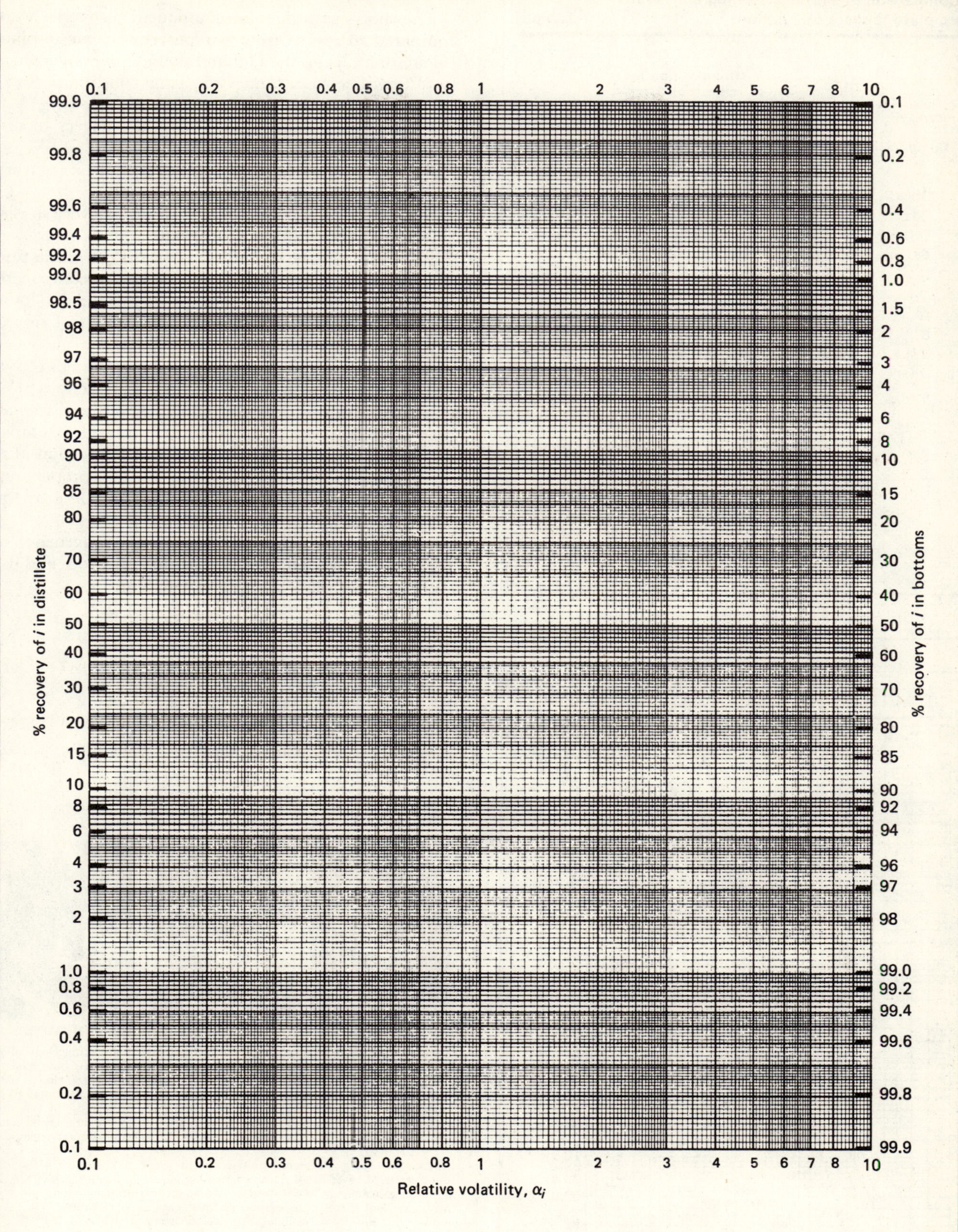

Multicomponent distillation nomograph for estimating component recovery in distillate and bottoms **Fig. 3**

Comparison of short-cut nomograph results vs. plate-to-plate calculations **Table II**

		Composition X_i			
		Distillate		Bottoms	
	Component	Nomograph	Plate to plate	Nomograph	Plate to plate
Case I	A	0.0901	0.0901	0.0002	0.0002
High	B	0.3588	0.3591	0.0026	0.0023
recovery	C (LK)	0.5197	0.5190	0.0269	0.0278
(16 trays)	D (HK)	0.0271	0.0271	0.5271	0.5271
	E	0.0041	0.0045	0.3314	0.3308
	F	0.0002	0.0002	0.1118	0.1118
Case II	A	0.0879	0.0880	0.0016	0.0012
Inter-	B	0.3466	0.3464	0.0128	0.0120
mediate	C (LK)	0.4814	0.4770	0.0683	0.0726
recovery	D (HK)	0.0668	0.0682	0.4839	0.4835
(13 trays)	E	0.0155	0.0187	0.3218	0.3187
	F	0.0018	0.0018	0.1116	0.1120
Case III	A	0.0866	0.0872	0.0034	0.0028
Low	B	0.3376	0.3395	0.0250	0.0227
recovery	C (LK)	0.4561	0.4552	0.1015	0.1027
(9 trays)	D (HK)	0.0844	0.0839	0.4606	0.4610
	E	0.0308	0.0295	0.3016	0.3031
	F	0.0045	0.0046	0.1079	0.1077

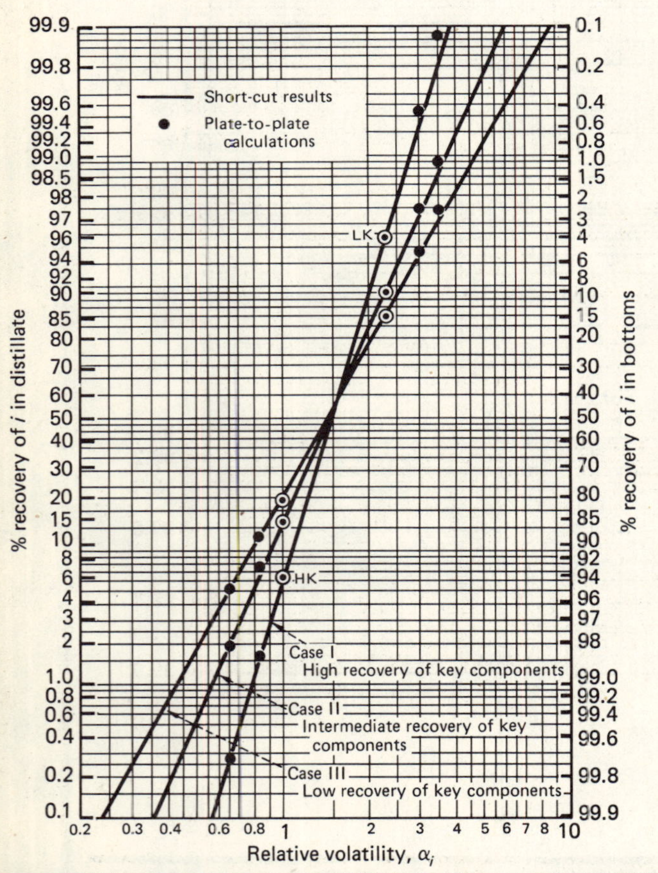

Comparison of short-cut results and plate-to-plate calculations **Fig. 4**

Comparison of results

The short-cut results for component recoveries were compared to results from computerized plate-to-plate calculations, using the Holland method of convergence [4,5] for three cases, under the same conditions as existed in the previous example.

- Case I—High Recovery: 96% LK recovery in distillate, 94% HK recovery in bottoms.
- Case II—Intermediate Recovery: 90% LK recovery in distillate, 85% HK recovery in bottoms.
- Case III—Low Recovery: 85% LK recovery in distillate, 81% HK recovery in bottoms.

Fig. 4 presents the comparison of short-cut results and plate-to-plate calculations for the three key-component recovery levels (I–high, II–intermediate and III–low). The comparison shows favorable agreement of results between the two methods.

Table II compares the results when, instead of recovery rates, we use composition in mol fractions. For example, in Case I, the recovery of A in the distillate, read from the nomograph, is 99.82%. The feed composition, X_{iF}, is 0.05. Therefore, the mol proportion of A in the distillate is 0.04991. Adding the mol proportions of all components in the distillate and dividing by the sum (55.4135) gives $X_{iD} = 0.0901$.

Table II summarizes the remarkable congruence of results obtained by the nomograph method and by computerized plate-to-plate calculations.

References

1. VanWinkle, M., "Distillation," McGraw-Hill, New York, 1967.
2. Hengstebeck, R. J., *Trans. AIChE,* 1946, pp. 309–329.
3. Geddes. R. L., *AIChE J.,* December 1958, pp. 389–392.
4. Holland, C. D., "Fundamentals and Modeling of Separation Processes," Prentice-Hall, Englewood Cliffs, N.J., 1975.
5. Holland, C. D., "Multicomponent Distillation," Prentice-Hall, Englewood Cliffs, N.J., 1963.

The authors

C. L. YAWS

C.-S. FANG

P. M. PATEL

Carl L. Yaws is associate professor in the Chemical Engineering Dept. of Lamar University, P.O. Box 10053, Beaumont, TX 77710. His industrial experience has been primarily in research and development and design with Exxon Corp., Ethyl Corp. and Texas Instruments Co. He received a B.S. from Texas A&I University, and an M.S. and Ph.D. from the University of Houston, all in chemical engineering. He has written many articles on mass transfer, distillation reaction kinetics, process scaleup and property data.

Cheng-Shen Fang is associate professor of the Dept. of Chemical Engineering, University of Southwestern Louisiana, Lafayette, LA 70504. In addition to teaching, he has been a participant in various research projects in the areas of energy conservation and alternative energy sources. He received his B.S. from National Taiwan University, and M.S. and Ph.D. from the University of Houston.

Prabodh M. Patel is a graduate student in Chemical Engineering at Lamar University, and is currently conducting, as a research assistant, process design and evaluation studies on energy-related processes. He received a B.S. degree in chemistry from Gujarat University, India, and B.S. and M.E. degrees in chemical engineering from Lamar University, Beaumont, Tex. He has publications in the areas of chemical engineering and property data.

Calculating actual plates in absorbers and strippers

The author develops simplified equations to determine the actual stages, using the Murphree efficiency. Compositions at each plate—ideal and actual—can also be determined. These equations often are both more accurate than typical graphical means and easier to use than most other formulas.

Hung Xuan Nguyen, Philip Morris U.S.A.

☐ There are several methods for determining the number of ideal stages for a plate absorber or stripper [*1-5*]; however, these methods are complicated. The conventional graphical technique is tedious and time-consuming, especially if a large number of plates are required. Also, stepping off stages at the ends of the curves is impractical unless considerable magnification of the chart is made; however, such a process often leads to inaccuracies.

Here, we will develop a generalized, simplified method to calculate the number of actual plates, employing the Murphree stage efficiency. This approach has also been used by the author to determine the actual number of stages in a constant-underflow, countercurrent-leaching cascade [*6*].

Operating line

Consider a countercurrent absorber having n plates (see Fig. 1). Let G represent the total vapor flow of the carrier gas in moles per unit time of composition y, and let L be the total moles per unit time of the inert carrier-solvent flow of composition x.

The material balance of the solute around the envelope in Fig. 1 is:

$$G_1 y_1 + L_i x_i = G_{i-1} y_{i-1} + L_1 x_1 \tag{1}$$

Assuming constant molal overflow, Eq. (1) may be rearranged to yield:

$$y_{i-1} = \left(\frac{L}{G}\right) x_i + \left(y_1 - \frac{L}{G} x_1\right) \tag{2}$$

This is the equation for the operating line with a slope of L/G.

Design equations

Assuming that (1) the gas and liquid leaving each plate are in equilibrium and that (2) a linear relationship exists between x and y:

$$y_i^* = m x_i + c \tag{3}$$

where y_i^* is the equilibrium concentration, m is the slope of the equilibrium line, and c is the y-intercept.

Eliminating x_i in Eq. (2) and (3) yields:

$$y_{i-1} = \left(\frac{L}{mG}\right) y_i^* - \frac{Lc}{mG} + y_1 - \left(\frac{L}{G}\right) x_1 \tag{4}$$

The vapor Murphree-plate-efficiency is:

$$E_V = \frac{y_i - y_{i-1}}{y_i^* - y_{i-1}} \tag{5}$$

or

$$y_i^* = \frac{y_i - y_{i-1}}{E_V} + y_{i-1} \tag{6}$$

Substituting y_i^* from Eq. (6) into Eq. (4), and then rearranging:

$$y_i - \left(1 + \frac{mGE_V}{L} - E_V\right) y_{i-1} = mE_V x_1 + cE_V - \left(\frac{mGE_V}{L}\right) y_1 \tag{7}$$

Eq. (7) is a linear difference equation and can be solved by a method given by Mickley, others [*7*].

Eq. (7) is in the general form:

$$y_i - k_1 y_{i-1} = k_2 \tag{8}$$

Therefore:

$$y_i = Ak_1{}^i + \frac{k_2}{1 - k_1} \tag{9}$$

$$y_i = Ak_1{}^i + \frac{mE_V x_1 + cE_V - \left(\frac{mGE_V}{L}\right) y_1}{1 - \left(1 + \frac{E_V mG}{L} - E_V\right)} \tag{10}$$

$$y_i = Ak_1{}^i + \frac{E_V\left(mx_1 + c - \left(\frac{mG}{L}\right) y_1\right)}{E_V\left(1 - \frac{mG}{L}\right)} \tag{11}$$

Originally published April 9, 1979

Nomenclature

- A Constant in Eq. (12)
- c Intercept of equilibrium line, Eq. (3)
- E Murphree plate efficiency
- E^0 Overall efficiency
- G Total vapor flow, moles per unit time
- L Total liquid overflow, moles per unit time
- m Slope of equilibrium line
- n Number of actual plates
- n' Number of theoretical plates
- x Mole fraction of solute in liquid phase
- y Mole fraction of solute in vapor phase
- y^* Equilibrium mole fraction of solute in vapor phase
- α As defined by Eq. (17)
- α' As defined by Eq. (26)
- β_L As defined by Eq. (25)
- β_V As defined by Eq. (16)
- λ L/mG

Subscripts:

- i ith plate of column
- 1 Bottom of column
- 2 Top of column
- L Liquid
- V Vapor

Cancelling E_V and putting in the value of k_1:

$$y_i = A\left(1 + \frac{mGE_V}{L} - E_V\right)^i + \frac{mx_1 + c - \left(\frac{mG}{L}\right)y_1}{1 - \frac{mG}{L}} \qquad (12)$$

Where A is a constant that is evaluated from boundary conditions.

When $i = 0, y_i = y_1$, and A is determined. The complete solution for n actual plates is:

$$\left(1 + \frac{mGE_V}{L} - E_V\right)^n = \frac{\left(1 - \frac{mG}{L}\right)y_n - mx_1 - c + \left(\frac{mG}{L}\right)y_1}{y_1 - mx_1 - c} \qquad (13)$$

Since the liquid composition at the bottom, x_1, is not often specified, it may be eliminated from Eq. (13). A material balance of the solute for the entire column yields:

$$x_1 = \left(\frac{G}{L}\right)y_1 - \left(\frac{G}{L}\right)y_2 + x_2 \qquad (14)$$

Combining Eq. (13) and (14), and simplifying, yields the final design equation for n actual plates:

$$\beta_V^n = \frac{y_1 + \alpha}{y_n + \alpha} \qquad (15)$$

where:

$$\beta_V = \frac{1}{1 + E_V\left(\frac{mG}{L} - 1\right)} \qquad (16)$$

and:

$$\alpha = \frac{y_2 - \left(\frac{L}{mG}\right)(mx_2 + c)}{\left(\frac{L}{mG}\right) - 1} \qquad (17)$$

Eq. (15) is used to compute actual compositions at each plate. In addition, when $E_V = 1$, then $\beta_V = L/mG$, which is the absorption factor. β_V in Eq. (15) may be called the modified absorption factor.

Furthermore, when $E_V = 1$, Eq. (15) reduces to Chen's equation for calculating the number of theoretical plates in absorption [5].

For calculating the number of actual plates, Eq. (15) is rearranged as:

$$n = \frac{\ln\left[\frac{y_1 + \alpha}{y_n + \alpha}\right]}{\ln \beta_V} \qquad (18)$$

and the number of ideal plates, n', is:

$$n' = \frac{\ln\left[\frac{y_1 + \alpha}{y_n + \alpha}\right]}{\ln \lambda} \qquad (19)$$

where:

$$\lambda = L/mG \qquad (20)$$

The overall efficiency, E^0, is calculated as follows:

$$E^0 = \frac{n'}{n} = \frac{\ln \beta_V}{\ln \lambda} = \frac{\ln\left[1 + E_V\left(\frac{mG}{L} - 1\right)\right]}{\ln\left(\frac{mG}{L}\right)} \qquad (21)$$

Eq. (21) is the same as that given by Treybal [8].

Liquid-phase equations

Similarly, the Murphree plate efficiency in terms of the liquid phase is:

$$E_L = \frac{x_{i+1} - x_i}{x_{i+1}^* - x_i} \qquad (22)$$

Combining Eq. (22) with those equations for the equilibrium and operating lines, the following difference equation is obtained for boundary conditions of: $x_i = x_2$ at $i = 0$; and $x_i = x_n = x_1$ at $i = n$.

$$x_{i+1} - \left(1 - E_L + \frac{E_L L}{mG}\right)x_i = \frac{E_L y_2}{m} - \frac{E_L L}{mG}x_2 - \frac{E_L c}{m} \qquad (23)$$

As before, the final design equation for n actual plates is:

$$\beta_L^n = \frac{x_n + \alpha'}{x_2 + \alpha'} \qquad (24)$$

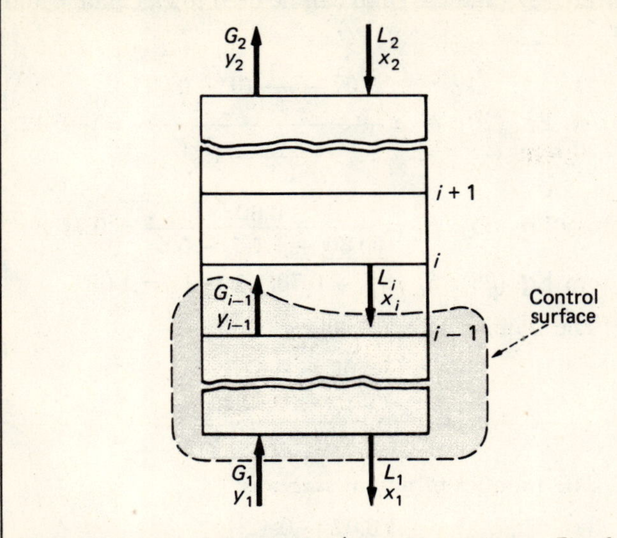

Material balance for stripping/absorbing column Fig. 1

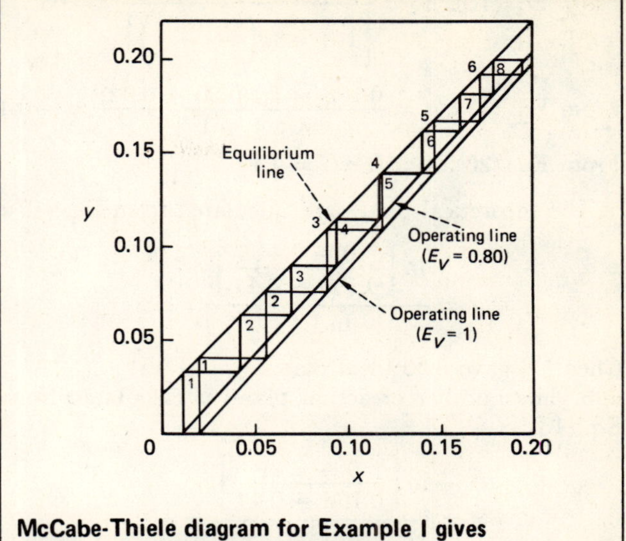

McCabe-Thiele diagram for Example I gives 6 theoretical stages, 8 actual stages Fig. 2

where:

$$\beta_L = 1 + E_L\left(\frac{L}{mG} - 1\right) \quad (25)$$

$$\alpha' = \frac{x_1 - \left(\frac{G}{L}\right)(y_1 - c)}{\left(\frac{mG}{L}\right) - 1} \quad (26)$$

Therefore:

$$n = \frac{\ln\left[\frac{x_n + \alpha'}{x_2 + \alpha'}\right]}{\ln \beta_L} \quad (27)$$

When $E_L = 1$, Eq. (27) becomes:

$$n' = \frac{\ln\left[\frac{x_n + \alpha'}{x_2 + \alpha'}\right]}{\ln \lambda} \quad (28)$$

From the last two equations:

$$E^0 = \frac{n'}{n} = \frac{\ln \beta_L}{\ln \lambda} = \frac{\ln\left[1 + E_L\left(\frac{L}{mG} - 1\right)\right]}{\ln\left(\frac{L}{mG}\right)} \quad (29)$$

This is the same equation developed by Chen [12]. Equating Eq. (21) and (29) gives the relationship between E_L and E_V:

$$\frac{\ln\left[1 + E_L\left(\frac{L}{mG} - 1\right)\right]}{\ln\left(\frac{L}{mG}\right)} = \frac{\ln\left[1 + E_V\left(\frac{mG}{L} - 1\right)\right]}{\ln\left(\frac{mG}{L}\right)} \quad (30)$$

Since $-\ln\left(\frac{L}{mG}\right) = \ln\left(\frac{mG}{L}\right)$:

$$-\ln\left[1 + E_L\left(\frac{L}{mG} - 1\right)\right] = \ln\left[1 + E_V\left(\frac{mG}{L} - 1\right)\right] \quad (31)$$

Or, since $\lambda = L/mG$:

$$1 + E_V\left(\frac{1}{\lambda} - 1\right) = \frac{1}{1 + E_L(\lambda - 1)} \quad (32)$$

$$E_V = \frac{\lambda}{(1-\lambda)(1 + E_L(\lambda - 1))} - \frac{\lambda}{(1-\lambda)} \quad (33)$$

$$E_V = \frac{E_L}{E_L + \frac{mG}{L}(1 - E_L)} = \frac{\lambda E_L}{1 + E_L(\lambda - 1)} \quad (34)$$

This equation is the same as that given by Smith [13]:

$$E_L = \frac{E_V}{E_V + \lambda(1 - E_V)} \quad (35)$$

Illustrations

Examples are given to show the use of the proposed equations, and results are compared with those from the literature.

Example I: It is desired to desorb a solute from a solution by using a gas. The solution enters a plate column at 20 mole % of solute A and leaves the bottom at 2 mole %. The entering gas is solute-free. The slope of the operating line is 1.1 and the equilibrium line is given by $y = x + 0.02$ (from Chen [5]).

Determine the number of theoretical plates and the vapor concentration at each plate. When the Murphree plate efficiency is $E_V = 0.80$, calculate the number of actual plates. Solution:

a. The number of theoretical plates:

The values given in the problem are: $y_1 = 0$, $x_1 = 0.02$, $x_2 = 0.20$, $m = 1$, and $L/G = 1.1$.

y_2 can be calculated by making a material balance around the entire column:

$$y_2 = \frac{L}{G}(x_2 - x_1) + y_1 = 1.1(0.20 - 0.02) + 0$$

$$= 0.198$$

Chen calculated y_2 to three significant figures. The same value is used here for comparison.

From Eq. (16), $\beta_V = \dfrac{1}{\left[1 + 0.80\left(\dfrac{1}{1.1} - 1\right)\right]} = 1.08$

From Eq. (17), $\alpha = \dfrac{0.198 - 1.1(0.20 + 0.02)}{1.1 - 1} = -0.44$

From Eq. (20), $\lambda = L/mG = 1.1$

The theoretical plates are calculated from Eq. (19):

$$n' = \frac{\ln\left[\dfrac{0 - 0.44}{0.198 - 0.44}\right]}{\ln 1.1} = 6.3$$

Chen [5] gave 6.30 ideal plates.

b. The number of actual plates is calculated from Eq. (18):

$$n = \frac{\ln\left[\dfrac{0 - 0.44}{0.198 - 0.44}\right]}{\ln 1.08} = 8.0$$

The overall efficiency is:

$$E^0 = \frac{n'}{n} = \frac{6.3}{8.0} = 0.79$$

c. The theoretical and actual compositions at each plate:

Ideal (from Eq. (19)):

$$y_{n'} = \frac{y_1 + \alpha}{\lambda^{n'}} - \alpha = \frac{-0.44}{1.1^{n'}} + 0.44$$

$$x_{n'} = y_{n'} - 0.02$$

Actual (from Eq. (18)):

$$y_n = \frac{y_1 + \alpha}{\beta_V{}^n} - \alpha = \frac{-0.44}{1.078^n} + 0.44$$

$$x_n = y_n - 0.22$$

Only calculations for theoretical compositions are shown here. Values are also obtained from a McCabe-Thiele diagram (see Fig. 2).

Theoretical vapor composition

n	From Eq. (19)	From Chen [5]	Graphical (Fig. 2)
1	0.04	0.04	0.04
2	0.08	0.08	0.08
3	0.11	0.11	0.11
4	0.14	0.14	0.14
5	0.17	0.17	0.17
6	0.19	0.19	0.19
7	0.21	0.21	0.21

Theoretical liquid composition

n	From Eq. (19)	From Chen [5]	Graphical (Fig. 2)
1	0.02	0.02	0.02
2	0.06	0.06	0.06
3	0.09	0.09	0.09
4	0.12	0.12	0.12
5	0.15	0.15	0.15
6	0.17	0.17	0.17
7	0.19	0.19	0.19

Eq. (27) and (28) also can be used to calculate n and n':

$$\text{From Eq. (26): } \alpha' = \frac{0.02 - \dfrac{1}{1.1}(0 - 0.02)}{\dfrac{1}{1.1} - 1} = -0.42$$

$$\text{From Eq. (35): } E_L = \frac{0.80}{0.80 + 1.1(1 - 0.8)} = 0.78$$

From Eq. (25): $\beta_L = 1 + 0.78(1.1 - 1) = 1.08$

The number of ideal stages is:

$$n' = \frac{\ln\left[\dfrac{0.02 - 0.42}{0.20 - 0.42}\right]}{\ln 1.1} = 6.3$$

The number of actual stages is:

$$n = \frac{\ln\left[\dfrac{0.02 - 0.42}{0.2 - 0.42}\right]}{\ln 1.08} = 7.8$$

Example II: A process for making small amounts of hydrogen by cracking ammonia is being considered, and it is desired to remove uncracked NH_3 from the resulting gas. The gas will consist of hydrogen and nitrogen in the molar ratio of 3:1, containing 3% ammonia by volume, at 2 atm and 80°F. Determine the number of actual plates if $L = 0.322$ lb-mole/s, $G = 0.145$ lb-mole/s and the equilibrium line is $y = 0.707\,x$ (from Treybal [8], p. 226).

The following values are calculated by Treybal: $x_2 = 0, y_2 = 0.000018, y_1 = 0.03$ and $E_V = 0.575$.

From Eq. (16):

$$\beta_V = \frac{1}{1 + 0.575\left(\dfrac{0.707 \times 0.145}{0.322} - 1\right)} = 1.645$$

From Eq. (17):

$$\alpha = \frac{0.000018}{\dfrac{0.322}{0.707 \times 0.145} - 1} = 0.00000841$$

From Eq. (18):

$$\text{n} = \frac{\ln\left[\dfrac{0.03 + 0.00000841}{0.000018 + 0.00000841}\right]}{\ln 1.645} = 14.1$$

The value given by Treybal is 14.

When determining the required stages in the very dilute and very concentrated regions at the tower ends, a McCabe-Thiele plot can lead to errors. Other methods, such as using a logarithmic scale [9,10] or a probability scale [11] are time-consuming. The proposed equations can replace these graphical methods. It should be noted that these equations are only applicable at the two far ends of a column, where the equilibrium lines are linear.

Example III: A continuous distillation column separates a mixture containing 55 mole % A and 45 mole % B into an overhead product of 99.95 mole % A and a bottoms product of 0.05 mole % A. The feed enters the tower at its boiling point; the average relative volatility is 6; the optimum reflux ratio L/G is 0.40, whereas the corresponding stripping ratio is $L/G = 1.49$. From a material balance, the ratio D/G is 0.60 and $W/G = 0.49$. Here, D = distillate, and W = bottoms.

This example is from Horvath and Schubert [9]. Calculations for the very dilute and very concentrated regions of this example were performed by Chen [12]. Here, we will check our equations with the results of Chen, as well as those of Horvath and Schubert.

First, calculate the number of ideal stages in the very dilute region of the stripping section for x_n ranging from 0.0005 to 0.03. The value of $x_n = 0.03$ was obtained by Chen from a standard McCabe-Thiele diagram for a point several stages above the tower bottom. Next, calculate the number of plates in the very concentrated region for x_n from 0.95 to 0.9995. The value of 0.95 is also taken from a McCabe-Thiele diagram by Chen.

The equilibrium line in the low-concentration region is calculated by Chen as:

$$y_n^* = 0.1666x_n + 0.8333$$

and the operating line is given by Horvath and Schubert as:

$$y_n = 0.40x_n + 0.5997$$

From these two equations, we have: $m = 0.1666$, $c = 0.8333$, $L/G = 0.40$, or $\lambda = L/mG = 2.40$. Points given are: $x_2 = 0.9995$, and $x_1 = 0.95$. Therefore:

$$y_1 = 0.40(0.95) + 0.5997 = 0.9797$$
$$y_2 = 0.40(0.9995) + 0.5997 = 0.9995$$

Using Eq. (17):

$$\alpha = \frac{0.9995 - (2.4)(0.1666 \times 0.9995 + 0.8333)}{2.4 - 1}$$
$$= -1.00$$

The number of ideal plates in this region is calculated from Eq. (19):

$$n' = \frac{\ln\left[\dfrac{0.9797 - 1.00}{0.9995 - 1.00}\right]}{\ln 2.4} = 4.2$$

Chen calculated 3.96 stages, and Horvath and Schubert gave about 4 ideal plates (graphically).

The equilibrium line for the dilute region is:

$$y_n^* = 6x_n$$

Thus, $m = 6$, $c = 0$.

Points given:

$$x_2 = 0.03,\ x_1 = 0.0005 = x_n = y_1$$

and

$$\frac{L}{G} = 1.49, \text{ or } \frac{L}{mG} = 0.248 = \lambda.$$

Using Eq. (26):

$$\alpha' = \frac{0.0005 - \dfrac{1}{1.49}(0.0005 - 0)}{\dfrac{6}{1.49} - 1} = 0.0000543$$

Then, from Eq. (28):

$$n' = \frac{\ln\left[\dfrac{0.0005 + 0.0000543}{0.03 + 0.0000543}\right]}{\ln 0.248}$$
$$= 2.9 \text{ ideal plates}$$

This agrees with the value obtained graphically by Horvath and Schubert, which is slightly less than 3 ideal stages.

Conclusion

From the above examples, it is shown that the proposed equations can predict the number of ideal and actual plates, and the actual compositions at each plate, for either the absorber or stripper in a plate column. Furthermore, these equations can also be used to determine the efficiency of the column, provided that other parameters are specified.

When the equilibrium line is not straight, the column can be considered to be composed of several short sections with linear equilibrium and operation lines in each. Thus, the proposed equations would be used to calculate the number of plates in each section, and the actual plates would be the sum of the values for each section.

References

1. Walker, W. H., others, "Principles of Chemical Engineering," 3rd ed., McGraw-Hill, New York, 1937, p. 488.
2. Souders, M., The Countercurrent Separation Process, *Chem. Eng. Progr.*, Vol. 60, No. 2, Feb. 1964, p. 75.
3. Souders, M. and Brown, G. G., Fundamental Design of Absorbing and Stripping Columns for Complex Vapors, *Ind. Eng. Chem.*, Vol. 24, 1932, p. 519.
4. Kremser, A., Theoretical Analysis of the Absorption Process, *Nat. Petrol. News*, Vol. 22, No. 21, May 21, 1930, p. 48.
5. Chen, Ning Hsing, Calculating Theoretical Plates in Absorbers or Strippers, *Chem. Eng.*, Vol. 71, No. 10, May 11, 1964, p. 159.
6. Nguyen, Hung Xuan, Calculating Actual Stages in Countercurrent Leaching, *Chem. Eng.*, Vol. 85, No. 25, Nov. 6, 1978, p. 121.
7. Mickely, H. S., others, "Applied Mathematics in Chemical Engineering," Chap. 4, McGraw-Hill, New York, 1957.
8. Treybal, R. E., "Mass Transfer Operations," McGraw-Hill, New York, 1955, p. 226.
9. Horvath, P. J., and Schubert, R. F., Find Distillation Stages Graphically, *Chem. Eng.*, Vol. 65, No. 3, Feb. 10, 1958, p. 129.
10. Robinson, C. S., and Gilliland, E. R., "Elements of Fractional Distillation," McGraw-Hill, New York, 1957, p. 129.
11. Lowenstein, J. G., Distillation Column Designing, *Ind. Eng. Chem.*, Vol. 54, No. 1, Jan. 1962, p. 61.
12. Chen, Ning Hsing, Basic Equation Solves Many Mass-Transfer Problems, *Chem. Eng.*, Vol. 74, No. 1, Jan. 2, 1967, p. 93.
13. Smith, B. D., "Design of Equilibrium Stage Processes," McGraw-Hill, New York, 1963, p. 599.

The author

Hung Xuan Nguyen is a research professional at Philip Morris U.S.A., Research Center, P.O. Box 26583, Richmond, VA 23261. Telephone: (804) 271-2000. He has worked for Allied Chemical Corp., where his responsibilities included solving technical problems and making process improvements in heat transfer, fluid dynamics and mass transfer. He holds two chemical engineering degrees—a B.S. from the University of Lowell and an M.S. from the University of Massachusetts. He is an associate member of AICHE.

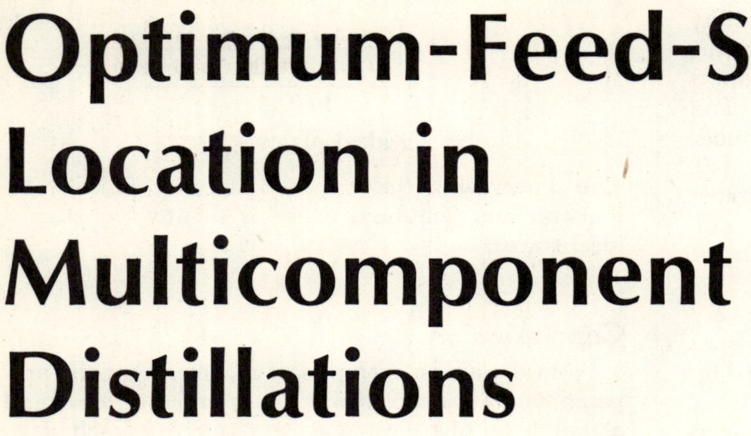

Optimum-Feed-Stage Location in Multicomponent Distillations

The simple rules described here facilitate determining the feed stage of a distillation column that will yield the greatest separation of a selected pair of key components.

JOSEPH H. MAAS, Consultant

A suitable feed location in a fractionating column is an essential requirement for effective distillation. The location is an optimum one if it results either in producing maximum separation from a fixed number of fractionation stages, or in a minimum number of total stages for some specified separation.

Multicomponent distillations can often be reduced to the separation of only two of the feed components in the mixture. Then, the Fenske equation is convenient to calculate separations between pairs of components.

The location of the optimum feed stage depends upon which two feed components (called *key* components) are selected for their maximum separation, and also upon distillation conditions. Fig. 1 shows how the optimum location changes with the component pair chosen. (For a given component pair, the feed location yielding the greatest number of Fenske minimum stages is the optimum location for the feed.)

Fig. 1 also shows that the optimum feed stage for a volatile key pair, such as *iso*-butane and *n*-butane will occur at a higher stage level in the tower than if a less volatile pair—such as *n*-butane and *iso*-pentane—were chosen as the key components requiring maximum separation. For this rule to hold true, however, the comparison must be made at some constant molar ratio of distillate to bottoms, with a constant reflux rate, and with other distillation variables—such as operating pressure—held constant.

Rules for Locating the Optimum Feed Stage

Rules regarding the location of an optimum feed stage are important whether or not multicomponent-distillation computer programs are used. For both types of calculations, the number of tower stages and the feed-stage location must be known.

When using computer programs, preliminary calculations are still necessary to determine the optimum location of a feed. Here, another rule applies to the approach taken to the optimum stage from data obtained from a trial calculation. This rule is illustrated by the data in Fig. 2, which is a plot of the mole fractions of a pair of key components selected for maximum separation in the debutanizer. Each curve corresponds to one feed inlet; the concentrations used are those of the stage liquors.

From the position of the feed inlet used in a trial computation, and from the slope of the curve just above and just below the feed, it is possible to predict the direction in which the optimum feed location lies. The rule is that the optimum feed location will be on that side of the feed stage that shows the most negative slope condition.

Only one stage above and one below the feed stage are necessary to determine the slopes. These are considered positive when the curve is upward and to the right. If both slopes are negative, the location of the optimum feed will be in the direction of the side with the larger negative slope value.

The optimum feed-stage location is confirmed in Fig. 2 as that giving the lowest ratio of key component concentrations in the reboiler liquid (extreme left of the curve) and the highest ratios at the distillate top (Stage 34). Notice how the shape of the curve tends to flatten as the optimum is approached near the center.

High reflux rates produce curves that are almost straight lines, on which increased separations—when near the optimum—are indicated by even lower concentration ratios at the bottom stage, and higher ratios at the top.

Simple plots of key-component concentrations would also illustrate the approach to an optimum feed stage.

Originally published April 16, 1973

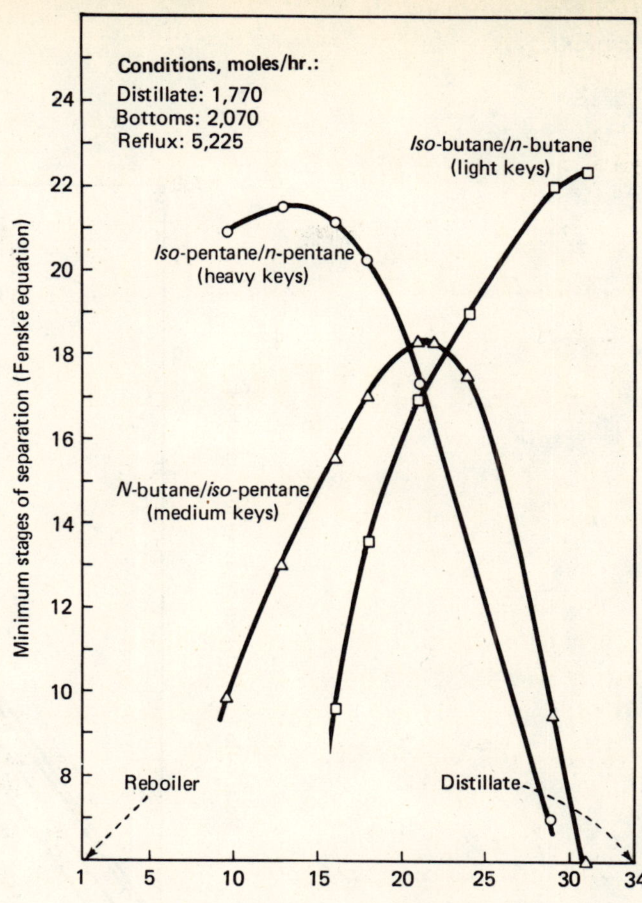

SEPARATION of three key pairs at various feed locations (stagewise) in debutanizer operation—Fig. 1

Here, the rule stating that the direction of the optimum stage lies towards the negative slope is easy to apply, if it is desired to revise existing distillation programs so that the optimum feed stage may be determined within them.

Physical conditions are responsible for the oddly shaped curves of Fig. 2, which indicate that the concentration of the light key is higher below the feed stage than it is at some stages above. The condition at Stage 25, for example, shows what happens when the feed enters too high up in a well-refluxed tower. Under such circumstances, the feed is chilled and is carried several stages downward before it begins to evaporate and yield its butane, which is carried upward with the vapor boilup.

A similar condition exists when the feed is at Stage 16, where the key components in the feed are being carried upward into a region of the tower where the stage liquors are rich in *iso*-pentane. The result of this is a temporary drop in the concentration ratio of the keys on the stages just above the feed (16th) stage.

The final placing of the feed at Stage 21 comes very close to placing it at the optimum stage, where the smallest amount of key components is recycled from the feed. This suggests that the optimum feed-stage location could be defined as that stage where entrance of the feed results in a minimum recirculation of the key components.

Sample Calculation

Determine the optimum feed stage for a multicomponent distillation whose key components are *n*-butane and *iso*-pentane.

When using the type of plot shown in Fig. 1, the Fenske minimum stages are calculated. For example, take a set of data such as shown in Fig. 2, and assume that the relative volatility (α) of the keys is 1.73.

The Fenske equation—used to calculate the number of minimum stages of separation between two components—is:

$$(n_{LK_D}/n_{HK_D})(n_{HK_B}/n_{LK_B}) = \alpha^{N_m}$$

where n_{LK_D} = moles of light key in distillate; n_{HK_D} = moles of heavy key in distillate; n_{HK_B} = moles of heavy key in bottoms; n_{LK_B} = moles of light key in bottoms; α = relative volatility of the keys, average over tower; and N_m = number of stages of separation when reflux is infinite.

Starting with Stage 21, the ratio of *n*-butane to *iso*-pentane (Fig. 2) in the distillate = 74.0; in the reboiler

---

Meet the Author

Joseph H. Maas is an independent chemical engineer in the Delaware Valley (104 Treaty Rd., Drexel Hill, PA 19026), who works on research and development projects. Before, he was a senior process engineer for Hydrocarbon Research, Inc., in West Germany. He has worked on both the design and startup operations of petroleum and chemical plants in Abadan, Wales, France, West Germany and the U.S. He holds a Ch.E. degree from the University of Cincinnati, and is a registered professional engineer in Pennsylvania and New Jersey. Recently, he was initiated as a Tau Beta Pi Eminent Engineer at Drexel University.

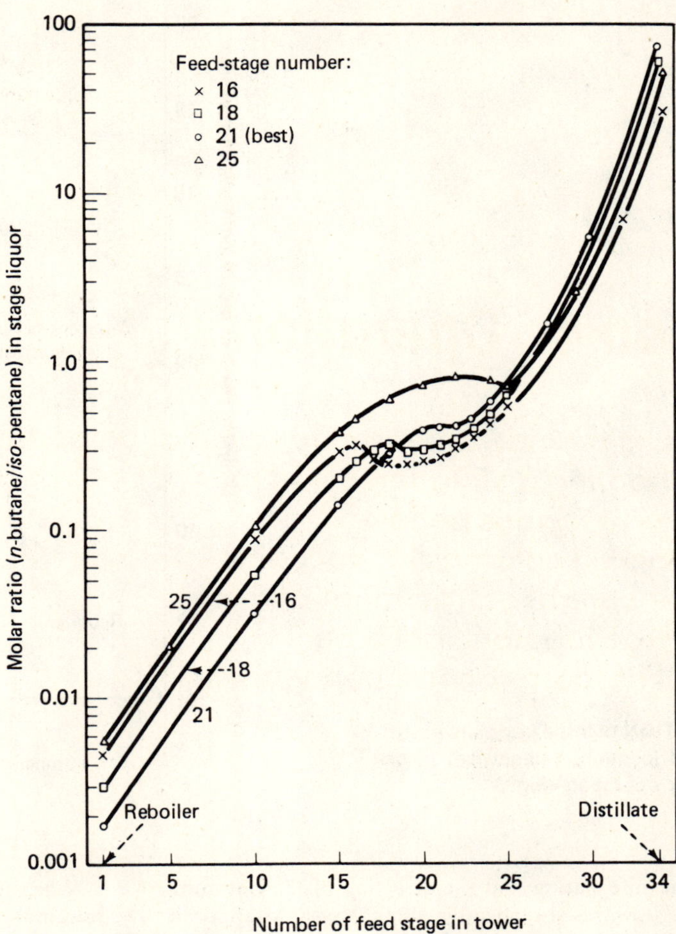

CONSTANT-FEED-LOCATION CURVES
show concentration ratio of key components in stage liquor—Fig. 2

bottoms, it is 0.00170. Since these ratios are equivalent to the mole ratios in the Fenske equation:

$$74.0/0.00170 = 1.73^{N_m}$$

where $N_m = \log(74.0/0.00170)/\log 1.73 = 19.5$

The next calculation shows how the results of computer runs are successively used to approach—in the direction of the largest negative slope—the optimum feed location.

To calculate the slope of the concentration ratio curves in Fig. 2, the following stage molar concentrations of *n*-butane and *iso*-pentane are taken into consideration, with Stage 21 as the target:

Component	Stage 20	Stage 21	Stage 22
N-butane	0.130816	0.128752	0.172266
Iso-pentane	0.334828	0.303372	0.431828
Molar ratio, *n*-butane/*iso*-pentane	0.39069	0.42440	0.39892
Ratio differential per tray (slope)	+0.03371		−0.02548

Since, according to the rule, the optimum feed stage will be found by moving the next trial calculation toward the 22nd stage, the next stage to be considered is the 25th:

Component	Stage 24	Stage 25	Stage 26
N-butane	0.200393	0.180123	0.273404
Iso-pentane	0.253670	0.245749	0.318066
Molar ratio, *n*-butane/*iso*-pentane	0.78997	0.73295	0.85958
Ratio differential per tray (slope)	−0.05702		+0.12663

This last trial calculation shows that the optimum feed stage lies below the 25th, possibly nearer to the 21st. A final trial at Stage 22 reveals that the optimum stage is between the 21st and the 22nd, possibly nearer to the 21st:

Component	Stage 21	Stage 22	Stage 23
N-butane	0.155016	0.140082	0.185121
Iso-pentane	0.301624	0.291048	0.399574
Molar ratio, *n*-butane/*iso*-pentane	0.51393	0.48130	0.46329
Ratio differential per tray (slope)	−0.03263		−0.01801

Because distillate and bottoms compositions are almost identical to the feed at Stage 21, their values are not plotted in Fig. 2.

Section 4 Operation

Simulations provide blueprint for distillation operation

An optimum strategy for operating a complex distillation system, which contains old and new columns in a combination of series and parallel arrangements, emerges from comprehensive simulation studies done before plant startup.

A. Chou, A. M. Fayon and B. L. Bauman, Mobil Oil Corp.

☐ The rapid rise in energy costs has given engineers a new reason for finding more opportunities to save energy in the chemical process industries. One such opportunity arose at Mobil Oil Corp.'s ethylene plant in Beaumont, Tex. As part of a plan to modernize the facility (built in 1961) and double ethylene capacity, the ethane/ethylene separation section was considerably revamped.*

The present system consists of an entirely new setup that contains previously existing and newly built distillation columns in a combination of series and parallel arrangements. There is also a heat pump or vapor-compression system in which overhead vapor is compressed to a temperature and pressure such that the heat of condensation can be used to supply reboiler heat.

Recognizing the complexity of these distillation-system configurations, digital-computer simulation studies were undertaken before plant startup in mid-1975. The studies were intended to identify unusual operating characteristics, and to provide guidelines for making specification product with a minimum consumption of energy. This article describes how the simulation studies were made, and what was learned about the optimum strategy for operating the distillation system.

Process background

The final distillation train of the ethylene plant is the ethane/ethylene separation section. This consists of seven columns (Fig. 1), including a rectifier section, topping still and C_2 splitters. Both the topping still and splitters produce a 99.9% ethylene-product stream. The

*Design and construction were contracted to Stone & Webster Engineering Corp.

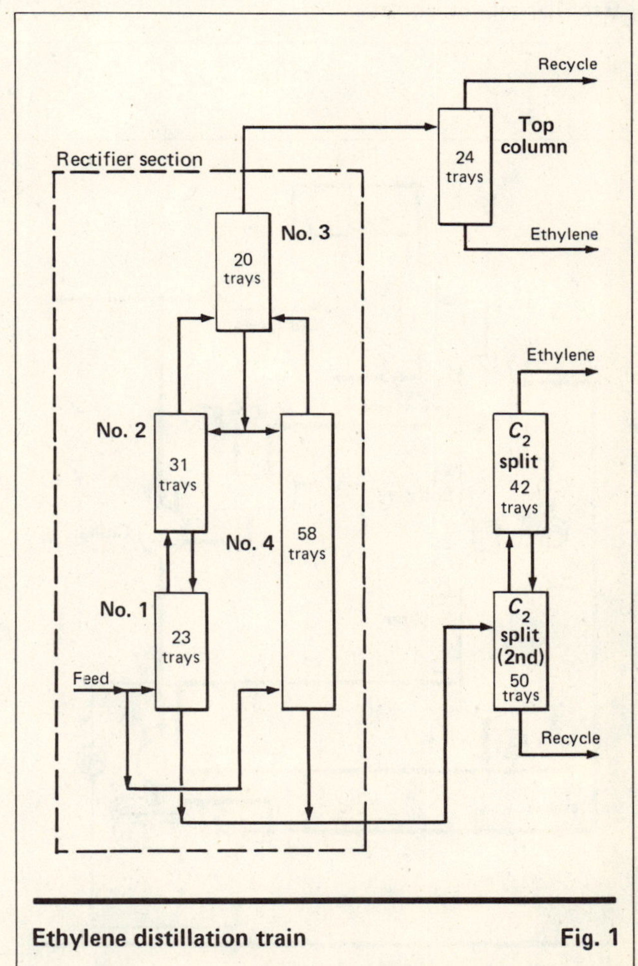

Ethylene distillation train **Fig. 1**

Originally published June 7, 1976

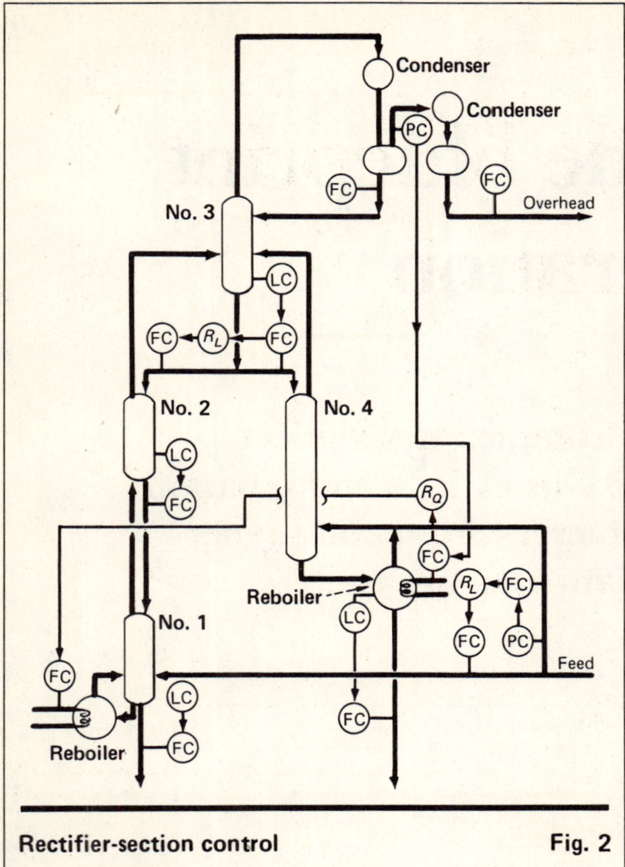

Rectifier-section control Fig. 2

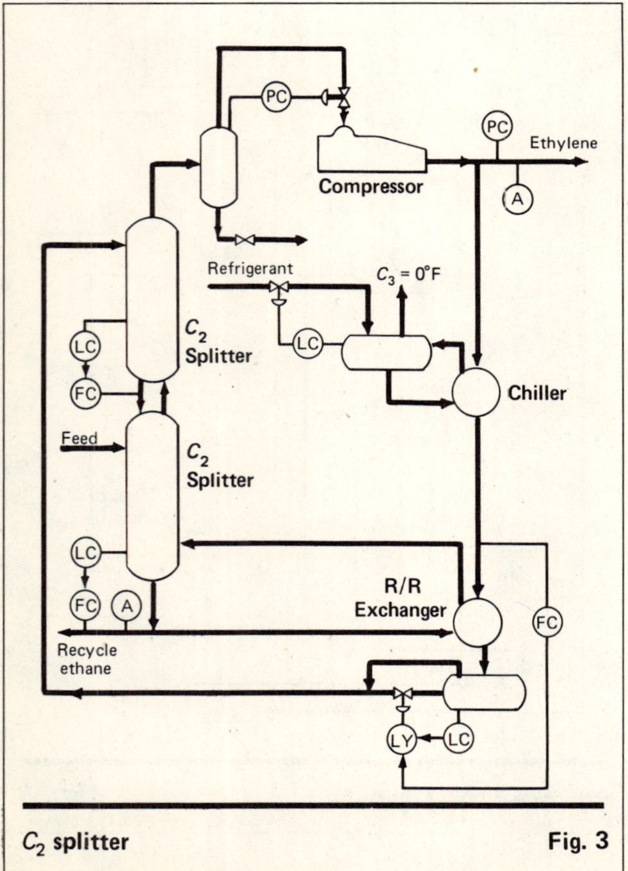

C_2 splitter Fig. 3

Nomenclature

C	Composition, mole %
H_I	Enthalpy of reflux to the C_2 splitter, Btu/h
H_O	Enthalpy of reflux vapor, Btu/h
L/V	Column internal liquid-to-vapor ratio, dimensionless
Q_{ch}	Heat removed by chiller, Btu/h
Q_{comp}	Heat input by compressor to C_2-splitter reflux vapor, Btu/h
Q_{cond}	Condenser duty, Btu/h
Q_r	Reboiler duty, Btu/h
Q_t	Total heat removed, Btu/h, as defined by Eq. (4)
R_f, R_l, R_q	Ratios of feedrates, intermediate reflux rates, and reboiler duties, respectively, for parallel towers of rectifying section, dimensionless
U	Overall heat-transfer coefficient, Btu/(h)(ft^2) (°F)

first unit removes the lighter-than-ethylene components, while the splitters remove the heavier-than-ethylene components. The rectifier section distributes the load between topping still and C_2 splitters for the most economical operation.

The rectifier section contains four columns. The No. 1 and No. 2 units, with 23 and 31 trays, respectively, operate as one column, and in parallel with the No. 4 column with 58 trays. Overhead vapors from the parallel rectifying columns are fed to the bottom of the No. 3 unit (20 trays), and the bottoms of the No. 3 column are introduced to the tops of these columns as refluxes, (referred to here as "intermediate" to avoid confusion with overhead reflux). Feed, as vapor, goes into the bottom of the No. 1 and No. 4 columns. Propylene refrigeration is used for the rectifiers and topping-still overhead systems.

The analog controls of the columns of the rectifier section are shown in Fig. 2. Overhead product from this section is on flow control, the bottom products are on level control, feed to the columns is on feedline pressure control, and the reboiler heat is on overhead pressure control. The distribution of feed, reflux and reboiler duty to the parallel columns is ratioed as set by operators. Fig. 2 shows that the main controls to the parallel rectifying columns are ratioed by analog instruments from the No. 4 column.

The new C_2 splitter contains two columns in series, as shown in Fig. 3. Overhead vapor, compressed and chilled with propylene refrigerant, then exchanges heat with column bottoms to supply reboiler heat. This saves considerable energy. At the same time, integration of the overhead condenser with the column reboiler by heat exchanging of these two streams, restricts the manipulations of two most-important variables in distillation-column control—reflux and boilup rates.

To provide an additional degree of freedom, a chiller is added to the reflux cooling train. This unit uses zero-degree refrigerant to precool the reflux vapor before it enters the reflux/reboiler (R/R) heat exchangers. By this means, the vapor and liquid traffic within the column can be manipulated. This series arrangement eliminates the need for a low-temperature refrigerant as used in a parallel trimmer (Ref. 1).

In the Beaumont setup, the two variables controlling the C_2 splitters are the chiller duty and the reflux vapor rate. The former is controlled by the refrigerant level in the accumulator, and the latter is on flow control. This study shows how these variables are related to the commonly used variables of reflux rate and reboiler duty in controlling product purities.

For the sake of clarity, discussion of the computer simulation study will be given in two separate parts, covering the rectifier section and C_2 splitters.

Rectifier section

The simulation model of the rectifier section, as shown in Fig. 4, was constructed with a Mobil computer program called QUICKBAL for steady-state simulation of process equipment. Ratios of feed, intermediate reflux and reboiler duties to the parallel rectifying columns are designated as R_f, R_l, and R_q, respectively.

The operation of the rectifier section is complicated by the configuration of two parallel sets of columns. In addition to the normal operating variables of feedrate, product rates and heat duties, operators need to specify ratios of feeds, intermediate reflux and boilups (R_f, R_l, R_q) to the parallel columns for optimal operation. The main purpose of this study is to provide guidelines in setting these ratios to conserve refrigeration.

Before proceeding with the simulation of the rectifier section in terms of manipulating R_f, R_l, and R_q, it is important to show the sensitivity of this distillation section to variations of heat duties. By keeping R_f, R_l, and R_q constant, the rectifier section performs much like a single column. The separation efficiency of the column can be varied by changing overhead reflux rate while keeping reboiler duties constant, or vice versa.

Fig. 5 shows the response of ethane content in the overhead product to the heat-duty variations. The design value is very close to the knee of the curve. In other words, increasing the separation by increasing condenser duty has little influence upon overhead purity. However, decreasing the condenser-duty/overhead-product ratio to below 16 will drastically increase ethane content. This type of response is typical of distillation columns, except that different columns are designed to operate at different portions of this curve. With R_f, R_l, and R_q properly selected and fixed, the rectifier section can be manipulated as a single column according to accepted distillation technology.

The effects of R_f, R_l, and R_q were studied as described below:

R_f and R_q were set constant at 0.79, based on the column diameters, and R_l was varied while maintaining overhead ethane content at 0.07 mole % and reboiler duties at a fixed value. The effects of R_l changes on overhead product rate and Q_c/D (condenser-duty/overhead-product) are shown in Fig. 6. An optimal R_l value of 0.8 was obtained where maximum overhead product was produced and minimum condenser duty or refrigerant was used. The shape of the curves was anticipated, but the sharpness in curvature was surprising. The data show that changing R_l from 0.8 to 0.75 will result in more than an 8% increase in condenser-duty/overhead-product.

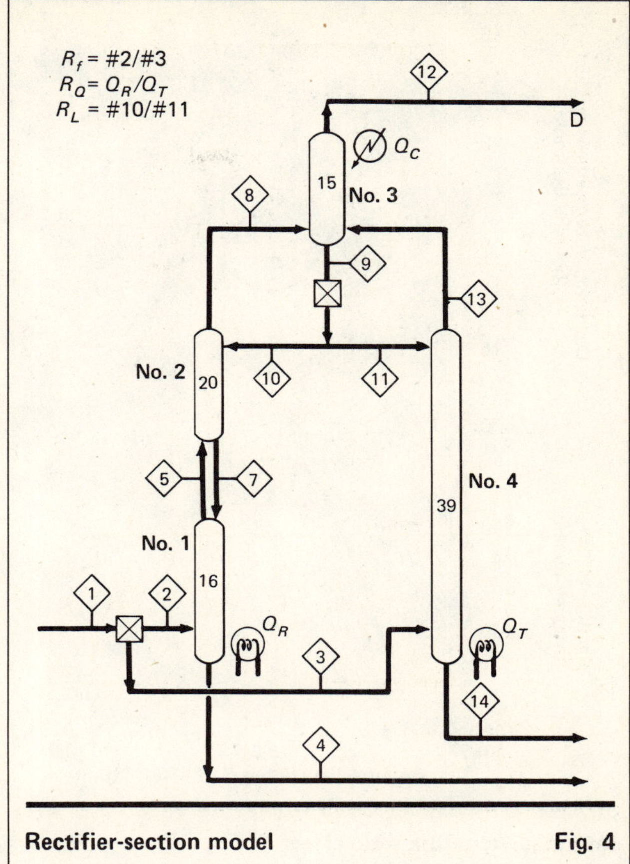

Rectifier-section model — Fig. 4

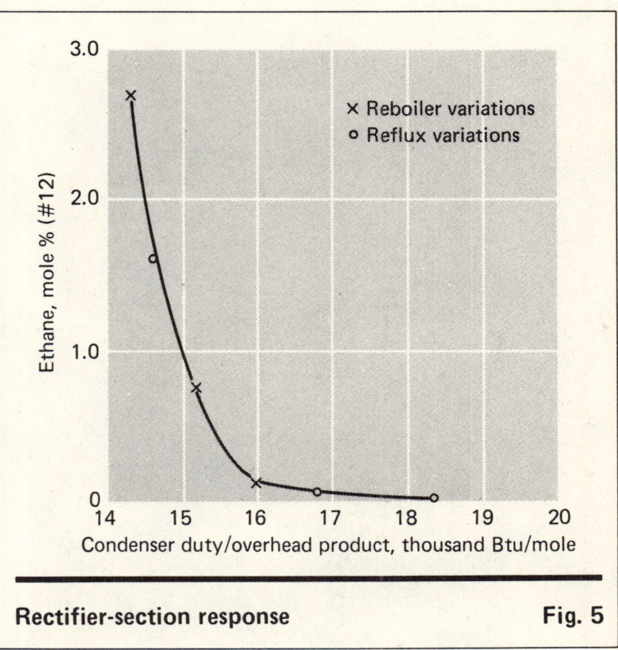

Rectifier-section response — Fig. 5

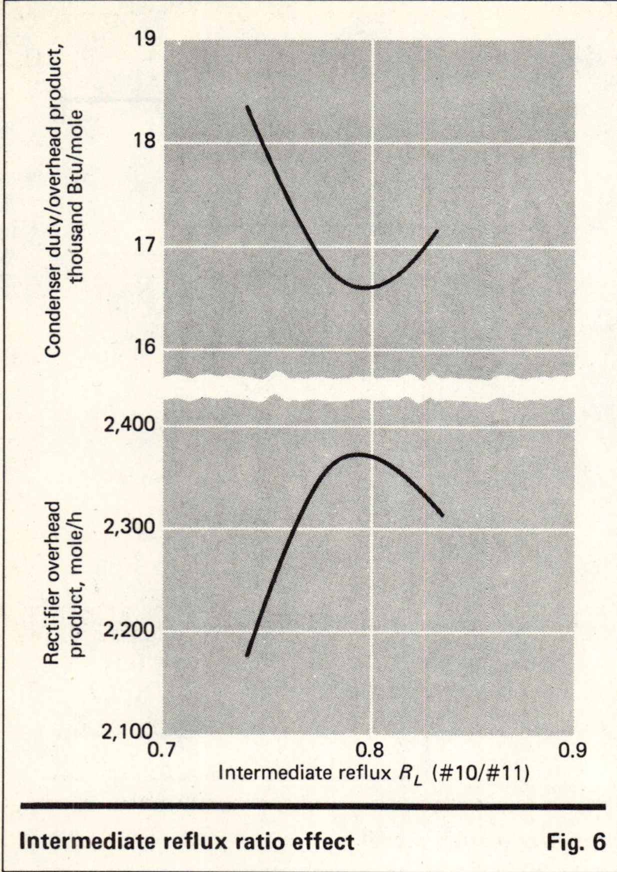

Intermediate reflux ratio effect **Fig. 6**

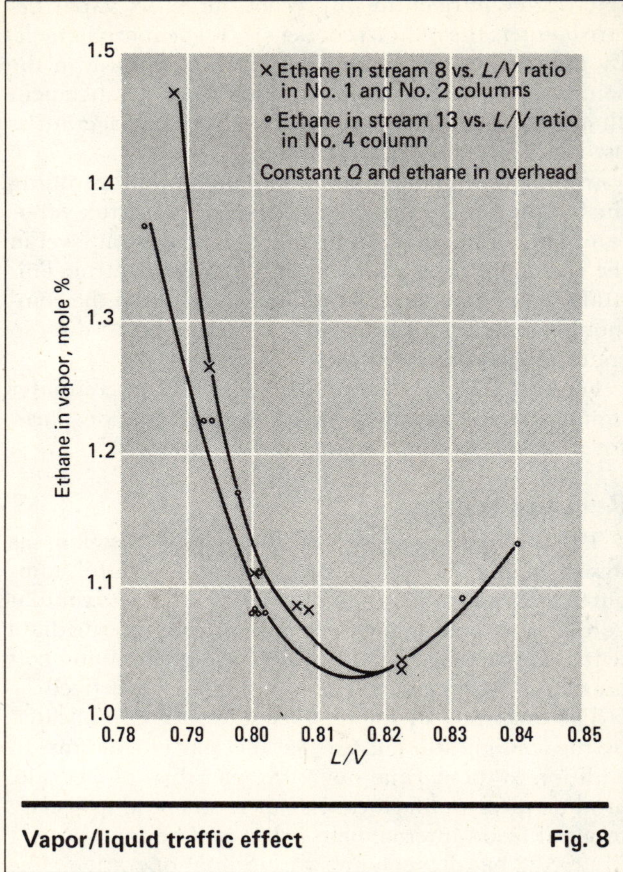

Vapor/liquid traffic effect **Fig. 8**

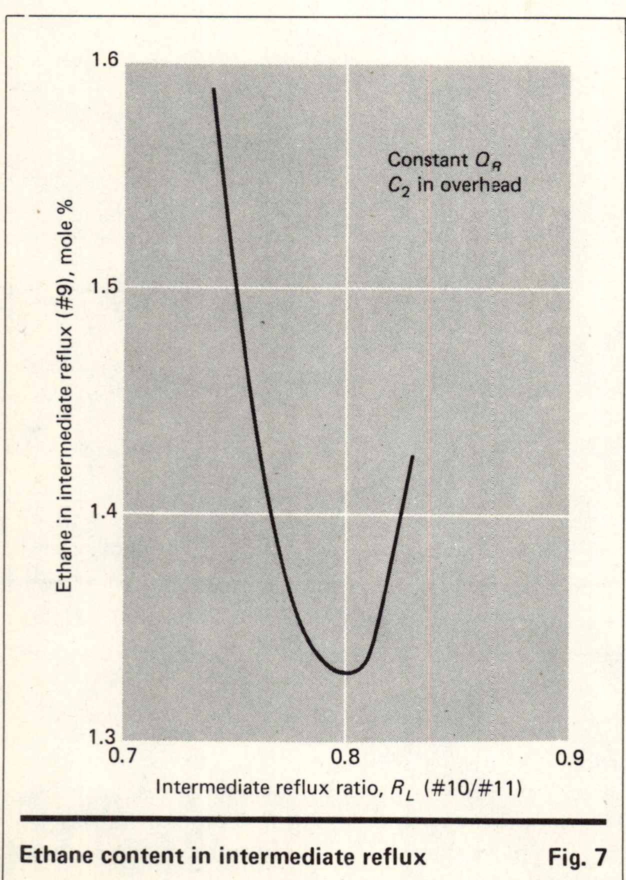

Ethane content in intermediate reflux **Fig. 7**

This phenomenon can be attributed to two causes. As shown in Fig. 5, the reflux effect is highly nonlinear. For a given amount of reflux from the second rectifier, there is an optimum distribution of the intermediate reflux between the two parallel columns in terms of separation. Wasting reflux on one column, with little improvement in purity, will result in a drastic reduction in purity for the other column. The combined overhead product from the parallel columns will show higher ethane content than it would if R_l were properly selected.

In addition, the parallel columns, as shown in Fig. 4, are not truly independent—i.e., the overheads are fed to a common column (No. 3). The intermediate reflux to the parallel rectifying columns is the bottom liquid from the No. 3 column. For given reboiler duties and overhead-product specification as defined previously, ethane content in the intermediate reflux decreases and then increases with increasing R_l values, as shown in Fig. 7. This is caused in part by the nonlinear characteristics of distillation columns, as was explained earlier.

A higher ethane content in the intermediate reflux results in poorer separation, or less product at a given purity. This is verified by the results plotted in Fig. 8, in which ethane content of the parallel rectifying columns' overhead vapor is plotted vs. L/V (liquid/vapor ratio) within the columns. The L/V ratio is equivalent to the condenser-duty changes, as shown in Fig. 5. The drastically different shapes of Fig. 5 and 8 illustrate the interaction of the two parallel columns at the conditions of simulation as given.

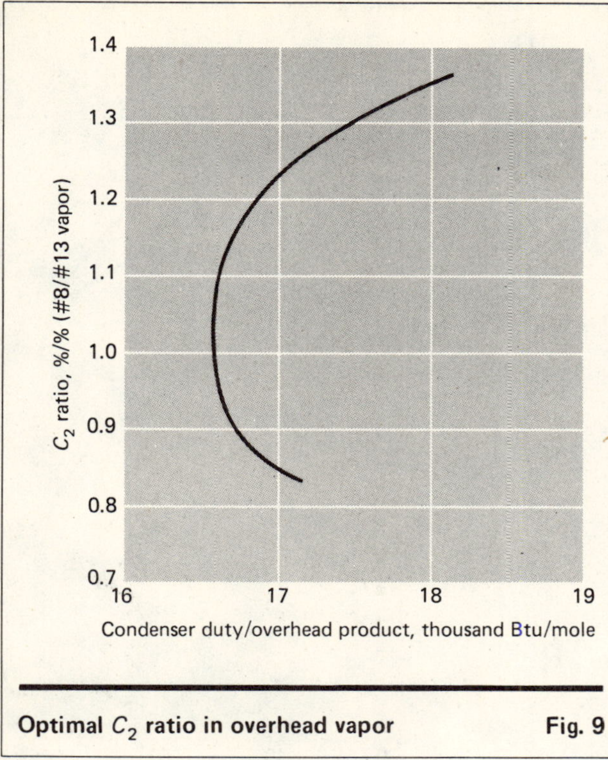

Optimal C_2 ratio in overhead vapor **Fig. 9**

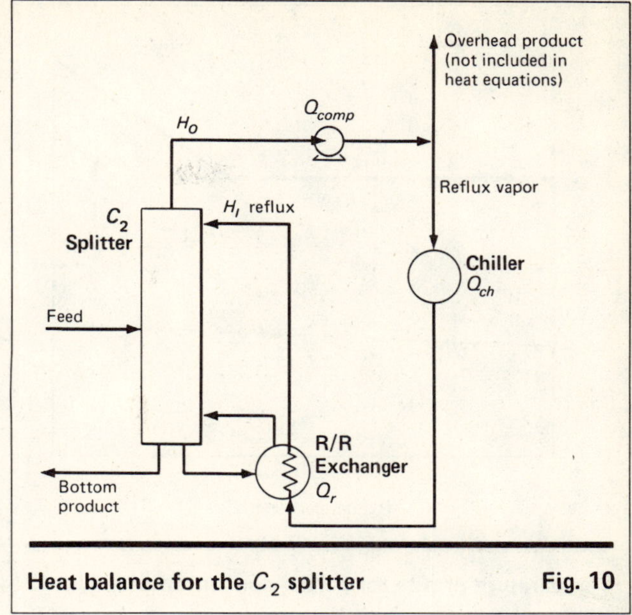

Heat balance for the C_2 splitter **Fig. 10**

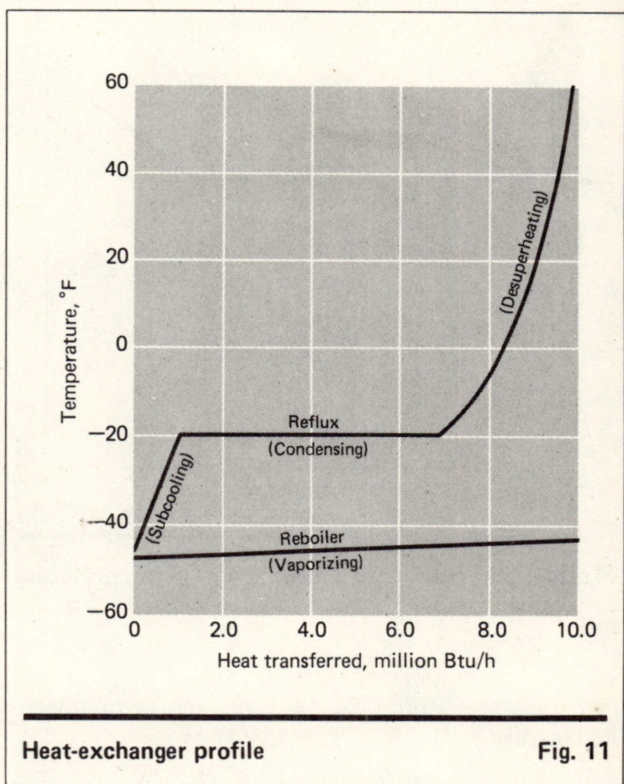

Heat-exchanger profile **Fig. 11**

At low L/V ratios, the separation effect dominates, as shown in Fig. 9, and normal distillation behavior prevails (same as Fig. 5 at low condenser-duty/overhead-product rate). At high L/V ratio, the benefit from a high reflux ratio is overshadowed by the additional ethane introduced by the intermediate reflux stream. Consequently, instead of a reduced ethane content (Fig. 5), the overhead streams show higher ethane content (Fig. 8).

It is felt that the combination of these two factors results in an extremely high sensitivity of the rectifier section to R_l changes. The optimal point of $R_l = 0.8$ indicates that slightly more reflux should be used in the "column" that has fewer theoretical trays (36 trays for the two-tower rectifier vs. 39 for the No. 4 column).

The model further shows that overhead vapor composition of the parallel columns can be used as a guideline to operate them at a minimum refrigeration consumption. Fig. 9 indicates that minimum condenser duty occurs when the overhead ethane content of each of the parallel columns is the same (or in terms of ratio, equals 1 in the figure). Hence, operators manipulate R_l so that the ratio of ethane content in the overheads is kept at 1 to make efficient use of refrigeration.

The C_2 splitters

To facilitate the understanding of C_2-splitter operation (Fig. 10), the heat balance of a distillation column with a heat pump is presented. A heat balance of the reflux vapor stream only can be represented by:

$$H_O + Q_{comp} - Q_{ch} - Q_r - H_I = 0 \qquad (1)$$

Since the overhead product leaves the distillation system as vapor and is not a part of this heat balance, the condenser duty (Q_{cond}) relating to the internal traffic of the column can be expressed by:

$$Q_{cond} = H_O - H_I \qquad (2)$$

Substituting Eq. (2) in Eq. (1), a relationship of condenser duty in terms of the heat duties in the heat-pump circuit is obtained:

$$Q_{cond} = Q_r + Q_{ch} - Q_{comp} \qquad (3)$$

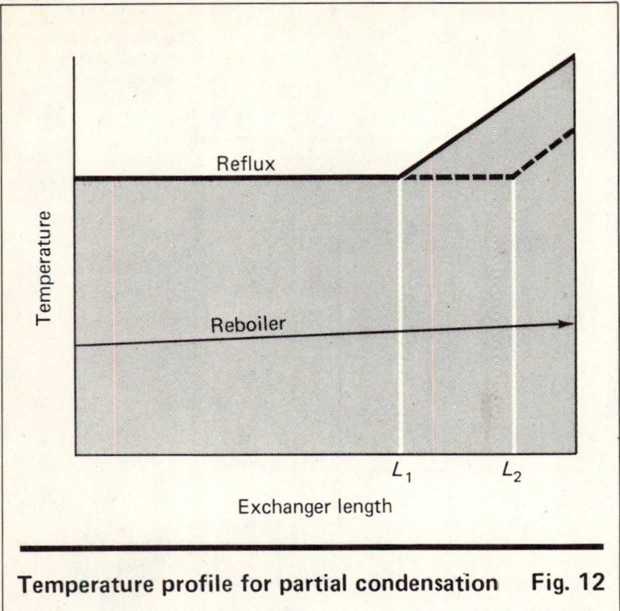

Temperature profile for partial condensation **Fig. 12**

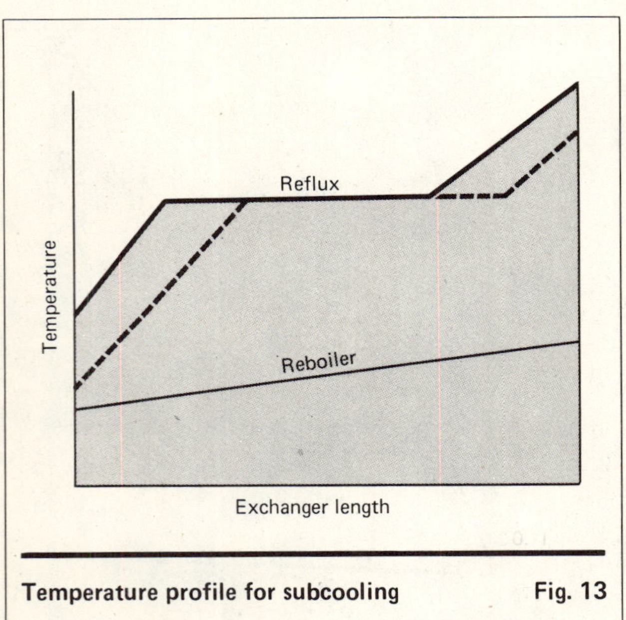

Temperature profile for subcooling **Fig. 13**

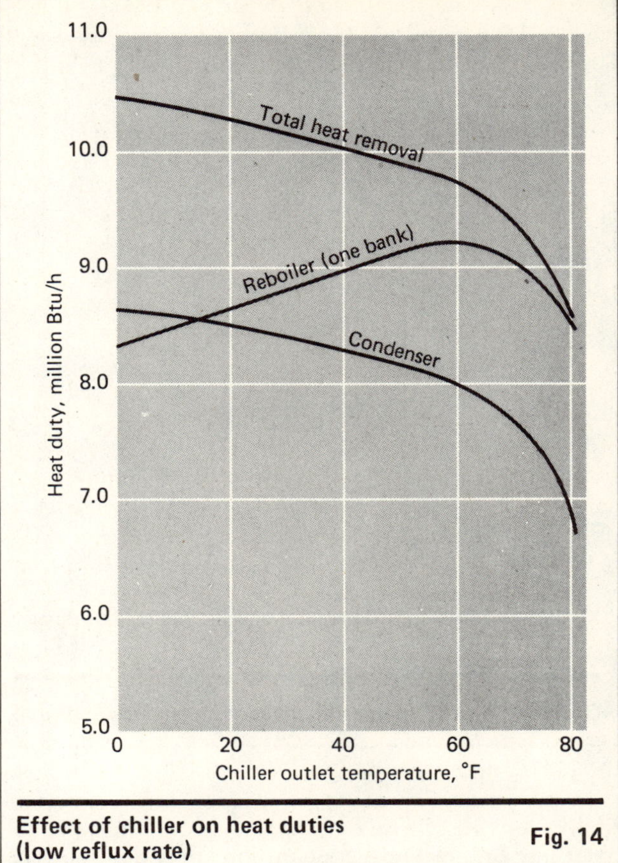

Effect of chiller on heat duties (low reflux rate) **Fig. 14**

A rearrangement of Eq. (3) leads to an equation for chiller duty:

$$Q_{ch} = Q_{comp} - (Q_r - Q_{cond}) \tag{3a}$$

For the convenience of future discussion, the term Q_t will be used to designate total heat removal:

$$Q_t = Q_r + Q_{ch} \tag{4}$$

Operation analysis

A good start for an analysis of C_2-splitters operation is the R/R exchanger. Fig. 11 shows a typical R/R exchanger profile. Since the reflux vapor is nearly pure ethylene (99.9%), it exhibits, during cooling a nonlinear temperature profile comprised of desuperheating, condensing and subcooling regions. On the reboiler side, the liquid contains mainly ethane; a nearly constant boiling point is obtained, as shown by the vaporizing curve in Fig. 11. This nonlinear temperature profile on the reflux-vapor side leads to very interesting behavior of the R/R exchanger.

To illustrate the effect of the R/R exchanger upon a distillation column with a heat pump, it is best to first explain the heat-exchanger performance in the partial condensation and subcooling regions in terms of reflux-vapor inlet (chiller outlet) temperature.

At partial-condensing reflux-outlet conditions, a reduction in reflux-vapor inlet temperature will result in an increase in the heat transferred. This is best explained with a schematic diagram of a heat-exchanger profile, as given in Fig. 12. Solid lines show the temperature profile for a given operation. If the reflux vapor is chilled, as shown by the dashed line, less desuperheating is required. The additional heat-exchanger surface can now be applied for condensation. Since the heat-transfer coefficient for condensation is considerably larger than for desuperheating, the overall heat transferred in the exchanger increases, despite the reduction in temperature driving force.

In the case of subcooling, the converse is true, as shown in Fig. 13. Lower reflux-vapor temperature will result in less desuperheating as before. Since the condensation area is fixed as a result of a fixed U and a nearly constant temperature difference between condensing vapor and boiling liquid, the additional exchanger area will be used for subcooling.

It is easy to visualize that for a high degree of subcooling, the outlet temperature of the reflux will approach the inlet temperature of the boiling liquid. The overall effect is less heat transferred as the reflux-vapor temperature is reduced.

To explain these heat-exchanger behaviors, look at Fig. 14, in which chiller temperature is plotted against reboiler duty, condenser duty and total heat removal. Condenser duty and total heat removal have already been defined. Condenser duty is used to explain distillation operation because of the general familiarity with the effects of condenser and reboiler duties upon distillation-column performance.

As expected, Fig. 14 shows that reboiler duty increases and then decreases as chiller temperature is reduced. Maximum reboiler duty is the total condensation point of the reflux vapor. Total heat removal and condenser duty increase as the chiller temperature is reduced. The trend indicates a leveling of these values at low chiller temperature.

Similarly, Fig. 15 contains results calculated for a reflux-vapor rate 20% higher than in Fig. 14. The trends in both figures are the same, except that the relative positions are shifted. For the convenience of later discussion, these data on reboiler duties (Fig. 14 and 15) have been combined in Fig. 16 to show the reflux-vapor rate effect upon reboiler duties at different chiller temperatures. Reboiler duty increases or decreases as reflux-vapor rate increases, depending upon whether the heat exchanger is operating in the subcooling or partial-condensation regions.

As a basis for relating heat duties to column fractionation performance, QUICKBAL simulations of the C_2 splitter were made. Fig. 17 shows the effect of reboiler duties with constant condenser duty upon the purities of overhead and bottom products. Increased reboiler duty will reduce ethylene content at the bottom and increase the ethane content overhead. Since condenser duty is fixed for this column simulation, increased reboiler duty also means decreasing condenser/reboiler duty ratio. Using this approach, Fig. 14 and 15 are replotted in terms of condenser/reboiler ratio (Q_c/Q_r) in Fig. 18. With the help of Fig. 16, the operation of the C_2 splitter can be described as follows:

Fig. 18 shows that lowering the chiller outlet temperature will increase Q_c/Q_r—i.e., improve overhead purity. Also, reboiler duty increases and then decreases as chiller temperature is lowered, as shown in Fig. 16. The maximum reboiler duties are achieved at total condensation points, as shown in both Fig. 16 and 18. Column simulation (Fig. 17) indicates that an increase in reboiler duty reduces the ethylene content in the bottom product.

Fig. 18 illustrates the manipulations required for controlling the C_2 splitters. Since the R/R exchanger of the C_2 splitters is designed to be operated flooded, this discussion is restricted to operation exclusively in the subcooling region. It is easy to visualize that if operation to include the total region, as described by the simulation, is ever required, control will be considerably more complex than it is in the following discussion.

Assuming the overhead purity is high in ethane, the operator will increase chiller duty to increase Q_c/Q_r to

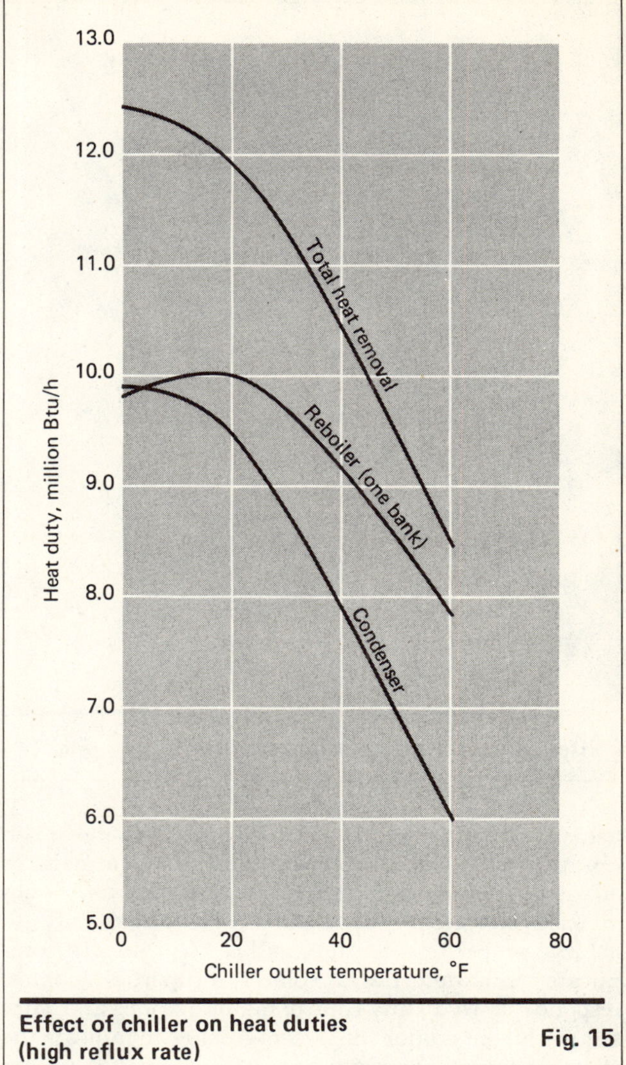

Effect of chiller on heat duties (high reflux rate) **Fig. 15**

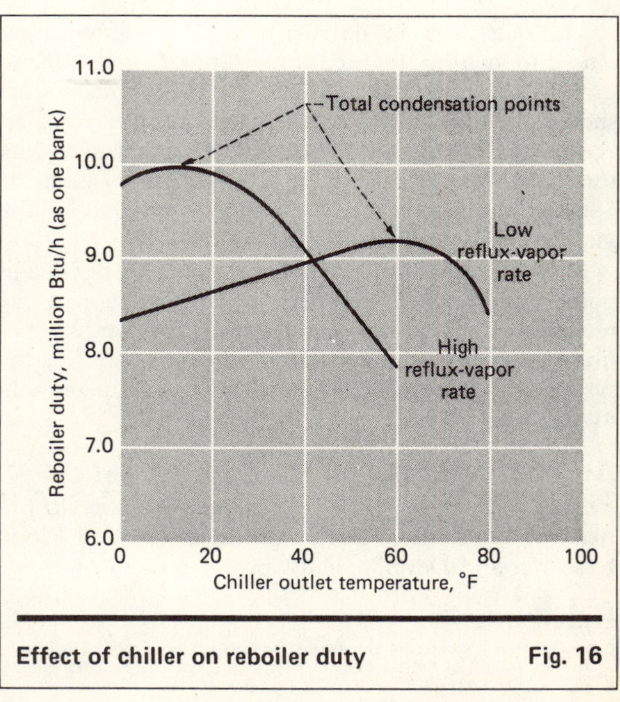

Effect of chiller on reboiler duty **Fig. 16**

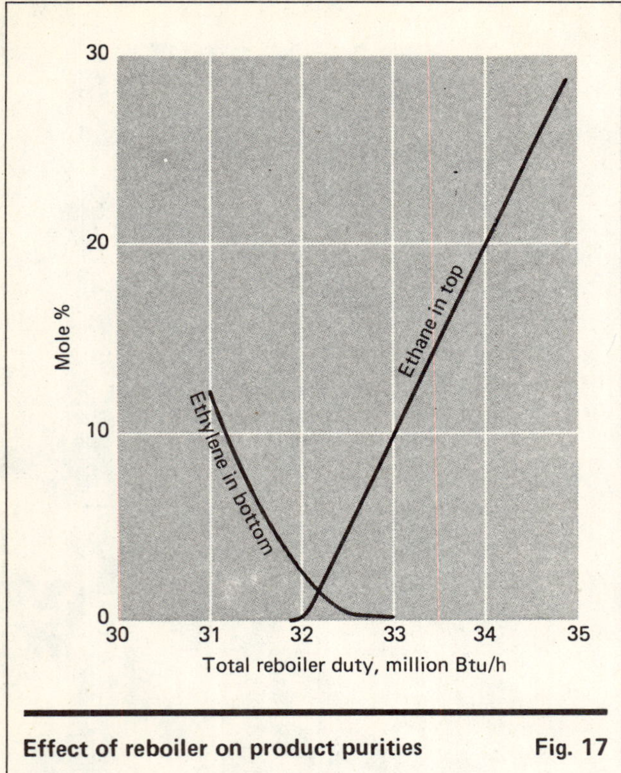

Effect of reboiler on product purities **Fig. 17**

achieve overhead specification. However, this move also results in a reduction of reboiler duty, and causes loss of ethylene to the bottom. Therefore, the operator should increase the reflux-vapor rate and chiller duty to achieve proper reboiler duty and Q_c/Q_r ratio. To complicate the matter, the C_2 splitter, a superfractionator, responds slowly to any control manipulations and, consequently, the product analyzers are slow to indicate the effect of the operator's manipulations. Even if proper operational procedures are known, it is difficult to gauge the responses in regulating the amount of control actions.

The described simulations of the C_2 splitters have served to identify the need for additional analog measurements and provide insight into how this system should be operated. With plant-operating experience, consideration can be given to the desirability of adding the model to an online process computer to assist the operator's control of the system in a closed-loop fashion via the computer if this is warranted.

In conclusion, simulation has resulted in significant potential savings in refrigeration in the case of the rectifier section. In the case of the C_2 splitters, simulation has explained some special characteristics of the system and provided solutions to potential operational difficulties.

Acknowledgements

The authors gratefully acknowledge the contributions by the staffs of Mobil Chemical Co. and Mobil Research and Development Corp.

References

1. Mosler, H. A., "Control of Sidestream and Energy Conservation Distillation Tower," Industrial Process Control Workshop, AIChE, Nov. 1974 Tampa, Fla.

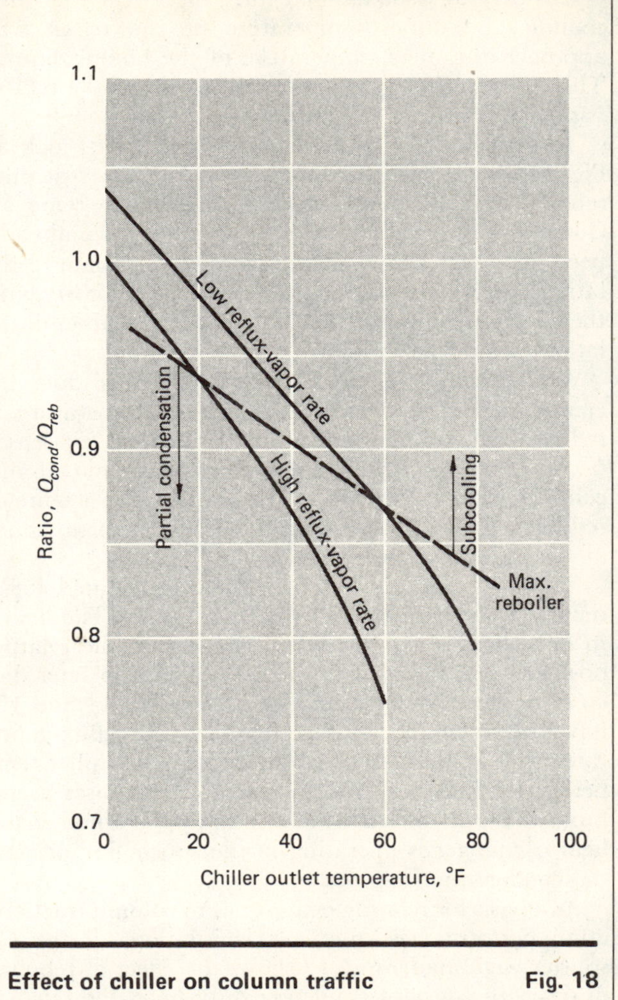

Effect of chiller on column traffic **Fig. 18**

The authors

Chou

Fayon

Bauman

Adam Chou is an associate engineer with Mobile Research and Development Corp. Besides broad experience in the petroleum and chemical industries, his recent experience includes process-system analysis and computer-control development and applications. He holds a Ph.D. from the University of Minnesota.

A. M. Fayon is a senior planning associate in the Petrochemical Div. of Mobil Chemical Co. His responsibilities include the evaluation of new projects, and the simulation of petrochemical processing schemes. He holds a B.S. in Chemical Engineering from the University of Istanbul, Turkey, an M.S.E. from Johns Hopkins University, and an Sc.D. from New York University.

Burton L. Bauman is a senior associate engineer in the Control Systems Section, Engineering Dept., Mobil Research and Development Corp. He leads a group responsible for developing and applying advanced analog and digital control systems in Mobil's worldwide operations. He has a B.Ch.E. from New York University, and an M.Ch.E. and Sc.D. from Stevens Institute of Technology.

Troubleshooting distillation columns

Here are practical methods for analyzing and correcting common problems that arise during the continuous operation of towers and their overhead and reboiler systems.

G. C. Shah, *Oxirane Corp.*

☐ Problems encountered during the operation of distillation equipment can often be simply explained. In this article, we will explore some of the common sources of problems that arise in the (1) overhead systems, (2) towers, and (3) reboiler systems.

Also, problems arising in one or more of the system components will result in problems in other areas as well. For example, excessive boilup in a reboiler will cause entrainment in the tower. Though the origin of the problem is in the reboiler, it manifests itself as a tower hydraulic problem.

Overhead systems

Conventionally, the main function of an overhead system has been that of pressure control of a tower. Fig. 1 shows some of the commonly encountered control systems. Although still limited to relatively close-boiling mixtures (small Δt through the tower), the heat-pump arrangement (Fig. 1e) is thermally efficient and is becoming economically attractive. For a brief comparison of various control systems, refer to Shinskey [*1*].

Pressure-control problems on a tower can be due to instrumentation, condensers, or vacuum systems.

Instrumentation problems

The overhead control system usually consists of pressure transmitter, controller and control valves. Although the process engineer does not get deeply involved with instrumentation problems, proper knowledge of the system is essential.

Problems with instrumentation originate from one or more of its equipment components, i.e., transmitter, controllers or control valves. The control system may be electronic or pneumatic.

Originally published July 31, 1978

Electronic pressure transmitters operate on straingage [*2*], reluctance [*3*], and capacitance bases, or their variations, while pneumatic transmitters use flapper-nozzle combinations to convert pressure to a proportional output signal.

Proper installation of the transmitters is important. For vapor service (as in overhead systems), a transmitter should be located above the elevation of the process tap. If the transmitter has to be located below the process line, we must allow for free drainage of condensate, as shown in Fig. 2. Condensate can cause the transmitter output to be erratic. Vibration-free mounting and watertight housing are just as important. The sensing lines should be insulated/traced where the potential for condensation exists.

Some common transmitter problems are: (1) Vibration, improper orientation, loose connections and/or shorting can produce erratic or fluctuating output; and (2) Improper span or zero adjustment will give an erroneous (too high or too low) pressure indication.

Electronic controllers incorporate numerous variations of *RC* (resistance/capacitance) circuits and amplifiers, while their pneumatic counterparts use a restriction orifice and bellows combination for their functional modes, i.e., proportional, reset and rate. Proper calibration and tuning of a controller are necessary to identify bad components and assure smooth response.

The output from an electronic controller (usually 4 to 20 mA d.c.) is fed to *I/P* (current to pneumatic) transducers to convert it to a pneumatic signal (3 to 15 psig) to operate a control valve. In the transducer, the current input is first converted into torque that, in turn, produces a pressure signal by a flapper-nozzle-type combination [*4*]. Zeroing and span-calibration of a trans-

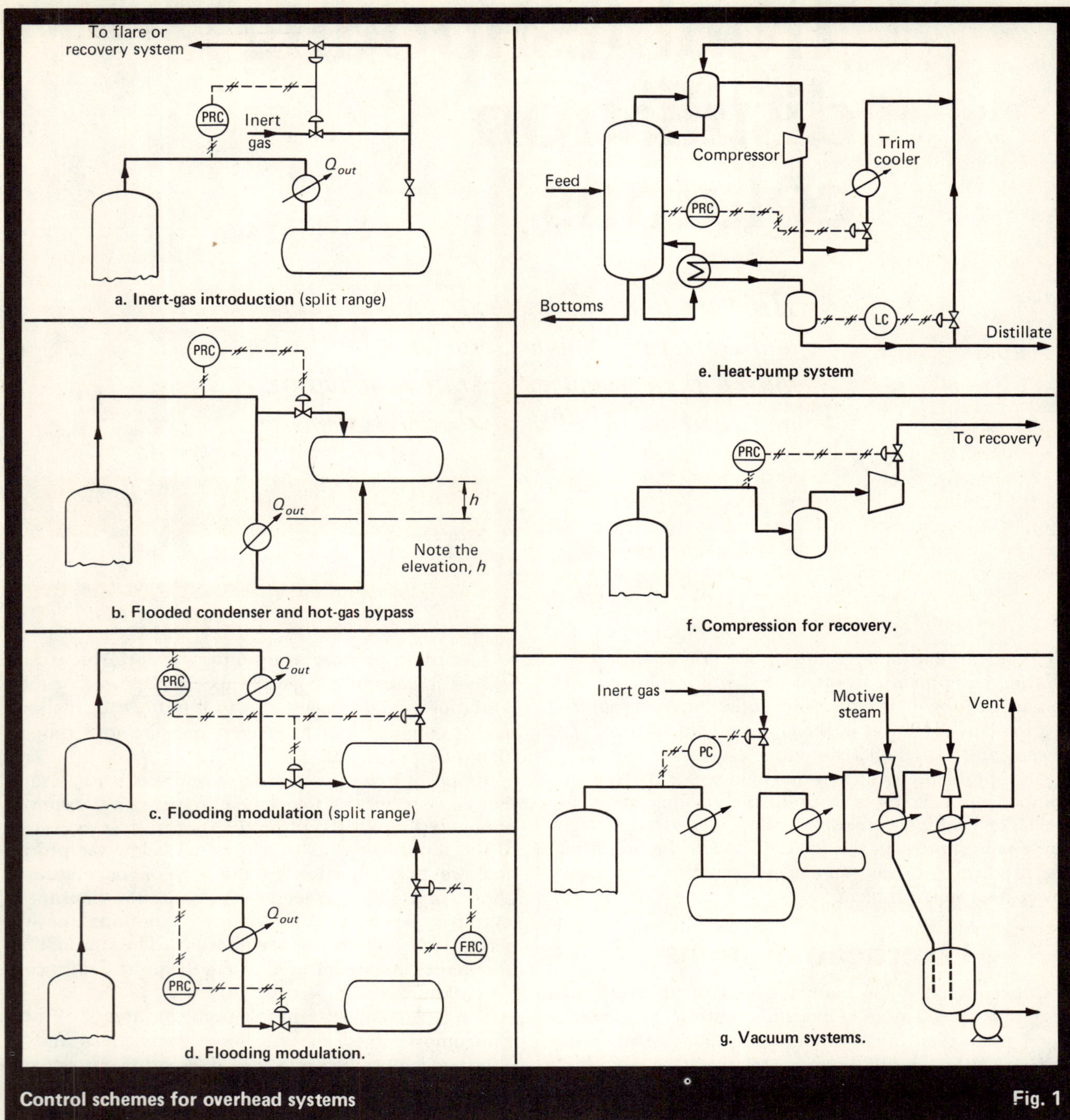

Control schemes for overhead systems Fig. 1

ducer are important for proper pressure control. Excessive vibration will cause the output to be erratic, while moisture may render the transducer inoperable.

The final control elements are control valves. Some of their problems are: (1) dirt in instrument-air supply, hindering the valve action; (2) sticky valve stem, packing gland too tight, or galling; (3) leaky or ruptured diaphragm; and (4) bad valve positioner, including faulty bellows, relay or nozzle, or other components.

Condenser problems

Common problems [5] with condensing of process fluids or coolants are:

1. Inert gases decrease heat transfer. Absence of properly located vents on condensers will reduce the effective surface available for condensation, and can result in high pressure. The inerts problem is most common during startups.

2. Inadequate condensate removal will likewise reduce heat transfer. Proper sizing and location of the nozzles is important.

3. Vapor binding on the coolant side is most commonly encountered during startups. All air must be purged from the coolant side (high-point vents); otherwise, the air will severely limit heat transfer.

4. Process-side fouling usually is slow and will begin to cause problems over a long period of time. If the overhead vapor contains small amounts of sticky mate-

rials (viscous), the arrangement shown in Fig. 3 may be considered. The viscous material that condenses first will foul the condenser unless properly washed by "lights."

5. Coolant-side fouling is also usually slow and will cause problems over an extended period. Periodically air-rolling the condensers helps reduce fouling, but the ultimate solution is proper coolant treatment. For air condensers, dirt and trash accumulated on the tube walls must be periodically removed.

Multicomponent condensers can be designed by using several techniques [*6,7*]. Usually, problems traceable to design are uncommon.

Vacuum systems

To minimize thermal degradation and optimize reboiler capacity, many fractionations are carried out under vacuum.

Although steam ejectors are poor users of energy, they are commonly employed to create vacuum because of their reliability and versatility. Their limited turndown ratio makes steam throttling impractical. The degree of vacuum required will dictate the number of stages. Multistage ejectors may be either the condensing or noncondensing type. To achieve greater reliability, many jets contain two elements.

For convenience, the problems of an eductor system can be grouped as either external or internal [*8,9*]. External problems are:

1. Low steam pressure or wet steam will adversely affect jet performance. Steam of too high a degree of superheat, or whose pressure is much higher than design, will likewise affect performance by overloading the ejector throats.

2. Hotwell flooding can be caused by pluggage or hotwell-pump problems. This will ultimately result in water hammer in the condensers, and loss of vacuum. A very short condenser barometric leg will also cause erratic operation.

3. High backpressure on the jet system will reduce its capacity.

4. Process upsets or changes will affect system performance. Too much air leakage, or an excess of components lighter than design molecular-weight, will quickly overload the system, e.g., waterbatching (molecular weight = 18) of the columns handling high-molecular-weight components.

Internal problems are:

1. Nozzle/throat pluggage and erosion or corrosion will severely limit jet capacity, with partial or total loss of vacuum.

2. Leaks in the steam chest decrease the vacuum.

3. Inadequate water supply, fouling (long term), or vapor binding on the water side (air must be vented during startups) cause condenser problems. By preferentially blocking steam to some of the jets, the condensing load can be reduced. Erosion or pluggage of the water nozzle for a contact-type condenser will also result in partial or total loss of vacuum.

4. For multistage ejectors, occasionally the dimensions of several stages look nearly identical. Mismatch will cause vacuum problems. Vendors' identifications should be carefully followed.

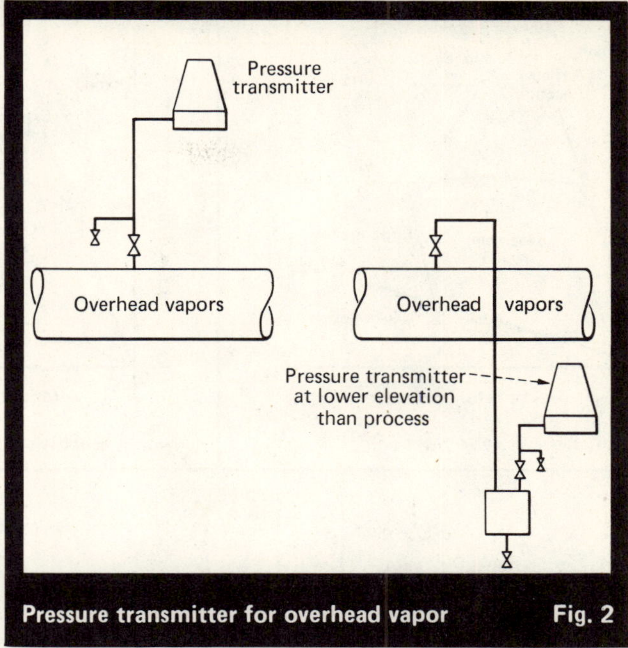

Pressure transmitter for overhead vapor **Fig. 2**

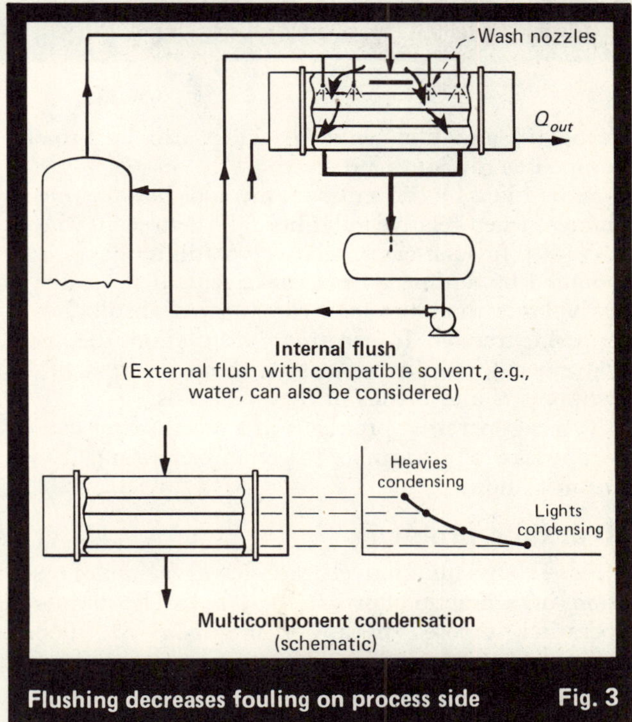

Flushing decreases fouling on process side **Fig. 3**

Problems in the tower

A distillation tower is a series of vapor-liquid contacting stages to effect component separation, based on volatility differences. Vapor-liquid mass transfer can be carried out in a plate tower or a packed tower.

Plate, or tray, towers accomplish separation by contact in discrete stages, whereas packed towers are characterized by continuous stagewise mass transfer. Cost, capacity and system properties (viscosity, fouling potential, corrosion, etc.) are some of the factors affecting the choice between a plate and a packed column. Different types of trays include bubble-cap, sieve, valve,

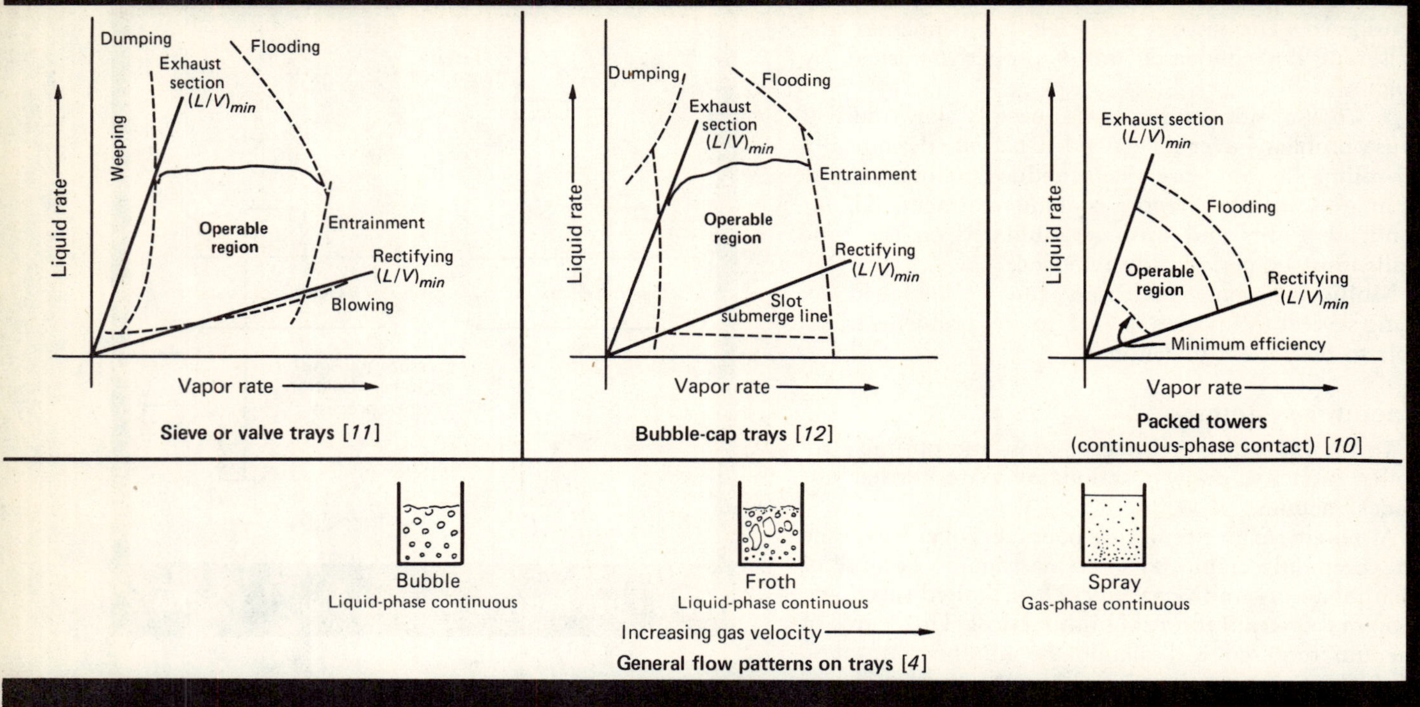

Performance diagrams (schematic) for fractionators **Fig. 4**

Turbogrid and film, while packings can be broadly grouped as random or stacked.

Separations by conventional methods become uneconomical when relative volatility falls between 0.95 and 1.05 [*10*]. In such cases, relative volatilities have to be modified by adding an external agent. If the agent is less volatile than the feed components, distillation is termed extractive. In azeotropic distillation, the agent (entrainer) forms an azeotrope with one or more of the components and distills in the overheads.

Typical operating problems in a fractionator can be grouped as (a) hydraulic, (b) equilibrium and (c) instrumentation.

Hydraulic problems

Fig. 4 shows qualitatively the various regions of operation in a fractionator [*10,11,12,13*]. The limits of operation imposed by tower internals, viz., flood/entrainment, weeping and dumping are well known and need no elaboration. Since these phenomena contribute to reduced contact efficiencies, towers must be operated in the range suggested by tray or packing manufacturers.

For packings, vapor-liquid traffic directly affects mass-transfer coefficients. For vacuum operations and towers of large diameters, loading significantly affects contact efficiency and, hence, product distribution. Hydraulic calculations for fractionators can be made by using several correlations [*12,14,15,16*]. Some of the key factors affecting hydraulics are:

1. Cold reflux will cause internal condensation and, thus, increase the liquid traffic in a column. Internal condensation makes the overhead product purer (fewer heavies). However, it can impair tower performance by overloading the tower (e.g., downcomer flood in the top section) for columns operating at high percent flood or for foaming systems. This can be calculated by a heat balance. For azeotropic and extractive distillations, solvent (entrainer) should be introduced at temperatures close to design.

2. For columns operating close to critical pressure of the system (density differences between liquid and vapor are small), inadequate foam factors will contribute to premature flood, and will severely limit operating capacity. Presence of impurities can contribute to foaming and poor performance (especially during startups). For fouling systems (presence of solids), washing with a compatible solvent is the easiest way to delay flood.

3. For vacuum columns, the liquid gradient on a tray is important from a hydraulic point of view. A high gradient can contribute to weeping. Ref. 17 shows that improvement in performance can be obtained by installing froth initiators. Vapor-liquid oscillations and their harmful effect on tray performance has been reported [*18*] for vacuum operations. Refer to Fig. 5 for explanation.

4. A thorough inspection of tower internals prior to startup will eliminate many problems. For example, numerous cases have been reported where missing tray manways or downcomer seal strips quickly resulted in entrainment problems. For packed towers, the distributors, hold-down plates, packing supports and proper packing are the relevant items to be inspected.

5. Proper vapor-liquid distribution is important for packed-tower performance. Liquid distributor pluggage (or improper orientation) will cause liquid maldistribution. Onstream cleaning of the distributor can be considered. Defoamers may have to be used because foaming also reduces efficiency. This is especially important

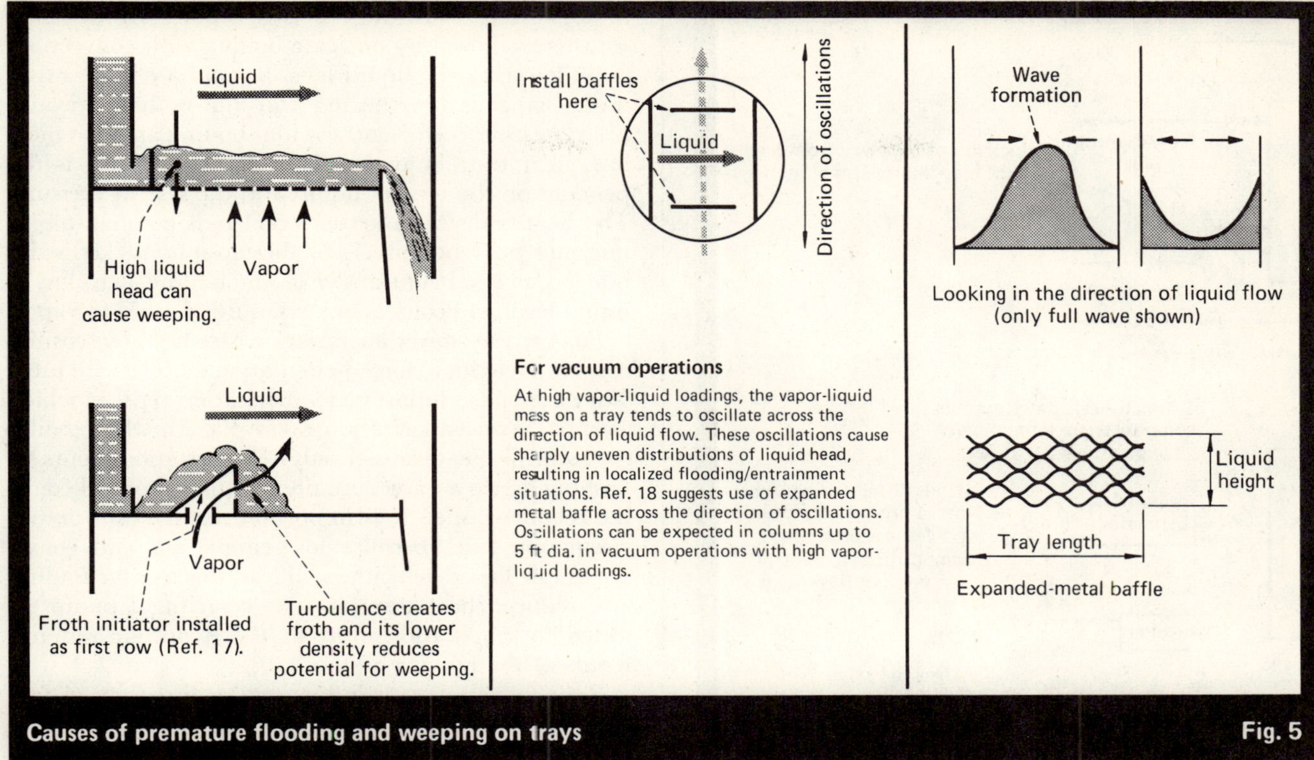

Causes of premature flooding and weeping on trays **Fig. 5**

during startup when a high concentration of impurities can quickly lead to foaming.

Equilibrium problems

The term *equilibrium problems* may be a misnomer, but the intent here is to discuss problems in distillation columns that relate to vapor-liquid equilibrium and mass-energy balances.

It is widely known that the reflux ratio, and hence boilup, directly govern product distribution for a given feedrate of fixed composition, feed enthalpy, feed-point location, number of stages, pressure of operation, and product withdrawal rates. For example, increasing the reflux ratio increases product purity. At low reflux ratios, the fractionator becomes increasingly sensitive to disturbances. Low reflux ratios often contribute to lower plate efficiencies for plate columns, and to higher HETP (height equivalent to a theoretical plate) for packed towers.

For multicomponent systems, it is difficult to assess the effect of reflux on product distribution and temperature profile in a tower. In such cases, a McCabe-Thiele-type diagram makes it considerably easier to visualize directional trends. Use of accurate K and H values (vapor-liquid equilibrium ratio and enthalpy) is of primary importance for a reliable prediction of tower performance. For extractive and azeotropic distillations, the solvent or entrainer must be maintained in adequately pure condition and concentration if the desired separation is to be achieved.

Instrumentation problems

The choice among a variety of control strategies, e.g., feedforward, direct or indirect material-balance control, heat integrated systems, etc., is made during design and will not be discussed here. Some pertinent comments regarding instrumentation are:

1. Since feed-tray zone temperatures are highly sensitive to variations in distillate or bottoms composition, temperature control in this area has often been used successfully for column control [i.e., a TRC (temperature-recorder-controller) controlling boilup]. This strategy proves unsuccessful, however, when the sensitivity is so high that fluctuations (following a temperature change) never cease, which makes operation unstable. Obviously, the TRC has to be relocated where the tray temperature is optimally sensitive to product compositions. Control may be improved by using multiple temperature measurements [*19*], as shown in Fig. 6a.

2. Startup of heat-integrated columns or a column train, where enthalpies are exchanged between several streams, requires extreme care and small incremental changes. Fig. 6b illustrates a case where problems could be experienced.

3. Interface controllers, especially the displacer type, can become inoperable unless properly purged with compatible liquid. This applies to displacers used for chimney trays for water draws.

Reboiler systems

Thermal energy required for distillation is supplied by reboilers. Before discussing the problems, let us briefly review reboiler systems encountered in industry, as shown in Fig. 7.

Depending on the mode of liquid circulation, reboilers can be grouped as natural circulation or forced circulation.

Natural-circulation reboilers, thermosiphon reboilers and kettle reboilers operate on density differences. For

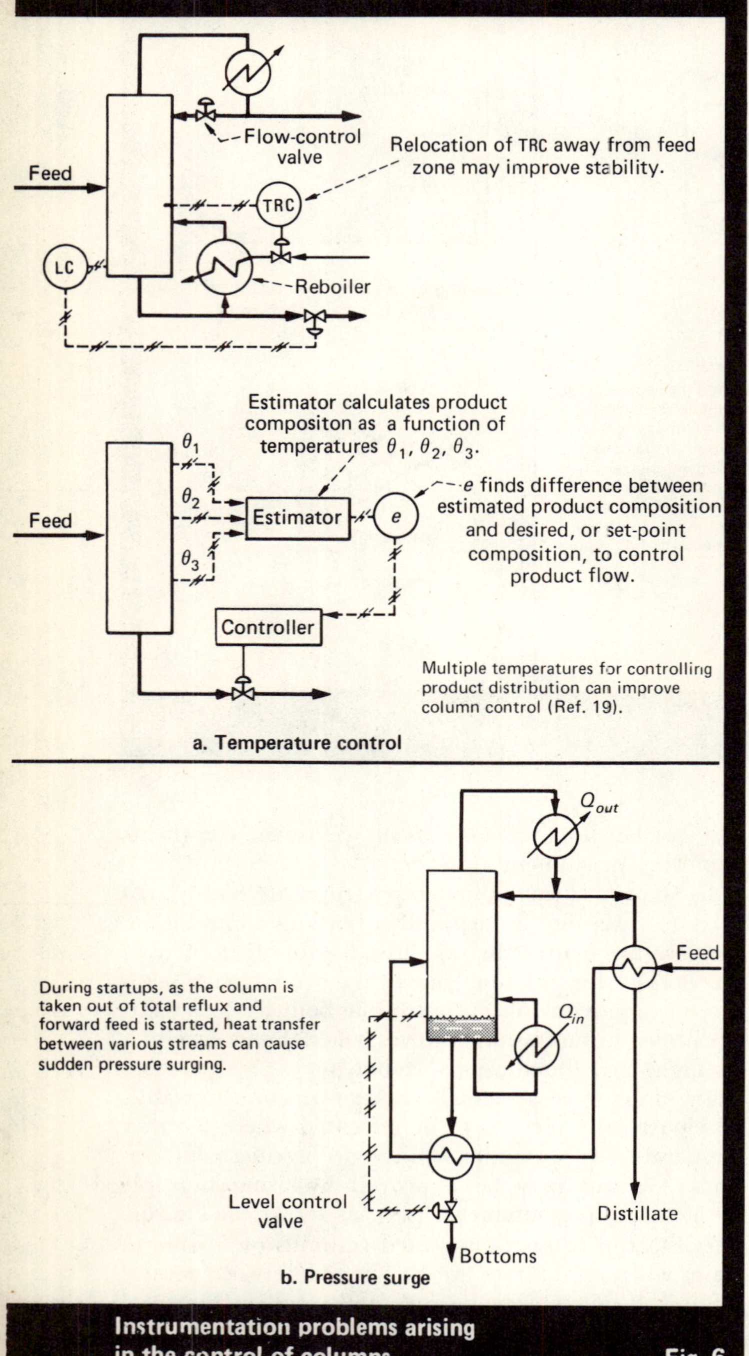

Instrumentation problems arising in the control of columns **Fig. 6**

example, the greater density of the liquid at the reboiler inlet than that at the outlet provides a driving force for circulation. These reboilers are suited for low-viscosity systems. Typical percent vaporizations in the thermosiphons are 5 to 25%, while for kettle reboilers, they are nearly 70 to 75%.

Kettle reboilers are prone to fouling because the greater vaporization reduces circulation rates and hence contributes to fouling. For thermosiphon reboilers (Fig. 7), liquid level is maintained near the top tubesheet. By lowering the liquid level, vaporization increases to a point, but beyond that there is a sharp drop in heat transfer. Although heat transfer in thermosiphons is complex, it can be viewed in a simplified form as essentially consisting of two zones—sensible heat transfer, followed by nucleate boiling with convection.

By lowering the liquid level beyond a certain minimum, vapor is superheated and film boiling prevails. Since transfer coefficients for film boiling are extremely low, heat transfer drops. This minimum level is dependent on the type of liquid and the system pressure. The heat transfer in kettle reboilers is comparable to nucleate pool boiling. Like thermosiphons, kettle reboilers can lose heat transfer in film boiling—i.e., loss of liquid level and consequent exposure of tubes to vapor.

For vacuum-tower operations where liquid viscosities are relatively high, forced-circulation reboilers are used. They are of the suppressed-vaporization type, in which essentially no vaporization takes place in the reboiler (i.e., sensible heat transfer only). Vaporization occurs by flashing across a valve, usually manually controlled, at the reboiler outlet. It is important to maintain proper circulation rates because low circulation and consequent low liquid velocities will lead to excessive fouling and reduced heat transfer. The centrifugal pump selected for circulation must fit the hydraulic requirements of the reboiler system [*10*].

Reboiler selection is influenced by a variety of factors, such as viscosity, fouling characteristics, maintenance and vaporization requirements. Ref. 21 gives a detailed description of these factors.

Troubleshooting reboilers

The following problems can be expected during the operation of reboiler systems:

1. If steam is used as a heating medium, the presence of inerts (CO_2, NH_3, etc.) and steam-trap malfunction are common problems during startups. Even though the steam-condensation coefficients are high [1,500 to 2,000 Btu/(h)(ft^2)(°F)], minor quantities of inerts, say 1 to 2%, can severely reduce these coefficients. Inerts must be vented. Trap pluggage due to sludge/trash and malfunctioning internals will likewise reduce heat transfer by condensate flooding. Proper installation of adequately sized condensate nozzles and traps is equally important.

2. For proper heat transfer in vertical or horizontal reboilers, liquid level is quite significant. In vacuum systems, the effect of level (head) is most profound because it strongly influences the boiling point of the liquid. A liquid level that is too low will cause loss of thermosiphon and heat transfer because of film boiling; while too high a level will flood the reboiler outlet (Fig. 7).

Level is also important for kettle reboilers. Exposing a significant portion of the tubes to vapor will result in film boiling, and consequent loss of heat transfer. To allow for proper liquid disengagement, the top tube row should not be higher than 60% of the shell diameter [*27*]. Proper sizing of the vapor-return lines is also important, to maintain the level in the kettle reboiler. Because of excessive frictional losses in too small a vapor line, the level in the kettle reboiler will be suppressed and affect its performance.

3. It is extremely important to maintain high circulation rates for fouling services or polymerizable materials, in order to avoid loss of heat transfer due to

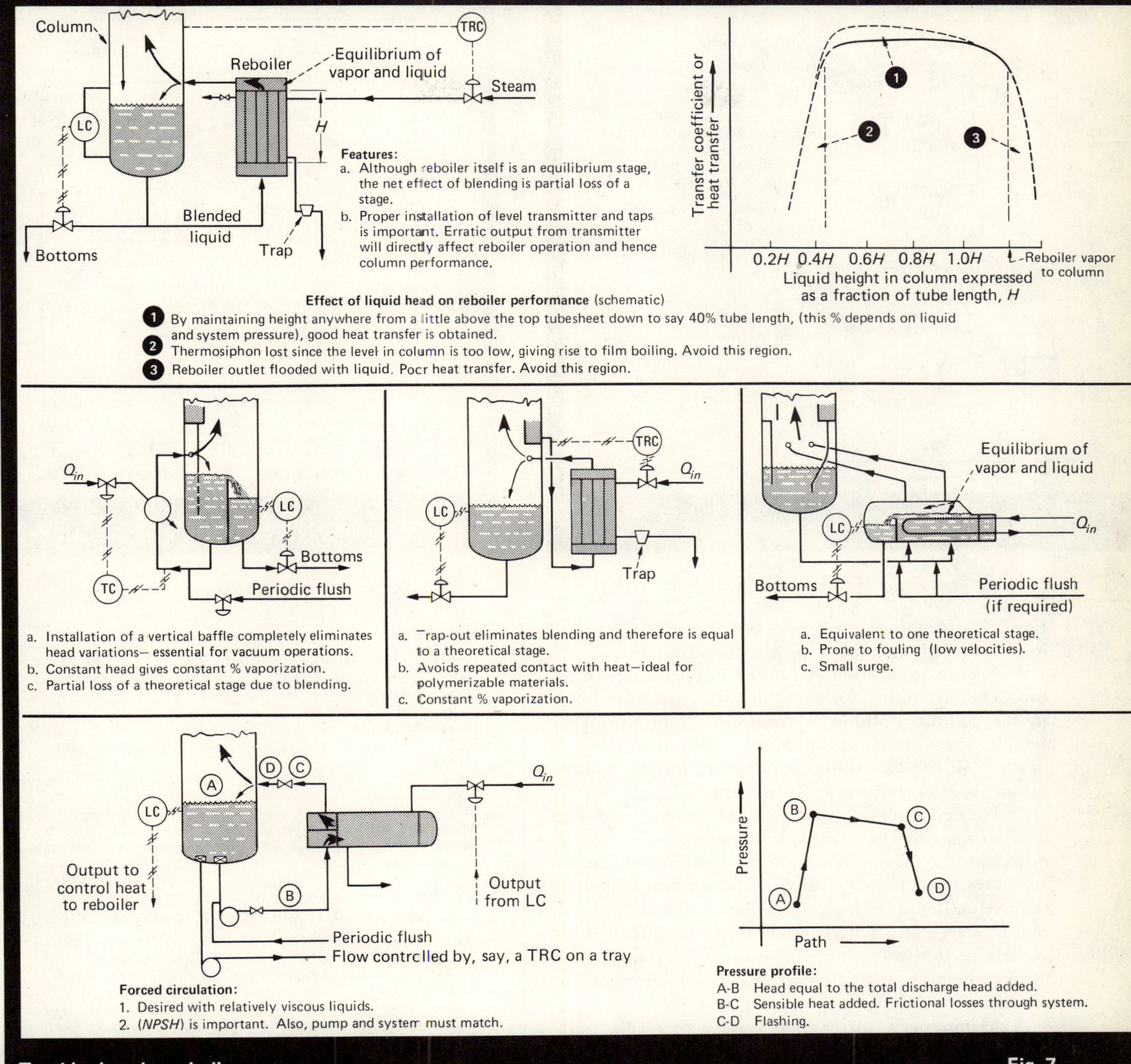

Trouble-shooting reboiler systems **Fig. 7**

deposits. A similar condition holds when the liquid contains predominantly high boilers. Low circulation in these cases will give rise to surging [*23*], as explained in Fig. 8. For fouling services, periodic flushes with "lights" or fouling inhibitors should be considered. These inhibitors, of course, should be compatible with downstream processing. Treatment systems such as caustic neutralization/wash, upstream of the distillation step, often contribute to fouling/pluggage.

4. An expression for heat flux is: $U\Delta t$. When the temperature-driving force, Δt, approaches small values, heat transfer will drop. In thermosiphon reboilers that are designed for a wide range of boiling mixtures, Δt may become small for a high degree of vaporization. The apparent inferior performance of a reboiler in such cases may mislead one to conclude that the reboiler is fouled. Ref. 22 suggests that performance can be improved by:

a. Sparging gas such as nitrogen at reboiler bottom.

b. Adjusting the liquid head in the column. This will change fractional vaporization and, therefore, Δt across the tube wall.

c. Installing a restriction (a valve or an orifice) on the liquid line to the reboiler (Fig. 9).

5. The U tube formed between a column and a reboiler can cause liquid oscillations. Unless dampened by friction, these oscillations may have an adverse effect on reboiler performance, especially in vacuum opera-

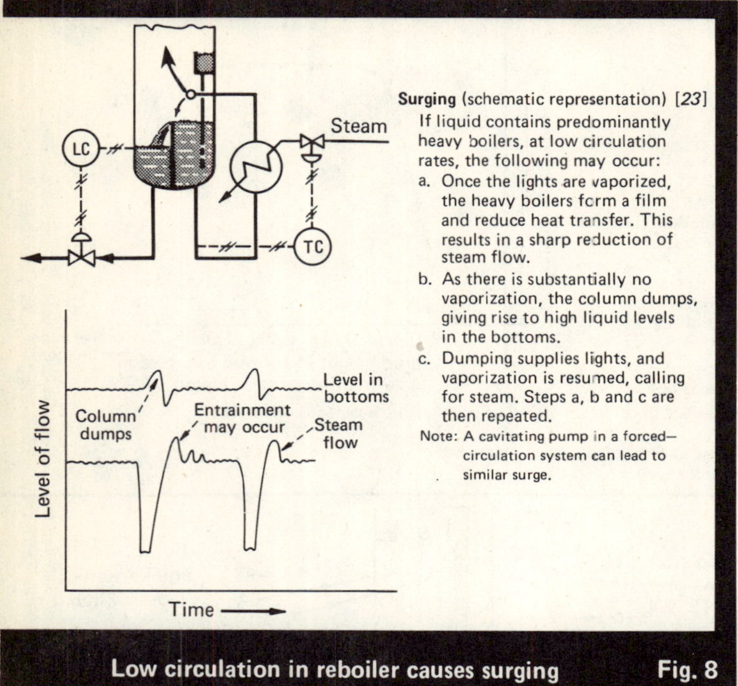

Low circulation in reboiler causes surging Fig. 8

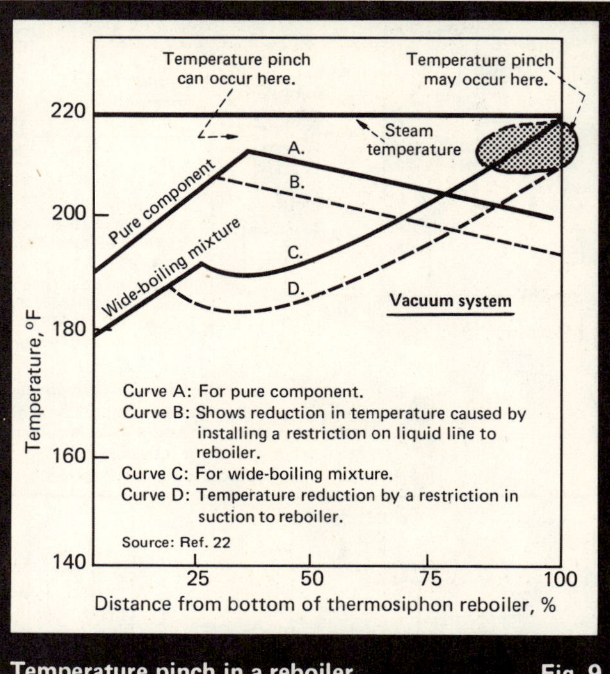

Temperature pinch in a reboiler Fig. 9

tions. The installation of some restriction in the liquid line to the reboiler will improve the situation [1].

6. Proper orientation of level or temperature instruments is important. For example, if a level transmitter in the column bottoms is continuously impinged by reboiler vapors or stripping steam, erratic operation will occur [24]. For direct-fired reboilers, proper flow distribution is important to avoid hot spots.

7. For forced-circulation reboilers, adequate liquid level in the column to provide the net positive suction head (NPSH) required for the pump is important. Liquid velocities from 10 to 15 ft/s should be maintained to reduce fouling [23]. In turn, this means that the valve at the discharge of a reboiler should not be throttled beyond a certain minimum. Of course, the pump curve must be compatible with hydraulic requirements of the reboiler system [20].

A design check on reboilers can be made by using several references [25–27].

The guidelines presented here do not cover all possible situations for continuous distillations, but will serve to streamline troubleshooting efforts in an operating environment.

The author

G. C. Shah is a senior field staff engineer with Oxirane Corp. (Houston, TX 77027), currently assigned to the Channelview plant (P.O. Box 580, Channelview, TX 77530). He has an M.S. in chemical engineering from Oklahoma State University, and is a member of AIChE and the American Management Assn. He is also a registered professional engineer in Louisiana.

References

1. Shinskey, F. G., "Distillation Control," McGraw-Hill, New York, 1977.
2. "Diffused Silicon PP/I Transmitter," Operators Manual 811-666, Honeywell Process Control Div., Fort Washington, Pa.
3. "Taylor Pressure Transmitters," Instruction Manual, Taylor Instrument Process Control Div., Rochester, N.Y.
4. "Electro Pneumatic Transmitters," Form 1783, Fisher Controls Co., Marshalltown, Iowa.
5. Steinmeyer, D. E. and Mueller, A. C., Why Condensers Don't Operate as They Are Supposed To, *Chem. Eng. Prog.,* July 1974.
6. Price, B. C. and Bell, K. J., Design of Binary Vapor Condensers Using the Colburn-Drew Equations, *AIChE Symp. Ser. 138,* 1974.
7. Ward, D. J., How to Design a Multiple Component Partial Condenser, *Petrochem. Eng.,* Oct. 1960.
8. "Design and Application of Steam Jet Ejectors," Bulletin E68A, Croll-Reynolds Co. Inc., Westfield, N.J.
9. Power, R. B., Steam Jet Ejectors—Part 10, *Oil Gas Equipment,* July 1966.
10. Van Winkle, M., "Distillation," McGraw-Hill, New York, 1967.
11. Chase, J. D., Sieve Tray Design—Part I, *Chem. Eng.,* July 31, 1967.
12. Smith, B. D., "Design of Equilibrium Stage Processes," McGraw-Hill, New York, 1963.
13. Ho, G. E., Mueller, R. L. and Prince, R. G. H., Characterization of Two-Phase Flow Patterns in Plate Columns, Int. Symp. on Distillation, Brighton (England), 1969.
14. Eckert, J. S., How Tower Packings Behave, *Chem. Eng.,* Apr. 14, 1975.
15. "Float Valve Design Manual," Nutter Engineering Co., Tulsa, Okla., Apr. 1976.
16. Bulletin 960, Koch Engineering Co., New York.
17. Winter, G. R. and Uitti, K. D., Froth Initiators Can Improve Tray Performance, *Chem. Eng. Prog.,* Sept. 1976.
18. Biddulph, M. W. and Stephen, D. J., Oscillating Behavior on Distillation Trays, *AIChE J.,* Jan. 1974.
19. Joseph, B., Brosilow, C. B., Howell, J. C. and Kerr, W. R. D., Multi-temps Give Better Control, *Hydrocarbon Process.,* Mar. 1976.
20. Gilmour, C. H., Trouble Shooting Heat Exchanger Design, *Chem. Eng.,* June 19, 1967.
21. Jacobs, J. K., Reboiler Selection Simplified, *Hydrocarbon Process.,* July 1961.
22. Smith, J. V., Improving the Performance of Vertical Thermosiphon Reboilers, *Chem. Eng. Prog.,* July 1974.
23. Lord, R. C., Minton, P. E. and Slusser, R. P., Design Parameters for Condensers and Reboilers, *Chem. Eng.,* Mar. 23, 1970.
24. Love, F. S., Troubleshooting Distillation Problems, *Chem. Eng. Prog.,* June 1975.
25. Fair, J. R., What You Need to Design Thermosiphon Reboilers, *Pet. Refiner,* Feb. 1960.
26. Hughmark, G. A., Designing Thermosiphon Reboilers, *Chem. Eng. Prog.,* July 1964.
27. Kern, D. Q., "Process Heat Transfer," McGraw-Hill, New York, 1950.

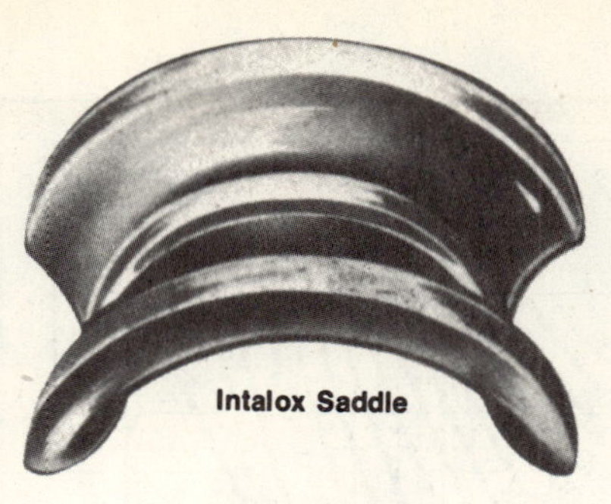

Intalox Saddle

Raschig Ring

Pall Ring

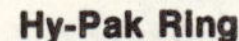

Hy-Pak Ring

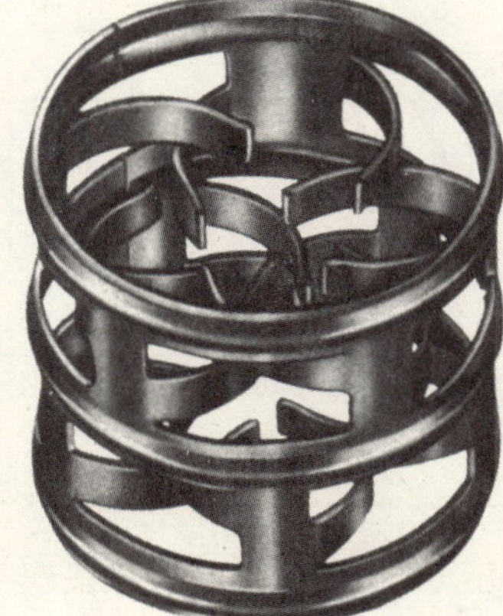

Berl Saddle

How Tower Packings Behave

Here is information on the pressure drop in tower packings, and the methods of measuring efficiency. Various packings are compared to show what one may expect in the way of performance and general behavior.

JOHN S. ECKERT, Norton Co.

Packed towers are widely used in gas absorption and desorption, distillation and liquid-liquid contacting. They are also used as entrainment separators and for the removal of dusts, mists and odors.

Such towers are useful mass-transfer devices that are available in a very great variety of materials of construction, are extremely versatile in their ability to give good performance over widely varying liquid and gas rates, are very predictable as to both capacity and efficiency, and usually are less costly to operate than other types of mass-transfer devices.

The vast majority of tower packings sold are of the ring or saddle type (Fig. 1). While the design approach used here will feature these two types of packing, the same technique may be used equally well for the lesser used packings, such as metal mesh and special shapes of injection-molded plastics, if their relative capacities or packing factors are known.

Various Types of Packings

Ring-type packings are commonly made of metal or plastic, except for Raschig rings, which are generally ceramic, seldom in plastic; saddle-type packings are commonly made from ceramic or plastic, almost never from metal (Fig. 1).

Originally published April 14, 1975

Ring-type packings, such as the Hy-Pak and Pall ring, lend themselves mostly to distillation because of their very good turndown properties, and availability in metals of all types that can be press formed. Saddles lend themselves largely to absorption and regeneration operations because of their good liquid redistribution and their availability in ceramic and plastic, which yield good corrosion resistance at very low cost. Rings and saddles actively compete with each other where injection-moldable plastic may be used.

Usually ring-type packings are used in handling organic materials when there are no major corrosion problems; saddle-type packings generally are specified for aqueous systems when corrosion is a major factor. These are only generalities—metals are often used in aqueous corrosive systems, and ceramics are used in nonaqueous noncorrosive systems. Though plastics are employed in both organic and aqueous systems, great care must be taken when solvents are present, when a system is oxidizing in nature, or when an elevated temperature must be considered.

Saddles are best for redistribution of liquid and for the correction of maldistribution. Rings, on the other hand, do not promote much redistribution; and the Raschig ring may occasionally actually promote maldistribution.

Mesh-type packings ordinarily do not effectively redistribute liquid, so they must have a very good primary distribution of liquid to be effective. Once this has been achieved, mesh-type packings provide low HETP (height equivalent to a theoretical plate).

Ring and saddle packings are much lower in cost than mesh packings, and thus enjoy by far the greatest part of the rapidly growing packings market, which has increased at least twentyfold over the past sixteen years.

Pressure-Drop in a Packed Column

If any type of packed tower is operated in countercurrent flow with the liquid rate constant and the gas rate allowed to vary (Fig. 1), the pressure drop across the bed will be roughly proportional to the square of the gas mass-velocity. Actually this will vary a little with different geometries of packing (due to special holdup variations). Therefore, the pressure will not actually vary with the square, but with somewhere between the 1.7 and 1.9 power of the gas rate. As the gas rate is increased through the bed, it holds back the flow of the liquid; this results in lower voidage and, consequently, a higher-than-normal pressure drop. For this reason, as the gas rate is increased through an irrigated bed, the pressure drop will increase faster than if dry packing is used.

Capacity of Packed Beds

Early investigators thought the holdup of liquid in a packed bed could be characterized by three separate realms, namely: below a lower loading zone, between a lower and upper loading zone, and above an upper loading zone. None of our laboratory work has ever detected these separate realms of flow. Instead, it has shown a steadily increasing slope of the pressure-drop curve with increasing gas rate, until the slope becomes infinite (classical flood). At this point, the liquid will be the continuous phase (in the interstices between the pieces of packing) and the gas will be bubbling through it. The bed will be in "operational flood" (non-steady state) sometime before classical flooding takes place. Packed beds are always designed for an operating pressure-drop well below these flood points.

The specific value of a pressure-drop chart (Fig. 2)—when various liquid rates are used as parameters against continuously changing gas rates—is to develop a log-log plot of gas-rate versus pressure-drop, and thus establish a capacity rating for the given packing. This rating is expressed as the packing factor, or *F*, and has been determined for all of the commonly used packings (Table I) [*1*]. The packing factor results from the a/ε^3 of the original Sherwood correlation [*2*] (Fig. 3). Lobo, et al. [*3*] found that the a/ε^3 of the Sherwood correlation would not properly express the capacity of all geometric shapes in a packed bed, and proposed a packing factor (*F*) to characterize this capacity, based on actual operation in a tower.

Until the papers of Leva [*4*], all capacity work on packed beds was done with a so-called flood line. The flood line is actually quite broad, being very sensitive to the foaminess of the system (Fig. 3). Leva introduced

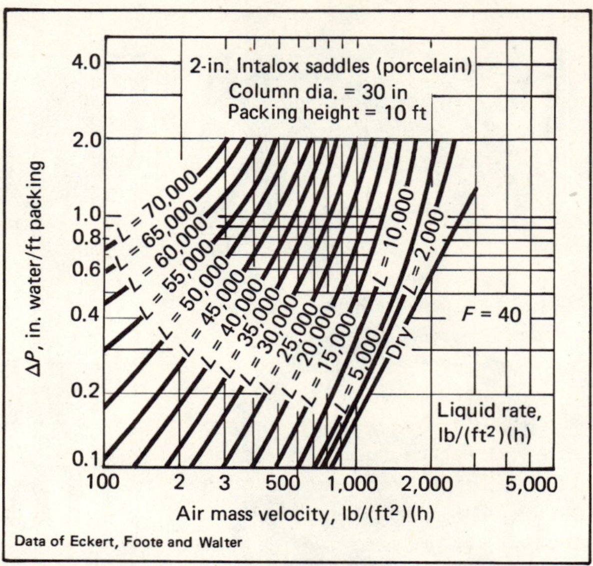

PRESSURE-DROP vs. gas rate in a packed column—Fig. 1

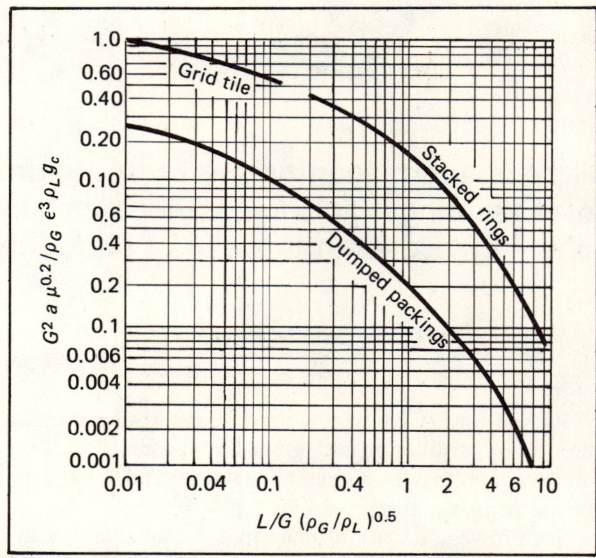

SHERWOOD flood correlation for packings—Fig. 2

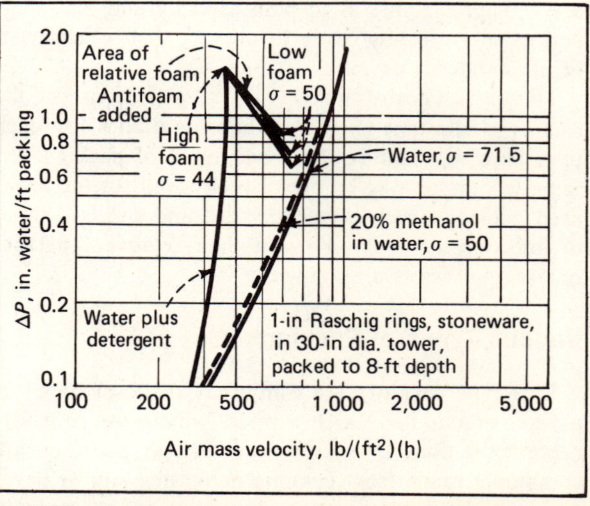

EFFECT of foam and surface tension on Δp—Fig. 3

parameters of constant pressure-drop, which have since been used as the design criteria for packed beds. Absorbers and regenerators are designed to operate at 0.25 to 0.50 in. of water-pressure-drop per foot of packed depth; stills and fractionators of the vaccum type, from 0.01 to 0.5 in. depending on the purpose of vacuum operation; and atmospheric and pressure stills and fractionators at 0.75 to 1.0 in. Packed beds should not be designed to operate at over 1 in. pressure drop per foot, since a very small increase in gas rate at this point (Fig. 4) will cause the bed to flood.

Packing factors have been determined for all of the commonly used packings (Table I), and the modified Sherwood correlation (Fig. 5) has come to be accepted as the best design approach for arriving at the required area of a packed bed (for a given process stream and duty). The sizing technique is very direct in that the value of the abscissa can be calculated directly from the requirements of the process stream. The desired pressure-drop sets the ordinate value. In any sizable installation where 2 and 3-in packings will be used, a packing factor of 20 (Table I) may be assumed (if the designer is not already committed on type and size of packing). The value of the gas rate, G', in lb/(ft^2)(s) can be calculated, and from the quantity of G' in the process stream the area of the required tower (hence its diameter) may then be determined. If the designer wishes to change packings, the change in gas duty at the same pressure-drop may be determined by the ratio:

$$\frac{G_2}{G_1} = \left(\frac{F_1}{F_2}\right)^{0.5} \tag{1}$$

which is obtained by inspection of the G^2F relationship of the generalized pressure-drop correlation, Fig. 5.

Nomenclature

a	Specific packing surface, ft^2/ft
F	Packing factor
G	Gas rate, lb/(ft)(h)
G'	Gas rate, lb/(ft)(s)
g_c	Acceleration of gravity, 4.17×10^8 ft/h^2
g_c'	Acceleration of gravity, 32.2 ft/s^2
HETP	Height equivalent to a theoretical plate, ft
K_Ga	Mass-transfer coefficient for gas, lb-moles/(h)(ft^3)(atm)
K_La	Mass-transfer coefficient for liquid
L	Liquid rate, lb/(ft^2)(h)
L'	Liquid rate, lb/(ft^2)(s)
N	Number of lb/moles of gas transferred per hour
P_1, P_2	Net partial pressures, atm
ΔP	Pressure drop, in. water/ft packing
ln Δp	Natural log-mean of net drive between bottom and top of bed, atm
V	Volume of bed, ft^3
ϵ	Bed voidage, dimensionless
μ	Liquid viscosity, cp
ρ_G	Gas density, lb/ft^3
ρ_L	Liquid density, lb/ft^3
σ	Surface tension, dynes/cm

System properties other than relative volatility have little or no effect on difficulty of separation by distillation. Hence, if the vapor-liquid equilibrium data are accurately known for a given system (and, therefore, the number of theoretical plates required to effect a given separation), the required bed depth can quickly be determined for a known packing size.

Open metal rings such as Hy-Pak and Pall rings tend

Packing Factors for Wet and Dump-Packed Packings—Table I

Type of Packing	Mat'l.	Nominal Packing Size, In										
		¼	⅜	½	⅝	¾	1	1¼	1½	2	3	3½
Super Intalox	Ceramic	—	—	—	—	—	60	—	—	30	—	—
Super Intalox	Plastic	—	—	—	—	—	33	—	—	21	16	—
Intalox saddles	Ceramic	725	330	200	—	145	98	—	52	40	22	—
Hy-Pak rings	Metal	—	—	—	—	—	42	—	—	18	15	—
Pall rings	Plastic	—	—	—	97	—	52	—	40	25	—	16
Pall rings	Metal	—	—	—	70	—	48	—	28	20	—	16
Berl saddles	Ceramic	900[×]	—	240[×]	—	170[h]	110[h]	—	65[h]	45[×]	—	—
Raschig rings	Ceramic	1,600[b,×]	1,000[b,×]	580[c]	380[c]	255[c]	155[d]	125[e,×]	95[e]	65[f]	37[g,×]	—
Raschig rings 1/32-in wall	Metal	700[×]	390[×]	300[×]	170	155	115[×]	—	—	—	—	—
Raschig rings 1/16-in wall	Metal	—	—	410	290	220	137	110[×]	83	57	32[×]	—
Tellerettes	Plastic	—	—	—	—	—	40	—	—	20	—	—
Maspak	Plastic	—	—	—	—	—	—	—	—	32	20	—
Lessing exp.	Metal	—	—	—	—	—	—	—	30	—	—	—
Cross partition	Ceramic	—	—	—	—	—	—	—	—	—	70	—

[a] 1/32 Wall [b] 1/16 Wall [c] 3/32 Wall [d] ⅛ Wall [e] 3/16 Wall [f] ¼ Wall [g] ⅜ Wall [h] Packing factors obtained in 16- and 30-in. I.D. towers. [×] Extrapolated

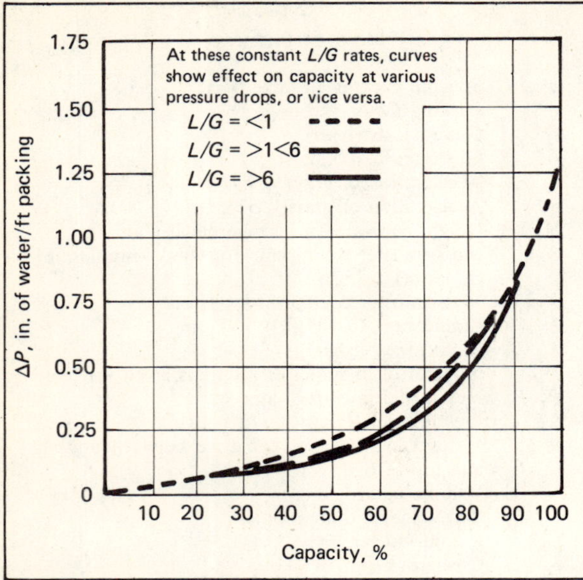

PERMISSIBLE operating rate with packed bed—Fig. 4

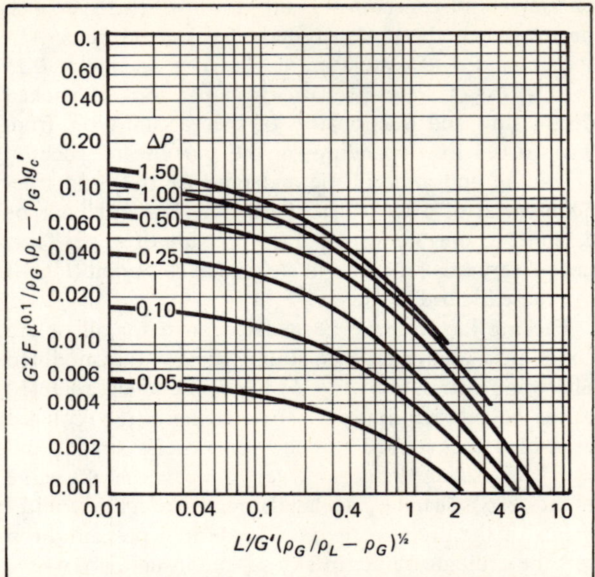

GENERALIZED pressure-drop correlation—Fig. 5

to have a very uniform efficiency over a wide range of loading (Fig. 6), while saddles are not quite as good (but are still considerably better than Raschig rings). The HETP that might be expected from an open-ring packing bears a relationship to its size, i.e., 1-in. packings may be expected to give an HETP of from $1\frac{1}{4}$ to $1\frac{3}{4}$ ft in industrial practice, while the $1\frac{1}{2}$-in. size will give an HETP of 2 to $2\frac{1}{2}$ ft, and the 2-in. size from $2\frac{1}{2}$ to 3 ft. Saddles will behave in the same manner, but only where the pressure drop exceeds 0.35 in.; Raschig rings need to be operated at about 0.5 in., or above, to obtain equal efficiency.

The packed bed has its greatest advantage in vacuum distillation, where it shows much lower pressure drop per theoretical plate than a trayed tower. Hy-Pak followed by the Pall ring are the packings to use because of the need to operate at very low pressure-drop. Mesh-type packings are good in this service; however, they must be considered a specialty packing because of their high cost per cubic foot and their special requirement for good distribution. They will give low HETP and at low pressure drop.

Designing Absorption and Regeneration Towers

The design of absorption and regeneration towers always presents a challenge to the engineer because of the many variables. A vast majority of distillations involve a gas that is totally soluble in a liquid in which no reaction takes place. Hence, equilibrium between gas and liquid may be quickly established, so long as the gas and liquid are kept in a turbulent condition. However, absorption and regeneration usually involve chemical reaction, heat of solution, and an inert gas along with a gas that will be only partially soluble in the liquid. Because of the complex liquid phase, absorbers and regenerators show a greater tendency to foam than do fractionators or stills. For this reason, they must be designed for a lower maximum pressure-drop.

A distillation column with bad distribution will operate at a lower reflux ratio than anticipated and will require a greater number of theoretical plates than expected for a separation (often interpreted as low packing efficiency). Bad distribution in an absorber or regenerator will show up quickly, either by a massive increase in breakthrough of the gas being absorbed or poor regeneration of the liquid. Fortunately, these devices are designed to operate at high liquid rates (as compared to distillation devices), so good distribution is not as difficult to obtain. However it is very advantageous to use a packing that will exhibit the maximum redistribution properties, because the effects of bad distribution in absorbers are more noticeable (when compared to those in distillation) since the breakthrough of absorbed gas will increase rapidly.

The complexity of the absorption/regeneration process is such that it does not lend itself to design by either the concepts of theoretical plates or transfer units, because these will change both in number required and in height within the same tower and over the same packed bed. Because of this complexity, it is necessary to invent

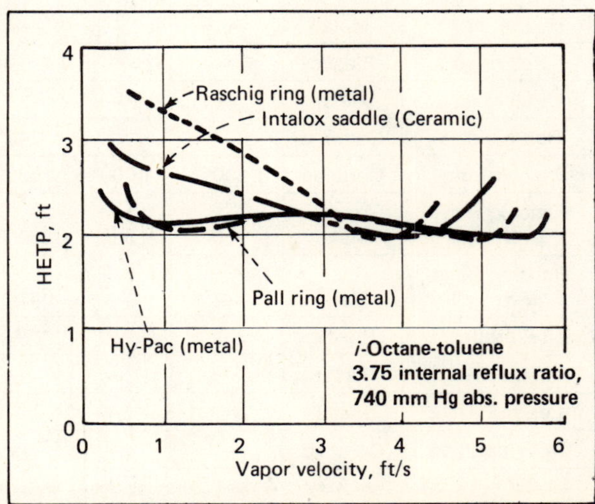

DISTILLATION performance (typical), 2-in packings—Fig. 6

Mass-Transfer Coefficient, K_Ga, of Various Packings—Table II

Packing	Size (Approximate) In: 1		1½		2		3		3½	
	L=5,000	L=10,000	L=5,000	L=10,000	L=5,000	L=10,000	L=5,000	L=10,000	L=5,000	L=10,000
Berl ceramic	2.5	2.8	1.8	2.2	1.45	1.8				
Cross partition rings, ceramic							1.05	1.35		
Heil Pak, plastic							1.15	1.2		
Hy-Pak, metal	2.85	3.30			2.1	2.6				
Intalox, ceramic	2.6	3.2	2.3	2.7	1.8	2.1	1.4	1.6		
Intalox Super, ceramic	3.2	4.0			2.1	2.45				
Intalox, plastic	3.2	3.5			2.0	2.3	1.5	1.7		
Lessing expanded, metal			1.35	1.45						
Maspac, plastic					1.7	2.1	1.15	1.45		
Mini-Ring #5 ceramic							1.15	1.30		
Mini-Ring #2, plastic					1.65	2.0				
Pall, metal	3.0	3.5	2.6	3.2	2.2	2.6	1.7	1.9	1.45	1.85
Pall, plastic	2.4	2.8	2.2	2.6	1.9	2.25			1.4	1.5
Raschig, ceramic	2.25	3.0	1.9	2.05	1.65	1.95				
Raschig, metal	2.4	2.8	1.95	2.15	1.5	1.65				
Tellerettes, plastic					1.9	2.1				

CO_2 1%, NaOH 4%; 25% conversion to carbonate, 75°F

Mass-Transfer Coefficient of Packing in Various Systems—Table III

Gas	Reagent	Hydrate	Ion		K_Ga*	Ionization Constant, K
Cl_2	$H_2O \cdot NaOH$	NaOCl	Na^+	OCl^-	25.0	—
HCl	H_2O	HCl	H^+	Cl^-	25.0	7×10^{-4}
NH_3	H_2O	NH_4OH	NH^+	OH^-	16.0	1.8×10^{-5}
H_2S	$H_2O \cdot MEA$	$MEA \cdot SO_3$	MEA^{++}	SO_3^{--}	8.0	—
SO_2	$H_2O \cdot Na_2SO_3$	$NaHSO_3$	NA^+	SO_3^{--}	7.0	—
H_2S	$H_2O \cdot DEA$	$DEA \cdot S$	DEA^{++}	S^{--}	5.0	—
CO_2	$H_2O \cdot KOH$	K_2CO_3	K^+	CO_3^{--}	3.10	—
CO_2	$H_2O \cdot MEA$	$MEA \cdot CO_3$	MEA^{++}	CO_3^{--}	2.50	—
CO_2	$H_2O \cdot NaOH$	Na_2CO_3	NA^+	CO_3^{--}	2.25	—
H_2S	H_2O	H_2S	H^+	S^{--}	0.400	1.1×10^{-7}
SO_2	H_2O	H_2SO_3	H^+	SO^{--}	0.317	5.6×10^{-8}
Cl_2	H_2O	HOCl	H^+	OCL^-	0.138	3.2×10^{-8}
CO_2	H_2O	H_2CO_3	H^+	CO_3^{--}	0.072	5.0×10^{-11}
O_2	H_2O	—	H^+	OH^-	0.0072	1.0×10^{-14}
KOH	$H_2O \cdot KOH$	KOH	K^+	OH^-	K_Ga above at 25% completion	>10
NaOH	$H_2O \cdot NaOH$	NaOH	Na^+	OH^-		10
MEA	$H_2O \cdot MEA$	$MEA \cdot OH$	MEA	OH^-		6.0×10^{-1}
DEA	$H_2O \cdot DEA$	$DEA \cdot OH$	DEA	OH^-		1.6×10^{-2}

* 1½-in. Intalox saddles

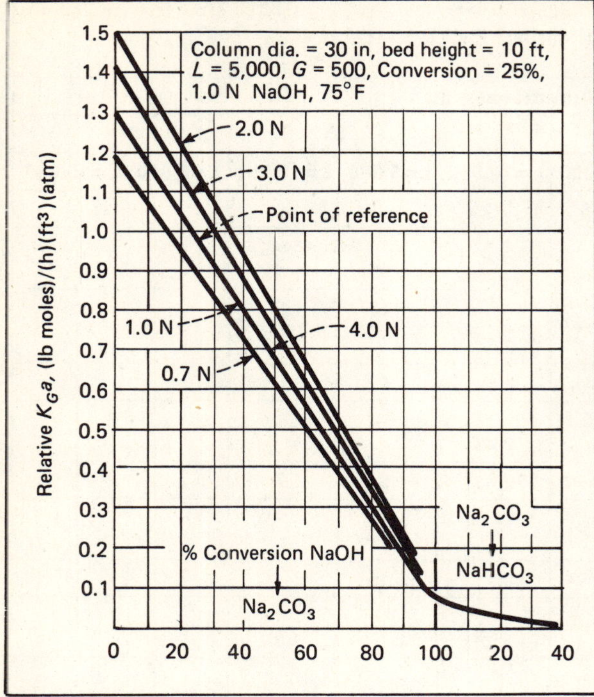

PERCENT conversion vs. K_Ga (CO_2, NaOH, H_2O)—Fig. 7

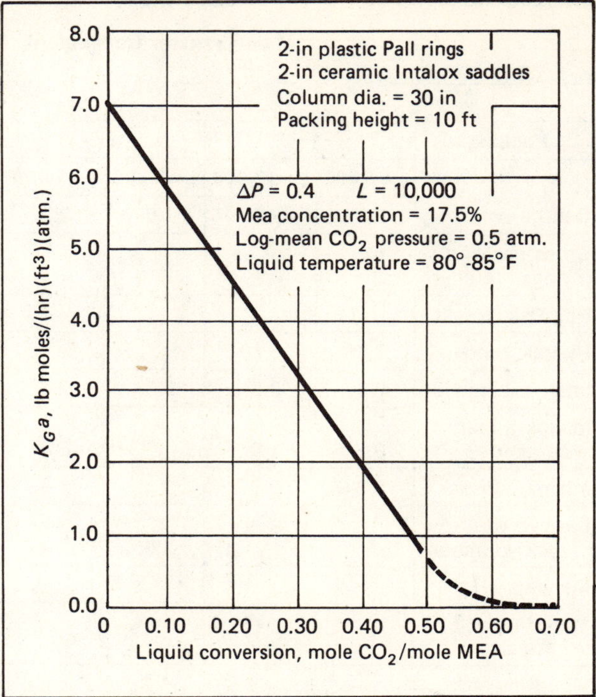

CONVERSION vs. K_Ga (CO_2, monoethanolamine)—Fig. 8

a "variable constant" to arrive at a design procedure—so we have the mass-transfer coefficient, or K_Ga, which is:

$$K_Ga = \frac{N}{V \ln \Delta p} \quad (2)$$

$$\ln \Delta p = \frac{P_1 - P_2}{\ln (P_1/P_2)} \quad (3)$$

K_Ga may be approximated using concepts of diffusional mass transfer and reaction kinetics, but coefficients in common use have been taken from operating equipment, either in a plant or an engineering laboratory. Once a mass-transfer coefficient has been determined for a certain system with a given packing, it can be converted to a different packing or size of packing; see Table II and take a simple ratio of the tabular mass-transfer coefficients to arrive at the new mass-transfer coefficient.

Certain laws governing the behavior of the mass-transfer coefficient must be understood:

■ Those systems having a high heat of reaction (and/or solution) will have a high K_Ga; those with a low level of heat release will have a low K_Ga. This is not a linear but a very complex function; therefore, when using this law to predict the K_Ga of an unknown system, interpolation should be used between systems approximating the unknown and having greater and lesser heat release. Table III gives the K_Ga of some of the common systems.

■ K_Ga will decrease as $\ln \Delta p$ increases, by approximately the 0.32 power of the log-mean drive:

$$K_Ga_2 = K_Ga_1\left(\frac{\ln \Delta p_2}{\ln \Delta p_1}\right)^{0.32} \text{ where } \ln \Delta p_2 \text{ is more than 1} \quad (4)$$

There is some evidence that K_Ga will approach infinity at driving forces of less than 50 ppm (5×10^{-5} atm), but so little is known about true vapor pressure, and therefore backpressure, in this region that this is more or less speculative.

■ As temperature increases, K_Ga increases about 1%/°C for carbon dioxide and sodium hydroxide systems. It has not been measured for other systems, but is probably about the same (for the same reaction order) as CO_2 and sodium hydroxide.

■ K_Ga will decrease as the reaction in the liquid phase nears equilibrium or completion. This decrease may be regarded as a linear function in systems of medium to low K_Ga—i.e., where $K_Ga = 7$ or less (Fig. 7, 8), but will follow a more-complex path where heat of reaction is high, as in the sulfur dioxide/sodium hydroxide system (Fig. 9). Most systems the engineer encounters fall into one or the other of these types. It should be noted that if the assumption is made in all cases that the decrease is a linear function, it will tend to give a resultant K_Ga on the conservative side.

■ Equilibrium partial-pressure of the gas over the liquid must be subtracted from the vapor-phase partial pressure to arrive at the true log-mean drive of the system. In most absorbers, as the liquid moves down the tower its temperature increases (as the reaction moves toward equilibrium or completion). Both of these factors result in the absorbed gas exhibiting an increasing vapor pressure over the liquid as it moves down the tower, and can very materially reduce the initial driving pressure in such towers as hydrochloric acid absorbers, sulfuric acid oleum absorbers, etc. This can also affect the exit drive on the system, in cases such as hydrogen sulfide absorbers, where the amount of absorbable gas in the exit gas must be kept very low. In cases of this type, consideration must be given to external heat-exchange and, sometimes, increased circulation rates.

Fortunately, most industrial processes operate within

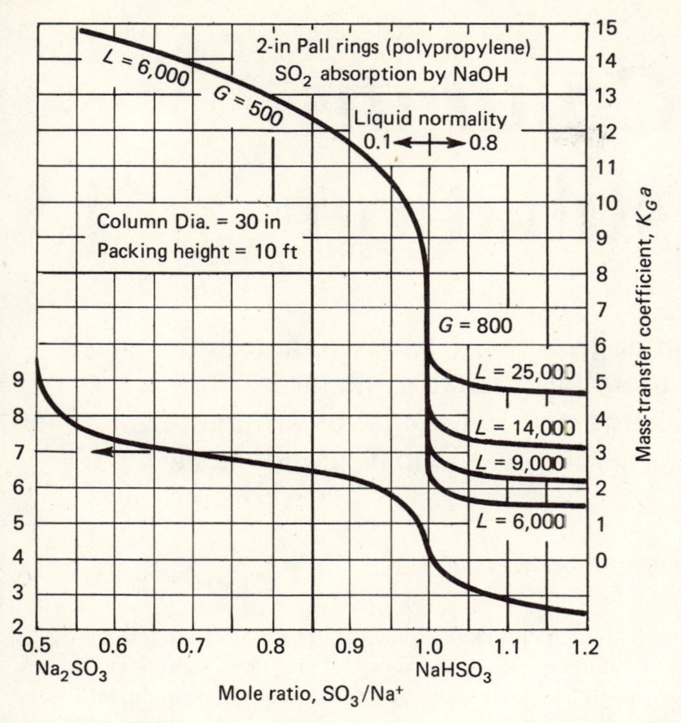

MASS transfer coefficient (SO_2, NaOH system)—Fig. 9

a relatively narrow range for each specific system, so experienced designers will have these mass-transfer coefficients already established for the processes they use.

It would be nice if we could say that in the same way that the mass-transfer coefficient (K_Ga) can be used in determining bed volume for absorbers, so K_La could be used to arrive at the volume required for regenerators. But, unfortunately, this is not true. The best regenerator designs are those based on experience—in other words, they are purely empirical. To understand this situation, one must understand just what happens during the regeneration of an absorption liquor.

Usually heat is exchanged with liquid coming from the bottom of the regenerator as a heat-conservation measure. Then the pressure is let off, and the absorbed gas is flashed to some lower pressure, either in a flash drum or at the top of the regenerator. The top of the regenerator is usually the best place to do this because here the liquid can be kept in a degree of turbulence sufficient for it to come to practical equilibrium. Also, there is usually a wash section at the top of the regenerator, where evaporated solution and mist may be caught and returned to the system. If the liquor is not kept turbulent during this period, the gas will not "spring" completely; hence, the top of the regenerator or the top of the upper packed bed will be overloaded and may flood. The rate at which the rest of the absorbed gas will be driven from the liquid now becomes an interrelation of time, temperature and pressure.

The amount of regeneration is a function of time in the regenerator and not one of K_La, height of a transfer unit, or height of a theoretical plate—however, any or all of these values can be "backed out" of an operating regenerator. These values will then change with a change in gas rate, liquid rate, pressure, and temperature.

What is really taking place in the regenerator is the thermal degradation of a compound of the absorbed gas in the liquid phase. Hence, increasing the pressure will increase the rate of this degradation because both the temperature and the backpressure on the system will be increased. However, within reasonable limits, increasing the pressure on the regenerator increases the rate of regeneration faster than the backpressure decreases it, and therefore results in a saving in heat input to the system. Attempts to arrive at design parameters based on this rate of degradation have not met with any great degree of success, because if the rate used is that predicted by a boiling vessel, such as that used by Ellis and Leachman [*5*], a packed bed will have to be twice as large (based on residence time) than those known to be successful. This is probably because the dissolved gas is released from the liquid while nearer to theoretical equilibrium (due to turbulent flow over the packed bed) than can be realized from a boiling vessel. Even though this situation is true, regenerators are still designed oversize so the liquid coming from them has attained equilibrium at some place up in the bottom bed. It is important to design the regenerator in this manner, because if acid gas is regenerated in the heat exchangers, it tends to corrode them very rapidly. Gas and liquid loadings are easily calculated for regenerators to establish the diameter of the vessel.

In conclusion, it might be said that enough knowledge has been gained over the past twenty years about the behavior of packed beds to say that we do know how to design them with confidence. They are used today in very large sizes and throughout the chemical, petrochemical and petroleum industries. Time and space do not permit a complete discussion of the theory and application of the packed bed; however, it is hoped that enough of the design concepts and principles have been covered here for the engineer to approach the design of gas-liquid-contacting devices using tower packings with some degree of confidence, and with an understanding of some of the mechanics of what is taking place.

References

1. Eckert, J. S., *Chem. Eng. Progr.*, **57,** No. 9 pp. 54–58, (1961).
2. Sherwood, T. K., others, *Ind. Eng. Chem.*, **30,** p. 765 (1938).
3. Lobo, W. E., others, *Trans. Am. Inst. Chem. Eng.*, **41,** 693–710 (1945).
4. Leva, Max, "Tower Packings and Packed Tower Design," The United States Stoneware Co. (1933).
5. Ellis, G. C., others, Allied Chemical Co. Release, April 1963.

Meet the Author

John S. Eckert is director of chemical engineering for the Chemical Process Products Div. of the Norton Co., P.O. Box 350, Akron, OH 44309. He is an authority on mass transfer and has developed several types of tower packings, improved methods of liquid distribution and tower-packing support; he holds numerous patents on tower internals and accessories. He is a graduate of Ohio State University, where he is presently an adjunct associate professor in the School of Chemical Engineering, lecturing on mass transfer.

Gauze-Packed Columns For Vacuum Distillation

Metallic gauze packings for distillation columns give better performance and lower costs for high-vacuum separations of thermally unstable mixtures. An evaluation and comparison of gauze packings and Pall rings for a number of different mixtures shows where each packing can be used most economically.

REINHARD BILLET, Badische Anilin- & Soda-Fabrik AG

The thermal instability of many mixtures to be separated by distillation or rectification is a frequent source of difficulties for design engineers and operators.

Since heat-sensitive high-boiling components may decompose or polymerize, there is an upper limit on the temperature, and this requires separating at a lower pressure. Here again, a limit is imposed by an economical condensation temperature at the head of the column and by an economically justifiable degree of vacuum.

Consequently, for rectifications of this nature, the temperature and pressure at the head and foot of the column are fixed, and with them, the maximum permissible pressure drop in the vapors flowing through the column.

Thus, the economics of vacuum rectification are governed not only by the prime costs but also by the number of theoretical trays that can be attained for a given pressure drop, i.e., on the energy requirements.

Gauze Packings

A new packing with a regular woven structure for use in separation equipment gives an extremely low pressure drop per theoretical tray.[1] It consists of stainless-steel strips with oblique serrations. They are arranged vertically and parallel to one another to form a cylindrical packing of 100 to 200 mm. height. Since the teeth of adjacent strips are mutually opposed, a criss-cross pattern of parallel flow channels with a triangular cross-section, and about 10 mm. side is obtained.

Thus, vapor and liquid flow counter-currently in a zigzag pattern through the column. By virtue of the large capillary action of the gauze surface, even comparatively small amounts of reflux are uniformly distributed to form a liquid film. The slope of the flow channels ensures intensive lateral mixing in the direction of the parallel layers. The packing, staggered at an angle of 90°, ensures that both media are also thoroughly mixed over the horizontal cross-section of the column. Hence, conditions are excellent for efficient separation.

Up to now, two different types of gauze packing have been used in industrial practice. The flow channels in these differ in the dimensions of the flow section and the angle of inclination. As a result of its geometrical structure. Type BX features an extremely low pressure drop per theoretical tray, and is preferred for vacuum separation of thermally unstable mixtures. The other, Type CY, has a very large specific surface and is particularly suitable for rectifications requiring many theoretical trays, e.g., for the separation of D_2O from water. Only Type BX will be dealt with in this article.

Normally, an annual allowance of 0.1 mm. is made for corrosion in technical equipment. However, this value is unacceptable for gauze packing because the thickness of the strips is only 0.16 mm. For this reason, the packing is made solely from corrosion-resistant metals such as stainless steel, phosphor-bronze, and Monel. The gauze packing cannot be cleaned mechanically.

Due to the regular structure of the packing and the thorough lateral mixing of the two phases, maldistribution such as occurs with ring packing is very slight if the liquid is evenly distributed. As a result, efficiency is independent of the diameter of the packing.

Evaluation of Performance and Costs

Owing to its low pressure drop per theoretical stage, the thin-wall metal Pall ring is often preferred to other packings for vacuum rectification on a technical scale. Therefore, it was obvious to compare the performance and costs of the gauze packing with those of metal Pall rings. The methods of assessment are described in detail elsewhere.[2] The steps in the calculation are shown qualitatively in Fig. 1 for two types of packing with different load characteristics. In Fig. 1, the calculation steps for optimization and comparison of mass-transfer columns are numbered. On the left-hand side are plotted: (1) efficiency, (2) pressure drop, (3) specific pressure drop and (4) specific column volume as function of the vapor-capacity factor. On the right-hand side, we find (5) cost per unit column volume (including the shell) as a func-

Originally published February 21, 1972

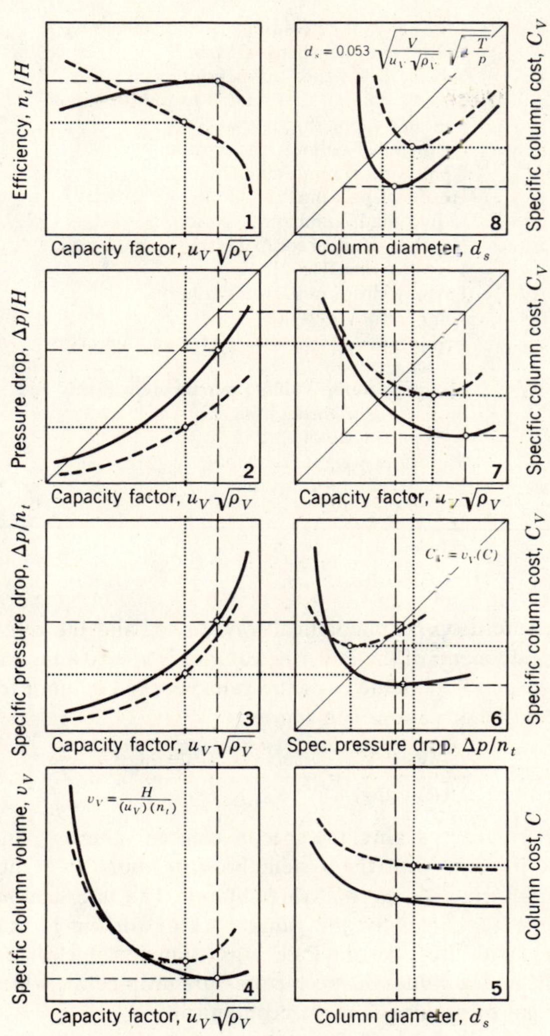

The evaluation procedure is carried out as follows:

1. Determine efficiency, in terms of the theoretical number of trays per unit height of packing, n_t/H, as a function of the vapor capacity factor, $u_V(\rho_V)^{1/2}$.
2. Determine pressure drop per unit height of packing, $\Delta p/H$, as a function of the vapor capacity factor.
3. Calculate specific pressure drop per theoretical tray, $\Delta p/n_t$, as a function of the vapor capacity factor.
4. Calculate specific column volume as a function of the vapor capacity factor. The specific column volume is defined as the specific volume, $v_V = H/u_V n_t$, required for unit vapor throughput and unit efficiency.
5. Quote the column costs per unit volume, C, as a function of the column diameter, d_s. The costs for the column shell must be included.
6. Calculate specific column costs, $C_V = v_V C$ as functions of the specific pressure drop, $\Delta p/n_t$, whenever this is of any significance.
7. Determine C_V as a function of the vapor capacity factor, $u_V(\rho_V)^{1/2}$, if the pressure drop is of no significance.
8. Calculate the specific column costs, C_V, as a function of the column diameter, d_s.

The dotted lines in the charts connect those points of the calculation steps that belong to the optimum conditions according to the minimum value of the specific column costs.

OPTIMIZATION of mass-transfer columns—Fig. 1

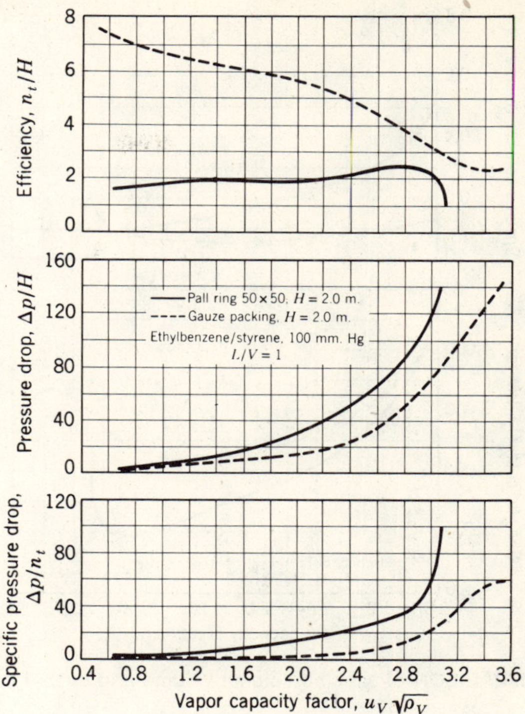

PERFORMANCE of gauze packing vs. Pall rings—Fig. 2

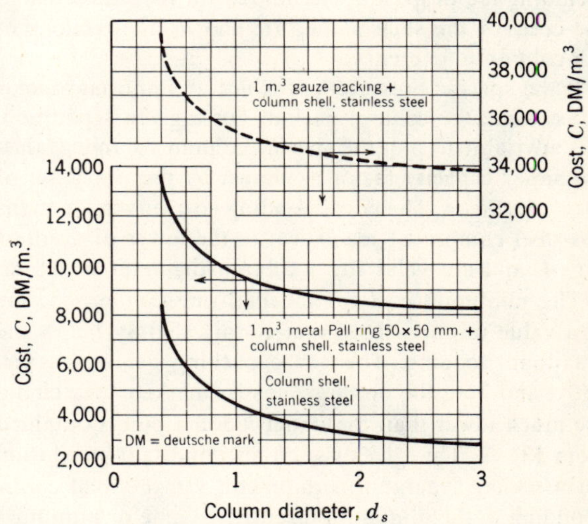

COSTS for gauze- vs. ring-packed columns—Fig. 3

tion of column diameter. Also, specific column cost related to unit of separation and vapor throughput is plotted as a function of: (6) specific pressure drop, (7) vapor-capacity factor, and (8) column diameter.

In order to determine the diameter of the cheapest column, the numerical value of $u_V(\rho_V)^{1/2}$ has to be found at which the product $v_V C$ is a minimum.

For example, in Fig. 2, the characteristic magnitudes for assessing the performance of the two columns (e.g., the efficiency, n_t/H, and the pressure losses $\Delta p/H$ or $\Delta p/n_t$) are plotted against the load, $u_V(\rho_V)^{1/2}$. These graphs show that much higher efficiencies and consid-

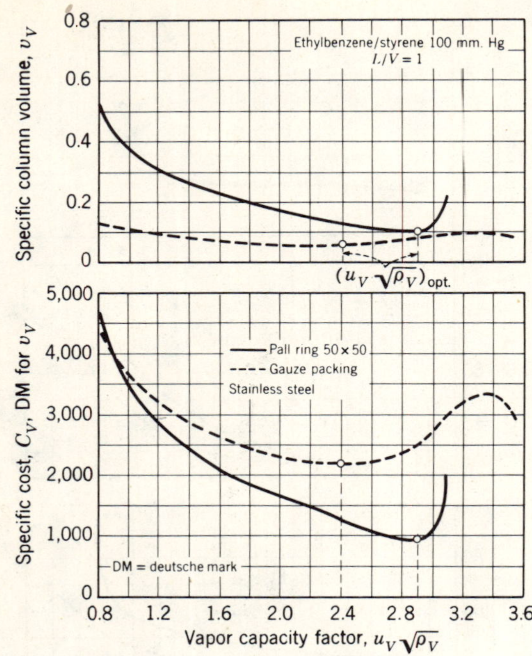

COMPARISON of specific column volume and cost—Fig. 4

erably lower pressure drops per theoretical tray can be achieved in the main loading range with gauze packing.

In Fig. 3, the costs C (DM/m.^{3} of column packing) including the proportion accounted for by the shell, and the costs of the shell alone, are shown as functions of the column diameter.

If the specific pressure loss is not an important factor in assessing the column packing for a given separation, it is advisable to plot the specific column volume against the vapor capacity factor as shown by the upper set of curves in Fig. 4. The corresponding cost curves for stainless-steel equipment are shown in the lower diagram of Fig. 4, and are valid for a column diameter of 0.8 m.

The minimum cost for the Pall-ring columns occurs at a value of vapor load that is only slightly below the maximum loading. The gauze packing behaves differently, and here the optimum loads (marked by a circle) are much lower than the upper loading points obtained from Fig. 2. This gives rise to an important conclusion for assessing separation equipment, viz, the load corresponding to the minimum installed volume or minimum costs need not necessarily be identical to the maximum loading.

Optimum column diameter, $(d_s)_{opt}$, can be calculated from the optimum load, $[u_V(\rho_V)^{1/2}]_{opt}$ from:

$$(d_s)_{opt} = 0.053\sqrt{\frac{V\sqrt{\mu(T/p)}}{(u_V\sqrt{\rho_V})_{opt}}} \quad (1)$$

where μ is the molecular weight, p is the absolute pressure and T is the absolute temperature of the vapor in the column.

The two columns were further investigated with different systems in order to determine the effects of various physical properties of the mixture on the assessment of performance and the cost optimization (see Fig. 5).

If the two columns are compared at the same specific pressure drop per theoretical tray, $\Delta p/n_t$, and the reference diameter of the Pall-ring column is $d_s = 0.8$ m., the corresponding diameter of the gauze-packed column for the cost calculations is given by:

$$\frac{d_s\text{ (gauze)}}{d_s\text{ (Pall ring)}} = \sqrt{\frac{u_V\sqrt{\rho_V}\text{ (Pall ring)}}{u_V\sqrt{\rho_V}\text{ (gauze)}}} \quad (2)$$

For gauze packing, the specific column values, v_V and C_V, increase with the system pressure, but this is not entirely true for the Pall-ring column. The intersections of the C_V curves for the gauze packing (drawn in thin lines) with those for the Pall rings (drawn in thick lines) indicate the values of specific pressure drop below which the gauze packing is more economical.

It is undeniable that gauze packing requires the least space. However, it has no economic advantage unless the specific pressure drop is below 10–20 mm. water column. The actual specific pressure drop depends on the system.

Application of Gauze Packing

The economically justifiable range of applications for gauze packing is mainly restricted to rectifications in which the pressure drop per theoretical tray has to be very low. This demand may be imposed either in the separation of thermally unstable systems, or in separations for which the pressure at the foot of the column may not exceed a maximum value that is dictated by a limitation in the saturated steam pressure for the column vaporizer and the available surface of the vaporizer.

If a large number of theoretical trays is required for a given separation, gauze packing may prove advantageous because the height of the packing is much lower than in conventional columns.

A column with gauze packing is quite capable of competing with other known equipment[2] from the point of view of price if the pressure drop per theoretical tray is less than 10 mm. water column. The equipment compared in the accompanying table has been arranged in the sequence of increasing specific cost factor, C_s, which

Nomenclature

C	Cost per unit of column volume (including equipment plus shell), DM/m.^{3}
C_S	Specific cost factors, dimensionless.
C_V	Specific column cost related to unit of separation and vapor throughput, DM.
d_s	Diameter of column, m. or mm.
$(d_s)_{opt}$	Optimum column diameter, m.
H	Height of packing, m.
L	Liquid load, kmol./hr.
n_t	Number of theoretical trays.
p	Pressure, mm. Hg.
Δp	Pressure drop, mm. water column.
T	Temperature, deg. K.
u_V	Vapor velocity per unit of free column cross-section, m./sec.
v_V	Specific column volume per theoretical tray and unit of vapor throughput, m.^{3}/(m.^{3})(sec.).
V	Vapor load, kmol./hr.
μ	Molecular weight, kg./kmol.
ρ_V	Density of vapor, kg./m.^{3}

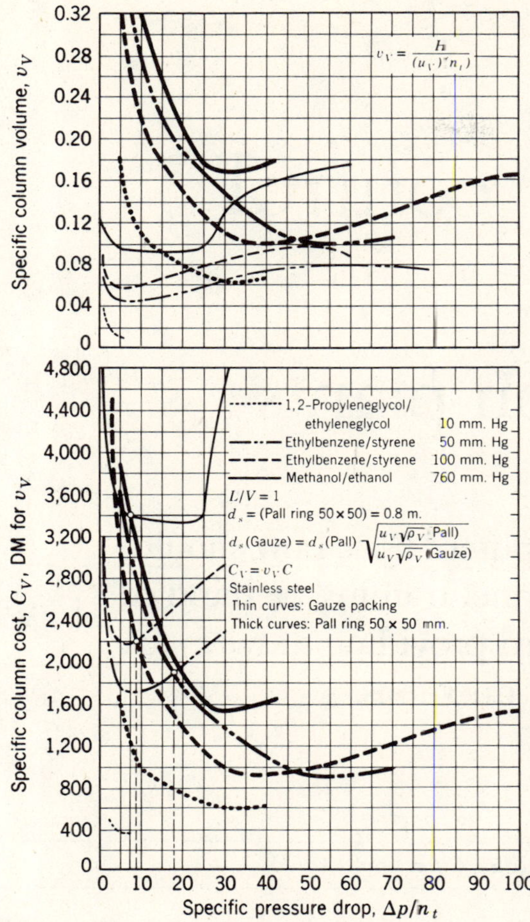

EFFECTS of different systems on column volume and cost between gauze- and ring-packed columns—Fig. 5

Comparison of Packings

Column Packing	Specific Cost Factor, C_S	Load Factor, $u_V\sqrt{\rho_V}$	Specific Column Volume, v_V
Gauze packing	0.71	2.40	0.057
Pall ring (50 × 50 × 1, mm.)	1.00	1.15	0.324
Pall ring (80 × 80 × 1.5, mm.)	1.25	1.25	0.406
Fine grid	1.25	1.30	0.429
Coarse grid	1.40	1.60	0.578
Diamond grid	2.00	1.30	0.428

Packing and column are stainless steel. Valid for the system ethylbenzene/styrene at 100 mm. Hg. $L/V = 1$, $\Delta p/n_t = 5$ mm. water column, and d_s (coarse grid) = 0.8 m. reference diameter.

is defined as the ratio of the specific costs of the column in question to those of a column packed with 50 × 50 mm. metal Pall rings for the equivalent separation, i.e.:

$$C_S = \frac{C_V\ (\text{packing} + \text{column})}{C_V\ (50 \times 50\ \text{mm. Pall rings} + \text{column})} \quad (3)$$

The C_S factors thus indicate how much the columns concerned are more expensive than a column packed with 50-mm. Pall rings. The columns headed load factor and specific column volume show that at the pressure drop mentioned, gauze packing allows the highest throughput and by far the smallest column volume. Thus, its liquid content is also low—a factor that may be a decisive advantage in the rectification of thermally unstable mixtures.

Fundamentally, gauze packings may be used in all sectors of the chemical and allied industries. Products for which gauze-packed columns have now been used are: menthol isomers, neraniol/geraniol, di- and triethanolamines, di-, tri-, and tetra-ethylene glycols, higher fatty alcohols, and heavy water.

Conclusions

A feature of the gauze packing discussed here is that equipment costs need not be expected to be a minimum if the column has been designed for maximum throughput. In fact, owing to the rapid decrease in specific efficiency with increasing load, the optimum load factor is equal to a vapor capacity factor that in the vacuum systems is less than 60 to 70% of the corresponding flood point (the actual decrease depending on the mixture). In the light of the investigations, the average is given by:

$$(u_V\sqrt{\rho_V})_{opt} = 2.4,\ (\text{m.}^{-1/2}\ \text{sec.}^{-1}\ \text{kg.}^{1/2}) \quad (4)$$

The associated specific efficiency of the packing is identical to:

$$n_t/H = 5\ \text{theoretical trays/m.} \quad (5)$$

Average pressure drop per theoretical tray is only about:

$$(\Delta p/n_t)_{opt} = 5\ \text{mm. water column} \quad (6)$$

Hence, the column costs are less than those of any other separation equipment designed for the same specific pressure drop. For this reason, and because the specific volume and time of residence are much less than for conventional equipment, much better performance and reductions in costs can be achieved by using gauze packing for the separation of thermally unstable mixtures under very high vacuum.

References

1. Swiss Patent 398,503, Sulzer A. G., Winterthur, Switzerland.
2. Billet, R., paper read at International Symposium on Distillation, Brighton, England, 1969. See also, *Chem. Eng. Progr.*, Jan. 1970, pp. 41–50.

Meet the Author

Reinhard Billet is a member of the planning department of Badische Anilin- & Soda-Fabrik, AG, 67 Ludwigshafen, Germany. He joined BASF in 1960 where he started in the chemical engineering research department. Most of his work has been in distillation and evaporation. He was awarded the VDI-Ring of Honor of the German Institution of Engineers (VDI). He received the degrees of Dipl.-Ing. and Dr.-Ing. from the Technical University of Karlsruhe, Germany.

Computer control of fractionation plants

Does it pay to install a computer? If so, how should installation, startup and personnel training be handled? Here is a useful guide toward providing answers to these and other questions.

***Daniel W. Kemp** and **Donald G. Ellis,** Cities Service Oil Co.*

☐ Economic savings and reliability are the two most powerful arguments in favor of computer control of fractionators. A project engineer contemplating such an application must first justify, at least on paper, the purchase of a computer and accessory hardware. He must also be aware of certain pitfalls concerning installation and operation of the system, and operator training.

In this article, the authors attempt to cover these aspects and others related to computer installation, such as location of sensors, valve-operating devices, sample systems, and flexibility to run without a computer.

Calculating a payout

The first step in the justification procedure is to compare how well the existing control system is operating in comparison with a theoretical ideal. This involves collecting basic operating data over a normal or average period of operation. In a typical fractionating column—e.g., a deethanizer—the data should include the following variables:

- Heavy key component in the overhead product
- Light key component in the bottom discharge
- Temperature, pressure, composition and rate of flow of the feed
- Energy required to run the column—usually, heat to the reboiler.

Deviations from the norm are determined by integrating the composition recordings and measurements of heat to the reboiler, and presenting them on a time-weighted basis. High and low deviations are selected at the points within the period that consumed 90% of the total time. These data define the actual operating circumstances in terms of purity vs. energy required.

Meanwhile, tray-to-tray calculations using feed analyses previously determined over a period of time are plotted as shown in Fig. 2. This graph establishes the relationship between the purity of the overhead product and the energy required at a specific bottom purity—in this case 6% ethane in the propane. In the same fashion, an entire family of bottom-purity curves can be developed—at a constant tray-efficiency, as in Fig. 3, or at varying tray-efficiencies (Fig. 4).

By comparing the actual operating data (Fig. 5) with the theoretical calculation (Fig. 6), a range of deviations can be obtained. This represents conditions previous to the application of advanced controls, and serves as a basis for estimating project payout.

Measuring the value of a computer

At this point, how can the project engineer determine whether investing in a computer is justifiable? Fig. 7, which shows the deviation from set point in relation to maximum product impurity, should provide a good indication. During normal operation, the set point must be set at considerably below the maximum impurity level, which is also that of minimum utility requirements. But if controllability could be improved

Originally published December 8, 1975

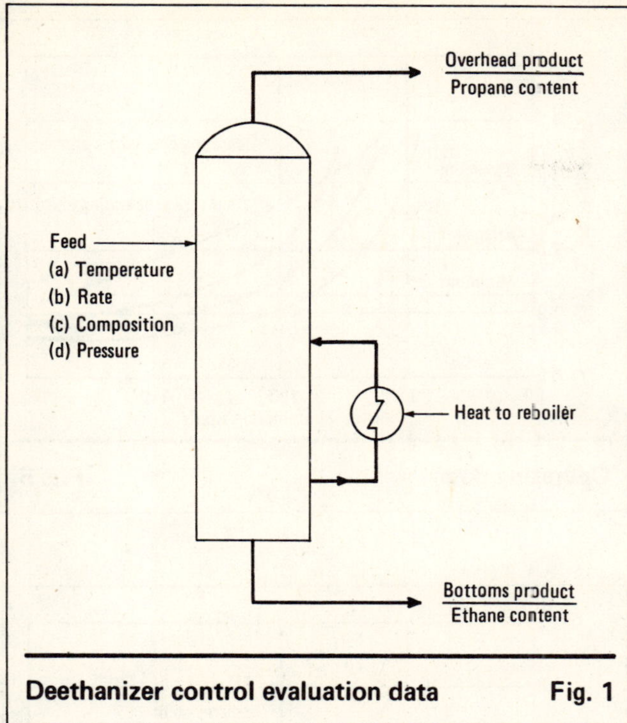

Deethanizer control evaluation data — Fig. 1

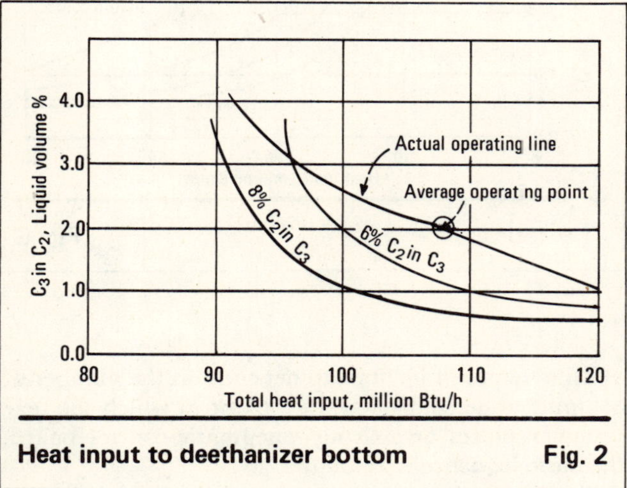

Heat input to deethanizer bottom — Fig. 2

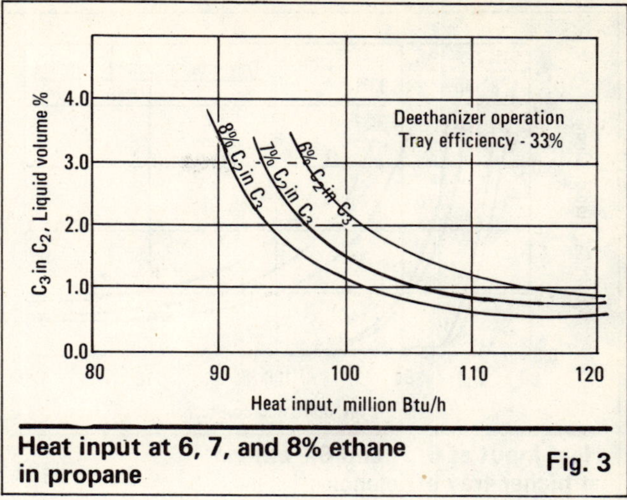

Heat input at 6, 7, and 8% ethane in propane — Fig. 3

to reduce the deviation to much lower limits—as Fig. 7 further illustrates—the set point could be moved much closer to the maximum impurity level.

The improvement is better seen in Fig. 6. The new set point here relates to a much lower energy demand, which leads to lower operating costs and a higher average product-quality. Improving the control at both ends of the column maximizes the light key component in the bottom discharge and the heavy key component in the overhead—all at a much lower energy requirement. Tray efficiency also increases with improved column controllability.

Fig. 8 illustrates the total reduction in utilities that can be achieved by installing advanced controls on a fractionator, as seen from the viewpoint of decreased control-point deviations and increased tray efficiency.

Reflections before buying

After justification has resulted in management approval, and the computer vendor has been selected, the project engineer is ready to proceed with the purchase. The following factors must first be determined:

- Process inputs to the computer
- Computer outputs for process control
- Power requirements
- Needed reliability

The vendor usually provides an answer to all of the above except for reliability, which is the customer's responsibility.

At the time of purchase, the project engineer should exercise care in choosing hardware for computer interfacing. The input sensors provide data upon which a computer performs its calculation, while the output transducers handle the control valves on a fractionation column. High quality in these items is of vital importance. And the engineer must make sure that the purchased equipment is compatible with the computer.

The computer vendor should always be provided with specifications of any equipment that his customer is buying, and he must approve the purchase.

Measurements provided to the computer should be taken at the best possible locations around the fractionator. Accuracy is important in determining the type of sensor needed. For example, a temperature change of 0.1°F in the overhead line from a deisobutanizer reflects a considerably different composition-shift than does the same change in the overhead line of a deethanizer. At any rate, responsibility for the complete sensor system, including analyzer, programmer, interfacing, sampling system and sampling point should be clearly defined, and preferably handled by one party.

In addition, the individual control points must be evaluated in terms of their effect on the new control package. Additional hardware, if needed, must be specified.

Electrical requirements for a computer installation in a fractionation plant may include main and alternate power supplies for the computer equipment; power for transducers, sensors, chromatographs and related units; and utility outlets. Plants that are required to recover automatically from short-term variations in

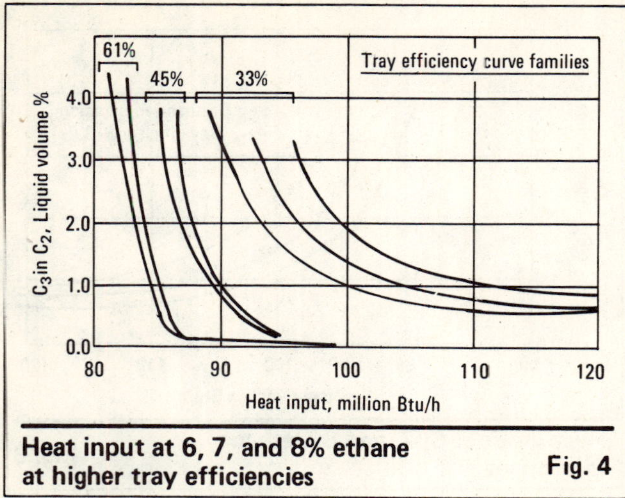

Heat input at 6, 7, and 8% ethane at higher tray efficiencies **Fig. 4**

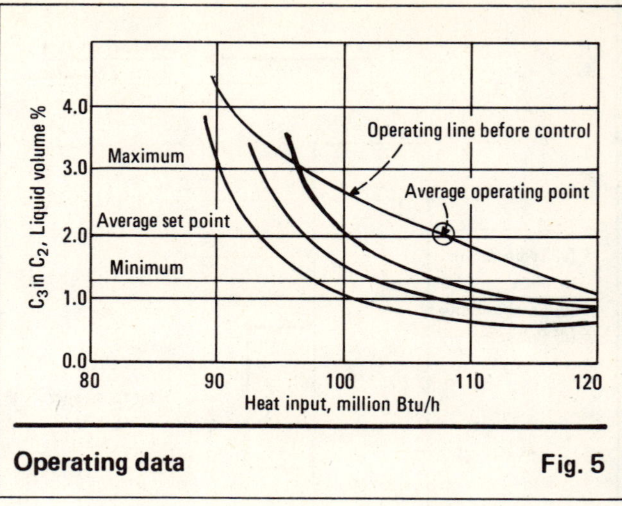

Operating data **Fig. 5**

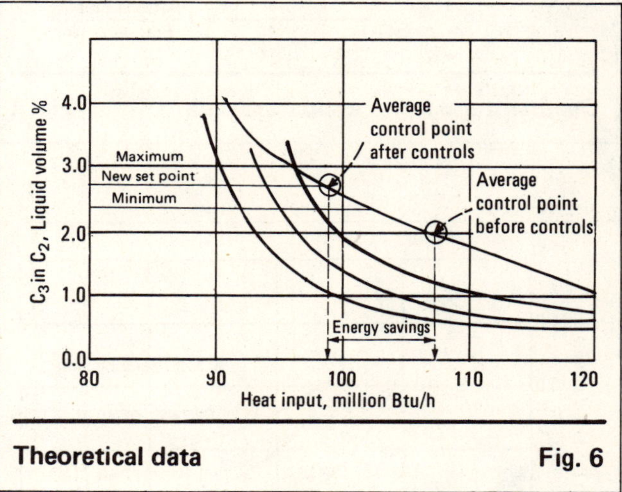

Theoretical data **Fig. 6**

power supply may need a safeguard mechanism, such as battery backup systems.

Installation and startup

It is essential to have special drawings made for the installation phase, because a great deal of the work involves electrical wiring. Mistakes can be time-consuming and expensive to correct. The engineer should therefore provide good drawings, amend them as the job proceeds, and ensure that it is the "as-built" versions that find their way into the plant files.

Actual installation comprises the computer and associated equipment, input/output hardware, power supplies, and signal and power wiring.

In addition to providing a proper environment for the entire system, the installation should proceed with an eye to operation and maintenance. The wiring, for instance, should be sufficiently flexible. Because installation is usually underway simultaneously with process-control-model development, the sensor inputs are defined at an early stage—usually according to past experience and similar applications. If the computer monitors all of the process variables, there is no need for input wiring spares. But if the installation calls for a limited number of sensors, it may be prudent to install accessories such as spare-signal and thermocouple-extension wires.

Output transducers should present no problem, since they are normally limited by the number of control valves.

Upon completion, the installation job should be checked by the project engineer and the computer vendor. Power supplies must be tested before the equipment is switched on. Input and output connections to the computer should be verified, and tie-ins to the proper process points should be checked.

Depending on the equipment and personnel available, input and output hardware can be calibrated either before or after installation. However, following the verification step, input devices should be run full range, and the measurement on the computer checked to ensure proper operation. Output devices should be similarly checked before startup. This may help detect early failures.

Time required for startup depends on the number of columns to be controlled. In a plant in which the new system replaces an existing pneumatic-control board, the basic sequence is as follows:

- Shift one control loop to the computer
- Tune the local loop on the computer
- Proceed to the next control loop and tune

Whether or not each control loop is left on the computer at this stage depends on the normal mode of operation. For instance, it may be desirable to return a cascade loop to the board after local tuning.

In general, loop tuning can take from ten minutes to several hours, depending on the type of loop and what is happening in the process. Personnel involved in this operation should not leave the console area until performance is satisfactory and the loop is on process control.

Testing the control models

Process-control models should be tested upon completion of control-loop tuning. If possible, independent functions should be tested one at a time. Transfer of controls to the computer should begin as early as possible during the working day to achieve maximum run-

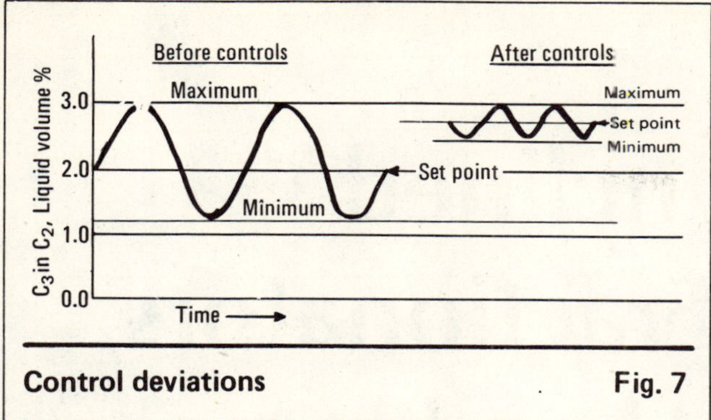

Control deviations Fig. 7

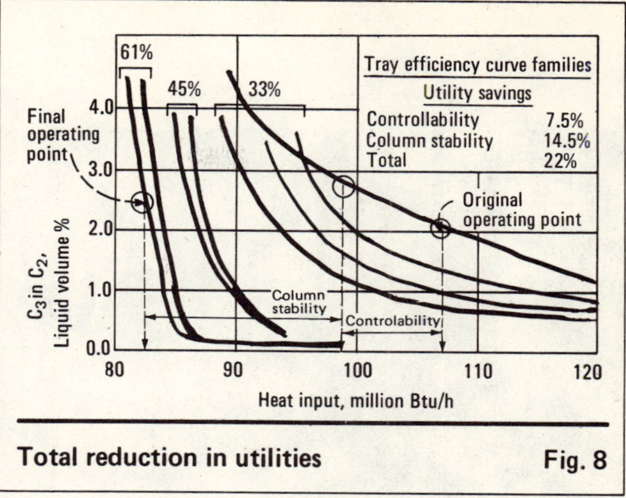

Total reduction in utilities Fig. 8

ning time on a given control-function while the startup personnel is at the plant.

Controls must be closely monitored to verify proper performance. It may be necessary to change model-parameters or override the controls if deficiencies are observed. The new controls should have performed satisfactorily in a "hands-off" manner for several hours before personnel is allowed to leave the computer-console area for any extended period.

Performance of each section of the control model requires at least a 48-h trouble-free run before a final decision is made. Of course, failures should not be too surprising. Models developed for one plant may not perform well in another due to differing physical systems. In a digital computer, failures due to deficient models are easily corrected by modifying the model and feeding a new program. However, a major modification of this kind may take several days.

When a system has been totally operational for several days, the project engineer collects the data required for performance evaluation. These are broken down and overlaid on the curves developed during the justification phase.

Meanwhile, the vendor prepares final documentation for the project. This consists of an update on the "as-built" operating models, and the final configuration of the equipment. The data, along with the manuals and information that the customer needs to make changes is then forwarded to the customer.

Thoughts on personnel training

Training of operating personnel should normally be accomplished in two phases. The first occurs after the process models are ready, but before startup begins. Operators should learn what the project goals are, how the models work, and how to use the system for control and information retrieval.

The second phase coincides with startup. Under the supervision of vendor representatives and the project engineer, the personnel should become familiar with the operation and use of the new control system.

As a rule, operators are skeptical of advanced control systems. They have to see actual improvements in process control before they accept the system. One of the worst mistakes the project engineer or vendor can make before the final stages of a project is to constantly tell operators how well the new system will work. Focus instead on what the system is supposed to accomplish. Openly discuss the possibility of control-logic errors, and disruptions that may occur while shifting over to the new unit. Once the new controls are running well, the fact will speak for itself.

The completion of an advanced control project in a fractionation plant should open a Pandora's box of questions from the operating group and the engineering staff. Here is a list of the most common ones.

1. How can the data-logging capability of the computer be used to upgrade the plant process capabilities?
2. Should the customer develop his own in-house capability for program modifications, or use the vendor's?
3. How many modifications can be added to the installation before system degradation occurs?
4. For future applications, what type of system should be used?
5. How can the rest of the engineering staff be educated in the areas of modern logic packages, sensors and existing advanced applications?

The authors

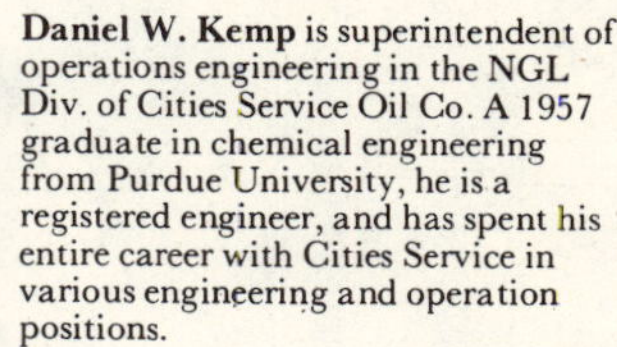

Daniel W. Kemp is superintendent of operations engineering in the NGL Div. of Cities Service Oil Co. A 1957 graduate in chemical engineering from Purdue University, he is a registered engineer, and has spent his entire career with Cities Service in various engineering and operation positions.

Donald G. Ellis, Staff Engineer, is responsible for evaluation and application of advanced controls in different areas of NGL recovery and fractionation at Cities Service. He obtained a B.S. from the Naval Academy in 1964.

Inclined Fractionators, Absorbers

For absorption, fractionation and stripping operations, an inclined contactor is more efficient and costs less to construct than a vertical tray column.

D. P. RAO, Birla Institute of Technology and Science

An inclined gas-liquid contactor can be fairly cheaply made in a plant shop for absorption, fractionation and stripping operations. The unit has these advantages over vertical tray columns: (1) higher throughputs per unit tray area; (2) greater efficiency due to less entrainment, higher vapor velocities, and shorter liquid paths; (3) use of nominal shell thickness;* and (4) greater accessibility for inspection and maintenance.

* For vertical columns, one must consider wind velocity and seismic forces in determining shell thickness.

Looking like a long rectangular box supported on an inclined structure (Fig. 1 and 2), the trays are arranged one after another as stairway steps, at 6 to 12-in heights, depending on liquid throughput. The trays are provided with a prismatic downcomer for the vapor to flow from the space above one tray to the space below the next. All kinds of trays equipped with liquid downcomers can be installed. The vapor rises through the plate orifices (Fig. 2), and exits down the downcomer (Fig. 1) to the space beneath the next plate. Entrained liquid falling through

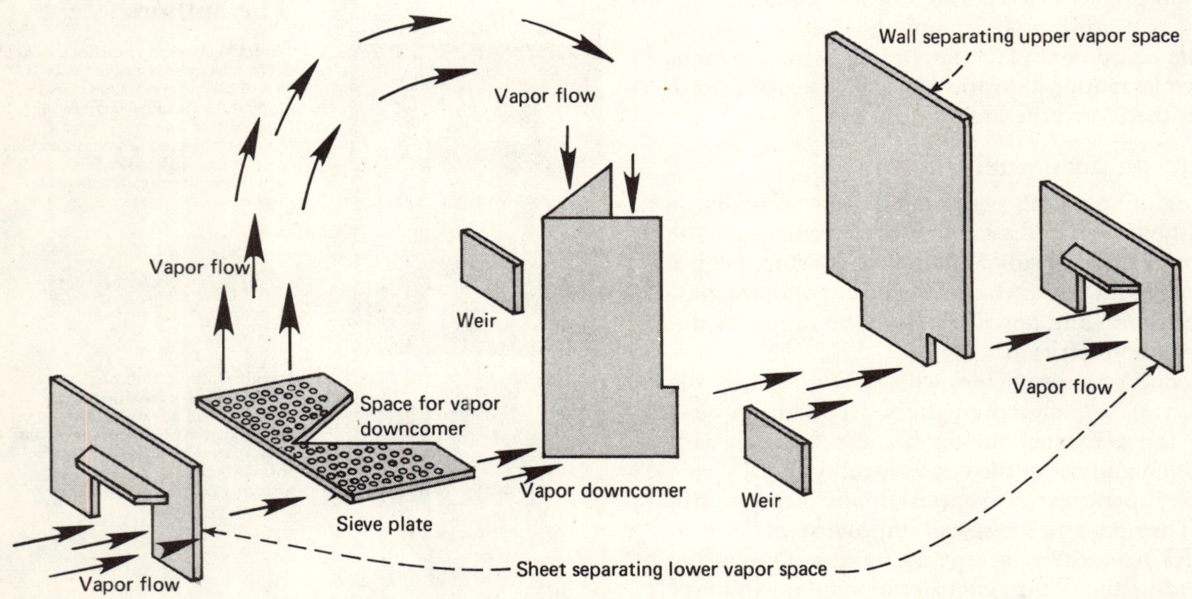

COMPONENTS of tray section. Note how vapor travels upward and then flows down through the downcomer—Fig. 1

Originally published April 28, 1975

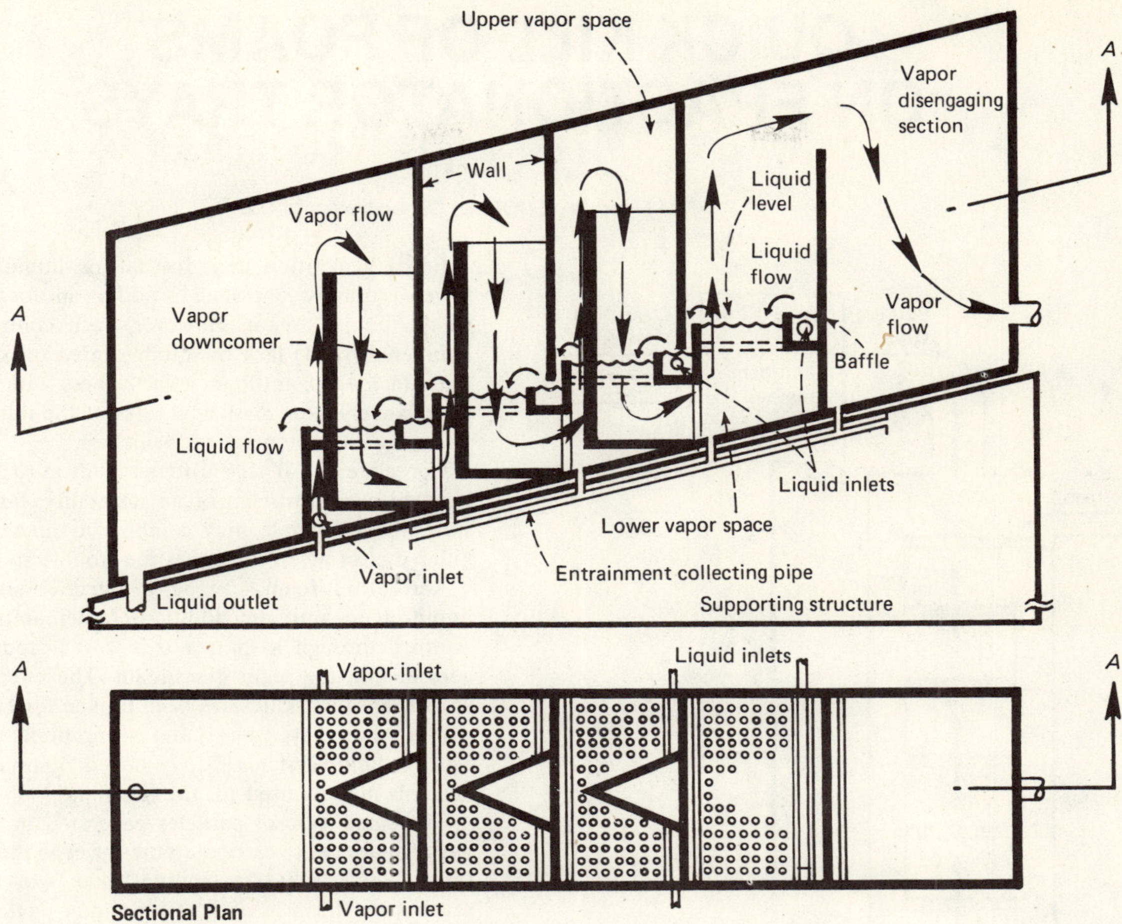

VAPOR-LIQUID CONTACTOR shows fluid inlets, direction of fluid travel, and partitions between trays—Fig. 2

the sieve plates is collected in a pipe running at the bottom of the structure.

The vapor downcomer can be installed in the center or at both ends of the tray. In Fig. 2, it is located at the center, being open at the top and in the lower portion to permit passage of vapor beneath to the next tray. As seen in the figure, the vapor enters the lowest part of the section containing the bottom tray, and travels upward—bubbling through the trays—until it reaches the top tray. There is no need on this tray for a vapor downcomer because the vapor passes over the baffle to reach the top part of the vapor-disengaging section. The liquid, on the other hand, is introduced on the top tray (Fig. 2) or on any other tray liquid-downcomer and travels downward to exit at the lowest part of the structure. Note that a weir is provided close to the sheet (wall) separating the upper vapor space. This is to prevent the vapor from bypassing the tray during startup or transient periods.

Because of high vapor spaces and shorter liquid paths, high vapor and liquid throughputs can be achieved without an appreciable increase in contactor cost. Tried on a pilot-plant operation, the unit has shown no appreciable entrainment. One disadvantage, though, is that more floor area is required than would be the case for a conventional vertical column.

For design purposes, known data for various trays can be used, because the flow patterns of the two fluids are similar to those in conventional columns. Compared to the total pressure drop across the tray, the vapor pressure-drop that is developed in the downcomers is negligible.

For optimum design, entrainment studies are needed—although to start, entrainment correlations for conventional columns can be used. Flooding can occur due to liquid backup on an upper tray. However, flooding conditions can be estimated based on vapor pressure-drop and liquid-gradient considerations. Thus, the height between immediate trays can be established.

Acknowledgments

The efforts of V. S. Bakshi and Ravi Kumar in building and operating the pilot unit to demonstrate this design are deeply appreciated.

Meet the Author

D. P. Rao is lecturer in the Chemical Engineering Dept., Birla Institute of Technology and Science, Pilani (Raj.) 333031, India. He holds the following professional degrees: B.Tech. from Andhra University, Waltair; M.Tech. from Indian Institute of Technology, Kharagpur; and Ph.D. from Birla Institute of Technology and Science, Pilani, India.

QUICK KILL OF FOAMS ON FRACTIONATOR TRAYS

C. F. PRATT and S. Y. HOBBS, General Electric Co.

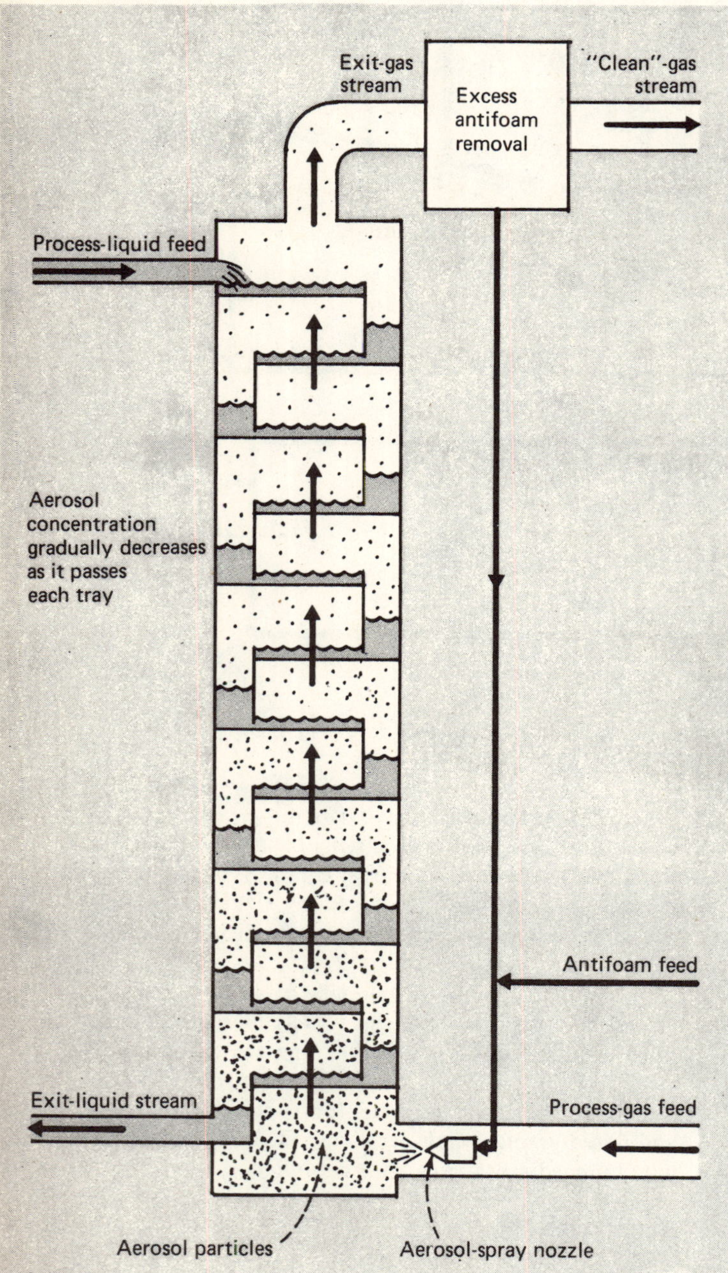

Foam generation in industrial gas-liquid separation towers is usually controlled by adding antifoaming agents to the liquid stream. However, such control may be hampered by (1) lack of suitable injection sites, (2) fatigue (a loss of antifoam effectiveness with time), and (3) relatively slow mass-flow rates of the liquid in comparison with superficial gas velocities.

Consequently, if an antifoam agent must be added a long distance upstream of an instability, fatigue and a slow liquid flowrate may combine to render the foam killer useless when it reaches the trouble spot.

Aerosol antifoams can be an effective alternative to liquid-phase antifoam addition. A defoaming agent is pumped through a spray nozzle and introduced as an aerosol into the main gas stream. The effectiveness of an aerosol foam killer has been proven for both aerated stagnant liquid systems [*1*] and conventional countercurrent gas-liquid systems [*2*]. The flow diagram depicts how aerosols may be used for the latter.

The small aerosol particles generated in the process gas feed are easily carried up the tower to the separation trays, where foam instabilities may exist. Antifoam reaches the trays in a short time, due to the relatively high superficial gas velocity. For small enough aerosol particles ($\sim 5\mu$), the antifoam will not be stopped by the first few trays, and every tray in the tower can be treated. The foam-killing capacity of the aerosols is as good as, or better than, that of conventional aqueous systems.

Costs are reduced in this system, because continuous antifoam addition is usually not necessary. Foam instabilities can generally be controlled with intermittent sprays of the aerosol. Should continuous use be desirable, cost reductions can still be realized by placing aerosol separation equipment (jet impactors, vane-type mist extractors, etc.) in the exit gas stream. Excess antifoam would simply be recycled to the aerosol-generating nozzle. This would have the added advantage of solving any exit-gas contamination problems.

References

[*1*] Evans, J., U.K. Pat. 1,091,199 (to Midland Silicones Ltd.), Oct. 21, 1966.

[*2*] Hobbs, S. Y., and Pratt, C. F., U.S. Pat. application (to General Electric Co.), on file.

Part II
DISTILLATION ALTERNATIVES

Section 5 Liquid-liquid Extraction

Liquid-liquid extraction: The process, the equipment

Liquid-liquid extraction (also known somewhat incorrectly as solvent extraction) is a unit operation that is growing in importance. Here is a review of the theory underlying the operation and a survey of the equipment that is used.

***P. J. Bailes, C. Hanson** and **M. A. Hughes,** University of Bradford (Great Britain)*

☐ Liquid extraction has long been a powerful separation technique for laboratory use. Its application for large-scale industrial separations dates from the early 1930s, when it answered the need for a method of removing aromatic hydrocarbons from the kerosene fraction during oil refining. Since then it has found ever-increasing application in a wide range of industries from copper production to the manufacture of antibiotics.

Liquid-liquid extraction is attracting a great deal of interest at the present time. This article will explain the basic principles involved and show something of the developments which are taking place. For further information, readers are referred to the standard text by Treybal [*1*] and the review of recent developments by one of the authors [*2*]. Much valuable information has been published over the past decade in the proceedings of a series of international solvent extraction conferences (the ISEC meetings) [*3–7*]. Regular reviews have also been published over recent years by the Society of Chemical Industry [*8*] and are a valuable source of information.

THE PROCESS

Equilibrium position

If a liquid solvent is added to a solution of some *solute, A,* in a second solvent, either immiscible or only partially miscible with that which is added, then the solute will distribute between the two liquid phases until equilibrium is established. The solute's concentrations in the two phases at equilibrium will depend on its relative affinities for the two solvents. Although it is inevitable that two solvents should be involved, it is conventional to refer to the added liquid as *the solvent.* The product of the desired solute in the solvent is *the extract,* while the residue leaving in the initial phase is *the raffinate.* At equilibrium, the ratio of the concentrations of solute in the extract, y, and raffinate, x, phases is called the distribution coefficient, D, and gives a measure of the affinity of the solute for the two phases:

$$D_A = \frac{y_A}{x_A}$$

Since considerations of thermodynamics demand that the activity (or chemical potential) of a solute should be equal in two phases at equilibrium, a distribution coefficient of other than unity implies that the solute must have different activity coefficients in the two phases. The origin of such a difference usually lies in the degree of interaction between the solute and the two solvents. While no sharp dividing line can be drawn, it is possible to distinguish two broad categories of solvent extraction processes, depending on whether the solute-solvent interaction is nonspecific or specific. The former implies some form of "physical" interaction such as might arise from polarity differences or hydrogen bonding. With the latter, a more definite "chemical" interaction can be distinguished, usually allowing a stoichiometric equation to be written. In this case there will be a limiting concentration of solute in the solvent corresponding to complexing of all the solvent available.

If only one solute is involved, such as in the recovery of an impurity from an effluent stream, only the distribution coefficient need be considered, and it is desir-

Originally published January 19, 1976

Confusion sometimes arises through use of the terms solvent extraction and liquid extraction for the same operation. This article is restricted to those applications of solvent extraction in which one or more of the components of a liquid solution are separated by extraction into a second liquid phase (the solvent), which is added and which is either immiscible or only partially miscible with the first liquid. While frequently called solvent extraction, the operation should strictly be called liquid-liquid extraction, because the term "solvent extraction" is also applied to the recovery of substances from solids by treatment with an organic solvent—typically recovery of natural oils from seeds. The technology of the latter operation has developed separately from that of liquid-liquid extraction and will not be given further consideration in these pages.

able for this to be as large as possible. In other cases, however, the aim is to achieve a separation between two solutes. While a large distribution coefficient for the solute being extracted is still desirable, consideration also has to be given to the selectivity of the solvent for solute A as against B. This is measured by the *separation factor* α, which is the analog of relative volatility in distillation:

$$\alpha_{AB} = \frac{D_A}{D_B}$$

Since the condition of equilibrium demands that:

$$\gamma_A^* y_A = \gamma_A x_A$$

it follows that:

$$\alpha_{AB} = \frac{\gamma_A \gamma_B^*}{\gamma_A^* \gamma_B}$$

where γ^* and γ are activity coefficients in the extract and raffinate phases, respectively. Where specific interactions are involved, the separation factor can vary significantly with concentration, as there will be competition between the solutes for the available solvent if this becomes limited.

It will be clear from the above that separations by solvent extraction are very dependent on chemical considerations and the technique tends to effect a separation on the basis of the chemical nature of a solute. This makes the technique complementary to distillation, in which volatility can be primarily correlated with molecular size.

Solvent systems

The solvent may be a single chemical species, but this is not always the case. Particularly when a specific interaction is involved, the most appropriate active reagent may not have suitable physical properties for direct use as the solvent. In such cases it can be used dissolved in another suitable liquid. The active reagent is then known as the *extractant* and the liquid in which it is dissolved as the *diluent*—the solvent, then, comprises the two together. The diluent is often dismissed as being inert and its choice not critical. This is oversimplification, as it is now realized that the diluent can have a considerable influence on both the position of equilibrium and the rate of extraction, as well as determine the maximum possible solute concentration in the solvent phase (through its solvency for the solute-extractant complex). The latter is a significant consideration. The maximum operable concentration (or loading) of a solute in a solvent determines the volume of solvent phase that has to be processed for a given production of the solute. In addition, if the solubility of the complex in the diluent is exceeded it will come out of solution, possibly as a solid but more usually as a third liquid phase—the situation being known as third-phase formation. This makes an industrial plant inoperable and must be avoided. If the solute-extractant complex has a low solubility in all convenient diluents, the limitation can often be alleviated by addition of a further component to improve the solvency. Such an additional component is known as a *modifier*.

The flowsheet

Extraction—The amount of solute that can be recovered from a particular feed solution by equilibration

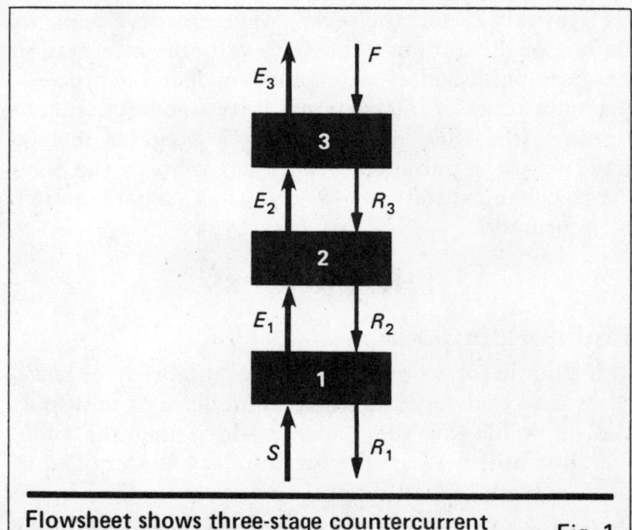

Flowsheet shows three-stage countercurrent liquid-liquid contacting — Fig. 1

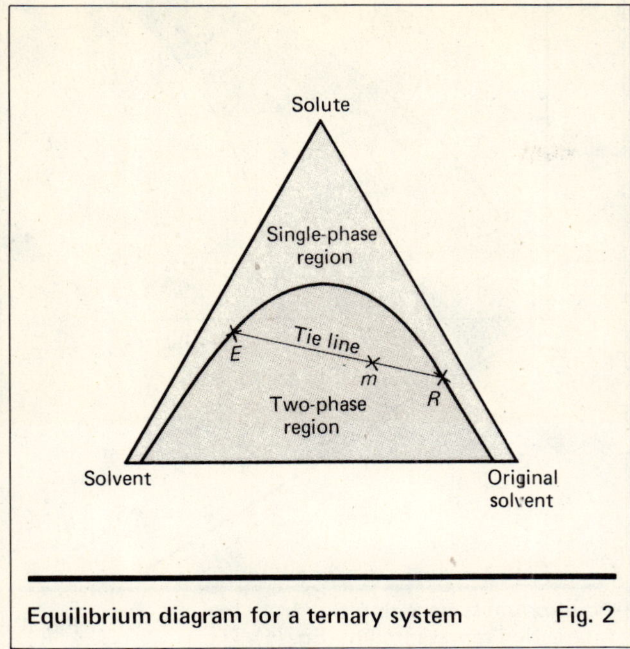

Equilibrium diagram for a ternary system **Fig. 2**

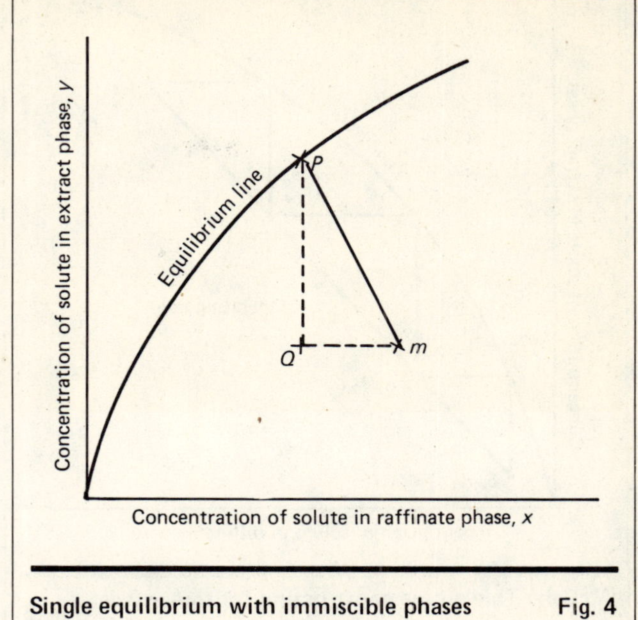

Single equilibrium with immiscible phases **Fig. 4**

with a solvent depends on both the distribution coefficient and the volumetric ratio of extract to raffinate phase. An increase in the relative amount of solvent gives a greater recovery of solute but produces a more dilute extract. In order to combine the criteria (which are usually sought) of a reasonably concentrated extract and a high recovery of solute from the raffinate phase, it is normally best to employ multistage countercurrent contacting, as illustrated in Fig. 1, this being the most effective way of obtaining multiple equilibrations on a continuous-flow basis.

The amount of separation achieved in a single equilibration can easily be calculated from the distribution coefficient of the solute and an overall mass balance for this component. Simultaneous solution of the distribution coefficient and mass-balance equations can also be used in a stage-to-stage calculation to find the number of equilibrium stages required to effect a particular separation in a multistage manner.* However, graphical methods are often more convenient.

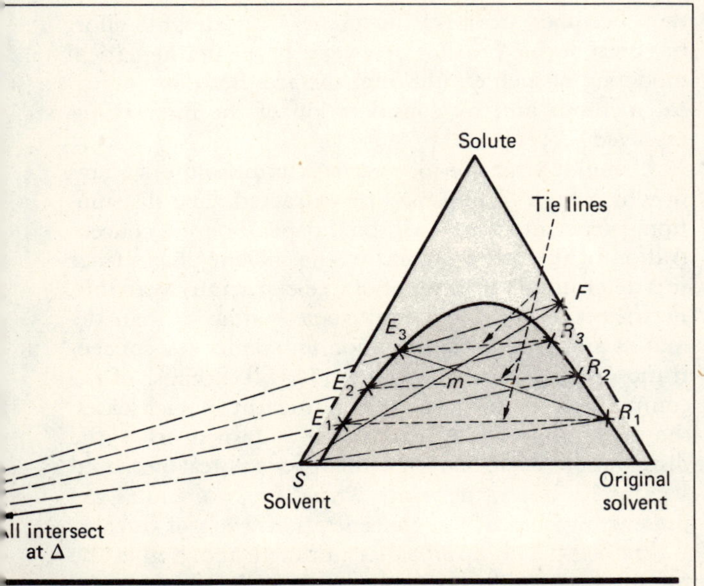

riangular diagram showing construction for he three-stage system of Fig. 1 **Fig. 3**

There is a considerable difference in approach depending on whether the relative miscibility of the two solvents varies appreciably as a function of the solute concentration. For systems involving physical interactions only, as are often found in the organic chemical industry, the miscibility usually changes with the concentration of solute. For a three-component system, this is most conveniently shown on a triangular diagram; a typical case being illustrated in Fig. 2. Systems having compositions enclosed by the binodal curve will all split into two liquid phases. The compositions of the conjugate phases (i.e. phases at equilibrium) will lie on the curve at either end of the tie line that passes through the average composition of the total system. Thus a mixture of overall composition corresponding to Point M will separate into two phases of compositions E and R. The relative amounts of the two phases can be found from the so-called "lever rule," with the amount of E to that of R corresponding to the ratio of the lengths MR to EM.

With a multistage contactor such as is illustrated in Fig. 1, the extract and raffinate phase compositions will again lie on the curve, but the line joining these points will not be a tie-line since these phases are not in equilibrium. The feed and solvent compositions can also be entered on the diagram, as in Fig. 3, and the lever rule then gives an easy method of determining phase flow ratios. Thus Point M is located on Line FS in such a way that the length ratio MS/FM equals the flow ratio of feed to solvent. The line connecting E_3 and R_1 must also pass through M to satisfy the requirement of an

*For ease of visualization, the discussion is based on equilibrium stages. For design of a differential extraction column, it might be desirable to work in terms of the number of transfer units required for a particular separation.

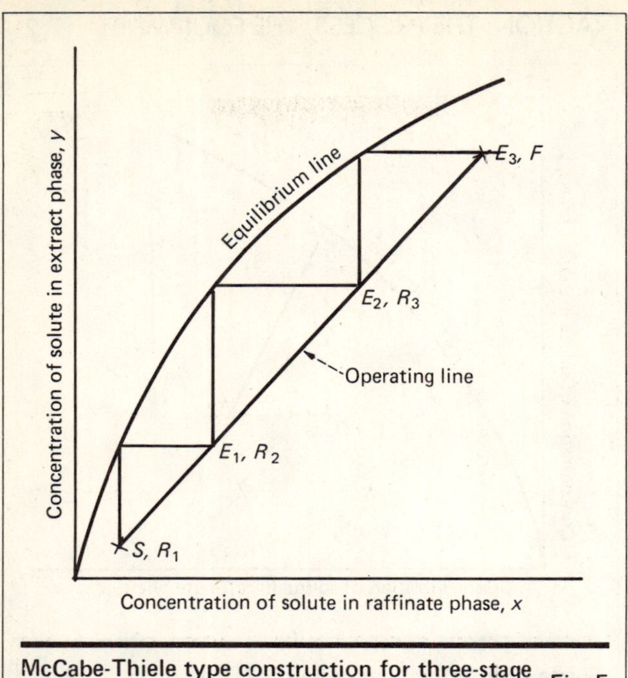

McCabe-Thiele type construction for three-stage system of Fig. 1, assuming phases are immiscible Fig. 5

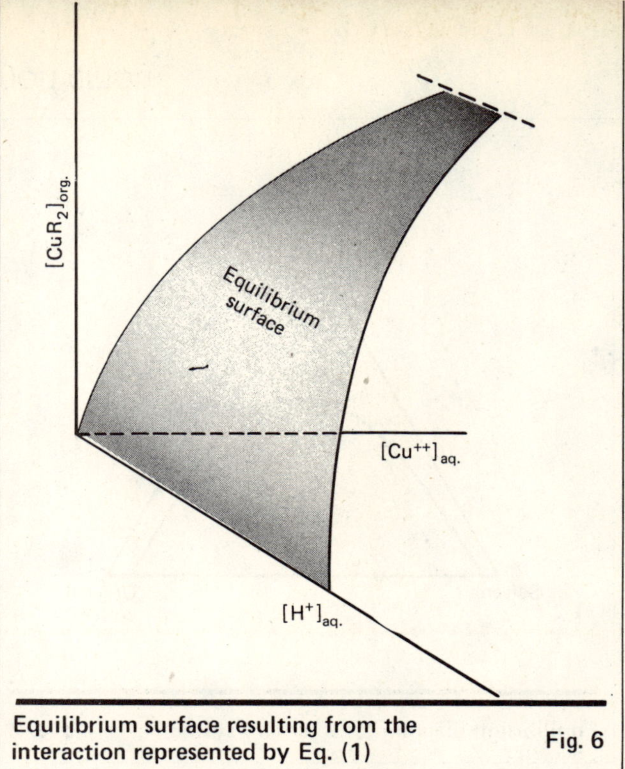

Equilibrium surface resulting from the interaction represented by Eq. (1) Fig. 6

overall mass balance. The length ratio R_1M/E_3M is the flow ratio of extract to raffinate. Such a diagram provides an easy way of picturing the consequences of changes in flow rates. Thus, a reduction in the flow of solvent relative to feed moves Point M towards F. If the concentration of solute in the raffinate (Point R_1) is fixed, say by effluent regulations or simple economics, E will move to a higher concentration of solute (but will need more stages to get there).

So far we have only been concerned with the overall position. The number of equilibrium stages required to effect the separation between E and R can also be found via the graphical construction using the difference point Δ, as shown in Fig. 3. The method is well documented in standard textbooks [*1*].

When a separation is achieved by a specific chemical interaction, the two solvents are often essentially immiscible, and the only volume changes taking place are due to transfer of the solute. In this case the situation can be represented conveniently in two dimensions on an *x-y* plot. Fig. 4 shows a curve representing the equilibrium compositions of the two phases. If a mixture of overall composition M is brought to equilibrium, the compositions of the conjugate phases will be given by a point such as P on the equilibrium line. The ratio of the amount of extract to that of raffinate is then the length ratio $QM:PQ$, which also equals the ratio of solvent to feed. Increase in the relative amount of solvent used decreases the concentration of solute in both phases, although it increases the recovery of solute in the extract.

The number of equilibrium stages required to achieve a particular separation in a continuous countercurrent contactor, the phases being immiscible, can be obtained graphically as shown in Fig. 5. This is often called a McCabe-Thiele construction, although this is not strictly accurate since the latter was derived for two-component distillation calculations and the similarity is only superficial in appearance. The method as applied to liquid-liquid extraction is described in more detail elsewhere [*9*].

In certain cases of metals extraction, the overall solute-extractant interaction is analogous to an ion-exchange process. Thus the extraction of copper into oximes may be represented by:

$$Cu^{++}_{aq} + 2RH_{org} \rightleftarrows CuR_{2org} + 2H^{+}_{aq} \qquad (1)$$

As would be expected from this equation, the distribution coefficient (governed by the position of equilibrium) depends on the concentration of free protons, i.e., the acidity. An equilibrium line as in Fig. 5, then, applies to one particular acidity only. To represent the full equilibrium position requires addition of a third axis for proton concentration in the aqueous phase, as shown in Fig. 6—all equilibrium positions lying on a surface in this diagram. The operating line of Fig. 5 then becomes an operating plane. Considerable effort has been devoted the last few years to the mathematical modeling of such equilibrium surfaces both by empirical methods and by consideration of the interactions involved [*10,11*].

A similar situation occurs with multisolute systems in which both solutes may be extracted. The distribution coefficient for one will then depend on the concentration of the other present in the solvent phase, since this determines the amount of free extractant available. Flowsheet design for a multistage contactor then demands an iterative computation to balance the concentration profiles for both solutes [*12–13*]. Because of the competition for the available extractant in such cases, the separation factor cannot be obtained from the distribution coefficients of the two solutes measured separately, but rather only when the two solutes are present together at the concentration levels of interest.

Scrubbing—The approaches outlined above give the number of equilibrium stages required to reduce the concentration of solute in the raffinate to a particular level. The use of multistage contacting for extraction

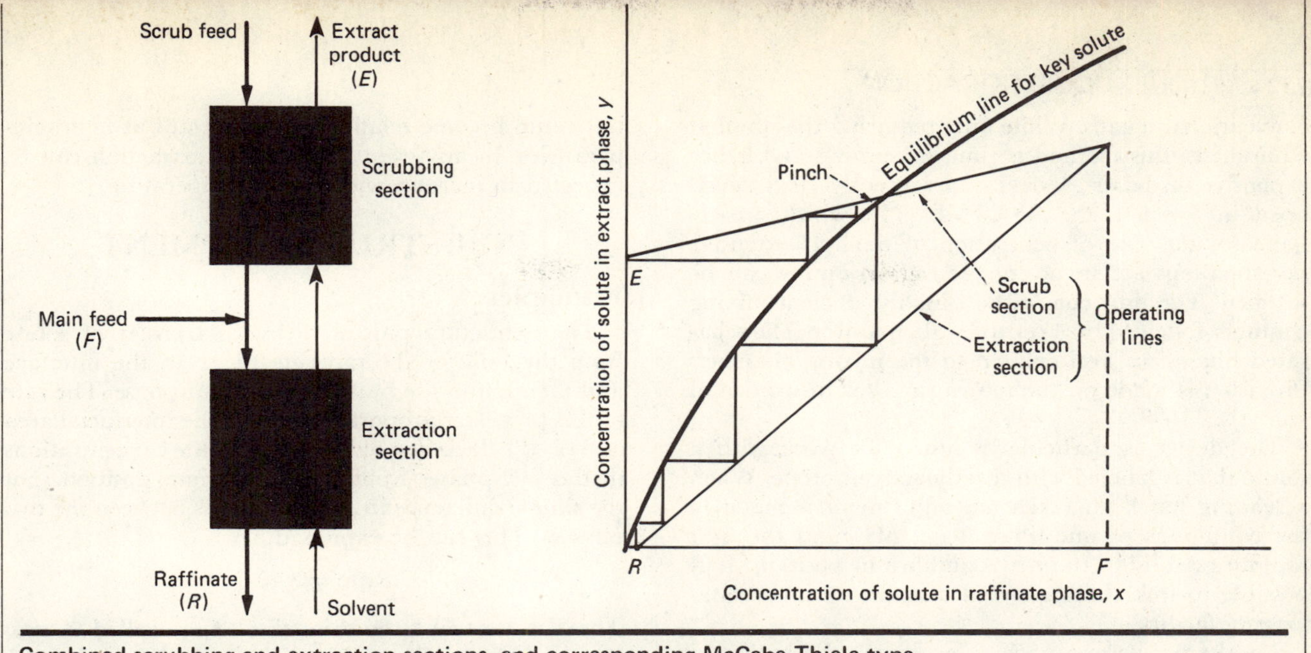

Combined scrubbing and extraction sections, and corresponding McCabe-Thiele type construction (assuming immiscible phases)

Fig. 7

has comparatively little effect on the concentration of impurities or secondary solutes in the extract as compared with what would arise from a single equilibration. Instead, the concentration depends on the selectivity of the solvent. Few solvents are perfectly selective for one solute, and so the extract often contains higher concentrations of certain impurities than can be tolerated in the final product. This can be overcome by *scrubbing* the extract with a second liquid phase, usually involving the same solvent as the original feed, under such conditions that the desired solute is largely retained in the extract while the impurities are washed out. Examples will be given in a future article on industrial applications of liquid-liquid extraction.

If it is necessary to employ multistage scrubbing, the number of stages required is calculated by using equilibrium data for the impurities, not for the desired solute. However, some of the latter will be washed from the extract unless its distribution coefficient is infinite. To recover this, the raffinate from the scrubbing section is usually added to the main feed to the extraction section, as shown in Fig. 7. Allowance for the effect of this on the feed concentration and its flow rate must be made when calculating the number of stages required in the extraction section. A graphical construction for the key solute for combined scrubbing and extraction sections is also illustrated in Fig. 7. Remembering that the number of stages in the scrubbing section is determined by the partition of the impurities, it is desirable to have a "pinch" for the key solute across as much of the scrubbing section as possible. This limits the buildup of key solute in the raffinate phase, and hence the additional load imposed on the extraction section in the solute's recovery. Similarly, there should ideally be a pinch with respect to the impurities within the extraction section. If several impurities are involved, the composition of the scrub feed is a parameter for ensuring their maximum removal [*12*].

It will be apparent from Fig. 7 that the concentration of key solute will be higher at some stages within the scrubbing section than in the extract product. Recognition of the existence of such a concentration peak can be significant, say when operating near the point of third-phase formation. A concentration peak also has to be considered during the reprocessing of nuclear fuels, when the concentration of plutonium must be kept down to a safe level to avoid establishment of a critical-mass situation [*14*]. Increase in the raffinate/extract-phase flow ratio in the scrubbing section, by an increase in the gradient of the operating line, increases the solute concentration at the pinch point or can even lead to loss of the pinch. Ultimately, of course, increase in the scrub feedrate will cause an increased loss of solute in the final raffinate.

Solvent Recovery—Extraction, plus scrubbing if necessary, separates the desired solute from the original solvent and any impurities present in the feed. However, it leaves the solute in the solvent. The final task is to separate the solute from the extract and recover the solvent in a suitable form for recycle.

Where the original solvent-extraction process involves nonspecific "physical" interactions, the final solute/solvent separation is usually by physical means, distillation being the most common. However, when extraction is based on a specific interaction, it is necessary to reverse this by chemical methods. Thus in the case of copper extraction, represented by Eq. (1), copper can be recovered from the organic product by reversing the reaction on contacting with relatively strong acid. Solvent recovery by reversal of the interaction in this way is known as *stripping*. Multistage countercurrent contacting is usually necessary. The process is essentially the reverse of extraction, and the number of stages required can be found by an analogous graphical construction.

Equilibrium data

Any flowsheet design and evaluation demands a knowledge of equilibrium data for the system. Almost invariably, this needs to be determined experimentally (by equilibration of the two phases, followed by their analysis). The simplest method, and still the most widely employed, is to shake the two phases in a separating funnel for some time, and then to analyze by conventional means. While only requiring the simplest equipment, this method is time-consuming and hence expensive on labor. A device has recently been developed in Sweden, the AKUFVE [*15*], which greatly increases the rate of generation of data. If extensive investigations are involved, the capital outlay can be justified. The unit comprises a highly efficient mixing chamber followed by a centrifugal separator. The separated phases are recirculated to the mixing chamber. Provision is made on the return lines for instrumental analysis.

The device is particularly suited for work with a solute that is labeled with a radioactive isotope. When a reading has been taken, an adjustment is made in the conditions of one phase (e.g., pH) and the unit rapidly establishes the new equilibrium position. It is possible to link the analytical instruments to a data-logging facility.

An alternative approach to overcome the high cost of data generation is the development of methods for predicting a whole equilibrium surface from very limited experimental data. Reference has already been made to this possibility for systems exhibiting specific interactions [*10,11,13*]. Our understanding of the thermodynamics of systems involving nonspecific interactions is also improving, and some data correlation is now possible, e.g. by use of the NRTL equation [*16*].

Temperature can have a marked effect on the equilibrium position, hence data should be obtained under isothermal conditions. Temperature dependence can sometimes be turned to advantage. Thus the extraction of uranium by tri-*n*-butyl phosphate is exothermic. Stripping therefore benefits from a high temperature, and this can be used to effect stripping with a smaller volume of aqueous phase, thus giving a more concentrated final product.

Extraction kinetics

The equilibrium position gives no guidance as to the rate of transfer of solute from one phase to the other. Liquid extraction has conventionally been considered a mass-transfer operation, and so its kinetics are normally discussed only in terms of diffusion between the two phases. The hydrodynamics of this process are discussed in the next section. However, it is now becoming apparent that in some processes involving specific interactions, the chemical kinetics of these reactions can also be important parameters. Such processes are really examples of mass transfer with simultaneous chemical reaction. They are characterized by having a chemical-rate dependence. The subject has been reviewed by two of the present authors [*17*]. Data have recently been published giving extraction kinetics for uranium into tri-*n*-butyl phosphate [*18,19*] and copper into LIX 64N (a commercial oxime extractant) [*20,21*].

Temperature dependence should provide a convenient method of distinguishing between processes that involve a kinetic resistance and those that are entirely diffusion-controlled, since temperature change has a much greater effect on the rate of a chemical reaction than on the rate of diffusion. For the same reason, chemical kinetics could play an important part in determining the rate of extraction at low temperatures but could become relatively unimportant at high temperatures. In any event, the rate of extraction can be expected to increase with rise in temperature.

INDUSTRIAL EQUIPMENT

Principles

The extraction process demands transfer of solute from the bulk of the raffinate phase to the interface and thence into the bulk of the solvent phase. The rate of this process is proportional to (1) the interfacial area and (2) the deviation of the actual solute concentrations in the two phases from the equilibrium position (not the simple difference in concentrations between the two phases). This can be expressed as:

$$\text{Rate} = kA\Delta C$$

The constant of proportionality, k, is called a *mass transfer coefficient*. Four such coefficients can be defined, depending on the measure used for ΔC. It is normal to seek to maximize rate since this should minimize equipment size. As will be seen below, there is some interaction between k, A and ΔC, and so it is important to maximize the product rather than the values of individual terms.

In all industrial solvent-extraction equipment, one phase is *dispersed* (in the form of drops) in the other, the *continuous phase*. The interfacial area per unit volume, A, depends on the fractional holdup of dispersed phase and the mean drop size. The latter is a function of the holdup, the physical properties of the phases and the degree of agitation. The higher the degree of agitation, the smaller the mean drop size, thus increasing A.

Considering the actual transport process, this may involve both molecular and eddy diffusion. The former arises from the general random movement of the molecules, which leads to a net movement in the direction of a concentration gradient (thus tending to give a uniform overall concentration). Eddy diffusion arises from bulk movement of the fluid, usually resulting from some form of turbulence. Molecular diffusion is a comparatively slow transport process. In a turbulent situation, eddy diffusion will be several orders of magnitude greater; thus, it is normally desirable to promote it.

If sufficient agitation is provided to create turbulent conditions, eddy diffusion is likely to predominate within the continuous phase, yielding a reasonably high component of k within that phase. However, agitation only affects the dispersed phase indirectly. When a drop is moving through a continuous phase, drag at the interface tends to set up internal circulation of the drop contents. Large drops also oscillate in shape. Both phenomena promote mixing of the contents of the drop, giving an effective eddy diffusion component and enhancing mass transfer in the dispersed phase. Whether this internal circulation occurs, depends on the velocity of the drop relative to the continuous phase, the drop diameter, and the physical properties. The smaller the drop, the smaller the drag forces relative to the forces of inertia. As a result, small drops behave as rigid spheres, and mass transfer within them is by the slow process of molecular diffusion.

Thus, increase in agitation will promote mass transfer

by decreasing drop size (thereby increasing A) and by increasing turbulence in the continuous phase. But beyond a certain point, further increase may start to inhibit the process by causing rigid-sphere behavior in the dispersed phase. The relative importance depends on the location of the principal resistance.

In the case of a dispersion, mean drop size does not give an adequate indication of the behavior to be expected from the dispersed phase. It needs to be supported also by knowledge of the drop-size distribution. Fig. 8 [*22*] shows two drop-size distributions in the region of minimum diameter for internal circulation. While the mean drop sizes for the two systems are not greatly different, it is obvious that most drops in System A will behave as rigid spheres but that the majority in System B will circulate.

One other process that promotes mass transfer in the dispersed phase is drop interaction, i.e. a drop coalescence-redispersion cycle. Again, this can be suppressed by excessive agitation.

It will readily be appreciated from the above that, while adequate agitation is desirable, excessive agitation can have a deleterious effect, particularly when one also considers phase separation.

The actual mass-transfer process can itself affect the magnitude of the mass transfer coefficient. High rates of mass transfer can lead to concentration gradients of solute along the interface which, in turn, cause interfacial tension gradients. These can result in interfacial turbulence, which promotes mass transfer by increasing the rate of interface renewal [*23*].

Returning to the other parameters in the above rate equation, the departure from equilibrium, ΔC, still has to be discussed. This is maximized when the contacting pattern for the two phases corresponds to perfect countercurrent plug-flow. In a stagewise contactor, the concentration profile must obviously change stepwise, but

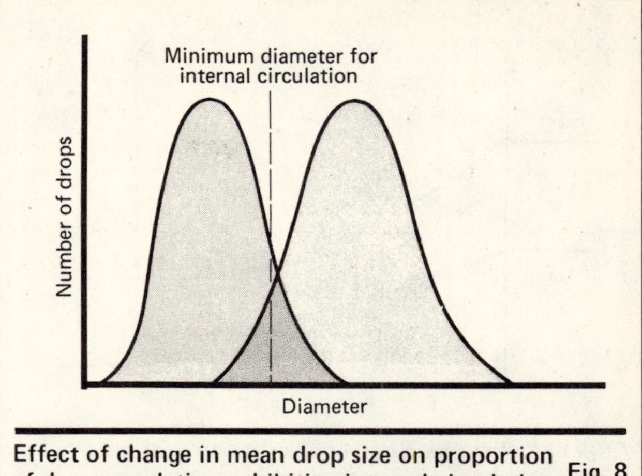

Effect of change in mean drop size on proportion of drop population exhibiting internal circulation Fig. 8

there is rarely any other deviation from the ideal. With differential contactors such as columns, however, it is possible to get considerable nonidealities in the flow pattern, causing an appreciable reduction in ΔC and, hence, of column performance.

All phenomena that lead to a distribution of residence times in the two phases are known collectively as *axial mixing*. The effect on ΔC is illustrated in Fig. 9. *Backmixing* comprises those phenomena that cause flow in the direction opposite to the bulk of the phase, while any giving a below-average residence time are termed *forward mixing*. Phenomena producing axial mixing of the continuous phase include entrainment behind the drops of dispersed phase, and a backmixing component, in many forms of agitation. Axial mixing of the dispersed phase arises from entrainment of drops in turbulent eddies of continuous phase and from the distribution of drop sizes.

Axial mixing can have a very deleterious effect on the performance of differential contactors. It has been stated that in some large-scale columns as much as 90% of the height has been necessary to compensate for axial mixing. It is obviously desirable to do everything possible to minimize such mixing at the design stage. However, it is equally important that this effect should be recognized, or serious underdesign can result. The subject has been extensively studied over the last decade, and many papers have appeared both reporting axial mixing results in specific types of contactor and proposing design methods allowing for the phenomena. Reviews are available [*24,25*] together with a useful monograph [*26*]. Mention should be made of the attractive design method recently proposed by Pratt [*27*].

One question that has to be asked is, which phase should be dispersed? Over a limited range of flow ratios (known as the region of ambivalence), it is possible to disperse either phase, depending on the startup conditions. Outside this region, the system will only be stable with the minority phase dispersed. It should be noted that the region of ambivalence varies for a given system according to whether mass transfer is taking place or not, while a flow system often shows results different from those of a batch mixer. If a process is operating in the region of ambivalence, there are a number of

Classification of industrial contactors

Force producing phase interdispersion	Differential contactors	Stagewise contactors
Gravity only	Spray column Packed column	Perforated plate column*
Pulsation	Pulsed packed column Pulsating plate column	Pulsed sieve-plate column* Pulsed mixer-settler
Mechanical agitation	Rotating disc contactor Oldshue-Rushton column Kühni column* Graesser contactor	Scheibel column* ARD extractor* Mixer-settlers
Centrifugal force	Podbielniak Quadronic DeLaval	Westfalia Robatel

*Could be considered intermediate between differential and stagewise

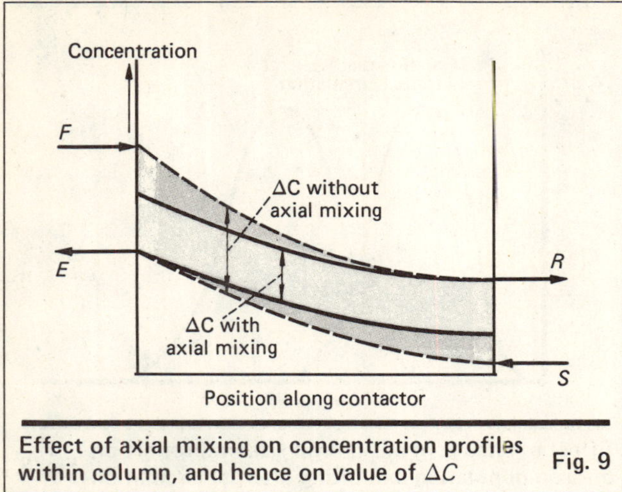

Effect of axial mixing on concentration profiles within column, and hence on value of ΔC Fig. 9

Spray column Packed column Perforated plate column

Three types of unagitated column contractors Fig. 10

parameters to take into account when deciding on the phase to disperse:

1. The direction of mass transfer should be chosen to promote interfacial turbulence in the phase having the maximum resistance.

2. Mass transfer from the dispersed to the continuous phase promotes coalescence.

3. Raffinate streams are best dispersed (certainly in the exit stage), as this reduces the danger of solvent loss by entrainment in the final raffinate. Such loss is an economic charge on the process and can pose an effluent disposal problem.

4. The two possible dispersions will usually have different settling rates. From this point of view, it is generally better to disperse the phase of higher viscosity. There is an optimum phase ratio for maximum rate.

In the case of stagewise contactors, such as mixer-settlers, it is possible to recycle one phase within a stage to achieve an optimum phase-ratio different from the overall flow-ratio. While this may allow a particular dispersion to be obtained, or an optimum phase ratio, the advantages must be weighed against the additional total throughput [*28*].

Contactor classification

Industrial solvent-extraction contactors can be classified into a number of groups. This is illustrated in the table for some common types. First, a division can be made into *differential* and *stagewise* contactors. The former provide conditions for mass transfer throughout their length, aiming to produce the ideal concentration profile shown in Fig. 9. Equilibrium should not be established between the phases at any point. As its name implies, a stagewise contactor provides a number of discrete stages in which the two phases are brought to equilibrium, separated, and passed countercurrent to the adjoining stages. It therefore reproduces the stagewise calculations illustrated in Fig. 3, 5 and 7. While such contactors do avoid the problems of axial mixing found with differential types, the necessity of separating the phases between each stage can add significantly to overall size and solvent inventory.

A further major division can be made according to whether the countercurrent flow is produced by the action of gravity on the density difference between the phases or by an applied force. Further subdivision follows from the means adopted for phase interdispersion.

Simple gravity-columns

These are the simplest possible contactors from a mechanical point of view, and three examples are shown in Fig. 10.

The *spray column* consists of a device to spray the dispersed phase through a column of the continuous phase. Either heavy (as illustrated) or light phase can be dispersed. While simple to construct, and offering the prospect of high throughputs per unit cross-sectional area because of the absence of any internal obstructions, the efficiency is low since axial mixing in the continuous phase is unchecked. The solute concentration in the continuous phase is often virtually constant throughout the length of the column, and values of HTU (height of a transfer unit) and HETS (height equivalent to a theoretical stage) are high.

Spray columns are usually employed only for simple separations requiring very few stages. They have been the subject of considerable study in recent years for possible use as contactors for direct liquid-liquid heat transfer in water desalting [*29,30*]. Axial mixing in the continuous phase can be reduced if it is possible to build up a dense packing of dispersed-phase drops throughout the column. However, it is difficult to guarantee the stability of such a dense-packed bed, particularly in a column of large diameter.

Continuous-phase backmixing can be substantially reduced by introduction of a suitable packing, resulting in the *simple packed column.* The packing should be preferentially wetted by the continuous phase; otherwise it will cause coalescence and reduce the interfacial area per unit volume. The packing will promote mass transfer by breaking up any large dispersed-phase drops and by promoting mixing in drops either by distorting them or producing a coalescence-redispersion cycle [*31*], as well as by reducing axial mixing. Presence of the packing reduces the cross-sectional area available for flow, and so the diameter of column required for a given throughput will be greater than with a spray column. Nevertheless, the reduction in HTU/HETS should be

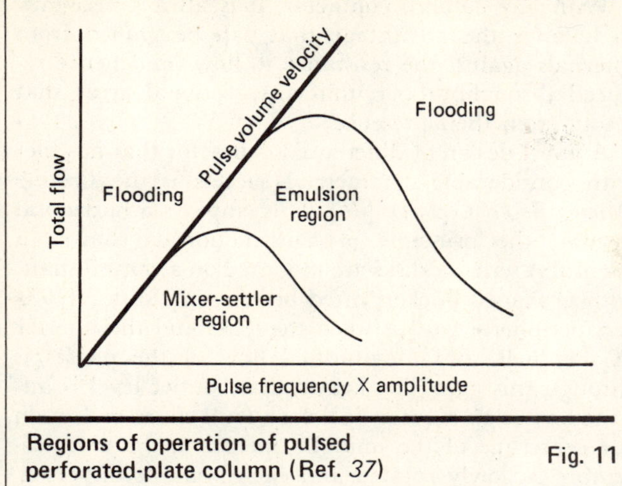

Regions of operation of pulsed perforated-plate column (Ref. *37*) Fig. 11

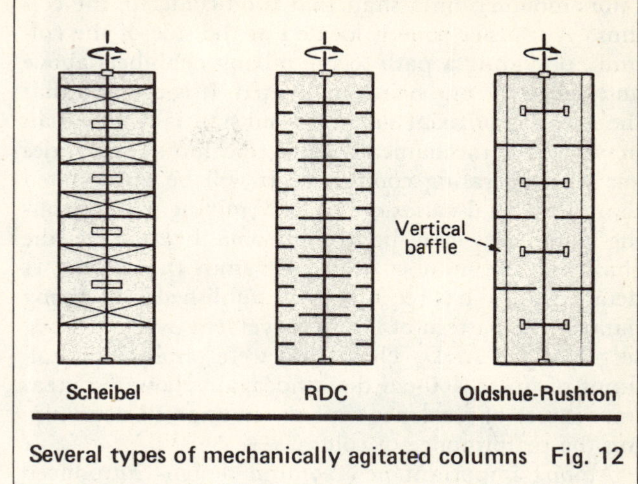

Several types of mechanically agitated columns Fig. 12

very substantial, providing a smaller holdup overall.

The *perforated-plate column* operates in an essentially stagewise manner, providing nominally vertical flow of the dispersed phase, and cross-flow of the continuous phase, within each stage. When the heavy phase is continuous, it is provided with downcomers as in Fig. 10. If the heavy phase is dispersed, the column is reversed and upcomers are used for the continuous phase. While their design is documented in the standard texts, few simple perforated-plate columns appear to be actually operating in commercial use at the present time.

Mechanically agitated columns

The provision of some form of mechanical agitation holds out the prospect of an improvement in performance through an increase in the interfacial area per unit volume. One method is to apply an oscillating pulse to the liquid. This can be done with a packed column and considerably reduces the height of column required for a given separation compared with an unpulsed unit. Mechanical difficulties with generation of the pulse were once expected to limit pulsed columns to comparatively small diameters. However, the construction of *pulsed packed columns* of up to 2.7m diameter has been reported [*32*], together with details of a suitable design approach [*33*]. Substantial strides have been made with pneumatic pulse generation [*34*], as well as with mechanical means [*34a*].

Pulsed perforated-plate columns have found considerable favor, particularly in the nuclear industry. They have been shown capable of handling solids in suspension and have been suggested for solvent-in-pulp processing in the minerals industry [*35*]. They have been reasonably well documented [36].

Several regions of operation can be distinguished, depending on the flowrates and degree of pulsation. These are illustrated in Fig. 11 [*37*]. In the so-called mixer-settler region, the two phases separate into discrete layers between the plates during each reversal of the pulse cycle. At higher pulsed-volume velocities, little or no coalescence takes place between the plates, and the column then behaves as a truly differential contactor.

Both pulsed packed and pulsed perforated-plate columns are open to the criticism that considerable energy is required to pulse the entire liquid contents of a column, particularly on a large scale. As an alternative, pulsing of the plates has been suggested, leading to the *reciprocating-plate column*. Research on this type of column has been reported in recent years, and some units have been built for industrial use [*38*].

Despite current interest in the various forms of pulsed column, and their promise for the future, the most important mechanically agitated columns at present employ some form of rotary agitator. There are three well-known designs: the *Scheibel* [*39*], *RDC* [*40*] and *Oldshue-Rushton* columns [*41*], which have been established for 20 or more years and are described in the standard texts [*1,2*] and previous reviews [*42*]. They are shown diagramatically in Fig. 12. Design improvements have continued and the Scheibel column, in particular, has appeared in a number of versions.

The last decade has seen the introduction of several other columns employing rotary agitators. In their design concept, these have aimed to minimize axial mixing, which can severely limit the performance of mechanically agitated columns.

The *Asymmetric Rotating Disk* (ARD) extractor, developed by Misek and coworkers [*43*], is semi-compartmentalized, its essential features being shown in Fig. 13. Phase interdispersion is by means of rotating-disk agi-

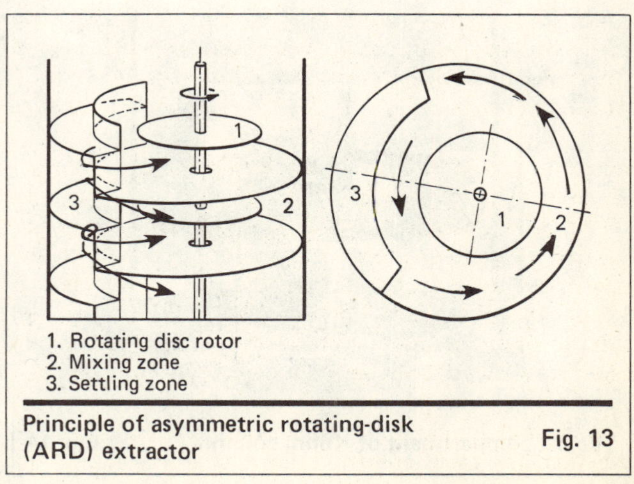

Principle of asymmetric rotating-disk (ARD) extractor Fig. 13

tators mounted on a shaft that is off-center in the column. A transfer zone is located at the side of the column, providing a path to the mixing chambers above and below the one being considered. It is claimed that the amount of axial mixing is substantially less than in some other mechanically agitated columns and varies less with operating conditions. It will be appreciated that substantial variations in axial mixing with operating conditions can be troublesome because of the change in the number of transfer units that results. A detailed study has recently been published [*44*] giving data on the pattern of liquid movement over the cross-section of an ARD. These data were obtained in columns of up to 400-cm dia. and again show the great importance of residence time distribution in determining the performance of columns.

Among important new column designs introduced commercially during the last few years is the *Kühni column*, developed in Switzerland. Details have been published of extensive studies of the droplet size distribution, mean droplet size and holdup characteristics for a wide range of liquid systems [*45a*] and of back-mixing characteristics for a 150-mm pilot-scale extractor [*45,45b*]. The main features of the column are the use of a shrouded turbine-impeller to promote radial discharge characteristics within the compartments, and the possibility of varying the stator-plate free-flow area by the use of differing plate geometries.

A typical compartment is illustrated in Fig. 14. The variable hole arrangement promotes flexibility of design for different column applications. It can also be exploited within a single column, since the compartmental geometry may be changed along the column to assure high uniform holdup and constant drop diameter, even when the physical properties of the two phases change appreciably. Standard columns of this type have been built with diameters up to 2.5 m. In one particular application in the petroleum industry, a column of 2.3 m dia. containing 48 compartments gave a performance equivalent to about 14 theoretical stages and had a combined phase throughput of about 180 m^3/h. Columns of up to 5 m dia. have been constructed. On this scale, they are provided with three agitators, with appropriate stator-plate design, distributed around each compartment.

Typical compartment of Kühni column — Fig. 14

With any column contactor, it is always necessary to balance the advantage that can be gained from internals against the resistance to flow (and hence reduced throughput per unit cross-sectional area) that results from their presence.

A novel design of differential contactor that has met with considerable commercial success is the *Graesser Raining Bucket Contactor* [*46*]. This employs a horizontal format; the principle of construction is shown in Fig. 15. A series of disks are mounted on a central shaft, with C-shaped buckets fitted between the disks. There is a peripheral gap between the disks and the interior of the shell, and longitudinal flow of the phases is through this gap. The heavy-phase outlet level is adjusted so as to give an interface level more or less on the centerline of the unit. In operation, the rotor assembly is slowly rotated and each phase is dispersed, in turn, in the other. The design is virtually unique in not having one phase dispersed and the other continuous throughout.

Dispersing action is very gentle and so this contactor is particularly attractive for systems having poor phase-separation characteristics. Because of the limited cross-sectional area for flow, the unit throughput is only modest. However, moderate cost and simplicity of operation have made the machine popular for medium-throughput applications. Data have been published on the axial mixing characteristics [*47*]. A modified version of the unit is available for countercurrent solid-liquid contacting.

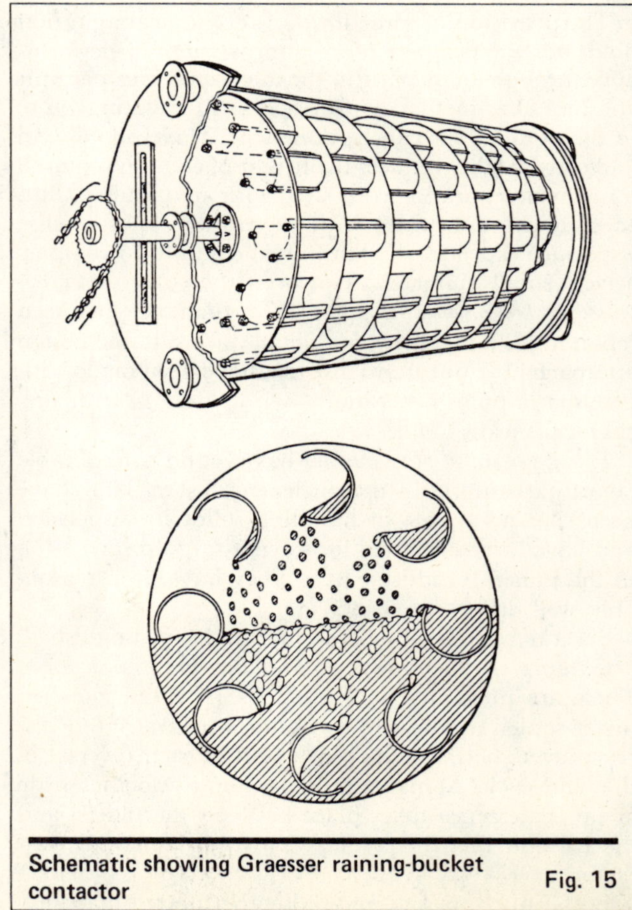

Schematic showing Graesser raining-bucket contactor — Fig. 15

Mixer-settlers

Some of the earliest commercial liquid-extraction plants employed mixer-settlers that involved separate mixing and settling vessels. The stages were usually staggered in height, so that one phase (normally the light) could flow from one stage to the next under gravity while the other was pumped. While unsophisticated, such equipment proved reliable in performance, and it is comparatively simple to design a unit that will function satisfactorily. The disadvantages are the obvious ones of space requirement and the inventory of material held up in processing. Considerable effort has therefore been expended in trying to improve design methods and produce a more compact unit.

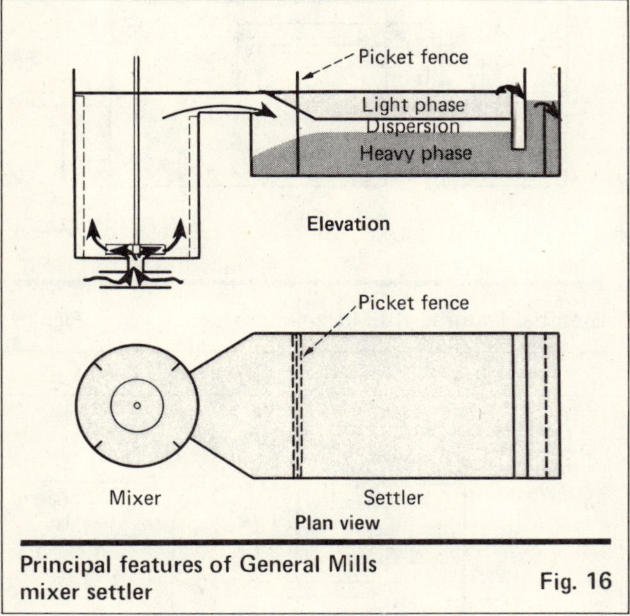

Principal features of General Mills mixer settler Fig. 16

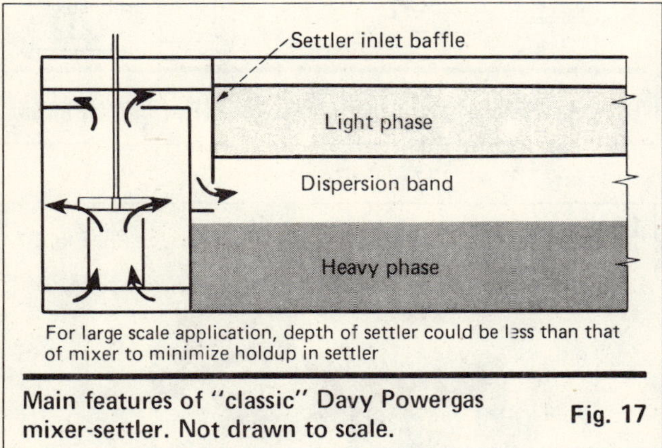

Main features of "classic" Davy Powergas mixer-settler. Not drawn to scale. Fig. 17

Davy Powergas mixer-settler under construction for copper extraction. Mixer in foreground, settler behind. Fig. 18

Photo: Nchanga Consolidated Copper Mines Ltd., Zambia

It was recognized that interstage pumping is not essential where density changes are modest. Since the mixed phase will have a higher average density than the light phase, there will be a positive driving force for flow from mixer to settler, provided that the interface in the settler is kept below the level of the entry port from the mixer. The same phenomenon produces a lower surface level in a mixer than in a settler, the difference providing a potential for flow from one stage to the next. This comparatively simple approach to the hydrodynamics of mixer-settlers led to the integral box mixer-settler, extensively used in the United Kingdom for nuclear fuel processing and often called the *Windscale mixer-settler* [*48,49*]. A multistage unit is built as one box, with partitions to create the different stages, thus eliminating all interstage piping. This design is simple, without mechanical complexity, and has proved very reliable in operation. However, although a major advance on early types, it does not attack the large inventory of materials held within the plant. The driving force producing interstage flow depends on the depth of the unit, inhibiting attempted size reduction.

Flow limitations of the simple gravity mixer-settler can obviously be overcome by interstage pumping. Rather than having the stages at different heights with the heavy phase pumped up hill between them, attention has been focused on some device that can be used to both mix and pump the two phases at the same time. The first widely recognized design to achieve this object was the *Pump-Mix extractor* [*50*]. This was primarily developed for nuclear fuel processing and has been widely used there. It was again an integral-box design rather than one having physically separate stages.

Some designs that incorporate a pumping action have the disadvantage that this can apply suction to the adjacent stages, thereby destroying hydraulic independence of the stages. This was avoided by use of a draught tube in a design first described by Hanson and Smith and subsequently developed further [*51*]. It has also been used as the basis for a small-scale laboratory mixer-settler constructed of glass [*52*].

The demand arising from the very large-scale applications of solvent extraction in nonferrous hydrometallurgy, and the purification of phosphoric acid for equipment able to handle high volumetric throughputs, has resulted in much development effort on mixer-settlers during the last decade. This has concentrated on various pump-mix designs, and the physical sizes involved have required physically separate stages.

The *General Mills unit* [*53*], designed by the manufacturers of the first commercially used copper extractant, has a baffled cylindrical mixer fitted in the base, with a top-shrouded turbine that both mixes and pumps the incoming phases. The dispersion leaves from the top of the mixer, as illustrated in Fig. 16, and flows via a trough into a shallow rectangular settler. The mixer is sized to give the residence time necessary for an acceptable approach to equilibrium. The shallow settler is designed to minimize holdup, on the premise that the throughput is primarily a function of the interfacial area available for coalescence.

Davy-Powergas [*54–56*] has also adopted a pump-mix approach. The mixer is a square-sectioned vessel. The feeds enter through a draft tube, as shown in Fig. 17. They are mixed and pumped by a shrouded impeller running immediately above the draft tube. Considerable effort has been reported on development of the impeller. The degree of internal circulation within the mixer can be varied by adjusting the clearance between the base of the impeller and the top of the draft tube. A horizontal plate with a central circular port is fitted in the top of the mixer as a vortex breaker. The dispersion flows off the top of the mixer and down a transfer channel into the settler, where it is led into the dispersion band to give optimum settler performance. The settler is rectangular in shape, with a small depth to minimize holdup. Extensive scaleup tests have been reported, and a very large-scale plant is in operation in Zambia for copper extraction. This is constructed of concrete lined with stainless steel.

The *IMI mixer-settler* [*57*], developed by Israeli Mining Industries Ltd., is known to be in commercial use for uranium extraction and phosphoric acid purification. This has many interesting features. The two feeds enter a cylindrical mixing compartment, from which they are drawn up a draft tube by an axial-flow pumping device. In contrast to the draft tube design of Hanson and coworkers [*51*], the pumping device is not required to act as the mixer. Instead, the two phases are dispersed by a separate impeller mounted on a shaft running coaxially with the drive to the pump. The two functions of mixing and pumping are therefore separated and can be optimized individually. Essentials of the design are illustrated in Fig. 19. The settler is also novel in being cylindrical in shape and comparatively deep. The dispersion is fed into the center of the settler and flows radially outward, the separated phases being removed through launders at the periphery. This has the effect of reducing linear velocities as the material moves through the settler. Antiturbulence baffles are provided.

The *Kemira mixer-settler* [*58*], developed in Finland and shown diagrammatically in Fig. 20, also employs pumping and mixing devices that are separate but mounted on the same shaft. The mixing phase is drawn from a point in line with the mixing impeller, the position recommended by Rushton [*59*] as yielding material most representative of the bulk of the mixer contents. The settler has a circular section with a conical base. Different stages are staggering in height so that the light phase flows by gravity, while the heavy phase is pumped between stages.

There are two quite distinct types of mixer-settler marketed by *Lurgi* of West Germany.

The horizontal type [*60*] uses an axial-flow impeller to both mix and pump the phases. If the rate of extraction is such that additional residence time is necessary, this impeller is followed by a suitable length of pipe in which turbulent flow is maintained. This is followed by a rectangular settler. The latter usually takes the form of a multi-tray settler, containing a series of horizontal trays to increase throughput.

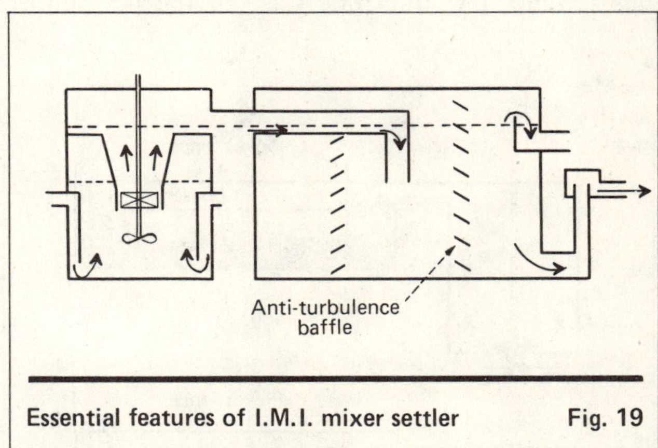

Essential features of I.M.I. mixer settler **Fig. 19**

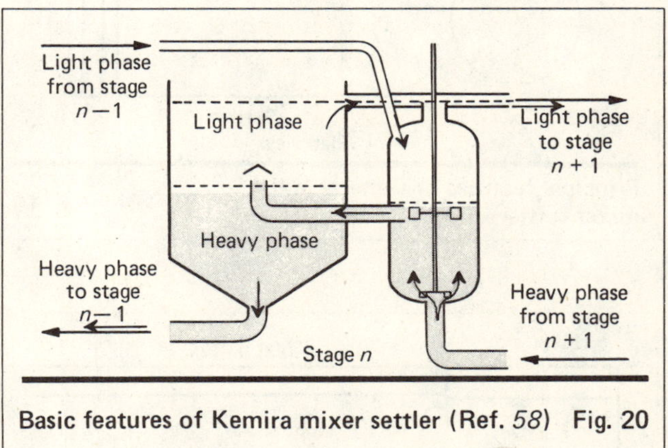

Basic features of Kemira mixer settler (Ref. *58*) **Fig. 20**

Lurgi tower extractor (Lurgi Mineralotechnik GmbH) **Fig. 21**

Photo: Werkfoto Lurgi

The other type comprises a vertical stack of stages and takes the form of a column [*60,61*]. It is extensively used in the NMP process for separation of aromatic from aliphatic hydrocarbons; an actual installation is shown in Fig. 21. Only the steeling compartments and interstage channels are within the column. Interstage flow and phase interdispersion are brought about by axial-flow pumps mounted external to the column. In suitable applications, electrostatic forces have been used to promote phase separation and minimize the size of settler required.

Finally, it should be noted that *Holmes and Narver Inc.* has disclosed [*62*] designs for mixer-settlers that, among other features, may incorporate multicompartment mixers.

With any of the designs that incorporate interstage pumping, it is comparatively easy to arrange recycle of one phase between settler and mixer if this is required to produce a particular phase continuity or holdup. As mentioned previously, phase ratio can be an important parameter in determining the performance of a mixer-settler [*28*].

While by no means representing an exhaustive list, the number and variety of the mixer-settler designs described above, that have appeared this decade shown the extent of commercial interest in this type of equipment.

Settler design

The processes taking place in a settler are quite complex and often interrelated. It is beyond the scope of this review to describe them in any detail, and readers are referred to other sources for a more comprehensive treatment [*63,64*].

When a single drop approaches an interface, it usually rests there for some time before coalescing with its homophase. This rest time at the interface is caused by a film of continuous phase caught between the drop and the interface when the drop first arrives there. The film has to drain radially outwards, under the action of gravity on the density difference between the phases, before coalescence takes place. This process explains some common observations. Thus coalescence is retarded by increase in viscosity of the continuous phase. This, in turn, is one reason why inversion gives dispersions that separate at different rates. Similarly, increase in temperature promotes coalescence by reducing viscosities. Decrease in interfacial tension retards coalescence because there is greater distortion of the interface, thereby producing a greater area of continuous-phase film that has to drain. This is quite separate from the effect of interfacial tension on the size of drops produced initially. Increase in density difference between the phases promotes coalescence by increasing the force acting to bring about drainage of the continuous-phase film.

Photo: Werkfoto Lurgi

Tray assembly for Lurgi multitray settler **Fig. 22**

In the case of a dispersion, there are three important processes to take into account. One is actual coalescence at the interface, as described above. Prior to this, the drops have to sediment to the interface. This is opposed by the dispersed phase released at the interface by the coalescence process, which has to flow counter-current to the drops. Finally, drop-to-drop coalescence may take place in the dispersion giving larger drops, which are able to sediment more rapidly.

The dispersion band in a settler has two apparent boundaries. One is the active interface at which the coalescence takes place. The other boundary is variously known as the passive or sedimentation interface. Within the dispersion band it is usually possible to distinguish two zones. Adjacent to the active interface there is a close-packed or dense layer in which the holdup of dispersed phase is much greater than in the feed. Between this region and the passive interface, the holdup is much less and there is a greater degree of motion amongst the droplets. In this zone the principal process is sedimentation, with limited interdroplet coalescence. As already mentioned, this is opposed by the release of continuous phase at the interface. There can be a classification effect that keeps small drops in suspension in the sedimentation zone until they can grow in size by drop-to-drop coalescence.

Small droplets can give considerable difficulty in practice as they are easily entrained in product streams, where they represent both a loss and an impurity. They can arise in two ways. One is in the mixer. Barnea and coworkers [*57*] have stressed the need for a mixer design that minimizes formation of fines. Small drops introduced in the feed to the settler can be particularly difficult as they may never be able to sediment into the dense region of the dispersion band. The other source is the coalescence process itself, which can give rise to small secondary droplets being formed in both phases [*63*]. There is a reasonable chance that small drops of dispersed phase formed in this way will be caught in the dense region of the dispersion band.

Measurement of the rate of collapse of a dispersion can be used to give some idea of mean drop sizes [*65*].

It is possible from the above picture to define some basic requirements for efficient operation of a settler. These include: (a) Minimize turbulence in the settler, as this will keep small drops in suspension. Entrance effects are a common cause of turbulence and can be countered by a picket fence or suitable baffles. (b) Minimize small drops in the original dispersion. (c) Keep linear velocities along the settler to a low value to avoid entrainment of small drops from the dispersion band.

When studying the performance of a continuous

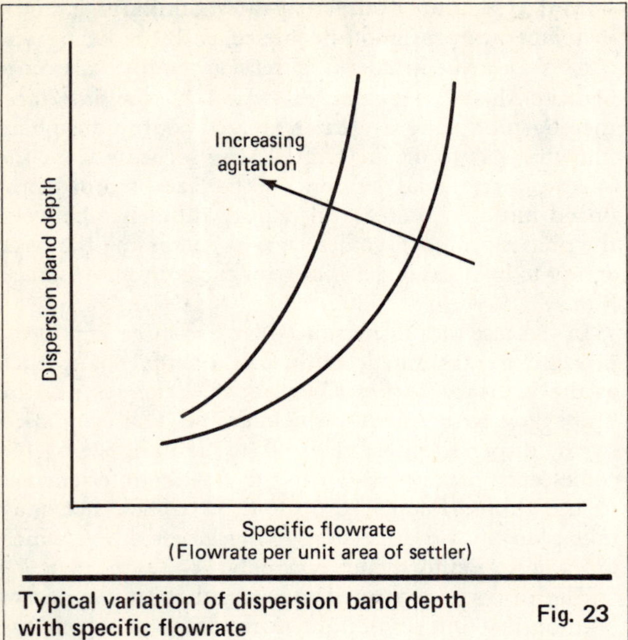

Typical variation of dispersion band depth with specific flowrate **Fig. 23**

settler, it is normal to plot dispersion-band thickness against specific throughput (i.e., throughput per unit area of interface). The relationship is often a power law giving the type of curve shown in Fig. 23. There are two important conclusions to be drawn from this. Firstly, there will be a safe, maximum specific throughput to use for design purposes above which quite small changes can lead to flooding. Secondly, the throughput *per unit volume of dispersion* decreases as the thickness of the dispersion band increases. One approach to the problem of increasing the throughput of a settler of given dimensions is therefore to divide the overall dispersion into a number of thinner bands. The Lurgi multi-tray settler [*60*] uses a system of trays with weirs to achieve this (a tray assembly is shown in Fig. 22). The IMI compact settler [*57*] has a similar aim using a series of horizontal plates.

Another approach to increasing settler throughput, and consequently reducing solvent inventory, is the introduction of some form of packing into the settler. The classical approach is to consider a packing preferentially wetted by the dispersed phase. To overcome the problems of possible inversion, changing wetting characteristics, etc., a woven-mesh packing incorporating materials that are both water and organic wetted [*66*] has been developed. Excellent results have been reported and the development is incorporated in a new mixer-settler recently announced by Davy Powergas [*56*]. This has a series of small settlers arranged around a central mixer. By using this approach, sections of settler can be closed for cleaning of the packing (if this proves necessary) without disruption of production. The overall unit is claimed to operate with much higher specific throughputs than a simple mixer-settler (more than an order of magnitude greater). The resultant saving in cost, particularly for a process involving an expensive solvent, should be much greater than the additional cost of the packing.

Apart from the application of an electric field [*67*], the other main approach to increased settling rates has been use of a centrifugal force.

Centrifugal contactors

Centrifugal contactors have been well established for many years. They offer short contact times and efficient phase separation and so have been favored for applications involving either chemically unstable (e.g., in the pharmaceutical industry) or slow-settling systems.

The *Podbielniak, De Laval* and *Westfalia* extractors are widely used industrially and their basic principles have been described in earlier reviews [*42*]. A useful discussion has recently been published [*68*] on factors determining the performance of centrifugal contactors of the Podbielniak type.

Reference should be made to the *Quadronic* design [*69*]. This is a differential machine incorporating the concept of gradient mixing energy. Also to appear recently has been the *Robatel* extractor [*70*]. This has essentially a stagewise contacting pattern. It is of general applicability, although it appears to have been considered particularly for the nuclear industry.

Contactor selection

It is impossible to give anything but the most general guide to contactor selection since a great many variables are involved. Any of the contactors described could be used for most simple processes, and a final choice will depend on an economic evaluation plus consideration of local factors, e.g., space available.

A chemically unstable system or one with bad phase separation characteristics will tend to favor centrifugal contactors. On the other hand, if a system is stable and cheap, and only a few stages are required, one would think in terms of a simple gravity column or a mixer-settler. If more stages are required, the simple gravity column would give way to a mechanically agitated column. Mixer-settlers are suitable for most chemically stable systems but require much more groundspace than columns. The holdup in mixer-settlers tends to be greater than in columns, although the difference need not be great. Mixer-settlers are better able to accept flow changes from stage to stage if these occur due to the transfer of material between the phases.

References

1. Treybal, R. E., "Liquid Extraction," 2nd ed., McGraw-Hill, New York, 1963.
2. Hanson, C. (ed), "Recent Advances in Liquid-Liquid Extraction," Pergamon Press, Oxford, 1971.
3. McKay, H. A. C., others (ed.), "Solvent Extraction Chemistry of Metals," Macmillan, London, 1965.
4. Dyrssen, D., others (ed.), "Solvent Extraction Chemistry," North Holland, Amsterdam, 1967.
5. Kertes, A. S. and Marcus, Y. (ed.), "Solvent Extraction Research," Wiley, New York, 1970.
6. Proceedings International Solvent Extraction Conference ISEC '71, Society or Chemical Industry, London, 1971.
7. Proceedings International Solvent Extraction Conference ISEC '74, Society of Chemical Industry, London, 1974.
8. Reports on the Progress of Applied Chemistry, vol. 53 (1968), vol. 55 (1970), vol. 57 (1972), vol. 59 (1974), Society of Chemical Industry, London.
9. Peterson, H. C. and Beyer, G. H., AIChE J., vol. 2, p. 38 (1956).
10. Robinson, C. G. and Paynter, J. C., Proceedings International Solvent Extraction Conference ISEC '71, vol. 2, p. 416 (Society of Chemical Industry, London, 1971).

11. Forrest, C. and Hughes, M.A., *Hydrometallurgy*, vol. 1, p. 25 (1975).
12. Wood, J. T. and Williams, J. A., *Trans. Inst. Chem. Eng.*, vol. 36, p. 382 (1958).
13. Burton, W. R. and Mills, A. L., *Nucl. Eng.*, 8, p. 248 (1963).
14. Hanson, C., *Chem. Proc. Eng.*, vol. 41, p. 445 (1960).
15. Reinhardt, H. and Rydberg, J., *Acta. Chem. Scand.*, vol. 8, p. 23 (1969).
16. Ashton, N., Soares, L. de J. and Ellis, S. R. M., Proceedings International Solvent Extraction Conference ISEC '74, vol. 2, p. 1815 (Society of Chemical Industry, London, 1974).
17. Hanson, C., Hughes, M. A. and Marsland, J. G., Proceedings International Solvent Extraction Conference ISEC '74, vol 3, p. 2401 (Society of Chemical Industry, London, 1974).
18. Baumgärtner, F. and Finsterwalder, L., *J. Phys. Chem.*, vol. 74, p. 108 (1970).
19. Farbu, L., others, Proceedings International Solvent Extraction Conference ISEC '74, vol. 3, p. 2427 (Society of Chemical Industry, London, 1974).
20. Flett, D., others, *J. Inorg. Nucl. Chem.*, vol. 35, p. 2471 (1973).
21. Whewell, R. J., others, *J. Inorg. Nucl. Chem.*, vol. 37, p. 2303 (1975).
22. Thorton, J. D., *Chem. Eng. & Fuel Science at Newcastle upon Tyne*, vol. 2, p. 25 (1966).
23. Sawistowski, H., Chap. 9 in "Recent Advances in Liquid-Liquid Extraction," ed. by Hanson, C., Pergamon Press, Oxford, 1971).
24. Misek, T. and Rod, V., Chap. 7 in "Recent Advances in Liquid-Liquid Extraction," ed. by Hanson, C., Pergamon Press, Oxford, 1971.
25. Ingham, J., Chap. 8 in "Recent Advances in Liquid-Liquid Extraction," ed. by Hanson, C., Pergamon Press, Oxford, 1971.
26. Hartland, S., "Counter-Current Extraction", Pergamon Press, Oxford, 1970.
27. Pratt, H. R. C., *Ind. Eng. Chem., Proc. Des. Dev.*, vol. 14, p. 74 (1975).
28. Rowden, G. A., others, Proceedings International Solvent Extraction Conference ISEC '74, vol. 1, p. 81 (Society of Chemical Industry, London 1974).
29. Kehat, E. and Sideman, S., Chap. 13 in "Recent Advances in Liquid-Liquid Extraction," ed. by Hanson, C., Pergamon Press, Oxford, 1971.
30. Kehat, E. and Letan, R., *AIChE J.*, vol. 17, p. 984 (1971).
31. Thornton, J. D., *Industrial Chemist*, vol. 39, p. 632 (1963).
32. Spaay, N. M., others, Proceedings International Solvent Extraction Conference ISEC '71, vol. 1, p. 281 (Society of Chemical Industry, London, 1971).
33. Ten Brink, G. and Simons, A., Proceedings International Symposium "Solvent Extraction in Metallurgical Processes" p. 59 (Technologisch Instituut K. VIV, Antwerp (1972).
34. Baird, M. H. I. and Ritcey, G. M., Proceedings International Solvent Extraction Conference ISEC '74, vol. 2, p. 1571 (Society of Chemical Industry, London, (1974).
34a. Brandt, H. W., others, *Verfahrenstechnik*, vol. 9. p. 383 (1975).
35. Ritcey, G. M., *Chemy. Ind.*, vol. 45, p. 1294 (1971).
36. Rouyer, H., others, Proceedings International Solvent Extraction Conference ISEC '74, vol. 3, p. 2339 (Society of Chemical Industry, London, 1974).
37. Logsdail, D. H. and Lowes, L., Chap. 5 in "Recent Advances in Liquid-Liquid Extraction," ed. by Hanson, C., Pergamon Press, Oxford, 1971.
38. Karr, A. E. and Lo, T. C., Proceedings International Solvent Extraction Conference ISEC '71, vol. 1, p. 299 (Society of Chemical Industry, London, 1971).
39. Scheibel, E. G., *Chem. Eng. Progr.*, vol. 44, p. 681 (1948).
40. Reman, G. H., Proceedings 3rd World Petroleum Congress, The Hague, Sect. III, p. 121 (1951).
41. Oldshue, J. Y. and Rushton, H., *Chem. Eng. Progr.*, vol. 48, p. 297 (1952).
42. Hanson, C., Chem. Eng., vol. 75, no. 18, p. 98 (1968).
43. Misek, T. and Marek, J., *Brit. Chem. Eng.*, vol. 15, p. 202 (1970).
44. Seidlová, B. and Misek, T., Proceedings International Solvent Extraction Conference ISEC '74, vol. 3, p. 2365 (Society of Chemical Industry, London, 1974).
45. Ingham, J., others, Proceedings International Solvent Extraction Conference ISEC '74, vol. 2, p. 1299 (Society of Chemical Industry, London, 1974).
45a. Fischer, A., *Verfahrenstechnik*, vol. 5, p. 360 (1971).
45b. Hody, D., *Chemische Rundschau*, vol. 28, p. 9 (1975).
46. Coleby, J., British Pat. 860, 880; 972,035 and 1,037,573.
47. Sheikh, A. R., others, *Trans. Inst. Chem. Engrs.*, vol. 50, 199 (1972).
48. Williams, J. A., others, *Trans. Inst. Chem. Engrs.*, vol. 36, 6 (1958).
49. Lowes, L. and Larkin, M. J.: *IChemE Symposium Series No. 26*, p. 111 (Inst. Chem. Engrs., London, 1967).
50. Coplan, B. V., others, *Chem. Eng. Progr.*, vol. 50, p. 403 (1954).
51. Hanson, C. and Kaye, D. A., *Chem. Proc. Eng.*, vol. 44, p. 27 (1963).
52. Anwar, M. M., others, *Chemy. Ind.*, p. 1090 (1969).
53. Ager, D. W. and Dement, E. R., Proceedings International Symposium "Solvent Extraction in Metallurgical Processes," p. 27 (Technologisch Instituut K. VIV, Antwerp, 1972).
54. Warwick, G. C. I., others, Proceedings International Solvent Extraction Conference ISEC '71, vol. 2, p. 1373 (Society of Chemical Industry, London, 1971).
55. Warwick, G. C. I. and Scuffham, J. B., Proceedings International Symposium "Solvent Extraction in Metallurgical Processes," p. 36 (Technologisch Instituut K. VIV, Antwerp, 1972).
56. Jackson, I. D., others, *IChemE Symposium Series No. 42*, Paper 15 (Inst. Chem. Engrs, London, 1975).
57. Mizrahi, J., others, Proceedings International Solvent Extraction Conference ISEC '74, vol. 1, p. 141 (Society of Chemical Industry, London, 1974).
58. Mattila, T. K., Proceedings Solvent Extraction Conference ISEC '74, vol. 1, p. 169 (Society of Chemical Industry, London, 1974).
59. Rushton, J. H., Paper 10.4 presented at AIChE–IChemE Meeting, London, 1965 (Inst. Chem. Engrs., London).
60. Stönner, H. M. and Wöhler, F., *IChemE Symposium Series No. 42*, Paper 14 (Inst. Chem. Engrs., London, 1975).
61. Mehner, W., others, Proceedings International Solvent Extraction Conference ISEC '71, vol. 2, p. 1265 (Society of Chemical Industry, London, 1971).
62. Exhibition during ISEC '74, Lyon (1974).
63. Jeffreys, G. V. and Davies, G. A., Chap. 14 in "Recent Advances in Liquid-Liquid Extraction," ed by Hanson, C., Pergamon Press, Oxford (1971).
64. Barnea, E. and Mizrahi, J., *Trans. Inst. Chem. Engrs.*, vol. 53, pp. 61, 70, 75 and 83 (1975) (published in 4 parts).
65. Slater, M. J., others, Proceedings International Solvent Extraction Conference, ISEC '74, vol. 1, p. 107 (Society of Chemical Industry, London, 1974).
66. Davies, G. A., others, *The Chemical Engineer* (London), no. 266, p. 392 (1972).
67. Brown, A. H. and Hanson, C., *Chem. Eng. Sci.*, vol. 23, p. 841 (1968).
68. Todd, D. B. and Davies, G. R., Proceedings International Solvent Extraction Conference ISEC '74, vol. 3, p. 2380 (Society of Chemical Industry, London, 1974).
69. Doyle, C. M., others, *Chem. Eng. Progr.*, vol. 64, no. 12, p. 68 (1968).
70. Miachon, J. P., *La Technique Moderne*, March/April 1971.

The authors

Philip Bailes is a lecturer in chemical engineering at the University of Bradford, Bradford, West Yorkshire BD7 1DP, England. He was educated at Heaton Grammar School, Newcastle-upon-Tyne, England, and then obtained a National Coal Board scholarship to read for his B.Sc. in chemical engineering at the University of Newcastle. This was followed by a Ph.D. at the same university and then by two years in a postdoctoral research appointment. He is a Chartered Engineer, a member of the Institution of Chemical Engineers, and an Associate of the Royal Institute of Chemistry.

Carl Hanson is a Professor and Chairman of the School of Chemical Engineering, Univ. of Bradford, where his principal research interest is in solvent extraction. He holds a B.Sc. in chemistry from the Univ. of London, and a Ph.D. in chemical engineering from the Univ. of Leeds. He edited "Recent Advances in Liquid-Liquid Extraction," (Pergamon Press, 1971), and has been involved with the International Solvent Extraction Conf. in 1971 and 1974. He is a Vice-President of the Soc. of Chemical Industry, and a member of the Council of the Institution of Chemical Engineers.

M. A. Hughes is a lecturer in chemical engineering at the University of Bradford, where his research interests are associated with the solvent extraction of metals. He holds the degrees of B.Sc., M.Sc. and Ph.D. in chemistry, and is a Fellow of the Royal Institute of Chemistry. He is also a member of the Institute of Fuel and is a Chartered Engineer. He has worked with the Gas Research Council Group at the University of Leeds, and with the U.K. Atomic Energy Authority, and has held several research consultancies concerning solvent extraction and aqueous pollution.

Liquid-liquid extraction: Nonmetallic materials

How liquid-liquid extraction applies to the separation of organic, pharmaceutical and inorganic substances. This review covers the processes, solvents, additives and operating conditions, and specifies suitable contacting equipment.

P. J. Bailes, C. Hanson and M. A. Hughes, University of Bradford (Great Britain)

☐ Liquid-liquid extraction (also known as solvent extraction) offers as a prime advantage a means of separating components according more to chemical type than to molecular weight. The theory underlying this unit operation, and a survey of the equipment used were presented in our earlier article [*1*]. Our purpose here will be to examine the industrial applications of liquid-liquid extraction in relation to separation involving nonmetals such as organic, pharmaceutical and inorganic materials.

Processes for separating organics

The need to separate mixtures of aliphatic and aromatic hydrocarbons provided one of the first large-scale applications for solvent extraction. The original Edeleanu process used liquid sulfur dioxide as a solvent for the removal of aromatics from lamp oil. While demand for this product has dwindled, there is still an ever-increasing need for aviation kerosene and lube oil having a low aromatics content. And an additional incentive to research new solvent systems has come through the use of high-purity aromatics such as benzene, toluene and xylenes (BTX) as feedstocks for the petrochemical industry.

The first major advance in this occurred with the development of the UDEX process, followed by processes based on solvents such as sulfolane, N-methyl pyrrolidone, dimethyl sulfoxide and N-formylmorpholine. The selectivity and capacity of these solvents (Fig. 1) with respect to aromatic hydrocarbons differ considerably [*2*]. Therefore, something can be gained by looking at the various processes in more detail.

The UDEX process [*3*] is most popular in the U.S. It uses mixtures of polyethylene glycols and water as the solvent. The original process used diethylene glycol in conjunction with water, but significant improvements in performance have been reported for the higher-molecular-weight solvent, tetraethylene glycol [*4,5*]. Apart from the increase in extraction capacity, the

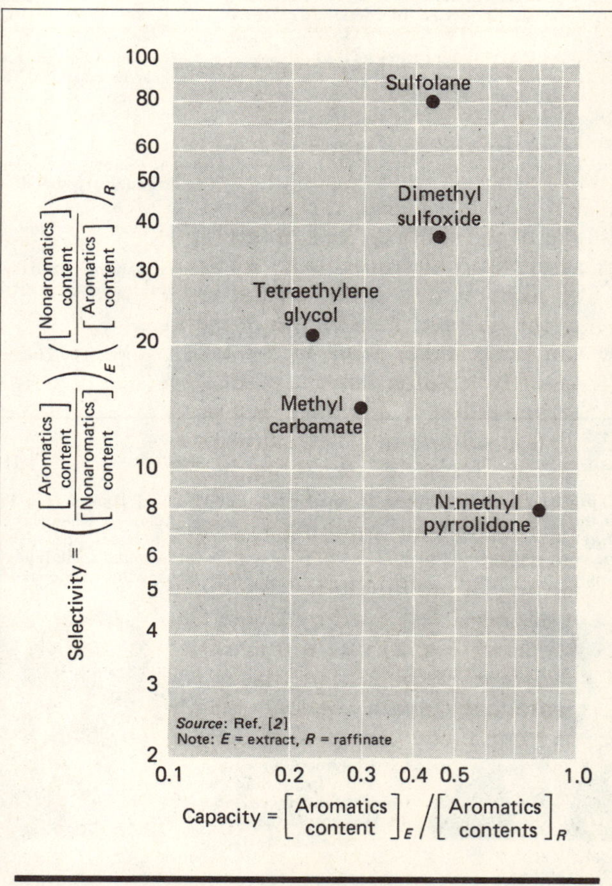

Capacity and selectivity of solvents for aromatics extraction from a hydrocarbon feedstock — Fig. 1

Originally published May 10, 1976

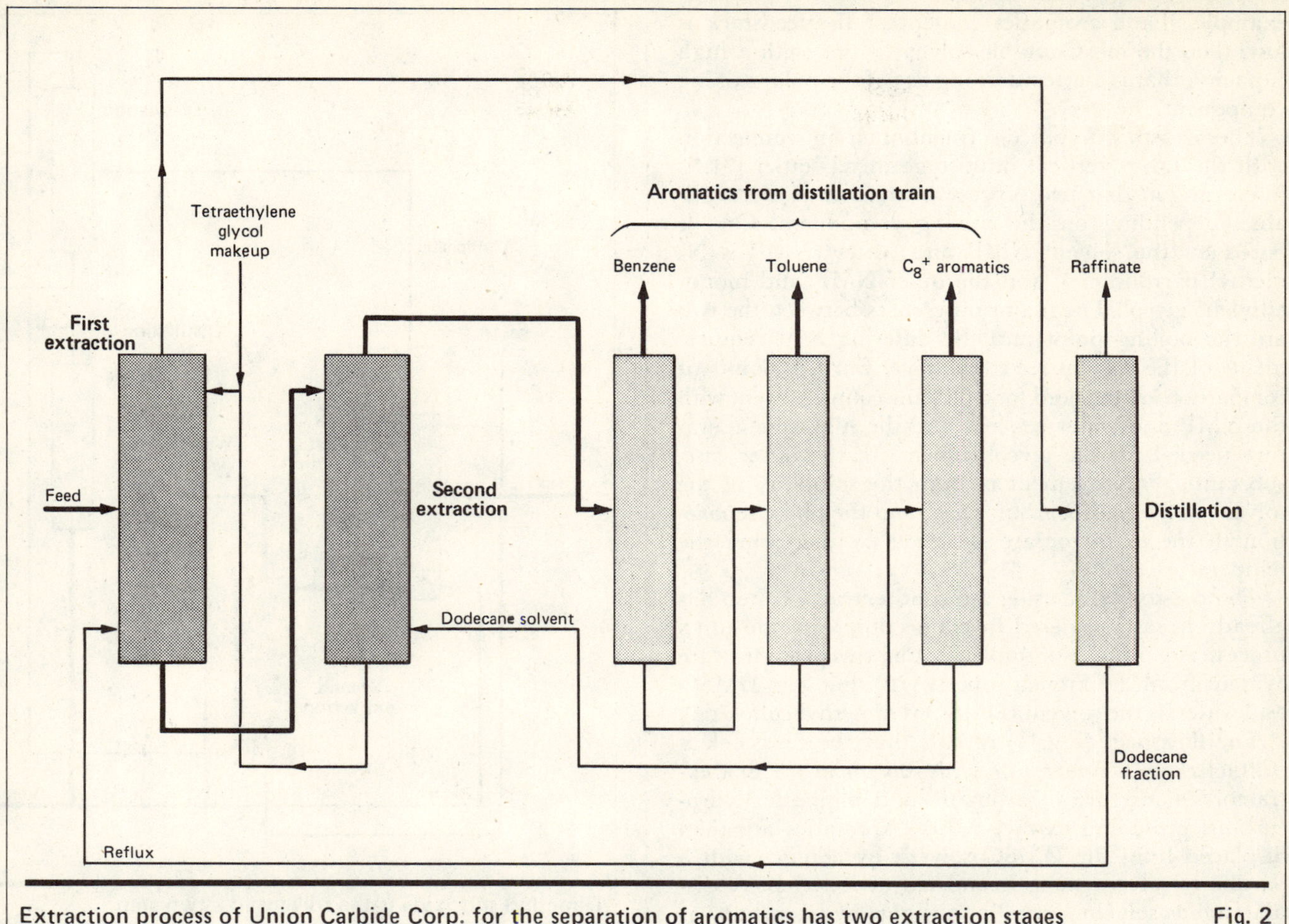

Extraction process of Union Carbide Corp. for the separation of aromatics has two extraction stages Fig. 2

latter solvent does not require the use of an antifoaming agent.

In a typical process, the feedstock enters near the middle of the extractor and progresses up the column, being depleted of aromatics by a tetraethylene glycol/ water solution that enters at the top and percolates down the column. Purification of the aromatics in the solvent phase takes place in the lower section of the extractor by counter current contact with a light hydrocarbon reflux. The extract then passes to a two-part distillation column in which extractor reflux and aqueous glycol solvent are independently separated from the aromatics. This stream and the raffinate from the extractor require water washing to recover any dissolved or entrained glycol.

One way of possibly improving this process arrangement has been developed by Union Carbide Corp. [*6*]. This scheme (Fig. 2) uses two extraction stages and a tetraethylene glycol solvent that is water-free. In the first extraction column, the contacting pattern is similar to the UDEX process except that there is no water, and the reflux is specified as a high-boiling-point aliphatic such as dodecane. The extract from the first extractor is fed to the top of a second extractor, where the aromatics are removed from the glycol by extraction with more dodecane. The distillation train separates aromatics from the dodecane and a small residue of glycol. Advantages are that no water is distilled, the aromatics are vaporized only once, and the bulk of the glycol does not contact high distillation temperatures. A disadvantage is that the raffinate from the primary extractor must be distilled to recover both glycol and dodecane.

The process arrangement used with sulfolane [*7*] is essentially similar to the UDEX, except that the extract is steam-stripped in two separate distillation columns. The reflux, together with some water, is distilled in the first column. The aromatics, together with some water, are separated from the sulfolane in the second column. Unlike the aqueous-glycol extraction, the resulting aromatics do not require water washing, although the raffinate stream does. In many respects, sulfolane has almost ideal solvent properties. Its superior selectivity and capacity are apparent from Fig. 1. Additional features are its high density, low heat capacity, and high boiling point. These minimize difficulties of phase separation, and ease solvent recovery.

Several competing commercial solvents that might compare unfavorably with sulfolane have had their selectivities modified by the addition of a polar mixing component. Notable examples are: N-methyl pyrrolidone [*2*], dimethyl sulfoxide [*8*] and N-formylmorpholine [*9*], where the addition of water has been shown to improve the selectivity at little expense to capacity. One advantage of this kind of mixed solvent is that the type and quantity of the mixing component can be varied to adapt the solvent to a greater range of duties. For example, if the aromatics content of the feedstock is low, then the most suitable solvent is one with a high capacity; that is, one containing less of the polar mixing component.

The AROSOLVAN process (mentioned in connection with the Lurgi vertical multistage mixer-settler [*1*]) is interesting in that two process arrangements are available, depending on the mixing component. One is based on the solvent NMP and water (NMP is N-methyl pyrrolidone), and the other NMP and monoethylene glycol. The major differences between the two are the boiling point and the differing heat requirements of the solvent-recovery stage. Further points of comparison are the need for a pentane countersolvent with the NMP and water process, and the higher temperature needed for the glycol solvent. In the latter case, substantial improvement in both the solubility of the solvent in the hydrocarbon phase and the phase separation in the extractor are achieved by increasing the temperature.

The possibility of using a second extraction step has already been considered in connection with the UDEX process. Advantages of implementing this procedure are evident from the French process [*10*] that uses DMSO and water as the solvent (DMSO is dimethyl sulfoxide).

The flowsheet (Fig. 3) reveals that the feedstock is contacted countercurrently with solvent in the first extractor—a mixture of aromatic and aliphatic hydrocarbons providing extract reflux. Aromatics are then displaced from the DMSO solvent by contact with a paraffinic solvent in a second extraction step. A displacement solvent, usually a light hydrocarbon, is chosen to achieve simple separation from the aromatics by distillation and is preferably the same as the paraffinic fraction used in the extract reflux to the first extractor. The advantage of this approach is that both extractions take place at ambient temperature, where the stability and selectivity of the solvent are at their best, and where carbon-steel corrosion is minimized. The nontoxic nature of DMSO could make it especially useful for providing hydrocarbon solvents to the food and drug industries.

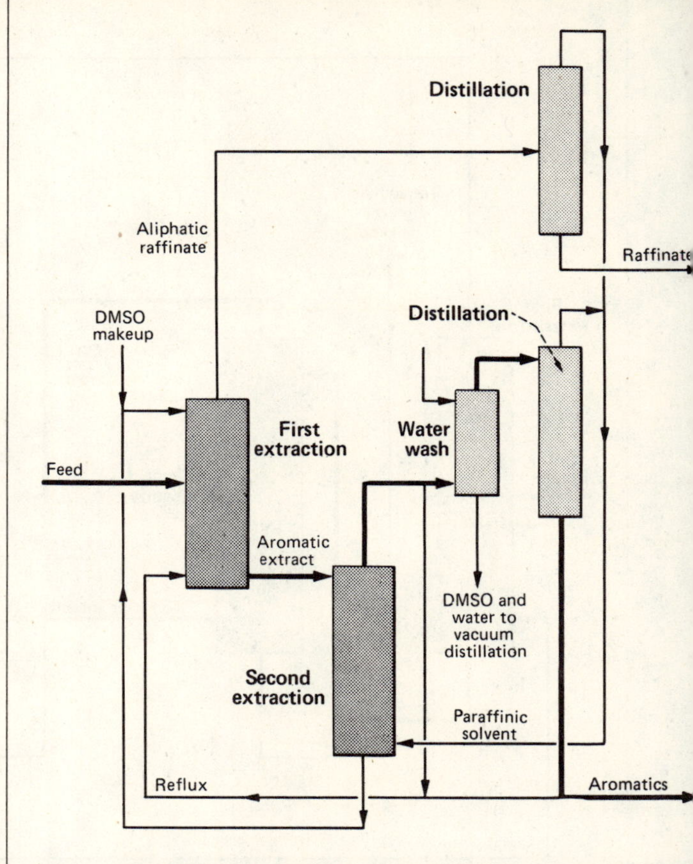

Dimethyl sulfoxide is the solvent in a two-step process for aromatics extraction from hydrocarbons Fig. 3

Other solvents for organic processes

Table I summarizes the main features of the processes described here. Other solvents for BTX recovery have been proposed, and many are covered in the review by Bailes and Winward [*11*]. Two pilot-scale tests that deserve mention involve liquid ammonia and methyl carbamate. The cost of producing aromatics from catalytically reformed naphthas has been shown to be less using ammonia extraction than with diethylene glycol as solvent [*12*]. Methyl carbamate has a high selectivity for aromatics even with respect to olefins. Pilot-plant data have been reported for the CARMEX process, which uses methyl carbamate as the solvent [*13*].

As previously mentioned, solvent extraction is extensively used in the refining of kerosene and lubricating oils. Contact with the correct solvent can improve the viscosity index, color, oxidation resistance, and reduce the carbon- and sludge-forming tendencies of lubricating oils. The solvents that are most widely used in this field are furfuraldehyde [*14*] and phenol [*15*], and the equipment ranges from packed columns to centrifugal contactors. Liquid propane [*16*] is often used to deasphalt vacuum residues from crude-oil distillation—the highly paraffinic extracts from this sort of operation being used as catcracker or hydrocracker feedstocks, or as high-viscosity-index lube oil.

Separations for other organics

In addition to the separation of aromatic-aliphatic hydrocarbon mixtures, there are many other applications of solvent extraction in the chemical process industries. Notable among these: processes for extracting caprolactam from ammonium sulfate solutions [*17*], acetic acid from aqueous solutions [*18*], acrylic acid from aqueous solutions [*19*], and *m*-xylene from a mixture of C_9 aromatic isomers [*20*]. This last process, developed by the Japan Gas Chemical Co., uses a potent mixture of hydrogen fluoride and boron trifluoride as both an extraction solvent and isomerization catalyst. The highly reactive solvent imposes its own restrictions but the new approach is claimed economically superior to conventional separation methods.

The coal-tar industry has also considered processes [*21,22*] for the separation of individual isomers. As yet, none has received commercial acceptance. Laboratory tests have shown that isomers can be separated from the mixtures of organic acids or bases typically found in coal-tar fractions by the method known as dissociation

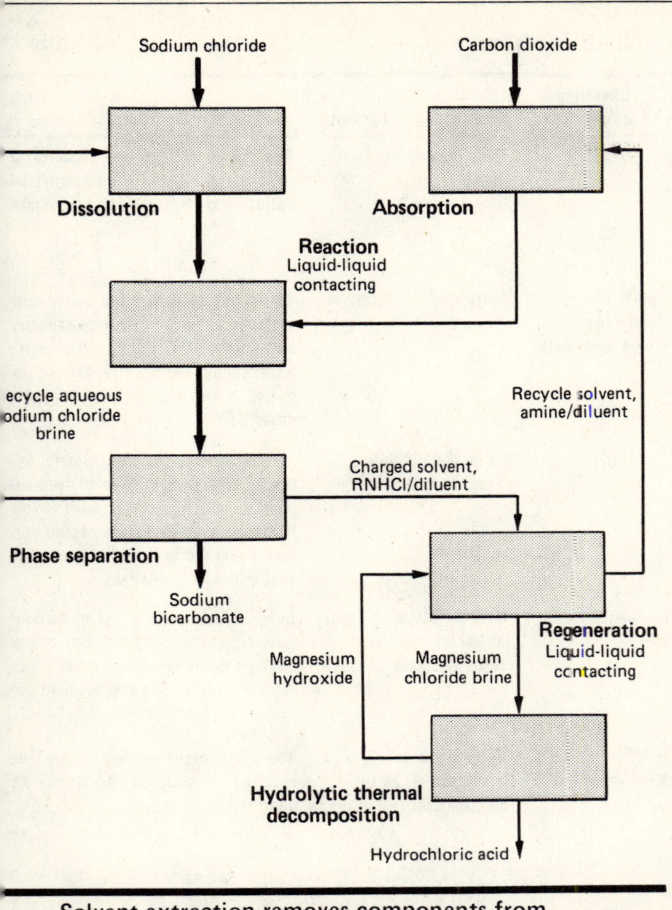

Solvent extraction removes components from reaction in manufacture of sodium bicarbonate Fig. 4

extraction [23]. This exploits the differences in dissociation constants of the components of the mixture. For example, in a mixture of *m*- and *p*-cresols, the stronger acid, *m*-cresol, reacts preferentially with a stoichiometric deficiency of an aqueous strong base to form a dissociated salt that is soluble in the aqueous phase. The weaker *p*-cresol loses the competition to react, and tends to remain undissociated in the organic solvent. In a multistage countercurrent process, separated components of high purity may be obtained. The aqueous product phase produces free *m*-cresol on acidification, and the *p*-cresol is recovered from the organic phase by distillation. In this example, the continuous consumption of strong acid and base can be avoided by using weak reagents in the manner described by Anwar, Hanson and Pratt [24].

The overriding requirement in the tar industry, however, is for large quantities of phenols, rather than specific isomers. For this reason, the favored extraction is still by means of caustic soda solution, which reacts chemically with the crude tar oil to remove most of the phenols.

Effluent treatment

An increasing number of commercial processes adopt solvent extraction as a means of treating "dirty" effluents. Perhaps the most common application is in the removal of phenols and related compounds from petroleum-refinery wastes [25], ammoniacal coke-oven liquors [26], and phenol-resin plant effluents [27]. For a comprehensive review of waste-water treatment by solvent extraction, see the article by Kiezyk and Mackay [28].

Pharmaceutical extractions

There is little information on the commercial use of solvent extraction in the manufacture of drugs and natural products. The process is widely used, however, because many pharmaceutical products are heat-sensitive, a feature that rules out separation methods such as distillation or evaporation. For example, solvent extraction is commonly used for the recovery of antibiotics from aqueous fermentation broths. In this connection, the production of penicillin is quite well documented [29,30] and mention should be made of bacitracin, erythromycin and the cephalosporins. A brief summary of the extraction procedure for penicillin typifies this kind of separation.

Penicillins are manufactured by a batchwise fermentation process. After filtration, the aqueous fermenter broth is passed forward continuously for concentration and purification by liquid extraction. There are different ways of accomplishing this.

A common method is to carry out a three-step extraction using *n*-butyl acetate as the solvent. The first step is carried out by adjusting the pH of the aqueous phase to between 2 and 2.5, and extracting the penicillin into butyl acetate. This extract is then treated with a buffer solution at a pH of about 6 to yield a penicillin-rich solution. At step three, the pH is again adjusted with acid, to a low value, and the penicillin reextracted into butyl acetate to yield a pure concentrated solution that, after further processing, gives the desired product. Adjustment of pH at each step ensures a favorable distribution coefficient.

Unfortunately, penicillin degrades rapidly as the pH is reduced. The data in Table II [31] give some idea of just how significant this stability factor is. Clearly, if the operation is to be performed at ambient temperature, it is necessary for the extraction at low pH to be accomplished as quickly as possible. A residence time as low as 5 s can be obtained in centrifugal contractors. This is one reason for their prevalence throughout the industry. An equally significant advantage of using such machines is their flexibility. The manufacture of drugs often involves small-scale batch extractions in which the same equipment is used to separate several different products and solvents. Centrifugal equipment, however, is not always essential. For example, a reciprocating-plate extractor has been used for removing a ketol (required in the manufacture of ephedrine) from a fermentation broth by means of butyl acetate [32].

Processes for separating inorganics

Organic processes rely on a purely physical preference that one type of solute molecule has over another for a particular solvent. The aim of the extraction train is to exploit this preference in order to achieve a separation of the components.

However, another function can be performed by solvent extraction whereby the solvent is used to remove

Solvents for the separation of benzene-toluene-xylene mixtures from light feedstocks **Table I**

Solvent	Process	Solvent additives and reflux conditions	Operating temperature	Contacting equipment	Comments
Sulfolane	SHELL process Licensor: Universal Oil Products	Sulfolane selectivity and capacity insensitive to water content caused by steam-stripping during solvent recovery. Heavy paraffinic countersolvent used.	120°C	Rotating-disk contactor, up to 4 m dia.	The high selectivity and capacity of sulfolane leads to low solvent/feed ratios, and thus smaller equipment.
Glycol/water mixtures	UDEX process Universal Oil Products	Solvent can be diethylene glycol and water, or a mixture of diethylene and dipropylene glycols and water, or tetraethylene glycol and water. Light hydrocarbon reflux.	150°C for diethylene glycol and water	Sieve-tray/extractor	Tetraethylene glycol and water mixtures are claimed to increase capacity by a factor of 4, and also require no antifoaming agent. The extract requires a two-step distillation to recover BTX.
Tetraethylene glycol	Union Carbide Corp.	The solvent is free of water. A dodecane reflux is used that is later recovered by distillation.	100°C	Reciprocating-plate extractor	The extract leaving the primary extractor is essentially free of feed aliphatics, and no further purification is necessary. Two-stage extraction uses dodecane as a displacement solvent in the second stage.
Dimethyl sulfoxide (DMSO)	Institut Français du Pétrole	Solvent contains up to 2% water to improve selectivity. Reflux consists of aromatics and paraffins.	Ambient	Rotating-blade extractor, typically 10 to 12 stages	Low corrosion allows use of carbon-steel equipment. Solvent has a low freezing point and is nontoxic. Two-stage extraction has displacement solvent in the second stage.
N-methyl pyrrolidone (NMP)	AROSOLVAN process, Lurgi	A polar mixing-component either water (12 to 20% by weight) or monoethylene glycol (40 to 50% by weight) must be added to the NMP to increase the selectivity and decrease the boiling point of the solvent. The NMP/water processes use pentane countersolvent.	NMP/glycol, 60°C NMP/water, 35°C	Vertical multistage mixer/settler, 24 to 30 stages, up to 8 m dia.	The quantity of mixing component required depends on the aromatics content of the feed.
N-formylmorpholine (FM)	FORMEX process, Snamprogetti	Water is added to the FM to increase its selectivity, and also to avoid high reboiler temperatures during solvent recovery by distillation.	40°C	Perforated-tray extractor FM density at 1.15 aids phase separation.	Low corrosion allows use of carbon-steel equipment.

one or more components of an equilibrium reaction. Such an extraction operation has been successfully applied in the manufacture of a number of inorganic compounds. The leader in this field is Israeli Mining Industries (IMI), which has developed commercial processes for potassium nitrate [*33*] and phosphoric acid [*34*], chiefly for use in the fertilizer industry.

The upgrading of potassium chloride to the nitrate exemplifies the type of process scheme involved:

$$KCl + HNO_3 \rightleftarrows KNO_3 + HCl$$

The reaction is forced to the right by selective extraction of the hydrochloric acid with isoamyl alcohol. The solvent is recovered and recycled, while the byproduct acid is subsequently used to treat phosphate rock in the first step of the IMI phosphoric acid process.

Recently, it has been suggested [*35*] that the production of sodium bicarbonate from brine and carbon dioxide could also be assisted if solvent extraction were used to confer a favorable bias to the main reaction:

$$NaCl + H_2O + CO_2 \rightleftarrows NaHCO_3 + HCl$$

The proposed process requires the removal of hydrochloric acid in such a way that the reaction proceeds to the right while the acid is recovered as a byproduct. At first sight, a good way of doing this seems to be by using a base such as magnesium hydroxide to remove the acid—the hydroxide and the acid being subsequently regenerated by thermohydrolytic decomposition of the magnesium chloride. Unfortunately, magnesium chloride reacts with sodium bicarbonate:

$$NaHCO_3 + \tfrac{1}{2}MgCl_2 \rightleftarrows NaCl + \tfrac{1}{2}MgCO_3 + \tfrac{1}{2}H_2O + \tfrac{1}{2}CO_2$$

Hence, it is necessary to interpose a second base to

Temperature and pH affect half-life of sodium penicillin G in aqueous solution **Table II**

pH	Half-life* at 0°C, h	10°C, h	24°C, h
2.0	4.25	1.30	0.31
3.0	24	7.6	1.7
4.0	197	52	12
5.0	2,000	341	92

*Time in hours to inactivate 50%.

avoid contact between Mg^{++} and HCO_3^-. The novelty of this process (flowsheet in Fig. 4) is that the base chosen as a vehicle for the chloride ions is a high-molecular-weight tertiary amine that forms a separate liquid phase. The reaction scheme is:

$$NaCl + CO_2 + H_2O + RN \rightleftharpoons NaHCO_3 + RNHCl$$
$$RNHCl + \tfrac{1}{2}Mg(OH)_2 \rightleftharpoons \tfrac{1}{2}MgCl_2 + RN + H_2O$$
$$\tfrac{1}{2}MgCl_2 + H_2O \rightleftharpoons \tfrac{1}{2}Mg(OH)_2 + HCl$$

In practice, the amine (denoted by RN) and its hydrochloride salt are present in an *n*-pentanol diluent. However, the process may be ultimately improved by using a different diluent such as nitrobenzene. Nitrobenzene causes a third liquid phase to form, in which the amine hydrochloride is much more soluble that the free amine. It is likely that this third phase could provide an additional means of controlling the system [*36*].

Another proposed extraction operation that uses an amine solvent is for the recovery and concentration of hydrofluoric acid from dilute aqueous solution. The process [*37*] has been developed in the U.K. at the Atomic Energy Research Establishment, Harwell. It provides one way of making anhydrous HF without interference from the troublesome high-boiling azeotrope that occurs at 38% HF. The trinonylamine "solvent," first contacted with the dilute acid feed, reacts with HF to form a predominantly bifluoride complex that incidentally increases the solvent viscosity considerably. The amine extract is then treated in a two-step distillation: first to dry it, and second, at a higher temperature, to recover the HF by dissociating the complex. The advantage of this procedure is that the most corrosive conditions are confined to the drying stage, whereas a single-step direct distillation of the wet extract would of necessity involve rectification of medium-strength aqueous acid under even more corrosive conditions. Of course, if the requirement is for a weaker acid of about 60%, there would be some advantage in removing both the water and HF in a single boiling step. It is envisaged that the process could be used to break feeds of azeotropic composition or to recovery HF from solutions containing only a few percent HF.

The waste solutions produced by stainless-steel pickling plants contain quantities of nitric and hydrofluoric acids, as well as various metals. One economic way of recoverying these pickling acids for reuse is to extract them with a suitable solvent.

Pilot-plant work in Sweden [*38*] has shown the merit of using 75% TBP (tributyl phosphate) in kerosene as the solvent in a pneumatically pumped sieve-plate column. The acids are stripped from the organic phase by contacting with pure water at a high temperature. The rate at which HF can be stripped is limited by slow kinetics. This factor, coupled with the need for temperature control, makes it desirable to use a high-residence-time contactor such as a mixer-settler for the stripping operation. In any event, the corrosive action of TBP and the acids necessitates equipment built from high-density polyethylene for which the temperature must not exceed 50°C.

References

1. Bailes, P. J., Hanson, C. and Hughes, M. A., Liquid-liquid extraction, *Chem. Eng.,* Jan. 19, 1976, pp. 86–100.
2. Müller, E., *Chem. Ind. (London)*, 518–522 (1973).
3. Read, D., Production of High-Purity Aromatics for Chemicals, presented at American Petroleum Institute meeting, San Francisco, May 1952.
4. Hoover, T. S., *Hydrocarbon Process. Petrol. Refiner,* Dec. 1969, p. 131.
5. Somekh, G. S. and Friedlander, B. I., American Chemical Soc. Div. of Petroleum Chemistry Preprints, Vol. 14, No. 4. D179 (1969).
6. Somekh, G. S., *Proc. Intern. Solvent Extraction Conference ISEC '71,* Vol. 1, p. 323, Soc. of Chemical Industry, London, 1971.
7. Beardmore, F. S. and Kosters, W. C. G., *J. Inst. Petrol.,* Vol. 49, 469 (1963).
8. Choffe, B., Raimbault, C., Navarre, F. P. and Lucas, M., *Hydrocarbon Process. Petrol. Refiner,* May 1966, pp. 188–192.
9. Cinelli, E., Noe, S. and Paret, G., *Hydrocarbon Process. Petrol. Refiner,* Apr. 1972, pp. 141–144.
10. *Hydrocarbon Process. Petrol. Refiner,* Sept. 1972, p. 185.
11. Bailes, P. J. and Winward, A., *Trans. Inst. Chem. Engrs. (London)*, Vol. 50, 240–258 (1972).
12. Barton, P., Fenske, M. R. and McCormick, R. H., *Brit. Chem. Eng.,* July 1970, pp. 921–923.
13. *Chim. Ind. (Paris) Genie Chim.,* Vol. 99, No. 7, 926 (1968).
14. *Hydrocarbon Process. Petrol. Refiner,* Sept. 1972, p. 191.
15. Fox, J. M., *Hydrocarbon Process. Petrol. Refiner,* Sept. 1963, pp. 2–7.
16. Bernard, J. D. T. and van Meurs, H. C. A., *Proc. Intern. Solvent Extraction Conference ISEC '71,* Vol. 1, p. 329, Soc. of Chemical Industry, London, 1971.
17. Coleby, J., in "Recent Advances in Liquid-Liquid Extraction," edited by Hanson, C., Pergamon Press, Oxford, 1971.
18. Lloyd-Jones, E., *Chem. Ind. (London)*, 1590 (1967).
19. British Patents 995,471; 995,472; 997,888 and 1,055,532.
20. Herrin, G. R. and Martel, E. H., *Chem. Engr.* included in *Trans. Inst. Chem. Engrs. (London)*, No. 253, 319 (1971).
21. Coleby, J., Symposium on Liquid Extraction, Inst. Chem. Engrs., Newcastle-upon-Tyne, 1967.
22. Anwar, M. M., Cook, S. T. M., Hanson, C. and Pratt, M. W. T., *Proc. Intern. Solvent Extraction Conference ISEC '74,* Vol. 1, p. 893, Soc. of Chemical Industry, London, 1974.
23. Wise, W. A. and Williams, D. F., Symposium on Less Common Means of Separation, Inst. Chem. Engrs., Birmingham, England, 1963.
24. Anwar, M. M., Hanson, C. and Pratt, M. W. T., *Proc. Intern. Solvent Extraction Conference ISEC '71,* Vol. 2, p. 911, Soc. of Chemical Industry, London, 1971.
25. "Manual on Disposal of Refinery Wastes," Chap. 10, American Petroleum Institute, Washington, 1969.
26. Carbone, W. E., Hall, R. N., Kaiser, H. R. and Bazell, G. C., *Blast Furnace Steel Plant,* May 1958.
27. Kirchgessner, N. H., *Sewage Ind. Wastes,* Vol. 2, No. 30, 191 (1958).
28. Kiezyk, P. R. and Mackay, D., *Can. J. Chem. Eng.,* Vol. 49, 747–752 (1971).
29. "Kirk-Othmer Encyclopedia of Chemical Technology," 2nd ed., Vol. 14, Wiley, New York, 1967.
30. Steel, R., "Biochemical Engineering," Heywood & Co., London, 1958.
31. Benedict, R. G., Schmidt, W. H. and Coghill, R. D., *J. Bacteriol.,* No. 3, 51 (1946).
32. Prochazka, J., Landau, J., Souhrada, F. and Heyberger, A., Report Institute of Chemical Process Fundamentals, Prague, 1967.
33. Baniel, A. and Blumberg, R., *Chim. Ind. (Paris)*, Vol. 4, 27 (1957).
34. Israel Patent 21,071 (1967).
35. Blumberg, R., Gai, J. E. and Hajdu, K., *Proc. Intern. Solvent Extraction Conference ISEC '74,* Vol. 3, p. 2787, Soc. of Chemical Industry, London, 1974.
36. Blumberg, R., *Israel Chem. Eng.,* Sept. 1973.
37. Harwick, W. H. and Wace, P. F., *Chem. & Process Eng.,* June 1965.
38. Kuylenstierna, U. and Ottertun H., *Proc. Intern. Solvent Extraction Conference ISEC '74,* Vol. 3, p. 2803, Soc. of Chemical Industry, London, 1974.

Liquid-liquid

First successful in extracting earths, liquid-liquid extraction i. metals such as copper. Here is a rundown

P. J. Bailes, C. Hanson and M. A. Hughes

☐ Liquid-liquid extraction for isolating metals on a large scale has experienced phenomenal growth in recent years. A greater variety of metals are now considered for treatment, and some recently installed (or designed) plants are large by any standards.

The hydrometallurgist/chemical engineer has been encouraged by the successful initial uses of liquid-liquid extraction (also called solvent extraction) for both the separation of uranium from its ore, and the subsequent treatment of spent reactor fuel to separate plutonium from uranium and its fission products. The pioneering work, in the 1940s, of the nuclear-materials industries in the U.S. and Britain led to the use of alkylamines to treat acid sulfate media [*1,2*], and tributyl phosphate to treat acid nitrate media [*3*].

Publications on the principles of these operations became available in the early 1950s. Soon, other reagents were being considered for uranium, so that the alkylphosphoric acids were screened [*4*]. It was realized that these derivatives had a wider use, in that other valuable cationic metal-species could be extracted, e.g., vanadium. Indeed, the value of the metal recovered was a primary driving force in developing the earliest solvent-extraction units, and so rare-earths [*5*], together with zirconium, hafnium [*6*] and vanadium [*7*] were extensively studied during 1959–1961. Gradually, the extractive metallurgist turned toward the less-valuable, but nevertheless important, transition metals such as cobalt and nickel [*8,9*], and by 1965 the commercial processing of ore bodies containing these transition metals was proposed.

It seemed that, apart from the nuclear materials and rare-earths, most early work had been associated with laboratory or pilot-plant studies; not until 1968 was solvent extraction of transition metals from concentrated leach liquors considered commercially viable [*10*].

Along with these developments, manufacturers of the commercial organic extractants had been active. Extensive work was reported in this period on alternatives to the alkylphosphoric acids, and so carboxylic acid [*11*] and a sulfonium chloride [*12*] were developed. However, the greatest step forward was when General Mills Inc. reported [*13*] the first in its series of hydroxyoxime LIX reagents, which were selective for copper. Other reagents, based on the hydroxyoxime structure, have followed, and it seems that there is commercial incentive for more-sophisticated reagents to be developed.

By 1974, the liquid-liquid extraction of copper on a large scale was a reality [*14*], and the growing values of metals on the world market suggested that even zinc could be economically recovered by hydrometallurgical routes involving solvents. Indeed, Reinhardt [*15*] reports the use of di-2ethylhexyl phosphoric acid (D2EHPA) to extract zinc from an effluent although, in this case, the penalties of pollution by the zinc if it were otherwise discharged spurred the application.

Now, there seems to be a variety of either proven or promising solvent-extraction processes for metals [*16,17*]. Commercial extractants are available and the future for these processes seems sure. The economics of a changing world will have a role to play—consequently, hydrometallurgical routes will supplant pyrometallurgical processes where the latter now prove uneconomical because of the financial burden of required antipollution equipment. Even the raw materials available for existing pyrometallurgical processes are changing, for low-grade ore must now be considered for treatment [*18*]; leaching to provide an aqueous feed to the solvent-extraction plant provides one way to solve this problem.

Interestingly, liquid-liquid extraction is being considered for the treatment of feeds derived from manganese nodules harvested from the sea bed [*19,20*].

Also, the increased cost of raw materials (and their diminishing supply) encourages the reprocessing of scrap; already, solvent extraction has made an impact on the isolation of metals for complex mixtures in solution, which are derived from waste residues, effluents and the like [*21*].

Despite this optimism, the technique has its problems. The loss of extractant and diluent to a raffinate stream presents two limitations. First, a major running cost incurred in these plants is the makeup of expensive organic reagent, and second, the extractant and diluent may make a major contribution to the BOD or COD of the final effluent stream [*22*]. Continuing problems exist with solids precipitating out in the hydrometallurgical circuits. Finally, the large-scale solvent-extraction process may be sensitive to the changing availability of world energy supplies, since the diluent and some of the extractants are derived from petroleum feedstocks.

Ritcey [*23*] has reviewed the economics of the recovery of metals by solvent extraction processing.

In the following section, the components of a solvent for the treatment of aqueous metal feeds are considered.

The solvents

The applicable solvent is often made up of an extractant dissolved in a diluent—many extractants are very viscous materials when "pure." Since "chemical

Originally published August 30, 1976

xtraction: Metals

uclear materials and rare
ow being used for more-mundane
n the solvents and techniques used.

niversity of Bradford (Great Britain)

energy" is spent in transferring the metal from the aqueous phase, it would be advantageous to use the extractant in a highly concentrated form. However, practical problems seem to suggest that 10–40% is often optimum. At higher loadings, the concentration-effect of solvent extraction is enhanced, but other features such as slow rates of mass transfer might then develop. Plants may be purposely operated at temperatures greater than ambient to overcome viscosity and mass-transfer problems, but diluents with sufficiently high flash-points are then required.

In some cases, a third component may be added to the organic phase to prevent "salt" precipitation or third-phase formation; such an additive is called a modifier. It would be a severe practical problem if third phases were formed in existing equipment, but interestingly this phenomenon also presents a challenge to the engineer because it is in this very phase where a maximum concentrating effect will be observed; new plant designs may yet use this feature to advantage.

The extractants

The more important commercial extractants, together with an indication of their area of application, are listed in Table I.

Several workers have sought to classify the chemicals into groups—the accepted classification is based on reaction type [*24*].

Inert extractants

Some metals may form essentially nonpolar species, e.g., germanium tetrachloride. Such species would then be extractable into "inert" solvents. Marcus and Kertes [*24*] have suggested that these solvents be graded according to (1) polarity, (2) polarizability and (3) hydrogen-bonding ability. It seems likely that the partition coefficients for most metals in this system will be low, which would not lead to commercial development.

Solvating extractants

If we employ the Lewis definition of a "base," then there are a series of basic extractants that extract metal species by means of an ability to solvate the hydrogen ion, e.g., the complex acid $HFeCl_4$. They might be graded according to their basicity—that is, their increasing ability to donate electrons:

$$R_3N > R_3PO > (RO)_3PO > R_2CO > R_2O$$

A range of important commercial extractants of this type exist. Of these compounds, the versatile tributyl phosphate molecule (TBP) has been much studied and employed. In the case of the extraction of uranylnitrate and nitric acid from certain leach liquors, the equations can be written:

$$2TBP_{org} + [UO_2(NO_3)_2]_{aq} \rightleftarrows [UO_2(NO_3)_2 \cdot 2TBP]_{org}$$
$$TBP_{org} + [HNO_3]_{aq} \rightleftarrows [TBP \cdot HNO_3]_{org}$$

Butex (Table I) is also in this class of extractant and has been used for fission-product treatment. However, in the presence of high concentrations of nitric acid, such extractant mixtures are prone to sudden and uncontrollable "fume offs," yielding oxidation products such as oxalic acid; this derivative would produce insoluble precipitates with plutonium, possibly creating severe criticality hazards!

Acidic extractants

A variety of commercial organic acids belong to this class, since they exchange the acidic hydrogen for the metal, thus forming a metal salt. Again, the acids may be graded according to their acidity, thus: dialkylphosphoric acids > carboxylic acids.

The phosphorus-based acid, di-2ethylhexyl phosphoric acid (D2EHPA) is important in the extraction of uranyl sulfate from sulfuric-acid leach liquors [*1*] and, in the form of its sodium or ammonium salt, it has been proposed as an extractant for cobalt and nickel [*25*]. But there is a complication in that the alkaline form "holds" appreciable quantities of water in the organic phase.

Naphthenic [*11*] and versatic [*26*] acids have been proposed as representatives of the carboxylic acids, for the extraction of the transition metals. Although some reports suggest that their appreciable solubility in the aqueous phase may deter their application at a commercial level, versatic 911 (V911) is used to extract rare earths and cobalt/nickel [*23*].

If these organic acids can be described by HR, then the extraction-reaction equation is:

$$M^{n+}_{aq} + nHR_{org} \rightleftarrows [MR_n]_{org} + nH^+_{aq}$$

Further treatment of this equation [*27*] demonstrates the use of the plot of %-extracted versus pH, since the extraction of a single metal must be dependent on the pH of the solution and the stability constant of the complex MR_n. One such plot for D2EHPA is given in Fig. 1. It can readily be seen that by a pH change in the feed to a solvent-extraction plant, one metal may be separated from another. Thus, as in Fig. 1, zinc is

(Text continues on p. 242)

Commerical extractants, their chemical structures and applications — Table I

Extractant classification and name	Structure	Known applications and sites of plants
1. Solvating Tributyl phosphate (TBP)	$(C_4H_9 \cdot O)_3P = O$	Uranium from nitric acid leach liquors, separation of uranium, plutonium and fission products, various countries Separation of Zr/Hf from nitric acid liquors, Canada Used in conjunction with amine circuit in separation of iron (as $HFeCl_4$) from nickel, cobalt. Falconbridge process, Norway
Methyl isobutyl ketone (MIBK)	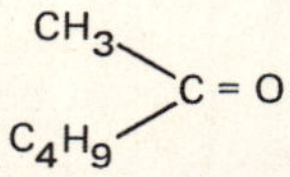 $CH_3(C_4H_9)C = O$	Extraction of niobium from H_2SO_4/HF mixtures, various countries Separation of Zr/Hf from H_2SO_4/ SCN^- mixtures, U.S.
Dibutyl carbitol (Butex, Hexone)	$C_4H_9 \cdot O \cdot CH_2 \cdot CH_2\ O \cdot CH_2 \cdot CH_2\ O \cdot C_4H_9$	Reprocessing of nuclear fuels, various countries Extraction of Au from HCl/HNO_3 mixtures, U.K.
Note: Alcohols may be used as solvating agents for magnesium species		
2. Acidic Naphthenic acids	Cyclopentane ring with substituents R′, R, R″, R‴ and $(CH_2)_2$ COOH	No known commercial application
Versatic acids	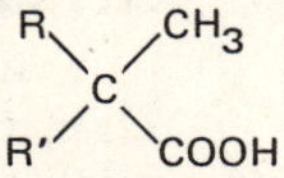 $R(R')C(CH_3)COOH$	V_{911} is used to separate Co/Ni, South Africa Extraction of rare earths, U.K.
D2EHPA (Di-2ethylhexyl phosphoric acid)	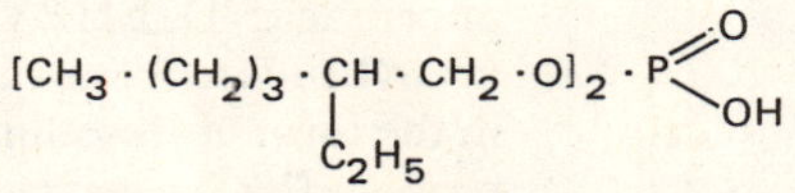 $[CH_3 \cdot (CH_2)_3 \cdot CH(C_2H_5) \cdot CH_2 \cdot O]_2 \cdot P(=O)OH$	Rare earths extracted from H_2SO_4; U.K., Norway, Canada Separation of Co/Ni, Canada Treatment of uranium in acid sulfate leach liquors, various countries Extraction of vanadium as vanadyl from reducing-acid-sulfate liquors, U.S.
3. Chelating LIX 63	$CH_3 \cdot (CH_2)_3 \cdot CH(C_2H_5) \cdot CH(OH) \cdot C(=N\text{-}OH) \cdot CH(C_2H_5) \cdot (CH_2)_3 \cdot CH_3$	
LIX 65N (anti form)	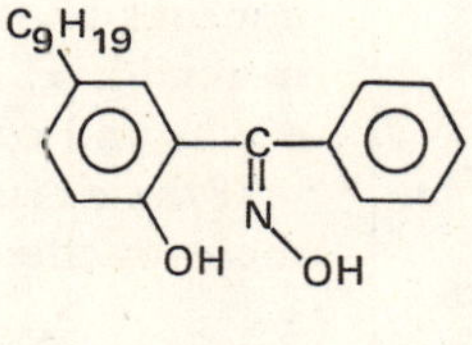 2-hydroxy-5-C_9H_{19}-phenyl–C(=N–OH)–phenyl	
LIX 64N	Mixture of 65N and 63	Extraction of copper from sulfuric acid or ammonia leach liquors, various countries
P17	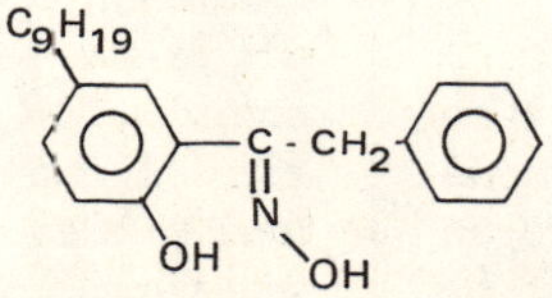 2-hydroxy-5-C_9H_{19}-phenyl–C(=N–OH)–CH_2–phenyl	Proposed alternative to 64N

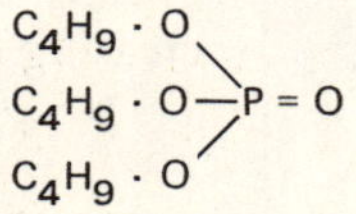

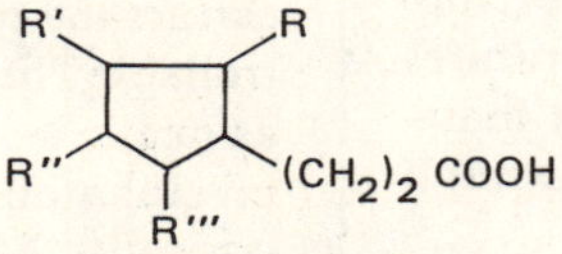

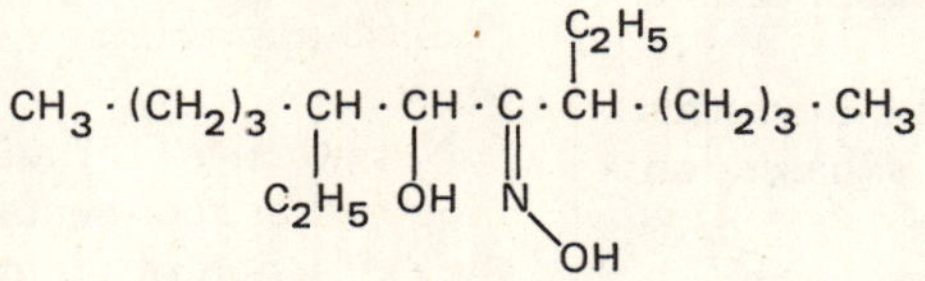

Commercial extractants, their chemical structures and applications (cont'd) **Table I**

Extractant classification and name	Structure	Known applications and sites of plants
P50	5-C_9H_{19}, 2-OH benzene ring with $-C(-H)=N-OH$	Proposed alternative to 64N
SME 529	5-C_9H_{19}, 2-OH benzene ring with $-C(-CH_3)=N-OH$	Proposed alternative to 64N
Kelex 100	8-hydroxyquinoline substituted with $CH_2=CH-CH(-CH_2-C(CH_3)_2-CH_2-(CH_3)_3)$	Proposed extractant for copper from leach liquors
Polyol	R–benzene ring–OH with X and CH_2OH substituents	Removal of boron from brines, U.S.

Note: A more detailed description of the LIX extractants, including LIX 70, 71 and 73, is given in Ref. *29*

Extractant classification and name	Structure	Known applications and sites of plants
4. Ionic		
(a) Primary amine Primene JMJ	$CH_3\cdot\underset{CH_3}{\overset{CH_3}{C}}\cdot\left(CH_2\cdot\underset{CH_3}{\overset{CH_3}{C}}\right)_4-NH_2$	
(b) Secondary amine LA-1, LA-2 Note 2	R_2^{IV} NH	
(c) Tertiary amines Alamine 336	R_3^{IV} N	Extraction of tungsten, U.S. Extraction of uranium, reprocessing of nuclear fuel, several countries
Adogen 336	R_3^{IV} N	Used in one circuit on the Falconbridge process for nickel
(d) Quaternary amines Aliquat 336	$(R_3^{IV}\cdot N\,CH_3)^+\cdot Cl^-$	Extraction of vanadium, U.S.
Adogen 464	$(R^{IV} = C_8 - C_{10})$	
(e) Sulfonic acid	$\left[R^V-S\begin{smallmatrix}CH_3\\CH_3\end{smallmatrix}\right]^+ Cl^-$ $R^V = C_{10} - C_{18}$	

Note 1 Alcohols have been used as solvating extractants to remove acids, e.g., phosphoric acid

Note 2 Secondary amines can be used to extract acid; they were also considered for uranium extraction

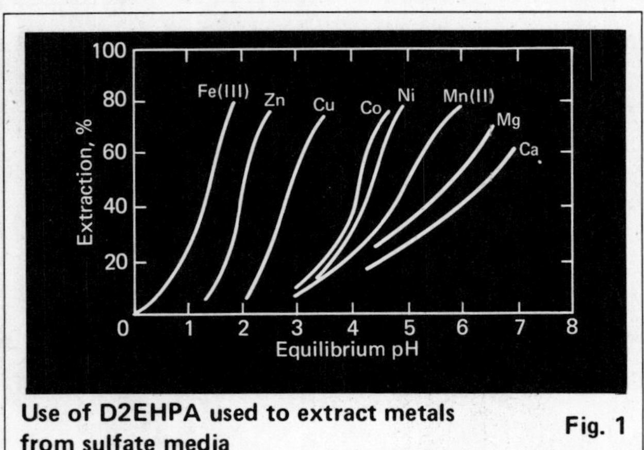

Use of D2EHPA used to extract metals from sulfate media **Fig. 1**

Approximate dates for the reporting of oxime reagents (see also Ref. 30) **Table II**

LIX 63	1963-64
LIX 64	1965
LIX 64N	1970
LIX 70	1971
LIX 65N	1971-72
LIX 71	1972
LIX 73	1973

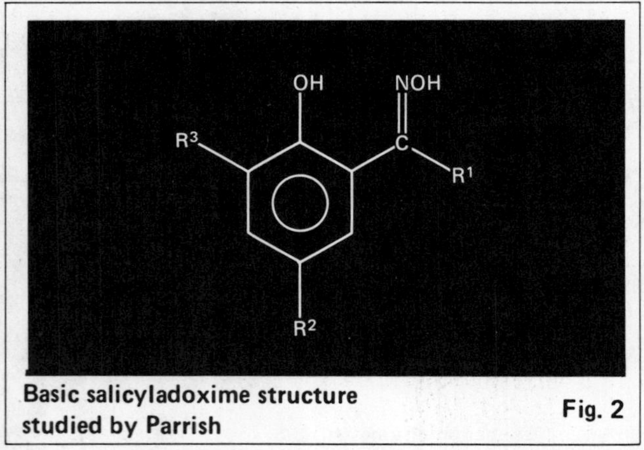

Basic salicyladoxime structure studied by Parrish **Fig. 2**

readily extracted at low pH and then could be separated from copper.

Chelating extractants

Several extractant molecules—which may be neutral or charged, or may contain zero or several displaceable hydrogen ions—show chelation behavior; that is, the formation of ring structures involving the extractant molecule as a ligand to the metal ion. Chelation has been much studied by coordination chemists [*28*] and it is agreed that the chelate structure shows additional molecular stability over the above equivalent structures containing the same total number of coordinate linkages and involving the same metal-ligand donor atom bond. Also, certain ring sizes, typically six- and seven-membered rings, are preferred.

The important LIX range of extractants belong to this class of compounds. It is of some interest to note the approximate dates when the LIX reagents were reported for hydrometallurgical application (see Table II).

During this development, the properties of derivatives of salicylaldoximes were investigated by Parrish in 1970 [*30*]; he studied a range of compounds with the basic structure shown in Fig. 2.

It should be noted that the above reagents could exist in a syn or anti form (see below), in which case only the anti form will be effective in the extraction of the copper. The existence of the two forms in LIX 65N is reported [*31*] and typical analysis suggests that at equilibrium the ratio is syn : anti : : 1 : 7.

C_9H_{19} ... C=N, OH, OH — *anti form*; C_9H_{19} ... C=N, HO, OH — *syn form*

Following 1973, companies other than General Mills reported oxime derivatives based on the above structure—P1 (later P50) and P17 are products of Acorga Ltd. [*32–34*] and SME 529 is a product of Shell Chemical Co.

Ashland Chemical Co. developed the substituted 8-hydroxyquinoline, Kelex 100, in 1968 [*36,37*]. Kelex 120 is a form of Kelex 100, diluted with nonyl phenol. It should be noted that this chemical extracts sulfuric acid as well as copper [*38*] and might be best used in a closed circuit where the sulfuric acid could be washed out and recycled. Interestingly, LIX 63 will also extract sulfuric acid from concentrated acid solutions.

At this point, the basic equation for the extraction process should be considered. Assuming that the final complex is the compound CuR_2 then we have:

$$Cu^{++}_{aq} + 2HR_{org} \rightleftarrows [CuR_2]_{org} + 2H^+_{aq}$$

This will be the case for several of the LIX reagents, the Kelex compound, and P50, P17 and SME 529. However, for LIX 63, there is evidence to show that the complex may be in the form $(CuR)_n$ (which may be polymerized) [*39,40*]. Of course, the reaction equation is much simplified and would be best applicable in situations near pH 5, and where small quantities of copper and reagents are involved; high concentrations of extractant in the organic phase would suggest aggregation, although the dimerization of the LIX 65N molecule is still under discussion [*41,42*]. In any event, the simplified structure of Fig. 3 for the complex between copper and the LIX 65N molecule would be appropriate, and it demonstrates the formation of two chelate rings.

The diagram shown in Fig. 4 demonstrates the ability of these reagents to extract copper from increasingly acid solutions.

The "impurities" in the hydroxyoxime-based chemicals must play an important role. Indeed, LIX 63 (about 1–2%) is purposely added to LIX 65N to produce the major commercial product LIX 64N, and the former acts as a "kinetic synergist." However, other materials left in from the manufacturing route may

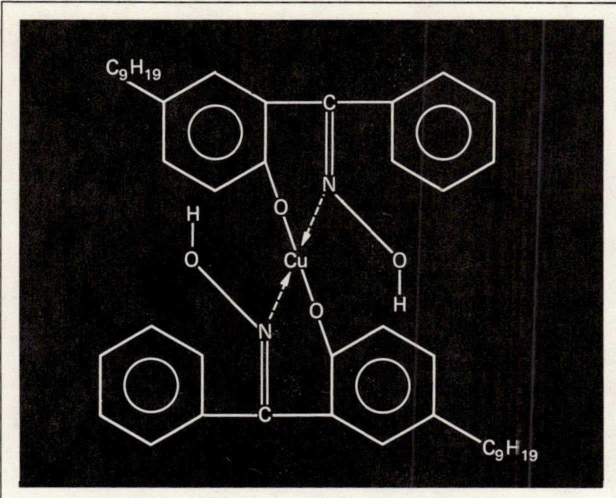

Basic structure for CuR_2, where R = LIX 65N Fig. 3

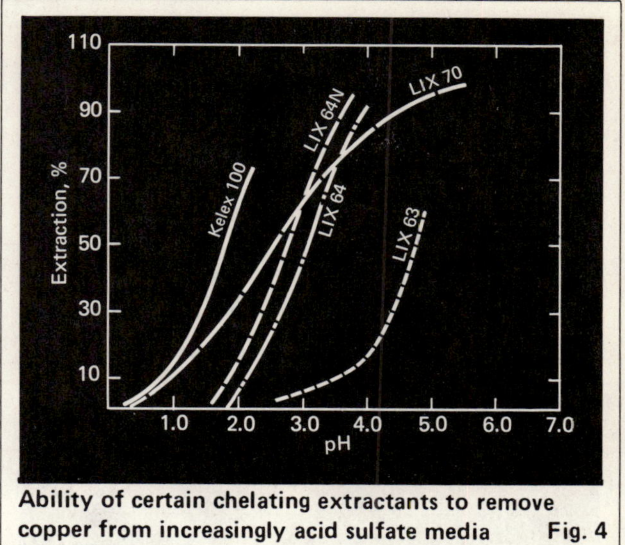

Ability of certain chelating extractants to remove copper from increasingly acid sulfate media Fig. 4

play an important role; thus nonyl phenol is a starting compound for the aromatic-based oximes, and is likely to be present in the final product [43] (also see [31]). Furthermore, the cracking of the nonyl side-chain has been reported [44] and this would lead to aqueous-soluble oxime derivatives. The additional degradation products may also be surface active and it is well known that such compounds may have an appreciable effect on the kinetics of solvent-extraction processes [45].

It seems likely that a number of combinations of these chelating reagents will be tried in the future to take advantage of any synergistic effect. Also, the synthesis of new purpose-made additives is in progress; already patents exist [46] that describe such additives.

The future is exciting for chelating reagents that are specific for transition metals, especially copper and nickel. Several companies are presently undertaking intensive research in this area, but the recent commercial reagents remain those based upon the β-hydroxy-oxime structure. Burkin and Preston [47] have reported the properties of dioximes and, in particular, 4,7-dimethyldecane-5,6-dionedioxime (see Fig. 5).

This reagent is related to the parent dioxime (dimethylglyoxime) used by analytical chemists to form the red nickel salt that is so insoluble in water that it is useful as a gravimetric analysis method. Not surprisingly, the reagent is selective for nickel where the $pH_{0.5}$ (the pH at the point where a given metal is extracted to 50% of its initial concentration) values [27] are:

Ni 0.88, Cu 1.61, Co 1.88, Fe^{+2} 3.85

However, the nickel is only slowly extracted, so the reagent offers another possibility for copper extraction, especially since it seems to offer some advantages where ferric iron is only very slightly extracted. The stripping conditions have yet to be thoroughly investigated and the reagent is still to be offered at a commercial level. If only a suitable additive could be found that would act as a kinetic synergist for nickel, this reagent might provide the method whereby nickel could be selectively extracted from acid sulfate media when cobalt is present.

Ionic extractants

These extractants are represented by those chemicals that carry within an ion pair a labile cation or anion that will exchange with the appropriate metal species in the aqueous phase. Their role may be regarded as a form of liquid ion-exchange. The several commercially available amines belong to this class. Such amines as primary, secondary and tertiary types react in a variety of ways [48] and of course are capable of extracting acids as well as metal species from leach liquors; their role in the treatment of uranium leach liquors has been evaluated by Lloyd and Oertel [49].

The quaternary amines, as exemplified by Aliquat 336 and Adogen 364, are interesting in that they are capable of reacting with strong alkali leach liquors, so that those metals capable of forming anionic species under such conditions, e.g., chromate and vanadate [50], molybdate and tungstate, can be extracted.

Of particular interest is the sulfonium compound reported by Spitzer [12], which can extract the transition metals in the form of the anionic chloride species formed in strong hydrochloric acid. The structure of the extractant is reported to be as shown in Fig. 6.

It is noteworthy that, according to the extraction curves given in Fig. 7, this reagent would separate cobalt and nickel. The reason lies not in any special

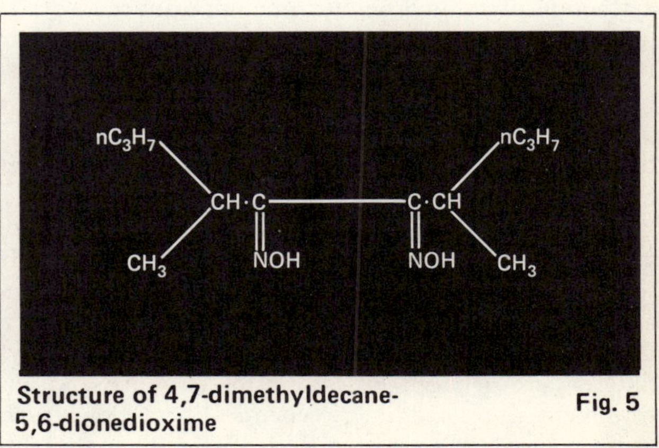

Structure of 4,7-dimethyldecane-5,6-dionedioxime Fig. 5

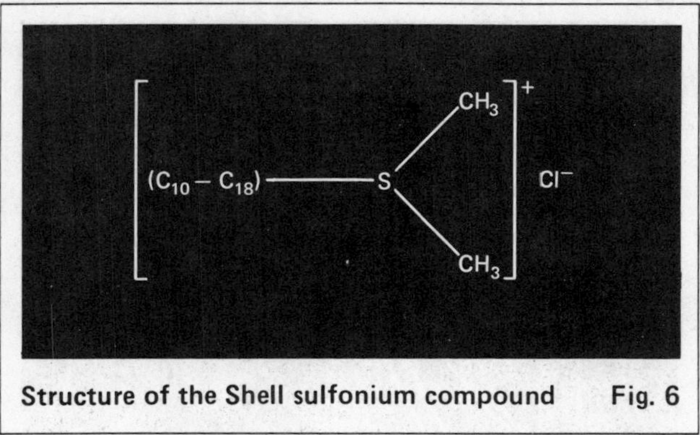

Structure of the Shell sulfonium compound Fig. 6

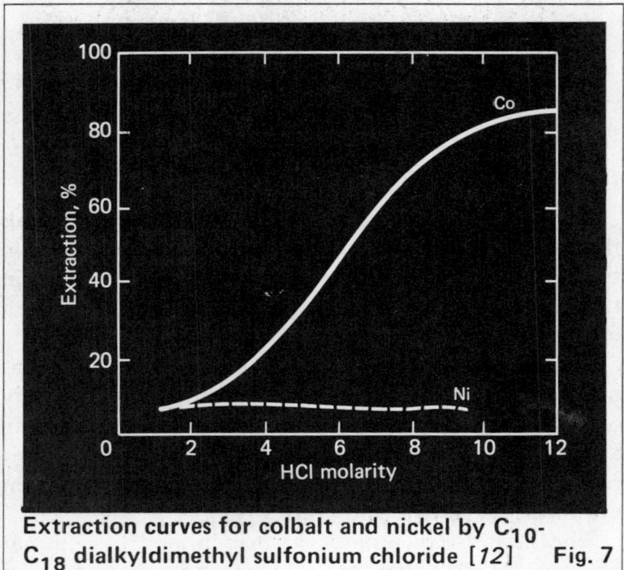

Extraction curves for colbalt and nickel by C_{10}-C_{18} dialkyldimethyl sulfonium chloride [12] Fig. 7

quality of the organic reagent but in the fact that nickel is loath to form tetrahalide anions of the form $[MCl_4]^=$ [51] It may be that operating hydrochloric acid circuits is not an encouraging prospect for chemical engineers because of corrosion problems. However, they might be encouraged by the successful Falconbridge process for nickel [52].

Extractant loss

The loss of extractant by evaporation, or by solubility and entrainment in the various streams of a solvent-extraction plant, will be of prime importance in determining the economic viability of the plant. In addition, the losses may cause pollution problems [22]. Most extractants are designed to have low solubility in the aqueous phase and, in terms of overall losses, it may be that the loss of diluent and modifier could exceed that of the extractant.

Appreciable losses of reagents have been reported, and Table III lists some values of Ashbrook [53]. It should be noted that the figures given are sometimes calculated by inventory rather than measured in a direct way; consequently, the results do not always differentiate between losses due to solubility and entrainment, and losses due to chemical breakdown such as hydrolysis and oxidation. Indeed, the ultimate solubility of some reagents may be very low where, for instance, Ashbrook reports [54] figures of $<$ 10ppm for certain LIX reagents in contact with an acid phase. The solubility is sensitive to the pH of the equilibrium phase and, not surprisingly, it is much higher when in contact with an alkaline aqueous phase [55].

Extractant losses **Table III**

Metal extracted	Extractant	pH at extraction	Loss, ppm
Cu	Kelex 100	1-2	10
Cu	LIX 64N	1.5-2.0	4-15
Co	V 911	7.7	100
Co	D2EHPA	5.5-6.5	30
Ni	Naphthenic acid	4.0	90
Ni	Naphthenic acid	6.5	900
Ni	V 911	7.0	900
			300
Hf	MIBK	1.5 M HCl	20,000
Rare earths	D2EHPA	2.0	7
U	Tertiary amines	1.5-2.0	4-15
U	TBP	2.0	25-40

(According to Ashbrook [53])

In some instances, the value of the metal recovered makes the economics of solvent loss a minor feature in the overall costing. Hence, with certain platinum metals, one might even consider a "burn off" of the solvent from the final concentrated organic phase. In the case of the extraction of hafnium by MIBK (see Table I), the extractant is indeed removed by thermal decomposition.

The diluent

The diluent, or carrier solvent, is mainly required as a diluting medium to lower the viscosity of the organic phase and facilitate contacting between the two phases. The additional important carrier properties are:

1. The ability to retain in solution both the complexed and uncomplexed extractant.
2. A low solubility in the aqueous phase.
3. A high flash point and a consistent low rate of evaporation.
4. High chemical stability over the range of conditions for plant operation.

Point (3) is especially important in plants erected in countries having high ambient temperatures, or where it is proposed to run the solvent-extraction process at a high temperature.

Apart from the features listed above, it is apparent that the carrier can influence both the rate of extraction and the position of equilibrium; thus the work of Murray and Boubouliais [56] shows the effects for a copper/oxime/diluent system. And Ritcey et al. [57] have attempted to relate the behavior of the diluents to their fundamental properties, such as dielectric constant, solubility parameter, dipole moment, and polarizability. It seems that an approach based on Hansen's analysis [58] of the individual contributions to the solubility parameter would be worthwhile. It is surprising to find that hydrometallurgists have previously thought of the diluent as being inert, inasmuch as the classical work of Irving and Williams [*59,60*] on the extraction of metals with dithizone and other chelates demonstrated the solvent effect as far back as 1949.

Many diluents have been considered and used by the extractive metallurgist, and some of these are listed in Table IV. It should be realized that the commercial diluents are "cuts" from certain processes adopted during the production of chemicals from petroleum feed-stocks and, as such, are complex mixtures.

It is noteworthy that, apart from the complex composition of commercial diluents, their aromatic-to-aliphatic content remains a major feature. In the extraction of copper by oximes, the increased aromatic content seems to favor the loading of the copper with respect to equilibrium, but increased aliphatics favor faster rates of mass transfer [*56*].

The choice of carrier also affects the settling characteristics of the mixed phase.

Little has been said of stripping, but the several points raised above would be applicable to this part of the process as well as to forward extraction.

Some commercial diluents for metals extraction **Table IV**

"High" aromatics	Sp. gr., 20°C	Boiling pt., °F	Flash pt., °F	Aromatic content, approximate %
Solvesso 100	0.876	315	112	99
Solvesso 150	0.985	370	151	97
HAN	0.933	357	105	89
Chevron 3	0.888	360	145	98
Chevron 25	0.875	316	115	99
Chevron 40L	0.886	360	141	78
Chevron 44L	0.893	366	154	70
"Medium to low" aromatics	**Sp. gr., 20°C**	**Boiling pt., °F**	**Flash pt., °F**	**Aromatic content approximate %**
Escaid 100	0.790	376	168	20
Napoleum 470	0.811	410	175	12
"Low" aromatics	**Sp. gr., 20°C**	**Boiling pt., °F**	**Flash pt., °F**	**Aliphatic content, approximate %**
Isopar L	0.767	373	144	93
Isopar E	0.723	240	<45	99.9
Isopar M	0.782	405	172	80
Norpar 12	0.751	384	156	98
Shell 140	0.785	364	141	45*
Shell MSB 210	0.783		165	
DX 3641	0.793	361	135	45*
Escaid 200	0.796	383	152	52*

*Contains appreciable quantities (ca. 50%) of naphthenes.

Notes: Solvesso, Isopar, Norpar, Escaid, HAN and DX, are products of Exxon. A high naphthenes content is desirable in some systems.

Although it would be advantageous to have a tailor-made hydrocarbon diluent, it seems unlikely that this will be available in the near future. The cost of blending and manufacturing such a sophisticated chemical is prohibitive. The major oil companies can at present only provide selected cuts from the standard processing of crude-oil feedstocks. The total volume of diluent sold to existing or projected solvent-extraction plants is very small in comparison with overall petrochemical production. Indeed, under some circumstances, the extractive metallurgist might find himself out on a limb with regard to any single diluent, and he would be wise to arrange his processing to allow for the use of a variety of similar diluents from several suppliers.

The use of chlorinated compounds as diluents seems to offer some advantages where they are heavier than the aqueous phase, e.g., perchloroethylene, and so evaporation losses of the diluent would be minimized. However, some of these solvents (e.g., chloroform and carbon tetrachloride) can be toxic to man, and many of them have to be stabilized to avoid breakdown to oxidation products, especially in the presence of light. Some chlorinated solvents are stabilized with small amounts of amines, which would be quickly removed on contact with acid liquors.

The modifiers

Additional chemicals are sometimes added to solvents to prevent third-phase formation, and long-chain alcohols or tributyl phosphate have been used in the past.

Experience shows that the alcohols have the disadvantage of being appreciably soluble in the aqueous phase and indeed may contribute a severe environmental problem. The work on the Dapex process [*61*] highlighted the losses of modifiers in the system involving D2EHPA under either acid or alkali conditions.

Although possible losses might affect the selection of a modifier, these losses may be outweighed by the other effects to be gained, so that certain modifiers enhance the rate of extraction and the final equilibrium position. Indeed, it is well known that tributyl phosphate acts synergistically with dialkylphosphoric acids. In other cases [*62*] tributyl phosphate might poison the system if it reduced the mass-transfer rate significantly.

The future

Future developments will be chiefly influenced by the changing characteristics of available world resources. Copper will be increasingly processed by hydrometallurgical routes because low-grade ores will have to be exploited. As metals become scarcer, then secondary sources such as scrap and sludge will have to be recycled. Solvent extraction has a role to play in such recycling. New processes will be developed where extractants selective for given metals are produced, and other metals not presently featured in the commercial liquid-liquid extraction processes will be considered—e.g., molybdenum, bismuth and rhenium are likely to be in short world supply in the years to come. Finally, aluminum presents a great challenge because of its worldwide abundance, albeit distributed in aluminosilicate form. New routes to this last metal are being sought by many research groups.

The chemical engineer has already developed the basic contacting equipment, and the mixer-settler continues to be preferred. Only faster extractants will encourage use of differential contactors. However, solvent loss through entrainment from any contacting equipment will require attention—thus solvent-recovery devices will be further developed. This solvent loss is of paramount importance, and the alternative solid-liquid ion-exchange processes for the treatment of certain feeds will continue to offer a strong alternative where they do not lose reagent to an effluent stream. However, such solid-liquid exchangers do degrade through poisoning and attrition. One alternative might be to absorb the organic extractants onto a solid support, so producing a type of macroporous resin.

It seems certain that, provided the chemistry of the processes can be developed, the chemical engineer can readily adapt such inventiveness for the winning of metals in a profitable way, at the same time making a contribution to solving the world's increasing requirements for these metals.

References

1. Wart, J. C., U. S. Patent No. 2,564,241, 1951.
2. Coleman, C. F., others, *Ind. Eng. Chem.,* Vol. 50, p. 1756 (1958).
3. Stoller, S. M. and Richards, R. B., eds., "Reactor Handbook," Vol. II, "Fuel Reprocessing," Interscience, New York, 1961.
4. Blake, C. A., others, *Ind. Eng. Chem.,* Vol. 50, p. 1736 (1958).
5. Arnold, W. D., Crouse, D. J., U. S. Atomic Energy Commission Report, ORNL, p. 3030, 1961.
6. Cerrai, E., *Testa. Energia Nucleare,* Vol. 6, p. 771 (1959).
7. Swanson, R. R., others, *Eng. Min. J.,* Vol. 162, No. 10, p. 110 (1961).
8. Good, M. L., and Bryan, S. E., *J. Inorg. Nucl. Chem.,* Vol. 20, p. 140 (1961).
9. Queneau, P., ed., "Extractive Metallurgy of Copper, Nickel and Cobalt," Interscience, N. Y., 1961.
10. "Advances in Extractive Metallurgy," publication of Inst. of Min. & Met., London, 1968.
11. Fletcher, A. W., and Wilson, J. C., *Trans. Inst. Min. & Met.,* Vol. 70, p. 355 (1961).
12. Spitzer, E. L. T. M., and Radder, J., "Advances in Extractive Metallurgy," publication of Inst. of Min. & Met., London, 1968.
13. Swanson, R. R., (LIX 63), U. S. Patent No. 3,224,873, 1965.
14. Holmes, J. A., and Fisher, J. F. C., "Advances in Extractive Metallurgy," publication of Inst. Min. & Met., London, 1971.
15. Reinhardt, H., *Chem. Ind.* (*London*), Vol. 5, p. 210 (1975).
16. Flett, D. S., International Symposium on Solvent Extraction in Metallurgical Processes, published by Tech. Inst. K. VIV, Antwerp, Belgium, 1972.
17. Flett, D. S., and Spink, D. R., Dept. of Industry, Warren Spring Laboratories, Report LR211(ME), 1975.
18. Kirby, R. C., and Barclay, J. A., *Min. Eng.,* Vol. 27, No. 6, p. 42 (1975).
19. Brooks, P. T., and Martin, D. A., U. S. Bur. Mines Rept. Invest., p. 7473 (1971).
20. Kane, W. S., and Cardwell, P. H., U.S. Patent No. 3,752,745.
21. Hughes, M. A., *Chem. Ind.* (*London*), p. 104, Dec. 1975.
22. Ritcey, G. M., others, Proceedings International Solvent Extraction Conference ISEC '74, Vol. 3, p. 2783 (Soc. of Chemical Industry, London, 1974).
23. Ritcey, G. M., *CIM Bulletin,* p. 85, Jan. 1976.
24. Marcus, Y., and kertes, A. S., "Ion Exchange & Solvent Extraction of Metal Complexes," Wiley-Interscience, London, 1969.
25. Ritcey, G. M., others, *CIM Bulletin,* Jan. 1975, p. 111.
26. Spitzer, E. L. T. M., others, *Trans. Inst. Min & Met.,* Section C, Vol. 75, p. 265 (1966).
27. Starcy, J., "The Solvent Extraction of Metal Chelates," Pergamon Press, Elmsford, New York, 1964.
28. Orgel, L. E., "An Introduction to Transition Metal Chemistry," Methuen, London, 1960.
29. Ashbrook, A. W., *Hydromet.,* Vol. 1, p. 5 (1975).
30. Parrish, J. F., *J. South African Inst. Chem.,* Vol. 23, p. 129 (1970).
31. Ashbrook, A. W., *J. Chromatog.,* Vol. 105, p. 141 (1975).
32. Anderson, B., German Patent No. 2,305,694, 1973.
33. Ackerley, N., and Mack, P. A., German Patent No. 2,334,901, 1973.
34. Anderson, B., German Patent No. 2,342,878, 1973.
35. Shell Chemicals Preliminary Technical Information Bulletin, INT, 74, M1.
36. Budde, W. M., and Hartlage, J. A., U.S. Patent No. 3,637,711, 1972.
37. Hartlage, J. A., and Cronberg, A. D., U.S. Patent No. 3,725,046, 1973.
38. Flett, D. S., *Trans. Inst. Min. & Met.,* Section C, Vol. 83, p. 30 (1974).
39. Chakravorty, A., *Co-ord. Chem. Rev.,* Vol. 13, p. 1 (1974).
40. Preston, J. S., *J. Inorg. Nucl. Chem.,* Vol. 37, p. 1235 (1975).
41. Laskorin, B. W., others, Proceedings International Solvent Extraction Conference ISEC '74 (Soc. of Chemical Industry, London, 1974).
42. Dalton, R. F., others, Soc. Chem. Ind. Symposium on Solvent Extraction, Bristol, U.K., 1975.
42. Hughes, M. A., Preston, J. S., and Whewell, R. J. Analyses carried out at the University of Bradford, England.
44. van der Zeeuw, A. J., I. Chem. E. Symp. Series, No. 42, 1975.
45. Hanson, Carl, ed., "Recent Advances in Solvent Extraction, Pergamon Press, Elmsford, New York, 1971.
46. Morin, E. A., and Peterson, H. D., U.S. Patent No. 3,878,286, 1975.
47. Burkin, A. R., and Preston, J. S., *J. Inorg. Nucl. Chem.,* Vol. 37, p. 2187 (1975).
48. Schmidt, V. S., "Amine Extraction," Israel Program for Scientific Translations Ltd., Jerusalem, Israel, 1971.
49. Lloyd, P. J., and Oertel, M. K., Unit Processes in Hydrometallurgy, Met. Soc. Conf., Dallas, Tex., ed by Wadsworth, M. E., and Davis, F. T., Vol. 24, p. 253 (1963).
50. Hughes, M. A., and Leaver, T., Proceedings International Solvent Extraction Conference ISEC '74, Vol. 2, p. 1147 (Soc. of Chem. Ind., London, 1974).
51. "Stability Constants Handbook," a publication of Chem. Soc., London.
52. Thornhill, P. G., others, *J. Metals,* July 1971, p. 2.
53. Ashbrook, A. W., *Min. Sci. Eng.,* Vol. 5, No. 3, p. 169 (1973).
54. Ashbrook, A. W., *Anal. Chim. Acta.,* Vol. 58, p. 115 (1972).
55. Hughes, M. A., work carried out at the University of Bradford, England.
56. Murray, K. J., and Bouboulias, C. J., *Eng. Min. J.,* July, 1974, p. 74.
57. Ritcey, G. M., and Lucas, B. H., Proceedings International Solvent Extraction Conference ISEC '74, Vol. 3, p. 2437 (Soc. of Chem. Ind., London, 1974).
58. Hansen, C. M., and Beerbower, J., Solubility Parameters, in "Kirk-Othmer Encyclopedia of Chemical Technology," 2nd ed., Supplement Volume, pp. 889–910, Wiley, New York, 1971.
59. Irving, H., and Williams, R. J. P., *J. Chem. Soc.,* III, p. 1841 (1949).
60. Irving, H., others, *J. Chem. Soc.,* p. 1847 (1949).
61. U.S. Atomic Energy Commission, report, Oak Ridge National Laboratories, p. 2172, 1955.
62. U.S. Atomic Energy Commission, report, Oak Ridge National Laboratories, p. 1903, 1955.

Selection criteria for liquid-liquid extractors

Based on the authors' operating experience, here are practical guidelines for making the initial choice of contacting equipment.

K.-H. Reissinger and *Jürgen Schröter, Bayer AG*

☐ Liquid-liquid extraction has become an important separation technique in modern process technology. This has resulted in the rapid development of a great variety of extractor types, in the evaluation of which the chemical engineer must primarily depend on manufacturers' literature. Comparative measurements obtained with selected, standardized test systems are available only in exceptional cases. Let us, therefore, present the conventional extractor types used, and follow this by attempting to establish selection criteria on the basis of available operating experience.

Industrial extraction apparatus

An overall review of industrial extractors and their relationship to different extraction processes is given in Fig. 1. The different types of apparatus will now be described briefly and looked at critically. Most of the evaluation is based on our own experience.

As a rule, industrial extractors can be designed with a maximum of 10 to 15 theoretical stages per unit. This small number of stages compared with distillation processes is due to a low stage efficiency caused by the impeded mass transfer between two liquid phases. As a result of the low extract concentrations present in most instances, most of the mass transfer takes place by diffusion. Improvement in mass transfer can be achieved by means of large and turbulent interfaces that are continuously renewed. This renewal is accomplished by different devices in the individual apparatus, such as internals; and the mechanical energy input is determined by the properties of the substances processed. Reduced efficiency is caused primarily by backmixing.

This article is based on a paper originally presented at the symposium on Alternatives to Distillation sponsored by the North Western Branch of the Institution of Chemical Engineers, Rugby, England, April 1978.

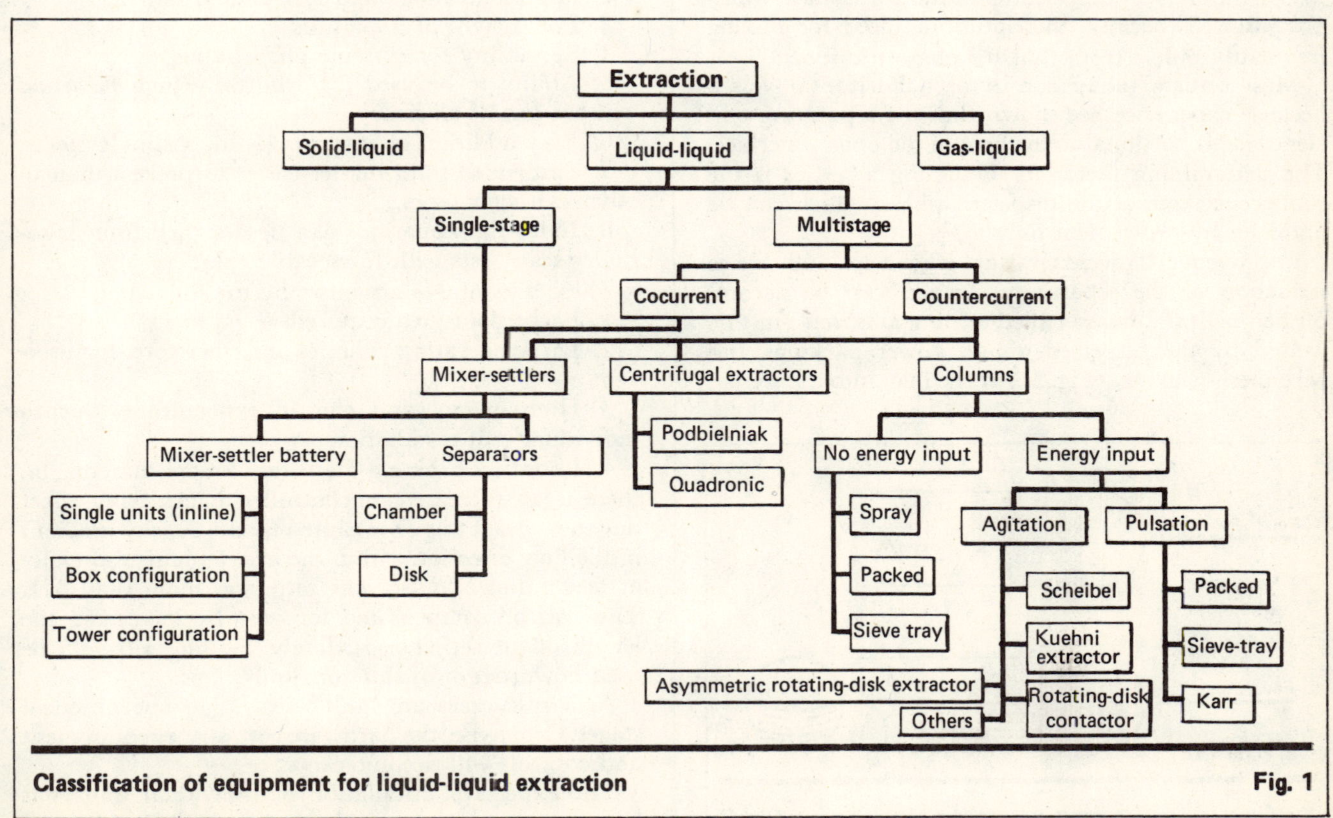

Classification of equipment for liquid-liquid extraction — Fig. 1

Originally published November 6, 1978.

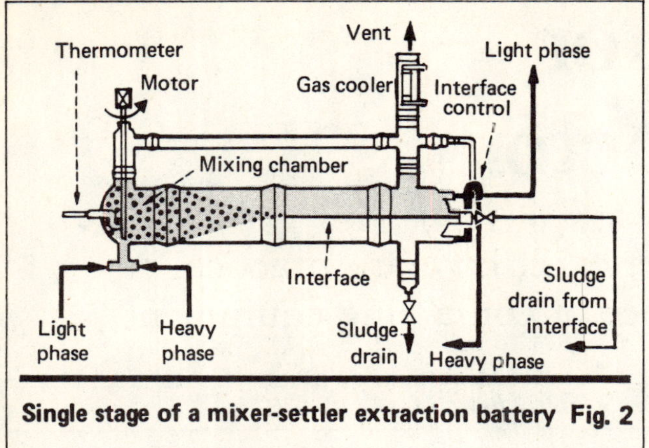

Single stage of a mixer-settler extraction battery Fig. 2

Mixer-settlers

The simplest type of apparatus is the mixer-settler battery (MSB) [2,3,4]. Fig. 2 shows a single stage of such a battery, consisting of a mixing chamber and downstream settler that can be either integral or separate. In an integrated system, the mixing and settling chambers are separated by a slotted baffle. Wherever possible, sludge drains must be provided in the separator at the interface level and at the bottom.

In the mixing unit, an adjustment of the optimum droplet size must be made, which consists of a suitable compromise between small droplets for good mass transfer and large droplets for short separation times. Agitated vessels, pumps, mixing nozzles and static mixers are used as mixing elements, depending on the individual conditions. It has proven advantageous not to use the mixer pumps for simultaneous phase transport as well, because the optimum speed for mixing frequently differs from that for phase transport.

Most settlers today consist of a horizontal vessel, because experience has shown that the separating efficiency is proportional to the area of the phase interface. The determining factor for designing a settler is the coalescence rate of the dispersed phase, which can be obtained by experiment only.

Improvements of separating efficiency, and hence reduction of the separating surface, can be accomplished by installing so-called settling aids that must be wetted by the dispersed phase. Tower packings and wire-mesh packings (Fig. 3a) at the inlet into the settler,

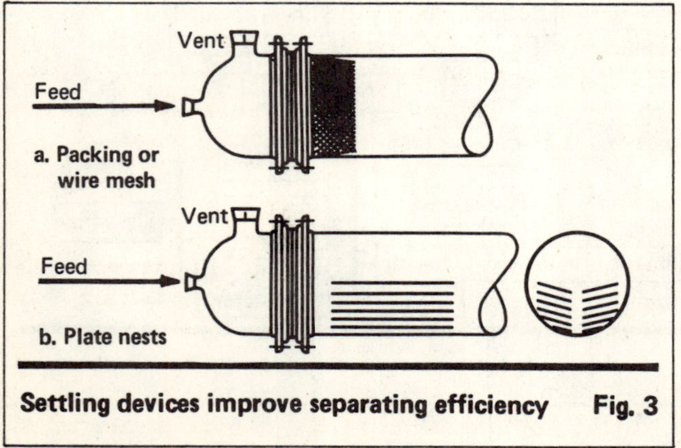

Settling devices improve separating efficiency Fig. 3

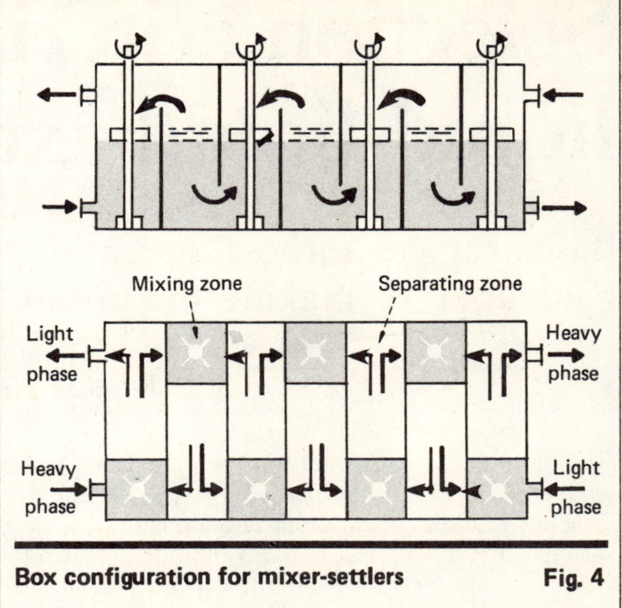

Box configuration for mixer-settlers Fig. 4

or oblique plate nests (Fig. 3b) in the forward section of the settler, are especially suited for this purpose.

These aids can be used to realize very good phase separation—achieving entrainment rates of the respective counter phase of less than 1%, so that stage efficiencies of 90% and more can be attained.

Other advantages of the MSB include:

1. Optimum selection of the degree of dispersion. This is possible since the mixers can be adjusted independently.
2. High load range and any desired flowrate.
3. Low height of apparatus.
4. Suitability for extreme phase ratios.
5. Ability to be used for arbitrarily high flowrates without loss of efficiency.
6. Easy addition of extra stages to existing systems.
7. Suspended solid matter easier to process than in other extractor types.
8. Industrial apparatus can be designed from laboratory-scale tests with foreseeable risk.

These advantages are offset by the following:

1. Large floor area required.
2. Large operating volumes and, therefore, high solvent costs.
3. High energy and control expenditures due to individual unit installation.
4. In settlers having a diameter of more than one m, there is an increased, uncontrolled backmixing effect due to recirculating flow caused by the density gradient in the flow direction within the heavy phase, especially at feed ratios S/R (solvent/raffinate) more than 5:1. This can be compensated for only by increasing the length of the settler accordingly. Settling aids can be used downstream of this zone only.

As a rule, MSBs are built with up to five practical stages. Otherwise, the hardware cost, space requirement and controls will become excessive.

The expensive installation of individual units has been circumvented by the box configuration of Denver Equipment Co. (Fig. 4). In this system, the mixing and

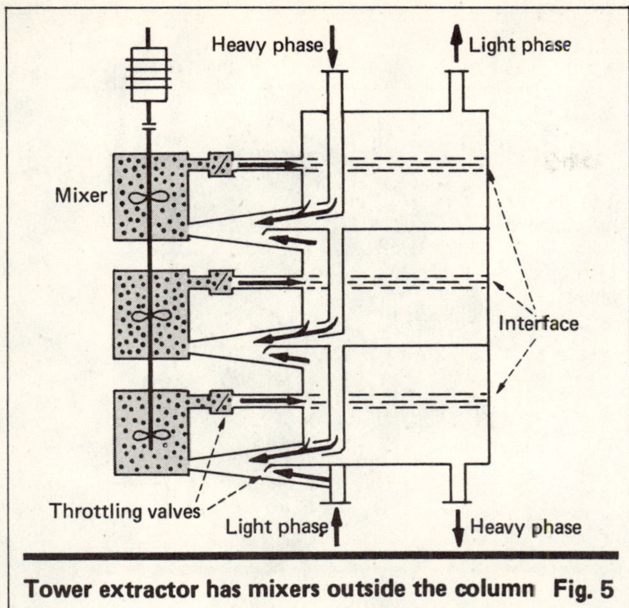

Tower extractor has mixers outside the column Fig. 5

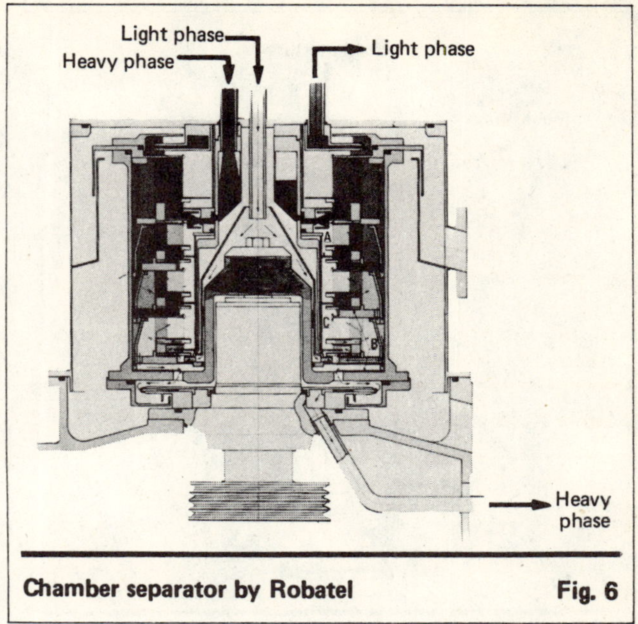

Chamber separator by Robatel Fig. 6

separating zones as well as the individual stages are separated by weirs. These weirs can considerably impair the phase separation so that stage efficiency can drop to values under 50%.

Phase transport within the apparatus is accomplished externally by means of the feed pumps. This means that feed fluctuations will immediately affect the location of the interfaces and, therefore, the entire operating behavior of the system. This configuration can be used for up to ten stages, but the space is still considerable.

A marked reduction in floor-space requirement is achieved by means of the tower configuration (Fig. 5). In a tower extractor developed by Lurgi Co., the mixers are located outside the column, which is used for phase separation only. The phases are mixed by centrifugal stirrers and are passed through a coalescent zone prior to entering the actual separator space.

Three to five mixer pumps are driven by one motor. This coupling effect and the transport characteristic of the mixing units, which cannot be ignored, result in an adaptation of mass flows for each stage. This has the effect that one of the two phases is recirculated across one stage and can be controlled externally by means of throttling valves. This internal cycle will increase the entrainment rate, thus impairing the stage efficiency. Due to flooded operation, perturbations in feed metering cause fluctuations throughout the column. Hence, the feed ratios to the individual stages, in addition to the internal circulation, are further changed compared with the external balance, and efficiencies reduced.

These tower extractors are popular for extracting aromatics (separation of aromatic and aliphatic hydrocarbons by N-methyl pyrrolidone).

Separators

A separator is an extractor operating on the mixer-settler principle, in which phase separation is accomplished by means of centrifugal force. Therefore, these units are especially suited for processing liquid pairs having small density differences ($\Delta\rho \geq 0.02$ g/cm^3). A distinction is made between two systems of different design, i.e., chamber separator and disk separator.

A typical apparatus of the chamber type is the Robatel extractor developed in France (Fig. 6). Up to eight stages can be accommodated in this apparatus, which, for example, is used for uranium extraction.

Disk separators are made, for instance, by Westfalia and Alfa-Laval. These extractors consist of a mixing chamber and a separator space in which a large number of conical disks are installed. These disks ensure very-thin liquid layers and, therefore, short paths for the separation of the droplets. This apparatus is popular for extracting essences and aroma substances, and antibiotics in the pharmaceutical industry (Fig. 7).

The Luwesta extractor, designed by Lurgi and Westfalia, is a multistage apparatus of this type. However, throughput is decreased as a result of the multistage design. Consequently, for higher throughput the advantages of the multistage configuration are lost because of the number of parallel units required. Also as a result of their sophisticated design, these multistage extractors are trouble-prone, so that in recent years they have only rarely been built. And, these separators generally have a very high first cost and require extensive maintenance. However, their small operating volume yields a decisive advantage—short residence times, and low material costs if the solvents are expensive. Recent experience has shown that pressure up to 100 bars may occur within the separator space. When the equilibria are pressure-dependent, this can cause undesirable re-extractions in some stages of multistage systems.

Centrifugal extractors

This type of apparatus, which includes the Podbielniak and the Quadronic extractors, occupies an intermediate position between the separators operating on the mixer-settler principle and the column extractors [*19*]. Because of its design, it must be considered, with reservations, to be a "rotating sieve-tray column."

In the Podbielniak extractor (Fig. 8), a number of

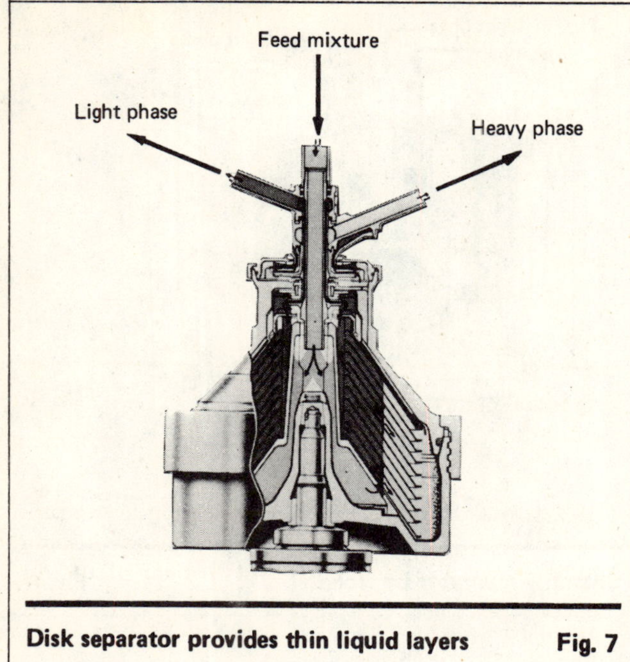

Disk separator provides thin liquid layers **Fig. 7**

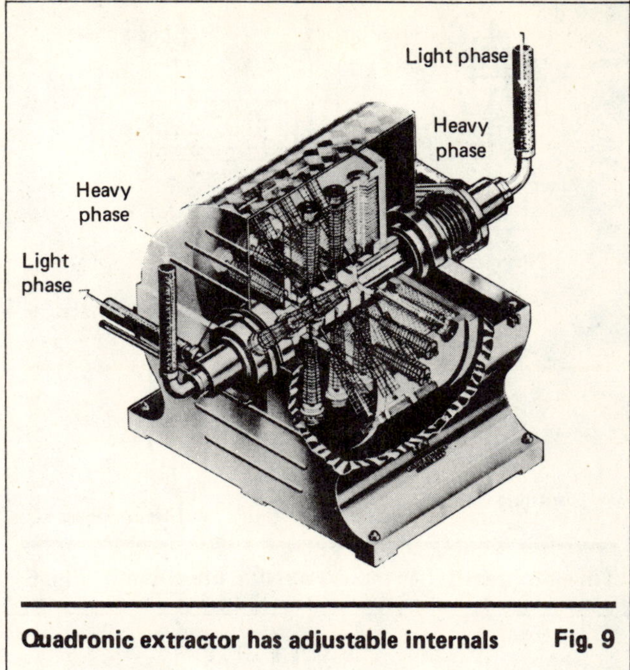

Quadronic extractor has adjustable internals **Fig. 9**

concentric sieve trays are located around a horizontal axis through which the two liquid phases flow in a countercurrent pattern. Liquid inlet and outlet are accomplished via the shaft. This requires an inlet pressure of 4–7 bars to overcome the pressure drop and the centrifugal forces.

In a Podbielniak extractor, three to five theoretical stages per machine can be achieved at throughputs up to 130 m^3/h. However, this extractor is not suitable for processing emulsions, and requires density differences $\Delta\rho \geq 0.05$ g/cm^3. This type of apparatus is also characterized by a low operating volume and, consequently, short residence time. Its susceptibility to trouble is primarily due to sealing problems. Again, in the outer range of the drum, pressure up to 70 bars will occur in the liquid. This extractor is frequently used in the pharmaceutical industry and, due to its great throughput, is also used to handle large flows.

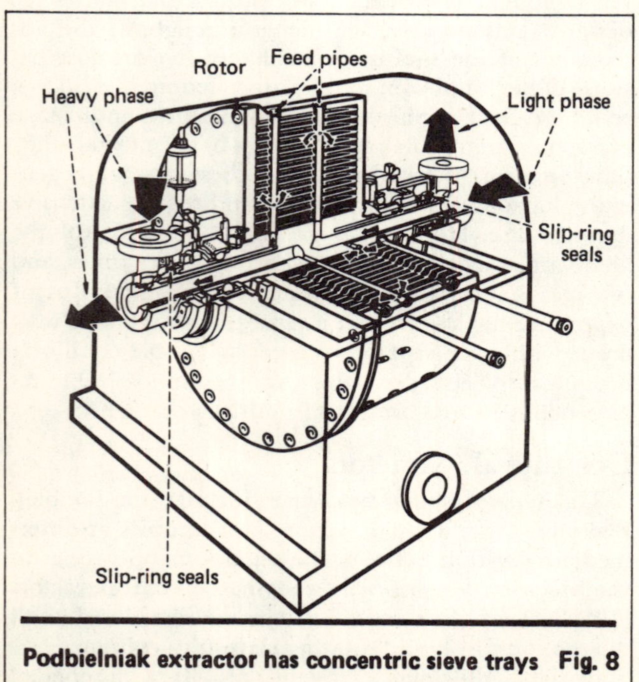

Podbielniak extractor has concentric sieve trays **Fig. 8**

A unit of fairly similar design is the Quadronic extractor (Fig. 9). The significant difference to the Podbielniak extractor is that the free cross-sectional area of its internals is adjustable, so that it can better handle different loads. According to the manufacturer's information, throughputs up to 150 m^3/h and approximately four to six theoretical stages can be achieved. The higher number of stages, compared with the Podbielniak extractor, is achieved by better cross-mixing of the two phases. Detailed operational experience is not available for this system.

Column extractors

The significant feature of column extractors is the vertical countercurrent flow of the two liquids, due to density difference and gravity. Since continuous interface renewal has a significant effect on mass transfer, it will consequently affect the necessary column height. In the different types of column, this is taken into account by using various means for energy input. On the other hand, the backmixing that reduces efficiency must be kept within acceptable limits.

Columns without energy input [*2,3,5*] include those operating without external influence on the liquid flow and droplet distribution. The efficiency of these units is fairly low because the droplet size distribution is unfavorable and interface renewal is poor.

The simplest type of column extractor is the spray-column (Fig. 10), consisting of a smooth tube without any internals. The droplets of the dispersed phases are generated only once—at the input and, in most instances, by means of annular spray nozzles. The throughput for these columns depends greatly on the

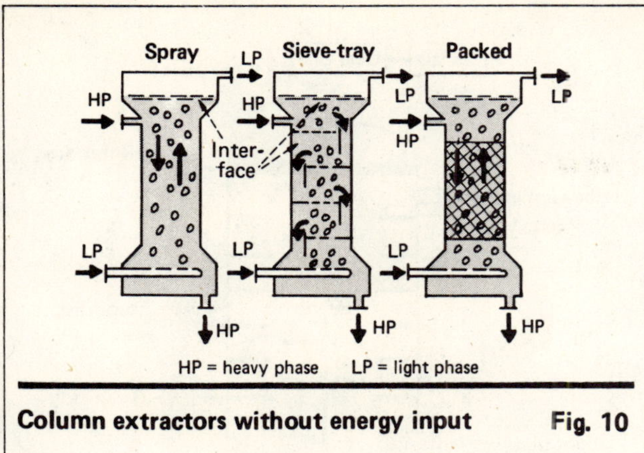

Column extractors without energy input **Fig. 10**

density difference and the viscosity of the two phases. Axial mixing of the continuous phase is high, and it increases greatly with increasing diameter-to-length ratio of the column. Therefore, these columns are rarely used—for instance, as neutralizing scrubbers for organic substances or for direct liquid-liquid heat exchange.

Sieve-tray and packed columns, on the other hand, are used more frequently. In the sieve-tray column [*6,7*], which is designed on the same principle as distillation systems using overflow weirs and downcomers, the droplets are reformed on every tray, since the dispersed phase will back up on the trays, coalescing into a continuous layer. By means of a suitable arrangement of drain pipes (such as several pipes distributed over the circumference, with outlet ports toward the center of the column), a considerable cross-flow is generated within the column, which results in an additional backup layer for the dispersed phase, thus improving efficiency. The throughput through these columns depends on the density difference between the two phases and the height of the backup layer under the trays. The efficiency is also affected by the height of the backup layer and the spacing of the trays. The load range is small.

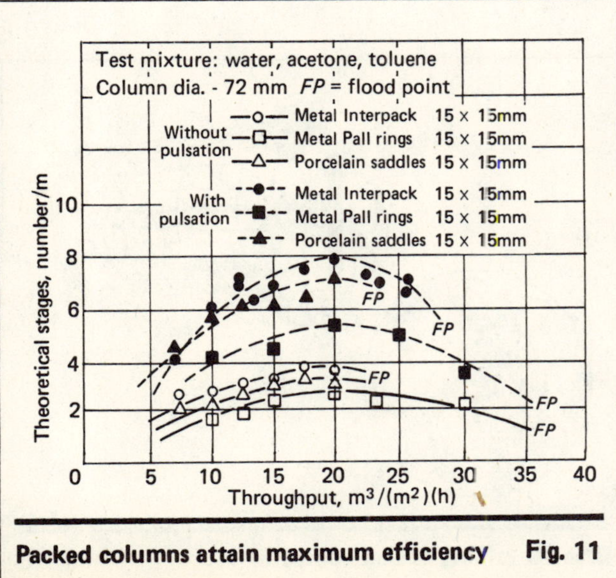

Packed columns attain maximum efficiency **Fig. 11**

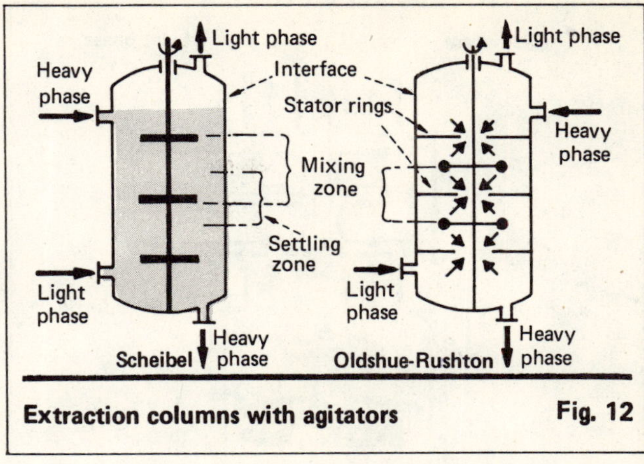

Extraction columns with agitators **Fig. 12**

Packed columns consist of a completely packed unit, without redistributors as used in distillation systems. Columns have a tendency to form streams of the dispersed phase that are no longer dispersed. The wettability of the packing by the continuous phase has a strong effect on efficiency. If the packing is wetted by the dispersed phase, coalescence will be increased and the mass-transfer area per unit of volume will be accordingly reduced.

The normal packing size for all column diameters is between 15 and 25 mm. At the present time, this size is still determined by scaleup calculations, since the same packing dimensions are used in the design experiments as are used for the industrial-size columns. An increase in the column diameter will decrease the efficiency per unit of height, due to increased backmixing.

The necessity of increasing the column height, however, will increase the probability of stream formation, thus further impairing efficiency.

Our measurements performed on a test system of water, acetone and toluene [*8*], using different types of packing, led to the conclusion that the maximum efficiency is approximately 2.5 to 3.5 theoretical stages per meter (Fig. 11) at a throughput per unit area of 20 $m^3/(m^2)(h)$, and an efficiency loss of just under 10% at 25% underload.

A marked improvement in efficiency, by a factor of two, can be achieved through pulsation of the liquid column. Raschig rings are unsuitable as a packing, since the droplets of the dispersed phase will easily adhere to them. These columns operate without problems due to their very simple design. They are especially suitable for extraction processes where the liquid properties do not change significantly in the extractor. Applications are found in the nuclear and petrochemical industries.

Column extractors with energy input

For columns having energy input, the mechanical energy will improve formation of new droplets and increase interfacial turbulence, resulting in greater efficiency. The required energy can be introduced via agitation or pulsation. Such columns represent a dependable method for achieving high mass transfer regardless of liquid properties but, on the other hand, they increase the susceptibility to trouble, as well as the cost of the apparatus.

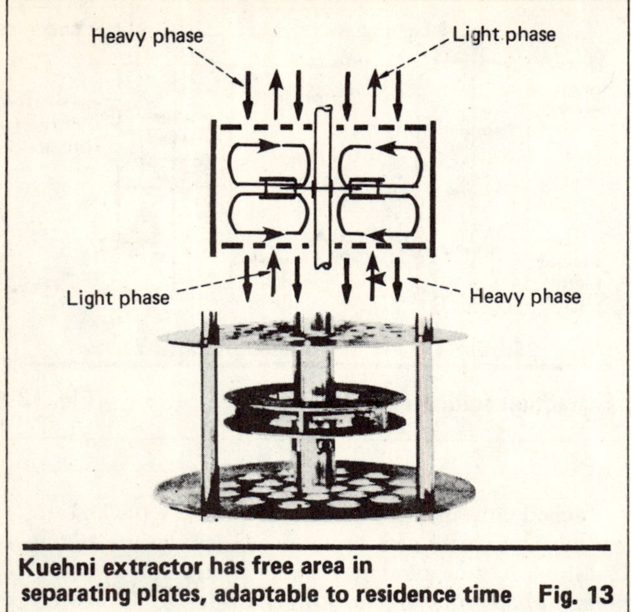

Kuehni extractor has free area in separating plates, adaptable to residence time **Fig. 13**

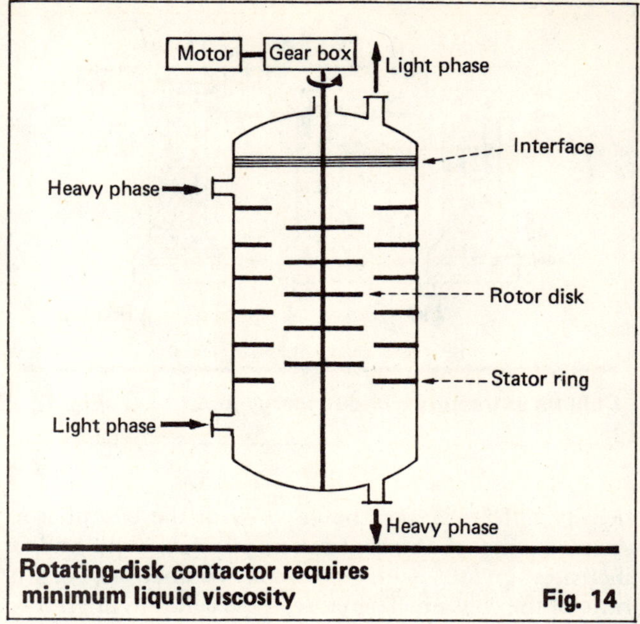

Rotating-disk contactor requires minimum liquid viscosity **Fig. 14**

Probably the oldest of columns having agitators is the Scheibel column [*2,3*] (Fig. 12). In this column, double-bladed agitators are mounted on a vertical shaft, at fixed intervals, in order to achieve phase mixing. Particle or wire-mesh packings are installed in the separation zones between the agitators to improve coalescence. These packings must be wetted by the dispersed phase and must be approximately one to three times the height of the agitating zones. The column operates on the mixer-settler principle. The good separation effect and the low backmixing result in efficiencies of approximately three to five theoretical stages per meter.

Disadvantages of this column include:

- Poor disassembly characteristics.
- Easy fouling of the separation zones.
- Need for several bearings on the shaft, thus increasing corrosion and wear.
- Poor efficiencies at column diameters over one m, as a result of a broad droplet spectrum caused by a circumferential velocity gradient between the inside and outside diameters of the agitator at low speeds.

The Scheibel column is rarely built today. Its deficiencies resulted in an advanced development, the Oldshue-Rushton column (Fig. 12), in which the packing zones are replaced by simple stator rings [*2,3*]. Residence time, throughput and efficiency can be controlled by the aperture size of these rings. However, backmixing is larger than for the Scheibel column, so that the efficiencies are lower. With narrow apertures, there can be achieved approximately 1 to 3 theoretical stages per meter, and with greater apertures only about 8 to 1. The configuration having larger apertures is less sensitive to fouling and easier to clean.

A column based on a very similar principle is the Kuehni extractor (Fig. 13), where the stator disks are made of perforated plates [*19*]. Centrifugal agitators are used for this system. The free cross-sectional area of the separating plates can be adapted to the desired conditions. Hence, this column can be adjusted to required residence times. As a result, it is especially suitable for

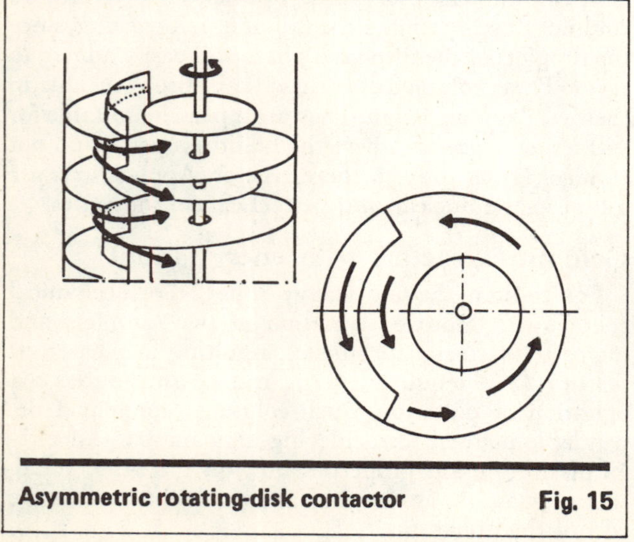

Asymmetric rotating-disk contactor **Fig. 15**

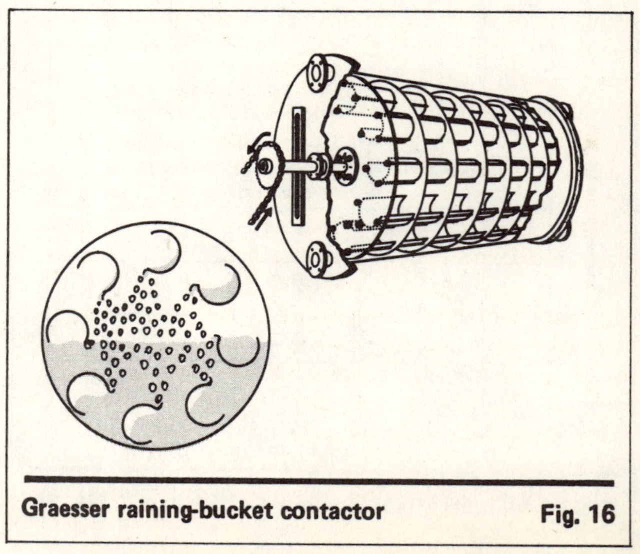

Graesser raining-bucket contactor **Fig. 16**

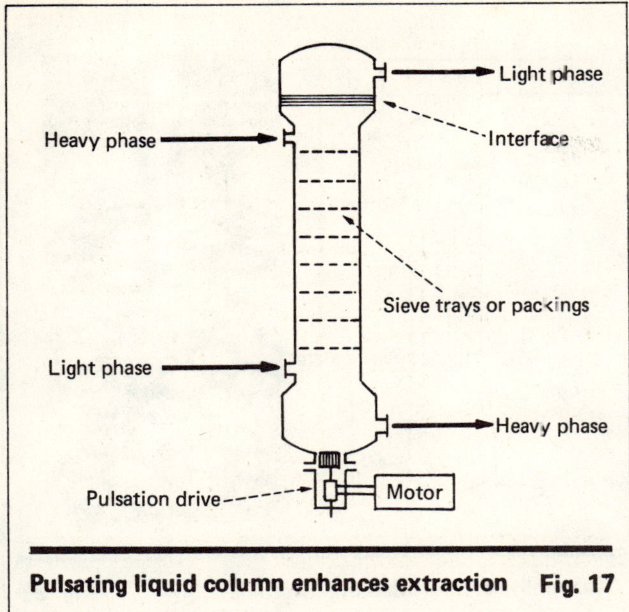

Pulsating liquid column enhances extraction Fig. 17

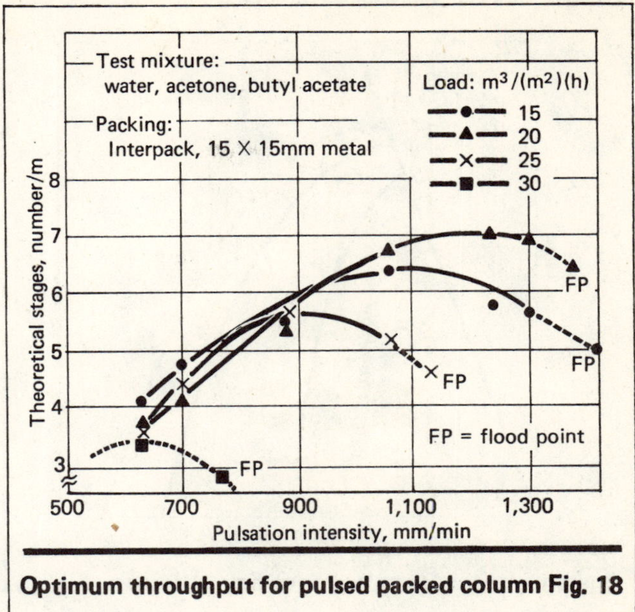

Optimum throughput for pulsed packed column Fig. 18

extractions combined with reactions and for extreme phase ratios. Throughput and efficiency depend greatly on the mixer speed and the free cross-sectional area of the separating plates. Up to ten theoretical stages per meter can be achieved.

For large diameters (over 2.5 m), three agitator shafts are used so as to achieve more-uniform circumferential velocities. The available design documentation is good. However, there are disadvantages in that the shaft requires many bearings, and the separating plates and agitators are difficult to install, requiring one manhole for each practical stage.

Probably the best-known agitated extractor is the RDC (rotating-disk contactor) (Fig. 14) [*9,10,11*]. In this system, horizontal disks are used as agitating elements, which are mounted on a centrally supported shaft. Offset against the agitator disks, stator rings whose aperture is greater than the agitator disk diameter are mounted at the column walls, simplifying installation.

Agitation of the dispersed phase by rotating disks requires shear forces that presuppose a certain minimum viscosity of the liquids. The throughput and efficiency are affected by the rotor speed. Thus, underloads down to 50% of nominal load can be processed without loss of efficiency by increasing the rotor speed. However, excessive speed will produce an excessively small droplet size that reduces efficiency as a result of increased backmixing.

Because of the RDC's inherently large backmixing effect, a maximum of approximately 0.5 to 1 theoretical stage per meter can be achieved. Therefore, this unit is preferred for high throughputs at low separating performance—for instance, in the petrochemical industry, for phenol extraction from waste waters or for solvent recovery. The RDC is ideal for processing dirty products.

For design and operational reasons, the device can be built with large diameters (6 to 8 m) but with only low height (10 to 12 m), since the shaft must be continuous without any coupling and since intermediate bearings cannot be used. Based on our experience, flooding would occur at these points. This extractor has been studied intensively by several operators so that comparatively good documentation is available.

An advanced development of RDC is the ARD (asymmetrical rotating-disk contactor), as shown in Fig. 15 [*19*]. In this system, the shaft, fitted with agitator disks, is located asymmetrically to the column axis. The agitating zones are separated by nonperforated plates.

Phase transport and separation take place in a settling zone that is separated from the extraction zone by a vertical wall. Again, the liquids must have a certain minimum viscosity.

As a result of the spatial subdivision of the column cross-section into agitating and settling zones at a ratio of approximately 2:1, the throughput per unit of area of the ARD is smaller than for the RDC. However, the settling space ensures better phase separation and, hence, better efficiencies of approximately one to three theoretical stages per meter. Again, the throughput and the separation efficiency can be influenced within certain limits by means of the rotor speed.

The ARD is probably the most expensive extractor based on the column principle. Compared with the RDC, it has the advantage that the rotor shaft can be made with couplings and can therefore be extended in existing systems. Intermediate shaft bearings are possible. Apparatus having large diameters (over 2.8 m) are fitted with two agitator shafts in order to achieve greater uniformity of the circumferential velocities of the rotating disks. Moreover, the shafts can be installed and removed without major effort through appropriate designs. This is another apparatus for which basic research is available, so that its design is quite feasible.

Another system to be classified as extractor with agitator is the Graesser contactor (Fig. 16) [*2,5,19*]. Unlike the other apparatus, it consists of a cylinder rotating about a horizontal axis. In order to improve mass transfer, the cylinder space is subdivided into chambers by means of vertical partitions, and dippers resembling rain gutters are used to disperse and mix the

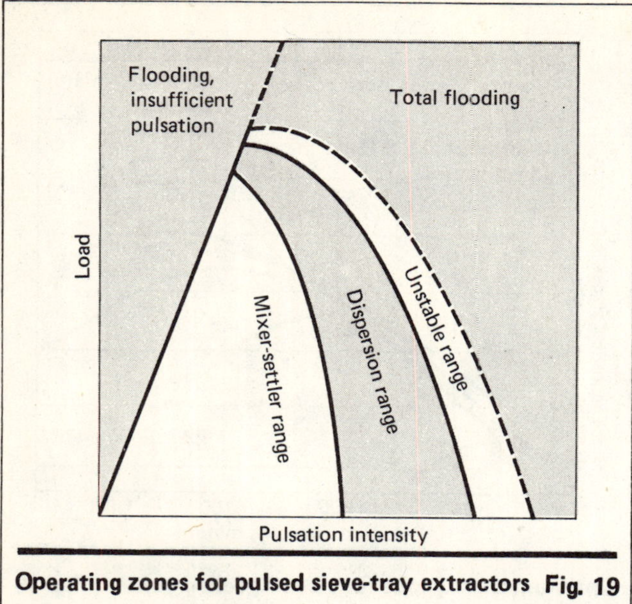

Operating zones for pulsed sieve-tray extractors Fig. 19

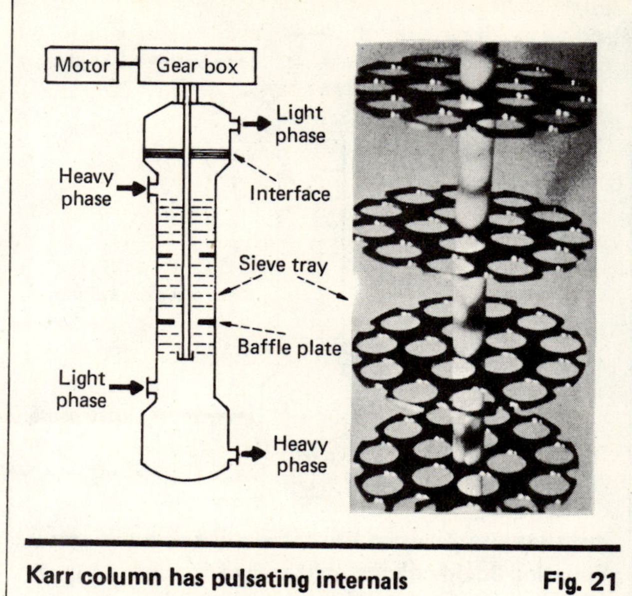

Karr column has pulsating internals Fig. 21

two phases. The liquids pass through the drum in a counterflow pattern, proceeding into the next chamber through gaps between the partition wall and the cylinder envelope. Length of a theoretical stage is approximately identical to the drum radius. Throughput is low, compared with other column extractors.

Pulsating packed columns

In columns having pulsation, the required energy can be introduced through pulsation either of the entire liquid column or of the built-in trays. The pulsating-liquid column has the advantage that the moving parts are located outside the column and, therefore, are more easily accessible. These columns are designed as packed or sieve-tray units (Fig. 17).

Extensions for phase settling are installed at the column head and column sump. The interface forms in

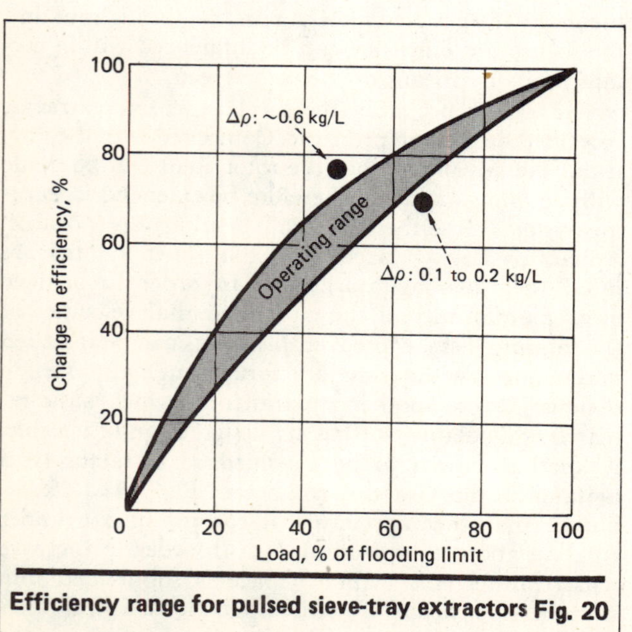

Efficiency range for pulsed sieve-tray extractors Fig. 20

these elements and may be on the top or the bottom, depending on the type of operation. Pulsation is generated from the column sump, using special-purpose pumps or compressed air.

The pulsation packed column [*12*] has been in operation for about ten years and is widely used throughout industry. However, it is a poor mixing column and the formation of new droplets is limited. But the pulsation prevents the formation of streams occurring, as in the simple packed columns.

As a result, the improved extraction effect represented in Fig. 11 is achieved. However, the pulsation will not permit the maximum to move toward greater loads, nor will it change the underload behavior of the column. In a test series using water - acetone - butyl acetate as the test system, an optimum throughput of 20 $m^3/(m^2)(h)$ was found (Fig. 18) [*13*]. If the diameters are increased, backmixing will occur at an increasing rate that is not susceptible to calculation, so that the efficiency can drop to approximately 0.5 theoretical stages per meter. As a result, the design of these columns is difficult. The pulsating packed column requires density differences of more than 0.08 g/cm^3 and is used for problems not imposing high requirements with respect to the number of stages (about 5 theoretical stages per meter) and the throughput. One disadvantage is the easy fouling.

In pulsed sieve-tray extractors (PSE), the entire column cross-section is occupied by the trays, so that both phases must pass through the holes, the lighter phase on the upward stroke and the heavier phase on the downward stroke. This will continuously create new interfaces, improving mass transfer. At high throughput per unit area of about 60 $m^3/(m^2)(h)$, the PSE will yield good stage efficiencies compared with other extractor types [*8,18*], provided that it is operated on the boundary between mixer-settler and dispersion mode (Fig. 19) [*14*]. This can be achieved at frequencies between 60 and 150 strokes/min and amplitudes under 10 mm.

A tray spacing of 100 mm has proved favorable. Closer tray spacing will result in lower throughput and design problems in industrial-scale columns. If the spacing is larger, the efficiency per unit of length will decrease.

The preferred hole diameter is 2 mm. This means that a maximum of 23% free cross-section area can be obtained. Greater aperture ratios are possible only with larger hole diameters. More-recent investigations have shown [15] that such increases are always worthwhile for systems having a low interfacial tension. At an intermediate interfacial tension, a reduction of the column diameter by means of using larger aperture ratios can be achieved only at the expense of greater height. Small aperture ratios are best for systems having high interfacial tensions.

The disadvantages of the PSE include its sensitivity to fouling, especially with sticky products, and a relatively small load range. In Fig. 20, the relative change or efficiency has been plotted as a function of the relative load change. Starting out from the design point of 80 to 90% of flooding load, the efficiency reduction with increasing underload is highly dependent on the density difference of the system. The smaller the density difference, the greater the efficiency reduction. Only part of this loss can be offset by changing the pulsation intensity so that the PSE operating at small density differences has a load range of about 50 to 95% of flood load. Major load variations, especially in the underload range, should be compensated for by artificially increasing the load, for instance by means of raffinate feedback. In industrial-scale columns, the safety surcharge to the column length used for the design has a significant effect on underload behavior.

Columns with pulsating internals include, for instance, the designs by Prochazka, Ziehl and Karr. From these types, the Karr column will be the most important one. It has sieve trays moving up and down, with free cross-section areas of 50 to 60%. As a result of these large aperture ratios, this column appears suitable for systems having intermediate to low interfacial tensions. In the experiments carried out by Karr [16], xylene - acetic acid - water was used as test mixture. However, with this system interface, turbulences must be expected, so that it is unsuitable for testing purposes. This column has been used predominantly in the U.S. and has been built up to 1 m dia. Further information is not available. Fig. 21 shows the Karr column schematically.

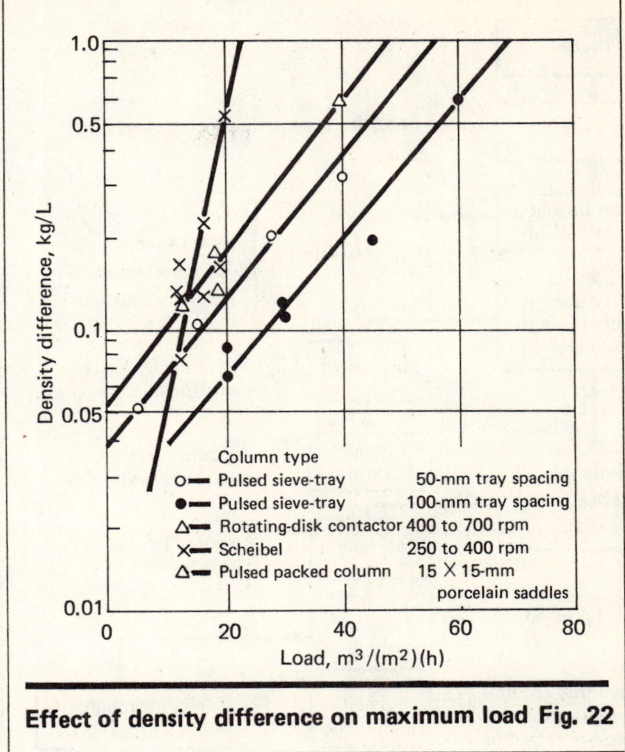

Effect of density difference on maximum load Fig. 22

Load comparison between column extractors

Any comparison and, therefore, evaluation of different extractor types on the basis of the data available today is almost impossible. The parameters of the materials being processed have highly different effects on the performance of the individual types of apparatus, so that classification requires determination of the charac-

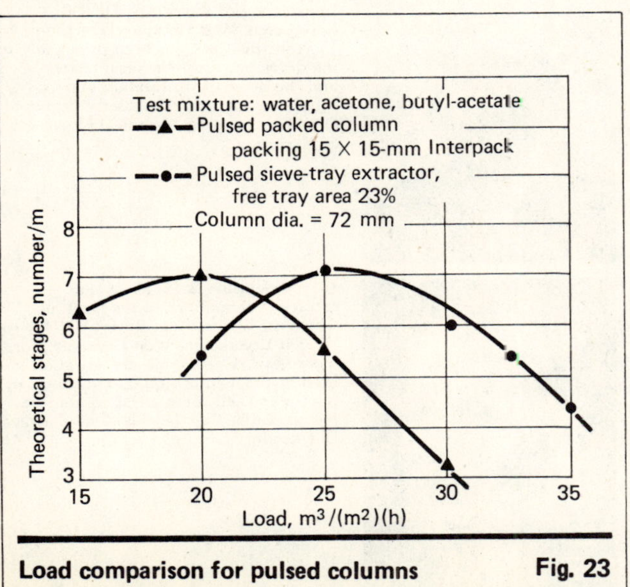

Load comparison for pulsed columns Fig. 23

Maximum loadings for extractors **Table I**

Column type	Maximum load $m^3/(m^2)(h)$	Maximum column dia., m	Maximum throughput, m^3/h
Graesser contactor	< 10	1.5	25
Scheibel	< 20	1.0	16
Asymmetric rotating-disk	≈ 20	4.0	250
Lurgi tower	≈ 30	8.0	1,500
Pulsed packed	≈ 40	2.0	120
Rotating-disk contactor	≈ 40	8.0	2,000
Kuehni	≈ 50	3.0	350
Pulsed sieve-tray extractor	≈ 60	3.0	420
Karr	80–100	1.5	<180

The foregoing data apply at:
High interfacial tension (30–40 dyne/cm)
Viscosity like water
Inlet ratio of the phases 1:1 parts by volume
Density difference ≈ 0.6 g/cm^3

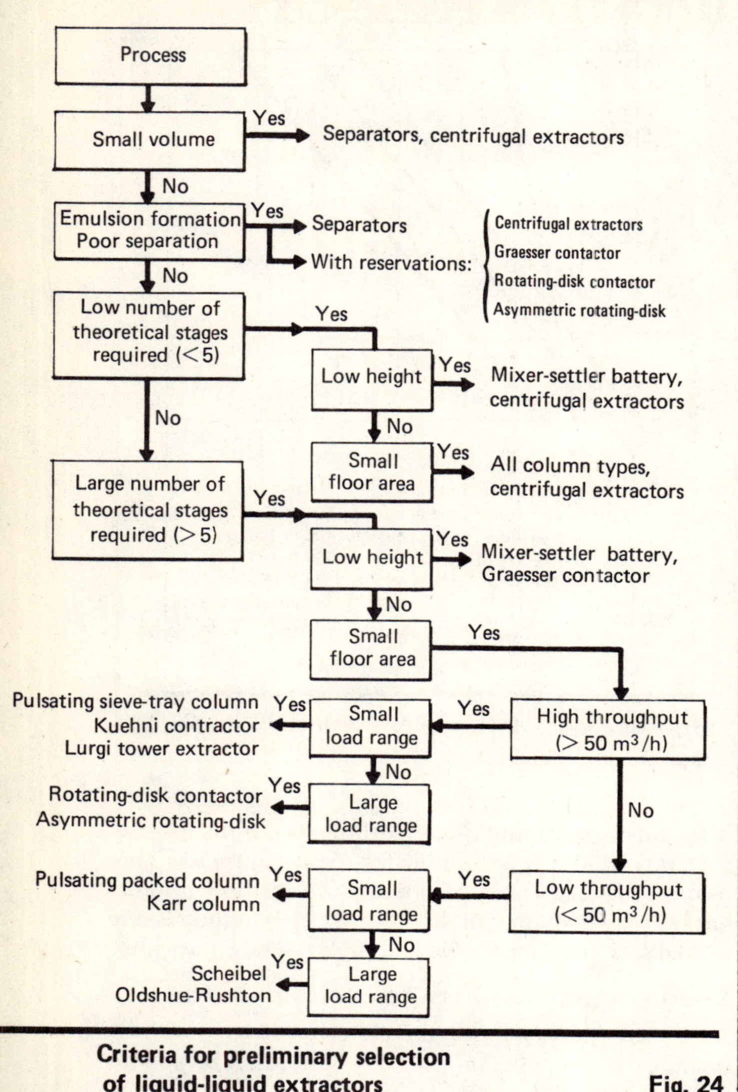

Criteria for preliminary selection of liquid-liquid extractors **Fig. 24**

teristic curves of the individual extractors. Experiments of this type are very expensive and time-consuming.

One possibility will be to compare the maximum load of the extractors, as shown in Table I. The effect of density difference on the load has been plotted in Fig. 22. Comparative measurements have been made for the pulsed packed column and PSE (Fig. 23) [*13*].

These tests have shown that both columns achieve nearly the same efficiency. However, the PSE is better by approximately 25% with respect to throughput and lowered pulsating-energy requirement. These values may be different for another test system.

Extractor selection

The establishment of selection criteria must be preceded by an evaluation and classification of the different extractor types. Since this is extremely difficult, it is impossible to present any clear-cut selection proposals.

Since most extractors react very differently to the various liquid properties and since these properties may be further modified by trace impurities for any separation problem (no matter how similar), final selection and design can be based only on experiments. Moreover, these experiments must be carried out with the original products. Even then, it is not uncommon for the experimental product to show a different behavior than the product processed later in an industrial system.

Based on available experimental data and operating experience, it is possible to make a certain preselection between apparatus for a new separation problem, which can then be used to implement specific design experiments. An attempt has been made to establish a selection procedure by means of a diagram (Fig. 24), where the most important parameters have been taken into consideration [*4,17*]. This scheme can be used as a rule-of-thumb only.

References

1. Perry, R. H. and Chilton, C. H., "Chemical Engineers' Handbook," 5th ed., McGraw-Hill, New York, 1974.
2. Mueller, E., "Fluessig-Fluessig Extraktion," Vol. 2, Verlag Chemie, Weinheim, Germany, 1972.
3. Treybal, R. E., "Liquid-Liquid Extraction," 2nd ed., McGraw-Hill, New York, 1963.
4. Hanson, C., "Neuere Fortschritte der Fluessig-Fluessig Extraktion," Saarlaender, Aarau, Switzerland, 1974.
5. Bailes, P. J., Hanson, C. and Hughes, M. A., *Chem. Eng.*, Jan. 19, 1976, pp. 86–100.
6. Mewes, D. and Kunkel, W., *Chem. Ing. Tech.*, Jan. 1977, p. 65.
7. Pilhofer, T. and Goedel, R., *Chem. Ing. Tech.*, May 1977, p. 431.
8. Schaefer, J. P., Thesis, Cologne Technical College, 1976/1977.
9. Marr, R., Moser, F. and Husung, G., *Chem. Ing. Tech.*, May 1974, p. 207.
10. Marr, R., Moser, F. and Husung, G., *Chem. Ing. Tech.*, Mar. 1977, pp. 203–212.
11. Moehring, D. and Weiss, S., *Chem. Tech. (Berlin)*, May 1977, pp. 262–265.
12. Simons, A., *Chem. Ing. Tech.*, May 1976, p. 487.
13. Straussfeld, E., Thesis, Cologne Technical College, 1977.
14. Brandt, H. W., Reissinger, K. and Schröter, J., *Verfahrenstechnik*, Aug. 1975, pp. 383–387.
15. Berger, R., Leukel, W. and Wolf, D., Paper read at the GVC (Gesellschaft fuer Verfahrenstechnik und Chemieingenieurwesen) symposium, Bad Kissingen, Germany, 1976; *Chem. Ing. Tech.*, July 1978, p. 544.
16. Karr, A. E. and Lo, T. C., *Chem. Eng. Prog.*, Nov. 1976, pp. 68–70.
17. Hanson, C., *Chem. Eng.*, Aug. 26, 1968, pp. 76–98.
18. Baecker, W., Thesis, Cologne Technical College, 1976.
19. Miscellaneous equipment manufacturers' literature.
20. Marr, R., *Chemie Technik*, Sept. 1976, pp. 373–376.
21. Weiss, S. and Wuerfel, R., *Chem. Tech. (Berlin)*, Aug. 1975, pp. 442–446.

The authors

Karl-Hermann Reissinger is group leader for liquid-liquid extraction and distillation in the Dept. of Chemical Engineering of Bayer AG, D-5090 Leverkusen, West Germany. For the past 15 years, his work has been principally in the development of chemical processes and the design of distillation and extraction columns. He has the degree of Dipl.-Ing. from the Technical University of Karlsruhe.

Jürgen Schröter is leader of the extraction laboratory in the Dept. of Chemical Engineering, Bayer AG, D-5090 Leverkusen, West Germany. For more than 20 years, his specialization has been in the field of liquid extraction for pilot-plant and commercial equipment. He has studied at the School of Chemical Engineering, Leipzig.

Multistage-leaching simulation

A material-balance model of multistage, multicomponent, countercurrent extraction and wash systems determines liquid-phase compositions or flowrates in each stage, as well as efficiencies.

J. C. Agarwal and *Ivan V. Klumpar,* Ledgemont Laboratory, Kennecott Copper Corp.

☐ Published computational methods for multistage, countercurrent leaching [*1* through *9*] are inadequate for the simulation of a general extraction and wash system for one or more of the following reasons:

- Complete extraction in the first stage is assumed (methods are derived for countercurrent decantations).
- Only the minimum number of stages is calculated.
- Some methods are limited to a countercurrent operation, i.e., multiple extraction (or wash-liquor inlets), and losses are not considered.
- Recycling is not included.
- Interaction between the liquid and solid phase is oversimplified—i.e., the same equilibrium is assumed in each stage and/or kinetics are neglected.
- Gas-liquid interactions are not accounted for.
- Residence-time calculations and equipment sizing are not included.

To overcome the above drawbacks, a flexible model for multistage, multicomponent, countercurrent (or partially crosscurrent) extraction and wash systems is proposed. The sequential combination of extraction and washing will be referred to as *leaching*. The model assumes variable solid-phase compositions, and solute buildup in recycled wash liquor. For each stage, liquid-phase compositions or flowrates—together with three different efficiencies—are calculated. Simple gas-liquid interactions can also be simulated. An extension of the model permits the incorporation of equilibrium and kinetic data for solid-liquid-gas systems, for residence-time estimation, and for equipment design.

The structure of the model [*10*] is described first, followed by the various optional features of the system. Then, the equilibrium and kinetic extension is outlined and, finally, the mathematics are provided.

Countercurrent leach system

Consider first the simple, multistage countercurrent system of mixer-settlers in Fig. 1. For the moment,

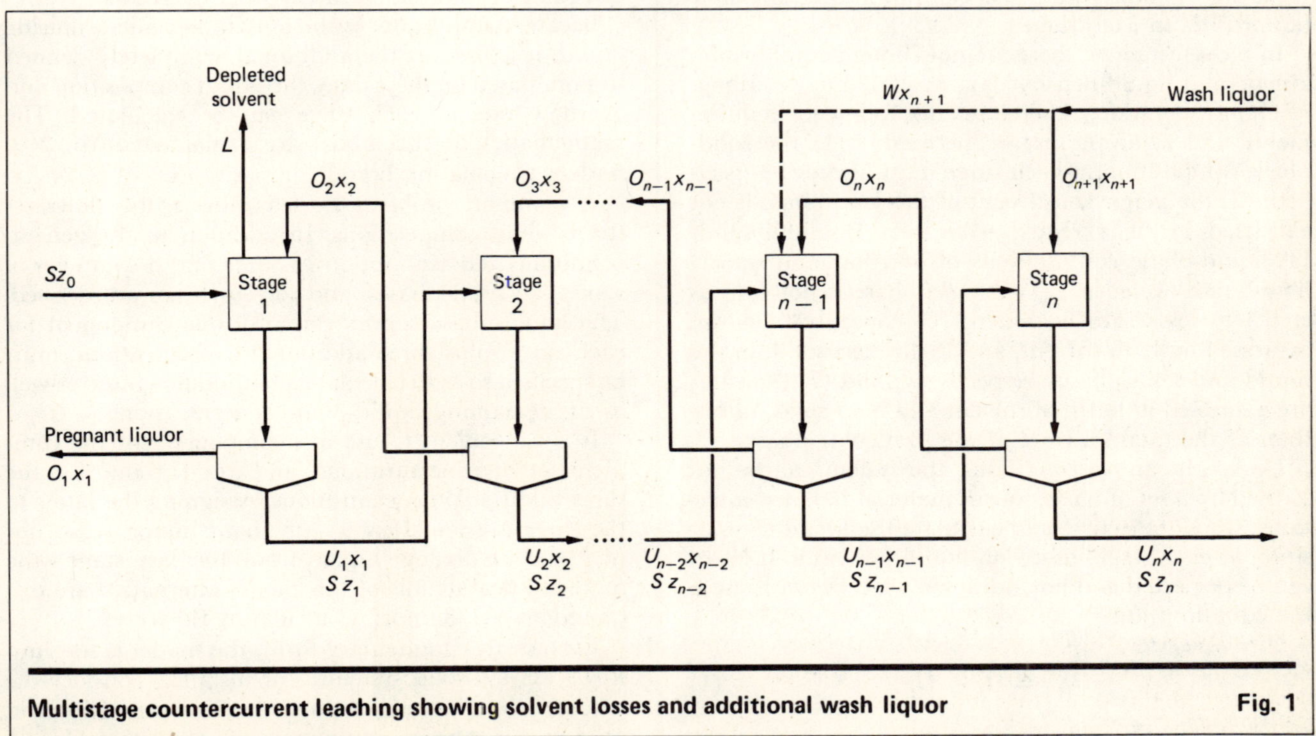

Multistage countercurrent leaching showing solvent losses and additional wash liquor **Fig. 1**

Originally published May 24, 1976

Nomenclature

A,B	Reactants
a,b	Stoichiometric coefficients, dimensionless
C	Liquid-phase concentration, (lb/moles)/ft^3 solution
D	Diffusivity, ft^2/h
E,F,G	Reactants and reaction products
e	Extraction efficiency, dimensionless
f,g	Stoichiometric coefficients, dimensionless
I	Area of gas-liquid interface, ft^2
j	Stage counter, dimensionless
k	Mass-transfer coefficient, (lb-moles)/[(h)(ft^2)(psi)], or ft/h
k_r	Reaction-rate constant, ft/h
L	Solvent losses, lb/h
l	Leach efficiency, dimensionless
M	Molecular weight, lb/(lb-mole)
m	Constant, dimensionless
N	Mass, lb-moles
n	Number of stages
O	Solvent overflow, lb/h
o	Fraction of pregnant liquor recycled
P	Partial pressure, psia
p	Number of particles
q	Constant, dimensionless
R	Rate, (lb-moles)/h
r	Radius of particle, ft
S	Inert-solid flowrate, lb/h
T	Residence time, h
t	Time, h
U	Solvent underflow, lb/h
u	Fraction of last-stage underflow recycled
W	Additional solvent wash-liquor, lb/h
w	Washing efficiency, dimensionless
x	Liquid-phase concentration, lb/lb solvent
Y	Makeup solvent, lb/h
z	Solid-phase concentration, lb/lb inert solid
π	3.1416
ρ	Density, lb/ft^3

Subscripts and superscripts

A,B	Components A and B
b	Bulk liquid
c	Core surface
E,F,G	Components E, F and G
g	Bulk gas
e	Equivalent
i	Interface
j	Stage counter
n	Last stage
p	Particle surface
r	Reaction
*	Equilibrium
0,1,2 . . .	Stage number

disregard the broken lines marked L and Wx_{n+1}. Because wash liquor is fed to the last stage, no recycling is involved. A typical problem would be: Given the underflow liquor-rate, solid throughout and solid composition in each stage, as well as the number of stages, what are the liquid-phase concentrations and overflow liquor-rates in each stage?

In most instances, there are not enough equilibrium, kinetic and liquid-density data available to describe a leaching operation. Therefore, the following requirements and assumptions are necessary: (1) the solid-phase composition in each stage must always be specified; (2) the major constituent of the solid phase is not extracted, i.e., it is treated as an inert, insoluble solid; (3) liquid-phase compositions of overflow and underflow liquor are equal; (4) the gas phase is not considered; (5) losses are neglected; (6) concentrations are expressed in lb/lb solvent, and lb/lb inert solid, in the liquid and solid phases, respectively; and (7) flowrates are expressed in terms of solvent and inert solid, rather than as the total flowrate of the particular phase.

Under the above conditions, the system can be described by a set of basic, linear, material-balance equations, whose objective is to calculate the liquid composition at each stage, based on liquid flowrates. If losses can be neglected and no additional wash liquor is used, the equations are:

Overall balance:

$$O_{j+1} = O_1 + U_j \quad (1)$$

Component balance:

$$O_{j+1}x_{j+1} + Sz_0 = O_1x_1 + U_jx_j + Sz_j \quad (2)$$

An analysis of the degrees of freedom indicates that the following variables must be specified: the flowrates of overflow or underflow solvent from each stage, and the flowrate and composition of one additional liquid stream. It has been arbitrarily decided to specify the flowrates of underflows rather than overflows.

Because wash liquor is the only independent outside liquid, it is used as the additional, completely defined stream. Based on these data, the liquid composition and overflow rate at each stage can be calculated. The mathematics of the model are explained on p. 260, under "Calculating liquid compositions."

A common problem is determining the flowrates from liquid compositions. In addition to the general conditions and assumptions, assume that the number of stages, all solid phases, and solvent losses are defined. The liquid-phase concentration of one component for each stage, plus three additional concentrations, must be specified so as to calculate all liquid flowrates, as well as the remaining liquid-phase concentrations.

It is convenient to use one component for the complete set of concentrations, and another one for the three additional concentrations. Assigning the latter to the three outside streams—pregnant liquor, wash liquor, and underflow liquor from the last stage—the mathematical details of this model alternative are discussed on p. 138 under "Calculating flowrates."

Even in its rudimentary form, the model is not limited to liquid-solid systems. For instance, consider the oxidative leaching of a reduced lateritic ore with air in an aqueous solution of ammonium carbonate [*11*]. A decreasing carbon dioxide content may be specified in the solid phase, and its concentration in the liquid phase calculated, assuming that no transfer to or from the gas phase occurs.

On the other hand, ammonia or oxygen content in the liquid phase is immaterial and need not be specified because the system is defined independently of NH_3 and O_2. This is possible because flowrates and concentrations are based on "carriers" (inert solid and solvent) rather than on total phases. Thus, the model is consistent despite possible variations in the total weights of the solid and liquid phases, due to the consumption of oxygen or absorption of ammonia from the gas phase.

Optional features

The model can account for losses at any point in the system and/or feed of additional wash (or extraction) liquor to any stage. The broken lines in Fig. 1 indicate a simple case based on the following assumptions: (1) solvent losses due to hydration of solids and evaporation are grouped together and are assumed to occur in the first stage; and (2) an additional wash-liquor stream is introduced to the stage preceding the last one.

The basic material-balance equations for this case appear on p. 261 under "Losses and additional wash liquor." Similar relationships may be derived for multiple losses and multiple wash (or extraction) liquor additions. Based on such equations, techniques to calculate either liquid compositions or flowrates can be developed along the lines indicated under "Calculating liquid compositions" and "Calculating flowrates," both discussed on p. 260.

Another optional feature of the model is the material balance of recycled material (Fig. 2). A solute might build up in the leaching system as a result of recycling a solution containing the solute. For example, the recycle might originate from the pregnant liquor—after the valuable components have been recovered—or from the last-stage underflow, when the solids have been separated. If water effluents from a metals-recovery section, and/or tailing-pond water from the separation section, are returned to the leaching section in a hydrometallurgical plant, impurities may gradually increase in the wash liquor.

For recycling calculations, additional input needed is (1) the fraction of solvent recycled from the pregnant liquor, and (2) concentrations of the solute components in the makeup stream.

By using the technique discussed under "Recycling," p. 261, the amount of makeup stream and the steady-state concentrations of the solute components in the wash liquor are determined.

The user of the model also has the option of calculating three types of efficiencies for each component: (1) total leach efficiency for the entire system; (2) wash efficiency for all washing stages; and (3) individual extraction efficiencies for each extraction stage.

Efficiencies are defined on p. 261. Leach efficiency equals the wash efficiency multiplied by the last-stage extraction value.

If the solid-phase composition reaches a constant level after several extraction stages, the required number of stages can be determined by using the model. As a rule, this number is higher than the number of extraction stages, due to the additional washing stages. The minimum number of stages can be computed by choosing an arbitrarily high number and running the calculations described in the previous section. This minimum number corresponds to the point where the wash liquor concentrations are reached by all components in the overflow.

Equilibrium and kinetic calculations

In the previous sections, it was assumed that the solid composition in each stage is specified. If the physico-chemical equilibrium between the solid and liquid phases can be expressed as an algebraic equation, the solid compositions can be computed for each theoretical stage. For example, it has been found that for a particular case of oxidative leaching of a copper ore, the metal content in the underflow solid is approximately a linear function of the metal concentration in the overflow liquid (see "Physico-chemical equilibria," p. 261.).

Kinetic calculations are performed to determine the residence time in each theoretical stage, i.e., the time required to reach physico-chemical equilibrium. This is the basic parameter for equipment design. The kinetics of leaching is a complex phenomenon, involving homogeneous and heterogeneous chemical reactions, as well as mass transfer between solid, liquid and, in some cases, gas phases [*12* through *16*]. To extend our model, rate equations are used, some of which are based on the shrinking-core concept. The basic assumptions are listed under "Kinetic calculations," p. 261.

As an example, assume that a gaseous Component A dissolves in the liquid phase and reacts with Solute B. The liquid-phase reaction Product E reacts, in turn, with solid Component F to form the final Product G, which again dissolves in the liquid (Fig. 3). The process can be imagined as consisting of the following steps: (1) transfer of A from the bulk gas to the gas-liquid interface; (2) transfer of A from the interface through a liquid-reaction zone, where A reacts with B to form E; transfer of E from the bulk liquid to the solid surface; (4) diffusion of E through the equilibrium zone of a solid particle to its unreacted core, which contains Component F (equilibrium zone and core are explained below); reaction of E and F at the surface of the core to form Product G; diffusion of G through the equilibrium zone; and (7) transfer of G to the liquid bulk.

Steps 1, 3 and 7 involve mass transfer in fluids, while steps 4 and 6 represent diffusion through a porous solid. Let us now examine the more complex steps 2 and 5.

The reaction between A and B starts at the interface at a relatively high rate [*13*]. As the unreacted portion of A moves away from the interface, both the mass-transfer rate and the reaction rate slow down, because of the decreasing concentration of A, until equilibrium is reached. The net rate of Step 2 is a function of a mass-transfer coefficient, a reaction-rate constant and component concentrations.

Each of the seven steps represents a resistance to mass transfer or to chemical conversion (see p. 261, "Kinetic calculations"). The residence time could be calculated as a function of all resistances in series, but the mathematics are so complex that doing this becomes unmanageable. Also, the individual equilibrium, the mass

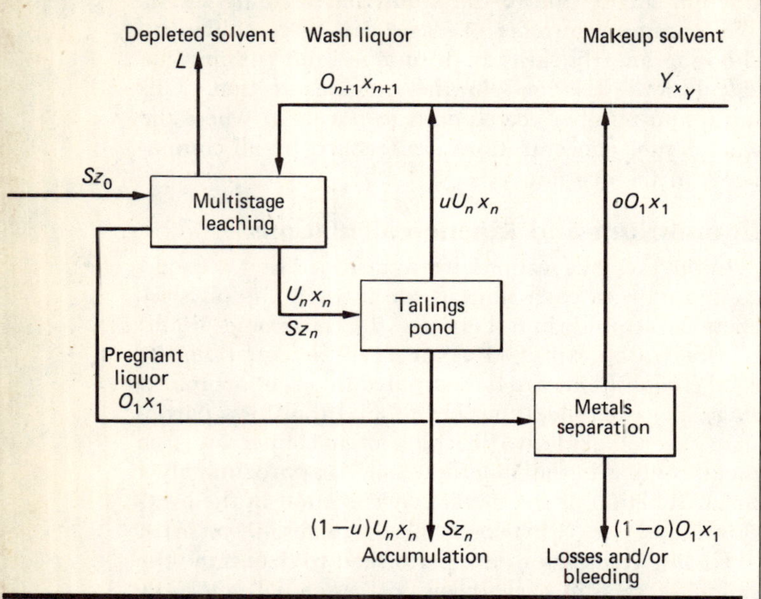

Recycling wash liquor can build up impurities in leaching system **Fig. 2**

transfer and the reaction data that are required for such calculations are difficult to obtain.

In most cases, the equations can be simplified and the resulting expressions directly related to the mass balance and equilibrium relationships (see p. 262, "Simplified kinetic calculations"). Thus, a system of equations is available that directly leads from a set of flowrates, or concentrations, and physico-chemical data to the residence times in each leaching stage.

Applications

The leaching model discussed here has been computerized, and its program written in FORTRAN IV. The typing of input data is facilitated by an interactive loading routine. The program can handle up to 12 components and 20 stages, with the results typed as two tables. The first is a material balance showing flowrates of overflow, underflow, and wash-liquor solvent, as well as the solid- and liquid-phase compositions for each stage. This table also indicates solid throughput, solvent losses, and recycling information, if any. The second table presents efficiencies for all components.

One application of the model could be copper-concentrate leaching, with a leach liquor recycle after tailings separation and copper recovery. If it is desired to calculate liquid-phase compositions in each stage, efficiencies, requirements for makeup water, and the number of stages, then copper concentrations should be included, as well as chlorides buildup.

The following variables must be specified: solid throughput, liquid flowrates, residual concentrations of copper and chlorides in the solid phase, chloride content in makeup water, losses, and the amount of recycle. Input data would look like the following:

Inert solids, lb	1,500
Overflow H_2O, lb	10,000
Underflow H_2O, lb	4,000
Lb Cu/1,000 lb inert solids:	
In feed	90
After Stage 1	30
After Stage 2	10
After Stages 3, 4, 5, etc.	5
Lb Cl/1,000 lb makeup water	0.1
Lb Cl/1,000 lb inert solids:	
In feed	10
After stages 1, 2, 3, etc.	1
Losses:	
H_2O, % leach liquor	5
Lb Cu/1,000 lb underflow	1.5
Recycle from pregnant liquor, %	80
Recycle from last-stage underflow, %	90

An inspection of the output data would show that washing continues after the extraction has been completed in the third stage. If, at Stage 7, the copper content drops to the permissible value of 1.5 lb/1,000 lb water, seven stages would be required. The buildup of chlorides could be surprisingly high, compared with their concentration in the makeup water. Makeup requirements could be 2,000 tons/d of water.

For the sake of simplicity, the above example shows specified solid compositions in each extraction stage. However, the model can be expanded to include equilibrium and kinetic relationships, if enough data on the interaction of the solid, liquid and gas phases are available. Based solely on boundary conditions and a set of flowrates (or compositions), the expanded model predicts solid and liquid compositions (or flowrates), and equipment sizes.

Calculating liquid compositions

If five stages are to be considered, all overflows are determined from Eq. (1), starting at the last stage and going backwards. For the last stage:

$$O_6 + U_4 = O_5 + U_5 \tag{1a}$$

Because leach Liquor O_6 and Underflows U_4 and U_5 are specified, Overflow O_5 can be calculated. Overflow O_4 (from the fourth stage) can then be computed by using an analogous equation for the second-to-the-last stage, and so on.

Next, the concentration of each component in every stage is determined by a set of equations, one per stage (Eq. 2). For five stages, there are five equations to compute $(x_1, x_2 \ldots x_5)$ for a particular component. For instance, for Stage 5:

$$O_6x_6 + U_4x_4 + Sz_4 = O_5x_5 + Sz_5 \tag{2a}$$

Overflow O_2 was calculated from Eq. (1a), but the following parameters are specified: solid throughput, S; solid concentration, z; underflows, U; leach-solution flowrate, O_6; and concentration, x_6. The only unknowns to be determined are the remaining liquid concentrations, x.

Calculating flowrates

The flowrates of pregnant liquor and wash liquor are determined by eliminating Underflow U from the modified material balances—Eq. (1) and (2)—around stages 1 through 5 for either component, and by combining

$$U_5 = O_6 - O_1 - L \tag{1b}$$

and

$$O_6x_6' + Sz_0' = O_1x_1' + U_5x_5' + Sz_5' \quad (2b)$$

where the prime mark denotes the specified concentrations of one of the selected components. Similarly, Eq. (1b) and (2b) are combined for the other component, which provides a set of two equations from which the two unknowns, O_1 and O_6, are calculated.

Next, flowrates for the remaining stages are computed successively in the same way, by using the component for which the complete set of concentrations has been specified. Only one combined equation is needed for each stage because the pregnant-liquor flowrate, O_1, has been calculated. Finally, the remaining liquid-phase concentrations are determined, as discussed in the foregoing section.

Losses and additional wash liquor

The two basic material-balance equations (1) and (2) can be modified as follows for stages 1 through $n-1$, assuming that solvent losses from the first stage and an additional wash-liquor stream, W, are going to the second-from-the-last stage:

Overall balance:

$$O_n + W = O_1 + U_{n-1} + L \quad (3)$$

Component balance:

$$O_nx_n + Wx_{n+1} + Sz_0 = O_1x_1 + U_{n-1}x_{n-1} + Sz_{n-1} \quad (4)$$

Recycling

The amount of makeup solvent, Y, is calculated from the overall material balance of the streams that constitute the wash liquor (see Fig. 2):

$$Y = O_{n+1} - oO_1 - uU_n \quad (5)$$

where o and u are the fraction of solvent recycled from the pregnant liquor, and the last-stage underflow liquor, respectively.

Based on the corresponding component balance, the solute buildup is then established. For losses from the first stage and a single-component solute:

$$x_{n+1}O_{n+1} = ox_1O_1 + ux_nU_n + x_yY \quad (6)$$

where x_y is the solute concentration in the makeup solvent.

As the computations are repeated, the solute concentration, x_6, in the wash liquor is increased until a constant value is reached.

Efficiencies

Assume that an additional wash-liquor stream is introduced to Stage $n-1$. Then, the total leach efficiency, l, of each component is the fraction of the original solid phase transferred into the pregnant liquor:

$$l = \frac{x_1O_1 - x_{n+1}(O_{n+1} + W)}{z_0S} \quad (7)$$

The wash efficiency, w, of each component is the fraction of the amount removed from the solid phase transferred into the pregnant liquor:

$$w = \frac{x_1O_1 - x_{n+1}(O_{n+1} + W)}{z_0S - z_nS} \quad (8)$$

The individual extraction efficiency, e, of each component at each stage is the fraction of the original solid phase removed from the solid in the particular stage and in the previous stages:

$$e = (z_0 - z_j)/z_0 \quad (9)$$

Physico-chemical equilibria

Assume the content of each component in the underflow solid, z, is a linear function of its concentration in the liquid phase:

$$z = mx + q \quad (10)$$

For each stage and each component, Eq. (2) can be rewritten as:

$$O_{j+1}x_{j+1} + Sz_0 = O_1x_1 + (U_j + Sm)x_j + Sq \quad (11)$$

The constants m and q can be determined, for example, by measuring overflow concentrations x_1, x_2 in Stages 1 and 2, respectively, and solving a pair of equations similar to Eq. (11) for m and q. Thereafter, liquid compositions or flowrates can be calculated as shown on p. 260. For instance, three equations similar to Eq. (11) can be written for each component in a five-stage system to compute x_3, x_4 and x_5.

General kinetic calculations

For kinetic calculations, these assumptions must be made:

1. In each stage, the process is in a pseudo steady-state condition. That is, the rates of all steps that constitute the process are equal.
2. The rate equations—developed for first-order irreversible reactions in both the liquid and solid phase—are also approximately valid for other cases.
3. The liquid phase is well mixed.
4. The solid phase consists of particles whose geometry can be reduced to uniform spheres by using mean particle-sizes and shape factors. (The equations presented in this article apply only to uniform spheres.)
5. The temperature of the particles and the adjacent liquid is uniform.
6. The density of the equilibrium zone equals the density of solid-phase-reaction Product G.

The process shown in Fig. 3 can be described in terms of the following realtionships:

Stoichiometric equations:

Liquid-phase reaction, $aA + bB = E$ (12)

Solid-phase reaction, $E + fF = gG$ (13)

Rate equations:

Definitions of process rate,

$$R = -a(dN_A)/dt = -b(dN_B)/dt = +(dN_E)/dt \quad (14)$$

$$R = -(dN_E)/dt = -f(dN_F)/dt = g(dN_G)/dt \quad (14)$$

Step 1, $R = ak_AI(P_{Ai} - P_{Ag})$ (15)

Step 2, $R = ak_eI(C_{Ai} - C_{Ab})$ (16)

The equivalent mass-transfer coefficient, k_e [13], is an empirical factor that covers both mass-transfer and chemical reaction. The coefficient may be expressed as a function of diffusivity, reaction-rate constant, and the width of the liquid-reaction zone, for a first-order reaction. For reactions of higher order, the function would also include the concentration of B and the order of reaction.

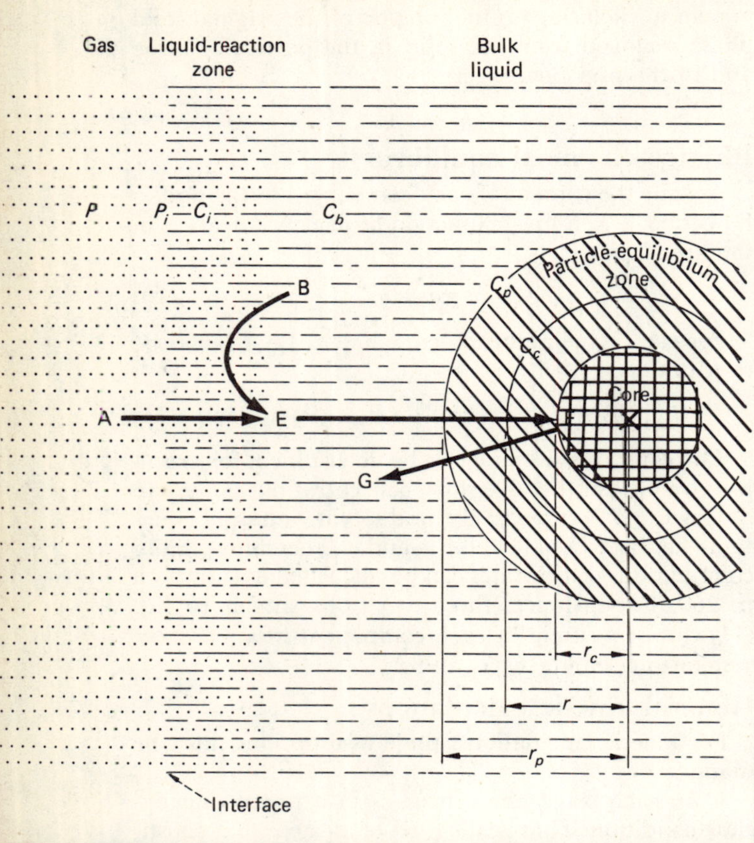

Leaching-process steps for kinetic calculations **Fig. 3**

$$\text{Step 3, } T = 4\pi r_p^2 p k_E(C_{Eb} - C_{Ep}) \tag{17}$$

$$\text{Step 4, } R = 4\pi r_c^2 p D_E(dC_{Eb}/dr)_c \tag{18}$$

where subscript c means differentiation is at $r = r_c$.

$$\text{Step 5, } R = 4\pi r_c^2 p k_r C_{Ec} \tag{19}$$

(See derivation of Eq. (17), (18), (19) in Ref. 12.)

$$\text{Step 6, } R = 4\pi r_c^2 p g D_G(dC_{Gb}/dr)_c \tag{20}$$

$$\text{Step 7, } R = 4\pi r_p^2 p g k_G(C_{Gp} - C_{Gb}) \tag{21}$$

Simplified kinetic calculations

Under certain conditions, the relationships of the preceding section can be simplified:

1. If no liquid-phase reaction occurs, k_e in Eq. (16) becomes a regular liquid-mass-transfer coefficient for Component A. Component E is equal to A or B, depending on which factors control the mass transfer and reaction rate in Steps 3, 4 and 5.

2. If no gas is involved in the process, or if the gas phase can be neglected, Steps 1 and 2 and Component A can be eliminated. Component E is equal to B. In a multicomponent system, E is again assigned to the solute component, which controls Steps 3, 4 and 5. If theoretical stages are considered, the liquid-phase concentration, x, in lb E/lb solvent, represents the equilibrium concentration that can be converted to the concentration, C_E, in lb-moles of E/ft³ solution as follows:

$$C_E = x\rho_b/M_E(1 + x) \tag{22}$$

In this manner, the kinetic equations can be directly related to the mass balance and equilibrium expressions discussed on pp. 258, and 260 through 261.

3. If the rate of the process is controlled by one step, relatively simple expressions for residence time can be derived [12]:

With the equilibrium zone controlling, the diffusion of E (Step 4) becomes:

$$T = \rho_F r_p^2/(6D_E f M_F C_{Eb}) \tag{23}$$

With the solid-phase reaction controlling (Step 5):

$$T = \rho_F r_p/(f M_F k_r C_{Eb}) \tag{24}$$

References

1. Boyadzhiev, L. A., *Dokl. Bolg. Akad. Nauk* (Sofia), Vol. 26, No. 3, p. 399, 1973.
2. Plachco, F. P., and Lago, M. E., *Chem. Eng. Sci.*, Vol. 28, No. 3, p. 897, 1973.
3. Plachco, F. P., and Lago, M. E., *Can. J. Chem. Eng.*, Vol. 50, No. 5, p. 611, 1972.
4. Kozlov, L. M., et al., *Zh. Prikl. Khim.* (Leningrad), Vol. 45, No. 9, p. 1993, 1972.
5. Hsieh, B. C. B., Paper 72-B-311, Soc. of Mining Engineers, Birmingham meeting, Oct. 1972.
6. Bruk, O. L., and Volodarskii, I. Kh., *Khim. Prom.* (Moscow), Vol. 47, No. 9, p. 704, 1971.
7. Chen, N. H., *Chem. Eng.*, Vol. 74, Jan. 2, 1967, p. 93.
8. Chen, N. H., *Chem. Eng.*, Vol. 71, Nov. 23, 1964, p. 125.
9. "Chemical Engineers' Handbook," 4th ed., Perry, J. H., ed., 17–2, 19–53, McGraw-Hill, New York (1963).
10. Klumpar, I. V., Paper A73-22, The Metallurgical Soc. of American Institute of Mining, Metallurgical and Petroleum Engineers, Chicago meeting, Mar. 1973.
11. *Eng. and Mining J.*, Vol. 158, No. 9, p. 82, 1957.
12. Smith, J. M., "Chemical Engineering Kinetics," 2nd ed., pp. 573–582, McGraw-Hill, New York (1970).
13. Hobler, T., "Mass Transfer and Absorbers," pp. 307–330, Pergamon Press, Oxford, England (1970).
14. Sato, S., and Sugita, J., *Kagaku* (Tokyo), Vol. 37, No. 1, p. 64, 1973.
15. Skrivanek, J., and Krzivsky, Z., *Zh. Prikl. Chim.* (Leningrad), Vol. 42, No. 2, p. 348, 1969.
16. Bretsznajder, S., and Piskorski, J., *Chem Stosow.* (Warsaw), Ser. B 3, No. 3, p. 275, 1966.

The authors

J. C. Agarwal is Director of Development at Ledgemont Laboratory, Kennecott Copper Corp., 128 Spring St., Lexington, MA 02173. Before, he was Chief, Process Analysis Div., U.S. Steel Corp. He holds M.S. and Ph.D. degrees in chemical engineering from Polytechnic Institute of New York. He has written over 80 articles on metals and minerals technology, and holds some 16 patents. He is a founding member of the American Assn. of Cost Engineers, a Fellow of AIChE and a member of the American Institute of Mining, Metallurgical & Petroleum Engineers.

Ivan V. Klumpar is a group leader at Ledgemont Laboratory, Kennecott Copper Corp., 128 Spring St., Lexington, MA 02173, where he does preliminary engineering and economic evaluations of metallurgical and related chemical processes. Before, he was with Diamond Shamrock Corp. He has M.S. and Ph.D. degrees in chemical engineering from Prague's Technical University. He is a member of AIChE, American Chemical Soc. and the American Assn. of Cost Engineers.

Section 6 Adsorption

Selecting and specifying activated-carbon-adsorption systems

The type of polymer and adsorbent treatment for removing hazardous and toxic chemicals from waste streams should be tailored to the specific waste through a series of laboratory and pilot tests. Here is a summary of the choices available and the tests necessary to make them.

F. E. Bernardin, Jr., Calgon Corp.

☐ Adsorption on granular activated carbon is one of the best commercially proven methods for removing toxic organic chemicals from waste streams. It has demonstrated, in full-scale plants, a strong affinity for some of the best known toxic materials, such as DDT, PCBs, Dieldrin, Aldrin and many others. Furthermore, the technology and equipment for applying activated carbon to such a service are well proven and available.

However, a given application to a specific wastewater still requires careful examination. An engineer must evaluate the objective of the treatment system and peripheral problems that may affect that system. Many variables affect the optimization of a granular activated-carbon system (see box: "Removing mixed toxic chemicals").

The components of an adsorption system, in order of increasing process-engineering involvement, consist of: (1) carbon-handling and storage facilities, (2) reactivation facilities, and (3) the adsorption equipment.

The carbon handling and storage is an essential part of any adsorption system. Granular activated carbon is shipped either dry or wet; it is transported and stored on-site in slurry form, since hydraulic transportation of carbon slurries is clean, rapid and economical. Such slurries can be handled by water-jet eductors, centrifugal pumps, and pneumatic or hydraulic blowcases.

Similarly, reactivation in a furnace is an essential part of economical carbon adsorption systems. Both multiple-hearth and rotary-kiln furnaces are used, depending on the exhaustion rate, economics, space limitations, flexibility requirements, and plant preference.

The furnace size is dictated by the relative ease of activation, as well as by the amount of carbon to be reactivated and the desired excess capacity. The excess

Removing mixed toxic chemicals

This chemical plant manufactures phenol, sulfuric acid, formaldehyde, pentaerythritol, sodium sulfite, sodium sulfate, and a number of synthetic resins and plastics. Its effluent, which averaged about 15 million gal/d, contained approximately 16,000 lb of BOD, 27,000 lb of COD, and 1,500 lb of phenol.

A target 90% reduction of these effluents was established in an agreement with the state regulatory agency; and a decision was made to install a clarification/adsorption system at a capital cost of about $2 million.

Prior to installing this system, the plant segregated its waste streams to reduce the flow of effluent for treatment to 400,000 gal/d. This collected effluent is adjusted for pH in an equalization basin and gravity-fed to a flocculation basin, where a nonionic polymer is added to enhance the formation of large floc particles. The flocculated wastewater then flows to a 40-ft^2 concrete clarifier, where essentially all suspended solids and floating material are removed.

The clarified waste is then pumped to either of two moving-bed adsorbers, each 12 ft dia. by 36 ft sidewall, made of steel, containing 124,000 lb of Filtrasorb granular activated carbon, and designed to accomodate a wastewater flow of 175 gpm.

This treatment has reduced waste to 1,450 lb/d of BOD, 2,675 lb/d of COD, 22 lb/d of suspended solids, and 0.5 lb/d of phenol. Approximately 70 gpm of the treated water is reused in the plant, and the rest is discharged to a river. The treatment plant has been operating over two years and is performing as designed.

Originally published October 18, 1976

capacity should be so limited as to have the furnace operating 80–90 percent of the time, since frequent shutdowns reduce refractory and arm life.

A hearth loading of 70–80 lb/(ft^2)(day) is generally required for multiple-hearth furnaces regenerating carbon for industrial wastewater applications. Depending on the application, this loading rate can range down to 50 and up to 120 lb/(ft^2)(day).

A 6% volumetric loading with 45 min at the reactivation temperature is generally used in countercurrent direct-fired rotary kilns for industrial wastewater carbons. Kilns can be designed with lifting flights to promote carbon drying and improve reactivation.

The adsorption equipment is the heart of the system. Three types of adsorbers are generally used in wastewater applications: downflow fixed-bed, packed moving-bed, and upflow expanded-bed.

Downflow fixed-bed adsorbers

These adsorbers offer the advantages of simple operation plus the ability to serve as a filter for simultaneous suspended-solids removal. Wastewater enters the top of the vessel and passes downward through the carbon bed, where the organics are adsorbed. The treated wastewater is collected in a header-lateral or tile underdrain at the bottom of the bed, and discharged to subsequent adsorbers or to the stream. The vessels can be operated either under pressure or by gravity flow.

If the bed is to serve as a simultaneous filter, 50% freeboard should be provided above the carbon to allow for expansion during backwash. Surface washers and an air scour are recommended for filtration applications. Backwash rates are generally 15–20 gpm/ft^2, and air scour rates 3–5 scfm/ft^2. As a rule of thumb, a carbon filter running at 5 gpm/ft^2 with a 50–75 mg/l suspended-solids loading will require backwashing for 10–20 min each day. Loadings are generally less than 5 gpm per ft^2 of cross-sectional area, if suspended solids are present, compared to rates up to 10 gpm/ft^2 in pressure systems with little or no suspended solids.

Fixed-bed adsorbers can be arranged for single-stage, multi-stage, or parallel operation. Single-stage adsorbers are used when the mass transfer zone is short (saturation of the carbon occurs shortly after initial breakthrough); two or more beds in series are used when an increased efficiency of carbon utilization is desired (see box: "Phenolic wastewater treatment"); two or more beds in parallel are used when the high flowrate would require excessively large vessel diameters.

Single-stage adsorbers are also used when the carbon exhaustion rate is low, and the cost of replacing or reactivating the carbon is a minor operating expense. Generally, when this occurs, the investment required for a multiple-column system cannot be justified by the lower carbon usage rate that would result.

When two or more stages are used, only one bed—that upstream—is removed at a time for reactivation; and the fresh carbon bed is put in the downstream, or polishing, position. When two or more beds are used in parallel, the effluents are blended, which is advantageous for reducing the carbon-exhaustion rate. In all cases, the entire vessel containing the exhausted carbon is emptied and refilled with fresh carbon.

Packed moving-bed adsorbers

The packed moving bed provides countercurrent operation, in that the fluid flows upward through a bed that is moved downward; portions of the bed are periodically removed from the bottom of the vessel, and replaced by reactivated carbon that is added at the top.

Capital investment for a moving-bed system is generally less than the investment for an equivalent fixed-bed system for treatments requiring large amounts of carbon. However, this type of bed generally requires the suspended solids to be less than 10 mg/l, which means that pretreatment is usually necessary.

Flowrates usually range from 3 to 7 gpm/ft^2. A height-to-diameter ratio of 3:1 is desirable to facilitate plug flow during carbon removal (pulsing). Pulse volumes range 2–10% of the total bed, so that the total bed is equivalent to 10–50 pulse-beds in series.

Adsorbers with total bed depths as great as 45 ft have been used without difficulty. Wastewater is fed peripherally through a ring header located on the bottom cone, while the treated water is collected in a ring header on the top cone. Screens or septa are submerged in the top of the carbon bed to retain the granules while passing the treated water.

Upflow expanded-bed adsorbers

If suspended solids in the effluent are not a concern, the upflow expanded-bed design should be considered. In this design, the wastewater enters the bottom of the beds at approximately 8–10 gpm/ft^2 and exits through the top of the vessel. The carbon near the top of the bed is expanded approximately 10%. A high rate of back-

Phenolic wastewater treatment

The plant is a producer of phenolic resins. Its wastewater contains floating oils, suspended solids, phenolics, and other organics. Flow is approximately 100,000 gal/d. Typical analyses are:

	Raw	Clarified	Carbon effluent
Suspended solids	220 mg/l	45 mg/l	$<$10.0 mg/l
Oil/grease	50 mg/l	25 mg/l	$<$5.0 mg/l
TOC	1,200 mg/l	650 mg/l	25.0 mg/l
Phenol	160 mg/l	130 mg/l	$<$0.1 mg/l

The treatment consists of separators for oil removal, coagulation utilizing a coagulant-aid and a polyelectrolyte, sedimentation, mixed-media filtration, and adsorption on activated carbon, which is supplied via an adsorption service that handles shipping, reactivation, and transfer of the material.

Adsorption is carried out in two fixed-bed adsorbers operated in series. Each vessel is 10 ft dia. by 11 ft sidewall and contains 20,000 lb of activated carbon.

The toxicity of this wastewater was evaluated by static tests in which bluegill sunfish were subjected to raw and final effluent samples. In the raw sample, all eight test fish died within the first 15 min of exposure. The fish exposed to carbon adsorber effluent showed no adverse reaction during the 10-day duration of the test.

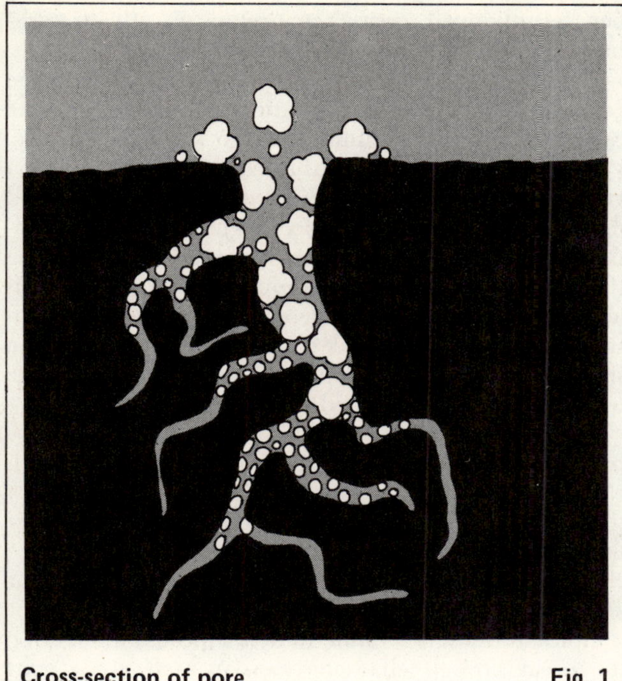

Cross-section of pore Fig. 1

wash with air scouring may be needed occasionally, depending on the nature of the solids in the wastewater and the extent of biological activity in the bed.

The upflow expanded bed can be arranged for single-stage, series, or parallel operation. Design of the inlet distribution system is critical, since the distributors must be able to pass suspended solids, but retain the granular activated carbon. The concept of upflow expanded beds is relatively new, and as a result the engineering development has not progressed as far as fixed-bed and moving-bed technology.

Wet carbon in static contact with carbon-steel incurs a galvanic type of corrosion. Consequently, adsorber vessels are lined carbon-steel, stainless steel, fiberglass-reinforced plastic (FRP), and concrete. The final choice among these materials depends on the corrosivity of the wastewater in the presence of granulated carbon.

When toxic chemicals can be adsorbed

Activated carbon can be made from any carbon-containing raw material, e.g. coal, wood, nut shells, sewage sludge, or petroleum residues. Many raw materials can produce activated carbon with an extensive internal surface area. The differences between these products are their residual ash content, their hardness, and their activity. Activated carbon derived from bituminous coal is preferred for wastewater treatment due to its high hardness—a characteristic needed to keep down handling losses during reactivation.

The activation process creates within the individual pieces of carbon selectively sized pores, which provide the internal surface area where adsorption takes place. Calculations show that one pound of high-quality activated carbon has an internal surface area equivalent to approximately 125 acres. The major part of this area exists on the walls of micropores toward the interior of the carbon particle as illustrated in Fig. 1.

A number of mechanisms govern the adsorption of organic molecules. First, molecules in the bulk of the solution must migrate to the carbon particle. Then they must migrate across a liquid film surrounding the particle, and thus into a pore. Thereafter they must migrate through the pore to finally come to rest on the ultimate adsorption site. When all of the available sites have been used up, the carbon particle is in equilibrium with the surrounding solution; and its capacity for further adsorption is exhausted.

Three principal factors affect the ease with which organic compounds are adsorbed: polarity, structure and molecular weight. Highly polar molecules are generally highly soluble; and vice versa. Since highly soluble molecules are adsorbed with difficulty (and vice versa), high polarity therefore reduces the ease of adsorption. Molecular structure affects the ease with which the molecules attach to the surface of the carbon particle; aromatic rings being conducive to adsorption.

Molecular weight affects ease of adsorption through two effects—solubility and surface attraction. The higher-molecular-weight compounds are generally less soluble, and consequently generally more easily adsorbable. Similarly, the surface attraction is generally greater for larger molecules, so that they are more easily adsorbed. However, this general rule holds only while the organic molecule is smaller than the pore size of the carbon. The major part of the pore surface area exists in pores ranging 10 Å to 1,000 Å, so that molecules larger than 1,000 Å are confronted with a severely limited surface-area capacity.

While these principles may afford some understanding of why a toxic chemical is more or less easily removed through adsorption, they do not help appreciably in estimating the equipment requirements of a specific treatment system. Such estimates require an adsorption isotherm and column-test results.

Adsorption isotherms

The equilibrium relationship between the concentration of an organic material adsorbed on the surface of carbon and the concentration of that material in the surrounding solution is given by the expression:

$$(x/m) = k\,C^{1/n}$$

where: (x/m) = concentration of the organic on the adsorbent, mg/g

C = concentration of the organic in the solution, mg/l

k, n = constants

Thus, if k, n and C are known, (x/m) can be calculated, and from that the theoretical equilibrium quantity of carbon. Since k and n vary with temperature, plots of the equation are known as adsorption isotherms. These are determined in the laboratory through tests on individual portions of activated carbon. A number of samples of weights ranging 0.1–20.0 g, for example, are placed in individual flasks. A given volume of the wastewater to be tested, e.g. 100 cm^3, is added to each of the flasks; and the flasks are then put into a shaker and left there long enough for the contents to reach equilibrium. Average time is about 2 h, since

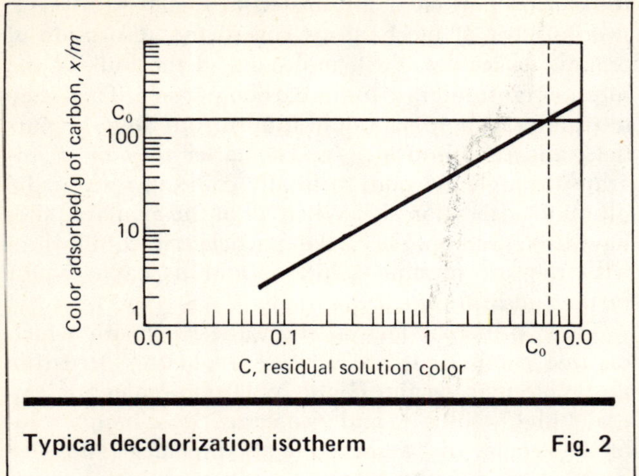

Typical decolorization isotherm **Fig. 2**

wastewater contaminants of primary concern are usually adsorbed relatively rapidly.

Subsequently, the carbon is separated from the treated solutions on 0.45-micron filters, and the purified solutions analyzed for a particular characteristic, such as total organic carbon (TOC) or chemical oxygen demand (COD). Since the conditions in each of the flasks represent a different relationship between the concentration on the adsorbent and the concentration in solution, the data can be used to determine k and n.

In actual practice, this is done through a plot on log-log paper (Fig. 2). The plot, when extrapolated a small distance in either direction, tells the maximum amount of a particular contaminant that can be adsorbed from any concentration of that contaminant—and consequently the minimum amount of carbon necessary to treat a particular concentration of wastewater. More specifically, the plot tells us:

- The relative affinity of a component for the carbon.
- The maximum concentration of the contaminant on the carbon.
- The minimum concentration that can be attained in the solution.
- The sensitivity of adsorption to concentration in the solution (indicated by the slope of the plot).
- The effect of pH (by controlling the pH of the tested solutions).

Note that all this information is qualitative, since it relates to equilibrium conditions. Since equilibrium rarely represents the most economical conditions, additional tests are necessary for sizing actual adsorption facilities. These tests are known as the "column tests."

Column testing

The three major parameters for designing adsorption systems are: (1) contact time, (2) carbon usage rate, and (3) pretreatment requirements.

The contact time indicates the amount of activated carbon required onstream at any given moment—thus the size of the equipment, and the capital cost.

The carbon usage rate indicates the rate at which the carbon must be replaced and reactivated, and this defines the operating costs of the system. The pretreatment requirements, which are dependent on pH, suspended solids, oil, grease, etc., relate directly to additional unit operations required for optimum treatment.

A wave-front or mass-transfer zone characterizes the gradual change in concentration of organic on the carbon during the continuous adsorption of the organic. When the solution moves through a carbon bed at a very rapid rate, this wave-front becomes long and drawn out. As a result, the bed begins to leak contaminant into the effluent very early, gradually increasing the concentration until complete breakthrough is achieved. If the solution moves through slowly, by contrast, the bed does not leak contaminant; and there is a sharp final breakthrough indicating a more complete utilization of the activated carbon (Fig. 3).

Contact time may be determined in the laboratory by a number of small carbon columns in series, with each column in the series representing a fixed contact time. Thus, if four columns each with 15 min contact time are put in series, it is possible to determine the effects of 15, 30, 45 and 60 minutes of contact time for any given volumetric throughput. By taking effluent samples from each column, data is collected to describe the breakthrough characteristics at each given contact time.

Fig. 4 shows breakthrough curves of an actual wastewater passed through five columns with contact times of 15, 40, 75, 105 and 240 min. The TOC in mg/l is plotted against the total volume throughput, in ml. As is shown, increasing the contact time from 15 to 40 min has virtually no effect on the breakthrough. However, an increase from 40 to 75 min results in the beginning of an effluent plateau at approximately 5,500 mg/l of TOC. A further increase to 105 min lowers this plateau and gives a significantly longer run before the average column has passed 4,000 mg/l.

Increasing the contact time to 240 min gives a more dramatic change in effluent quality. At this point, the additional requirement for equipment and carbon to achieve 240 min of contact should be worth the effort to achieve 4,000 mg/l in the effluent, because the amount of water that can be processed is much larger, considerably lowering operating costs. This is shown by a comparison with the curve for 75 min of contact.

At 75 min, about 400 cm^3 were passed prior to exceeding 4,000 mg/l in the effluent, compared with over

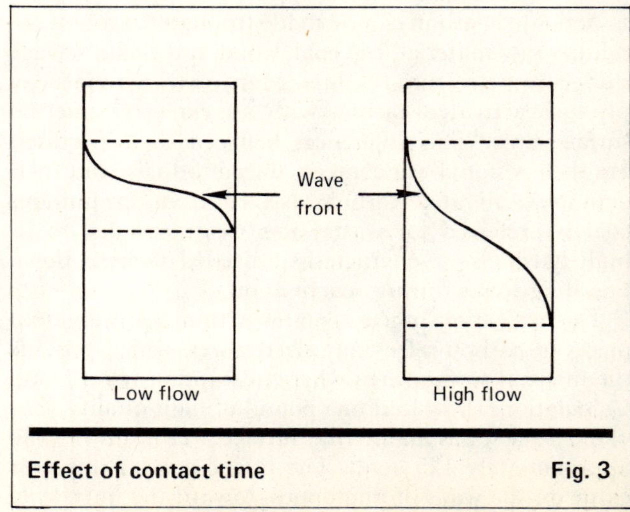

Effect of contact time **Fig. 3**

2,100 passed at 240 min. This represents a 5.3-fold increase in wastewater handled by a column that is by volume of carbon only 3.2-fold larger. This example is a rather extreme case of a contact-time-dependent adsorbate. In the majority of cases, a contact time of 60 min or less is adequate for good utilization of the carbon.

Once the contact time has been determined, the second of the three major parameters, carbon usage rate, can also be determined through a column test. Data can be obtained by operating two columns in series until the effluent of the second bed exceeds the effluent limitations. At this point, the first bed in the series is removed and a fresh bed of carbon placed in the final position. This is known as staging. By repeating the staging procedure a number of times and analyzing effluent concentrations, a series of curves can be developed.

These curves should begin to approach an equilibrium condition, as exhibited by a duplication of the shape of the breakthrough curve. At this point of duplication, the volume throughput per unit weight of carbon can be derived and should permit an accurate estimate of carbon usage for treating the wastewater.

Data from an actual wastewater passed through four operations illustrate this principle (Fig. 5). Columns A and B were placed in series service. When the effluent from B exceeded 4,000 mg/l TOC, column A was removed, and column C placed in the polishing position after column B. When the effluent from C reached 4,000 mg/l, B was removed from service and D installed after C—and so on through column E.

In reviewing this data (Fig. 5), the shapes of the breakthrough curves for columns C, D and E take on a characteristic shape. Accordingly, the volume throughput between equivalent points on the curves determines the volume throughput per unit of carbon. It can be seen from the figure that this was about 2½ liters for columns C and D and just over 2½ liters for column E.

The operating performance of this system is thus approximately 2½ liters capacity for the quantity of carbon contained in each column (in this case 25 g). It is this figure—2.5/25 = 0.1 liter/g, or 1,000 gal of wastewater per 80 lb of carbon—that determines the rate at which carbon must be replaced and reactivated.

Wastewater pretreatment

Although granular carbon can be used to filter out suspended solids and oil, as well as to adsorb soluble compounds, it is sometimes more economical to pretreat a wastewater to remove such solids prior to adsorption.

In general, solid particles are held suspended in water because of surface charges, so that it is necessary to remove or reduce the surface charges in order to separate the solids. This can be done in a number of ways through the use of polymers. Polymers, which can be either positively or negatively charged, or have an overall net charge equal to zero, can be used (1) to link suspended particles, (2) to neutralize the charged particles, or (3) to both link together and neutralize the suspended particles. The objective in any case is to reduce the stability of the suspension, so as to allow separation of the particles by subsequent treatment.

In some cases (i.e., suspended solids and oil) dissolved-air flotation may be the most applicable separation technique. Polymers added to such a wastewater will produce a floc particle consisting of both solid and oil, and with a density nearly equal to or perhaps slightly less than that of water. Such a floc is quite amenable to dissolved-air flotation.

In other cases, neutralization plus linking the suspended particles may result in a sediment that is best removed through some settling technique.

In either case, the maximum removal of particles calls for an additional step, such as filtration, to polish the effluent. Filtration may also be applied directly to a wastewater in cases where the particle concentrations are suitable, as when colloidal particles can be destabilized inline and removed directly by filtration.

Sometimes a primary coagulant may be needed to begin the capture of colloidal particles and prepare them for treatment by a polymer. An inorganic primary coagulant such as alum, lime or ferric chloride in combination with a polymer may produce separations superior to the mere addition of the effects of using the two treatments separately. Also, clay can be used to add weight to the floc, so as to help settling, or to provide a point on which the materials in suspension can aggregate, thereby making coagulation with a polymer somewhat easier. The chemical structure of the polymers can be designed to provide the required charges, whether positive, negative or neutral (Fig. 6).

In the laboratory, the application of polymers to a wastewater is usually evaluated in a small-scale batch

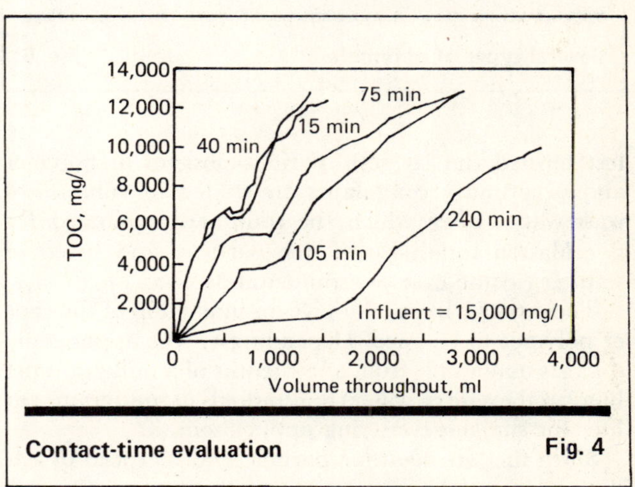

Contact-time evaluation **Fig. 4**

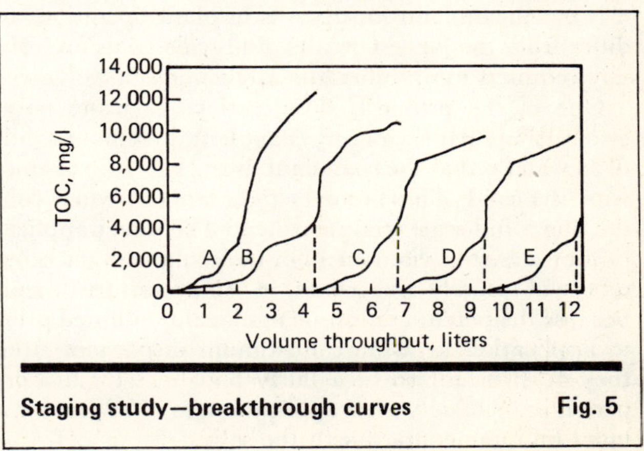

Staging study—breakthrough curves **Fig. 5**

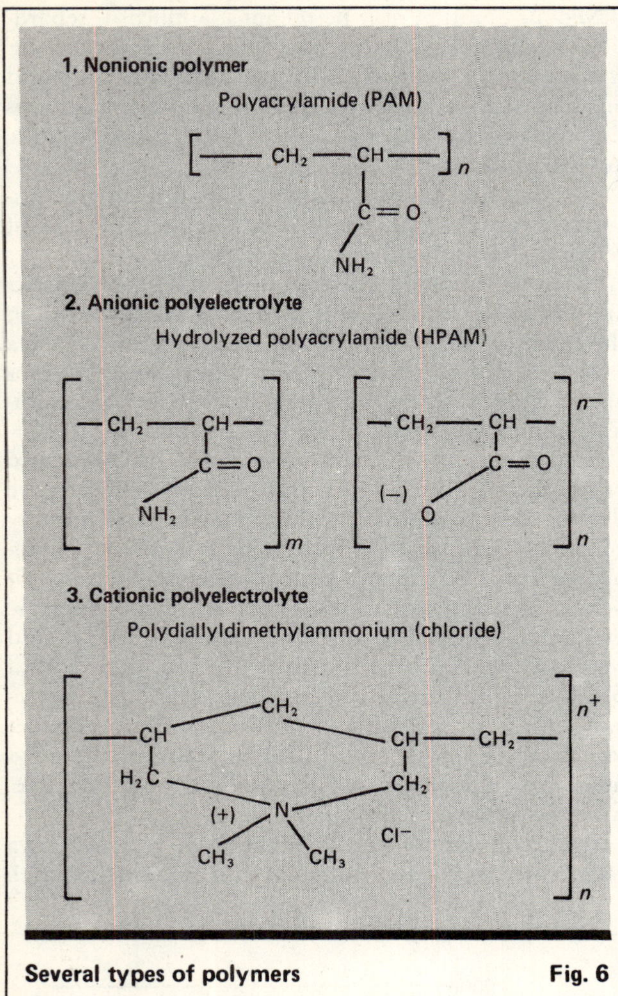

Several types of polymers **Fig. 6**

test termed the jar test. Various dosages of polymer and/or primary coagulants are added to volumes of wastewater, after which the solids are evaluated for flocculation time, size of floc particle, and speed of settling (in the case of sedimentation).

These tests will usually give an indication of the type of polymer to be tried. Depending on the application, dosages may range from a few tenths of a milligram per liter (in the case of water) to hundreds of milligrams per liter for sludge-dewatering applications.

Since the size of a floc particle and its speed of settling are somewhat dependent on the equipment used, it is not uncommon for dosages in plant operations to differ from the jar-test results. Full-scale trials are usually required to optimize the application of polymers.

One of the primary considerations of using polyelectrolyte polymers, or any coagulant, to remove solids from water is that the coagulant must come into contact with the solid. This becomes most evident when conducting a full-scale trial treatment. The direct application of a highly viscous high-molecular-weight polymer will usually not result in optimum treatment, because the polymers must be significantly diluted prior to application to achieve maximum dispersion. Also, they must be mixed at a fairly high rate for a short period immediately after application, so as to provide a more intimate contact with the solids.

Usually a separate mixing chamber is used, although in plants constructed without such mixing facilities Parshall flumes or weirs may be used to promote mixing. These latter are of particular interest when upgrading treatment plants that were not necessarily designed for the addition of polymers.

With the application of good coagulation and subsequent solid/liquid separation, the wastewater is ready for treatment by adsorption through contact with granular activated carbon.

Field pilot studies

In many cases it is necessary to run field pilot studies to verify the laboratory conclusions. Such tests demonstrate the process on a continuous basis under actual plant conditions of flow and concentration variations. They are generally desirable if suspended-solids removal is required or biological activity is expected. Granular carbon can be used as a filter for suspended solids and oil removal as well as an adsorber. Also, in the case of dilute, biodegradeable wastewaters, biological degradation of some of the organics can occur due to the growth of microorganisms on the surface of the carbon.

Generally, a pilot plant consisting of four $5\frac{1}{2}$-in. by 6-ft Plexiglas columns connected in series is used. This unit is designed to operate at a downflow rate of $\frac{1}{4}$–$\frac{1}{2}$ gpm.

The first column in the series will contain 3 ft of granular carbon, and will serve as a filter as well as an adsorber, so that series operation permits the study of bed depths and contact time provided by 3, 8, 13 and 18 ft.

The pressure drop across the pilot columns is monitored, and the lead column is backwashed when an excessive pressure drop is observed. Wastewater samples of the feed and effluent from each column are collected and analyzed. Based on this data, the design carbon-exhaustion rate, contact time, and filtration rate can be verified.

Finally, the spent carbon from the pilot plant should be subjected to a reactivation study to determine: (1) the spent-carbon characteristics, (2) the degree of difficulty in reactivating the carbon, (3) the existance of any air pollution problems, and (4) any potential for corrosion problems. Generally, batch laboratory reactivation tests are sufficient for generating this information, but in special cases, small-scale test furnaces can be used.

The author

Frederick E. Bernardin, Jr., is Technical Service Manager of the Adsorption Service group at Calgon Corp., Calgon Center (Box 1346), Pittsburgh, PA 15230, where he is responsible for providing technical service for the process performance of the firm's granular-activated-carbon adsorption service installations. Joining Calgon as a chemist in 1967, he was made group leader of the Activated Carbon R&D dept. before accepting his present position. He holds a B.S. in chemistry from Waynesburg College and an M.S. in sanitary engineering from the University of Pittsburgh.

Designing Fixed-Bed Adsorption Columns

There are two basic fixed-bed adsorption systems: adsorbent regenerated in place, or in a kiln. This article covers the design of both, including determination of column size, cycle time and heat required for regeneration.

W. A. JOHNSTON, Consultant

Free-energy adsorption can produce a filtrate of extremely high purity. However: the substance to be removed must be a minor constituent, usually less than 1% by volume; the substance must be polar or have bonds that can be polarized; the adsorbent must be able to accept the molecule to be adsorbed; the energy in the system must be below that at which chemisorption takes over.

Economics Dictate Equipment

The equipment for fixed-bed adsorption is determined by economics, which dictate that the adsorbent be regenerated and reused repeatedly. For continuous operation, a minimum of two, or possibly more, columns are required, depending on the relative proportions of the total cycle time consumed in operation and regeneration.

The columns are always vertical, cylindrical shells, closed at the top with a dished head, which includes a manhole. Bottom closure is dictated by the type of regeneration. Column fabrication is in accord with the ASME Code for Unfired Pressure Vessels.

The two most practical types of column are shown in Fig. 1, which also illustrates adsorbent bed supports. Bed supports other than those shown are used, but most of these are more costly to fabricate; the bed is more difficult to dump; and the pressure drop through some is high.

The support of the upper column in Fig. 1 is relatively inexpensive to fabricate and maintain, and it provides low pressure drop. Its disadvantage is the quantity of adsorbent at the manhead that has to be removed by hand. This is acceptable when the life of the absorbent bed is such that it need be changed only two or three times a year.

Although the head of the lower column in Fig. 1 is expensive to fabricate, it is the only type to consider when the adsorbent bed must be changed frequently. The column's low maintenance costs and small pressure drop quickly amortize the greater capital expense.

The most common feeding arrangements are the

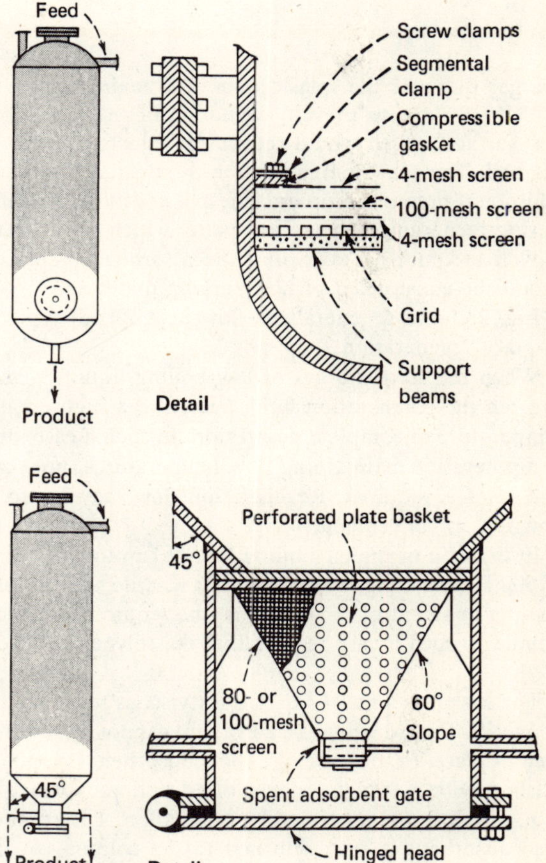

TOP COLUMN regenerates adsorbent in place—Fig. 1

This article is confined to adsorption in fixed-bed columns, wherein a feed fluid passes downward through an adsorbent bed under set conditions of flowrate, temperature and pressure. Cocurrent (contact) adsorption is not treated, and neither is moving-bed adsorption.

Originally published November 27, 1972

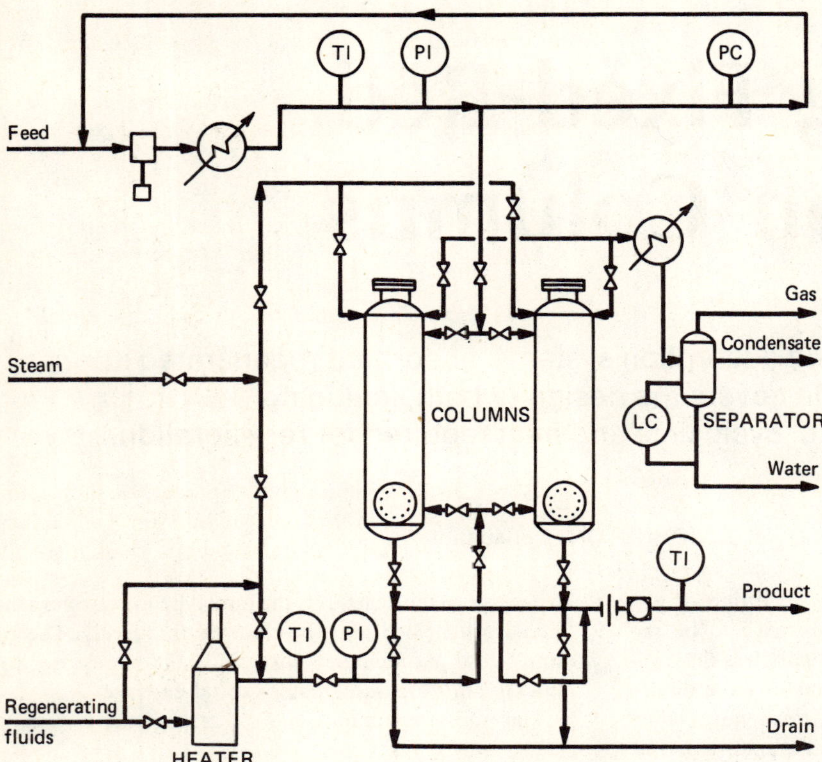

INSTRUMENTATION that is necessary is shown; with constant pressure feed and stable operating conditions, the replaceable orifice can control product flow adequately, but a flow controller may be used if it is desired—Fig. 2

sparger ring and the splash plate. The main purpose of a feed nozzle is to prevent cavitation of the adsorbent-bed surface, which would reduce the effective height of the bed. The sparger ring has the disadvantage of some additional pressure drop and the possibility of plugging during recharging. The splash plate, which must be removed and put back every time the adsorbent is replaced, should be constructed of nonsparking material.

Fig. 2 shows a generalized flowsheet for all types of in-place regeneration.

When the feed is a gas or low-boiling liquid, it itself is often the regeneration fluid, heated to a high enough temperature to complete desorption. In such a case, only equipment for maintaining flow temperatures, pressures and rates is required. Regeneration flow may be up or down, as is convenient.

In the case of higher-boiling liquids (up to end-points of 500 F.), draining, washing with a volatile solvent, plus heating or steaming, or both, may be required. The economics in such a case require that the solvent be recovered.

The lower column of Fig. 1 is for viscous feeds boiling above 500 F., and when the adsorbent cannot be regenerated in place. In this case, the spent adsorbent is drained, washed with a solvent, steam dried and removed to a kiln, where it is regenerated by calcination. The regenerated adsorbent is then returned to the column. In this case, a kiln, conveyors, elevators, hoppers, bins and chutes are required in addition to the equipment shown in Fig. 2. Also needed is a distillation unit to recover solvent and wash liquids.

Common Adsorbents

Beds in the type of equipment being discussed generally consist of the following adsorbents:

Molecular sieves (zeolitic-type alumino-silicates) are unique in that their synthesis produces a pore volume almost 100% of a single size. The pore-size distribution of other adsorbents ranges considerably wider.

Sieves for fixed-bed adsorption are spherical pellets or beads of uniform size. Grade designation is based on the nominal pore diameter in angstroms: 3Å adsorbs H_2O and NH_3, and is used for dehydrating unsaturated hydrocarbons and polar liquids; 4Å adsorbs H_2O, NH_3, H_2S, CO_2, SO_2, C_2H_4, C_2H_8 and C_3H_6, and is used to dehydrate and purify saturated hydrocarbons; 5Å adsorbs *n*-paraffins up to chain lengths of 20 carbon atoms, and offers the only means of separating *n*-paraffins from iso-paraffins and cyclic compounds; 8Å adsorbs iso-paraffins, olefins and C_6H_6, and is used to separate aromatic hydrocarbons; 10Å is used for drying liquids and gases and for sweetening natural gas.

Bone chars (natural and synthetic) are sized granules used almost exclusively for decolorizing sugar.

Activated carbons are sized granules, derived from the pyrolosis of wood or coal, used to remove noxious gases from air and other fluids, to purify water and light hydrocarbons, and to decolorize syrups.

Silica gel is a synthetic, porous, noncrystalline gel formed into spherical pellets and sized granules, used for drying gases and liquids, separating mononuclear aromatics and purifying hydrocarbons.

Activated alumina is a synthetic, porous crystalline gel formed into spherical pellets and sized granules, used for dehydrating air, organic fluids and other liquids, and purifying, decolorizing and refining petroleum oils and waxes.

Activated bauxite is a naturally porous crystalline alumina contaminated with kaolinite and iron oxides in varying proportions, depending on the place of origin. A sized granule, its uses are the same as those for activated alumina, and it has been experimentally shown to quantitatively remove most aerobic and anaerobic bacteria.

Activated clays are porous crystalline alumino-silicates derived from the minerals attapulgite, montmorillonite or sepiolite. They are used to dehydrate, purify, refine and decolorize the entire range of petroleum liquids, and are also known to remove quantitatively many aerobic and anaerobic bacteria from hydrocarbon liquids.

Many of the adsorbents listed above may be used for similar purposes. However, because they may vary considerably in product yield, quality and price, they should be selected after an overall comparison of these three factors.

Data for Design

The sizing and overall design of an adsorption system depend on the properties of both the feed fluid and the adsorbent. The aim is to produce a system in which the flow during adsorption is downward through the adsorbent bed under substantially isothermal and isobaric conditions. (Regeneration flow may be downward or upward.) The main consideration is to maintain a compact, stable bed of adsorbent.

The properties of the fluid that need to be known for design purposes are:

1. Chemical composition, including the content of inorganic solids. The makeup of many of the more viscous, complex fluids can only be generalized, and laboratory analyses should verify compatibility with the adsorbent, product yields and quality. Compatibility is important because adverse catalytic effects may occur with some of the adsorbents at temperatures between 200 and 300 F. (i.e., the operating temperature range for the more viscous fluids).

2. Fluid density at operating temperature and pressure (D_f, lb./cu.ft.), which can be calculated from standard data.

3. Fluid viscosity (μ, centipoise or lb./ft. × sec.) at operating temperature and pressure.

The properties of the adsorbent required for design are:

1. Volume-weight relations. As applied to stable beds of particle adsorbents, observable bulk volume (V_b) is the sum of the apparent particle volume (V_a) and the void volume (V_v), so that $V_b - V_a = V_v$ and $V_v/V_b = \varepsilon$, the void fraction. Manufacturers normally report these as reciprocals: bulk density (D_b) and apparent density (D_a), both in lb./cu.ft.

A leveled, dump-packed adsorbent bed produces minimum values of D_b and maximum values of ε. Tamping, which is impractical in commercial columns, produces maximum D_b and minimum ε. Both maximum and minimum values should be known for design. (The value of D_b also varies with particle size, increasing with decreasing size.)

2. Particle size, which is usually reported as a mean equivalent particle-diameter (D_p) in feet. In the case of uniformly shaped pellets, D_p is easily calculated from pellet dimensions.

In the case of granules and nonuniform pellets, D_p is calculated from a screen analysis by means of Eq. (1):

$$D_p = (ad_{p1} + bd_{p2} + cd_{p3} + \ldots + nd_{pn})/12 \qquad (1)$$

In Eq. (1), D_p = mean equivalent particle dia., ft.; $a, b, c, \ldots n$ = weight fraction of adsorbent lying between two adjacent screens; and $d_p = (d_1 \times d_2)^{1/2}$, with d_1 and d_2 the rated openings in the adjacent screens, in.

For most commercial granules suitable for fixed-bed adsorption, D_p values range from 0.0005 to 0.0260 ft. The larger the value of D_p, the slower the maximum permissible flow, and the lower the pressure drop through the bed.

3. Pore data, which are important because they permit elimination from consideration of adsorbents whose pore diameter will not admit the desired adsorbate molecule. Pore volume makes possible the estimation of capacity for adsorbate. Pore diameters or radii (or both) are reported in angstroms. Because only molecular sieves have uniform pore sizes, the pore-size ranges of other adsorbents are shown as a plot of the increment of pore volume divided by the increment of pore radius versus the pore radius. Radial measurements are also in angstroms.

4. Hardness, which indicates the care that must be taken in handling adsorbents to prevent the formation of undesirable fines. Most commercial adsorbents have a hardness of less than 2 on the Mohs scale.

5. Adsorbent life, which cannot be predicted with any certainty until the feed content of inorganic materials, and whether or not the regeneration process will affect pore structure, are known. The point at which the adsorbent is replaced is normally about when the onstream flow and regeneration times become equal. More often than not, this is when the cycle product yield has dropped to 35–40% of its original value. Under average conditions, when regeneration is in place, 60 to 100 cycles can be expected; with kiln regeneration; this can be 20–35 cycles, if the kiln temperature-gradient is well controlled.

Column Design

Design of the adsorption column is the key to the entire system. The methods of vessel design are based on two concepts: the height of an equivalent transfer unit (HETU) and maximum permissible space rate at which full adsorption takes place (Rx). The data on which these concepts are based are empirical.

The HETU unit, in ft., is based on the variation in the void fraction (ε) with variations in mean equivalent particle-dia. (D_p) and packing density. It reaches a maximum of 0.38 ft. when ε is 0.50 and D_p is 0.0260 ft. How HETU for any adsorbent may be estimated is illustrated in Fig. 3.

The minimum adsorbent-bed height for maximum product quality is 55–60 HETU. This normally results in minimum bed heights (L) of between 10 and 20 ft. Bed

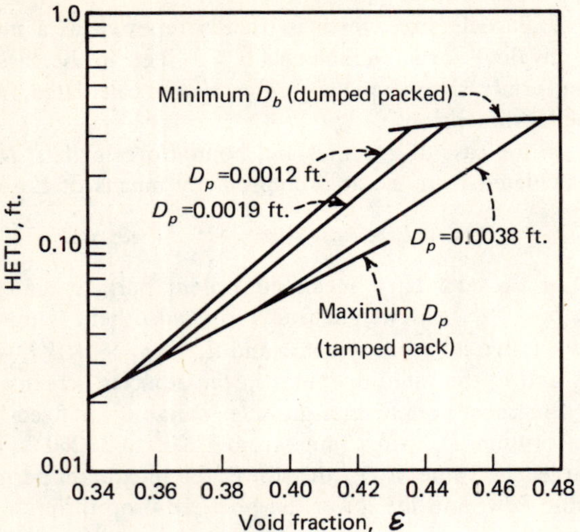

BED DEPTH should be minimum 50-60 HETU—Fig. 3

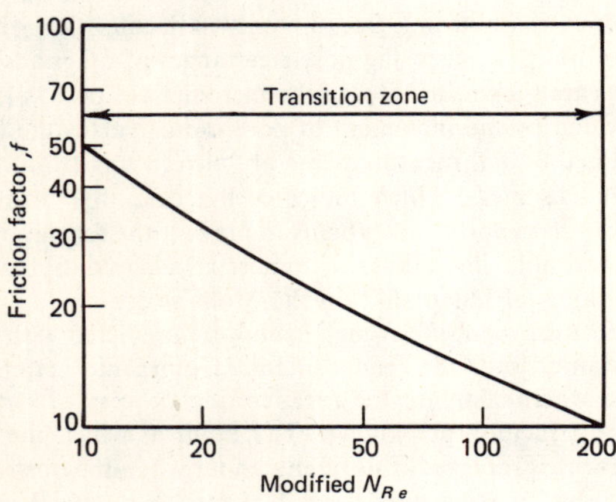

VARIABLES for pressure-drop equations—Fig. 4

heights greater than minimum are often desirable economically because of the longer onstream life achieved. Column heights above 30 ft. should not be used because of difficulties in maintaining stable packing.

The maximum permissible space rate (Rx) is expressed in cu.ft. of feed/cu.ft. adsorbent/hr., and is a function of the absolute viscosity of the feed (μ) at the temperature and pressure of operation and D_p. Rx may be computed with Eq. (2) or (3):

$$\log Rx = 2.0896 - 0.0387\mu + \log (0.0101 - D_p) \quad (2)$$

$$Rx = 101.2 - 3080\mu + 2286\mu^2 - 720D_p \quad (3)$$

Eq. (2) applies to values of μ greater than 0.5, and values of D_p less than 0.01; in other words, it applies to the viscous-liquid range. Eq. (2) shows that flowrates become too slow to be economical when μ is higher than 20 cp.

Eq. (3) applies to the range of μ from 0.005 to 0.5 cp., and values of D_p up to 0.0260 ft.

Eq. (2) produces values that are ±5%, and Eq. (3) ±10%, of the mean value. (Note that Rx determines the volume of the column.) When μ exceeds 20 at operating conditions, the flow viscosity can be reduced by dilution with a low-boiling inert solvent.

The I.D. of columns fabricated from plate is limited to about 36 in. by difficulties in manufacturing. Columns of less than 36 in. I.D. can be made from pipe fitted with blind collars. Because of handling difficulties, the height of pipe columns are usually limited to about 15 ft. Plate column diameters are generally no larger than 12 ft. because of the difficulty of obtaining stable adsorbent packing.

Pressure Drop

From a design viewpoint, flowrates may also be limited by pressure drop. Pressure drop (ΔP) through packed beds is estimated by relations similar to those for computing pressure drop in pipes:

$$\Delta P/L = 2fG^2/D_p g\rho \quad (4)$$

In Eq. (4), ΔP = pressure drop, lb./sq.ft.; D_p = mean particle dia., ft.; $g = 32.2$; G = mass velocity, lb./sec./sq.ft.; L = depth of the adsorbent bed, ft.; f = friction factor, dimensionless; ρ = fluid density at operating temperature, lb./cu.ft.; and μ = fluid viscosity at operating conditions, lb./(ft.)(sec.). The same factors for computing pressure drop in pipe apply: $N_{Re} = D_pG/\mu$; for laminar flow ($N_{Re} < 10$), $f = 480/N_{Re}$; for flow in the transition zone ($N_{Re} > 10 < 200$), see Fig. 4; for turbulent flow ($N_{Re} < 200$), $f = 60.3/(N_{Re})^{0.339}$.

The foregoing equations cover only the bed in the cylindrical shell. The following relations deal with pressure drop through bed supports and any adsorbent that may be held in the bottom of the cone.

For the type of support in upper Fig. 1, when free space is 40% or more:

$$\Delta P_b = 1.5\Delta P_b/L \quad (5)$$

For type of support in lower Fig. 1, when the ratio of d_1/d_2 is not more than 4:

$$\Delta P_s = 2.5\Delta P_b/L \quad (6)$$

For a 45-deg. cone, when L/d_1 and d_1/d_2 both equal 3:

$$\Delta P_c = 0.45\Delta P_b/L \quad (7)$$

For other dimensional relations in a cone, ΔP must be computed at small intervals of slope and integrated between d_1 and d_2 (d_1 is the I.D. of the cylindrical shell and d_2 is the I.D. of the cone at the point of truncation).

Of course, the total pressure drop through the column is:

$$\Delta P_t = \Delta P_b + \Delta P_s + \Delta P_c$$

A limitation of upflow regeneration occurs when the pressure of the flowing fluid exceeds the static head of the adsorbent bed. The point at which this expansion of the bed occurs may be calculated by Eq. (8):

$$\Delta P = [V_t(1.0 - \varepsilon)(\rho_s - \rho_f)]/A_t \quad (8)$$

In Eq. (8), V_t = vol. of adsorbent, cu.ft.; A_t = cross-sectional area of column, sq.ft.; and the subscripts s and f refer to the adsorbent and fluid, respectively.

Heat for Regeneration

The total heat input required for regeneration in place is found with Eq. (9):

$$H_b = H_1 + H_2 + H_3 + H_4 \tag{9}$$

In Eq. (9), H_1 = weight of steel in vessel × specific heat $(T_2 - T_1)$; H_2 = weight of adsorbent in bed × specific heat $(T_2 - T_1)$; H_3 = weight of retained fluid in the bed × specific heat $(T_2 - T_1)$; and H_4 = weight of retained fluid in bed × enthalpy.

The total heating requirement will be:

$$H_b = H_h + H_t \tag{10}$$

Eq. (10) allows for hysteresis, $H_h = 0.15H_b$, and thermal loss, $H_t = 0.1\ H_b$.

In these calculations, T_1 is the temperature of the bed at the start of heating, and T_2 the temperature after regeneration is complete (both in °F.). The heat removal required for cooling the bed back to operating temperature obviously involves only the heat in the vessel and the adsorbent bed. Because the bed is often cooled only to about 20 F. above the operating temperature, the heat-removal temperature will obviously be less than $T_2 - T_1$.

To determine the regeneration time-cycle, let $Q = H_b/(T_2 - T_1)$ and $C = (H_{T2} - H_{T1})/(T_2 - T_1)$; also let X stand for the lb. of regenerating fluid, and θ the time in hours desired for the cycle:

$$\theta = \int_{T_2}^{T_1} (Q/XC)(d_t/T_2 - T_1) \tag{11}$$

Obviously, the total regeneration time may be forecast by forming a similar integral for cooling and equating the sum of these integrals against θ.

To maintain substantially isothermal flow during flow to product, the column should be well insulated. Where severe weather may be expected, the columns should also be wrapped with steam coils. Steam coils should always be used when the feed fluid is normally solid.

Auxiliary Equipment

Other major pieces of equipment that may be required are heaters, kilns, solvent-recovery units, coolers, condensers, tanks, bins, conveyors and elevators.

Heaters for the regeneration fluid may be of any type capable of producing an outlet temperature of 500 F. Usually, direct-fired upshot heaters are most compact and economical.

Kilns for calcination regeneration should be capable of being operated in three zones of temperature. The upper third, receiving spent adsorbent, should maintain a gradient of 700 to 900 F. In this zone, no combustion should occur but the high-molecular-weight adsorbate should be cracked and partially vaporized, and the combustibles reduced to 4–5%. The gradient in the second third should be 900–1,100 F. and combustion should reduce the organic matter to zero. The last third's gradient should be 1,100–800 F. At the outlet, there should be a cooler capable of cooling the adsorbent from 800 to 150 F.

Residence time in each section should be 15 min., the total residence time 45 min. Hourly throughput should be set in relation to the total cycle.

It is most economical if the kiln can be operated continuously. Note that the maximum permissible adsorbent temperature is 1,100 F. Adsorbent life is improved if the complete removal of combustibles can be made at lower temperatures.

If the kiln is direct fired, the temperatures at the adsorbent surface must not exceed 1,500 F., to avoid adsorbent deterioration. Fuel requirements for direct firing may be estimated from the specific heat of the adsorbent (taking the combustion efficiency of the fuel as 45%).

Storage and Other Accessories

Bins are normally required only when regeneration is by calcination. With other types of regeneration, the adsorbent is discarded so infrequently that adsorbent storage bins are seldom justified. A means of hoisting palletized loads of bagged adsorbent to the upper manhead, and a platform from which they may be emptied, are all that is required.

For the calcination system, bins are required for fresh, spent and regenerated adsorbent. These bins should be circular with 60 deg. cone bottoms leading to circular chutes also 60 deg. to the vertical. Flows from the bins are controlled by orfice slide-valves in the chute.

The fresh adsorbent bin should have capacity for two or more carloads. The spent adsorbent bin should have a capacity of at least two column charges, as should the regenerated adsorbent bin. All bins should be elevated so as to feed back by gravity.

Because all commercial adsorbents are soft and will generate undesirable fines if roughly handled, they should only be carried on dished-bed and chain-belt elevators. Conveyors are required to bring fresh adsorbent makeup to the spent adsorbent going to the kiln, to carry spent adsorbent to the kiln, and to carry regenerated adsorbent back to the columns. The spent-adsorbent conveyor is usually equipped with a small movable bin to prevent overflow from the belt.

Elevators are required to lift fresh adsorbent from the truck or railcar to the storage bin, spent adsorbent to the kiln feed bin, and regenerated adsorbent to the bin charging the columns. The latter bin is usually arranged to serve four or more columns.

Piping should be designed for full safety under the operating conditions. Care should be taken that in manifolding no cross-contamination will occur. All valving should be double, with a drip drain between valves. Feedlines should be a loop with a backpressure relief valve returning at least 20% of the feed fluid back to the suction of the feed pump or compressor. All lines should be insulated, and those carrying materials that are solid at normal temperature should be steam traced or otherwise heated. Look boxes may be installed in product lines if safe and practical.

Pumps may be of any desired type available that give

adequate capacity. However, rotary or centrifugal pumps produce much less line pulsation.

Calculation Using HETU, Rx and ΔP

The problem is to design a fixed-bed adsorption unit to finish 17,200 metric tons/yr. (350 operating days/yr.) of a paraffin wax derived from crude to the following specifications:

Steam emulsion (ASTM 157)	30-sec. separation, maximum
Saybolt color (ASTM 156)	+21 minimum
TAPPI oxidation 650	pass 8 hr.
Odor and taste	none

The feed wax is produced by vacuum distillation, furfural extraction and methyl-ethyl-ketone (MEK) dewaxing and has the following characteristics:

Fusion point	50–54 C. (122–129 F.)
Density	285 lb./bbl.
Viscosity (at 210 F.)	3.8 centistokes
Saybolt color	+20
Oil content	not given

It is assumed that the oil content does not exceed 0.5% and the wax is free from furfural and MEK.

The stability, odor and taste requirements indicate that activated bauxite (20/60 mesh, 6% volatile matter) should be the adsorbent. To meet other requirements, it is also certain that product color will have to be +30 Saybolt.

The operation will be an equilibrium type, with a 2% makeup per cycle. With this type of feed, such an equilibrium operation should give an equilibrium efficiency of 65% of the new bauxite. Yields at +30 Saybolt color for the new bauxite are 220 lb./2,000 lb., and for the equilibrium bauxite, 143 bbl./2,000 lb.

Properties of the 20/60 mesh bauxite are:

	New	Equilibrium
D_b, lb./cu.ft. (dump packed and leveled)	57.5	62.5
D_p, ft.	0.0019	0.0013
Volatile matter, %	6.0–7.0	1.2–1.3
Void fraction, ε	0.426	0.37
Pore fraction, γ	0.21	0.14
Specific heat, Btu./lb.	0.2	0.2
HETU, ft.	0.326	0.316

The feedrate, converting 17,200 metric tons/yr. for 350 operating days, is 89.1 cu.ft./hr.

Operating at the lowest viscosity (highest temperature) will give the best flowrate for maximum yield. Because 210 F. is within normal plant steam temperatures and well below the 250-F. critical point, 210 F. is used as the operating point. The maximum permissible flowrate of feed (cu.ft./hr. per cu.ft. of adsorbent) determined by Eq. (2) is 0.766. The minimum adsorbent required is therefore 116.2 cu.ft. (89.1/0.766).

For the desired finished properties, a bed depth of 60 equivalent transfer units will be required, so the tower height will have to be 20 ft. (60 × 0.316 = 18.96 ft.). The cross-sectional area will be 5.81 sq.ft. (116.2/20), which gives a diameter of 0.846 ft.

Because such a diameter is too small for practical fabrication, a diameter of 3.0 ft. should be chosen. This will give a cross-sectional area of 7.06 sq.ft., and an adsorbent volume of 141.2 cu.ft. With this volume, *Rx* comes out to be 0.63, which is below the *Rx* previously calculated but still acceptable for maximum yield and product quality.

Pressure drop in lb./sq.in. per ft. is calculated with Eq. (4):

$$\frac{\Delta P}{L} = \frac{(2)(4{,}620)(0.177)^2}{(144)(0.0013)(32.2)(50.6)} = 0.953$$

The pressure drop through the bed, support and cone is then found with Eq. (5), (6) and (7) to be, respectively, 19.06 (20 × 0.953), 2.46 (2.5 × 0.953) and 8.56 (0.45 × 1,906) lb./sq.in. Summing these gives a total column pressure-drop of 30.08 lb./sq.in.

The column should be safely designed for a pressure of 75 psi., and the flow must be downward because the critical pressure, as calculated with Eq. (8), is equal to 1.13 lb./sq.in., which means that an upward flow would disperse the bed.

References

1. Allen, H.V., Pressure Drop for Flow Through Beds of Granular Adsorbents, *Petr. Ref.*, July 1944.
2. Eagle, S. and Scott, J.W., Liquid Phase Adsorption Equilibria and Kinetics, *Ind. Eng. Chem.*, **42**, 1287, 1950.
3. Gamson, B.W., Heat and Mass Transfer-Fluid Solid Systems, *Chem. Eng. Prog.*, **47**, 19, 1951.
4. Leva, M., Weintraub, M., Grummer, M., Polchik, M. and Storch, H.H., Fluid Flow Through Packed and Fluidized Systems, *U.S. Bureau of Mines Bull. No. 504*, 1951.
5. Mair, J.B., Westhaver, J.W. and Rossini, F.D., Theoretical Analysis of Fractionating Process of Adsorption, *Ind. Eng. Chem.*, **42**, 1279, 1950.
6. Spencer, M.M., and Stephens, O.F., "Petroleum Refining Processes," 4th ed., 1958; College of Minerals Industries, Pennsylvania State University.
7. Strain, H.H., Distribution in Adsorption Columns, *Ind. Eng. Chem.*, **42**, 1307, 1950.
8. White, G.E., Method and Chart for Number of Transfer Units in Diffusional Operations, *Chem. Eng. Prog.*, **46**, 363, 1950.

Meet the Author

Willis A. Johnston has been a consultant specializing in fixed-bed adsorption (610 U.S. 250 By Pass, Charlottesville, VA 22901) since his retirement from Attapulgas Clay Co. (which had been merged into Mineral & Chemicals Phillip Corp.). He finished his career in engineering sales. Before joining Attapulgas, he was assistant to the vice-president and general manager for Bradford Penn Refining Corp., responsible for research and development and engineering. He received his degree in chemical engineering from the University of Pennsylvania, and has been a member of ACS.

Adsorption Systems

Part I: Design by Mass-Transfer-Zone Concept

Both rate and capacity considerations confront the engineer designing a fixed-bed adsorption system. The mass-transfer-zone concept, which is fully discussed, permits a simpler calculation of column height and diameter.

GEORGE M. LUKCHIS, Union Carbide Corp.

In fixed-bed, dynamic adsorption systems, a gas or liquid feed that is rich in the sorbable component flows through a bed, or zone, containing adsorbent particles. Because the contact time between adsorbent and fluid is limited by the rate of flow and the geometry of the bed, the designer is confronted with both rate and capacity considerations.

Rates of adsorption in fixed-bed, dynamic systems may be described in either of the following two ways:

When the designer uses *mass-transfer coefficients*—which might be considered a "micro" approach—he considers each of the resistances encountered, as molecules transfer from the bulk fluid to a film on the adsorbent particles, and eventually to the adsorbed phase. Mass-transfer coefficients provide insight into the mechanism by which adsorption occurs, but they can be difficult to determine and tedious to use in design problems.

With the *mass-transfer-zone concept*—a "macro" approach to mass transfer—a total resistance is expressed in terms of an amount of unused adsorbent, which must be added to the adsorbent equilibrium requirement. The mass-transfer-zone concept—originally suggested by Michaels for fixed-bed ion-exchange columns[2]—provides a simple and effective method for considering rate phenomena in fixed-bed adsorption systems. The mass-transfer-zone concept is particularly amenable to rapid determination and correlation of rate data and to simple design procedures. We will direct our attention solely to the mass-transfer-zone concept, because it is widely applicable.[1]

Mass-Transfer Front

Consider an adsorbent bed with feed fluid entering at the top, and effluent—lean in sorbable component—leaving at the bottom. (The mathematical treatment

These articles are based on parts of a comprehensive manuscript on adsorption engineering written by Union Carbide's Molecular Sieve Dept. As Marketing Communications Manager for the department, Mr. Lukchis was the editor of this extensive work. The manuscript will appear as a series of brochures copyrighted by Union Carbide's Molecular Sieve Dept.

Originally published June 11, 1973

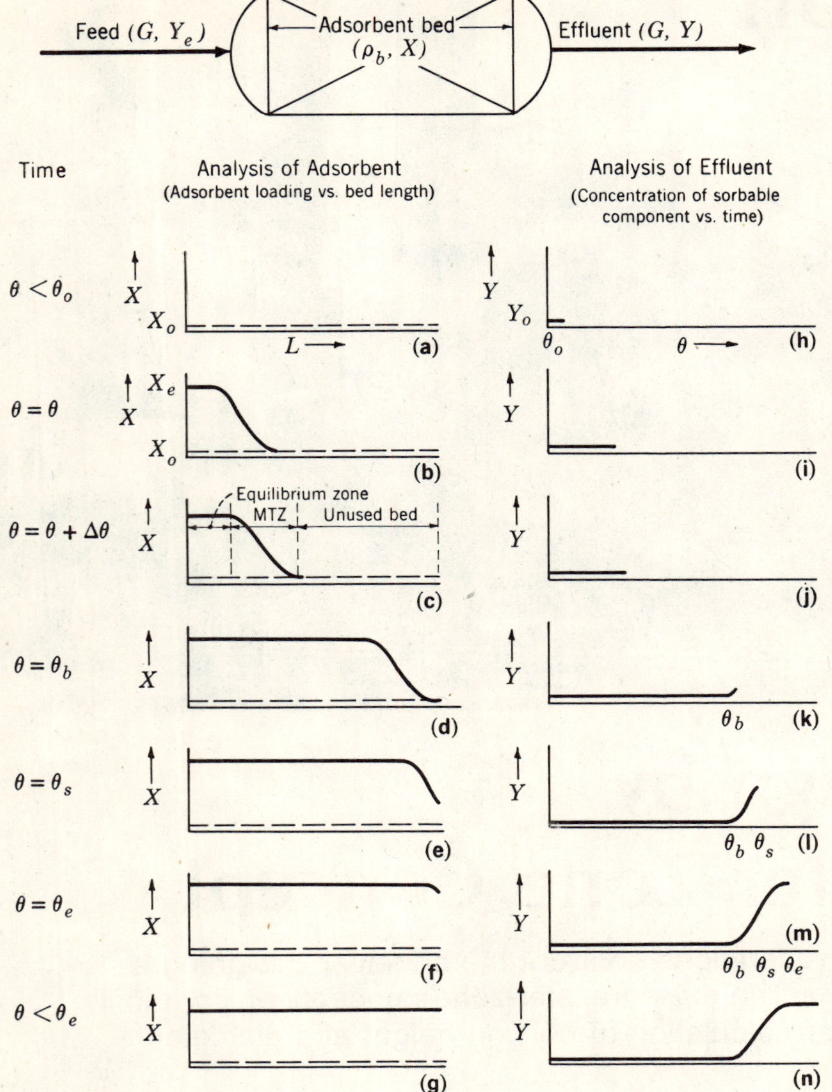

MASS-TRANSFER-ZONE concept—a "macro" approach to mass transfer—provides a simple and effective means for considering rate phenomena in fixed-bed adsorption columns; it facilitates rapid determination and correlation of rate data and makes possible the use of simple design procedures—Fig. 1.

would be identical if flow direction were reversed.) To facilitate our discussion, we will show the adsorbent bed resting on its side (as in Fig. 1), recognizing, however, that flow in a vertical direction is important for proper flow distribution in commercial adsorbers. Directly below the diagram of the adsorbent bed is a series of graphs. In each graph, the adsorbate loading, X, is plotted as ordinate, and distance from the inlet end of the bed, L, as abscissa.

Initially, the adsorbent is highly activated, and the initial adsorbate loading (designated X_o) is low. (This is indicated in Fig. 1a). At some starting time, which we shall designate θ_o, feed at a mass flow rate, G, with uniform concentration of sorbable component is admitted to the activated adsorbent bed. As the fluid passes through the bed, the sorbable component is attracted to and adsorbed on the adsorbent.

Flow continues under steady-state conditions. At some later time, θ, an analysis of uniformly spaced samples taken from the adsorbent bed would show an adsorbate loading curve similar to that shown in Fig. 1b.

At the inlet end of the bed, the adsorbate loading would be X_e, the equilibrium loading corresponding to the concentration of sorbable component in the feed. That portion of the bed whose adsorbate loading is equal to X_e is defined as the *Equilibrium Zone.*

Toward the effluent end of the bed, the adsorbate loading would be X_o, essentially the same as it was when the fluid was first admitted to the bed. That portion of the bed whose adsorbate loading is equal to X_o is defined as unused bed.

In some intermediate zone, the adsorbate loading changes from saturation, X_e, to the initial adsorbate loading, X_o, and this is represented by the "S-shaped wave" in Fig. 1b. It is in this zone that the sorbable component is being transferred from the bulk fluid to the adsorbate phase. That portion of the bed in which the adsorbate concentration transition occurs is defined

Nomenclature

A_x	Cross-sectional area of adsorbent bed, sq.ft.
D	Diameter of adsorbent bed, ft.
G	Superficial mass velocity of fluid, lb./(hr.)(sq.ft.)
L_o	Total length of adsorbent bed, ft.
L_s	Position of stoichiometric front, distance from inlet end of bed, ft.
LES	Length of equivalent equilibrium section, ft.
LES_b	Length of equivalent equilibrium section at breakthrough, ft.
LUB	Length of unused bed, ft.
ΔL	Increase in length of the equivalent equilibrium section, ft.
U	Front velocity, $(G\Delta Y/\rho_b\Delta X)$, ft./hr.
w	Weight of adsorbate, lb.
W_{SB}	Weight of adsorbent in a short bed, lb.
W_{MTZ}	Weight of adsorbent in a steady-state mass-transfer zone, lb.
WES	Weight of adsorbent in equivalent equilibrium section, lb.
WUB	Weight of unused bed, lb.
X_e	Equilibrium zone, or upstream, adsorbate loading, lb./100 lb. of activated adsorbent.
X_o	Initial, or downstream, adsorbate loading, lb./100 lb. of activated adsorbent.
ΔX	Delta loading, $X_e - X_o$, lb./100 lb. of activated adsorbent.
ΔX_{MTZ}	Average delta loading in steady-state mass-transfer zone, lb./100 lb. of activated adsorbent.
ΔX_{SB}	Average delta loading in a short bed, lb./100 lb. of activated adsorbent.
Y_e	Inlet, or upstream, concentration of sorbable component, lb./lb. of feed.
Y_o	Effluent, or downstream, concentration of sorbable component, lb./lb. of feed.
ΔY	Change in concentration of sorbable component in fluid, $Y_e - Y_o$, lb./lb. of feed.
θ	Time, hr.
θ_b	Breakthrough time, hr.
$\theta_{b_{SB}}$	Breakthrough time for a bed whose length is shorter than a steady-state transfer zone, hr.
θ_s	Stoichiometric time, hr.
ρ_b	Bulk density of the adsorbent, lb./cu.ft.

as the *mass-transfer zone*, abbreviated *MTZ*. The S-shaped wave in an *X-L* plot is defined as the *X-L mass-transfer-wave* or *front*.

Bed Breakthrough

As flow continues, the mass-transfer wave moves through the bed in a steady-state fashion, at uniform velocity. If adsorbent samples were analyzed at time $\theta + \Delta\theta$, the *X-L* mass transfer wave would be identical to the wave obtained at time θ, but displaced a distance ΔL in the L direction as shown in Fig. 1c.

Breakthrough is said to occur when the leading edge of the mass-transfer wave just reaches the effluent end of the bed. This is shown in Fig. 1d.

It is important to note, however, that *breakthrough concentration* is arbitrarily defined. It is usually taken as either the *minimum detectable* or the *maximum allowable* concentration of sorbable component in the absorber effluent.

The time at which breakthrough occurs is called the *breakthrough time*, and designated θ_b. If flow is continued, the time at which the trailing edge of the mass-transfer wave reaches the effluent end of the bed is called the *equilibrium time*, and designated θ_e. This is shown in Fig. 1f. At time θ_e, all of the adsorbent in the bed is at equilibrium with the concentration of sorbable component in the feed; the adsorbent capacity is completely spent; and the bed is said to be at *equilibrium*. No discernable changes occur if flow continues beyond time θ_e (as shown in Fig. 1g).

Analyzing Column Operation

Analyzing the adsorbent samples taken from various positions in the bed presents a difficult, although not insurmountable, problem. A simpler method of determining the shape of the mass-transfer front involves measuring the concentration of sorbable component in the effluent. Often, this can be accomplished with a continuous analyzer-recorder, as shown schematically in Fig. 1h through 1n. A curve similar to that shown in Fig. 1n results. The concentration of sorbable component in the fluid, designated Y, is plotted as the ordinate, and time, θ, as the abscissa.

Initially, the concentration of sorbable component in the effluent is equal to Y_o, a low value relative to the concentration, Y_e, in the feed. Y_o is the equilibrium concentration corresponding to X_o, the adsorbate loading at the effluent end of the bed.

From time $\theta = \theta_o$, until the breakthrough time, θ_b, the concentration of sorbable component in the effluent remains equal to Y_o. At time θ_b, when the leading edge of the *X-L* mass-transfer wave reaches the effluent end of the bed, the concentration of sorbable component in the effluent begins to increase.

From time θ_b until time θ_e, the concentration of sorbable component in the effluent increases until it is equal to the Y_e. The smooth, S-shaped wave in the Y-θ plot is defined as the *Y-θ mass-transfer wave* or *front*.

If the bed length were increased by some amount ΔL, and the experiment repeated with all other parameters held constant, the Y-θ mass-transfer wave would be identical to that previously obtained, but displaced in the θ direction. A comparison of the *X-L* and Y-θ mass-transfer waves shows that they are similar but inverted.

In a steady-state system, X and Y are related by equilibrium requirements, and it can be shown that time onstream and bed length are equivalent parameters.

Resistance to Mass Transfer

The fact that real mass-transfer waves are S-shaped is evidence of resistance to mass transfer. The greater the resistance, the longer is the wave. As mass transfer resistance decreases, the wave becomes shorter. In the ultimate, "ideal" case, the curve becomes a vertical, straight line (zero length). The "ideal" mass-transfer wave is called the *stoichiometric wave* or *front*.

An adsorber bed with a stoichiometric front is comprised solely of an equilibrium zone and unused bed, because the length of the mass-transfer zone is zero. At breakthrough under ideal conditions, the concentration

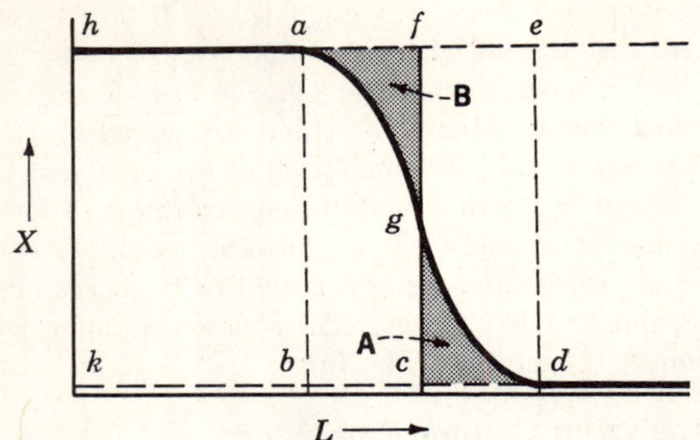

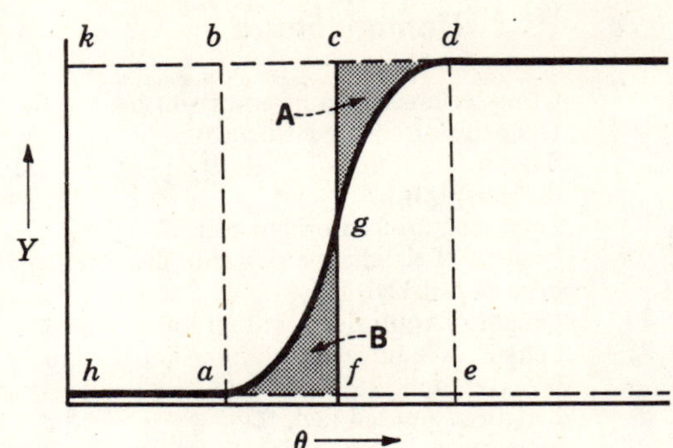

STOICHIOMETRIC FRONT, geometry of which is shown, can be used to analyze real mass-transfer fronts—Fig. 2

of sorbable component in the effluent would change instantaneously from Y_o to Y_e, and the entire bed would be loaded to equilibrium capacity. Breakthrough under the ideal conditions of no mass-transfer resistance is defined as the *stoichiometric breakthrough*. The time at which stoichiometric breakthrough occurs is called the *stoichiometric time,* and designated θ_s.

Analyzing Mass-Transfer Waves

All real mass-transfer waves can be analyzed in terms of an equivalent stoichiometric front. Consider the geometry of the transfer waves in Fig. 2. The area under the *X-L* wave (*a,g,d,b,a*) represents used adsorbent capacity; the area above the wave (*a,g,d,e,a*), unused adsorbent capacity. Thus, the ratio of the area under the wave to the total area of the rectangle (*a,b,d,e,a,* which just encloses the mass-transfer wave) is the fraction of used adsorbent in the mass-transfer zone.

In a *Y-θ* diagram, geometric considerations are reversed. The area below the wave (*a,g,d,e,a*) reflects unused adsorbent capacity. The ratio of the area above the wave (*a,g,d,b,a*) to the total area of a rectangle (*a,b,d,e,a,* which just encloses the wave) is the fraction of used adsorbent in the mass-transfer zone.

In either plot, a vertical straight line drawn through the mass transfer wave at *g*—such that the area *a,f,c,b,a,* is equal to the area *a,g,d,b,a;* and the area *f,e,d,c,f* is equal to the area *a,g,d,e,a*—yields an equivalent stoichiometric front for the particular system. Then the entire rectangle *h,f,c,k,h* corresponds to adsorbent at its equilibrium loading Y_e, and is defined as the equivalent equilibrium section. It is specified in terms of the *length of equivalent equilibrium section, LES,* or *weight of equivalent equilibrium section, WES:*

$$WES = LES\,(\pi D^2/4)\rho_b \tag{1}$$

The area *f,e,d,c,f* corresponds to adsorbent at its initial loading, X_o, and by definition, it is *equivalent unused bed,* or briefly called *unused bed.* It is specified in terms of the *length or weight of unused bed, LUB* or *WUB:*

$$WUB = LUB\,(\pi D^2/4)\rho_b \tag{2}$$

Thus, in accordance with this model, an adsorbent bed at breakthrough is actually comprised of an equilibrium zone and a mass-transfer zone (*MTZ*); or it is hypothetically comprised of an equivalent equilibrium section (*WES* or *LES*) plus equivalent unused bed (*WUB* or *LUB*), and this is tantamount to a bed with a stoichiometric front. The *WES* can be calculated directly from equilibrium data and the quantity of sorbable component to be removed from the feed. The *WUB* is developed from dynamic data obtained in a fixed-bed adsorption system.

Assumptions of the Model

Some important assumptions are implied in the development of this model: (1) the bed is uniformly packed; (2) the adsorbent temperature and initial adsorbate loading are uniform throughout the bed; (3) the feed rate, feed temperature and feed composition are constant; (4) there are no radial temperature, concentration or flowrate gradients; (5) the temperatures of both the fluid and adsorbent are essentially equal; (6) the fluid does not undergo a phase change; (7) adsorption heat effects are negligible; and (8) chemical reactions do not occur.

Caution should be exercised if this model is applied to systems involving chemisorption reactions, high concentrations of sorbable component (i.e., bulk separations), large heat effects, or non-steady-state conditions.

Mathematics of the Model

A mathematical treatment will help to make the model clear. Consider the adsorber shown schematically in Fig. 3. Part **a** of Fig. 3 depicts a stable mass-transfer zone moving at uniform velocity, *U,* through an adsorber vessel. In Part **b**, the bed is shown with stoichiometric fronts only, and in Part **c**, the stoichiometric transfer front is superimposed over the actual transfer front.

The total amount of sorbable component, *w,* removed from the bulk fluid between time θ and time $(\theta + \Delta\theta)$ is expressed by Eq. (3):

$$w = G(Y_e - Y_0)\,A_z\Delta\theta \tag{3}$$

As *w* lb. of sorbable component is adsorbed on the bed, the shape of the mass transfer wave is unaltered although its position changes. Thus, the tangible effect of adsorbing *w* on the bed is an increase in length, ΔL.

of either the equilibrium zone or equivalent equilibrium section, *LES,* as expressed by Eq. (4):

$$w = (\Delta L)A_x\rho_b(X_e - X_0) \tag{4}$$

Therefore,

$$(\Delta L)\,A_x\rho_b(X_e - X_0) = G(Y_e - Y_0)A_x\Delta\theta \tag{5}$$

Rearranging,

$$\Delta L = (G\Delta Y/\rho_b\Delta X)\Delta\theta \tag{6}$$

Or,

$$\Delta L = U\Delta\theta \tag{7}$$

During steady-state operation, the values of G, ρ_b, ΔX and ΔY are fixed. Thus, the expression $(G\Delta Y/\rho_b\Delta X)$ is a constant having units of length-time, and it is seen to be the velocity at which the equilibrium zone expands. It is also the velocity at which the mass transfer wave moves through the bed.

It is noteworthy that the velocity of a steady-state mass-transfer wave depends only on the fluid mass velocity, the bulk density of the adsorbent and the terminal conditions. Local mass transfer rates within the front have no effect on the front velocity, although they do affect the shape of the front.

At any time, θ, the position of the stoichiometric front is given by Eq. (8):

$$L_s = U\theta \tag{8}$$

By definition, at time θ_b:

$$L_s = LES \tag{9}$$

Therefore,

$$LES = U\theta_b \tag{10}$$

By definition, at time θ_s:

$$L_s = L_o \tag{11}$$

Therefore,

$$L_o = U\theta_s \tag{12}$$

However,

$$LUB = L_o - LES \tag{13}$$

Or, combining Eq. (10), (12) and (13):

$$LUB = U(\theta_s - \theta_b) \tag{14}$$

Combining Eq. (12) and (14), by eliminating U:

$$LUB = L_0(\theta_s - \theta_b/\theta_s) \tag{15}$$

Thus, simple breakthrough curves provide the data required for the correlation of rate data in dynamic systems. Eq. (15) is the basic equation used for analyzing breakthrough data.

Analyzing Commercial Units

When analyzing the performance of commercial units, operation past breakthrough is not usually permissible. Combining $U = (G\Delta Y/\rho_b\Delta X)$ (the velocity front) with Eq. (12) gives:

$$\theta_s = L_o/U = L_o(\rho_b\Delta X/G\Delta Y) \tag{16}$$

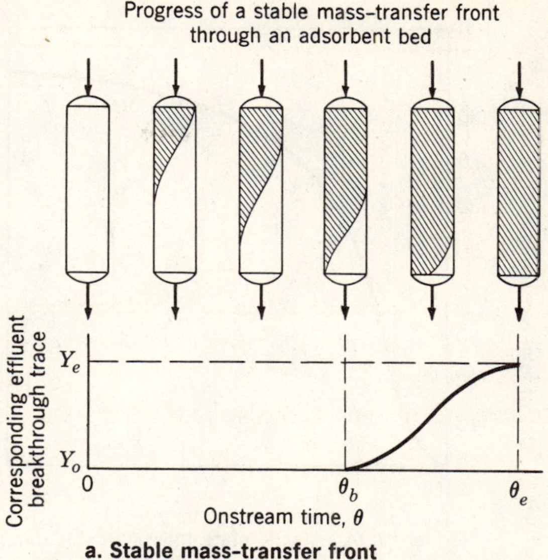

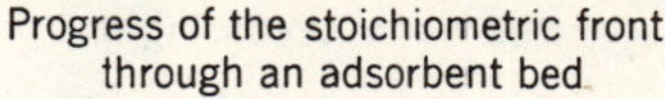
a. Stable mass-transfer front

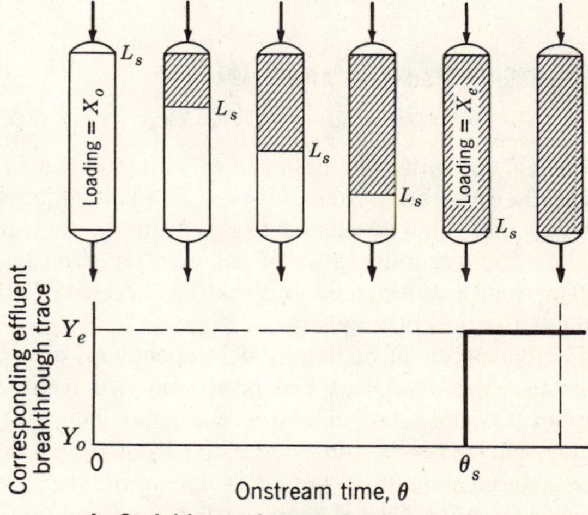

b. Stoichiometric transfer front

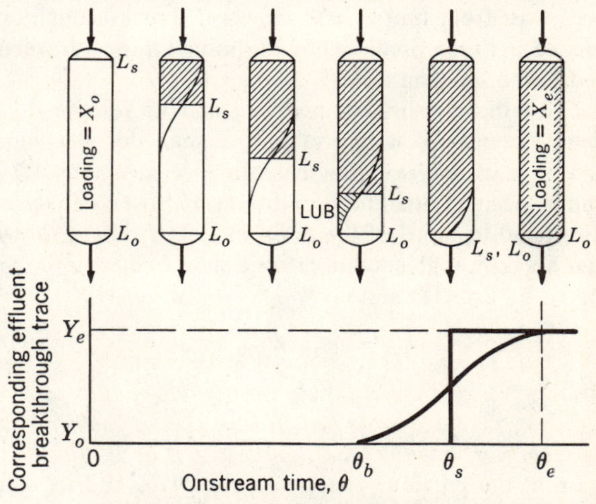

c. Stable front relative to stoichiometric front

TRANSFER FRONTS shown moving through adsorber—Fig. 3

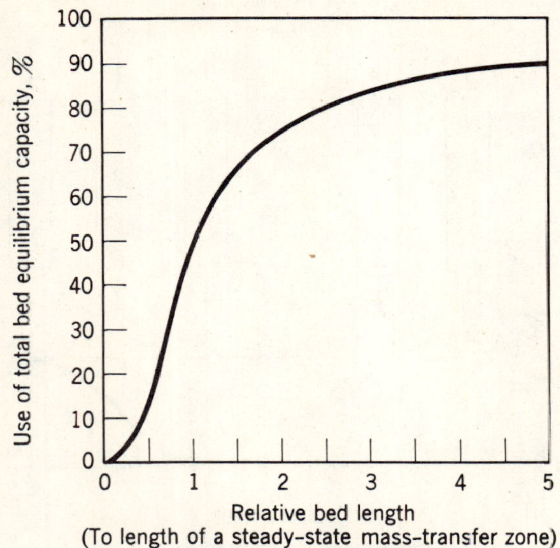

BED USE is compared to increasing bed length—Fig. 4

Then, combining Eq. (14) and (15) gives:

$$LUB = L_0 - (G\Delta Y/\rho_o \Delta X)\theta_b \tag{17}$$

Eq. (17) permits the designer to estimate the *LUB* when the adsorption step must be stopped at breakthrough. The *LUB* so calculated is sensitive to the value of ΔX. (An accurate value of ΔX is often difficult to obtain from a commercial unit that has "cycled" for an extended period of time.)

The development of this model depends on assumptions that the adsorbent bed is of sufficient length to contain a steady-state mass-transfer zone, and that a steady-state mass-transfer zone exists from time $\theta = 0$. The fact that some finite period of time at the beginning of an adsorption step is required for the formation of the steady-state *MTZ* does not complicate this model, provided that the bed is long enough to contain a steady-state *MTZ*. An equivalent stoichiometric front does exist from time $\theta = 0$. However, breakthrough occurs earlier than predicted by this model if the adsorbent bed is shorter than an *MTZ*.

Thus, the question of bed size and its relation to a steady-state *MTZ* is important. We shall define a short bed as one that will not contain a steady-state *MTZ* under specified operating conditions. If Rosen's analysis[3] is restated in terms of the *MTZ* model, it can be shown that breakthrough conditions for a short bed are approximated by Eq. (18) and (19):

$$\theta_{b_{SB}} = (W_{SB}/W_{MTZ})^2 \theta_{MTZ} \tag{18}$$

Or,

$$(\Delta X)_{SB} = (W_{SB}/W_{MTZ})^2 \, (\Delta X)_{MTZ} \tag{19}$$

From the previous analysis of the *MTZ* and its relationship to *LUB*, it is readily apparent that the average adsorbate delta-loading in the *MTZ* is approximately half that of the equilibrium zone; that is, adsorbent use in the *MTZ* is approximately 50%. Eq. (19) suggests that adsorbent use falls off rapidly as bed size is reduced below that required to contain an *MTZ*.

Increasing the bed size beyond the length of the *MTZ* adds equilibrium capacity—adsorbent that provides 100% use. An approximate relationship between capacity and bed length is shown in Fig. 4. Bed length is expressed in terms of the length of an *MTZ*. Capacity is expressed in terms of overall, average adsorbent use, in which equilibrium zone ΔX is defined as 100%.

If adsorbent cost is a major factor influencing the design of an adsorbent bed, one would deduce from Fig. 4 that the preferred bed length would be approximately 1.5 to 3 times the size of the *MTZ*.

Much of the adsorption breakthrough data that has been reported had been obtained in short columns. In some instances, the conclusions based on these data led to downgrading the adsorption process because of apparent low adsorbate loadings. Had the data been analyzed in light of Fig. 4, or Eq. (19), the adsorption process might have appeared more attractive.

To obtain reliable breakthrough data, the laboratory column should be at least twice the length of the steady-state mass-transfer zone, and the column diameter at least 10 times the diameter of the largest adsorbent particle.

Heat Effects and Space Velocity

Typically, the steady-state wave is inversely symmetrical. A skewing (or tailing) of the wave is caused principally by the heat generated during adsorption. When the concentration of the sorbable component is high, the heat generated can become large enough to grossly distort the mass-transfer wave. In extreme cases, the heat can even prevent formation of a stable mass-transfer wave.

Space velocity appears to be a poor design-parameter for adsorption systems that operate with steady-state mass-transfer zones. Space velocity is defined as: (vol. of feed fluid/hr.)/vol. of adsorbent bed. The definition implies that the entire volume of adsorbent is actively involved in the adsorption process at all times, which contradicts observations of mass-transfer zones.

References

1. Collins, J. J., *Chem. Eng. Prog.*, Symp. Series No. 74.
2. Michaels, A. S., *Ind. Eng. Chem.*, **44,** *8,* 1952, pp. 1922–30.
3. Rosen, J. B., *Ind. Eng. Chem.*, **46,** *8,* 1954, pp. 1590–94.

Meet the Author

George M. Lukchis is Marketing Communications Manager of Union Carbide Corp.'s Molecular Sieve Dept. (270 Park Ave., New York, NY 10017). He has been with Union Carbide for seven years, during which time he has been involved with a variety of business areas dealing with the chemical process industries. Previously, he had been a copywriter with McGraw-Hill Inc. and American Standard Co., and had written numerous technical abstracts as a freelance writer. He graduated from Upsala College.

Adsorption Systems

Part II—Equipment Design

Should flows pass up or down through adsorbers during adsorption and regeneration? What is the right bed length and diameter? How can the problem of fluid holdup be handled best? What are suitable internal arrangements?

GEORGE M. LUKCHIS, Union Carbide Corp.

A question frequently debated is whether fluids should flow up or down through adsorber beds. There is ample commercial experience to demonstrate successful operation for either flow direction. Nevertheless, there are factors that may influence the choice.

As a rule, flow through adsorber beds should not be in a horizontal direction. Bed settling would permit flow over the top of the bed, bypassing it. There would also be a high probability of particle stratification—in planes parallel to the direction of flow—because the forces associated with gravity and fluid flow would be acting at right angles to the bed of adsorbent. So, flow channeling would likely be a problem.

Flows in the Vertical Direction

In liquid-phase adsorption and vapor-phase regeneration systems, a downward purge-gas flow facilitates the removal of holdup liquid. When liquid filling is from bottom to top, holdup gas is displaced uniformly by the piston-like interface. When, however, liquid filling is from the top, the interface still moves upward, but now it will form gas pockets unless a method of venting is provided. Entrapped gas pockets could cause flow channeling problems. However, both flow schemes are being commercially used successfully.

Desorption flow direction relative to adsorption flow

These articles are based on parts of a comprehensive manuscript on adsorption engineering written by Union Carbide's Molecular Sieve Dept. As Marketing Communications Manager for the department, Mr. Lukchis was the editor of this extensive work. The manuscript will appear as a series of brochures copyrighted by Union Carbide's Molecular Sieve Dept.

Originally published July 9, 1973

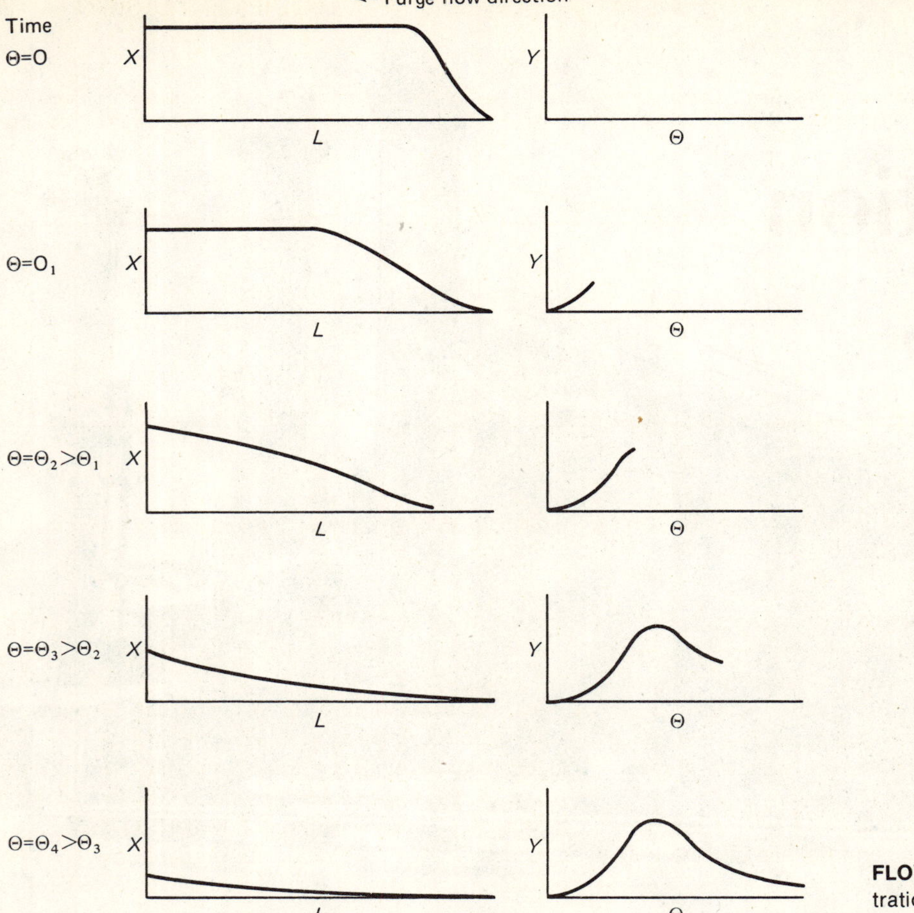

FLOW DIRECTION affects concentration front—Fig. 1

direction influences product purity, heat and purge requirements, and mechanical and operational simplicity. If desorption is carried out cocurrently to adsorption flow, all of the adsorbate must move through the entire bed, including the unused section, in each cycle. Some of the fluid initially desorbed from the equilibrium section will be readsorbed in the unused section, where it must be desorbed a second time. This condition is essentially equivalent to one in which the entire bulk of adsorbent is initially loaded to equilibrium.

If, on the other hand, desorption flow is countercurrent to adsorption flow, the unused section may absorb some heat but not otherwise influence desorption. Unused bed will never contact primary adsorbate.

Countercurrent desorption with a stripping medium can be schematically depicted on a series of *X-L* or *Y-θ* diagrams (Fig. 1). The end of the bed into which the desorption fluid is fed is desorbed first. The concentration of the sorbable component in the desorption fluid increases as the fluid flows through the bed. This gives rise to a residual loading gradient, which is always at its lowest concentration at the end of the bed at which the desorption fluid enters.

If the flow is cocurrent, the residual gradient would be reversed. A uniform, equilibrium residual gradient throughout the entire length of the adsorbent bed would require very large volumes of purge fluid, and is economically and technically unrealistic.

Two qualitative conclusions are: countercurrent stripping is more efficient because the unused section of the bed is not desorbed, and the effluent end of the bed in adsorption becomes the most highly activated part when desorption flow is countercurrent (this favors a lower concentration of sorbable component in the product fluid in the subsequent adsorption step).

Thermal-Swing-Cycle Regeneration

In thermal swing cycles, an additional consideration involves the direction of flow during the cooling step. Purge flow during the heating step can be cocurrent or countercurrent to the adsorption flow. For each of these two cases, the cooling flow can be either cocurrent or countercurrent. A set of *X-L* diagrams that show qualitatively the general shape of the residual loading gradients after the cooling step for each of the four cases is presented in Fig. 2.

Some qualitative observations are:

1. *Countercurrent heating and cooling* can provide the lowest residual loading at the effluent end of the bed in the adsorption step if the cooling fluid is free of sorbable component. Countercurrent cooling would be highly detrimental to subsequent adsorption if the cooling fluid contained significant quantities of sorbable component, because the sorbable component from the cooling purge would be adsorbed as the bed was cooling. This would increase the residual loading, which would be highest at the end of the adsorber that becomes the adsorption-step effluent.

2. *Cocurrent heating and cooling* always leaves a relatively high residual loading at the effluent end of the bed. Both in cocurrent heating and cooling and counter-

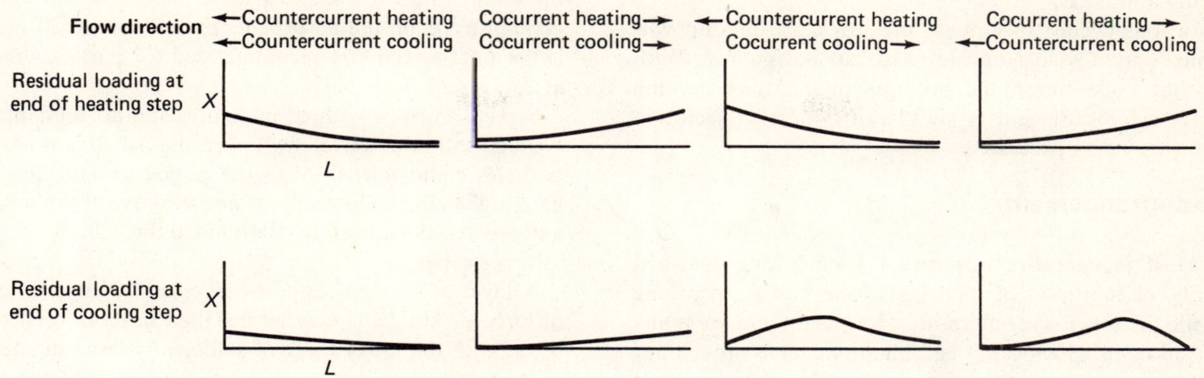

SHAPE of residual loading gradient depends on whether purge and cooling flows are cocurrent or countercurrent—Fig. 2

current heating and cooling, desorption continues during the cooling step, because the movement of concentration and thermal gradients is not reversed when cooling commences. The cooling fluid, after being heated upstream, acts as a hot purge and continues to do some desorbing as it flows through downstream sections of the bed.

3. *Countercurrent heating and cocurrent cooling* makes the residual low at the effluent end of the bed in the subsequent adsorption step. This is a preferred mode of operation in many drying and purification operations because adsorption is not afterwards harmed by the sorbable component in the cooling fluid. In many commercial processes, the sorbable component may be present in the regeneration fluid during startups and periods of emergency operation.

4. *Cocurrent heating and countercurrent cooling* is not a common mode of operation. It is used occasionally when feed fluid is the closed-cycle hot purge, and purified product is the cool purge—for example, to minimize dilution of the desorbed component.

Flow Velocity

Typical superficial linear velocities through beds of molecular sieves are on the order of 1 ft./sec. for gases, and 1 ft./min. for liquids. Although these rates are often presented as typical, the actual velocities in commercial units range broadly. For example, a superficial linear velocity of 0.1 ft./sec. in a gas-adsorber system is not unusual.

Flow velocity, pressure drop and bed diameter are all related. When any one of these parameters is fixed, all three are. A limitation on pressure drop is usually the key parameter, and is generally the basis for fixing the other two.

Bed Diameter

Vessel costs tend to increase dramatically with diameter. This becomes more significant as the operating pressure (wall thickness) goes up. The minimum diameter for an adsorber bed is set by pressure drop limitations. A pressure drop analysis is required for each of the steps in the adsorption cycle, including the pressuring and depressuring steps.

Maximum pressure drop in commercial systems usually occurs during regeneration, when the bed is at its highest temperature or lowest pressure. For thermal swing cycles, these conditions occur at the end of the heating step or the beginning of the cooling step.

Length-to-diameter ratio of adsorber vessels is a poor correlating parameter for molecular sieve systems because wide variations in ratios are possible. In any case, careful attention to inlet and outlet flow distributors is necessary.

Fluid Holdup

During the normal step-changes of an adsorption cycle, fluid from the previously completed step is present in macropore voids, in the voids between particles and in the vessel head spaces. This fluid is referred to as holdup. If improperly processed and accounted for, this holdup fluid can contaminate product streams, upset heat balances, and adversely affect regeneration.

In thermal swing cycles, liquid holdup may grossly distort regeneration heat requirements. Liquid that is not drained, or otherwise removed, from a bed must be vaporized during regeneration. The latent heat requirements can represent a sizable heat load on the regeneration system.

Time should be allowed in the regeneration step for thorough drainage in liquid systems. Vapor purge or padding with gas pressure in the down direction can increase the drainage rate and remove liquid holdup more completely. A minimum drainage period of a half hour is recommended in liquid-phase drying and purification. Even after thorough drainage, some liquid is still retained in the macropore structure of the adsorbent.

If the holdup fluid is a reactive liquid, it can be displaced by washing with a nonreactive liquid before normal regeneration is started. For example, olefins should be removed before a thermal regeneration to prevent excessive polymerization. One regeneration procedure involves a displacement washing with butane, followed by a normal draining, then by a thermal-swing regeneration with natural gas.

In long-cycle vapor-phase processes, the holdup volume is usually insignificant and can be ignored. With shorter cycle times, the holdup volume may become more significant, and is always an important factor in pressure-swing processes.

Bed Arrangement

Most regenerative adsorption systems are dual-bed units, or multiples of dual beds; one bed is adsorbing while the other is regenerating. However, there are many instances in which other combinations have proven advantageous. Three, four, five or more bed units are not uncommon.

For example, three-bed units are sometimes used when mass-transfer zones are long. Each bed "moves" sequentially from (1) *trim*, or downstream, adsorption to (2) *lead*, or upstream, adsorption and then to (3) regeneration. A bed spends a third of its cycle time in each of the three positions. The trim bed is long enough to contain a mass-transfer zone, and it guards against breakthrough. In the lead position, nearly all of the adsorbent becomes loaded to equilibrium capacity.

Three-bed units are also used in thermal swing cycles, when there is a shortage of regeneration gas. For example, in CO_2 removal- or natural-gas-sweetening units, each bed "moves" from (1) adorption to (2) heating and then to (3) cooling. A bed spends a third of its cycle time in each of the three positions. In one arrangement, clean purge gas flows first to the bed to be cooled, next to a heater, then to the bed to be heated and desorbed, and finally to discharge.

Many bed combinations are possible. They are limited only by the imagination of the designer and the economics of the application.

Equipment Considerations

Simplicity is the key to the design of commercial adsorber beds. Adsorbers seem to perform best when the quantity and complexity of vessel internals are kept to a minimum.

The internals of a typical adsorber consist of inlet and outlet distributors and a support for the bed. An inert ballast may be installed on top of the bed. Intermediate supports may be added if the bed is deep.

Adsorbent particles may be supported on a mechanical grid or on inert catalyst support-balls. Mechanical grids may consist of catwalk grating or a similar material, supported on I-beams. A series of screens, or a 4-to-8-in. layer of inert catalyst support-balls, resting on the mechanical grid provides the final support for the adsorbent.

An alternate approach to the mechanical grid involves completely filling the lower head space with inert catalyst support-balls. Normally, the balls are graded in layers, with 1 to 2-in.-dia. ones at the bottom, and gradually smaller sizes on top, to perhaps ¼-in. dia. The adsorbent rests directly on the support-balls.

Screens may be used to retain the adsorbent above a mechanical grid. A set of screens is customary because the screens must be stepped down in size to one small enough to retain the adsorbent. U. S. Standard 20 mesh is the final screen size recommended for particles averaging 1/16 or ⅛ in. dia.

Screens must be without punctures and must maintain contact with the vessel wall, because small adsorbent particles could flow through the punctures and around gaps at the edges. Contact between the edge of the screen and the vessel wall can be maintained through the use of asbestos rope.

A layer of catalyst support-balls may be used instead of screens. The balls may be installed in more than one layer, with the lower layer of sufficiently large diameter to be retained on the mechanical support, and the top layer's diameter small enough to retain the adsorbent. In general, the average particle diameter can be stepped up or down by a factor of 2-to-4 without an excessive accumulation of smaller particles in the void space of the larger-diameter balls in the lower layer.

The top layers of an adsorbent bed are sometimes subjected to liquid splashing or high-velocity fluid impingement. This occurs, for example, when liquid is initially added to the bed after a vapor-phase regeneration. It also occurs with high-pressure gas systems, with improperly designed or installed distributors, and during repressuring steps.

A layer of support-balls on top of the adsorbent bed serves as a fluid buffer and prevents movement of adsorbent on the surface caused by high local velocities. When the balls are placed on top of an adsorbent bed, the layer in direct contact with the adsorbent should have a diameter approximately two to four times larger than the average diameter of the adsorbent. A second layer of balls of larger diameter is often used.

If large-diameter balls are placed directly on top of the adsorbent, they tend to migrate to the bottom of the bed. Migration is promoted by the cyclic nature of adsorption processes, and is most likely to occur if the adsorbers are located in an area of high vibration—for example, in the vicinity of reciprocating compressors. A floating screen between the inert balls and the adsorbent is recommended.

Poor flow distribution at either the inlet or outlet of an adsorber may cause flow channeling through the bed. One of the simpler methods of assuring good flow distribution is to provide ample plenum space above and below the bed. Plenum space can be free volume, or filled with coarse support-balls.

Simple baffle plates are often preferred for inlet and outlet distributors. They function well over a wide range of fluid velocities, and are not subject to plugging or fouling. If the vessel heads are packed with support balls, the baffle plates must be covered with strong screens, and the balls in direct contact with the baffle plate should be of a large diameter.

All adsorber beds that operate at elevated temperatures should be insulated. This includes all systems that are thermally regenerated, because heat losses add to regeneration purge requirements. The insulation should cover all hot lines and the vessel heads, not merely the vessel walls. Even in well-insulated systems, heat losses

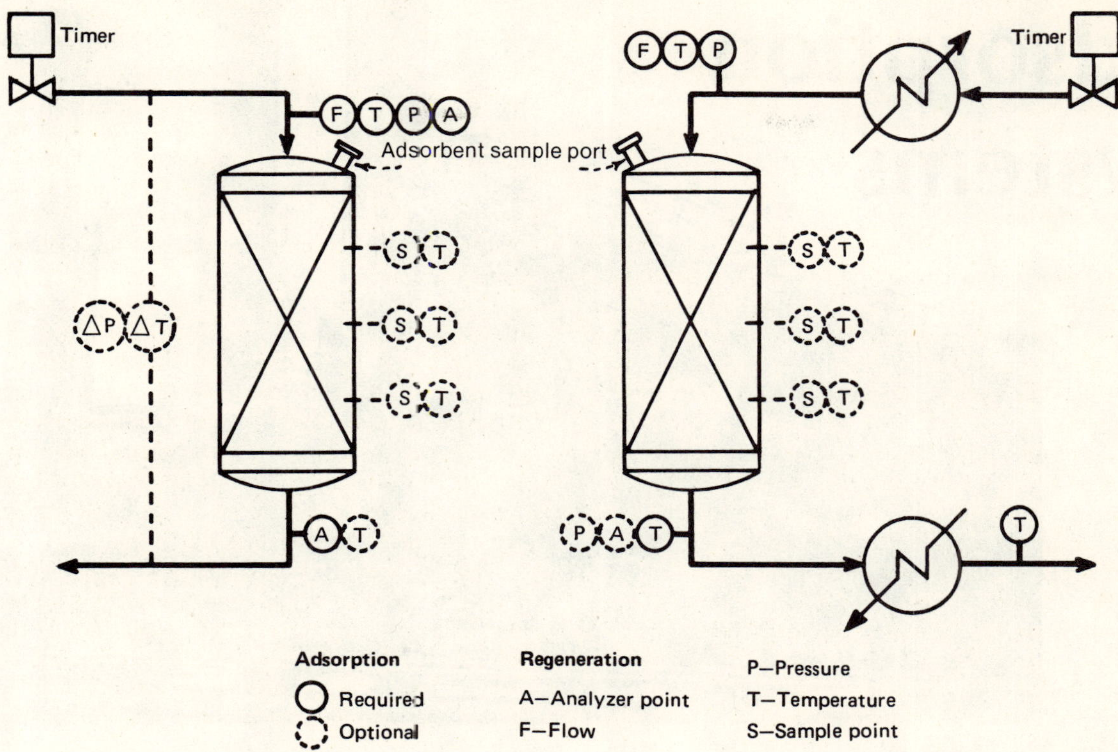

TIMER-CONTROLLED VALVES primarily control the operation of adsorber systems. Valve sequencing may be automatic or semi-automatic. Noted is whether particular instrumentation and sample taps are required or optional—Fig. 3

in large commercial adsorbers generally limit the steady-state temperature approach between heater outlet and bed effluent to a δT of approximately 50 F. External insulation is usually sufficient. Internal insulation can be more effective, because it eliminates heat loss to the vessel. It is especially recommended when the availability of regeneration gas is limited.

Design of the desorption steps is usually based upon a constant flowrate for the purge or displacement fluids. For a thermal swing cycle, the time assigned to heating and cooling is directly proportional to the energy balances associated with these two steps. Equipment requirements are simplified when the heating and cooling fluids flow in the same direction. Ordinarily, this results in a simpler manifold and fewer valves than when the cooling fluid flows countercurrently to the heating fluid. Bed movement is also minimized when all flows are in the same direction. Downward flow is preferred because larger pressure drops can be tolerated. Pressuring and depressuring steps should preferably be done in the down direction, because this minimizes particle movement that can result from rapid pressure changes.

Operating Control

The operation of adsorber systems tends to be simple. Most adsorbers operate on automatic or semi-automatic valve sequencing, with timer mechanisms providing primary control. Performance evaluation is based upon flow, temperature, pressure and composition measurements. Analytical instrumentation is recommended for most systems. Fig. 3 schematically depicts the instrumentation suggested.

Vessel entry ports for taking adsorbent samples can be most helpful. Periodic analyses of bed samples are useful in measuring a unit's performance and in projecting when the adsorbent will have to be changed. Adsorbent analyses also help to identify secondary adsorbates and adsorbent poisons.

An example of the kind of problem that can arise when instrumentation or provisions for analyses are inadequate is illustrated by the performance of a purifier upstream to a plant for liquefying natural gas. The molecular-sieve system had been designed for a CO_2 feed concentration of 0.55%. The unit did have a CO_2 analyzer on the effluent stream, but not one for measuring CO_2 in the feed.

Shortly after startup, freezeups occurred in the cold box. It took considerable investigation to establish that the CO_2 content in the natural-gas feed was oscillating widely, dropping to as low as 0.2% and rising to considerably higher than 1%. The gas supplier consistently, and accurately, reported the daily average CO_2 concentration at less than 0.5%. Of course, it was the instantaneous concentrations that were important.

The CO_2 content is now analyzed several hundred miles upstream in an incoming pipeline and the data are relayed directly to the control room. This alerts the operators to make necessary adjustments several hours before a sharp CO_2 concentration peak arrives.

Adsorption Systems

Part III: Adsorbent Regeneration

Desorption is a key step in the performance of regenerative adsorption systems. It involves simultaneous mass and heat transfer passing through a packed bed of adsorbent under carefully controlled conditions.

GEORGE M. LUKCHIS, Union Carbide Corp.

Regenerative adsorption systems can be divided into two categories, which differ primarily in terms of the concentration of sorbable component in the bulk fluid:

Drying and purification includes applications in which a contaminant at concentrations of about 3% or less is removed from a carrier fluid. The removal of water, odorants and colorants represent typical examples. The adsorbate in these applications is generally low valued and is frequently removed for disposal.

Bulk separations involve applications in which the sorbable component represents a significant fraction of a bulk fluid, generally on the order of 3% to 50%. A bulk separation process recovers both the carrier fluid and the adsorbate.

In many bulk separations, the adsorbate represents the higher-valued product, and its efficient recovery at high purity is essential to the economical operation of the process. The separation of normal paraffins from kerosene is one example. In other bulk separation processes, the carrier fluid is the higher-valued product. The upgrading of a refinery off-gas to produce a hydrogen product of 95% to 99% purity is a typical example. Most bulk separation processes operate on rapid cycles, with relatively small delta loadings.

These articles are based on parts of a comprehensive manuscript on adsorption engineering written by Union Carbide's Molecular Sieve Dept. As Marketing Communications Manager for the department, Mr. Lukchis was the editor of this extensive work. The manuscript will appear later as a series of brochures copyrighted by Union Carbide's Molecular Sieve Dept.

Originally published August 6, 1973

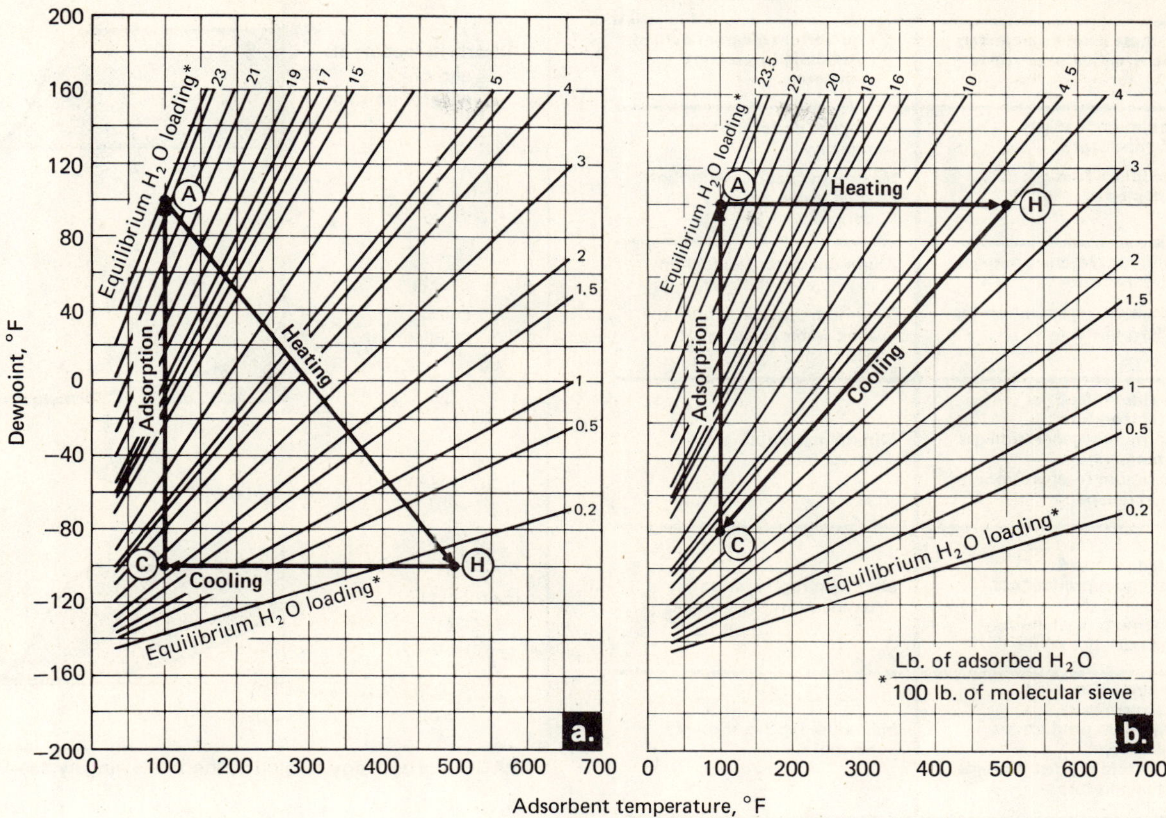

ISOSTERES for water on molecular sieve; activation conditions are 350 C. and less than 10 microns Hg.—Fig. 1

Desorption Step Classifies Cycle

A regenerative adsorption cycle may be classified according to the manner in which desorption occurs.

A *thermal-swing cycle* is defined as one in which desorption takes place at a temperature higher than that of the adsorption step, and in which the temperature differential is the major factor causing desorption. Cooling is usually required before the adsorbent is ready for a subsequent adsorption step.

A *pressure-swing cycle* is one in which desorption is carried out at a pressure lower than that of the adsorption step, and in which the pressure differential is the major factor bringing about desorption.

A *purge-gas-stripping cycle* is one in which desorption occurs primarily through the stripping action of a purge gas that is not adsorbed. The gas reduces the partial pressure of the sorbable component in the fluid phase and thereby affects desorption.

A *displacement cycle* is one in which an initial adsorbate is desorbed by the action of a second sorbable fluid, which displaces all or part of the initial adsorbate. Displacement fluids may be more strongly or less strongly sorbed than the initial adsorbate.

Equilibrium Limits of a Cycle

The equilibrium limits achieved at the end of each operating step of an adsorption-desorption cycle can be depicted as points on equilibrium diagrams, if the assumption is made that equilibrium conditions exist at the end of each step. This is not actually the case with most commercial units because of rate limitations. However, equilibrium diagrams permit one to consider limiting performance.

An operating cycle for a thermally regenerated, fixed-bed dehydrator can be shown by plotting an isostere diagram. (The isostere diagram is chosen, in this instance, because each of the steps can be depicted as straight lines, but any equilibrium diagram could be used.)

The procedure is as follows:

The equilibria corresponding to the operating conditions at the end of each step of a complete cycle are plotted. Then the steps that make up the cycle are depicted as lines connecting the points, in sequence as ordered by the process cycle. Although the actual path between the points may not be known, the steps can be represented as straight lines.

Consider a simple air dryer. Assume that design conditions stipulate that feed air, available at 100F. and water saturated, is to be dried to a dewpoint of – 100 F., and that regeneration is to be accomplished with a slipstream of dry product air heated to 500 F. The foregoing is sufficient to establish thermodynamic limits relative to the performance of the dryer.

Fig. 1a is an isostere diagram for water on molecular

Design Information from Equilibrium Diagrams

If these design parameters are specified or assumed:	Equilibrium diagram defines these design/operating parameters:
Maximum feed-gas temperature Maximum feed-gas dew-point	Maximum desiccant water capacity Maximum cyclic equilibrium-zone water capacity of desiccant
Minimum regeneration-gas temperature Maximum regeneration-gas dewpoint	Minimum water residual on regenerated desiccant Maximum cyclic equilibrium-zone water capacity of desiccant
Maximum feed-gas temperature Minimum regeneration-gas temperature Maximum regeneration-gas dewpoint	Minimum product-gas dewpoint
Maximum feed-gas temperature Maximum product-gas dewpoint Maximum regeneration-gas dewpoint	Minimum regeneration-gas temperature
Maximum feed-gas temperature Maximum product-gas dewpoint Minimum regeneration-gas temperature	Maximum regeneration-gas dewpoint
Maximum feed-gas temperature Maximum product-gas dewpoint	Combination of minimum regeneration gas temperature and maximum regeneration gas dewpoint

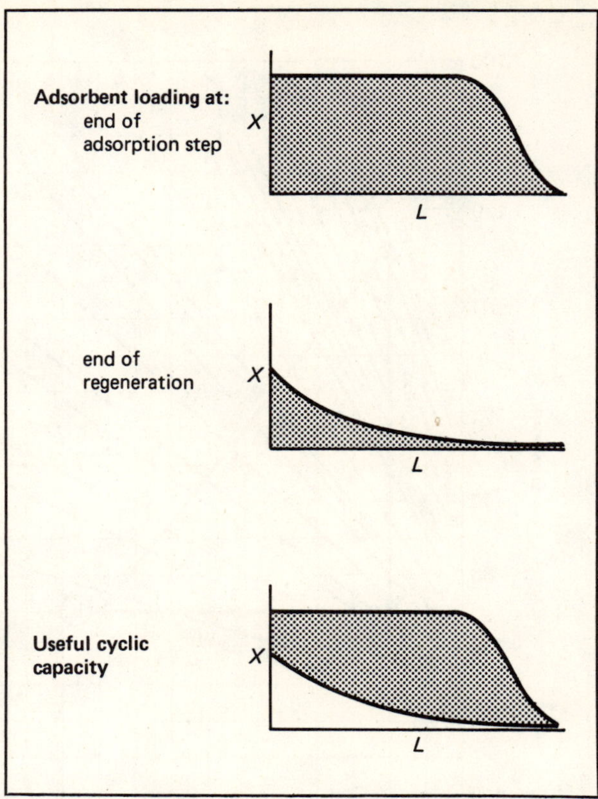

CAPACITY of bed regenerated by thermal-swing cycle—Fig. 2

sieve Type 4A. Point A depicts the equilibrium condition at the end of the adsorption step, assuming equilibrium had been attained. We will refer to these conditions as pseudo-equilibria. Point H depicts the psuedo-equilibrium condition at the end of the heating step, and Point C the same at the end of the cooling step (only if impurities in the cooling gas are sufficient to bring the total bed to the higher residual level).

Thus, at Point C the adsorbent bed is at a temperature of 100 F., in equilibrium with gas having a dewpoint of −100 F., and the equilibrium residual water loading is equal to 2.5%. (This assumes that sufficient cooling gas is used so that the quantity of contained water is adequate to bring the total bed residual loading from 0.2% to 2.5%; because this is usually not the case, the line H–C exhibits a slight slope to an even lower dewpoint.)

As adsorption proceeds, the water loading on the adsorbent increases to 23.5%, which is the loading at 100 F. for molecular sieve in equilibrium with sorbable component whose concentration is expressed as a dewpoint of +100 F. Because the adsorption process is essentially isothermal, the adsorption step is depicted as a vertical (constant temperature) line from Point C to Point A.

During the heating step, the adsorbent temperature rises from 100 F. to 500 F., when contacted with regeneration purge fluid having a dewpoint of −100 F. Point H would be the equilibrium condition if the adsorbent were maintained at a temperature of 500 F. in contact with a purge fluid having a dewpoint of −100 F.

Again, the actual thermodynamic path, depicting the mechanism by which this step occurs, is not a straight line; nonetheless, the heating step can be shown as a straight line from Point A to Point H. The cooling step is carried out along the horizontal line (constant dewpoint) Point H to Point C, reflecting the decrease in adsorbent temperature while the bed is continuously contacted with a fluid whose dewpoint is −100 F.

Now suppose that some of the process conditions were slightly different. Specifically, consider the case when regeneration is carried out at 500 F. with wet feed gas having a dewpoint of +100 F., and when the cooling step is conducted with a closed-cycle purge.

The process conditions for this case are shown in Fig. 1b. Point A depicts the pseudo-equilibrium condition at the end of the adsorption step, Point H at the end of the heating step, and Point C at the end of the cooling step.

As feed gas contacts the activated adsorbent, the water loading on the adsorbent increases to 23½%, and the adsorption step is again depicted as the constant temperature line from Point C to Point A.

During the heating step, the adsorbent is now continuously contacted with a fluid having a dewpoint of +100 F. Thus, the heating step is shown as a horizontal (constant dewpoint) line from Point A to Point H.

During closed-cycle cooling, it can be assumed that no additional water is admitted to the system and that the quantity of it in the interstitial gas is small (this is equivalent to saying that the cooling is performed isosterically).

Nomenclature

a_v	External surface area per unit volume of packing, sq.ft./cu.ft.
A_x	Cross-sectional area of adsorbent bed, sq.ft.
B_H	Overall particle heat-transfer coefficient, Btu./(hr.)(cu.ft.)(°F.)
c_{pf}	Specific heat of fluid, Btu./(lb.)(°F.)
c_{pg}	Specific heat of gas, Btu./(lb.)(°F.)
c_{ps}	Specific heat of solid, Btu./(lb.)(°F.)
E	Reduced temperature, dimensionless
F	Volume of adsorbate-rich feed gas, std.cu.ft.
G	Superficial mass velocity, lb./(hr.)(sq.ft.)
h_g	Fluid-phase particle heat-transfer coefficient, Btu./(hr.)(cu.ft.)(°F.)
h_s	Solid-phase particle heat-transfer coefficient, Btu./(hr.)(cu.ft.)(°F.)
k	Thermal conductivity of fluid, Btu./(hr.)(ft.)(°F.)
L	Height of packing or distance from bed entrance, ft.
N_{Re}	Modified Reynolds Number ($G/a_v\mu\psi$), dimensionless
P_1	System pressure during adsorption step, psia.
P_2	System pressure during stripping step, psia.
P_{a1}	Partial pressure of sorbable component in feed, psi.
P_{a2}	Partial pressure of sorbable component in purge effluent, psi.
Q	Actual quantity of heat supplied or required, Btu.
Q_{min}	Minimum heat requirement, Btu.
R_{min}	Minimum volume of clean purge fluid required to strip adsorbent to its original condition, std.cu.ft.
t_f	Temperature of fluid at distance, L, from the inlet end of bed, °F.
t_f^o	Initial or inlet temperature of fluid, °F.
t_s^o	Initial temperature of adsorbent, °F
V_f	Volume of sorbable component in feed gas, std.cu.ft.
V_p	Volume of sorbable component in effluent, std.cu.ft.
W	Weight of adsorbent, lb.
θ	Time, hr.
μ	Fluid viscosity, lb./(hr.)(ft.)
ρ_b	Bulk density of adsorbent, lb./cu.ft.
ψ	Particle shape factor: 1.00 spheres, 0.91 for cylinders, and 0.86 for flakes

The cooling step, then, is depicted as a line, essentially parallel to the adjacent isostere lines, from Point H to Point C. Note that the residual loading is 4% on the regenerated and cooled adsorbent, and the water content of a gas in equilibrium with this adsorbent represents a dewpoint of −80 F. This is the lowest dewpoint possible with the adsorbent regenerated with this cycle. Thus, the effect of using wet feed gas for regeneration, rather than dry product gas, is a reduction in delta-loading and in the quality of the product gas from the subsequent adsorption step.

Many variations of the basic cycle could be shown in such equilibrium diagrams. Certain limits would be immediately defined for each case, and important design information could be determined directly. The table presents a summary of design information that can be determined directly after plotting a cycle on an equilibrium diagram.

Information obtained from equilibrium diagrams can also be used to quickly estimate the effect of operational changes and upsets on the performance of commercial dehydrators. If one assumes that the mass-transfer zone is a fixed fraction of the adsorber bed, and that this fraction is not grossly altered by process changes or upsets, the equilibrium diagrams can serve as a qualitative guide to the effects of operational changes.

General Desorption Considerations

The performance of a cyclic adsorption process is strongly dependent upon: (1) the condition of the adsorbent at the end of the bed that serves as the adsorption step effluent; and (2) the average delta loading. Both factors are related to the residual loadings following the regeneration step.

In dynamic processes, the adsorbent at the effluent end of the adsorption step serves as the final contacting stage of a multistage process. The equilibrium associated with the residual loading at the effluent end determines the minimum possible concentration of sorbable component in the effluent fluid.

The average delta loading is the difference in loading between the adsorption and desorption steps, and it fixes the relationship between adsorbent requirement and cycle time. This is shown schematically in the *X-L* diagrams of Fig. 2. It may be noteworthy that the residual at the end of the regeneration step need be neither uniform nor zero throughout the length of the bed.

Drying and purification processes usually operate on long cycles. This follows because the amount of sorbable component per unit time is relatively small, giving rise to a correspondingly small adsorbent requirement. Often times, too, the minimum adsorbent inventory is dependent upon mass-transfer zone requirements essential to prevent breakthrough. And, finally, low residuals may be required to attain nearly complete removal of the sorbable component.

In bulk separation processes, the sorbable component per unit time may be quite large. The adsorbent inventory required tends to become very large as the cycle time increases. Because the breakthrough of small amounts of sorbable component is acceptable, low residual loadings are not necessary. Thus, bulk separation processes tend toward short cycle times in order to minimize adsorbent inventory. Short regeneration periods may, in turn, result in relatively small delta loadings.

Regeneration strongly influences the efficiency of cyclic adsorption processes and tends to be the major factor in the overall economics.

Thermal-Swing Desorption

Thermal-swing desorption results in low residual loadings and, hence, large delta loadings. Equilibrium considerations associated with low residual loadings make it possible to almost completely remove a sorbable component from a fluid. For this reason, thermal-swing-cycle operation is generally preferred for drying and purification applications. Such regeneration is simple, and equipment requirements not elaborate.

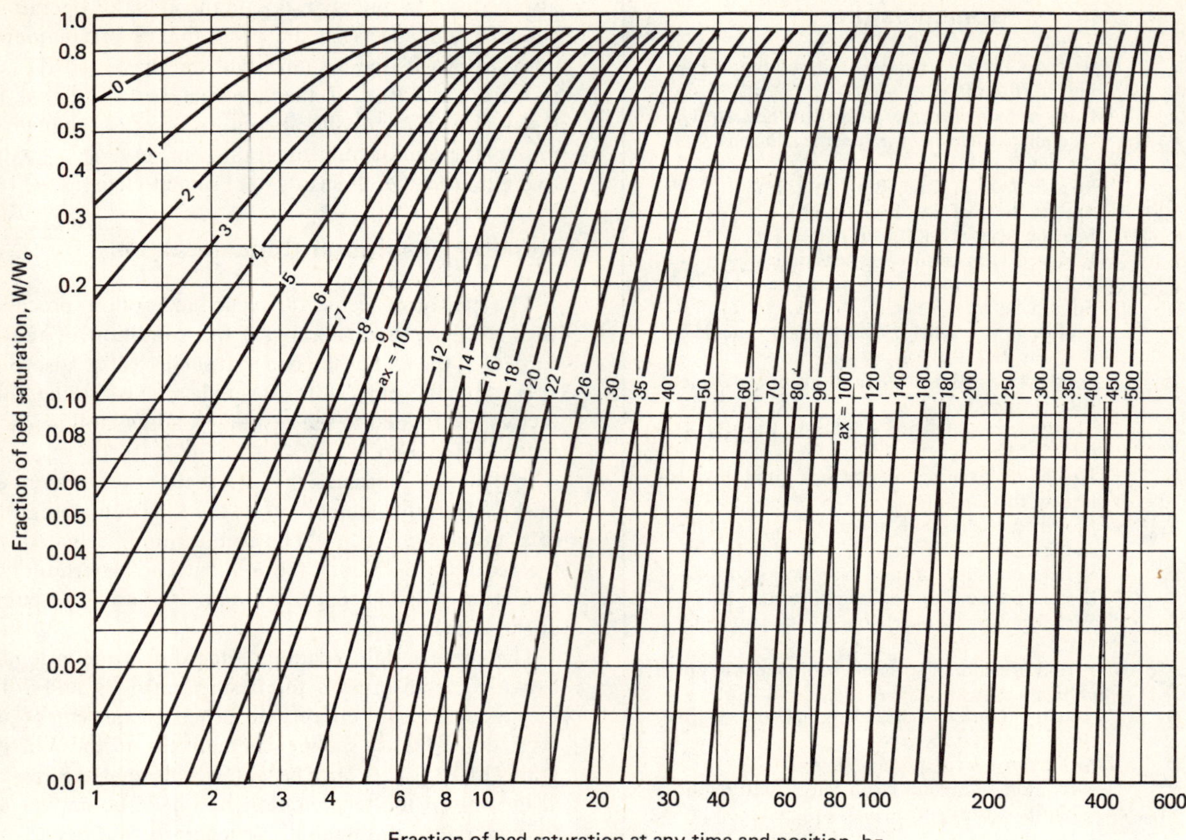

SHAPE of heat-transfer front at any time can be determined as a function of parameters *ax* and *bt*—Fig. 3

Thermal-swing cycles tend to be inefficient in the sense that a relatively large amount of time is required for regeneration, and a nonproductive cooling step consumes a significant amount of time. Although some thermal-swing regenerations can be finished in less than 1 hour, most designs specify a regeneration period of 8 to 24 hours. This long regeneration time makes thermal-swing desorption techniques unattractive for use in bulk separation processes.

The application of heat alone is seldom sufficient to effect suitable regeneration; a purge or vacuum is needed to remove the sorbable component. Without the purge or vacuum, the sorbable component can be readsorbed when the system cools. However, in some cases, the pressure-volume relationship may permit escape from the system of a suitable fraction of the desorbed material.

Direct heating with a hot purge fluid affects desorption in two ways. The hot fluid provides the heat necessary to shift the system equilibrium, and it acts as a stripping medium to carry away the desorbed material.

Thus, two limiting conditions should be recognized. With low-pressure purge gases, desorption is often limited by the rate of heat addition to the system. And this rate of addition depends on the rate of fluid flow, which, in turn, may be limited by pressure-drop considerations. With high-pressure purge gases, desorption may be limited by the stripping action of the purge fluid. Thus, heat balances alone are not always a sufficient criterion for designing a thermal regeneration system.

Indirect heating complicates the mechanical design of an adsorber bed and tends to increase the probability of flow-channeling problems. It also results in a nonuniform addition of heat.

However, indirect heating does provide an attractive method for heat addition when there is a shortage of purge fluid or when dilution of the desorbed component must be limited.

Direct Heating With a Purge Gas

During direct heating or cooling with a purge gas, a transient temperature front is formed within a packed bed. Heat-transfer efficiency depends on the shape of the front. Experiments with Type 5A, ⅛-in.-dia. pellets in gas-phase operation indicate that resistance to heat transfer frequently exists in both the fluid and solid phase. The overall particle heat-transfer coefficient, B_H, is:

$$1/B_H = 1/h_s + 1/h_g \qquad (1)$$

The fluid-phase coefficient, h_g, can be calculated by means of the Colburn factor,* j_H, an empirical correlation defined as:

$$j_H = 0.91\,(N_{Re})^{0.51}\,\psi \quad (N_{Re} < 50) \qquad (2)$$

$$j_H = 0.61\,(N_{Re})^{0.41}\,\psi \quad (N_{Re} < 50) \qquad (3)$$

$$j_H = \left(\frac{h_g}{c_{pf}G}\right)\left(\frac{c_{pf}\,\mu}{k}\right)^{2/3} \qquad (4)$$

* Bird, R. B., Stewart, W. E. and Lightfoot, E. N., "Transport Phenomena," Wiley, New York, 1962, p. 411.

Values of h_s have been calculated from Eq. (1) with B_H determined experimentally, and h_g determined by j_H factors for beds heated and cooled with air. The values of h_s varied between 2,500 and 8,000 Btu./(hr.)(cu.ft.)(°F.). For conservative estimation of heat requirements, a value of h_s equal to 2,500 Btu./(hr.)(cu.ft.)(°F.) is suggested for ⅛-in.-dia. particles.

The rate of temperature rise for pellets and spheres is inversely proportional to the square of the radius; hence, for typical commercial particles, h_s is inversely proportional to the square of effective particle diameter.

The method of Hougen and Marshall can be used to estimate the shape of the heat-transfer front.† This calculation applies to the ideal case involving no phase changes, chemical reactions, heat of adsorption or heat losses.

With the aid of Fig. 3, the shape of the wave at any time, θ, is determined as a function of two parameters: ax and $B\tau$, which are defined as:

$$ax = B_H L / c_{pf} G \quad (5)$$

$$b\tau = B_H \theta / c_{ps} \rho_b \quad (6)$$

At any time, θ, the temperature at any position, L, can be readily determined.

The minimum amount of heat necessary to change the temperature of W lb. of adsorbent from t_s^o to t_f^o is:

$$Q_{min} = W c_{ps} (t_s^o - t_f^o) = A_x L \rho_b c_{ps} (t_s^o - t_f^o) \quad (7)$$

The rate of heat input to the bed is:

$$q_i = G A_x c_{pf} (t_s^o - t_f^o) \quad (8)$$

And at any time, θ, the actual amount of heat supplied to the bed is:

$$Q = G A_x c_{pf} (t_s^o - t_f^o) \theta \quad (9)$$

If the cooling or heating front were infinitely short and if it did not widen as it passed through the bed, the flow of purge gas could be stopped just as the front reached the effluent end of the bed, and under these conditions:

$$Q = Q_{min} \quad (10)$$

However, the wave is known to spread, and the actual amount of heat required is greater than Q_{min}. An efficiency factor, η, can be defined as the ratio of these terms:

$$\eta = Q_{min}/Q \quad (11)$$

Combining Eq. (7) and (9):

$$\eta = \frac{Q_{min}}{Q} = \frac{A_x L \rho_b c_{ps} (t_s^o - t_f^o)}{G A_x c_{pf} (t_s^o - t_f^o) \theta} = \frac{L \rho_b c_{ps}}{G c_{pf} \theta} \quad (12)$$

But, from Eq. (5) and (6):

$$ax/b\tau = L \rho_b c_{ps} / G c_{pf} \theta \quad (13)$$

Therefore:

$$\eta = ax/b\tau = Q_{min}/Q \quad (14)$$

Thus, the actual amount of purge gas required can be estimated by using the ratio of ax to $b\tau$, from Fig. 3.

†Hougen, O. A. and Marshall, W. K., *Chem. Eng. Prog.*, 43, 1947, p. 197.

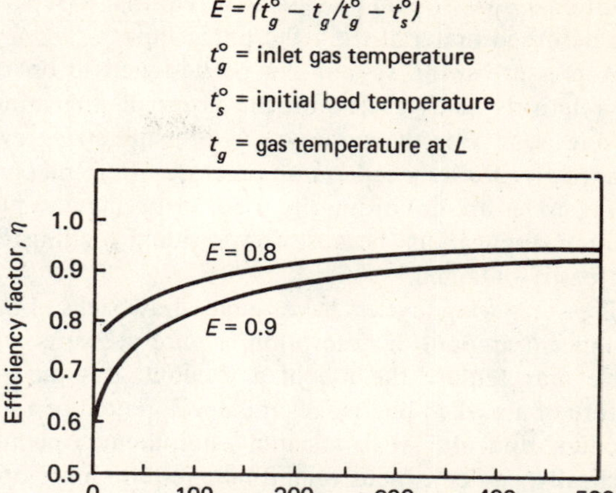

EFFICIENCY FACTORS for heating, cooling fronts—Fig. 4

In most commercial units, fluid and adsorbent temperatures approach each other, but are seldom equal. A reduced temperature, E, relates the temperature approach and is defined as:

$$E = (t_f^o - t_f)/(t_f^o - t_s) \quad (15)$$

A plot of efficiency, η, as a function of $B_H L / c_{pg} G$ for several values of reduced temperature, is presented in Fig. 4. Other experience factors may be incorporated to account for heat generation, heat losses and various equipment characteristics.

In considering the regeneration of an adsorbent bed, the foregoing procedures must be modified to include additional heat terms. As a first approximation in preliminary designs, Eq. (7) should include the sensible heat change for the absorbent bed, the heat of vaporization of liquid holdup, the sensible heat of the adsorber vessel, plus the latent heat change for the adsorbate.

The minimum regeneration temperature varies for each adsorbate-adsorbent combination, and depends on the residual loading to be achieved. Minimum practical regeneration temperatures for water desorption from molecular sieves are on the order of 375 F. There is little incentive for exceeding 650 F. If the cycle is not limited by stripping rate, desorption is complete when the effluent temperature reaches the prescribed level.

In large commercial adsorbers, it is common practice to design for a ΔT of approximately 50 F. between the heater outlet and the bed effluent. Thus, if the regeneration purge gas is heated to 450 F., regeneration is usually complete when the purge effluent temperature reaches 400 F., except when stripping limitations are particularly severe.

Pressure-Swing Desorption

A principal advantage of pressure-swing cycles is that the desorbed material can be recovered directly as high-purity product. With most other desorption techniques, additional downstream processing is required to separate the desorbed material from the purge fluid.

A pressure-swing system can be designed to operate on relatively short cycles, on the order of one minute to one hour. For these reasons, a pressure-swing cycle is attractive for bulk separation processes. Pure pressure-swing cycles are not ordinarily used for drying and purification applications because low residual loadings are not easily obtained.

Pressure-swing cycles have some drawbacks. For a significant amount of desorption, a pure pressure-swing cycle may require the use of a vacuum. But the very nature of a packed bed makes the development of a high vacuum difficult. Also, vacuum equipment, operating cyclically, can be expensive and may require more operating attention than other types of regeneration equipment. An additional consideration involves control over the rate of pressure change so as to avoid particle movement.

Vacuum desorption can be used successfully in systems involving a weakly adsorbed material. Alternatively, an operating temperature level can be chosen that is sufficiently high that desorption will readily occur upon a moderate reduction in pressure. Pressure-swing cycles can be considered for adsorbate systems when the isotherms exhibit steep slopes at relatively high adsorbate pressure levels.

Purge-Gas-Stripping Desorption

Purge-gas-stripping cycles usually operate adiabatically and at moderate pressure ratios between the adsorption and desorption steps. This permits the use of short cycles and a relatively small quantity of adsorbent. For applications in which the effluent purge containing the desorbed fluid can be discarded, auxiliary equipment requirements are minimal.

Purge-gas stripping becomes more efficient as the desorption temperature is increased or the pressure decreased. When significant temperature increases occur, the cycle is generally referred to as thermal swing. Few purge-gas-stripping cycles operate isobarically because the purge-volume requirement becomes excessive.

The process commonly referred to as "heatless adsorption" provides a classic example of a purge-gas-stripping cycle. In a heatless-adsorption cycle, the bulk fluid contacts the adsorbent at some relatively high pressure, say P_1. The desorption is affected by flowing a slipstream of purified carrier fluid countercurrent through a second adsorber bed at some lower pressure, P_2.

The beds are switched from the adsorption step to the desorption step before the thermal front associated with adsorption reaches the effluent end of the onstream bed. In this manner, the beds retain sufficient heat to compensate for the endothermic heat of desorption. This generally dictates rapid cycling. Cycles on the order of 10 sec. to approximately 5 min. are typical.

Except for the minor temperature changes associated with heat of adsorption, the process is essentially isothermal and adiabatic. Desorption is dependent strictly on the ability of the purge gas to strip the adsorbate at the lower pressure, P_2. Purge-gas-stripping cycles are used in small air dryers that contain only a few grams of adsorbent, and in very large hydrogen upgraders that contain more than 50,000 lb. of adsorbent.

Consider the case in which a volume of bulk fluid, F, contacts adsorbent at a uniform flowrate and at some initial relatively high pressure, P_1. The partial pressure of sorbable component is P_{a1}. As bulk fluid flows through a bed, a mass-transfer front develops, and the partial pressure of sorbable component decreases to lower values, and finally to $(P_{a1})_e$ in the effluent.

If equilibrium conditions are assumed, the adsorbent loading at the inlet end is X_a and at the effluent end $(X_a)_e$. Across the mass-transfer zone, the adsorbent loadings are intermediate values.

During desorption, a slipstream of effluent, after pressure reduction to P_2, becomes the purge fluid. The volume of purge fluid is R. The partial pressure of sorbable component in the entering purge fluid is $(P_{a2})_{e'}$:

$$(P_{a2})_e \leqq (P_2/P_1)(P_{a1})_e \tag{16}$$

But, because $P_2 < P_1$, it follows that $(P_{a2})_e < (P_{a1})_e$.

If equilibrium conditions were approached during the adsorption step, equilibrium considerations would favor desorption during the purge step. This follows because the partial pressure of sorbable component in the purge fluid is less than the equilibrium partial pressure associated with the adsorbent loadings.

As desorption occurs, P_{a2} becomes the partial pressure of sorbable component in the effluent purge fluid. If we consider very small movements of the mass-transfer zone, and if a close approach to equilibrium is obtained during the adsorption and purge steps, the partial pressure, P_{a2}, will approach a value equal to $P_{a1'}$ and at the limit will equal P_{a1}.

Assuming ideal gases, the volume of sorbable component in the feed is:

$$V_f = F(P_{a1}/P_1) \tag{17}$$

Let R_{min} be the minimum volume of clean purge gas required to restore the bed to its original condition, and let V_p equal the volume of sorbable component in the purge effluent:

$$V_p = (R_{min} + V_p)(P_{a2}/P_2) \tag{18}$$

But V_p is usually small relative to R_{min}, so Eq. (18) can be approximated by:

$$V_p = R_{min}(P_{a2}/P_2) \tag{19}$$

Also, if equilibrium conditions were attained:

$$V_f = V_p \tag{20}$$

Or:

$$F(P_{a1}/P_1) = R_{min}(P_{a2}/P_2) \tag{21}$$

And so:

$$R_{min} = F(P_2/P_1) \tag{22}$$

These conditions apply when a steady-state mass-transfer zone is made to oscillate back and forth over short distances within an adsorbent bed. There are two requirements: (1) a steady-state mass-transfer condition must be established, and this may require an initial period of nonsteady-state cycling, and (2) the beds must be switched from the adsorption to the desorption step before the thermal front associated with adsorption breaks through the effluent end of the bed.

Within a reasonable range of operating conditions for a particular system, a fixed volume is required to effect desorption to a specified level; the desorption is nearly independent of purge rate.

Eq. (22) can be used to estimate the minimum purge requirement of any adsorption system. Actual purge requirements for commercial systems will be larger.

Purge requirements for systems that deviate from the isothermal, adiabatic operations outlined may differ grossly. If heat is added during stripping, purge requirements are reduced. However, Eq. (22) and Eq. (7), modified for latent heat effects, are two useful design tools. They define the minimum purge requirements for ideal systems. When applied to actual systems, efficiency terms and experience factors must be added.

Displacement Desorption

Displacement desorption is widely used in bulk separation processes. Most displacement cycles operate with a displacement fluid exhibiting sorption characteristics, including heat of adsorption, similar to those of the primary adsorbate. Thus, mass transfer depends primarily on mass action. The transition from the adsorption to the desorption step is further simplified by operating at nearly isothermal and isobaric conditions. A disadvantage of a displacement cycle is that downstream separation facilities must usually be provided to separate the displacement fluid both from the adsorbate-free carrier fluid and from the desorbed product.

The economics of large bulk-separation processes depend on a capability to adsorb and desorb large quantities of adsorbate on relatively small amounts of adsorbent. The mechanical and thermal efficiency associated with the use of displacement fluids permits operation with very short cycles, on the order of 1 to 5 min. Under these conditions, a relatively small delta loading is justified. A delta loading of 1 to 2% of the weight of adsorbent, representing approximately 10% of the maximum possible delta loading, is common.

When a more strongly sorbed fluid is used as purge, it immediately displaces the primary adsorbate. Little or no stripping action occurs. Generally, the more strongly the displacing agent is adsorbed, the more complete is the desorption of the previous adsorbate, and the smaller is the displacing fluid requirement. However, the problems are shifted to the subsequent adsorption step. If the bed is switched to adsorption without first desorbing the strongly held displacement fluid, the useful capacity of the bed may be reduced or the recovery of sorbable component decreased.

Desorption with a less strongly adsorbed displacing fluid is accomplished both through displacement and purge-gas stripping. A less strongly adsorbed fluid generally entails higher displacement fluid circulation rates and larger downstream separation equipment than is required for more strongly adsorbed ones.

Process Design

A simplified process design procedure for regenerative adsorption processes can be summarized as follows:

1. Determine the fresh adsorbent equilibrium, X_e, for the sorbable component from an equilibrium diagram. Special data is required for multicomponent systems.
2. Select a regeneration procedure; then determine a corresponding residual loading, X_r, for the sorbable component.
3. Calculate the fresh-bed equilibrium-zone delta loading ($\Delta X = X_e - X_r$).
4. Calculate the quantity of sorbable component, w_a, in the carrier fluid that is to be removed during each adsorption step. An assumption concerning cycle time is required.
5. Calculate the weight of fresh adsorbent, *WES,* required in the equilibrium section ($WES - W_a/\Delta X$).
6. Determine the weight of unused bed, *WUB,* from mass-transfer-zone data obtained for the systems.
7. Obtain an equilibrium section adsorbent life factor, f_1, based upon the required calendar life, cycle time and number of cycles for the particular application. Calculate the cycled equilibrium section adsorbent requirements ($WES_c = WES_1/f_1$).
8. Obtain an adsorbent mass-transfer zone life factor, f_2, based on required calendar life, cycle time and number of cycles, then calculate cycled mass-transfer section adsorbent requirements ($WUB_c = WUB_1/f_2$), and next the cycled adsorbent requirement ($W_c = WES_c + WUB_c$).
9. Incorporate other experience factors that may apply. For example, some systems are found to contain traces of various adsorbent poisons or secondary adsorbates (such as very heavy oils, certain chemical decomposition products), and some additional adsorbent may be included to compensate for capacity loss attributable to these causes.
10. Fix a detailed regeneration procedure. Determine flowrates and other regeneration operating parameters. The following should be included: temperatures, pressures, flowrates, flow directions, time for each step, and heat duty for heaters and coolers.
11. After fixing the regeneration conditions, verify that the residual loading determined in Step 2 above is obtained. If not, repeat Steps 2 through 10, using the new residual loading.
12. Assume various bed configurations, then calculate pressure drops for each of the steps. Fix a bed configuration that avoids excessive pressure drop.
13. Size the vessels and auxiliary equipment.

The design of multicomponent systems is similar but considerably more complex.

New Method Simplifies Design of . . .

Activated-Carbon Systems

With bed-depth/service-time analysis, designers can calculate effects of varying feed concentrations, flowrates or effluent compositions in activated-carbon adsorption systems.

ROY A. HUTCHINS, ICI America, Inc.

Activated-carbon adsorption systems, widely used in the chemical process industries for several decades, are now playing an important role in cleaning up plant effluents and municipal wastewaters. Design of granular-carbon systems, however, can be difficult and time consuming in some cases. The usual approach involves a four-step progression—studying powdered-carbon adsorption isotherms, running laboratory column tests, followed by pilot-scale tests, and finally designing the commercial unit.

Now there is a new calculation method, called bed-depth/service-time (BDST) analysis, that should speed up the design process by reducing the amount of preliminary testing. Basically, it is a means for predicting the effects of different feed concentrations, flowrates or effluent compositions.

BDST analysis applies to granular-carbon systems, which are not in equilibrium. Powdered carbons are usually found in batch operations that operate at or near equilibrium. In such cases, somewhat rigid mathematics can be applied for computing the optimum carbon dosage.

When liquid flows through a granular-carbon column, there is no sharp demarcation between purified liquid and the feed. Rather, there is an adsorption zone of some length, in which the concentration of adsorbable impurities will vary from a maximum at the back to zero at the front. This situation is particularly pronounced when the feed to the column contains a number of different impurities.

When the adsorption zone is entirely within the column, the effluent will be almost completely purified.

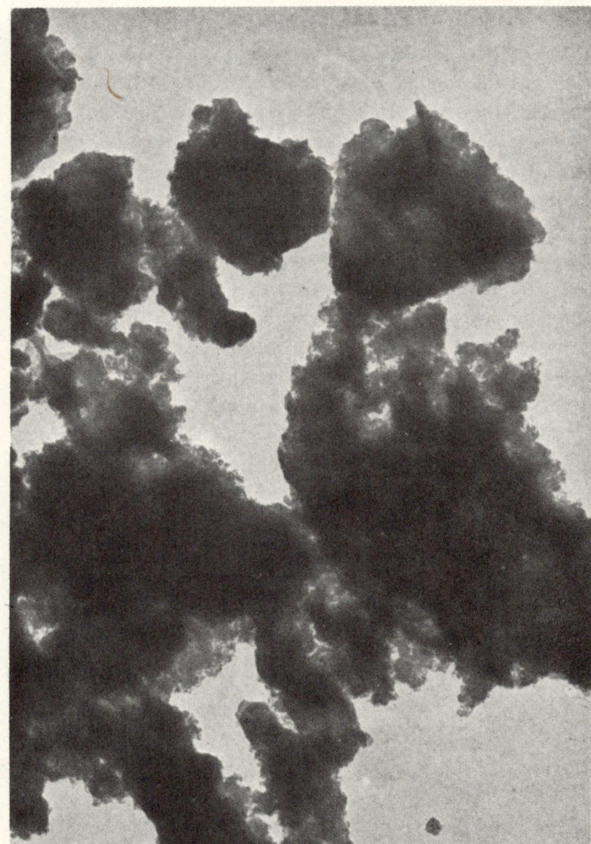

LARGE SURFACE AREA of activated carbon—450 to 1,800 sq. meters/gm.—is shown in this photomicrograph of granular carbon enlarged 25,700 times.

Originally pub sihed August 20, 1973

However, once the front of the zone arrives at the effluent end, quality begins to drop off. As the adsorption zone continues moving forward, the concentration of impurities in the effluent increases and eventually equals that in the feed. At this point, the carbon is saturated—it has reached maximum loading under the particular set of operating conditions.

Design Parameters

Contact or residence time is the major design parameter for the adsorption portion of the system. That is, the optimum residence time determines the size of the adsorbers and volume of the carbon bed. Regeneration equipment is sized according to the rate of carbon consumption required to maintain an acceptable product.

Minor design and operating parameters include: linear flowrate, impurity concentration and composition in feed and product, pH, temperature, and viscosity. Additional factors are properties of the adsorbent such as particle size, adsorptive capacity, pore-size distribution, and chemical nature of the adsorbent's surface.

Normal variations in the above parameters have only minor effects on investment and operating costs. Some of these factors are interrelated, however, and could cause longer residence times and higher regeneration rates. Therefore, both the adsorption and regeneration sections of the system should be larger than the optimum size based on estimated residence time and carbon consumption.

Isotherms and Column Tests

As mentioned earlier, adsorption isotherms have been used traditionally for preliminary screening before running more-costly tests. These procedures are well known, and give an indication both of the effectiveness of carbon adsorption for removing specific impurities, and the maximum quantity that can be adsorbed by a unit of carbon.[1,2] But isotherms cannot give accurate scaleup data for granular-carbon systems because:

- Adsorption in a granular-carbon column is not at equilibrium.
- Granular carbon rarely becomes totally exhausted in commercial processes.
- Powdered carbon has a significantly greater adsorption rate than granular material.
- Effects of carbon recycling cannot be studied, and isotherms cannot predict chemical or biological changes occurring in the adsorber.

Granular-carbon column tests are often run in ways that give unreliable results. For example, testing at unrealistic flowrates or residence times; columns less than 1-in. dia.; short runs; or unrepresentative feed. Obviously, column tests should be based on the same conditions as those expected in the actual plant system.

Bed-Depth/Service-Time Method

The BDST method is based on work by Bohart and Adams,[3] Klotz,[5] and Dole and Klotz.[4] Others[6,7,8] have developed similar methods, but none of these appear to be as useful.

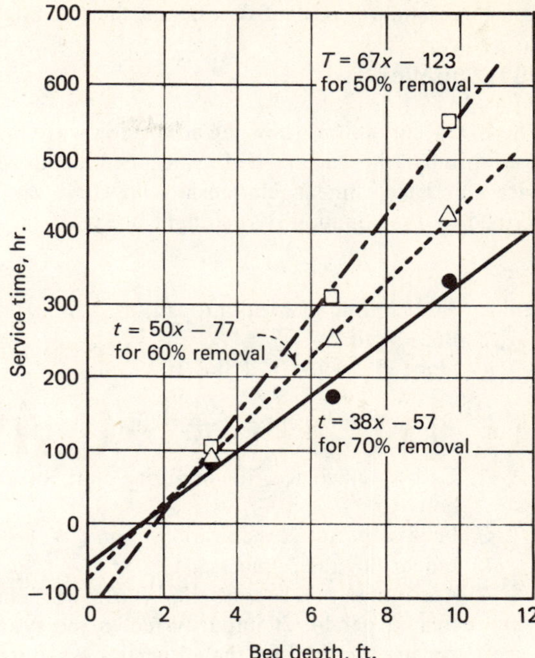

BDST curves show linear relationship between bed depth and service time—Fig. 1

A typical test system consists of at least three granular-carbon columns operating in series. If the liquid is to be pretreated in the plant system, it should be similarly processed before entering the test columns. The total carbon depth and linear flowrate should approximate those expected in the commercial plant. Usually, either upflow or downflow is acceptable. Linear flowrate as well as concentration and composition of feed impurities should be reasonably constant during the test.

As liquid flows through the system, samples of the first-column feed and effluent from each column are taken and analyzed at regular intervals. These data are plotted as percent impurity remaining vs. service time for each column, to give curves called "column exhaustion curves." Then, breakthrough points can be determined. This is the place on each column exhaustion curve where the impurity concentration becomes unacceptable. Typical breakthrough data are in the following table. These are based on a refinery waste containing 129 ppm. COD, flowing at 2 gpm./sq.ft.

Column Number	Cumulative Bed Depth, Ft.	Service Time, Hr., at Breakthrough for % Removals		
		70%	60%	50%
1	3.3	80	90	105
2	6.6	170	250	310
3	9.9	330	420	550

The breakthrough points, expressed as service time, are plotted against bed depth to give the BDST curves in Fig. 1. The service time at breakthrough should be a linear function of bed depth if the linear flowrate and the concentration and composition of impurities in the feed are reasonably constant throughout the test.

BDST Equation

The BDST curve shows how the adsorption wave front moves through the carbon bed. Since bed depth and service time have a linear relationship, the curve can be described by the equation of a straight line:

$$t = ax + b$$

where t Service time at breakthrough, hr.

x Bed depth, ft.

a Slope $= 1990N_o/C_oV$, hr./ft.

b Ordinate intercept $= \dfrac{16.018}{KC_o} \ln\left(\dfrac{C_o}{C_B} - 1\right)$, hr.

N_o Carbon efficiency, lb. impurity/cu.ft. of carbon.

C_o Feed-impurity concentration, ppm.

V Linear flowrate, gpm./sq.ft.

K Adsorption rate constant, cu.ft. of liquid treated per lb. of impurity fed to the system per hr., required for the adsorption wave front to move through a bed depth equivalent to the critical bed depth.

C_B Impurity concentration in effluent at breakthrough, ppm.

The preceding is the BDST equation. The numerical constants will vary depending on the units used to develop the data. For example, APHA (American Public Health Assn.) color units per 100 ml. could replace ppm. as a measure of impurities, and the breakthrough point could be expressed in days rather than hours.

The slope, a, of the line is the time in hours required to exhaust 1 ft. of carbon bed under the test conditions. Stated another way, it is the amount of time needed for the wave front to move through 1 ft. of carbon.

The reciprocal of the slope, then, is the rate at which the carbon bed is spent. Multiplying this value by the adsorbent's apparent bulk density gives an estimate of the carbon-use rate (or rate at which carbon must be regenerated) to continuously produce acceptable product (in this example, wastewater).

The intercept of the abscissa (X axis) in Fig. 1 is the critical bed-depth, X_o, defined as the minimum depth for obtaining satisfactory effluent at time zero under the test operating conditions. The intercept (b) of the ordinate is a measure of the adsorption rate and is defined as the time required for the adsorption wave front to pass through the critical bed-depth.

Calculating New Flowrates

After developing a BDST equation from column tests at one linear flowrate, the designer can calculate equations for other rates by multiplying the original slope (a) by the ratio of the original and new rates. This gives a valid equation for the new condition because changing the linear flowrate has little or no effect on the ordinate intercept (b).

For example, Fig. 2 shows BDST equations obtained from testing phenol removal from water at linear flowrates of 0.75 and 1.5 gpm./sq.ft. In both cases, phenol concentration was reduced from 100 ppm. to 1 ppm. The slope for the higher rate calculated from data at 0.75 gpm./sq.ft. is 33 hr./ft. This agrees very well with the slope of 32 hr./ft. obtained in actual tests. We have found similar good correlations during tests on a number of complex mixtures.

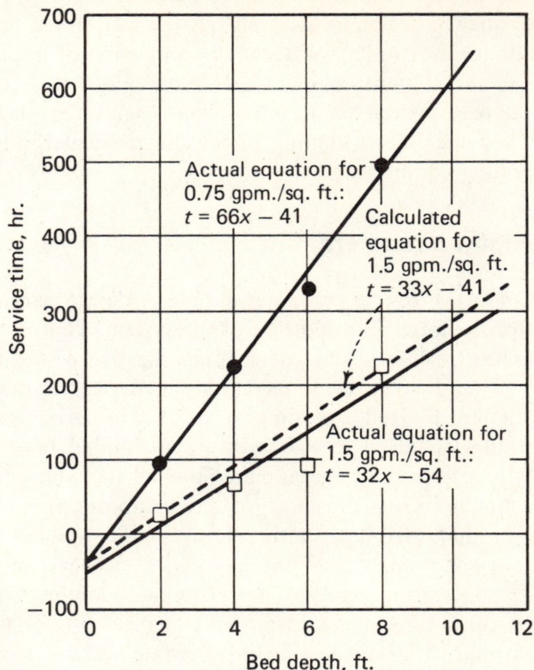

ACTUAL versus calculated BDST curves for new flowrate show good agreement—Fig. 2

With this procedure, laboratory tests can be reliably scaled-up to other flowrates without running pilot tests in larger columns. One precaution is that tests in 1- to 1.5-in. laboratory columns usually must be run at less than 1.5 to 2 gpm./sq.ft. with liquid viscosities about that of water. Otherwise, wall effects will cause misleading results. We have found that tests in 1.5-in., 4-in. and 5-ft.-dia. columns yield equivalent results when wall effects are eliminated.

Changes in Feed Concentration

The designer can also calculate equations for other impurity concentrations, although running additional tests yields the most reliable data. Fig. 3 shows actual BDST curves and equations for removing phenol from water at 100 and 200-ppm. feed concentrations. The figure also contains an estimated curve for the higher concentration calculated from the BDST equation. The slope of the curve for the higher concentration was estimated by multiplying the slope at 100 ppm. by the ratio of the two feed concentrations:

$$a' = 66 \times 100/200 = 33$$

The intercept b for 100-ppm. feed is shown on the graph as −41. An estimated value for the intercept at

200 ppm. was calculated from the relationship:

$$b' = -41(100/200)\frac{\ln[(200/1) - 1]}{\ln[(100/1) - 1]}$$

$$b' = -23$$

In this case, the difference between actual and calculated equations is within experimental error. This method for estimating the effect of changing feed concentration appears to work well when removing one-component impurities from water. Additional work must be done to determine if it can be applied to more-complex mixtures in process streams and wastewaters.

Types of Systems

Fixed-bed adsorption systems may consist of a single column with another on standby, or several columns arranged in parallel or series. In parallel systems, all columns receive the same feed. Usually, the columns are "staggered" so they are at varying levels of carbon exhaustion, and then the desired product is made up by blending effluents.

With columns arranged in series, the effluent from the last column is the product. When carbon in the lead column is spent, the column is shut down and a fresh one put onstream at the end of the series. The second column then becomes the lead column.

Systems may also be arranged so that the carbon moves opposite to the liquid flow. Spent carbon is removed periodically from the bottom of the unit while an equal volume of fresh or regenerated material is added to the top. Moving-bed systems are, in effect, the same as a number of fixed-bed columns operating in series.

Modifying the Equation

The BDST equation, calculated at instantaneous breakthrough points, describes how an adsorption-wave front moves through a single fixed-bed column. However, the equation may be varied to cover series, parallel and moving-bed systems as well.

In parallel systems, whose design procedures will not be covered here, a column is not removed as soon as it begins producing unacceptable effluent. By contrast, with series and moving-bed operations a unit is removed or fresh carbon added when the adsorption-wave front begins to leave the system. Therefore, the rate at which the wave front moves will be the rate for determining when the next column should be taken offline or when a "pulse" of carbon must be removed.

Conducting actual tests is the most reliable way of determining the wave front rate. However, a good estimate can be made by applying the BDST equation, with an assumed average concentration of adsorbable impurities in the feed, and the appropriate factors from Table I.

The values in the table were developed by assuming that the moving bed is pulsed or a column removed just as the wave front begins to exit from the system. This is the point at which the carbon is exhausted, i.e. in equilibrium with the feed concentration. The data also assume that the impurity concentration in the adsorption

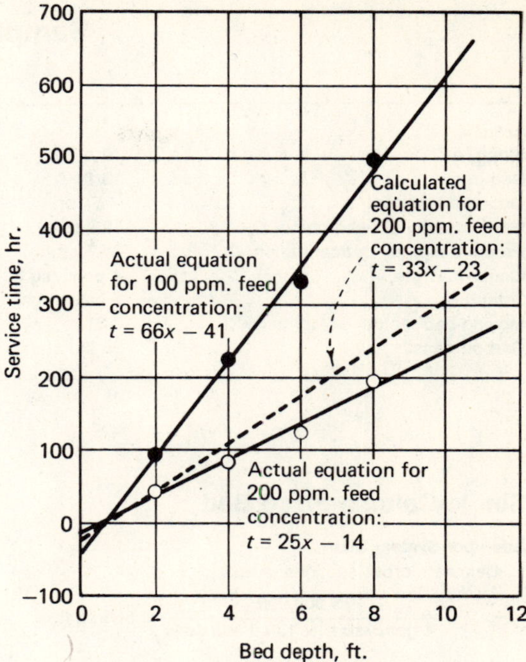

EFFECT of new feed concentration may also be estimated with BDST equation—Fig. 3

zone is an exponential function of the bed depth in the column.

As the wave front moves through fresh carbon, the concentration of impurities entering the freshly added carbon increases. The average concentration fed to the freshly added carbon is the average of that fed from the time the adsorption-wave front enters the freshly added carbon until it begins to exit. The factors in Table I are the fraction of the original impurity concentration that is assumed to be the average fed to freshly added granular carbon.

Consider a moving-bed system operating at 4 gpm./sq.ft., with a wastewater impurity of 129 ppm. Acceptable effluent is 38.7 ppm. (70% removal). The BDST equation developed from a column test was:

$$t = 19x - 57$$

If we assume that the composition of adsorbable impuri-

Factors for Modifying the BDST Equation—Table I

	Required Removal of Adsorbable Impurities, %					
	50	60	70	80	90	95
Moving-bed factors, pulsing 5% of bed.	0.51	0.41	0.31	0.21	0.107	0.054
Series factors for number of columns.						
2	0.70	0.63	0.54	0.44	0.31	0.22
3	0.59	0.50	0.40	0.30	0.18	0.11
4	0.56	0.46	0.37	0.26	0.144	0.080
5	0.55	0.44	0.35	0.24	0.133	0.072

Sample Calculations

Assume:

Slope, a	19 hr./ft.
Bed depth, x	13 ft.
Ordinate intercept, b	−57 hr.
Feed impurity concentration, C_o	129 ppm.
Product impurity concentration, C_1	38.7 ppm.
Linear flowrate, v	4 gpm./sq.ft.
Total flow	5 million gal./day
Moving-bed factor	0.31
Carbon density	23 lb./cu.ft.

Single-Column Fixed Bed

Adsorber-System Size

1. Calculate cross-sectional area:

$$\frac{5{,}000{,}000}{4 \text{ gpm./sq.ft} \times 1{,}440 \text{ min./day}} = 868 \text{ sq.ft.}$$

2. Calculate volume of carbon required:

$$868 \text{ sq.ft.} \times 13 \text{ ft.} = 11{,}284 \text{ cu.ft.}$$

3. Calculate adsorber volume (allow 50% expansion):

$$11{,}284 \text{ cu.ft.} \times 1.5 = 16{,}926 \text{ cu.ft.}$$

4. Make a preliminary cost estimate for an adsorption system with this volume.

Regeneration-Furnace-System Size

1. Convert carbon volume to weight:

$$11{,}284 \text{ cu.ft.} \times 23 \text{ lb./cu.ft.} = 260{,}000 \text{ lb.}$$

2. Use the BDST equation to calculate service time:

$$(19 \text{ hr./ft.} \times 13 \text{ ft.}) + (-57 \text{ hr.}) = 190 \text{ hr.}$$

3. Calculate carbon-use rate:

$$260{,}000 \text{ lb.}/190 \text{ hr.} = 1{,}368 \text{ lb./hr.}$$

4. Make a preliminary cost estimate for a regeneration furnace system of this capacity or the next largest available size.

Moving Bed

Adsorber-System Size

1. Cross-sectional area same as fixed bed.
2. Volume of carbon same as fixed bed.
3. No allowance needed for bed expansion. Absorber volume thus equals 11,284 cu.ft.
4. Make a preliminary cost estimate for an adsorption system with this volume.

Regeneration-Furnace-System Size

1. Use the slope a to calculate a new slope, a', for the adsorption wave front passing through fresh carbon:

$$a' = (C_o \times a)/(f \times C_o)$$

$$a' = (129 \times 19)/(0.31 \times 129) = 61.3 \text{ hr./ft.}$$

2. Calculate the rate at which the adsorption-wave front passes through the fresh carbon:

$$1/61.3 \text{ hr./ft.} = 0.0163 \text{ ft./hr.}$$

3. Calculate the carbon-use rate:

$$0.0163 \text{ ft./hr.} \times 23 \text{ lb./cu.ft.} \times 868 \text{ sq.ft.} = 325 \text{ lb./hr.}$$

4. The carbon-use rate is a minimum value. Make a preliminary cost estimate for the furnace system of the next largest available size.

Estimating New Slopes and Intercepts

Linear Flowrate:

$$a' = (V \times a)/V'$$

where a and a' are the old and new slopes and V and V' are the old and new linear flowrates.

Intercept, b, is not affected by linear flowrate.

Feed Concentration:

$$a' = (C_o \times a)/C_1$$

$$b' = b \times (C_o/C_1) \times \frac{\ln[(C_1/C_F) - 1]}{\ln[(C_o/C_B) - 1]}$$

where C_o and C_1 are the old and new feed concentrations, and C_B and C_F are the old and new effluent concentrations.

ties at the wave front compares to a diluted feed with average concentration of 0.31 C_o (from Table I), and that the Y axis intercept (−57) is not significant in a series or moving-bed system, we can estimate the slope of a BDST equation describing the movement of a wave front (38.7 ppm.) through fresh or regenerated carbon.

$$a' = (C_o \times a)/(f \times C_o)$$

$$a' = (129 \times 19)/(0.31 \times 129) = 61.3 \text{ hr./ft.}$$

This example is shown in the sample calculations. Generally, the reciprocal of a' can be used to estimate the carbon-use rate for series and moving-bed operations, although this gives liberal rather than conservative values of carbon consumption.

Cost Estimates and Sizing

After defining the basic design equation, cost estimates can be made to optimize the size of the adsorption and regeneration units. Adsorption-system size depends on carbon-bed depth and linear flowrate. These two factors determine residence time, which is the major design parameter. Size of the regeneration system, of course, is based on the carbon-use rate, a function of the BDST equation. Estimates for these separate operations—adsorption and regeneration—determine total investment and operating costs.

Examples of using the BDST equation for sizing are presented in the sample calculations. The initial estimate should be based on reasonable operating parameters, usually selected from prior experience. Then other situations are examined to find the best balance between investment and operating costs.

The relationships between residence time, investment and operating costs (direct and indirect) are shown in Fig. 4 and 5. These curves are based on a fixed-bed system and the operating conditions listed in the sample calculations. Usually, residence time at minimum total

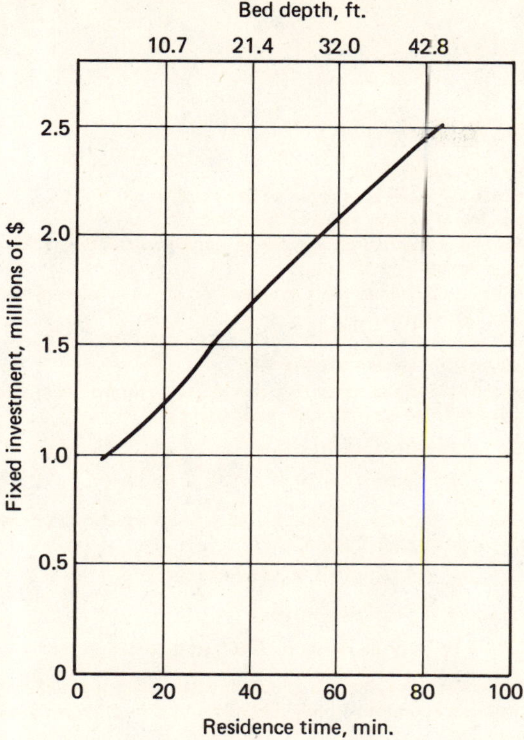

INVESTMENT in a fixed-bed carbon system varies directly with residence time—Fig. 4

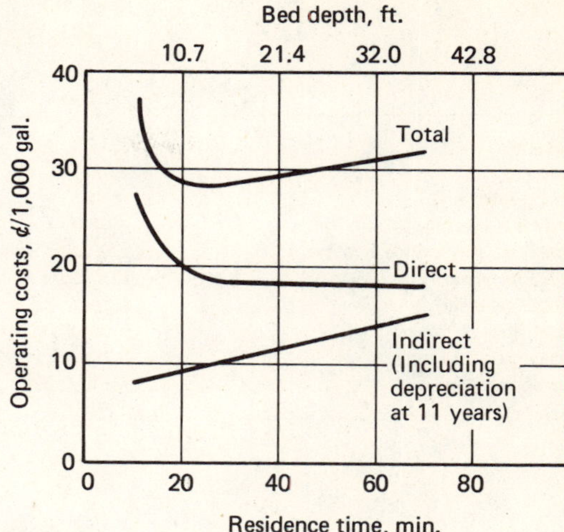

RESIDENCE time at minimum operating cost usually determines best system size—Fig. 5

operating cost fixes the optimum system size. In the example, this is 25 min., equal to a bed depth of 13 ft.

The curve for total operating cost will generally have a minimum point at a certain residence time when working with single-column or parallel systems. This may not be true for series or moving-bed installations because the carbon-use rate is more a function of the reciprocal of the slope of the BDST equation than of the entire equation. In these cases, system selection must be based on the relative importance of capital and operating costs in the particular company or municipality.

Limitations

As with all data evaluation techniques, the BDST equation will have limited value if the column feed is not representative of the normal plant system. Also, if there is chemical or biological activity in the test system, results may not be equivalent to those found in the plant-scale unit.

For the most accurate results, the adsorption-wave front must move through the column at a constant rate. Conditions for this requirement—reasonably constant impurity concentration and composition, and uniform linear flowrate—are not always met during testing. Care should be taken to run tests at the flowrate and bed depth most likely to be used in the final plant. Then, continued testing until unacceptable effluent is consistently obtained in the last column will give an estimate of the carbon-use rate even under poor test conditions.

Even with very good BDST data, design should be tempered with experience and judgment. For example, it is usually advantageous—particularly in wastewater applications—to have some overdesign. This will allow for:

- Unexpected flowrate surges or high impurity concentrations.
- Changes in feed or required product quality.
- Increased production demands.
- Possible decrease in activated-carbon performance, caused by numerous regenerations.

The BDST equation is by no means the ultimate method of analyzing data from granular-carbon column tests. But to our knowledge it is the best technique available. And, of course, we and others are continually trying to improve existing methods and develop new ones.

References

1. Evaluation of Granular Carbon for Chemical Process Applications, D-116, ICI America Inc., 1971.
2. Adsorption Isotherms of Granular Carbon for Wastewater, PC-2, ICI America Inc., May 1972.
3. Bohart, G. S., and Adams, E. Q., *Journal of American Chemical Society*, **42;** 523–544, 1920.
4. Dole, M., and Klotz, I. M., *Industrial and Engineering Chemistry*, **38;** 1289–1297, 1946.
5. Klotz, I. M., *Chemical Reviews*, **39;** 241–268, 1946.
6. Eckenfelder, Jr., W. W., and Ford, D. L., "Laboratory and Design Procedures for Wastewater Treatment Processes", 84–91, University of Texas at Austin, EHE-10-6802, CRWR-31, 1968.
7. Erskine, D. B., and Schuliger, W. G., *Chemical Engineering Progress*, **67;** 41–44, Nov. 1971.
8. Evaluation of Granular Activated Carbon in Columns for Wastewater Treatment, PC-3, ICI America, Inc., May 1972.

Meet the Author

Roy A. Hutchins is an applications supervisor in the pollution-control-venture department of ICI America Inc., (formerly Atlas Chemical Industries) Wilmington, DE 19899. His duties include supervising granular carbon testing programs for industrial and municipal wastewater applications. Most of his industrial experience has been with activated carbon, including research, process improvement, technical service, and process design. Mr. Hutchins received a B.S. in chemical engineering from North Carolina State University in 1960 and has done graduate work at the University of Delaware.

Section 7 Other Separation Methods

Membrane separation processes
Ultrafiltration: An emerging unit-operation
Design factors in reverse osmosis
How to design settling drums
Freeze crystallization

Membrane Separation Processes

ROBERT E. LACEY, Southern Research Institute

Although membrane processes have been studied for more than a century, they have only recently become of interest for industrial separations. This interest stems primarily from a basic advantage of membrane processes: they allow separation of dissolved materials from one another or from a solvent, with no phase change.

In contrast with evaporative or crystallization processes, membrane processes do not require the energy represented by the latent heat of vaporization or crystallization. Since the energy cost represents a sizable portion of the total operating costs for most separations, the possibility of effecting worthwhile economies in energy cost by using membrane processes is attractive.

However, to reduce the total operating cost, flux through the membranes has to be high enough to permit use of reasonably small areas of membranes, which results in a low equipment cost.

There are many different kinds of membrane processes, but all have certain features in common. In all of them, a fluid containing two or more components is in contact with one side of a membrane that is more permeable to one component (or a group of like components) than to other components—a *selective* membrane. The other side of the selective membrane is in contact with a fluid that receives the components transferred through the membrane. To cause the transfer of components, there must of course be a driving force of some kind. Such a force may be transmembrane differences in concentrations, as in dialysis; electrical potential, as in electrodialysis; or hydrostatic pressure, as in reverse osmosis, ultrafiltration and microfiltration. Transmembrane differences in temperature have not yet been used as a driving force in practical processes, mainly because heat transfer is faster than mass transfer, and an impractical amount of thermal energy had to be added.

This article is taken in part from "Industrial Processing With Membranes," ed. by R. Lacey and S. Loeb, Wiley, New York, 1972.

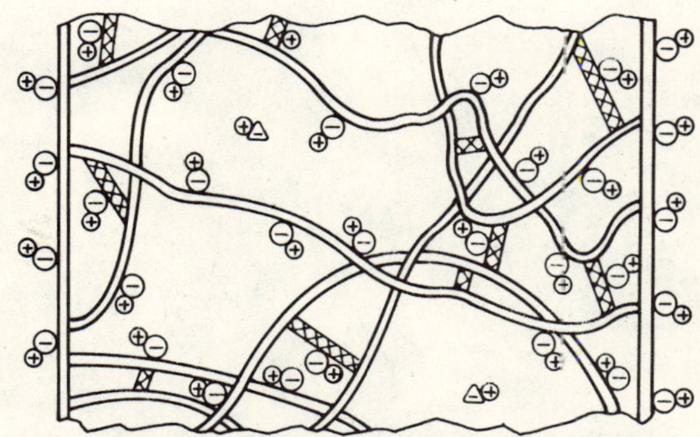

⊖ Fixed negatively charged exchange site, i.e. SO_3^-
⊕ Mobile positively charged exchangeable cation, i.e., Na^+
═ Polystyrene chain
Divinylbenzene crosslink

REPRESENTATION of cation-exchange membrane–Fig. 1

Visualizing a Membrane

It is convenient to picture a membrane as being similar to a jumble of wriggling worms. The worms represent the polymer chains, the void spaces between the worms represent the interstitial volume in a polymer through which transferring species pass, and the wriggle of the worms represents the thermal motion of polymer segments. In transfers through polymers with short interchain distances (i.e., close packing of the worms), the transferring species must often push polymer segments apart to slide past them. Highly crystalline or highly crosslinked polymers are of this type. Other polymers with less interchain attraction or that are less crystalline or less crosslinked have wider spaces between the polymer chains, or longer polymer segments that are more free to move aside. The resistance to transfer through such polymers is lower than that through polymers with very high interchain attractive forces, or though polymers that are highly crystalline or highly crosslinked. In the discussion to follow, some of these ideas about the general nature of membranes will be combined with other ideas to explain the source of selectivity of membranes.

Ion-Exchange Membranes

Ion-exchange membranes are selective in that they are permeable to positively charged ions (cations) but not to negatively charged ones (anions), or vice versa. In Fig. 1,

Originally published September 4, 1972

Processes for separating dissolved materials from fluids by use of membranes are now economically important. Update your knowledge of electrodialysis, reverse osmosis, ultrafiltration, as well as some important though lesser-known techniques.

polymer chains are shown that have negatively charged groups chemically attached to them. The polymer chains are intertwined and also crosslinked at various points. Positive ions are shown freely dispersed in the voids between the chains. However, the fixed negative charges on the chains repulse negative ions that try to enter the membrane, and exclude them. Thus, because of the negative fixed charges, negative ions cannot permeate the membrane, but positive ones can. If positive fixed charges are attached to the polymer chains instead of negative fixed charges, the opposite type of selectivity is achieved. This exclusion as a result of electrostatic repulsion is termed Donnan exclusion, and is an important phenomenon in Donnan dialysis, discussed later.

Selectivity by itself is not enough to make an ion-exchange membrane that is practical for low-cost processing. In addition, the resistance of the membrane to ion transfer must be low. To decrease the resistance, the degree of crosslinking is decreased so that the average interchain distances and the lengths of polymer segments that are free to move are increased. However, if this opening up of the void spaces between polymer chains is carried too far, it can result in volumes in the center of the voids that are not affected by the fixed charges on the chains (the repulsion effect of fixed charges falls off rapidly with distance). Volumes that are unaffected by the fixed charges result in ineffective repulsion of the undesired ions and lowered selectivity. For this reason, a compromise between selectivity and low resistance must usually be made. Happily, in membranes now available, it has been possible to combine excellent selectivity with low resistance, high physical strength, and long lifetimes.

Reverse-Osmosis Membranes

The selectivity of cellulose-acetate reverse-osmosis membranes stems from an entirely different mechanism. Cellulose acetate is a highly organized polymer having groups that can hydrogen bond to water or other solvents subject to hydrogen bonding, such as ammonia or alcohols. Water molecules or ammonia molecules can hydrogen bond to the carbonyl groups in cellulose acetate, as shown in Fig. 2, but ions and non-hydrogen bonding substances cannot enter the organic matrix. The water molecules that enter the polymer by hydrogen bonding

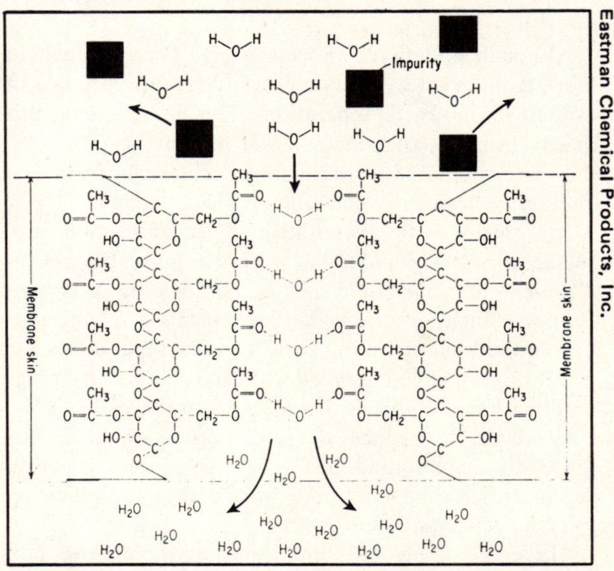

Eastman Chemical Products, Inc.

WATER transfer in cellulose acetate membrane—Fig. 2

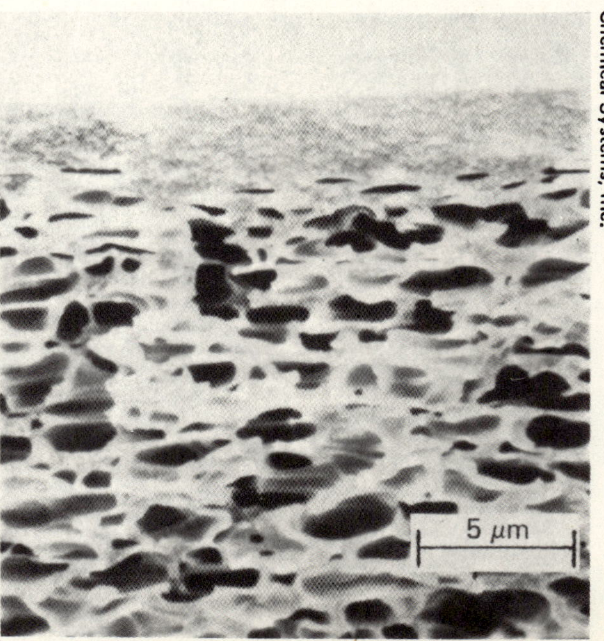

Chemical Systems, Inc.

REVERSE-osmosis membrane, electron micrograph—Fig. 3

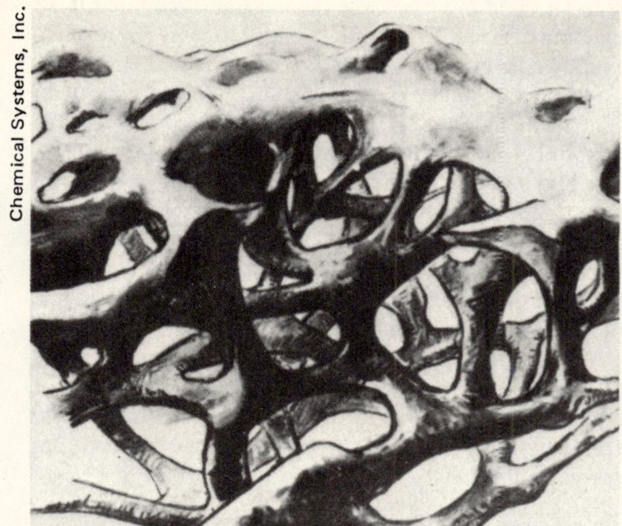

ULTRAFILTRATION membrane structure—Fig. 4

to it can move from one set of hydrogen-bonding sites to another and thus be transported through the polymer if there is a driving force to cause transfer. This type of transfer requires the making and breaking of hydrogen bonds and can only be accomplished with polymers that have the right combination of chemical groups in macromolecules that assume a highly organized structure.

The polymers must also be excellent film formers because even extremely tiny mechanical flaws in the film are enormously larger than the diameter of waterlike solvent molecules (3 to 5 Å). Transfer of species through such highly organized "tight" membranes is similar to the previously mentioned transfer in which the moving species pushes aside polymer strands. Therefore, the resistance to transfer is quite high. However, high fluxes through such materials have been achieved by making the effective thicknesses of the membranes extremely small. In fact the reverse-osmosis process did not appear to be economically practical until the late 1950's, when Loeb and Sourirajan found a method of casting anisotropic cellulose acetate films that had an extremely thin layer or "skin" on top of a thicker layer having an open cellular structure that had little resistance to transfer of water or other solvents. Fig. 3 shows the thin skin of a reverse-osmosis membrane with closed-cell pores just beneath it and open-cell pores still further down. Since that time, methods of making anisotropic membranes for ultrafiltration have been developed, and this process also can take advantage of the high fluxes available with such "skinned" membranes.

Other Membranes

The source of selectivity in ultrafiltration and microfiltration membranes lies in the size of the passages through the thin skin that may be tailored to have apparent diameters from about 20 Å to more than 200 Å. With one exception, these membranes do not have straight pores that extend through the membrane. Instead, they have pores in the surface skin, and tortuous channels that extend through the thickness of the membrane, as indicated in Fig. 4. The one microfiltration membrane that does have straight-through pores is made by irradiating a polycarbonate film with focused beams, and etching the portions that have received radiation damage until holes are formed that range from 0.1 to 8.0 microns dia.

Still other membranes derive their selectivity from preferential solubility of one component of a liquid or gaseous mixture in the polymeric material of the membrane. It may seem odd that liquids or gases are soluble in the polymers, but remember that polymers are merely extremely large molecules entangled one with another. The solubility of substances in polymers follows the same general principles as the solubility of one liquid in another. Polar liquids are usually more soluble in polar polymers, such as cellulose acetate, than in nonpolar polymers, such as polyethylene, and vice versa. Hexane, for example, is more soluble in polyethylene than is methyl alcohol, because both the hexane and the polymer are nonpolar, whereas methyl alcohol is polar. This type of membrane usually derives part of its selectivity from differences in the resistance of the polymer to the transfer of molecules with different shapes. For example, *n*-octane would diffuse more rapidly through polyethylene than would a branched-chain octane isomer because the isomer has a larger cross-sectional area and would

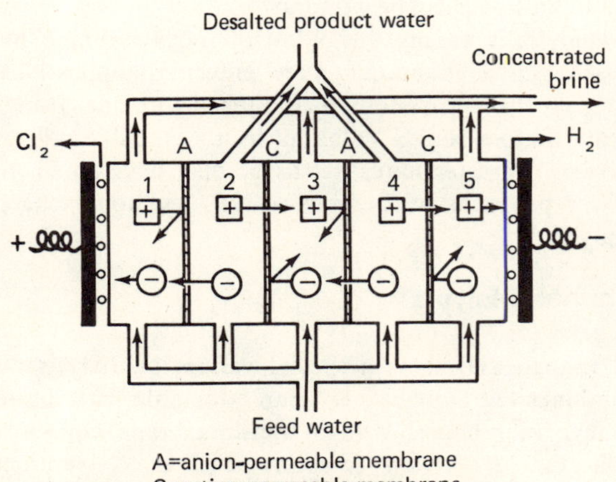

ELECTRODIALYSIS process shown diagrammatically—Fig. 5

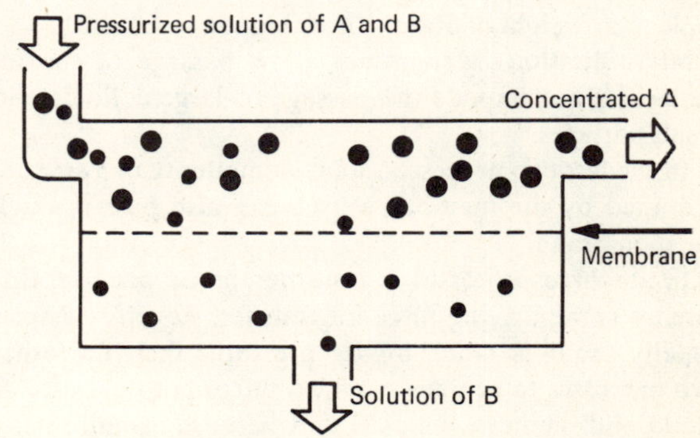

REVERSE osmosis, ultra- or microfiltration—Fig. 6

thus have greater frictional interactions with the polymer chains. Membranes with selectivity stemming from solubility and resistance characteristics are generally used for membrane permeation processes, such as gas permeation or pervaporation.

Selective membranes of one sort or another are used in many types of membrane processes, including electrodialysis, reverse osmosis, ultrafiltration, microfiltration, dialysis, Donnan dialysis, ion-exchange dialysis, gas permeation, and pervaporation. Each of these processes will be described briefly.

Electrodialysis

In electrodialysis, cation-exchange membranes are alternated with anion-exchange membranes in a parallel array to form thin solution-compartments (0.5 to 1.0 mm. thick). The entire assembly of membranes is held between two electrodes as shown in Fig. 5. A solution to be treated is circulated through the solution compartment. With the application of an electrical potential to the electrodes, all positive ions (cations) tend to transfer toward the negative electrode (cathode), and all negative ions (anions) tend to move toward the positive electrode (anode). The ions in the even-numbered compartments can transfer through the first membranes they encounter (cations through cation-exchange membranes, anions through anion-exchange membranes), but they are blocked by the next membranes they encounter as indicated by the arrows in the diagram. Ions in the odd-numbered compartments are blocked in both directions. By this mechanism, ions are removed from the solution circulating through one set of compartments (even) and transferred to the other set of compartments (odd). Ion depletion is accomplished for one solution, and ion concentration is accomplished for the second solution.

Reverse Osmosis, Ultra- and Micro-Filtration

The solution-flows and mechanical arrangements of reverse osmosis, ultrafiltration, and microfiltration are so similar the three processes will be described together. The main difference between the three is in the type of membrane used.

Reverse Osmosis—Membranes allow passage of water (or other hydrogen-bonding solvents) but impede the passage of salts and small molecules.

Ultrafiltration—Membranes allow passage of solvent molecules but impede the passage of molecules with a molecular weight of about 100 or higher.

Microfiltration—Membranes allow passage of solvent molecules but impede the passage of large colloids and small particles.

In the last two processes, solutes smaller than a size determined by the membrane used can also pass through the membrane.

In all three processes, a transmembrane pressure difference is the driving force for transfer. Reverse osmosis usually requires much higher pressures than the other two processes to achieve practical fluxes both because of the "tight" membranes used and because osmotic pressure must be overcome.

In all three processes, the membrane is mounted in an apparatus so that two solution compartments are formed, as shown in Fig. 6. The solution to be treated is pressurized and circulated through the high-pressure solution compartment. Some of the solvent is forced through the membrane into the low-pressure solution compartment. The material that permeates the membrane is withdrawn from the low-pressure solution compartment and either saved as product or discarded as waste, depending on the application. The concentrated material on the high-pressure side of the membrane can also be either the product or the unwanted material.

These pressure-driven processes are extremely simple methods of concentrating solutions or, with the proper type of membranes, of separating small solutes that will permeate the membranes from large ones that will not permeate through.

Dialysis Processes

Dialysis, Donnan dialysis, and ion-exchange dialysis are processes in which a transmembrane concentration difference is the driving force. Similar membrane equipment is used in all three processes, although they differ in certain respects. Therefore, the basic principles of the dialysis process will be presented first, and then the principles of Donnan dialysis and ion-exchange dialysis will be discussed.

In the dialysis process, many membranes are arranged parallel to each other to form thin solution-compartments in an apparatus similar to a filter press. The membranes used are selected to allow passage of salts or small molecules, but retain larger molecules and colloids. A solution containing a mixture of large and small solutes (for example, hemicelluloses and sodium hydroxide) is circulated through the odd-numbered compartments. Water or a dilute solution of the smaller solute is circulated through the even-numbered compartments. Because of the transmembrane differences in concentration, both the large and small solutes tend to transfer through the membranes into the dilute stream. The small solute can and does transfer. However, the membranes block the transfer of the larger solute molecules, and a solution of pure small solute molecules can be recovered. Simultaneously, a solution of the larger solute virtually free of the small solute can be withdrawn.

Dialysis is among the oldest of the membrane processes, but has found only a few industrial applications—notably the separation of sodium hydroxide from a hemicellulose-NaOH solution in the textile industry. However, greater industrial usage may develop for the newer processes of Donnan dialysis and ion-exchange dialysis.

Donnan Dialysis

Donnan dialysis is a special variant of dialysis that combines the Donnan exclusion attainable with ion-exchange membranes with a transmembrane concentration-gradient driving force as used in conventional dialysis. Its primary use is for recovering valuable cations or anions from dilute solutions. In the process, membranes are arranged in a parallel assembly to form solution compartments as in an electrodialyzer (Fig. 5), except that no electrodes are needed. However, the membranes used are either all cation-exchange or all anion-exchange membranes. A dilute solution that contains a valuable cation or anion (for example, $CuSO_4$) is circulated through the odd-numbered compartments. A relatively concentrated solution of an acid that is cheaper than the material to be recovered is circulated through the even-numbered compartments.

Because of the transmembrane difference in acid concentration, there is a tendency for both H^+ ions and the acid anions to transfer through the membranes into the solution in the odd-numbered compartments. The cation-exchange membranes allow the passage of the H^+ ions, but block the anions because of Donnan exclusion. The H^+ ions would enter the $CuSO_4$ solution in the odd-numbered compartments, but for this to happen either SO_4^{-2} ions must also enter those compartments or Cu^{+2} ions must leave the compartments to maintain electroneutrality. Since the SO_4^{-2} ions are prevented from transferring through membranes by Donnan exclusion, only the Cu^{+2} ions can transfer (from the odd-numbered compartments into the acid in the even-numbered compartments). With this process, valuable cations can be recovered from dilute solutions and concentrated in the acid stream. For example, copper has been recovered in the author's laboratory from a stream containing 30 ppm. Cu^{+2} and concentrated to 3,000 ppm. in an H_2SO_4 solution. Further stages would effect further concentration. Donnan dialysis has also been used to recover and concentrate uranyl ions from dilute uranyl nitrate solutions.

Some Uses of Membrane Processes—Table I

Electrodialysis

Commercial: Concentration of seawater for brine;[3,4] Electrohydrodimerization;[7] De-ashing whey;[1,2,9] Desalting iron-complexed dextrin.[6,7] *Pilot plant:* Regeneration of SO_2-scrubbing solution;[5] Recovery of acid from spent pickle liquor;[8,9] Recovery of values from pulping wastes.[10]

Reverse Osmosis

Commercial: Concentration of whey;[11-16] Treatment of municipal sewage.[18,19] *Pilot plant:* Recovery of values from plating wastes;[17] Concentration of orange juice;[20,21] Concentration of maple sap;[22,23] Concentration of coffee.[24]

Ultrafiltration

Commercial: Recovery of protein from whey;[11-14] Recovery of values from electrocoating painting operations.[25-27] *Pilot plant:* Recovery of protein from soy whey;[26,27] Concentration of egg white;[15] Hydrolysis of starch to sugars.[28]

Dialysis

Commercial: Separation of NaOH from hemicellulose.[29]

Donnan Dialysis

Pilot plant: Concentration of uranyl ions.[30]

Gas Permeation

Pilot plant: Separation of He from natural gas;[32] Separation of O_2 from air.[32]

Pervaporation

Laboratory: Separation of *o*-xylene from *o*-xylene - *p*-xylene mixtures.[33]

Ion-Exchange Dialysis

Ion-exchange dialysis takes advantage of a phenomenon often considered to be a disadvantage of ion-exchange membranes. Anion-exchange membranes are not very effective in blocking the transport of H^+ ions, although they block other cations effectively. Thus, when using anion-exchange membranes to block cations and transfer anions in a highly acid medium, the efficiency of using the electrical current can be as low as 20% to 30%, because 70% to 80% of the current through the anion-exchange membranes can be carried by H^+ ions that "leak" through them. Hydroxyl ions "leak" through cation-exchange membranes in a similar manner when the membranes are used to treat highly basic solutions. This leakage of ions would normally be a disadvantage for most processing but is used to advantage in ion-exchange dialysis.

Ion-exchange dialysis has been used most often to separate acids from metal salts. The arrangement of the membranes to form solution compartments is identical to that for Donnan dialysis. However, when used to separate acids from metal salts, all of the membranes are anion-exchange membranes instead of cation-exchange membranes. The mixed solution of acid and metal salts, for example, H_2SO_4 and $NiSO_4$, is circulated through the odd-numbered solution compartments, and water or a dilute solution of H_2SO_4 is circulated through the even-numbered ones. Because of the difference in concentration of H_2SO_4 and $NiSO_4$ in the two sets of compartments, both H_2SO_4 and $NiSO_4$ tend to transfer through the membranes from the odd-numbered compartments. Since the anion-exchange membranes leak H^+ ions, but block Ni^+ ions, the only cation that transfers (or leaks) is the H^+. The SO_4^{-2} anions transfer readily, of course. Thus, H^+ and SO_4^{-2} ions transfer to the dilute stream, but Ni^{-2} ions are left behind in the odd-numbered compartments. A stream of pure acid can be withdrawn from the even-numbered compartments, and a stream of purified $NiSO_4$ solution can be withdrawn from the odd-numbered compartments.

Ion-exchange dialysis has been used commercially to separate and recover metal salts from waste streams in the metal plating industry, and for separating impurities from acids so the acids can be reused.

Gas-Permeation Processes

Gas-permeation processes are pressure-driven processes in which one component can be separated from a mixture of gases. Selective membranes are arranged to form two sets of compartments, much as in the membrane processes previously described. The mixture of gases is circulated through one set of compartments at a high pressure. Membranes are selected that will allow one component of the mixture to permeate at a greater rate than others. The source of membrane selectivity is usually either preferential solubility of the desired component in the polymer, or a lower resistance to transfer for one component than for others, or both. With a driving force of a transmembrane hydrostatic-pressure difference, the preferred component is transferred more rapidly than other components, and the gases that permeate the membrane are enriched in the preferred component.

Pervaporation

Pervaporation is a membrane permeation process in which one component of a liquid mixture diffuses more rapidly through selective membranes than other components; the permeate, enriched in the one component, is removed from the opposite side of the membrane by evaporation into a stream of warm vapors and constitutes the product. The equipment used is a membrane stack, similar to previously described membrane units in that many compartments are formed by a parallel array of selective membranes. The liquid mixture to be treated is circulated through alternate compartments, and warm vapors are circulated through the other compartments. The warm vapors furnish the latent heat of vaporization needed to evaporate the permeate. After the vapors and permeate are withdrawn from the membrane permeator, a portion of the vapors is condensed as product. The remainder is reheated and recycled to the membrane permeator to furnish the latent heat for more permeate. Pervaporation has not yet found wide acceptance in industry.

Table I shows some applications of various membrane processes. Selected references are also given for readers who may wish to study given applications and processes further.

Selected books for readers who want to further study the theory and engineering principles of various membrane processes are given in the paragraph preceding the references.

Membrane Processes vs. Other Separations

Some of the methods of separating components from mixtures or solutions are evaporation, distillation, drum drying, spray drying, freeze drying, freeze concentration, centrifugation and filtration. Membrane processes have certain general advantages and disadvantages when compared with these other methods.

In general, membrane processes are gentler than evaporation or distillation. Membrane processes can separate heat-sensitive materials without damage, whereas evaporation or distillation cannot. For small to medium-sized plants, membrane processing is usually cheaper than evaporation or distillation, but for large plants the reverse is usually true. Evaporation can generally produce higher concentrations of solutes than membrane processes, and distillation is a highly selective fractionation process whose principles are well known.

Membrane processes are much gentler methods of separation than drum drying or spray drying, and can treat heat-sensitive materials much better than those two processes. In addition, membrane processes are usually considerably cheaper. Of course, the membrane processes cannot concentrate to dryness. But a combination of a membrane process to concentrate a solute followed by a drying process to remove the remaining solvent will often prove much cheaper than the drying process by itself.

Processing Costs for Competing Separation Processes — Table II
(Cost for removing 1,000 gal. of water)

Process	Cost, $
Evaporation and vacuum evaporation	0.20-5.00
Drum drying	25.00
Spray drying	17.00-50.00
Freeze drying	200.00-300.00
Freeze concentration	0.50-45.00
Centrifugation	0.30-10.00
Ultrafiltration and reverse osmosis	0.20-5.00
Electrodialysis	0.20-5.00

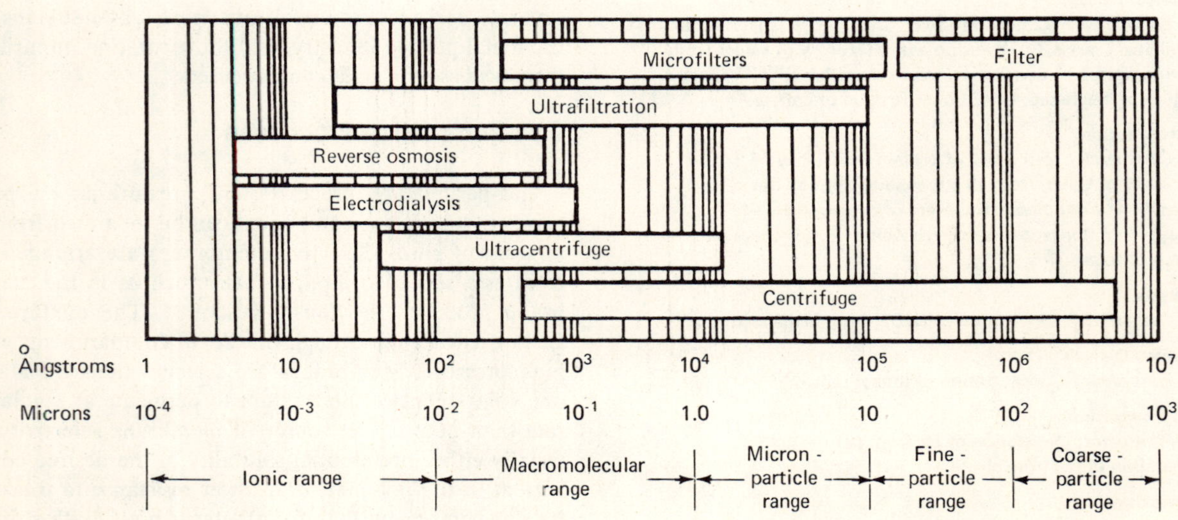

USEFUL RANGES of separation processes—Fig. 7

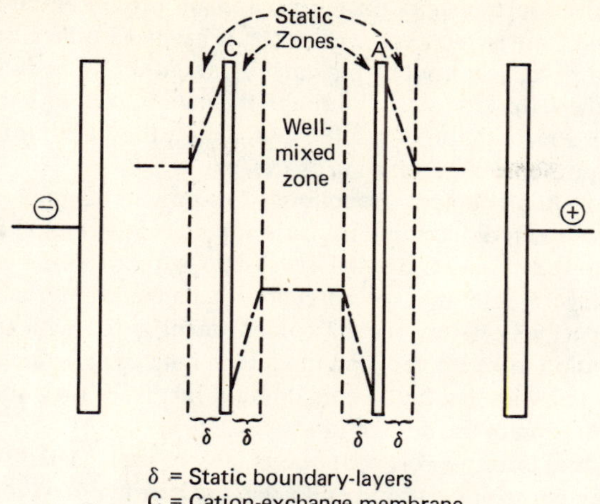

CONCENTRATION gradients in electrodialysis—Fig. 8

Table II shows costs of separating 1,000 lb. of water from various types of solutions or slurries by several methods of separation, as published in various articles in the literature.

Fig. 7 shows the size ranges of solutes, colloids and particles that can be effectively treated by various separation processes.

The membrane processes of greatest present utility in industry are electrodialysis, reverse osmosis, ultrafiltration and microfiltration. The most widespread use of microfiltration is to remove yeast cells from beer and wine and thus stabilize them. Strictly speaking, it is more a filtration operation than a membrane process. Accordingly, even though microfiltration will be discussed to some degree in the remainder of this article, the emphasis will be given to electrodialysis, reverse osmosis, and ultrafiltration.

ELECTRODIALYSIS

The nature of ion-exchange membranes has been shown in Fig. 1, and the way in which ion-exchange membranes are used to effect demineralization and concentration in electrodialysis was shown in Fig. 5. However, the detailed nature of ion transfer through a system of solutions and ion-exchange membranes warrants further description because it not only is the source of the depletion and concentration that occurs in electrodialysis but is also the source of excessive *concentration polarization*, which causes several difficulties in electrodialytic processing.

Fig. 8 shows a cation- and an anion-exchange membrane mounted between two electrodes. A solution of an electrolyte flows through the compartments formed by the two membranes. With the passage of an electrical current through the system of the membranes and solutions, anions (negative ions) transfer toward the anode (positively charged electrode), and cations transfer toward the cathode. These ion transfers are the way electrical current is carried in an electrolytic medium.

The cations and anions do not necessarily have to carry the same proportion of the current. When the electrolyte contains tiny cations but larger anions (for example, H^+ cations and Cl^- anions), the tiny cations travel at higher velocities through solutions than the larger anions, and thus carry more of the electrical current. For an electrolyte like KCl, in which the diameters of the hydrated cation and anion are almost equal, the velocities of the two types of ions are nearly equal and the fraction of the electrical current carried by each is nearly equal. The fraction of the current carried by an ionic species is termed its *transference number.* As an approximation, for KCl the transference number of K^+ and Cl^- are both 0.50.

Because of the flow of solution through the center compartment formed by the two membranes in Fig. 8, there is a zone of relatively well-mixed solution near the center of the compartment. The velocity of the solution and thus the degree of mixing diminish as the surfaces of the membranes are approached. In the figure, the so-called Nernst idealization is used for simplification. This assumes a completely mixed zone in the center of the compartment, and completely static zones of solution in boundary layers adjacent to the membranes. In the static boundary layers, ions are transferred only by electrical transfer and diffusion, but in the mixed zone ions transfer electrically, by diffusion and by physical mixing.

If we assume that the electrolyte is KCl, the transference numbers of K^+ and Cl^- are 0.50 in solution. However, because of the selectivity of the membranes, the transference number of K^+ is essentially 1.0 in the cation-exchange membrane and 0.0 in the anion-exchange membrane. Similarly, the transference number of Cl^- is 1.0 in the anion-exchange membrane.

Considering now only the anion-exchange membrane and its boundary layers, we find that the Cl^- ion carried only 50% of the electrical current in solution but 100% of the current in the membrane. If one Faraday of electricity were passed through the membrane and boundary layers, 0.5 g. eq. of Cl^- would be transferred to or away from the membrane surface, and 1.0 g. eq. of Cl^- would be transferred through the membrane. There would be a depletion of ions at the left-hand surface of the anion-exchange membrane in Fig. 8 and a concentration of ions at the right-hand surface. Obviously, this transient state of affairs could not continue long, before all ions were depleted at the left surface. In steady-state operation, concentration gradients are established in the static boundary layers, as indicated by the dashed lines in Fig. 8, so that the ions that are not electrically transferred to or from the membrane surface are supplied by diffusion through the boundary layers. These differences in transference numbers between solutions and membranes are the source of the depletion and concentration that occurs in electrodialysis, but they can also be a source of troubles.

If the current density (A./sq. ft.) is increased through the system of membranes and solutions in Fig. 8, the rate of electrical transport increases. Therefore, the rate of diffusional transport must also increase to supply the extra ions needed at the surfaces of the membranes. The only way diffusional transport can increase is for the thicknesses of the boundary layers to increase, or for the concentration gradients in the boundary layers to increase. If the solution velocity past the membranes stays constant, the thickness of the boundary layers also remains constant and the concentration gradients increase. The interfacial concentrations on the depleting sides of the membranes decrease; those on the concentrating sides increase.

If the current density is increased still more, a density will be reached at which the concentrations of electrolytes at the membrane interfaces on the depleting sides will approach zero. At this density, usually called the *limiting current density*, H^+ and OH^- ions from ionization of water will begin to be transferred through the membranes. The transfer of H^+ ions through the cation-exchange membranes is usually not particularly harmful, but transfer of OH^- ions through the anion-exchange membranes can cause increases in pH in the concentrating compartments that cause pH-sensitive salts like $CaCO_3$ or $Mg(OH)_2$ to precipitate in and on the membranes. In addition, the OH^- ions can cause dimensional changes in the anion-exchange membranes. Also, the presence of the layers of almost pure water at the membrane surfaces causes the resistance of the membrane cells to be high, and the energy requirements of the process to be high.

The thicknesses of the boundary layers decrease with increasing solution velocity. Therefore, it is important to use the highest practicable velocities, to have equal velocities in each solution compartment, and to have uniform velocities at all points along the membrane surfaces (i.e., have no points of solution stagnation).

Another problem encountered in electrodialysis is the fouling of ion-exchange membranes by charged organic materials. Organic anions can be driven to the surfaces of anion-exchange membranes, but if the organic portion of the molecule is much larger than the simplest organic anions (e.g., formate, acetate, etc.), the usual membrane will not allow the transfer of the organic portion, and a layer of negatively charged organic material is formed on the surface of the positively charged anion-exchange membrane. This negatively charged layer acts as though it were a cation-exchange layer adhering to the anion-exchange membranes. A similar phenomenon can occur with positively charged organic anions at negatively charged cation-exchange membranes. However, most of the difficulties that have been encountered in practice have been at the anion-exchange membranes.

Types of Separation Possible

Some typical classes of separations that are possible with electrodialysis are discussed below.

Demineralization—Since ions are transferred through the membranes in electrodialysis, but noncharged materials are not, electrodialysis can be used to separate ionized from nonionized components in solutions. Demineralization of sugar solutions and of whey are examples of this class of separation.

Concentration of Electrolytes—Concentration of electrolytes by electrodialysis usually accompanies demineralization, and the degree of concentration is limited only by the extent to which solvent transport accompanies ion transport. In the production of concentrated brine from seawater, for example, it has been difficult to achieve concentrations higher than about 3.5 N because of the water transported through the membranes with the ions. Precipitation of insoluble salts may also be a limitation in some concentration processes.

Ion-Replacement Reactions—Continuous anion replacement or continuous cation replacement may be achieved in an electrodialysis system with all anion-exchange membranes, or all cation-exchange membranes, respectively. Every second compartment is fed with the solution to be treated, and the intervening compartments are fed with a solution containing a relatively high concentration of the desired ion.

One example of continuous anion replacement is sweetening of citrus juices. In this process, hydroxyl ions furnished by a caustic solution replace the citrate ions in the juice. The hydroxyl ions are neutralized by hydrogen ions in the juice to produce water and are, therefore, unavailable for further transfer through the membrane system. The overall result is the removal of citric acid.

An example of continuous cation replacement is the removal of radioactive strontium from milk. In this process, a synthetic electrolyte solution having a cationic composition similar to milk furnishes nonradioactive cations to the milk to replace radioactive strontium.

Continuously controlled pH adjustment of solutions without the direct addition of acids or bases is another possible application of continuous ion replacement by electrodialysis.

Double-Decomposition Reactions—If the diluting compartments of a conventional electrodialysis system with alternating anion- and cation-exchange membranes are fed with electrolyte solutions of different compositions, double decomposition reactions are readily achieved. For example, photographic emulsions of AgBr have been produced by this technique from NaBr and $AgNO_3$ with a gelatin emulsion being fed into the concentrating compartments. The emulsion thus produced is free of extraneous electrolytes.

Separation of Electrolysis Products—When the prod-

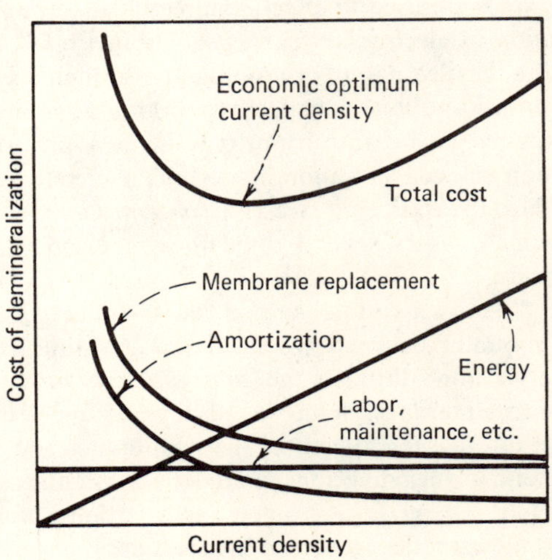

COST FACTORS in electrodialysis treatment—Fig. 9

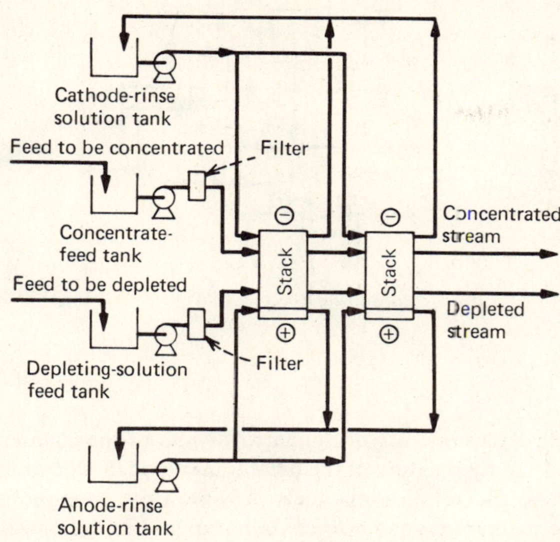

ELECTRODIALYSIS-plant flow diagram for concentrating or depleting electrolytes in solution—Fig. 10

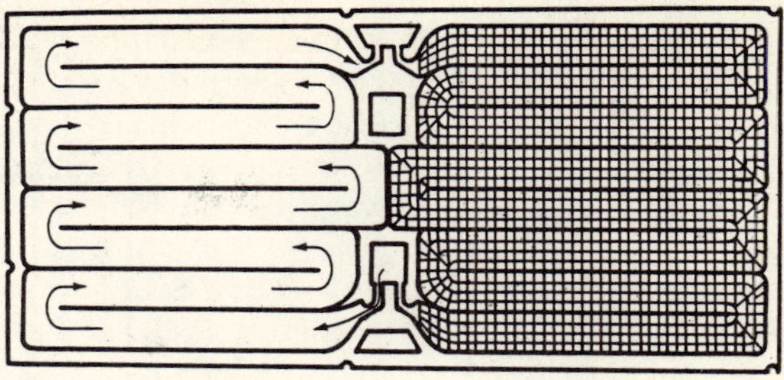

DIAGRAM of tortuous-path spacer—Fig. 12

ucts of electrolysis at one electrode, or the reactants fed to one electrode, must be kept separated from the reactants or products at the other electrode, ion-selective membranes sometimes offer an advantage over porous separator materials. For example, ion-exchange membranes have been used as cell separators in a process for hydrodimerizing acrylonitrile.

Fractionation of Electrolytes—In addition to separating electrolytes from solutions on the basis of their charge sign, membrane systems can be used to separate ions of like charge on the basis of their different rates of transport through membranes. When brackish waters are demineralized in most electrodialysis systems, for example, Ca and Mg ions are transported more rapidly than Na ions. The demineralized water is thereby softened to a greater degree than would be expected from the proportional reduction of all ionic components.

There are three categories of costs that contribute to the total cost of electrodialysis: costs that vary directly with current density (for electrical energy), costs that vary inversely with current density (for membrane replacement and amortization of capital investment), and costs that are essentially invariant with current density (operating labor, costs for chemicals, etc.). Fig. 9 shows typical variations of these cost items. Equations and methods for determining the optimized costs for electrodialysis have been published.[34-37]

Costs of desalting whey,[2] recovery of values in pulping liquors,[10] de-ashing glycerol,[39] and of concentrating seawater for use as chlor-alkali brines[3] are available in the open literature.

Other information on costs, is available from the manufacturers of membranes and equipment.

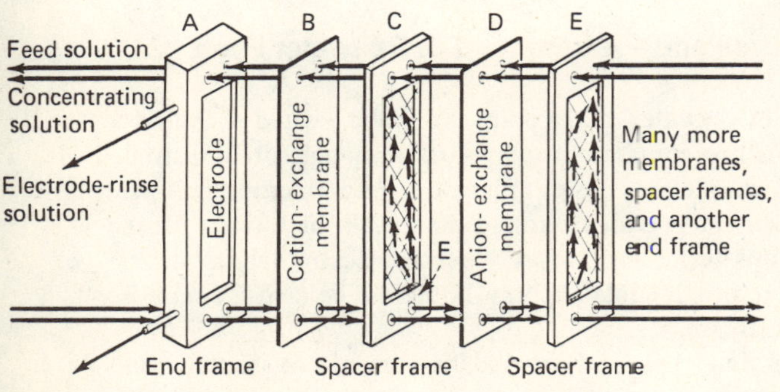

COMPONENTS in an electromembrane stack—Fig. 11

Electrodialysis Equipment

A flow diagram for a two-stage electrodialysis plant for concentrating and depleting electrolytes in solution is shown in Fig. 10. Solution to be fed to the concentrating compartment of the electrodialysis stacks is stored in the concentrate-feed tank. It is pumped from the feed tank through a filter and into the concentrating compartments of the first-stage electrodialysis stack, and from there through the concentrating compartments of the second-stage stack. The flowpath for the depleting stream is similar except that the depleting solution is pumped from its feed tank through the two stacks in series.

Anode- and cathode-rinse solutions, which may be a portion of the solution being treated, or specially prepared solutions of various kinds, are pumped from the electrode-rinse solution tanks through the electrode compartments of the stack and back to the proper tank. The pH of these solutions is usually kept at a value that will counteract the bases produced by electrochemical action at the cathode, and the acids produced at the anode, so the H^+ and OH^- ions do not enter the concentrating and depleting compartments of the electrodialysis stacks.

Several other methods of processing with electrodialysis stacks have been described by Mintz.[39] These other methods include batch-recirculation processing, and feed-and-bleed processing.

Fig. 11 is an exploded view of an electrodialysis stack that shows the main compartments. Component *A* is one of the two end-frames, each of which has provisions for holding an electrode and introducing and withdrawing the depleting, the concentrating, and the electrode-rinse solutions. The inside surfaces of the electrodes are usu-

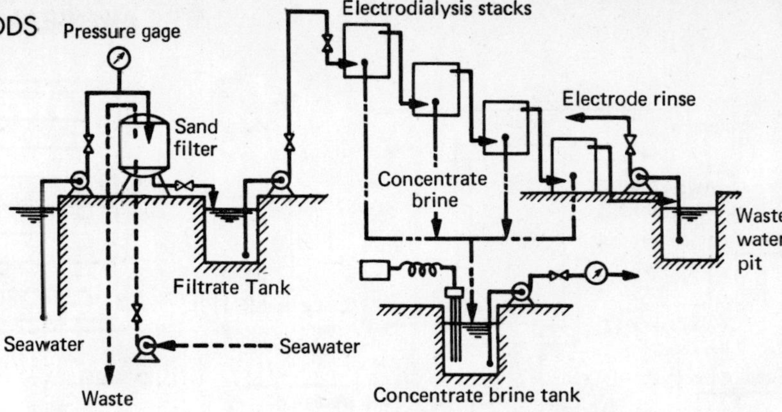

SEAWATER concentration-plant flowsheet—Fig. 13

ally recessed, as shown, so that an electrode-rinse compartment is formed when an ion-exchange membrane, *B*, is clamped in place. Components *C* and *D* are spacer frames. Spacer frames have gaskets at the edges and ends so that solution compartments are formed when ion-exchange membranes and spacer frames are clamped together.

Usually, the manifolds for the various solutions are formed by matching holes in the spacer frames, membranes, gaskets and end-frames. Each spacer frame is provided with solution channels (*E* in Fig. 11) that connect the manifolds with the solution compartments. The spacer frames have mesh spacers, or some other device, in the compartment space to support the ion-exchange membranes. A complete electrodialysis stack usually has many repeated sections, each consisting of components B, C, D and E, with a second end-frame at the other end.

There are three general types of electrodialysis stacks: tortuous-path stacks, sheet-flow stacks and unit-cell stacks.

In the tortuous-path type of electromembrane stack, the solution flowpath is a long narrow channel as illustrated in Fig. 12, which makes several 180-deg. bends between the entrance and exit ports of a compartment. The right half of the spacer gasket in Fig. 12 shows the individual, narrow, solution channels and the cross-straps used to promote turbulence, whereas the individual channels have been omitted in the left half of the figure so the flowpath could be better depicted. The ratio of the solution channel length to its width is high, usually greater than 100:1. Solution flow in a sheet-flow type of stack is approximately in a straight path from one or more entrance ports to an equal number of exit ports of a compartment, as was illustrated in the exploded view in Fig. 11. Thus the solution flows through the compartment as a sheet of liquid. The ratio of the solution channel length to width in the sheet-flow type of stack is much lower (about 1:2) than in the tortuous-flow type.

Solution velocities in the sheet-flow stacks are typically in the range of 5-15 cm./sec., whereas solution velocities are generally 30-50cm./sec. in the tortuous-path stacks. The drop in hydraulic pressure through a tortuous-path stack is generally higher than that through a sheet-flow stack because of the higher solution velocity, the changes in the direction of the flowing solution, and the longer path.

Spacer screens are usually not needed to support the membranes in the solution compartments of tortuous-path stacks because of the narrow width of the channels and the rigid nature of the membranes used. Small straps across the solution channels provide some support for the membranes and also cause turbulence in the flowing solution. This turbulence causes mixing of the depleted or enriched solution near the membrane surfaces with the bulk of the solution, and reduces the thickness of the boundary layers at the membranes' surfaces.

Spacer screens are generally used in the solution compartments of sheet-flow stacks, both to support the membranes and to produce turbulence in the flowing solution.

The unit-cell type of stack is described later in connection with concentration of brines.

SALT FROM SEAWATER

Because of the lack of native salt deposits in Japan, and because solar evaporation is expensive and undependable in Japan's wet climate, electrodialytic concentration of seawater is used to produce sodium chloride brines and, with further concentration by evaporation, to produce solid salt. The electrodialytic concentration process has been described by Nishiwaki,[3] and much of the material presented here is from that reference.

A flowsheet of a seawater concentration plant is shown in Fig. 13. The seawater feed is first filtered and then pumped through four stages of electrodialytic concentration. The concentrated brine is withdrawn and stored as product. The seawater, somewhat depleted in electrolyte content, is sent to a wastewater pit, and eventually to the wastewater outfall. The electrode-rinse streams are pumped to the electrodialysis stacks from the wastewater pits.

Source and Pretreatment of Seawater Feed

The seawater intake point should be located so the water is not contaminated by land drainage, or industrial and municipal wastes. Plankton, and microorganisms, which are prevalent around the mouth of a river, cause serious problems if they enter an electrodialysis stack. Therefore, the intake point should not be near the mouth of a stream.

A plant that produces 100,000 tons of salt annually requires a large amount of seawater (about 5,300,000 gal./hr.) Accordingly, care should be taken to locate the intake point and the outfall for depleted seawater to prevent mixing.

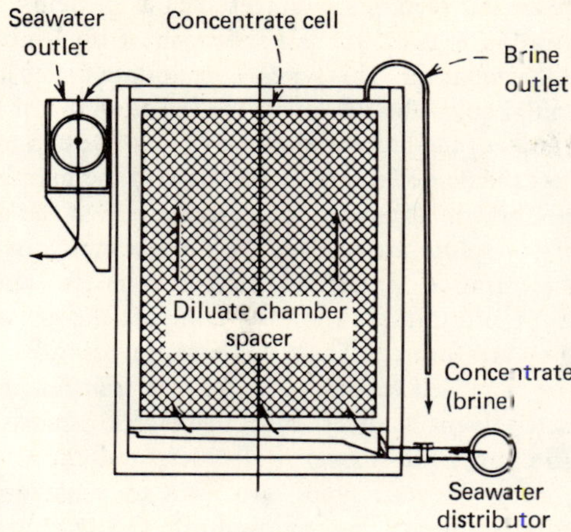

UNIT-CELL type electrodialysis stack—Fig. 14

Electrodialysis Stacks Used in the Process

One type of stack used is the conventional sheet-flow stack described previously.

However, there are usually two important differences between the stacks used for seawater concentration and others that have been described. Since the maximum possible degree of concentration is desired, no seawater is fed to the concentrate compartments. Only the water that transfers with the ions by osmosis and electro-osmosis is withdrawn as concentrated product. The other difference is that as many as 2,000 membranes are used in each stack, whereas for most other operations only 200 to 500 membranes are used.

Another type of stack is employed for seawater concentration—the unit-cell stack. In the unit-cell stack, each concentrating compartment consists of one cation-exchange membrane and one anion-exchange membrane sealed at the edges to form an envelope-like bag. These concentrate compartments are placed between electrodes with screen-like spacers between them to form compartments for the flow of seawater. Concentrated brine overflows each concentrate cell through small tubes attached to the unit cells, as shown in Fig. 14. More than a thousand concentrate cells may be assembled into a unit-cell stack. A central anode with cathodes at either end is often used to keep the total applied voltages to a tolerably low level.

In both the conventional sheet-flow and the unit-cell stacks, it is important to minimize the thickness of boundary layers to prevent excessive concentration polarization because seawater contains pH-sensitive salts, $CaCO_3$ and $Mg(OH)_2$, that will precipitate if water-splitting and zones of high pH occur. The Japanese companies that developed the process have each originated specially woven spacers of plastic monofilament that do a good job of minimizing boundary layers, with low hydraulic-pressure drops. The stacks developed for seawater concentration also have specially designed solution ducts that distribute seawater from the manifolds to the individual solution compartments. The purpose of the special designs is to minimize the flow of electrical current through the solution in the manifolds since that current bypasses the stack and does no useful work of transferring ions.

Compositions of Brines From Electrodialytic Concentration Plants — Table III

Ion	Concentration in seawater	Concentration of Brine From Plant A	Plant B	Plant C	Plant D
Na^+	0.4430	3.030	2.883	3.170	3.220
K^+	0.0095	0.105	0.087	0.150	0.090
Ca^{+2}	0.0198	0.075	0.104	0.080	0.100
Mg^{+2}	0.1080	0.199	0.304	0.120	0.260
Cl^-	0.523	3.402	3.366	3.500	3.658
SO_4^{-2}	0.0544	0.008	0.022	0.020	0.014

All concentrations are shown in equivalents/liter.

Ion-Exchange Membranes Used

If seawater is concentrated by electrodialysis with conventional ion-exchange membranes, the concentrated solution becomes saturated with $CaSO_4$ long before the $NaCl$ is concentrated to a desirable level. Since precipitates of $CaSO_4$ would damage electrodialysis stacks, it was necessary to find a way of overcoming this problem. Univalent-selective ion-exchange membranes were developed that transfer univalent ions in preference to di- or multivalent ions. Therefore, the Na^+ and Cl^- ions are transferred in preference to Ca^+ or SO_4^{-2} ions. Such membranes are made by either preparing homogenous membranes with resins that have low permeability for divalent ions, or coating conventional ion-exchange membranes with a thin film of a resin with low permeability for divalent ions. Membranes prepared by the coating method have much lower electrical resistance than those prepared by the homogenous-membrane process, and are therefore preferred for use in seawater concentration.

Brines, Compositions and Costs

Published data[3] on the composition of brines produced at four commercial electrodialysis concentration plants are summarized in Table III.

It is notable that the Ca^{+2} and SO_4^{-2} ions in the seawater were concentrated to a much lower degree than the univalent ions. In fact, the concentration of SO_4^{-2} ions is less than that in the original seawater. This low degree of concentration of divalent ions is a result of the preferential transport of univalent ions over divalent ions through the univalent-selective membranes. The brines, which are about 18% to 21% solids by weight and very low in sulfate, need only slight further concentration to serve as a raw material for the chlor-alkali industry. However, if solid table salt is produced for sale (as it is in Japan), further concentration and crystallization is necessary.

The first commercial electrodialytic concentration plant was erected in 1961 and had a capacity equivalent to 10,000 tons of salt per year. The size of the plants has grown since that time until plants now being constructed have capacities that range from 100,000 to 300,000 tons/yr. These large modern plants have captive electric-power facilities that are integrated with the salt-concentration facilities to produce electricity for the electrodialytic concentration, and low-pressure steam for the subsequent evaporation.

The total production cost of concentrated brine (including investment amortization) decreases from about $8/ton of salt to about $6/ton as the size of the plant is increased from 50,000 to 300,000 tons/yr. The energy consumption ranges from 250 to 300 kwh./ton of salt.

In this article on seawater concentration, Nishiwaki[3] also describes other applications of the electrodialytic concentration process, such as the concentration of sodium acetate solutions, and the concentration of copper ions from copper-plating wastes.

REVERSE OSMOSIS, ULTRAFILTRATION, AND MICROFILTRATION

These three processes have many points of similarity and will be discussed together. All three use membranes that can be said to have selectivity, although microfiltration membranes can also be considered closely akin to conventional filters. In all three processes, a difference in hydraulic pressure is the driving force. Moreover, the physical arrangement of the components of membrane permeators is similar whether the membrane used is a reverse-osmosis, ultrafiltration or microfiltration membrane.

The anisotropic nature of these types of membranes was shown in Fig. 3 and 4, and the basic features of the processes were discussed previously and shown in Fig. 6. As pointed out in the prior discussion, solvents (and perhaps small solutes) transfer through the membranes, whereas solutes are blocked by them. This passage of one component and blockage of another results in increased concentrations of solute at the high-pressure interfaces of the membranes. Concentration gradients are established in the solutions adjacent to those interfaces. These concentration gradients are similar to those discussed previously for electrodialysis in that the gradient exists mainly in static or semistatic boundary layers next to the membranes, and that the rate of diffusive transport through these layers sometimes controls the total rate of transport and sometimes limits the degree of concentration that can be achieved.

In reverse osmosis, thick boundary layers result in high solute concentrations at the membrane surfaces, which result in high osmotic pressures. High osmotic pressures reduce the total driving force, since the driving force is the difference between the applied pressure and the osmotic pressure. In both reverse osmosis and ultrafiltration, the high concentration of the solutes can result in precipitates of materials that exceed their solubility limits in the boundary layers. These layers of precipitates act as an added resistance to transfer and can dramatically reduce the fluxes through the membranes. In reverse osmosis, sparingly soluble salts, such as $CaSO_4$, precipitate on the membrane.

In ultrafiltration, the materials that precipitate are high-molecular-weight substances such as proteins, since ultrafiltration is used to treat solutions of high-molecular-weight materials. This type of precipitate forms layers of colloidal gels. The gel layers will build outward from the surfaces of the membranes and reduce the flux of solvent until the flux of solute carried toward the membrane with the solvent is reduced to a value equal to the diffusive flux of solute away from the membrane caused by the concentration gradients. Increases in the applied pressure in ultrafiltration tend to compact the gel layers once they are formed. Thus, increases in pressure often result in increased resistance to solvent transfer, not in increased solvent transfer. In both reverse osmosis and ultrafiltration, high solution-velocities, and turbulence-promoters of various types, are used to achieve good mixing of solutions and to diminish the thickness of boundary layers. The reduced thicknesses result in increases in the diffusive fluxes.

In tubular reverse-osmosis systems, a method of physically removing precipitates has been developed in which a pith ball, having the same diameter as the inside of the reverse-osmosis tubes, is introduced into the tube, and is moved through it by the applied pressure, scraping precipitate off as it goes.

Another method of removing precipitates, useful in all types of reverse-osmosis equipment, is to periodically reduce the applied pressure to zero and rinse the membrane compartments with water. A method of cleaning ultrafiltration equipment that has been found useful for many applications involves periodically reducing the pressure to zero and rinsing the membrane compartments with water adjusted in pH to a value that dissolves the proteins or other colloids of which the gel layer is formed. This method should be considered for applications in which the solutes being treated are materials that ionize at certain values of pH and become more soluble (i.e., ampholytes).

Types of Membrane Separations Possible

The general types of separations possible with these pressure-driven processes depend primarily on the membranes used. With microfiltration membranes, small solid particles and collodial solids can be separated from solutes. With ultrafiltration membranes, solvents can be separated from higher-molecular-weight solutes. With reverse-osmosis membranes, low-molecular-weight solutes can be separated from solvents

Ultrafiltration and reverse osmosis are often used in combination. A first stage of ultrafiltration is used to retain high-molecular-weight solutes while allowing a lower-molecular-weight solute and solvent to pass through the membranes. The permeate is then treated by reverse osmosis to concentrate the solution in the low-molecular-weight solute.

Costs of Reverse Osmosis and Ultrafiltration

Several studies of the costs of reverse osmosis and of methods of optimizing costs have been published[42-48] for desalting of water. The amortization schedules and other

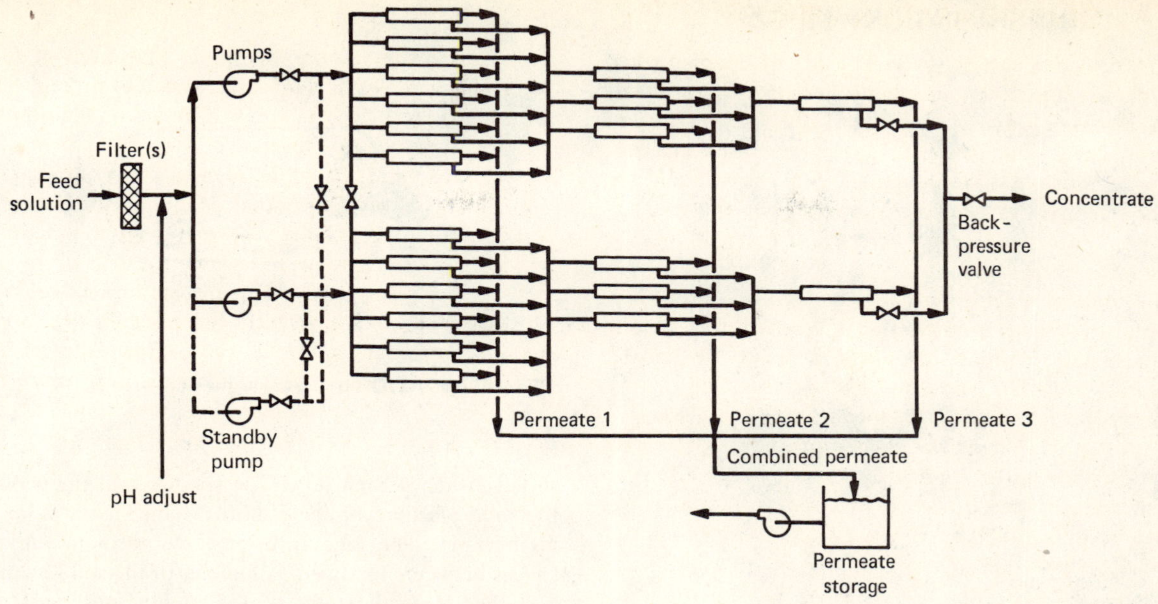

FLOW diagram of a pressure-driven membrane separation plant—Fig. 15

assumptions made for these cost studies are not directly applicable for industrial processes. Nevertheless, they are of value to the engineer in industry because they show the effects of the interaction of process variables on costs.

Other studies have been published on the costs of industrial processing by reverse osmosis and ultrafiltration. A Danish equipment supplier has issued a cost analysis (in English)[49] that gives a method and equations for comparing the cost of processing by reverse osmosis or ultrafiltration with the costs for competing processes, such as evaporation. Morrison[13] and Goldsmith[14] have published intensive studies of the costs of recovering protein from cottage-cheese whey. Goldsmith's data were for a plant treating 250,000 lb./day of whey. Morrison presents data showing the costs for several methods of processing whey by membranes to recover various kinds of products.

Other data have been published on the costs of treating industrial wastes,[50,51] and the costs of reclaiming chromium and other valuable metals from various kinds of plating wastes.[52,53]

Less-extensive cost data are available for recovery of paint solids from rinse waters in the electrocoating process,[54-57] recovery of protein from soy whey,[55] and the reclamation of a mineral oil used to treat textiles.[57]

Equipment for Pressure-Driven Processes

Fig. 15 shows a simplified flow-diagram of a pressure-driven membrane process plant for concentrating a solution. In a typical installation, the feed solution is filtered to remove gross particulate matter. Then, the pH of the feed is adjusted to a desirable value, if membranes of a cellulose ester are used. The feed is pressurized and sent to the membrane permeators. In Fig. 15, twelve modules of the first-stage permeators are shown in parallel. The concentrate from the first stage is sent to six modules in the second stage, and two modules in the third stage, since there are progressively smaller volumes of solution to be treated in those stages. The permeates from all three stages are combined and stored (if the permeate has value), or sent to waste. The pressure on the concentrated solution from the third stage is reduced with a backpressure valve and saved as a product.

The processing scheme shown in Fig. 15 is a simple straight-through flow plan, but the membrane permeators are versatile and can be used in various types of flow schemes by themselves or in combination with other separation processes, such as ion-exchange or evaporation.

The filters, tanks, pumps and piping used are conventional items used in the chemical process industries. The only special requirement is the capability for use at the pressures involved (50 to 1,000 psi., depending on the application).

Membrane Permeators

There are several basic types of membrane permeators: plate-and-frame, tubular, helical tubes, spiral-wrapped and hollow-fiber.

Plate-and-Frame—In a plate-and-frame membrane permeator, thin plastic support-plates shaped like a phonograph record are covered on both sides with a

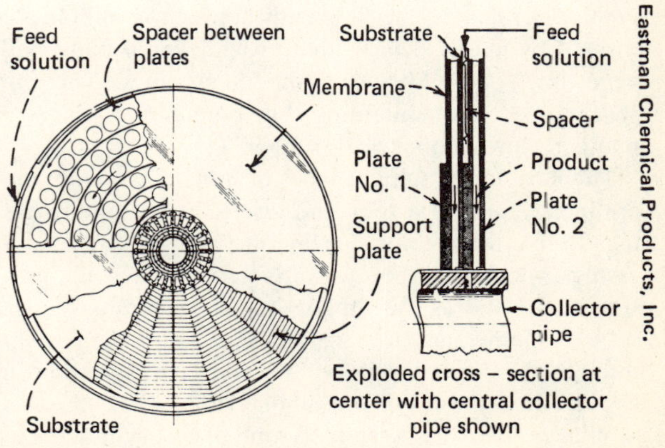

Eastman Chemical Products, Inc.

PLATE-AND-FRAME membrane permeator—Fig. 16

De Danske Sukkerfabriker

PLATE-AND-FRAME reverse osmosis units—Fig. 17

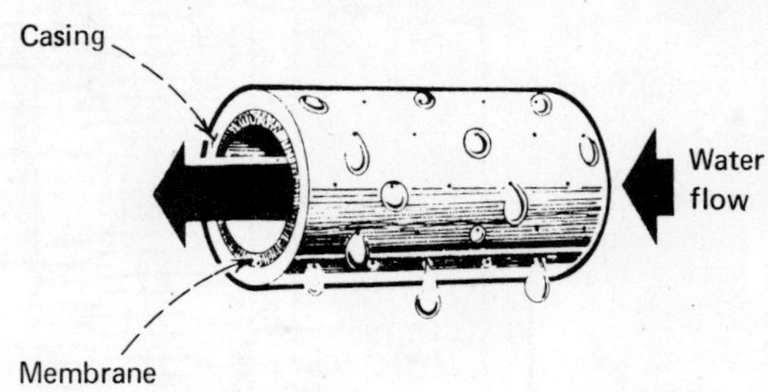

Eastman Chemical Products, Inc.

REPRESENTATION of a tubular membrane—Fig. 18

porous substrate and a membrane, and the membrane is cemented around the edges and at the center, as shown in Fig. 16. In one type of plate-and-frame unit, the center of each plastic support-disk contains a hole, and has shoulders that provide both a seal between the disks and also a flow channel between the membranes (when the disks are stacked on a central pipe with spacers between the membrane surfaces). The stacked structure is placed in a cylindrical pressure vessel. Pressurized feed solution enters the pressure vessel, and flows between the closely spaced membranes. Part of the solvent permeates the membranes and flows inward through thin grooves in the plastic support-disks toward the central pipe. The central pipe is provided with holes that receive the permeate, and the permeate leaves the membrane permeator through the pipe.

Another type of plate-and-frame unit is similar to the above-described unit except that the plastic support-disks have shoulders to effect pressure seals on their outer periphery as well as around the central pipe, and the grooves in the support disks lead the permeate to collector tubes at the outside of the stack of disks. This type of unit, pictured in Fig. 17, needs no cylindrical pressure vessel, since the pressure-sealed shoulders on the stack of disks form the pressure vessel.

Spiral—In the spiral-flow thin-cell plate-and-frame type of permeator, the flow path between membranes is formed by plastic devices that provide spiral paths with very thin spacing between the membranes. The thin spacings aid in maintaining thin boundary layers with relatively low volumetric flowrates.

Tubular—In the tubular types of units, the membranes are in the form of a tube and inserted into tubular casings so that the membranes line the casings. The tubular casings, which have porous walls, serve both as the pressure vessel and as the support for the membranes. The solvent permeates the membranes and then seeps through the porous walls of the casings, as indicated in Fig. 18. In a variant of the tubular unit called "thin-cell tubular permeators," a splined solid plastic cylinder is inserted inside of each tube. The splines hold the walls of the solid plastic cylinder a short distance away from the membranes. The feed solution flows through the thin spaces between the outside diameter of the solid cylinder and the inside diameter of the membrane-lined tube. High solution-velocities through these thin spacings result in high shear at the membrane surfaces, which aids in minimizing the "gel layers" in ultrafiltration.

A number of individual tubes are attached to headers, much like a shell-and-tube heat exchanger, and encased in a light housing. The permeate that seeps from the individual casings is collected in the housing and led off to storage. A photograph showing a number of the housings of a tubular unit used for treatment of primary sewage is given in Fig. 19. The pumps may be seen at the far end of the bank of tubes.

The tubular type of unit is also made with tubes in the shape of helices, as shown in Fig. 20. In these units, each helical tube segment is encased in a light plastic housing that collects permeate. The helical tube segments are designed so that a number of them can be manifolded together in a wide variety of parallel-series arrangements.

Spiral-Wound—Spiral-wound reverse-osmosis units are made with planar membranes. As shown in Fig. 21, they are fabricated by sandwiching a porous supporting material between reverse-osmosis or ultrafiltration membranes, sealing the edges of the membranes to each other and to a central tube, and then wrapping membranes, porous support, and a mesh feed-side spacer into a spiral

Calgon-Havens Systems

RAW SEWAGE treatment with reverse osmosis—Fig. 19

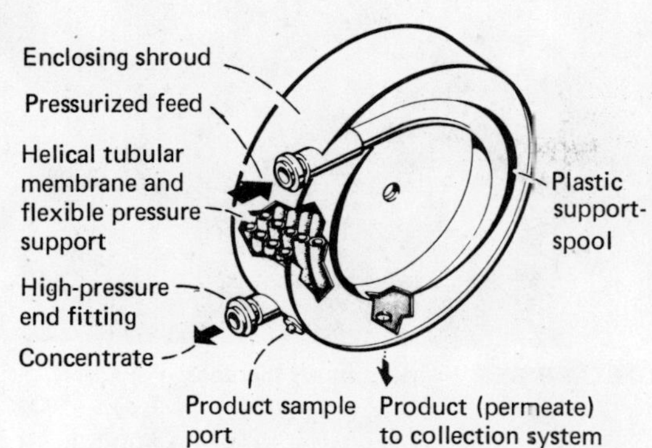

HELICAL-TUBE segment type permeator—Fig. 20

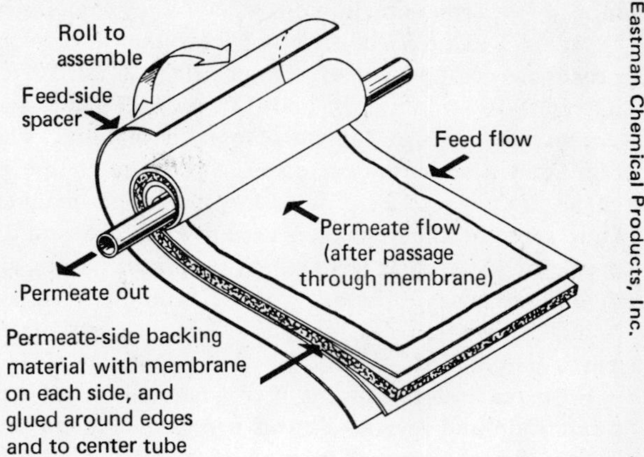

FABRICATION of spiral-wound module—Fig. 21

around the central tube. The spiral of membranes, spacer, and porous support is slipped into a cylindrical casing capable of withstanding high pressures. The pressurized feed solution is fed into the casing so that it flows through the mesh feed-side spacers and along the surfaces of the membranes. Part of the solvent permeates the membrane and flows through the porous support material into the central tube, from which it is collected.

Hollow-Fiber—In hollow-fiber reverse-osmosis or ultrafiltration units, the membranes are spun into hollow fibers. These are almost as fine as a human hair (100 to 200 microns) but are hollow, with walls about 25 microns thick. Cylinders this thin need no supporting structure to withstand large pressures without collapsing. For this reason, and because the fibers can be spun at low cost, it is possible to achieve vast areas of membrane surface cheaply—one of the main advantages of the hollow-fiber devices.

A bundle of long fibers is arranged in a U-shape, and the open ends of the U-shaped bundle are potted in plastic. The bundle with its potted ends is arranged in a cylindrical pressure shell as shown in Fig. 22. Feed solution is introduced into the middle of the bundle of fibers through a porous tube at one end of the cylinder, and the concentrated product leaves from an effluent tube at that same end. The permeate transfers through the fibers and flows through each fiber to the potted end, where it leaves the cylinder.

Fig. 23 shows a number of hollow-fiber modules arranged in parallel for producing extremely pure water used to rinse delicate electronic components.

Some of the above types of membrane permeators are more suitable for certain kinds of applications than others. The features of the various types are compared in Table IV.

For any particular application, it is prudent for the engineer in industry to seek advice from the manufacturers of the permeators.

Michaels[58] and Lonsdale[59] have published information about the characteristics of many of the commercially available membranes.

TREATMENT OF WHEY

The production of cheese creates large volumes of whey; milk is coagulated and the curd (cheese) is re-

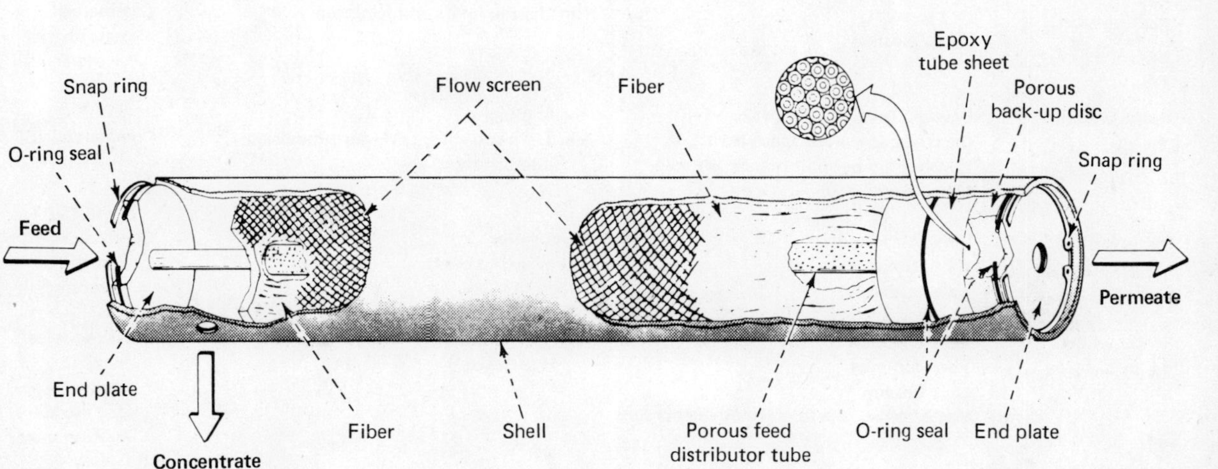

PERMASEP permeator, cutaway drawing—Fig. 22

moved—the remaining liquid is whey. Whey contains between 5.5% and 6.5% of dissolved solids that have an approximate composition of 12% protein, 1% fat, 70-75% lactose, 8-10% inorganics, and 0.1% to 1.0% lactic acid. Because of the high content of organic material, whey represents a serious waste-disposal problem (if the organics are not recovered or destroyed in some manner). Until recently, cheese-makers used several means to dispose of whey, such as paying high sewerage charges, giving the raw whey to farmers for use as animal feed, and drying the whole whey for sale to livestock growers for use as animal feed.

However, with the advent of commercially feasible ultrafiltration and reverse-osmosis processes, proteins and lactose can be recovered from whey as a profitable operation. At the same time, the recovery of these two products solves the cheese-makers' waste-disposal problem.

A flowsheet of a typical two-stage membrane process for recovering a protein concentrate and a lactose concentrate from whey is shown in Fig. 24.

Raw whey from the cheese-making operation is centrifuged to remove particles of cheese and globules of fats. The particles could cause blockage of some of the passages in the membrane equipment if not removed. If the fatty material were not removed, it could form a hydrophobic film on the membranes that would drastically reduce the flux of the watery portion of whey through the membranes.

HOLLOW-FIBER reverse osmosis modules—Fig. 23

The whey, which usually comes from the cheese-mak-

Comparison of Types of Membrane Permeators — Table IV

Types	Advantages	Disadvantages	Status
Plate-and-frame (conventional)	Low holdup per unit membrane area Much operating experience Low floor space per sq. ft.	Can plug at points of solution stagnation May be hard to clean acceptably for food uses Expensive at present (ca. $100/sq. ft.)	Commercial
Plate-and-frame (thin channel)	Low floor space per sq. ft. Very good for highly viscous solutions Very low holdup per unit membrane area High conversion per pass achievable	Meager large-scale operating experience at present Can plug Can be hard to clean acceptably Expensive at present Membranes in some designs hard to replace	Large-scale equipment just now on market
Tubular (conventional)	Easily cleaned; accepted for processing food products Much operating experience Individual tubes replaceable	High holdup per unit membrane area Relatively expensive (ca. $10-$20/sq. ft. Requires moderately large floor space per sq. ft. although tube modules can be placed separately in and around existing equipment in special cases	Commercial
Tubular (helical-tubes)	Moderately low floor space required per sq. ft. Easily cleaned	Meager operating experience at present High holdup per unit membrane area	Commercial new on market
Spiral-wound	Low in cost (ca. $3/sq. ft.) Compact; low floor space per sq. ft. Low holdup per unit membrane area Long operating experience	Easily plugged Hard to clean acceptably for processing food products	Commercial
Hollow-fiber (shell-side feed)	Very low in cost Very compact Low holdup	Plugs easily Very hard to clean	Commercial—for reverse osmosis only
Hollow-fiber (feed inside fibers)	Very low in cost Very compact Low holdup May be amenable to cleaning membranes	Just now on market—meager operating experience	Fiber modules new on market—for ultrafiltration only

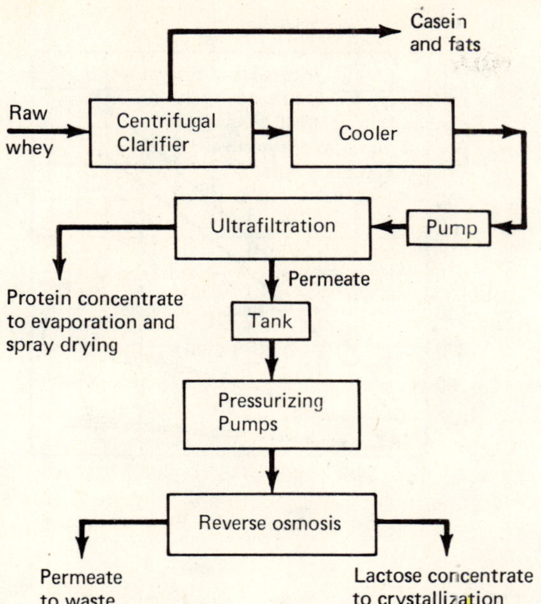

TWO-STAGE whey treatment membrane process—Fig. 24

ing operation at an elevated temperature, is cooled below 130 F. (with a few exceptions, the membranes that are commercially available at present can only be used at less than 130 F.).

The heat exchangers normally used are of the plate-type used for food products, which is designed to be easily cleaned.

Whey Ultrafiltration

Clarified and cooled material is pumped to the ultrafiltration units, usually the tubular type of membrane permeator, because tubular units are easily cleaned and are not as susceptible to plugging as some of the other types. Pressures are in the range of 30 to 100 psi. Ultrafiltration membranes are used that will retain the proteins in whey but will transfer the lactic acid, inorganic materials, water, and much of the lactose.

Solution is pumped past the membrane surfaces at relatively high velocities to minimize the thickness of boundary layers. Therefore, the amount of solvent removed in any one pass is small relative to the amount of whey pumped through the tubes, and the whey is recirculated through the system until a desired amount of water has been removed.

The amount of lactose removed from the protein depends on the amount of water removed from the whey. Although it has been shown in pilot-plant studies that products can be made that contain 80% or more of protein (on a dry solids basis) by removing 90% to 95% of the water in the whey, the commercial products usually contain 40% to 50% protein (on a dry basis), with the remainder being primarily lactose. The concentrated solutions produced contain from 12% to 15% solids. It has been found that there is a large demand for products in that range of protein content, and they can be produced and sold profitably.

The concentrated solution retained on the high-pressure side of the ultrafiltration membranes may find uses as a protein-containing solution, or it may be evaporated to a higher solids concentration and spray dried to give a dry product.

Reverse Osmosis of Whey Permeate

The permeate from the ultrafiltration unit is stored in interstage tanks, and then sent to the reverse-osmosis stage. Here, it is pressurized and pumped through the membrane permeators. Multistage, high-pressure centrifugal pumps and tubular-membrane permeators are usually used. Pressures are in the 600 to 800-psi. range. At these pressures, permeates from ultrafiltration have been concentrated from about 5.0% solids to more than 25% solids. However, for commercial operations, the concentrated product is usually in the 15% to 20% solids range because difficulties with fouling, plugging, and cleaning of the equipment are often encountered when solutions containing more than 20% solids are produced. The concentrated product is usually further concentrated by evaporation and the lactose is crystallized for sale.

Ninety-eight to ninety-nine percent of the chemical-oxygen demand in the feed is retained in the concentrated product. Therefore, the permeate from the reverse-osmosis stage is usually a permissible waste.

For treatment of whey, the entire equipment is designed so that it can be cleaned in place, since sanitation of any equipment used for food products is of prime importance. Clean-in-place techniques for both the ultrafiltration and reverse-osmosis stages have been developed that meet the stringent requirements of the dairy industry.

Whey-Treatment Costs

Capital costs and operating costs for treating whey both are dependent on the degree of concentration desired and on local factors, such as the availability of space and the cost of labor. However, generalized data relating costs to plant capacity have been published,[13, 14, 56, 60] and the data for the graphs in Fig. 25, 26 and 27 were compiled from these published sources.

Relationships of typical capital costs to plant capacity is shown in Fig. 25. The costs of individual sections of a plant are included, as well as a category labeled miscellaneous. This category includes the costs of items such as storage tanks, heat exchangers, piping, and clarifiers, and for the plants producing dry products, the cost of baggers, forklift trucks, and pallets. The cost of buildings and warehouses is not included.

The total capital costs of plants both to produce fluid concentrates and dry products are also shown. Data for plants producing dry products include the cost of evaporators, spray dryers, bagging and storing equipment, in addition to the costs of the fluid-concentrate plants.

Fig. 26 shows the relation of typical operating costs to plant capacity for the production of fluid concentrates. Although ultrafiltration can remove a pound of fluid more cheaply than reverse osmosis, the costs shown for the protein concentrate (on a dry basis) are considerably more than those for the lactose concentrate because there is much less protein in whey than lactose.

Fig. 27 shows the relation of operating costs for removal of fluid from whey by ultrafiltration, reverse osmosis, and evaporation. Ultrafiltration is slightly cheaper than evaporation.

Because of the low cost of removing fluid by ultrafiltration, it has been found that production of protein concentrates or of dry protein products is highly profitable. However, the market for lactose is depressed at present, and the costs of evaporation and crystallization add considerably to the costs indicated in Fig. 27. For that reason, the production of lactose is not as attractive at present as the production of protein products.

LOOK TO THE FUTURE

Electrodialysis, reverse osmosis, and ultrafiltration have already proved to be valuable industrial separation processes for a number of applications, and it is anticipated that more applications will be found.

For example, five years from now, ultrafiltration may be in daily use concentrating and desalting enzymes, concentrating egg albumin, and recovering valuable protein products from soybean whey and packinghouse blood. Or it might find application in dewatering biological sewage sludges or in treating wastes from metal-finishing operations.

In the near future, fruit-juice concentrates with improved taste might be produced by reverse osmosis more cheaply than by competing processes. On the other hand, reverse osmosis may make possible the recovery of valuable inorganic materials like metals and brighteners in the metals-plating industry, or it may find uses in removing and recovering materials from some of the dilute wastes in the pulp and paper industry.

Electrodialysis might be in use to preconcentrate solutions of electrolytes before spray or drum drying so that less water need be removed in drying. Perhaps it will find uses to separate and recover univalent and multivalent ions in certain dilute waste streams. On the other hand, electrodialysis might be used to de-ash organic products like glycerol or sorbitol.

When one considers possible improvements in membranes and equipment for membrane processes, the extent of man's imagination and ingenuity may be the only limitations for applying these processes.

Meet the Author

Robert E. Lacey is head of the chemical engineering section at Southern Research Institute, 2000 Ninth Ave. South, Birmingham, AL 35205, where he has performed and directed research and development on membrane processes for over 16 years. He has been a research chemical engineer at Monsanto Co. and has taught at Oklahoma State University—where he received his B.S. and M.S. in Ch.E. He has recently edited "Industrial Processing with Membranes," published by Wiley.

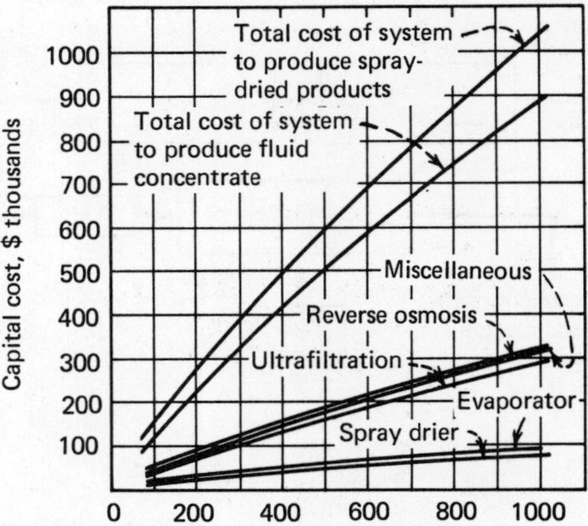

CAPITAL COST and plant capacity relationships—Fig. 25

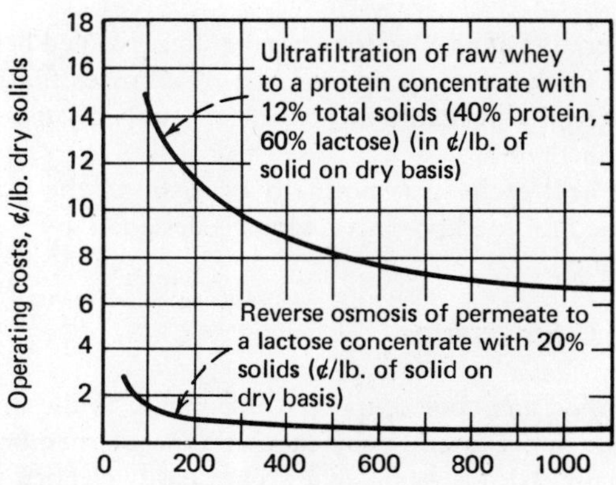

OPERATING cost and plant capacity relationships—Fig. 26

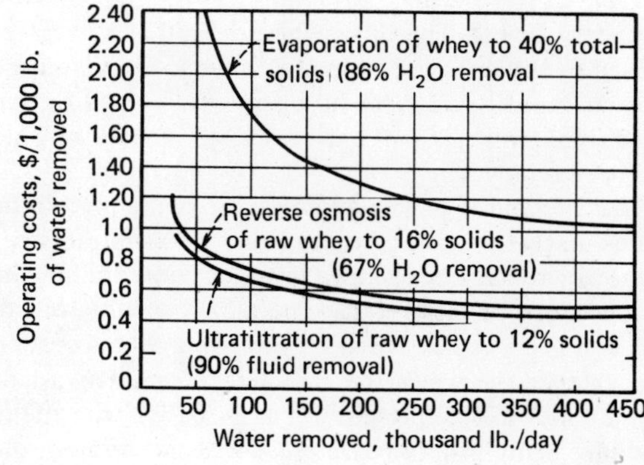

COST COMPARISONS of removing fluid by ultrafiltration, reverse osmosis and evaporation—Fig. 27

Books for Further Reading

Electrodialysis: Ref. 61, 62. Reverse osmosis: Ref. 63, 64. General (including untrafiltration, dialysis, and membrane permeation processes): Ref. 29, 62, 65, 66, 67. Reverse osmosis, ultrafiltration and microfiltration membranes: Ref. 68.

References

1. Royer, K. W., "Whey Processing Procedures," Proceedings of the Whey Utilization Conf., June 2 and 3, 1970, U. of Maryland, College Park, Md.
2. Ahlgren, R., "Electromembrane Processing of Cheese Whey," Chapter IV, in "Industrial Processing with Membranes," R. Lacey, and S. Loeb, editors, Wiley, New York, 1972.
3. Nishiwaki, T., "Concentration of Electrolytes with an Electromembrane Process," Chapter VI in "Industrial Processing with Membranes," R. Lacey and S. Loeb, editors, Wiley, New York, 1972.
4. Seko, M., Large-Scale Brackish Water Electrodialysis Plant at Webster, South Dakota, *Dechema Monographien*, **47**, p. 575, 1962.
5. Anonymous, Remove SO_2 at a Profit, *Electrical World*, Jan. 26, 1970.
6. Anonymous, Drug Output Spurts Tenfold via Electrodialysis Route, *Chem. Process.*, Dec. 1966.
7. Ioconelli, W. B., "Applications of Membrane Techniques," Bulletin from Ionics, Inc., Waterton, Mass., 1970.
8. Calmon, C., U.S. Pat. 3,394,068.
9. Lacey, R., "An Electromembrane Process for Regenerating Acid from Spent Pickle Liquor," Final Report on Contract 12010EQF, Environmental Protection Agency, Mar. 1971.
10. Ahlgren, R., "Electromembrane Processes for Recovery of Constituents from Pulping Liquors," Chapter V in "Industrial Processing with Membranes," R. Lacey and S. Loeb, editors, Wiley, New York, 1972.
11. McDonough, F. E., "Concentration and Fractionation of Whey by Reverse Osmosis," "Proceedings of the Whey Utilization Conference, June 2 and 3, 1970," U. of Maryland, College Park, Md.
12. McDonough F. E., Mattingly, W. A., *Food Technol.*, **24**, p. 88 (1970).
13. Morrison, M. C., "Economic Analysis of Acid Whey Protein Recovery," Brochure from Calgon-Havens Systems Div., Calgon Corp., Pittsburgh, Pa., 1970.
14. Goldsmith, R. L., others, "Membrane Processing of Cottage Cheese Whey for Pollution Abatement," Proceedings of National Symposium on Food Processing Wastes, Mar. 23-26, 1971, Denver, Col. Available from Environmental Protection Agency, Washington, D.C.
15. Porter M. C., Michaels, A. S., *Chem. Technol.*, **1**, p. 248 (1971).
16. Nielsen, I. K., others, "Concentration and Separation of Dairy Products with DDS Ultrafiltration-Reverse Osmosis System," Brochure from De Danske Sukkerfabrikker, Langebrogade-5, DK-1001, Copenhagen, Denmark.
17. Rozelle, L. T., "Ultrathin Membranes for Treating Metal Finishing Effluents by Reverse Osmosis," Environmental Protection Agency Report 12010DRH 11/71.
18. Hinden, E., others, "Water Reclamation by Reverse Osmosis," Federal Water Quality Administration, U. S. Dept. of the Interior, Washington, D.C., 1968.
19. Conn, W. M., "Raw Sewage Reverse Osmosis," presented at 69th Annual Meeting of A.I.Ch.E., Cincinnati, Ohio, May 17, 1971.
20. Merson, R. L., Morgan, A. I., *Food Technol.*, **22**, p. 631 (1968).
21. Elata, C., "Development of Concentrator of Orange Juice by Reverse Osmosis," Progress Report 352-1, Hydranautics-Israel, Ltd., Jan. 16, 1969.
22. Underwood, J. C., "Concentration of Maple Sap" U. S. Dept. of Agriculture Report ARS-74-51, 1969.
23. Willetts, C., *Food Technol.*, **21**, p. 24 (1967).
24. Underwood J. C., Willets, C. O., *Food Technol.*, **23**, p. 787 (1969).
25. Goldsmith, R. L., others, "Industrial Ultrafiltration," Brochure from Abcor, Inc., Cambridge, Mass.
26. Porter, M. C., *Ind. Water Eng.*, June/July 1971.
27. Porter, M. C., Michaels, A. S., "Applications of Membrane Ultrafiltration to Food Processing," presented at 3rd International Congress of Food Science and Technology, August 9-14, 1970, Washington, D.C., available from Institute of Food Technologists, Chicago, Ill.
28. Wang, D. I., others, "Enzyme Processing with Ultrafiltration Membranes," in "Membrane Science and Technology," J. E. Flynn, ed., Plenum Press, New York, 1970.
29. Tuwiner, S. B., "Diffusion and Membrane Technology," Reinhold, New York, 1962.
30. Wallace, R. M., *Ind. Eng. Chem. Process Design Develop.*, **6**, p. 423 (1967).
31. Nishiwaki T., Itoi, S., *Japan Chem. Quarterly*, **V-1**, p. 36 (1967).
32. Stern, S. A., "Gas Permeation Processes," Chapter 13 in "Industrial Processing with Membranes," R. E. Lacey and S. Loeb, editors, Wiley, 1972.
33. Choo, C. Y., "Membrane Permeation Processes" in "Advances in Petroleum Chemistry and Refining," K. A. Kobe, ed., Interscience, New York, 1962.
34. Cowan, D. A., Interaction of Technical and Economic Demands in the Design of Large Scale Electrodialysis Demineralizers, Advances in Chem. Ser., **27**, American Chemical Soc., Washington, D.C., 1960, p. 224.
35. Tribus M., Evans, R. Thermo-economic Considerations in the Preparation of Fresh Water from Sea Water, *Dechema Monograph*, **47**, Deutsche Gesellshaft Für Chemisches Apparatewesen, e.v., Frankfurt am Main, Germany, 1962.
36. Lacey, R. E., others, "Economics of Demineralization by Electrodialysis," Advances in Chem. Ser., **38**, American Chemical Soc., Washington, D.C., 1963.
37. Mattson, M. E., "Determining the Costs of an Electrodialysis Desalination Plant by Parametric Equations," Proceedings of the First International Symposium on Water Desalination, Washington, D.C., October 3-9, 1965, Vol. 3, U.S. Govt. Printing Office, Washington, D.C.
38. Rickles, R., "Membranes, Technology and Economics," Noyes Development Corp., Park Ridge, N.J., 1967.
39. Mintz, M. S., Optimize Electrodialysis Process Design, *Ind. Eng. Chem.*, **55**, p. 18 (1963).
40. Huffman E., Lacey R., "Engineering and Economic Considerations in Electromembrane Processing," Chapter 3 in "Industrial Processing with Membranes," R. Lacey and S. Loeb, editors, Wiley, New York, 1972.
41. Shaffer L., Mintz, M., "Electrodialysis," Chapter 6 in "Principles of Desalination," K. S. Spiegler, ed., Academic Press, New York, 1966.
42. Sourirajan, S., "Engineering Developments in Reverse Osmosis," Chapter 8 in "Reverse Osmosis" by S. Sourirajan, Academic Press, New York, 1970.
43. Menzel, H. F., U.S. Office of Saline Water Research and Devel. Rept. 236 1967.
44. Bray, D. T., "Engineering of Reverse Osmosis Plants," Chapter 6 in "Desalination by Reverse Osmosis," U. Merten, ed., M.I.T. Press, Cambridge, Mass., 1969.
45. Kaiser Engineering Co., U.S. Office of Saline Water Research and Devel. Rept. 509, 1970.
46. Lacey, R. E., "The Costs of Reverse Osmosis," Chapter IX in "Industrial Processing with Membranes," R. Lacey and S. Loeb, editors, Wiley, New York, 1972.
47. Shields, C. P., "Reverse Osmosis for Municipal Water Supply," Technical Bulletin from E. I. du Pont de Nemours & Co., Wilmington, Del., 1971.
48. Westbrook, G. T., "The Emergence of Membrane Technology," Technical Bulletin from the Dow Chemical Co., Walnut Creek, Calif.
49. Anonymous, Technical Bulletin No. 71-03, from De Danske Sukkerfabrikker, Langebrogade 5, DK-1001, Copenhagen, Denmark.
50. Okey, R. B., "The Treatment of Industrial Wastes by Pressure-Driven Membrane Processes," Chapter XII in "Industrial Processing with Membranes," R. Lacey and S. Loeb, editors, Wiley, New York, 1972.
51. Anonymous, "Economic Analysis of Industrial Waste Treatment Using Calgon-Havens Reverse Osmosis," Engineering Report from Calgon-Havens Systems, Pittsburgh, Pa., 1972.
52. Spatz, D., "Reverse Osmosis Reclamation Systems for the Plater," Technical Bulletin from Osmonics, Inc., Minneapolis, Minn.
53. Spatz, D., "Chromium Waste Treatment," Technical Bulletin from Osmonics, Inc., Minneapolis, Minn.
54. Anonymous, "Ultrafiltration Systems for Dewatering and Purification of Electrocoating Paints," Publication No. 1010A from Amicon Corp., Lexington, Mass.
55. Porter, M. C., "Ultrafiltration of Colloidal Suspensions," Technical Bulletin from Amicon Corp., Lexington, Mass., 1971.
56. Goldsmith, R. L., others, "Industrial Ultrafiltration," Technical Bulletin from Abcor, Inc., Cambridge, Mass., 1971.
57. Spatz, D., "Industrial Waste Processing with Reverse Osmosis," Technical Bulletin from Osmonics, Inc., Minnapolis, Minn.
58. Michaels, A. S., "Ultrafiltration" in "Progress in Separation and Purification," Vol. 1, E.S. Perry, ed., Interscience, London, 1968.
59. Lonsdale, H. K., "Theory and Practice of Reverse Osmosis and Ultrafiltration," Chapter VIII in "Industrial Processing with Membranes," R. Lacey and S. Loeb, editors, Wiley, New York, 1972.
60. Anonymous, "Osmotik Processes for the Dairy Industry," Calgon-Havens Systems Technical Bulletin, 1971.
61. Wilson, J., ed., "Demineralization by Electrodialysis," Butterworths, London, 1960.
62. Lacey, R. E., Loeb, S., editors, "Industrial Processing With Membranes," Wiley, New York, 1972.
63. Merten, U., ed., "Desalination by Reverse Osmosis," M.I.T. Press, Cambridge, Mass, 1966.
64. Sourirajan, S., "Reverse Osmosis," Academic Press, New York, 1970.
65. Lacey, R. E., ed., "Membrane Processes for Industry," Southern Research Institute, Birmingham, Ala., 1966.
66. Flinn, J., ed., "Membrane Science and Technology," Plenum Press, New York, 1970.
67. Bier, M., ed., "Membrane Processes in Industry and Biomedicine," Plenum Press, New York, 1971.
68. Kesting, R. E., "Synthetic Polymeric Membranes," McGraw-Hill, New York, 1971.

Ultrafiltration:

An emerging unit-operation

Ultrafiltration separates materials on the basis of molecular size, without change of state. The technique has potential in many processes—some commercial ones are described, together with what you need to know to consider it for your processes.

P. R. Klinkowski, *Dorr-Oliver, Inc.*

☐ Ultrafiltration has gained acceptance as a valuable tool for a wide variety of industrial processes. These include the stabilization of electrocoating baths, recovery of oil from waste effluents, concentration of heat-sensitive proteins for food additives, and water purification. Ultrafiltration can make difficult macromolecular separations, often replacing conventional change-of-phase methods, and performing where no practical alternative technology exists.

Ultrafiltration techniques are based on the ability of pressure-driven filtration membranes to separate multicomponent solutes, or solutes from solvents, according to molecular size, shape, and chemical bonding. Substances below a preselected molecular size are driven through the membranes by hydraulic pressure while larger molecules are held back. This is a simple physical process, requiring no chemical reaction, phase change or dilution steps.

In ultrafiltration, selective porous membranes accomplish separation in cases where the solute particles to be removed are at least one to two orders of magnitude larger than the solvent molecules. This includes the separation of large dissolved molecules (categorized as macromolecules), as well as dispersions, colloids and suspended particles. Ultrafiltration separations generally range from 10 to 100 angstroms (1 to 10 nm). The current state of the art allows tailoring of membranes to provide greater than 90% retention of a particular molecular-weight solute while permitting complete passage of solutes whose molecular weights are only one half that of the retained species.

Another membrane process, reverse osmosis, encom-

Originally published May 8, 1978

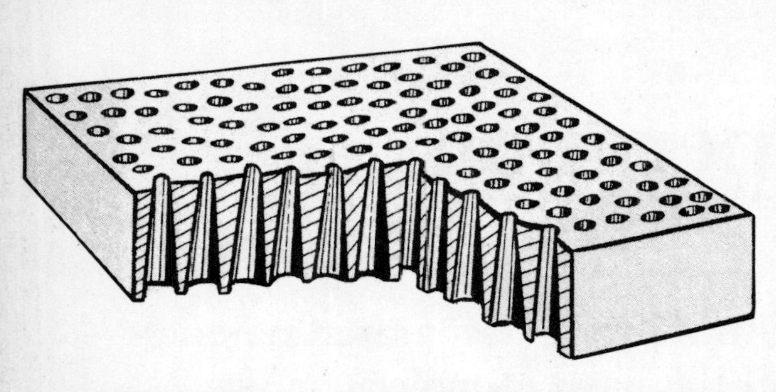

Schematic of a graded-pore microporous membrane **Fig. 1**

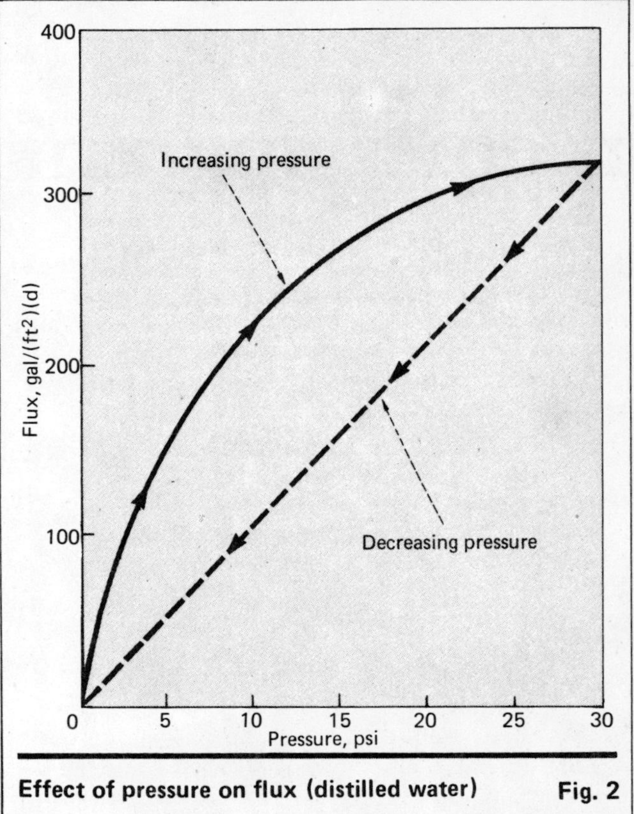

Effect of pressure on flux (distilled water) **Fig. 2**

passes the separation of inorganic salts and simple organic compounds—molecules whose size is of the same order of magnitude as that of the solvent. While reverse-osmosis membranes are continuous gels, ultrafiltration membrane materials are relatively porous media with little or no ability to reject salt. In ultrafiltration, separations are made by physical screening, and fluid transport is by pore flow. Osmotic pressures normally are insignificant and typical operating pressures vary from 10 to 100 psi.

How ultrafiltration works

The emergence of ultrafiltration as an industrial process can be traced back to the development of an asymmetric membrane by Loeb and Sourirajan at the University of California (Los Angeles) about 17 years ago. An electron-microscope examination of the asymmetric membrane cross-section shows an ultrathin surface layer of extremely fine pore structure supported on a thicker, more-porous layer.

Separations take place at the surface boundary. Material passing into the fine pores is readily transported through the membrane via the open-celled, spongelike structure of the support layer. Pores in the support layer are several orders of magnitude larger than those of the surface layer. One ideal membrane structure that has been developed is called graded-pore material (Fig. 1). It provides a continually increasing pore size through the structure, which prevents internal fouling or plugging.

Although early asymmetric membranes were prepared from cellulosic materials, modern ultrafiltration membranes are made from a variety of noncellulosic synthetic polymers (e.g., nylon, vinyl chloride - acrylonitrile copolymers, polyvinylidene fluoride, polysulfone, etc.) that are more resistant to thermal and chemical degradation.

The amount of material that will pass through a given area of ultrafiltration membrane in a given period is commonly called the membrane flux, which may be expressed by the equation:

$$J = Kp/t$$

where: J is membrane flux, e.g., gal/(ft^2)(d); K is permeability of the membrane; p is pressure difference across the membrane; and t is thickness of the membrane.

The above equation applies at low pressures, but due to compaction, a nonlinear flux response is obtained at higher pressures. The effect is irreversible (Fig. 2). Once the membrane has been conditioned at a high pressure, a linear flux-to-pressure response is reestablished, but at lower flux. For this reason, a membrane should never be exposed to a pressure exceeding the planned operating pressure.

While membrane composition, structure and thickness, as well as pressure across the membrane, are important to throughput or flux, other process factors, such as concentration polarization, shear rate, and temperature, also have a pronounced effect on ultrafiltration performance.

As the solvent in a mixture flows through the membrane, retained species are locally concentrated at the membrane surface, thereby adding resistance to permeate flow. When processing a solution, this localized concentration of solute normally results in precipitation of a solute gel onto the membrane. When processing a suspension, the solids collect as a porous layer atop the membrane surface. In both cases, this concentration polarization produces a layer of material that has a significantly lower permeability than the membrane itself. The permeate rate is then controlled by the rate of transport through the polarization layer rather than by membrane properties.

To maintain a high membrane flux, the concentration-polarization film thickness on the membrane surface must be limited. A high velocity of fluid flow across the membrane surface will tend to shear off the polarization layer, establishing an equilibrium film-thickness. As this scouring velocity increases, the membrane flux also increases (Fig. 3). At very high shear rates, the membrane approaches its theoretical throughput but, in practice, a balance must be struck between membrane throughput and pumping costs.

Still another factor affecting membrane flux is system temperature. In general, membrane flux will increase in direct proportion to temperature.

In summary, ultrafiltration throughput depends on physical properties of the membrane, such as permeability and thickness, as well as process and system variables, such as feed composition, feed concentration, system pressure, system velocity and system temperature. The design of practical ultrafiltration systems must consider the interacting effects of all these variables, while providing large, well-supported membrane areas in modular configuration that is capable of future expansion.

Making membrane modules

A number of schemes have been devised for supporting large areas of thin ultrafiltration membranes. These range from banks of membrane-coated porous tubes that are contained in pressure vessels, to bundles of small capillary tubing formed of membrane materials, to adhesive-sealed envelopes of material spiral-wound on a central support. Each scheme has its proponents. The system described here, a commercially available ultrafiltration cartridge, was specifically designed to provide a readily changeable, expandable module for process applications.

The design of this cartridge provides 18 to 20 ft^2 of active membrane surface distributed on 30 porous supports inserted in a plastic collection header (Fig. 4). The membrane is bonded to, and completely covers, all surfaces of the porous supports. Fluid flow parallels each of the membrane leaves, so that surface velocity and applied pressure are equal on all surfaces. Material passing through the membrane (permeate) enters the porous support and is conducted inside it to the collection header and then to a permeate channel. The cartridge acts as a self-cleaning filter, with feed continually scouring the membrane surface while permeate is being withdrawn through the porous support-medium. Individual cartridges are then assembled into an industrial module (Fig. 5).

The module consists of a fiberglass-reinforced housing that will accept three 20-ft^2-area replaceable cartridges. A clean-in-place (CIP) module for use in food and drug applications comprises a stainless steel housing and four special 18-ft^2-area cartridges designed for ease of cleaning and sanitation. Housings can be connected together in series or in parallel arrays, to build systems of various configurations and surface areas (see photograph on p. 324).

The number of modules that can be placed in series depends on the flow desired across the membrane (which generates a pressure drop) and on the maximum pressure allowed in the system. High recirculation-rates are required to promote turbulence and scouring of the membrane surfaces. Each application will have an optimum recirculation rate, depending on a number of factors, including: pumping costs, response of flux to velocity, fouling effects, product characteristics, fluid velocity, fouling effects, other product characteristics, and maximum allowable stresses in the system.

Process considerations

Since ultrafiltration systems operate at nominally high-velocity flowrates, single-pass, nonrecirculating configurations generally are impractical. Two basic ultrafiltration flowsheets are the open and closed loop.

The open-loop system (Fig. 6a) circulates feed over the membrane surface and then returns it to the feed reservoir. When practical, this system assures operation of the membrane loop at the same solids concentration as the feed reservoir and, in addition, eliminates the need for a separate system pressurizing-pump.

In the flowsheet consisting of a closed loop with a bleed (Fig. 6b), feed is pumped from the storage reservoir to a closed ultrafiltration recirculation-loop. A

Effect of flow velocity on flux through membrane **Fig. 3**

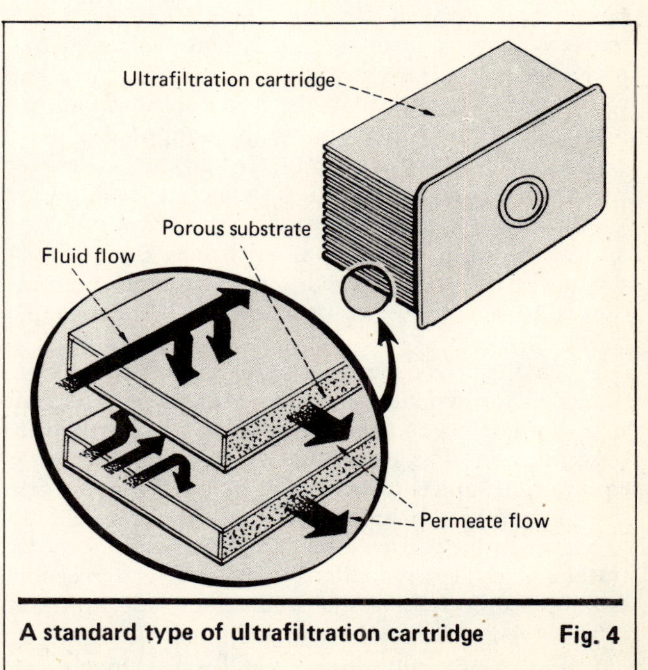

A standard type of ultrafiltration cartridge **Fig. 4**

Photo of cartridges installed in module Fig. 5

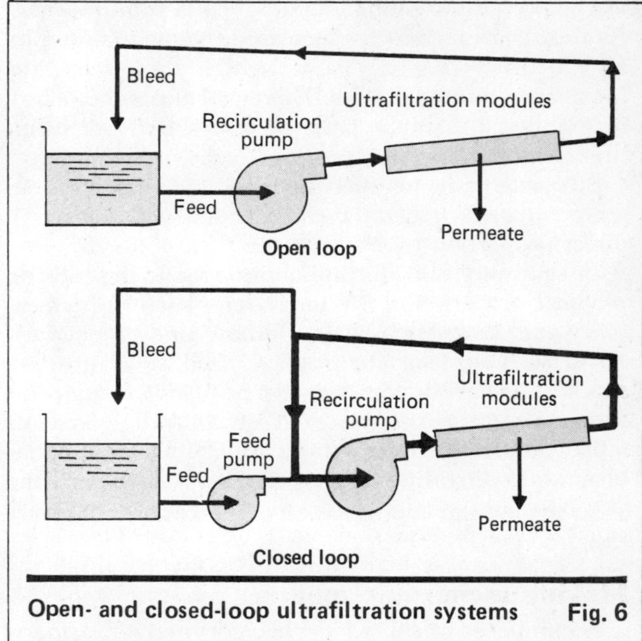

Open- and closed-loop ultrafiltration systems Fig. 6

separate feed-pump pressurizes the membrane loop and provides feed makeup to balance the permeate and bleed losses. Feed is pumped to the suction side of a larger recirculation pump that provides the needed turbulent flow across the membrane surfaces.

To prevent a buildup of concentrate in the recirculation loop, a bleed stream continuously removes concentrate from the loop. In a batch process, the loop concentration is continually increasing and the bleed is returned to the feed tank. In continuous processing, the loop operates at the desired bleed concentration. The bleed is transported to the next operation elsewhere in the plant.

Since the system pressure and recirculation rate are controlled independently, the closed loop with bleed offers increased flexibility. In addition, it requires less power than the open loop, since only the makeup feed must be pressurized to the full system pressure. However, the solids concentration of the material in the closed loop is higher than the feed concentration, resulting in decreased membrane throughput.

If a steady stream of concentrated product is desired, it is generally more economical to operate a multistage system in cascade configuration (Fig. 7). If a sufficient number of stages are used—usually more than three—the membrane area required to perform the concentration is not much greater than that needed for simple batch operation.

When a tenfold concentration is desired from a four-stage cascade system, for example, the bleedrate from stage four is regulated to one tenth of the feedrate concentration. The system is self-adjusting, and lower equilibrium concentrations are achieved in the first three stages, thereby improving flux rates as compared to operating the entire system at the final bleed concentration. If no flux reductions occur due to irreversible fouling or other effects, the equilibrium concentrations in the stages will remain constant.

Commercially, ultrafiltration still is a relatively new unit operation. Laboratory and pilot testing are recommended before planning and sizing a full-scale system. Since separation occurs at the molecular level, ultrafiltration rates are very sensitive to the nature of the fluid, contaminants, and many other subtle variables. With careful preliminary membrane screening and process development, scaleup from laboratory to pilot-plant to full-scale operation will proceed smoothly.

Some ultrafiltration applications

One of the early uses for the ultrafiltration process was in the treatment of sewage and wastewater. Some sewage treatment installations utilized a unique activated-sludge-plus-ultrafiltration process that produced an effluent with no suspended solids, no coliform bacteria, and less than 5 ppm biochemical oxygen demand (BOD). Several of these systems are in operation but, for most applications, the process is more effective than is required, and lower-priced less-efficient systems will suffice.

Oil-emulsion waste treatment

Many industrial wastewaters are amenable to treatment with the ultrafiltration process, particularly oil-water emulsions. At the present time, only about 50 million gal of the 2.5 billion gal of industrial oils sold annually in the U.S. are reclaimed. Of the oil that is not consumed during use, about 1.2 billion gal are dumped or burned. Obviously, this situation needs examination in the light of today's energy projections.

Since much industrial-oil use involves the generation of oil-water emulsions—particularly in large manufacturing operations—reclamation of both the oil and water is of environmental as well as economic importance. Ultrafiltration can be used to recover and concentrate emulsified oil from water and is particularly applicable in situations where the effluent cannot be treated by conventional oil/water separating devices.

The process will produce clear water for plant reuse or discharge; the concentrated oil can be recycled or utilized for its fuel value.

The most common method of processing is to collect all washings and dumps containing emulsified oil into a large storage tank. Prior to ultrafiltration, the material is skimmed to remove free-floating oil, filtered to remove grit and other foreign matter, and treated with an emulsifier, if necessary, to prevent destabilization.

Ultrafiltration is performed in either a straight-batch mode or continuously-fed-batch mode. The permeate generally contains less than 10 ppm of oil. The maximum attainable oil concentration ranges between 30 and 60%. During the process, the volume of feed material is reduced 20- to 30-fold in the recirculation loop, depending on the initial feed concentration. When this point is reached, the batch is dumped, the ultrafiltration unit is cleaned with a dilute nonionic detergent, and the next batch started. An average flux of about 20 gal/(ft^2)(d) can be maintained over the oil concentration range of 0.1 to 50%.

An alternative scheme of operation is to use an ultrafiltration system to continuously side-stream-filter an oil-water emulsion from a phosphating or degreasing tank (Fig. 8).

In this manner, contamination in the tank is maintained below a desired level. This is particularly useful when a small ultrafilter is attached to each cleaning station, since the cleaning tank needs to be dumped much less frequently. Clear permeate from the ultrafilter is recycled back to the tank, and the effluent is limited to a small, concentrated bleed-stream.

Electropaint recovery and tank control

Electrophoretic coating—electropainting—is the process by which an organic coating or paint film is applied to a metallic surface by passage of an electric current through an aqueous dispersion. With the ability to deposit uniform thin coatings on difficult-to-cover, recessed, or creased surfaces, electropainting can provide superior corrosion resistance and a saving of up to 50% in paint consumption, compared with other coating procedures.

Careful control of the electropaint solution is required for economical, efficient operation. Being an electrophoretic process, electropainting is adversely affected by ionic impurities that raise the bath conductivity. Even with careful control, contamination of the paint tank due to accumulation of soluble salts is a problem, arising from a number of sources, including the normal electrolytic decomposition of the paint resin, and the carryover of materials from pretreatment stages. In particular, solubilizer material accumulates in the paint tank while resin and pigment solids are deposited on the metallic part. The paint tank becomes progressively more alkaline, upsetting the process and producing inferior coatings. Bath solids are difficult to maintain, since the paint is deposited at 90+% solids and feed is 30 to 60% solids.

Ultrafiltration is used to process the paint by retaining the resin and pigment solids, while allowing inorganic salts, water, excess solubilizer, and solvent to permeate through the membrane. Organic solvent, deionized water and paint solids then are added to the paint tank in order to maintain optimum concentrations in the bath.

In an associated problem, washwater used to remove excess paint, or dragout, from coated parts causes pollution problems in sewer lines. Dragout losses range from a low of 10% for simple geometric forms to as high as 50% for large complex assemblies such as automotive bodies. Passing the coated part through a series of countercurrent water-spray stations typically produces an effluent flow of 10 to 30 gal/min at solids loadings ranging from 0.5 to 2.0%.

Ultrafiltration can sidestep conventional waste treatment schemes for recovering the lost paint solids and returning them to the paint bath. Often, the saving in recovered paint pays for the capital cost of the ultrafiltration installation in less than 6 months.

In this process, the paint bath is directly ultrafiltered, thereby providing permeate for the rinse (Fig. 9). Mem-

Four-stage cascade-system flowsheet **Fig. 7**

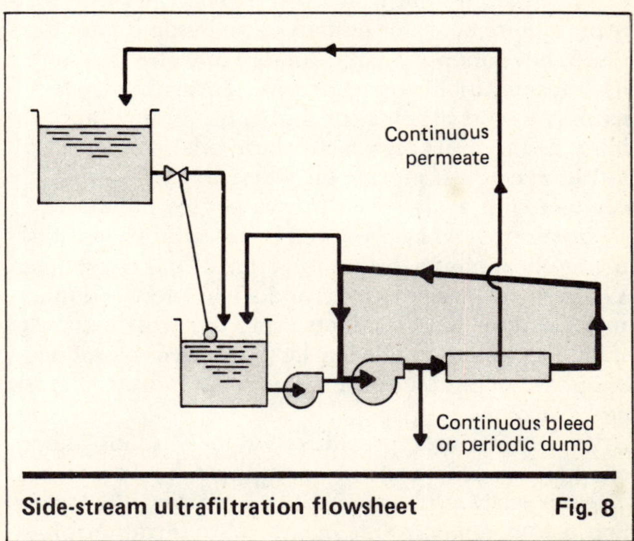

Side-stream ultrafiltration flowsheet **Fig. 8**

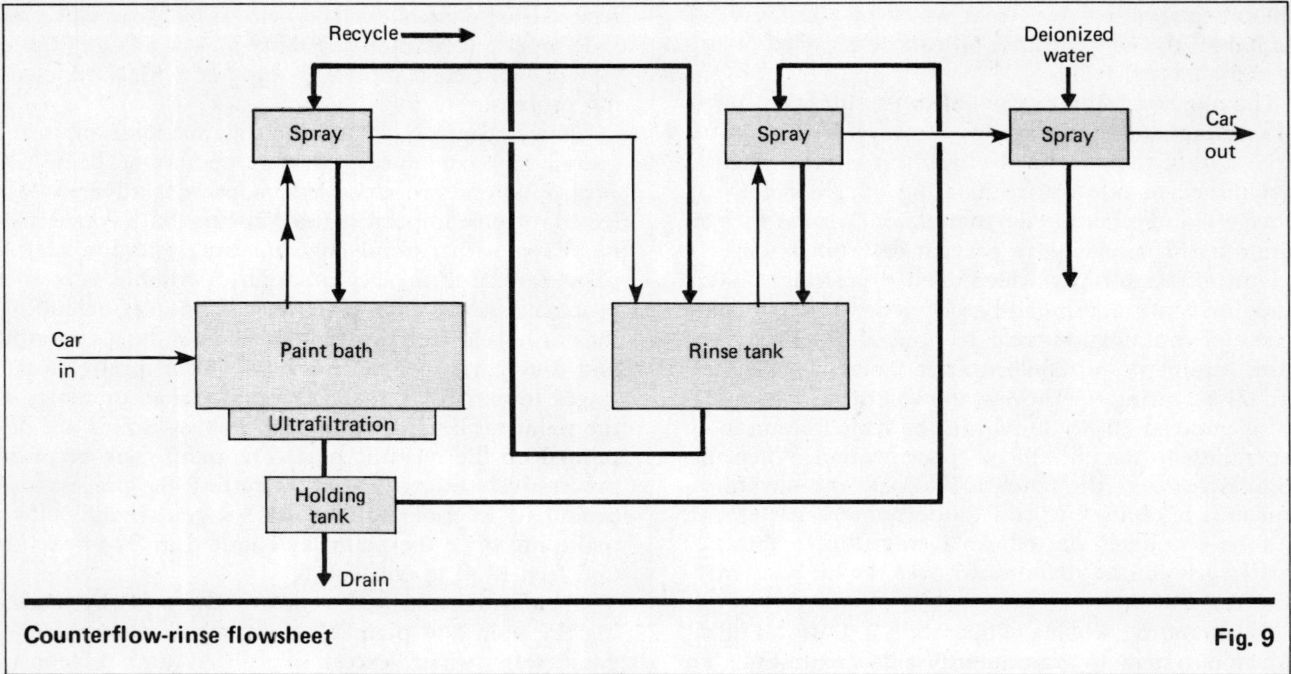

Counterflow-rinse flowsheet **Fig. 9**

brane-retained paint solids are returned to the paint bath. The permeate is transferred, through a holding tank, to a spray chamber where the workpiece is rinsed.

The used rinse fluid is collected in a tank and is recirculated to provide a high-volume rinse on workpieces entering the rinse station. A low-volume bypass is bled back to the paint tank to prerinse the workpiece as it emerges from the paint bath. Solubilizers in the permeate stabilize the paint solids in the rinse, allowing their reuse without affecting quality.

This counterflow, closed-cycle rinse can increase paint utilization up to 50%, while substantially reducing plant-effluent BOD loadings (Fig. 10).

Ultrafiltration in the dairy industry

In the production of cheese—one of the few dairy products in increasing demand—about 9 times more whey is produced than curd. Regarded as a difficult-to-handle wastewater, the greenish-yellow whey has an extremely high BOD, which can exceed even the treatment requirements for human waste in some communities. Whey contains a large amount of water that makes its transportation very expensive. Spray-dried whey is used as a food supplement, but it has limited applicability owing to its high lactose and ash content. Ultrafiltration can concentrate the whey proteins to obtain a product with a high nutritive value (see table).

Protein-rich whey produced by ultrafiltration is finding applications in the manufacture of many products. Whey yogurt, whey tortillas and a whey-fortified macaroni are three food products using the protein concentrate. Fermentation of whey lactose is used to produce a "pop" wine, and liqueur, as well as industrial-grade ethyl alcohol.

Whey contains approximately 6 to 7% solids—about one half the amount found in milk prior to processing. The raw whey solids consist of 13% protein, 72% lactose, and 8% inorganic solids or ash. Ultrafiltration physically separates the protein from the lactose and ash, enhancing the value of both the protein and lactose product streams. The solids in a typical ultrafiltered whey concentrate will have up to 50% protein.

Three modes of operation are currently employed. In the batch concentration mode, whey is pumped in a continuous loop from a feed storage tank. The continuously-fed batch mode operates similarly, except that raw whey is continuously added to the feed storage tank. In the third mode, cascade operation, whey is fed through a series of ultrafiltration units until the required degree of concentration is obtained in the final stage. Bleed from the first stage becomes the feed for the second stage, and so on. The final-stage bleed is sent elsewhere in the plant for spray drying and packaging.

During operation, the ultrafiltrate flux is affected by many factors, including membrane retention. A membrane designed for whey ultrafiltration will reject material having a molecular weight greater than 24,000, retaining 95 to 99% of the protein. The retention of lactose and ash with this membrane is essentially zero. Other factors affecting the ultrafiltrate flux include temperature, pressure, flow velocity, and pretreatment conditions.

Ultrafiltration also can be used to concentrate proteins in skim milk to about 5.2% liquid weight for the production of a low-fat yogurt that offers excellent consistency and taste, while reducing or eliminating the need for gelatin to set up the product. An enriched yogurt with 6% protein and 30% fat also has been prepared, as has a protein-fortified skim-milk product with 5% protein.

Other dairy applications for ultrafiltration include the concentration of milk prior to cheese production, which results in increased production capacity with existing equipment. Work being done in France shows the advantage of using ultrafiltration concentration to permit continuous cheese production.

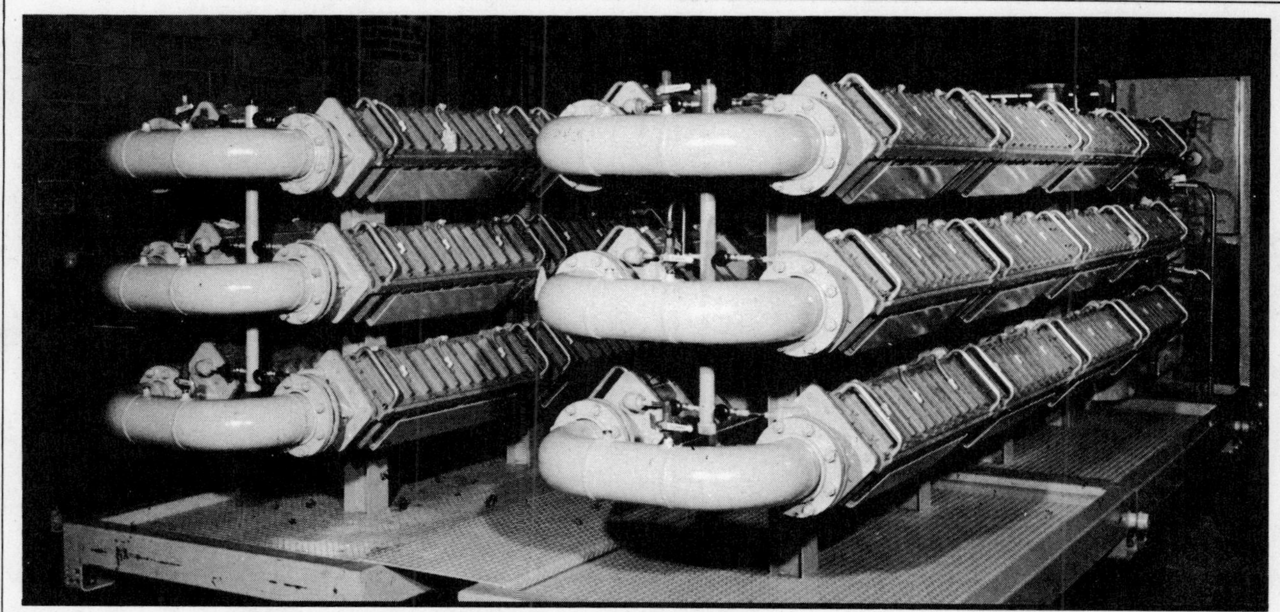

Ultrafiltration system used for electropainting Fig. 10

Other applications

Ultrafiltration has been used or proposed for many other industrial product streams, including enzyme and pharmaceutical preparations, polyvinyl chloride and other polymer lattice production, pulp-mill wastes, indigo-dye rinses, and synthetic-polymer desize wastes.

Most ultrafiltration installations now in use are of modest proportions. Advances in membrane technology and process design will foster the use of larger-scale equipment. The economies involved in scaleup will then open new application areas.

Economic considerations

Discussions dealing with ultrafiltration economics are often misleading. Membrane processes typically involve fractionation in which one feed stream is divided into two streams of altered composition. Each stream must be processed further to obtain the desired final products. The potential user needs to know the total processing costs involved, whereas the equipment manufacturer often can provide only economic data concerning a specific unit operation.

Prior to selecting ultrafiltration as a unit operation, some insight into the potential economics of this technology is needed. A convenient criterion for economic feasibility is the process flux required for a one-year payoff based on the value of the recovered product. Assume that the cost of a membrane system is $70/ft^2 = \text{income}/(ft^2)(yr)$. This reduces to:

$$\text{Flux required for one-year payoff} = \frac{2.3}{(\$/lb)(\%\text{total solids})}$$

where the increase in product value is expressed in terms of $/lb of dry solids, and the concentration of the retained species in % total solids.

Making some rudimentary assumptions concerning membrane-system economics and life, and method of depreciation, permits development of simple expressions for the cost of producing 1,000 gal of permeate.

Assumptions:

Installed cost of membrane system	$\$70/ft^2$
Membrane replacement cost	$\$10/ft^2$
Membrane replacement period	1 yr
Power cost	$0.02/kWh
Linear depreciation period	5 yr

Installation cost (amortization): $/1,000 gal = 38.4/flux

Membrane replacement: $/1,000 gal = 27.4/flux

Power: $/1,000 gal = 12.2/flux

Total operating cost: $/1,000 gal = 78/flux

The above relationships provide a quick method for determining the potential economics of a particular ultrafiltration application.

The use of membrane equipment for certain applications is a commercial reality. Fig. 11 and 12 develop capital and operating costs (less depreciation) for whey protein recovery. Similarly, Fig. 13 and 14 show cost implications for the concentration of skim milk using ultrafiltration. Each application is attractive from a payback standpoint because of the high value of the recovered product.

Electropainting economics

Similarly, electrocoating applications show attractive economics. For an electropaint system producing 24 gal/min of permeate, the capital cost will be about $200,000. For 1,000 gal of permeate, the operating cost component for membrane replacement will be $2.11 and the power cost $1.50, for a total operating cost of $3.61. Depreciation will add another $3 to the cost, for a total of $6.61 per 1,000 gal of permeate.

Using the expression previously developed for total

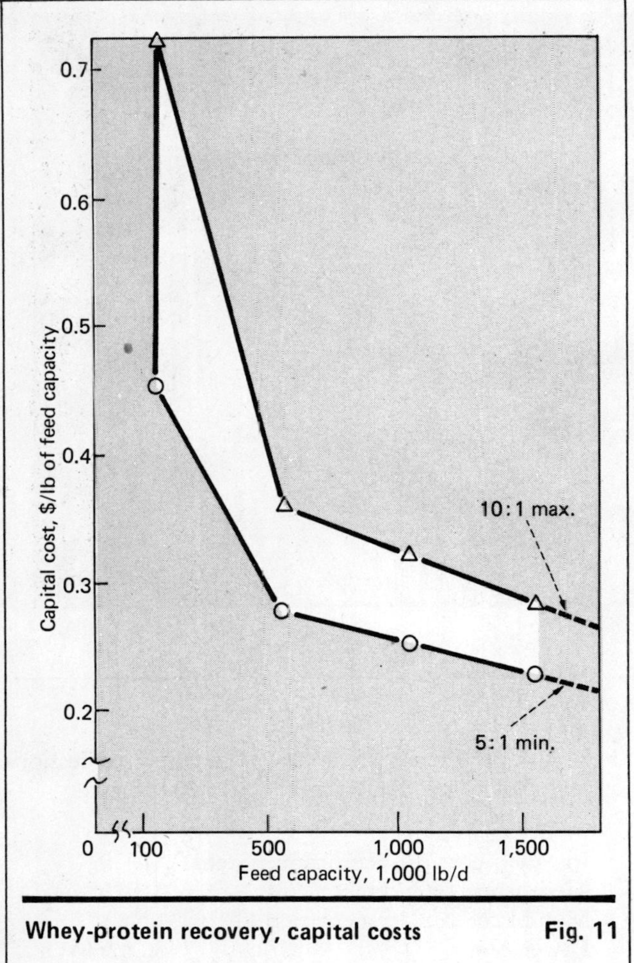

Whey-protein recovery, capital costs Fig. 11

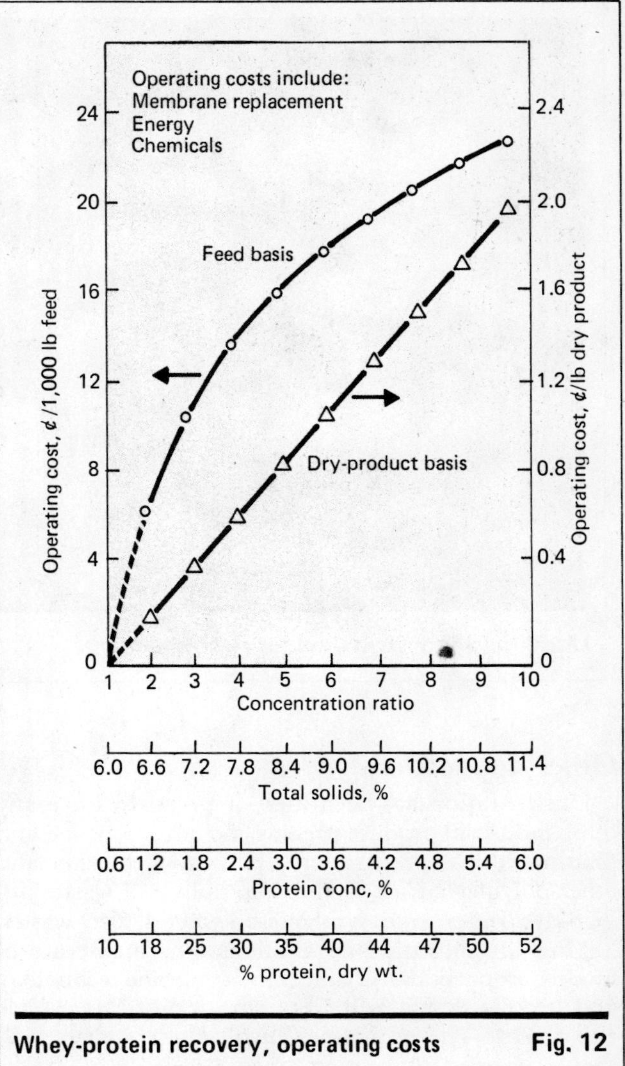

Whey-protein recovery, operating costs Fig. 12

operating costs, and a flux of 12 gal/(ft^2)(d), puts the total operating cost at:

$$\frac{\$}{1{,}000\ \text{gal}} = \frac{78}{12} = \$6.50$$

which substantially agrees with the initial cost estimate.

An exciting potential exists for the use of membrane equipment in sewage treatment and effluent polishing. The economics of the applications are quite different. A membrane sewage-treatment process where raw waste is converted to a high-quality effluent is expensive when compared with conventional package treatment-plants.

Approximate content of major nutrients in powdered-milk derivatives

Nutrient	Nonfat dry milk, %	Whole dried whey, %	Ultrafiltration dried-whey, % (typical)
Protein	36.0	13.0	50.0
Lactose	52.0	72.0	42.0
Ash	8.0	8.0	3.0
Fat	0.7	1.0	1.0
Water	4.0	4.5	4.0
PER*	3.0	2.5	3.2

*Protein efficiency ratio = wt gain/lb protein fed

However, if needed additional process steps are added to a package plant to produce an equivalent effluent quality, the cost difference often is reduced.

Installed membrane systems have demonstrated their ability to produce consistently high-quality effluents. Installed costs for small sewage plants (less than 15,000 gal/d) have been in the range of $3 per gallon of capacity. Operating costs are about $3.50 per 1,000 gal of treated water. Recent tests have shown that operation at fluxes in the range of 16 to 20 gal/(ft^2)(d) is possible on high-solids mixed liquors. This reduces effluent costs to about $1.10 per 1,000 gal.

Although ultrafiltration cannot produce dry products directly, it is a very effective way of supplementing thermal dewatering. That is, ultrafiltration can be used for preconcentration prior to evaporation.

For example, in removing water by evaporation, the food industry currently uses about 318 Btu/lb water evaporated. By applying optimum current evaporator art and add-on effects such as thermal recompression, the energy consumption can theoretically be reduced to 140 Btu/lb water removed.

Assuming a modest flux of 20 gal/(ft^2)(d), ultrafiltration—which does not rely on change of state to de-

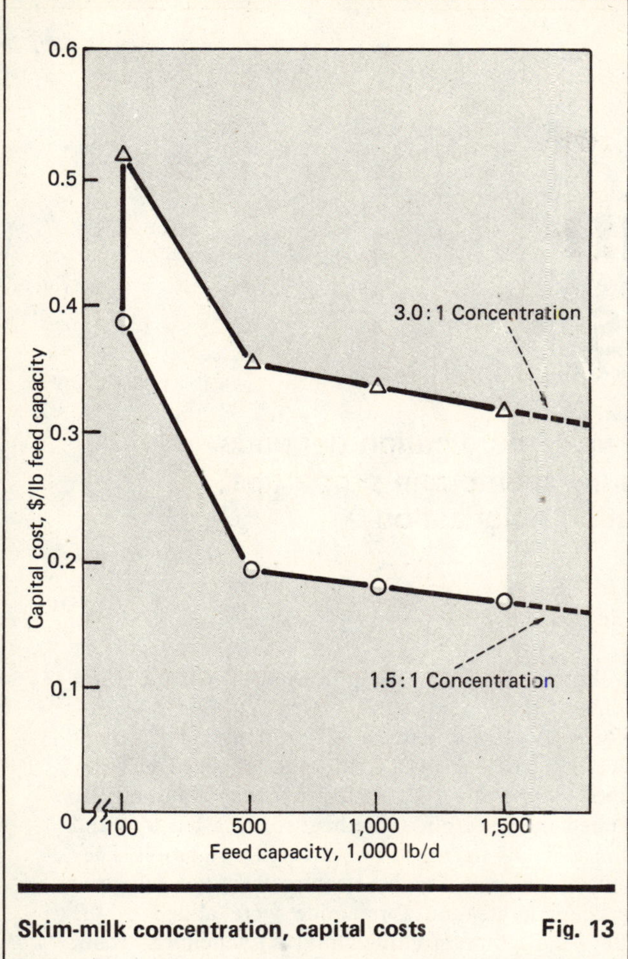

Skim-milk concentration, capital costs **Fig. 13**

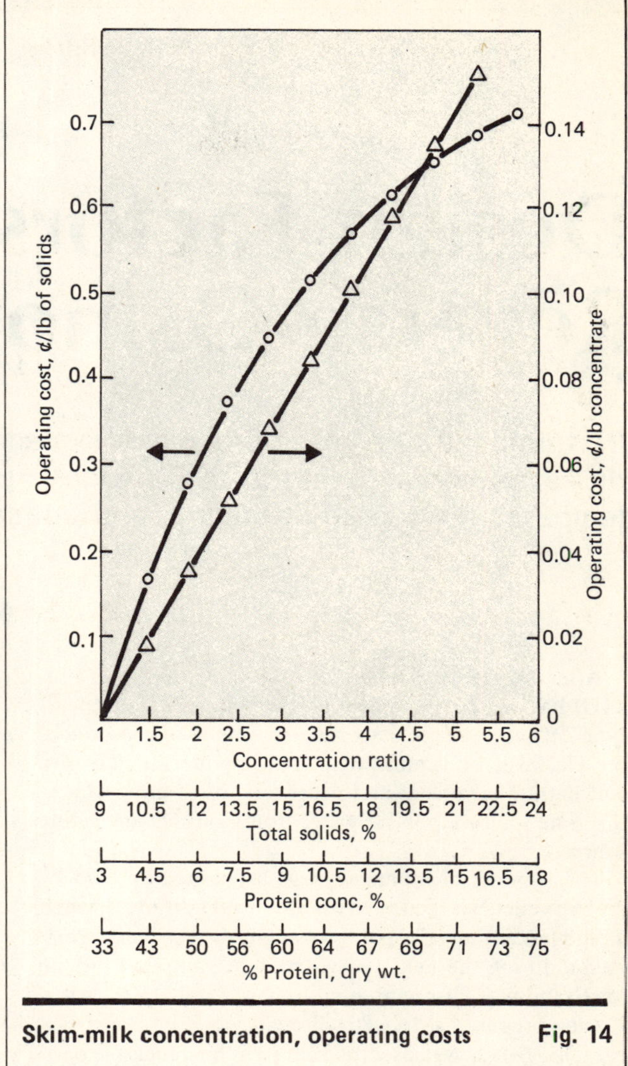

Skim-milk concentration, operating costs **Fig. 14**

water—uses the equivalent of about 13 Btu/lb of water removed. This is an order-of-magnitude less energy.

Since ultrafiltration system design can be tailored easily to a wide assortment of process requirements, it is possible to greatly influence not only the economics of the ultrafiltration stage but also of the entire processing plant. This is where the true economic potential for membrane processing can be seen.

Areas for improvement

In the area of membrane development, present work is designed to produce membrane materials with higher throughputs and increased selectivity. Combined with greater resistance to chemical and thermal extremes, new ultrafiltration membranes will also extend the useful life of ultrafiltration cartridges.

As presently fabricated, membrane products generally must be stored in a damp state to preserve their properties. Sometimes, rewetting is difficult, and full restoration of design flux and retention is not always possible. Because of the potentially high rate of bacterial and fungus growth during wet, damp or quasidamp storage, microbiocidal material must be included in the storage solution. Improved membrane products capable of dry storage will eliminate the dangers associated with the loss of desired properties now encountered under storage conditions.

As with all new unit operations, ultrafiltration will need to prove its effectiveness to the chemical engineer before it is fully accepted as an alternative separation process. Carefully controlled laboratory evaluations and small-scale pilot-plant studies are the rule before scaling up for full production facilities. As its inherent advantages are realized, and its applicability to processes presently impossible to perform with conventional change-of-phase separation techniques are proved, ultrafiltration will take its place among the common unit operations.

The author

Peter R. Klinkowski, the technical manager for ultrafiltration/membranes at Dorr-Oliver, Inc., Stamford, Ct. 06904, has spent nine years in the membranes division with design, applications, and systems responsibility. A graduate of the University of Michigan, with a B.S. in chemical engineering, he holds several U.S. patents pertaining to membranes, and has authored technical papers and presentations for technical meetings.

Design Factors in Reverse Osmosis

Performance of a reverse-osmosis system for water purification depends on several engineering factors—osmotic pressure, membrane properties, temperature, degree of fouling, and concentration polarization

EDGAR C. KAUP, Burns and Roe, Inc.

After 200 years as a laboratory curiosity, reverse osmosis (RO) has come into its own as a commercial unit operation. RO can be used to recover valuable byproducts, recycle wasted reactants, help reduce polluting effluents, concentrate beverages and drugs, and most important at this time produce potable water from brackish and saline sources.

The osmotic phenomenon was first noted in 1748 by Abbe' Nollet, who found that water would diffuse through a pig bladder into alcohol spontaneously. Further experiments during the next 100 years were hampered by the unreliability of animal membranes until 1864, when Traube prepared a "selective" membrane of copper ferrocyanide. This was used by Pfeffer in his precise experiments with sucrose and other solutions that linked osmotic pressure with temperature and solute concentration. His data are still referenced in physical chemistry texts as an illustration of osmotic phenomena.[1] Later, J. H. van't Hoff developed a mathematical relationship for osmotic pressure.

Interest in the osmotic process waned after the turn of the twentieth century because reliable membranes were not available. In the 1950's, amid predictions of water scarcity, the U.S. Dept. of Interior established its Office of Saline Water (OSW) to evaluate water purification methods. Reverse osmosis was attractive because of its simplicity and, theoretically, low energy requirements—no energy-wasting phase change takes place. All that was needed was a strong, corrosion-resistant, reliable, cheap, selective membrane.

Cellulose Acetate Studies

In 1958, Professor C. E. Reid[2] at the University of Florida suggested seeking a suitable membrane from commercially available films. In the ensuing program, he and J. E. Breton demonstrated that secondary cellulose acetate (CA), one of the first plastic films manufactured in this country, had the desired selectivity. However, CA had disappointingly low water transport and a very short productive life.

Sidney Loeb and associates,[3] working on a similar project at the University of California, enhanced the water permeability or "flux" of CA membranes and substantially extended their usefulness by heat treating the film and adding swelling agents to the casting formulation. They found that when hydrophilic compounds such as potassium perchlorate and formamide were added to the casting "dope," the resulting film was swollen with water bound to the polymer chains and had 10 to 20 times the water transport rate.

This work demonstrated that the technology necessary to purify water by reverse osmosis was in hand. From this point on, the major advances have been development, engineering and marketing of RO systems, although much investigation and argument still goes on concerning mode of solvent transport, membrane formulations and operating conditions.

Today's Applications

Reverse-osmosis systems for water treatment are now commercially available for any moderate need, ranging from a 2.5-gal./day unit for home drinking supply, to a 150,000-gal./day auxiliary source for municipal water, a 350,000-gal./day plant for vacation resorts and 800,000-gal./day plants for industrial water. Such units have been sold, with quality and quantity of water guaranteed by the manufacturers.

While most applications today are for treating brackish water, i.e., 1,000 to 10,000-ppm. total dissolved solids (TDS), the scope of RO systems is rapidly being extended. Seawater desalting is OSW's next target. This involves problems with high operating pressures, calcium sulfate scaling, and organic fouling. OSW is testing a two-stage plant. The first stage reduces the 35,000-ppm. seawater feed to brackish water with 3,500-ppm. salt, and the second stage yields 350-ppm. TDS product.

Originally published April 2, 1973

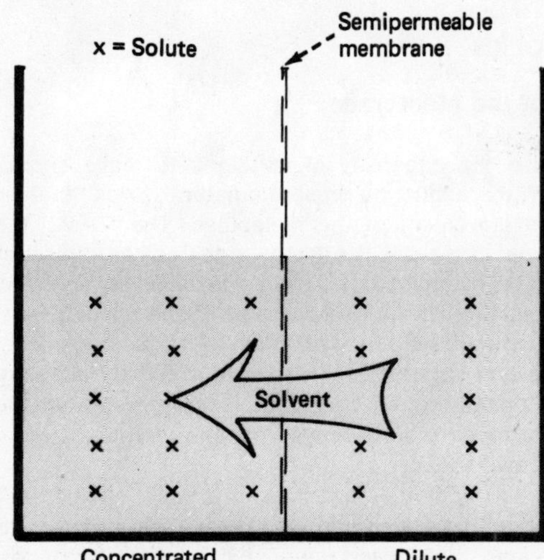

DIRECT OSMOSIS: solvent flows spontaneously through semipermeable membrane—Fig. 1

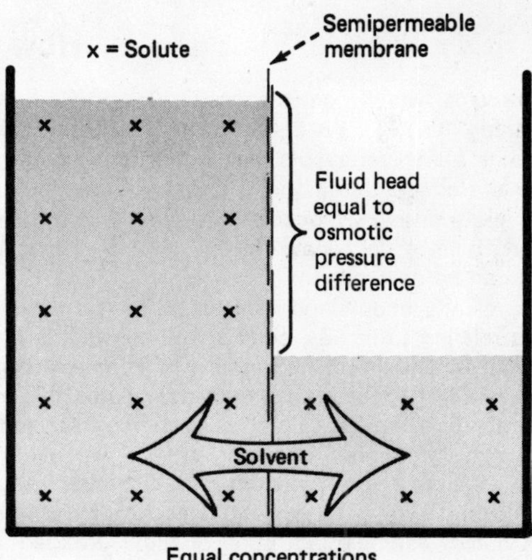

SOLVENT FLOW stops when the osmotic pressures of two solutions are equal—Fig. 2

Semiconductor fabricating plants, which use great quantities of ultrahigh-purity water for washing and rinsing, have reduced the cost of operating demineralizers 90% by pretreating feedwater with a RO unit. Demineralizing costs are determined by regenerant cost; regenerant consumption, in turn, is directly related to ionic composition of feedwater. Therefore, lowering the feed TDS to one tenth by RO pretreatment reduces operating cost proportionally. Some other applications are:

- Food processing—recovery of protein from cheese whey; concentration of maple sap, fruit juices, coffee and tea; concentration of drugs and biological products.
- Pollution control—removing chromate from cooling tower blowdowns; removal of sulfates from acid mine drainage; retrieval of gold, silver, platinum and other precious metals from electroplating solutions and rinses.
- Water reclamation—treating of secondary sewage effluent. An RO plant is operating at the San Diego municipal disposal plant, successfully reducing phosphate in the discharge.

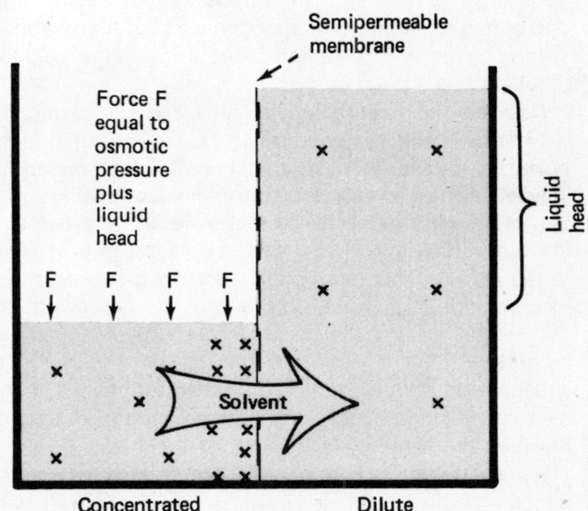

REVERSE OSMOSIS requires applied force equal to osmotic pressure plus liquid head—Fig. 3

Calculating Osmotic Pressure

Direct osmosis, the spontaneous passage of a solvent from a dilute solution through a semipermeable membrane into a more concentrated one, is depicted in Fig. 1; the opposite process, the moving of a pure solvent through the membrane by an external force on the concentrated solution, is represented in Fig. 3; and the intermediate step, osmotic equilibrium, is shown in Fig 2.

At equilibrium, the external force plus the osmotic pressure of the solution equals the osmotic pressure of the solvent on the right side of the membrane. Therefore, the total force necessary to stop the spontaneous flow of pure solvent into a solution is equivalent to the osmotic pressure of that solution. Osmotic pressure is stated relative to the pure solvent, but in actual practice it is the effective pressure between two solutions. Its magnitude is proportional to the amount of dissolved substances in the solution and to the temperature of the solution, and is completely independent of the membrane. In 1886, van't Hoff formulated an equation to calculate osmotic pressure, based on data for sugar solutions and the similarity of dilute solutions to ideal gases:

$$\pi = \frac{n}{v} RT$$

where: π = Osmotic pressure, atm.

$\frac{n}{v}$ = Ionic concentration, $\frac{\text{Moles}}{\text{liter}}$

T = Temperature, °K.

R = Proportionality factor, 0.083 $\frac{\text{atm. liter}}{\text{moles °K.}}$

This equation follows the ideal gas law so closely that the value of the gas law constant is used with reasonable

How Osmosis Works

Osmosis (which comes from the Greek word osmos, meaning "to push") is the tendency of a solvent to flow through a semipermeable membrane from a dilute solution to a concentrated one. It is a simple dilution operation that continually occurs in nature, but has always been a bit mystical because of the strong hydraulic forces that can be developed.

To clearly understand osmosis, a review of physical laws relating to liquids, solutes and solvents is helpful. All matter tends to distribute itself uniformly throughout the universe. Applied to solvents and solutes, this means that an unrestrained solute will disperse itself (diffuse) uniformly throughout a solvent and the two will in time form a homogeneous solution. The distribution rate depends on the solute (or solvent) concentration difference.

A homey example is a cube of sugar dropped into a cup of coffee. Sugar molecules diffuse rapidly from the saturated solution near the solid surface of the cube, throughout the sugar-less, "pure" coffee. Neglecting thermal effects that agitate the solution, the diffusion begins spontaneously, and the highest rate occurs at the beginning, when there is "pure" coffee in one place and pure sugar in another. The rate decreases as the sugar and coffee mix to form a homogeneous solution. Without stirring, uniform sweetening takes a long time because diffusion is a slow process.

One way of describing this effect is by saying the difference in sugar concentration causes the sugar molecules to diffuse. A more-fundamental explanation involves the free energies of the individual species. The partial free energy of the solvent molecule is greatest in the pure state, and is reduced in solution in proportion to the amount of solute added. As more solute is added to a solution, the partial free energy of the solvent becomes smaller and the partial free energy of the solute increases. Thus, if a pure solvent (or a weaker solution) is separated from a solution of that solvent by an impermeable membrane, there is a finite energy difference between the two liquids.

When the barrier is removed, the two species spontaneously begin to diffuse, seeking lower energy levels. Mixing is rapid at first because the energy difference is greatest, but it slows logarithmically as the concentration differences, and therefore the energy differences, throughout the solution decrease. Mixing stops when partial free energy throughout the "new" solution is uniform.

Role of the Membrane

To use the capability of solutions to move spontaneously, there must be present a natural "trick" that will harness movement of the molecules. The "trick" that makes osmosis work is a film with the ability to discriminate between molecules. This film is called semipermeable because it allows free passage of one species, such as water, while holding back dissolved substances such as salts and sugars. Such films are common in nature, e.g., in plant and tree cells and plasma membranes, but not until recently were any permanent man-made membranes available.

Back to our homey example. If a sack made from a semipermeable membrane is partially filled with a saturated sugar solution and then put into a cup of sugarless coffee, we will find that the water from the coffee will spontaneously pass into the sugar solution. Since the membrane is permeable only to water, it will allow water into the sack but not allow the sugar out. Furthermore, if the coffee supply is unlimited and the membrane is without leaks, the water level will continue to build up until the sugar solution is infinitely dilute, i.e., no osmotic pressure difference, or until some restraining force is placed on the sugar side.

Movement of solvent through a semipermeable membrane under the influence of a difference in solution concentrations is called osmotic flow. The process can be stopped at any time by applying a force equal, in our example, to the difference in osmotic pressures of the coffee and the sugar solutions.

When excessive pressure is exerted on the sugar solution, pure water moves into the coffee side of the system and, since the membrane is a barrier to the high-molecular-weight sugar molecules, the sugar solution is concentrated. This situation in which solvent (water) is transferred by an outside force from a solution with high solute concentration (a lower free energy level for the solvent) to a solution with low solute concentration (a higher free energy level for the solvent) is called reverse osmosis.

As water is "osmoted," the sugar solution becomes more concentrated, and more force is required to push out each drop. Carried to the extreme in an ideal system using an infinite amount of force, all the water could be squeezed out, leaving crystals of sugar on one side of the membrane and weak coffee on the other.

---

success. The similarity of the van't Hoff equation to the ideal gas law is so obvious that to lessen confusion the Greek Pi is used, instead of *P*, for pressure. Over a broader range and especially at higher solute concentrations, the following logarithmic expression gives more-accurate pressure values:

$$V = RT \ln \frac{P_0}{P_1}$$

where: P_0 = vapor pressure of the pure solvent
P_1 = vapor pressure of the solvent in solution

This relationship is not as convenient to use, since vapor-pressure measurements of "practical" solutions are not usually available. Also, since most RO applications are limited to weak, aqueous solutions (brackish water contains 0.1% to 1.0% dissolved solids, and seawater, 3.5%), it is usually adequate to roughly estimate the osmotic pressure from simple chemical analyses of the feed. The exact-pressure value is of little concern to the operation of a RO system. Therefore, refinements to the van't Hoff equation will not be discussed any further because for practical applications these improvements are inconsequential.

The osmotic pressure is only the barest minimum force required to obtain pure water from a solution. In current work, the operating pressures of RO systems are many times the osmotic pressure. For example, the osmotic pressure of the 2,000-ppm. brackish-water feed used for testing RO units at OSW's Roswell Test Facility in New Mexico is 32 psi., but the operating pressures of the units range from 250 to 800 psi. Table I lists osmotic pressures for several pure solutions and Roswell brackish waters calculated by van't Hoff's equation.

Types of Semipermeable Membranes

Many natural materials have semipermeable characteristics. Animal and plant membranes mentioned earlier are, of course, the best known examples. Collodion, cellophanes, porous glass frits, finely cracked glass, and inorganic precipitates such as copper ferrocyanide and zinc and uranyl phosphates have been used historically. All, however, have the common faults of leaks, short-lived selectivity and poor reproducibility.

Since Reid and Breton's discovery[4] that cellulose acetate possessed good ion selectivity, CA has become the most universally studied membrane material and the most successful. It is now the standard against which other developments are measured.

Numerous polymers have been found selective. Their suitability as a membrane is being investigated along with all sorts of modifications to cellulose acetate. Most of the research and development is sponsored by the U.S. Dept. of Interior's OSW, with the expectation that private industry will exploit the uses once feasibility is demonstrated.

One development that has received very little federal support until recently is Du Pont's Permasep membrane. Originally, Du Pont developed a system based on a nylon derivative that had moderate semipermeability. The membrane was spun in a unique form of a hollow, hairlike fiber to realize enough surface area to make it reasonably productive. This membrane was not successful because, even with its large surface area, water transport was very low (0.1 gal./day/sq.ft.) and rejection of the chloride ion was poor. Du Pont developed a second generation of Permasep membranes in 1969 with a new aromatic polyamide fiber that largely overcame the past failings. The system worked well enough so that in 1971 the company received a contract to work on a single-pass membrane system for producing potable water from seawater. To date, the published results have been quite optimistic.

One serious restriction of today's semipermeable membranes is their loss of ion selectivity (ability to reject specific ions) when operating above room temperature. The practical lower temperature limit is 40 to 45 F., a range in which diffusion is very sluggish. Increasing the temperature from 50 to 77 F. raises productivity 50%. Above 85 F., most membranes become unstable, show poor discrimination and lose strength.

General Electric is investigating polyphenylene oxide and sulfonated phenylene oxide polymers that show moderate selectivity at higher temperatures. The maximum operating limit has been increased to about 130 F. This is still not very hot, but it yields a three- or four-fold productivity increase.

Higher productivity is promised by dynamic membranes, which are loosely-structured polymeric mats with semipermeable characteristics. They are held in place on a filter support by the force of the liquid flowing through them. Dynamic membranes, however, have little cohesion and dissipate into the feed when flow stops.

Dynamic-membrane systems are operated continuously until cleaning is necessary. Then flow is stopped, the membrane dissipates and is flushed away. A new membrane is formed by reacting a mixture of hydrous zirconium oxide and polyacrylic acid in place. Production rates are very high, with fluxes on the order of 100 to 200 gal./day/sq.ft.

Properties of CA

Cellulose acetate, with an established technology gained through 60 years of experience in casting massive quantities of film, has good selectivity, dope formulations amenable to variation, good availability of raw materials and relatively low cost.

Chemically, cellulose acetate is a hydroxylic polymer made up of long chains of β-glucoside units (30,000–60,000) that have been acetylated with acetic anhydride and then hydrolyzed to reduce acetylation to about 40%. Most often, the CA powder supplied for RO membrane casting is in this partially acetylated form, which is known as 2.5 cellulose acetate. RO membranes are also cast from the fully acetylated form (triacetate). These perform equally well.

When cast as RO membranes, CA is a film about 4-mils thick. It is asymmetric—that is, the film has a thin, dense layer of about 0.25 microns above a thick, porous layer. Water passes easily from the dense layer through to the porous one, but with difficulty the other way. Unlike the thick, amorphous underlayer, the dense layer on top of the membrane is made up of tightly packed and organized chains of CA polymer that attract and hold water. Thus, water and solute are separated because the water molecules can form hydrogen bonds with the acetyl groups on the polymer, while many other species can not.

One simplified explanation of the separation process is: The CA polymer chains within the dense layer are highly organized because the membrane receives an annealing treatment that shrinks the film and crystallizes the polymer. The long molecules are somewhat separated and relatively immobile because of the interplay of van der Waals' forces. In this state, the polymer chains are close enough to crosslink with water molecules in the casting formulations and the annealing media.

These water molecules bridge across adjacent chains by forming strong hydrogen bonds with the acetyl groups. In this way, the voids between chains are filled with bound-water molecules and no foreign substance can pass. If the polymer chains were not densely packed, water molecules would bond to acetyls on the same chain and leave voids through which foreign ions could pass.

Water molecules move through the membrane by an applied pressure that pushes the water from a bond with one acetyl group to the next. Only a moderate force is necessary because the bonds are transferred, not broken. Dissolved ions or molecules that do not hydrogen bond cannot enter into attachments with bonding sites (acetyl groups) and are left to concentrate at the membrane surface.

Water Flux

Two parameters for characterizing an RO system are (1) production per unit area of membrane and (2) product quality. Production is measured by water flux, defined as the amount of product recovered per day from a unit area of membrane. English units (gal./day/sq.ft. are used in field work, while metric units (g./sec./sq.cm.) are used in the laboratory.

The flux through a particular membrane is determined by its physical characteristics (e.g., thickness, chemical composition, porosity) and by the conditions of the system (e.g., temperature, differential pressure across the membrane, salt concentration of solutions touching the membrane, and velocity of the feed moving across the membrane). In practice, the properties of the membrane and the solutions are relatively constant, and water flux becomes a simple function of pressure. Quantitatively, it is described by:

$$F_{H_2O} = A(\Delta P - \Delta \pi)$$

where: A = a permeation coefficient for a unit area of a membrane. This term includes the physical variables of the membrane and is relatively constant.
$\Delta P = (P_F P_p)$ = pressure exerted on the feed solution, P_F, less the pressure on product, P_p.
$\Delta \pi = (\pi_F - \pi_p)$ = osmotic pressure of the feed solution, π_F, less the osmotic pressure of the product, π_p.

The equation can be simplified for high pressures (above 600 psi.) by lumping all the constants into the coefficient A'. Flux for brackish water at constant temperature is very approximately:

$$F_{H_2O} \sim (A')(P_{Feed})$$

Temperature Factor

Flux is also affected by feed temperature. Water permeability of the membrane increases about 1½% per °F. The flux of a membrane is usually given by the manufacturer for 75–77 F., and a correction factor is applied at other temperatures. The correction can be derived on theoretical grounds from diffusivity and viscosity values, but experimentally determined corrections are held in more confidence.

The correction factor is applied to the area computed for feed at 25 C. (77 F.). This multiplier is obtained for the lowest expected operating temperature from a curve such as the flux/temperature correction curve in Fig. 4. This curve was developed by the Gulf Environmental Systems Co. for modified cellulose-acetate membranes. Application of the data is illustrated in the following membrane-area calculation:

It is desired to specify the membrane area for a 100,000-gal./day RO system to treat brackish water. Records show the lowest water temperature expected for any prolonged period is 68 F. (20 C.). The cellulose acetate membrane chosen for this system is expected to have an average flux of 15 gal./day/sq.ft. over its two-

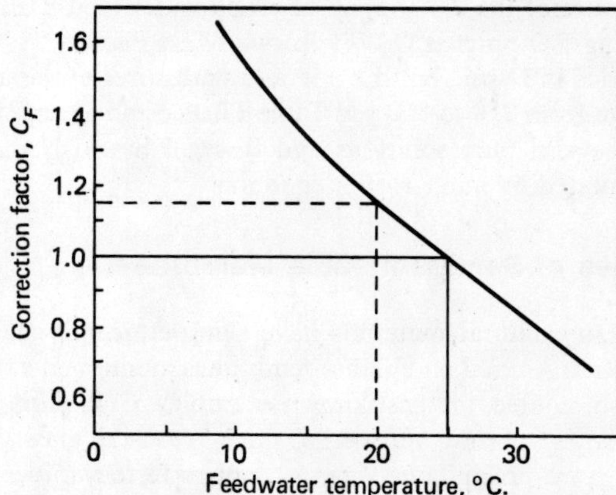

FLUX/TEMPERATURE correction curve—Fig. 4

year-life, operating at 600 psig. The average flux was determined at a base temperature of 77 F. The required membrane area is:

$$\text{Membrane area} = \frac{\text{Water production, gal./day}}{\text{Water flux, gal./day/sq.ft.}}$$

at 77 F.

$$\text{Area} = \frac{100{,}000}{15} = 6{,}667 \text{ sq.ft.}$$

The correction factor, C_F, for feed at 20 C., determined from Fig. 6, is 1.15.

$$\text{Area at 68 F.} = 6{,}667 \times 1.15 = 7{,}667 \text{ sq.ft.}$$

If the additional product is not desired or necessary when temperature is above 68 F., the operating pressure of the system can be reduced.

Flux Decline

Flux falls off during the life of the membrane for several causes. First, there is an initial decrease because capillary water is squeezed out as the operating pressure collapses the porous layer of membrane. CA has some "memory" and if the pressure is released before the membrane is permanently set, it will swell and regain some of the capillary water.

Following this initial compaction, there is a long-term compaction that slowly reduces the water flux. This results from densification of the thin, air-dried, membrane layer and probably corresponds to narrowing of pores through which water must pass. As the channels narrow, flow decreases.

Another cause of flux decline is the hydrolysis of acetyl groups that continually occurs during the life of the membrane.* This reaction results in a loss of hydrogen bonding sites, which further reduces the water transport. The reaction is also a source of salt leakage because there

*RO membranes are limited to a pH operating range of 3 to 7, outside of which rapid hydrolysis and membrane degradation occur. Presently, the optimum range is believed to be pH 5–6.

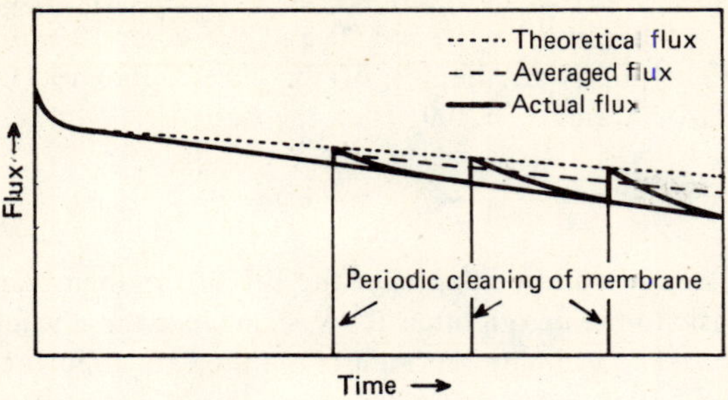

EFFECT of membrane cleaning on flux—Fig. 5

are fewer water bridges blocking the passage of foreign materials through the pores.

Loss in productivity happens to every membrane; it occurs slowly and is permanent. Chemical rejuvenation and low-pressure operation to relax and swell the matrix have been tried without success; the membrane simply ages and flux decreases, until economics dictate replacement.

Fouling: Causes and Cures

Fouling or temporary flux reduction is caused when foreign materials coat the membrane surface and interfere with inward movement of water. Materials accumulate because only hydrogen-bonding substances (water, ammonia) pass through the membrane's discriminating pores. Nonbonding materials are left in the quiescent film known as the liquid boundary layer.

The composition of deposits in the boundary layer reflects the composition of the feedwater. As expected, the most common constituents are calcium carbonate and sulfate scales, hydrates of iron oxides and aluminum, silicates, miscellaneous particulates, and biological growths. Most can be minimized by pretreating the feed to remove iron and control pH; by limiting the process to nonscaling concentrations of waste; by filtration; and by injection of small amounts of biocide.

Low concentrations of chlorine have been used to prevent organic growths on CA membranes, without discernable adverse effects. But chlorine definitely is not recommended for the polyamide type. Its use with Permasep membranes voids all guarantees.

Unfortunately, no matter how thorough the protection, fouling always occurs. The usual cleaning procedure is to first flush the membrane with feedwater at reduced pressure and two or three times normal velocity; the turbulent action of the fluid loosens the foulants and carries them away. Water flushing is a preliminary step in every cleaning operation.

Hardness scales (carbonate and sulfate salts of calcium and magnesium) can sometimes be removed by simply soaking the unit in distilled water for long periods. But this is time-consuming, and most often a warm solution of 1–2% citric acid is vigorously circulated through the unit. The acid dissolves large amounts of metallic ions and keeps them in solution by chelation.

Microbiological growths occur in most natural waters and are a particular problem when treating secondary sewage. They are often removed by recirculating washes of enzyme detergents. Organic fouling almost always reoccurs, and periodic cleanings are necessary to keep flux at a high level.

Flux degradation by fouling is an additional loss superimposed on the permanent losses. As seen in Fig. 5, the actual flux curve follows a declining, saw-tooth pattern when the membrane is cleaned periodically. Without cleaning, flux would follow the lowest curve, a projection of the initial smooth decline. The flux from a membrane that is never fouled is shown by the upper line, the theoretical flux curve that touches only the peaks of the saw-toothed curve.

Prediction of Flux

Unlike other unit operations in which production improves after an initial break-in period, water output from an RO system begins to decline as soon as the membrane is pressurized, and continues to degrade slowly throughout the membrane's life. The loss is irreversible, and if more flux is required the feed pressure must be increased. This alternative is somewhat self-defeating since additional pressure, while producing more water, also compresses the membrane further and hastens flux decline. Normal practice is to overspecify membrane area slightly and to keep operating pressure constant as long as possible, resorting to additional pressure late in the life of the membrane.

For a desalting system to be useful, its water production must, of course, be reliable. The output of a membrane is predictable because, as suggested above, the decline in productivity from a unit area of membrane is quite uniform and can be projected.

A linear plot of flux vs. operating time at a particular pressure yields a curve with an initial steep descent, followed by a prolonged, moderate decline. On exponential paper, this curve becomes a straight line and is reasonably constant for one and probably two years. Because of this constancy, flux can be predicted for the life of a membrane once the initial flux and curve slope are known.

Manufacturers provide initial flux values and estimates of curve slopes at various operating pressures. These slopes are experimentally determined for various simple salt solutions or local tap water. They can be applied only as guides for other waters.

Initial flux is the production for the first 24 hr., divided by the membrane area in the test unit. The decline slope is computed or graphically determined from flux values taken at time intervals such as 10, 100 and 1,000 hr. The decline rate is given as the slope of a log-log plot of flux in gal./day/sq.ft. vs. operating time in hours. The equation for the slope of the flux curve is:

$$m = \frac{\log F_i - F_x}{\theta_i - \theta_x}$$

where: F_i = initial flux in gal./day/sq.ft.
F_x = flux at time x hr.
θ_i = operating time, hr., for initial flux
θ_x = operating time, in x hr.

Some Reasons for Solute Rejection

- **Large nonelectrolytes are not permeable.** Water-soluble molecules that are too large to fit through the pores in the membrane are rejected whether they hydrogen bond or not. The historical example is, of course, Pfeffer's osmosis experiments with sucrose and dextrose. Blunk estimates the radius of an average pore to be about 3.6 A. and suggests sucrose is rejected because its radius is 4.4 A. Sourirajan showed that, besides large sugar molecules, sorbital glycerol, pentaerythritol are also well rejected. Molecular weight is sometimes used as a *rough* measure of permeability; nonelectrolytes with molecular weights over 200 are well rejected, those below are not.

- **Small, hydrogen-bonding, nonelectrolytes are permeable.** Simple straight-chain organics (four carbons or less) pass through membranes if they possess hydrogen-bonding abilities. Examples of molecules that permeate by hydrogen bonding are alcohols, aldehydes, acids, amines. Others are hydrogen peroxide, urea, acetamide and nitrates. Rejection of organic materials improves as the molecules become large, sterically complex and/or polyfunctional. Although organic acids and amines permeate easily in their free state, they are well rejected when neutralized to the salts. For example, acetic acid readily permeates CA membranes as the free acid but is 98–99% rejected when combined into the ionic sodium salt.

- **Dissolved gases such as carbon dioxide, sulfur dioxide, oxygen, chlorine and hydrogen sulfide are permeable in their free state.** Gases that are dissolved in water under conditions amenable to combinations with dissolved molecules are rejected. Carbon dioxide is poorly rejected from solutions with pH 4 or below but is 80–98% rejected when the acidic gas is neutralized to sodium bicarbonate. A similar permeability is found for sulfur dioxide at pH values below 3. Chlorine is poorly rejected below pH 7, but its permeability can be used to advantage when producing potable water with cellulose-acetate-membrane RO systems. Small amounts of chlorine added to the feed of a RO unit not only protect the membrane from biological growths but also purify the product.

- **Rejection of electrolytes increases with valences of their ionic species.** Di- and trivalent ions are more thoroughly rejected, no matter how strong their hydrogen-bonding tendencies, than monovalent ions. For instance, the divalent cations calcium and magnesium are more strongly rejected than the monovalent sodium and potassium. Similarly, the sulfate anion is much more strongly rejected than the monovalent chloride ion. A few small univalent species readily permeate membranes under most conditions because of their strong hydrogen-bonding tendencies. Hydronium, ammonium and fluoride ions and boric acid, phenol and acetic acid are examples of species that permeate easily. Acetic acid and phenol permeate so readily that they have shown some indications of concentrating in the product.

Time intervals in multiples of 10 are conveniently selected because log 10 = 1.00 and the computation is thus simplified to the log of the flux difference. Let $\theta_i = 10$ hr. and $\theta_x = 100$. Then the slope is:

$$m = \frac{\log F_i - F_x}{(1.0 - 2.0)} = -\log (F_i - F_x)$$

Determination of flux over the life of the membrane is basic to the design of an RO system since these values are used to estimate the membrane area required for a desired plant capacity. An average of initial and final flux can be chosen as a reasonable compromise to obtain an average area. But when such a choice is made, it is expected that the flux will be augmented during the later life of the membrane by increasing operating pressure. A second design approach is to specify membrane area from the final or smallest flux value. In this case, although fixed costs are highest, lower system pressures will decrease operating costs.

Another alternative is to select the initial flux as the design basis. This minimizes membrane area and capital costs but raises operating costs because pressures will have to be increased to maintain production. This is a logical choice for intermittent and short-term projects in which initial costs must be minimized.

Salt Permeability

Product quality from a semipermeable membrane is measured by the amount of solute or salt in the product. This depends on selectivity of the membrane and its imperfections. Theoretically, the salt passing through the membrane, or simply salt flux, is a function of the membrane's permeability coefficient, and the difference between the salt concentration in the feed and the product. (The coefficient is based on the physical characteristics of the membrane, such as thickness, salt diffusion, and the distribution of solute between the membrane and the solution.) The amount of salt passing through a unit area of membrane is:

$$F_{\text{salt}} = (C_H - C_L) = \beta \Delta C \approx \beta C_H$$

where: F_{salt} = salt flux, g./sq.cm./sec.
β = salt permeability coefficient, cm./sec.
C_H = concentration of solute on high-pressure side of membrane, e.g., concentrate side, g./cc.
C_L = concentration of solute on low-pressure side of membrane, e.g., product side, g./cc.

The equation says that salt passage is the result of the difference in salt concentrations across the membrane. It is believed that most salt moves by diffusion through the same membrane channels as water.

From the equation, normal salt flux (unlike water flux) is independent of pressure. Theoretically, if the pressure of the RO system is increased, the salt will diffuse at a constant rate, while the water flow will increase. The result will be greater production of purer water. Careful experimental work shows that these principles are correct, but often improvements are not seen because imperfections in the membrane leak substantial amounts of salt into the product. The problem of leakage is discussed later.

Rejection Factor

The measure of membrane selectivity is salt rejection, the ratio of solute rejected by a membrane to the solute in the feed. It is the most common method of evaluating a membrane's ability to separate salt, because the determination is simple and can be done as accurately in the field as in the laboratory.

Salt rejection is computed by dividing the difference in salt concentrations between the high- and low-pressure sides of the membrane by the high side. The result is expressed in percent. The solute concentrations can be obtained by chemical analyses or total dissolved solids (evaporation to dryness) but usually are determined by conductivity. This is a rapid, reliable method that requires a minimum of equipment and skill.

There are two possible bases for computation. One is the concentration of the feed solution as it is fed to the unit, which is an accurate representation if the membrane area is small. If, however, the membrane area is large, and the path between the inlet and outlet is long, there can be a considerable difference in feed concentration by the time the feed emerges from the discharge point. Since the latter portions of the membrane system are "seeing" a more concentrated solution, it follows that salt flux near the discharge of an RO unit will be greater than it was at the inlet.

To compensate for this increase in solute concentration, and give a more accurate description of the membrane's selectivity, most manufacturers use the feed-brine average between inlet and outlet as the basis for calculating salt rejection. This method always "improves" the salt rejection by a few percent over the simpler inlet-feed method. Field operations usually use the inlet basis because the calculation is simpler and the evaluation is of an RO system, not the membrane alone. Both methods of computing salt rejection are valid. Choice should be influenced by whether the system or membrane is being evaluated.

Influences on Salt Rejection

The ability of a membrane to reject salt is a complicated problem that seems to depend on combinations of physicochemical characteristics of the solute, the membrane and the water. The properties of the solute that have the most influence on the rejection of individual species are:

- Valence charge: Rejection increases with size of charge on ion.
- Molecular size: Rejection increases with physical size.
- Hydrogen bonding tendency: Permeation increases with strength of hydrogen bonding.

The membrane itself restricts solute, by limiting the physical size of the species that will pass through the pores, and by showing preference for solutes that hydrogen bond. The water molecules present in the membrane can also affect rejection. It appears that when water molecules become strongly bonded, as in a direct "bridge" between two acetyl groups, the water loses its solvating ability and rejects salts that it formerly dissolved as a free molecule.

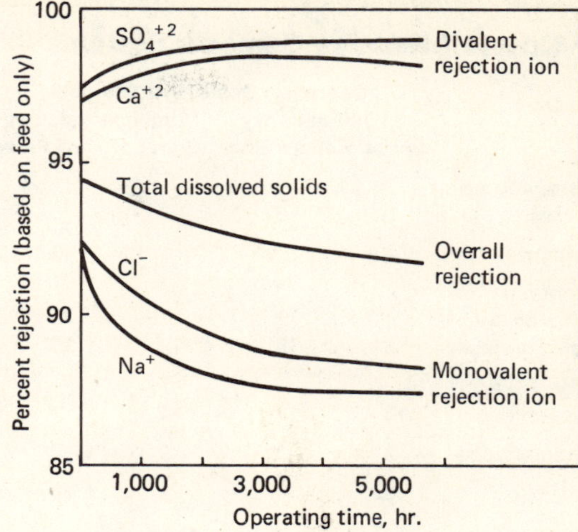

SALT REJECTION varies with time—Fig. 6

Also, water in some instances may be able to discriminate. It is believed that water molecules can form weak, secondary bonds to those that are strongly attached elsewhere. The weakly-bonded molecule will retain part of its solvency and be able to dissolve small, univalent ions. Thus, weakly-bound water molecules could act as a vehicle for some solutes and "carry" them through the membrane as part of the water flux.

The ability of a membrane to reject salts decreases with time. The decrease is noted first for small, univalent ions such as sodium and chloride. These are normally among the most permeable, show the lowest initial rejection, and have the highest rate of decline. Divalent cations such as calcium and magnesium, and anions such as sulfate, are initially rejected very well and show a very slow rate of decline.

Membrane units tested in Webster, S.D., show a constant overall rejection of 98% because the water is very hard and contains low salinity (sodium chloride). In another location, where both salinity and hardness are high, the rejection of divalent ions was at the same level but the overall rejection was only 91%. Because of the high saline content of the feed, rejection was initially lower and declined at a rate approaching that of sodium chloride.

The variation of rejection with time is illustrated in Fig. 6. Note the decline in rejection of monovalent ions as compared with the constant (and even improving) rejection for divalent ions. The overall salt rejection tends to follow the monovalent curve. The water in this test contained 950 mg./l. sodium chloride, 800 mg./l. hardness as $CaCO_3$, and total dissolved solids of 2,000 mg./l.

The progressive decrease in salt rejection may be caused by hydrolysis of the membrane, with subsequent loss in bonding sites. Another cause may be an increase in pore size due to membrane swelling. Most probably, it is the result of both effects.

At the present time, there are no quantitative laws governing the overall rejection of mixed species. Qualitative statements are made based on extensive experiments.

Costs of Reverse Osmosis Systems

	Operating Costs, ¢/1,000 gal.			
	50,000 gal./day		1 million gal./day	
	Spiral	Hollow Fiber	Spiral	Hollow Fiber
Power, 400 psi., 1¢/kwh.	8	8	8	8
Chemicals	2	2	1	1
Membrane replacement, including labor	18	18	10	10
Labor, $8/hr.				
Operations	18	18	1	1
Maintenance	3	12	2	8
Total	49	58	22	28

Data from Gulf Environmental Systems. Based on surface water with a residual feed turbidity of 1–5 Jackson units, and 75% recovery. Chemical costs based on conditioning with sulfuric acid at 1 lb./1,000 gal. Maintenance costs assume cleaning twice a week for hollow fiber units and once every two weeks for spiral systems.

Data have been reported by workers such as Pfeffer,[1] Reid and Breton,[2] Loeb and Sourirajan,[3] Lonsdale,[4] Keilin,[5] Glueckauf,[6] Hodges,[7] and Blunk.[8] Their results often agree, but often do not; sometimes for explainable reasons but sometimes not. The qualitative statements in the box concerning rejection of specific solutes by cellulose acetate membranes were distilled from the experiences of this author and many other investigators.

Membrane Leakage

Calculated salt rejection is always higher than the experimental values, even those that are determined under the most careful conditions. For example, the theoretical rejection of sodium chloride for a modified CA membrane is calculated as 99.7%, but experimental results show no better than 97–99% exclusion. The difference is believed caused by imperfections in the membrane (minute holidays) through which the pressurized brine can flow and contaminate the product water.

All membranes have imperfections; these are probably not manufacturing faults but a "property" of the membrane that must be adjusted or optimized to suit a particular service. Hence, very porous CA membranes are used to screen out large molecules (20–500 A. dia.) and very small particles for ultrafiltration applications; less-porous membranes are selected for high water flow and moderate salt-rejecting service in "saltless softeners"; and membranes heat-treated to low porosity are used for applications requiring high rejection.

Fortunately, most imperfections are small and easily plugged. Distribution runs from numerous holes with diameters of 100 A. and smaller, to a few with 1-micron diameters. The major source of product contamination results from salt passage through the larger holes because, by Poiseuille's law for viscose flow, salt leakage increases as the fourth power of the pore diameter. In addition to the heavy salt flow through large pores, considerable salt must also pass through the numerous small holes because the small-sized sodium and chloride ions always appear in the product, no matter how minor their concentration in the feed.

A number of ways have been tried to stop membrane leakage. Heat treating the membrane and modifying the dope formula seem to be the only permanent ways to improve semipermeability. Membranes for seawater must be highly heat treated to yield a film that can reject 99.5% of the salt in the feed.

Rejection can be temporarily improved by adding to the feed certain chemicals whose molecular size is large enough to plug up the leaking pores. Loeb[9] discovered that trace amounts of aluminum salts that occur naturally in Los Angeles tap water plugged the leaking pores of test membranes and improved salt rejection. Zephiran, a tetralkyl ammonium chloride, was used by Loeb and Manjikian[10] to gain a similar and more reproducible effect. Other materials that will improve rejection include polyvinyl methyl ether and Dowfax.

Unfortunately, leak-stopping additives have serious drawbacks that limit their use: They are more effective on low-flux, standard CA membranes than on the modified ones now used almost exclusively; they dissipate quickly and must be regularly replenished; most of them reduce water flux as they reduce salt leakage; and worst of all, they are expensive. These additives are useful only in special situations in which cost is not a factor.

Concentration Polarization

In all desalting processes, there are dynamic, localized concentrations of solute at those points where the solvent leaves the solution. In distillation processes, there are small increases in solute concentration at the solution surface when the water vapor leaves, but this is soon dispersed by bubble and thermal agitation. In membrane processes, however, the solute accumulates at a fixed boundary (the membrane) in a relatively stable layer (the boundary layer) and brings about an effect called concentration polarization. This is defined as the ratio of the solute concentration at the membrane wall relative to that in the bulk of solution.

Initially, the boundary layer is at the same concentration as the bulk solution, but since the wall (membrane) is permeable to solvent (water), the boundary layer becomes heavily populated with solute as solvent leaves through the channel wall. The layer grows thicker and more concentrated because the rate of solute diffusion away from the membrane cannot keep up with solvent flow through the membrane.

The layer at the wall can be reduced by higher feed velocities and turbulence. The extra flow through the unit at higher feed velocity results in overall lower product recovery, i.e. smaller ratio of product to feed. This increases power consumption and the amount of waste produced.

Turbulence promoters are a more attractive way to reduce the boundary layer. Tubular membrane units have small plastic balls in the annular space to break up the smooth flow of the feed solution. RO units with sheet membranes often place waffle-like polyethylene screens in the flowpath. With turbulance promoters, good filtering of the feed is necessary so that particulates or precipitates do not get lodged in narrow flow spaces and plug up the membrane element.

Effects of Polarization

When feed salinity increases, the product shows a corresponding increase in salt content. This effect is noted from the equation for salt flux, which shows that the salt or solute passed is almost entirely dependent upon feed concentration. Thus, the concentration of the solution nearest the membrane governs the amount of salt in the product. Unfortunately, the highest concentration is at the wall. For this reason, feed flow is adjusted to give the least concentrated solution in the boundary layer. The usual concentration polarization ratio is 1.2 to 2.0, which means that the concentration at the wall is 1.2 to 2.0 times that in the bulk of the feed.

Water flux will fall off when the salt content of the feed-brine solution increases, because the amount of salt affects the driving force of the flux. (Net driving force is the difference between the net pressure and the net osmotic pressure, and the salt content of the solutions determines the osmotic pressure.) The solution adjacent to the membrane is the most influential in the system because the osmotic pressure of this solution sets the driving force of the system.

As an example, let us examine what can occur in an RO unit treating seawater at 1,000 psi., when the product recovery is increased. Assume the seawater (35,000 mg./l. TDS) has an osmotic pressure of 350 psi. and that the water recovery of the unit is normally 17%. If the concentration polarization factor is 1.2, the salt concentration at the wall of the membrane is 42,000 mg./1. TDS, and the osmotic pressure at the membrane becomes 420 psi. If the feed is slowed so that half the water in it is recovered, the solution at the wall is concentrated 100%, and the local osmotic pressure is 700 psi. As a result of the increased concentration at the wall, water flux (production) will fall off almost 50% because the driving force has dropped from 580 psi., (1,000 − 420 = 580), to 300 psi., (1,000 − 700 = 300).

Thus, the power requirements of an RO system vary with concentration polarization. When the salt concentration at the membrane increases, either the operating pressure must be raised to compensate for higher osmotic pressure, or the pumping rate must be increased to gain more turbulence and thereby dilute the concentrated salt solution.

When dealing with natural brackish waters containing 2,000 to 10,000-ppm. dissolved solids, the concentration polarization effect is small because the net pressure is large and the osmotic pressure is low. The operating pressures of brackish-water units usually range from 250 and 800 psi., and the osmotic pressures are from 30 to 100 psi.

A very common problem caused by high concentration polarization is the plugging of membranes by sparsely soluble salts. If the solution at the wall becomes supersaturated, insoluble salts precipitate. These deposits can stop water flux and cause permanent membrane deterioration. The most common scales are calcium carbonate and calcium sulfate, since they occur in almost all natural waters. They have low solubilities, 14 mg./l. and 1,500 mg./l. respectively, which are easily exceeded if natural waters are overconcentrated.

Some of the precautions taken to prevent scaling are: (1) pH adjustment by adding acid to the feed, which breaks down the bicarbonate ion and prevents formation of carbonates in the boundary layer; (2) keeping the product recovery, i.e. product-to-feed ratio, low (less than 50% for most natural waters) to assure that a saturated calcium sulfate solution is not formed in the boundary layer; and (3) adding antiscale substances such as sodium hexametaphosphate to prevent deposits. Deposits can be dissolved by chelating agents described earlier; but after cleaning, the membrane always shows permanent damage.

Editor's Note

The cost information in the table is provided as a general guide for two types of RO systems. For additional information on membrane processes and equipment see Membrane Separation Processes, p. 305.

References

1. Glasstone, Samuel, "Textbook of Physical Chemistry," D. Van Nostrand Co., New York, pp. 651–683.
2. Reid, C. E., and Breton, J. E., *Journal of Applied Science,* 1959, pp. 1–133.
3. Leob, S., and Sourirajan, S., UCLA Department of Engineering Report 60-60, 1960.
4. Lonsdale, H., et al, Reverse Osmosis for Water Desalination, Office of Saline Water Research and Development Report No. 111, Pb. 18-1696, 1964.
5. Keilin, B., Mechanism of Desalination by Reverse Osmosis, Office of Saline Water Progress Report No. 117, Pb. 166395, 1964.
6. Glueckauf, E., Proceedings of First International Desalination Symposium, Paper SWD/1, Wash., D.C., Oct. 3–9, 1965.
7. Hodges, R. M. Jr., Polymer Films As Reverse Osmosis Membranes For Sea Water Desalination, Ph.D. Thesis, M.I.T. Dept. of Chemical Engineering, Cambridge, Mass., 1964.
8. Blunk, R. W., Study of Criteria for the Semipermeability of Cellulose Acetate Membranes to Aqueous Solutions, UCLA Dept. of Engineering Report 64-28, 1964.
9. Leob, S., Seawater Desalination by Means of a Semipermeable Membrane, UCLA Dept. of Engineering Report.
10. Leob, S., and Manjikian, S., Brackish Water Desalination by an Osmotic Membrane, UCLA Dept. of Engineering Report 63-37, 1963.

Meet the Author

Edgar G. Kaup is a senior chemical engineer at Burns and Roe, Inc., Oradell, NJ 07649. He is presently field manager for an acid-mine drainage-control project in Hawk Run, Pa., and previously was resident engineering manager at Roswell, N.M., where he evaluated membrane systems for the Office of Saline Water. Mr. Kaup has a B.S. degree in chemistry from Lehigh University, and B.S. and M.S. degrees in chemical engineering from Newark College of Engineering. He is a registered engineer in Pennsylvania and New Jersey.

How To Design Settling Drums

Gravity settlers, or decanters, for separating two immiscible liquids are often designed by trial-and-error. Here is a simple mathematical way to do it.

B. SIGALÉS, Instituto Petrolquimíca Aplicada, Barcelona, Spain

Horizontal drums are generally used for the separation of the liquid phases from a mixture of two immiscible liquids, that is, settling. In the most general case, two continuous phases are formed, each one containing unemulsified droplets of the other phase.

Separation occurs under the influence of gravity, owing to the difference in density between the two liquids. The rate of settling is influenced by viscosity and drop size.

Stokes law is used to determine the limit of the settling velocity, v, under the low Reynolds numbers that are encountered:

$$v = 8.3 \times 10^5 \frac{d^2(\rho_d - \rho)}{\mu}, \text{ in/min} \quad (1)$$

For a normal droplet size of $d = 0.005$ in, this gives:

$$v = 20.8 \frac{(\rho_d - \rho)}{\mu}, \text{ in/min} \quad (1a)$$

In practice, though, a maximum settling velocity of 10 in/min should be used, even if the equation gives a higher value [1].

The drum size is calculated from the settling and flow rates. The drum volume must provide enough holdup time for a drop to move from the top (or bottom) of the drum to the interface of the two liquid phases. Packing or horizontal baffles can be used to reduce the required drum volume, since the droplets then require only enough time to reach the baffle, rather than the interface.

Two settling volumes must be provided, but in most circumstances only the light phase must be recovered; the other (heavy) phase is recirculated. Therefore, only the recirculated phase must be removed completely from the product that is withdrawn. However, we will also cover the case in which both phases must be completely separated.

Design When the Heavy Phase Is Recirculated

Basically, the design consists of determining the drum dimensions and also the high and low positions of the interface (needed to fix instrument connections).

The drum's dimensions are determined for the case where it is operating at the high-interface level, because this is the most unfavorable condition for a droplet to reach the interface (the drum cross-section segmental area increases at a higher rate than the segment height—see Fig. 1).

The minimum required settling time will be found by dividing the distance between the top of the drum (or segment) and the interface level, by the settling velocity. To determine the length of the drum, multiply the flow rate of the light phase by the settling time and divide the result by the cross section of the drum above the interface level.

Use these rules in designing the drum (Fig. 2):

1. A minimum space for the heavy phase in the bottom of the drum should be set by fixing the low interface level at 20% of the drum diameter above the bottom (equivalent to an area of about 15% of the cross-section), or 8 in., whichever is greater.

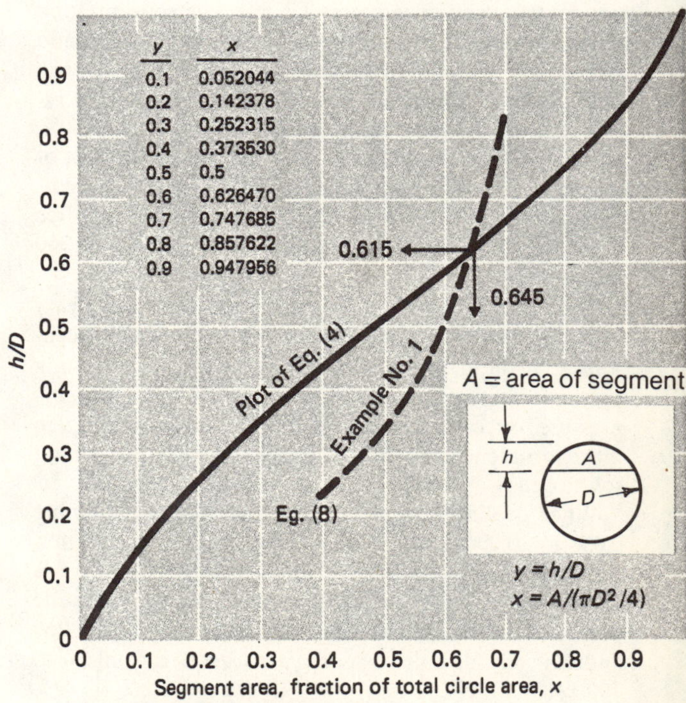

GRAPHICAL determination of settling-drum size—Fig. 1

Originally published June 23, 1975

Nomenclature

- A Area parameter, defined by Eq. (6), ft^2
- B Volume parameter, defined by Eq. (7), ft^3
- d Droplet diameter, in
- D Drum diameter, ft
- L Drum length between tangent lines, ft
- M Liquid flowrate, lb/h
- P Phase residence time, min
- r Economic drum proportions, L/D
- R Phase residence time, min
- S Settling time, min
- v Settling velocity, in/min
- x Passage-segment cross-sectional area as a fraction or total drum cross-sectional area
- y Ratio of passage segment height to drum diameter (see Fig. 1)
- μ Viscosity of continuous liquid (with subscripts l and h of light and heavy liquids, respectively), cp
- ρ Density of continuous liquid (with subscripts, the density of the liquid described by the subscript), lb/ft^3 (except in Eq. (1) and (1a), where it is g/cm^3)

Subscripts

- d Droplet
- h Heavy phase
- l Light phase

2. The minimum distance between the high and low interface level should be enough to provide 2-min holdup time, or 14 in. (required for level instrumentation).

The sizing of the drum has been traditionally done by trial-and-error. However, if the volume in the drum's dished heads is neglected, a graphical solution is possible.

Considering what we have already covered, the following equations can be established:

$$S = \frac{x(\pi D^2/4)L}{M_l/\rho_l(60)} = h/v \qquad (2)$$

$$R_h = \frac{(0.85 - x)(\pi D^2/4)L}{M_h/60\rho_h} \qquad (3)$$

$$y = y(x) = h/D \qquad (4)$$

This is graphed in Fig. 1. Note that y is the *ratio* of the section height of the passage segment to the total drum height.

$$L/D = r \text{ (economic drum proportions)} \qquad (5)$$

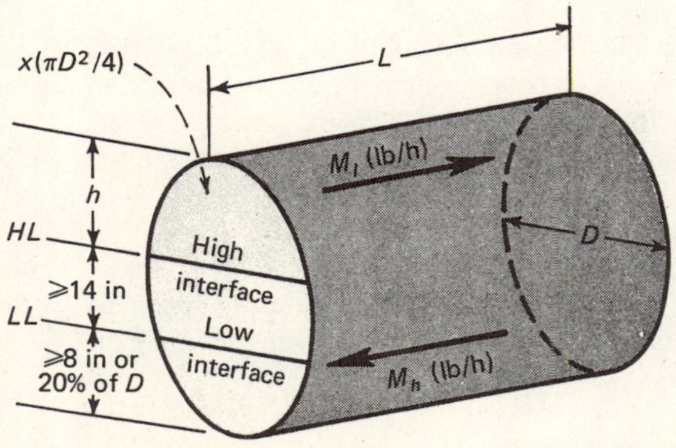

SETTLING drum for recirculated heavy phase—Fig. 2

From Eq. (2) and (5):

$$x = \frac{M_l}{15\pi v \rho_l DL}(h/D) = Ay/D^2 \qquad (6)$$

where:

$$A = \frac{M_l}{15\pi v \rho_l r}$$

From Eq. (3) and (5):

$$D^3 = \frac{M_h}{7.5\pi r \rho_h (0.85 - x)} = \frac{B}{0.85 - x} \qquad (7)$$

Where:

$$B = \frac{M_h}{7.5\pi r \rho}$$

Substituting Eq. (7) in (6), and solving for y:

$$y = \frac{x}{A}\left[\frac{B^2}{(0.85 - x)^2}\right]^{1/3} \qquad (8)$$

Eq. (8) can be plotted in Fig. 1, and its intersection with the plot of Eq. (4) will give the values of x and y.

From Eq. (6):

$$D = \frac{Ay}{x} \quad \text{and} \quad h = yD \qquad (9)$$

To meet the two rules for design that were given earlier, use either D from Eq. (9), or,

$$h + 8/12 + 14/12 = h + 1.84 \qquad (10)$$

whichever is greater.

Example No. 1

Assume the following data for a light-hydrocarbon and caustic-solution settling drum (caustic being recirculated):

Light-hydrocarbon load, gpm	95
Light-hydrocarbon density at operating conditions, g/cm^3	0.52
Caustic solution recirculated, gpm	26
Cautic solution density at operating conditions, g/cm^3	1.10
Light continuous-phase viscosity, cp	0.1
Economic L/D ratio, r	3.4

Solution—According to Eq. (1a), the theoretical settling velocity will be:

$$v = 20.8\frac{1.10 - 0.52}{0.1} = 120 \text{ in/min}$$

Since this value is greater than 10 in/min, we will use the latter value (10 in/min = 0.833 ft/min).

With this value, and the given data:

$$A = \frac{(95)(60)/7.48}{15\pi(3.4)(0.833)} = 5.71 \text{ ft}^2$$

$$B = \frac{(26)(60)/7.48}{7.5\pi(3.4)} = 2.60 \text{ ft}^3$$

From Eq. (8):

$$y = (x/5.71)[6.76/(0.85 - x)^2]^{1/3}$$

Solving for the following values of x:

x	y
0.60	0.50
0.70	0.82
0.64	0.547
0.85	∞

Plotting these data on Fig. 1 ("Example No. 1"):

$$x = 0.645;\ y = 0.615$$

therefore:

$$D = [(5.71)(0.615)/0.645]^{1/2} = 2.333 \text{ ft}$$
$$h = 0.615\ (2.333) = 1.435 \text{ ft}$$
$$h + 1.84 = 3.275 \text{ ft}$$

Since the latter is the larger, we will accept it and round it to 3.5 ft O.D.

$$L = 3.4(3.5) = 11.9; \text{ use } L = 12 \text{ ft}$$

High-interface level above bottom: 3.5 − 1.435 − [2 (shell thickness)] = 2.605 − 2(thickness), ft, say 24 in.

Low-interface level above bottom: 2.065 − 2(thickness) − 1.17 = 0.895 − 2(thickness), ft, say 10 in.

If it is desired to obtain an optimum-dimensions L-D plot (as used in Ref. [2]), it is necessary to solve Eq. (4) and (8) for x and y, using several values of r (three values will suffice) to obtain pairs of values of D and L that fulfill all requirements.

Designing for Complete Phase-Separation

When the two phases must be completely separated from each other, both the heavy- and light-phase droplets must reach the interface before the continuous phases leave the drum.

The basic conditions can be written (refer to Fig. 3):

$$\frac{x_l(\pi D^2/4)rD}{M_l/60\rho_l} = h_l/v_l, \text{ min}$$

(h_l measured from top of drum)

$$\frac{x_h(\pi D^2/4)rD}{M_l/60\rho_l} = h_h/v_h, \text{ min}$$

(h_h measured from bottom of drum)

$$h_l/D = y(x_l)$$
$$h_h/D = y(x_h)$$
$$D \geq h_l + h_h + 14/12$$

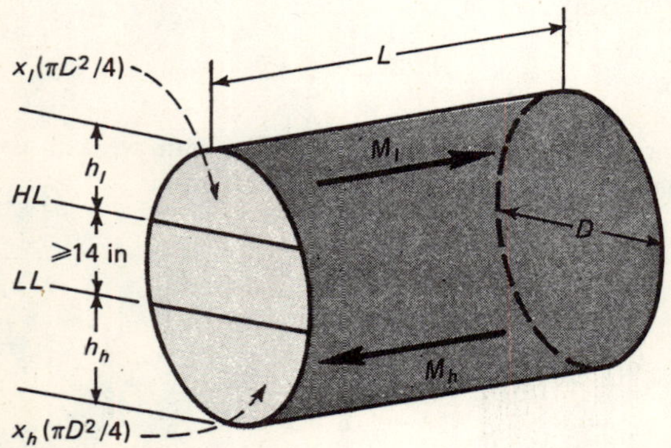

SETTLING drum for complete phase-separation—Fig. 3

The above relationships can be put in the form:

$$h_l/D = x_l D^2/A_l = y_l$$
$$h_h/D = x_h D^2/A_h = y_h$$
$$y_l = y(x_l)$$
$$y_h = y(x_h)$$
$$D = h_l + h_h + 1.17, \text{ ft (minimum dia.)}$$

where:

$$A_l = M_l/15\pi v_l r\rho_l \quad \text{and} \quad A_h = M_h/15\pi v_h r\rho_h$$

The diameter of the vessel will be:

$$D = (A_l y_l/x_l)^{1/2} = (A_h y_h/x_h)^{1/2} \tag{11}$$

The system reduces to:

$$A_l y_l/x_l = A_h y_h/x_h \tag{12}$$
$$1 = y_l + y_h + 1.17(x_l/A_l y_l)^{1/2} \tag{13}$$
$$y_l = y(x_l) \tag{14}$$
$$y_h = y(x_h) \tag{15}$$

Here, we have four equations, Eq. (12), (13), (14), (15), in four unknowns (x_l, y_l, x_h and y_h). This cannot be solved by using Fig. 1 as in the previous case, because two relationships are graphical, and it is impossible to obtain an algebraic equation with only the x_l, y_l pair, or the x_h, y_h one. The function $y = y(x)$ is transcendent and impractical to handle. An algebraic relation that approximates it can be obtained by regression analysis. With the nine points shown in the table on Fig. 1, plus the origin, one can obtain:

$y = 4.6727x^5 - 12.1145x^4 + 12.0254x^3 - 5.63107x^2 + 2.02853x$, having an "index of fit"* of 0.999908; or

$y = 1.00629x^3 - 1.57907x^2 + 1.54587x$, with an "index of fit" of 0.999553; and $y = 0.9626x$, with an "index of fit" of 0.994905.

For hand calculation, only the linear regression is practical, although the "index of fit" is not too good. Nevertheless, in practice, values of h lower than 15% of D ($y < 0.15$) are not used, and values of h higher than 80% of D are absurd. Inspection of Fig. 1 shows that within the limits $0.1 < y < 0.9$, the straight-line regression fits $y = y(x)$ rather well, and for $0.2 < y < 0.8$, it fits even better.

In effect, taking values of x and y ranging from $y = 0.2$ to $y = 0.8$, as shown in Fig. 1, we obtain:

$$y = 0.826103x + 0.0869497 = \alpha x + \beta \tag{16}$$

Eq. (16) has an "index of fit" (for $0.2 < y < 0.8$) of 0.9999, which is better than for any polynomial of degree 5, or lower.

Using Eq. (16), Eq. (14) and (15) become, respectively:

$$y_l = \alpha x_l + \beta$$
$$y_h = \alpha x_h + \beta$$

In addition, note that because of the symmetry of the function with respect to the point (0.5, 0.5) in Fig. 1:

$$1 - y = \alpha(1 - x) + \beta$$

*The "index of fit" equals:

$$1 - \left(\frac{\Sigma(\text{error})^2}{\Sigma y^2}\right)\left(\frac{\text{number of points} - y \text{ origin}}{\text{number of points} - \text{polynomial grade}}\right)$$

Therefore, in Eq. 13 we can write:

$$0 = 1 - y_l - y_h - 1.17(x_l/A_l y_l)^{1/2} = \alpha(1 - x_l) + \beta - \alpha x_h - \beta - [1.369x_l/(A_l(\alpha x_l + \beta))]^{1/2}$$

Dividing by α:

$$1 - x_l - x_h - \left[\frac{x_l}{(A_l\alpha^2/1.369)(\alpha x_l + \beta)}\right]^{1/2} = 0 \quad (17)$$

Eq. (12) becomes:

$$A_l(\alpha x_l + \beta)/x_l = A_h(\alpha x_h + \beta)/x_h$$

or:

$$x_h = \frac{x_l}{(A_l/A_h) + \alpha/\beta[(A_l/A_h) - 1]x_l} \quad (18)$$

Substituting Eq. (18) in (17):

$$1 - x_l - \frac{x_l}{(A_l/A_h) + \alpha/\beta[(A_l/A_h) - 1]x_l} - \left[\frac{x_l}{(\alpha^2 A_l/1.369)(\alpha x_l + \beta)}\right]^{1/2} = 0$$

and with $\alpha = 0.8261$ and $\beta = 0.0869$:

$$1 - x_l - \frac{x_l}{(A_l/A_h) + 9.506[(A_l/A_h) - 1)]x_l} - \left[\frac{x_l}{0.4118A_l x_l + 0.0433A_l}\right]^{1/2} = 0 \quad (19)$$

The above Eq. (19) can be solved graphically, yielding x_l and x_h (analytically). With these values, we use Fig. 1 to obtain y_l and y_h, and using Eq. (11), we find D.

Finally, with y_l and y_h, h_l and h_h are determined. If h_l or h_h are smaller than 8 in, use $h = 8$ in (increasing D by the same amount to keep the difference between the high interface level and low interface level equal to 14 in. (in Fig. 3, $HL - LL = 14$ in).

Also, solving with different values of r, the minimum-dimensions L-D curve can be plotted and used as in Ref. [2].

Example No. 2

Assume the following data:

Light-phase load, gpm	95
Heavy-phase load, gpm	39
Light-phase Stokes-law settling velocity, ft/min	0.417
Heavy-phase Stokes-law settling velocity, ft/min	10
Economic L/D ratio, r	3.4

Solution—The light-phase settling velocity = 10 ft/min = 120 in/min. This is greater than 10 in/min, so use:

$$v_l = 10 \text{ in/min} = 0.833 \text{ ft/min}$$

Heavy-phase settling velocity = 0.417 ft/min. (Use this velocity since it is less than 0.833 ft/min.) Therefore:

$$A_l = [(95)(60)/7.48]/15\pi(3.4)(0.833) = 5.71 \text{ ft}^2$$
$$A_h = [(39)(60)/7.48]/15\pi(3.4)(0.417) = 4.68 \text{ ft}^2$$

and:

$$A_l/A_h = 1.22; \quad A_l/A_h - 1 = 0.22$$

Substituting in Eq. (19):

$$1 - x_l - \frac{x_l}{1.22 + 2.09x_l} - \left[\frac{x_l}{2.351x_l + 0.247}\right]^{1/2} = 0$$

To solve this equation graphically, arrange your calculations this way:

	Eq. 19			
x_l	0.4	0.2	0.3	0.29*
$1 - x_l$	0.6	0.8	0.7	
$-x_h = \frac{-x_l}{1.22 + 2.09x_l}$	(0.195)	(0.122)	(0.162)	$x_h = 0.16$
$-\left[\frac{x_l}{2.351x_l + 0.247}\right]^{1/2}$	(0.580)	(0.528)	(0.561)	
Total	**(0.175)**	**0.150**	**(0.023)**	*(Graphically)

From Fig. 1: $y_l = 0.33$ and $y_h = 0.22$

From Eq. (11):

$$D = [(5.71)(0.33)/0.29]^{1/2} = 2.549$$
$$= [(4.68)(0.22)/0.16]^{1/2} = 2.54 \text{ ft}$$

and

$$h_l = (2.54)(0.33) = 0.838 \text{ ft} \cong 10 \text{ in}$$
$$h_h = (2.54)(0.22) = 0.559 \text{ ft} = 6.7 \text{ in}$$
$$HL - LL = 2.54 - 0.838 - 0.559 = 1.143 \text{ ft (O.K.)}$$

Taking $h_l = 10$ in, and $h_h = 8$ in, the diameter will be:

$$D = 10 + 8 + 14 = 32 \text{ in} = 2.67 \text{ ft}$$
$$L = (3.4)(2.67) \cong 9 \text{ ft}$$

Therefore, use a 3-ft-O.D. × 9-ft-long drum.

Note that this drum handles the same quantity of light phase and 50% more heavy phase than the one in Example No. 1 but has smaller dimensions. The reason is that the practical rules for design call for an oversize vessel—the theoretical dimensions of the drum of Example 1 were 2.33 × 7.9 ft, though the practical ones were 3.5 × 9 ft. Even with recirculation, using the method of calculation for complete settling of both phases can give more-economical designs. (But remember, if one h comes out less than 15% of D, the design is not good—i.e., Eq. (11) is not met.)

Acknowledgement

I should like to express my appreciation to Felix E. Saltor, Dr. I. I. (Dr. of Engineering) of IBM Espanola, for his help and comments concerning this article.

References

1. Happel, John, "Chemical Process Economics," Wiley, New York, 1958, p. 251.
2. Sigales, B., How to Design Reflux Drums, *Chem. Eng.*, Mar. 3, 1975, p. 157.

Meet the Author

B. Sigales is a professor at Instituto Petrolquimica Aplicada, Polytecnic University of Barcelona, Av. Generalisimo 647, Barcelona-14, Spain, where he specializes in crude-oil evaluations and workup and process-vessel design. He is also a professor at Centro Perfeccionamiento del Ingeniero, the National Assn. of Spanish Engineers for Industry (Barcelona and Madrid), and he is Chemical Participations Coordinator, Chemical Companies Board Advisor, Banco Industrial de Catalunya, Barcelona. He has a doctorate in engineering from Escuela Tecnica Superior Ingenieros de Industriales de Barcelona.

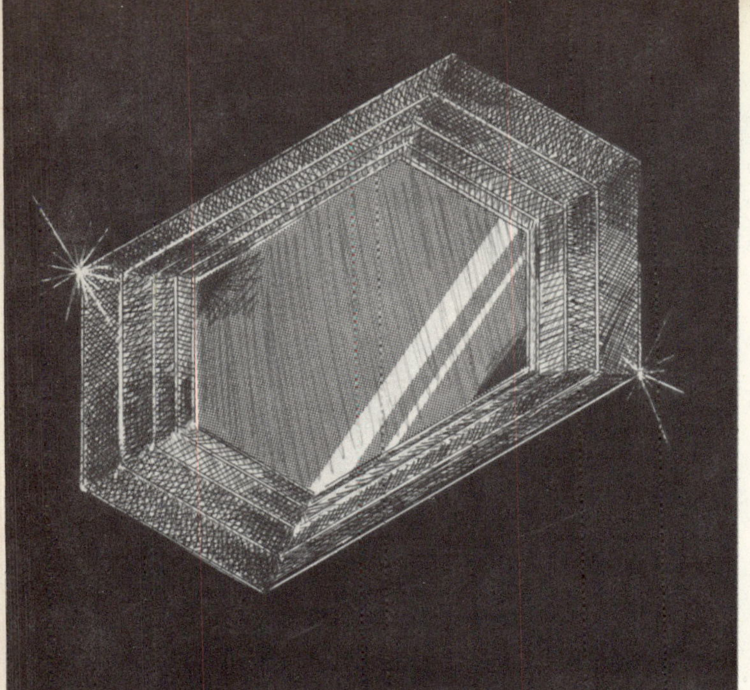

Freeze crystallization

Fractionation of components of a solution by freeze crystallization can be more energy efficient than are conventional separation processes. This article presents the theory and practice of the process, and ways of performing economic evaluations.

James A. Heist, *Concentration Specialists, Inc.*

☐ Although the vapor-liquid equilibrium is generally employed to separate the components of a solution, use of the solid-liquid equilibrium should also be considered—it may be cheaper. Although freeze crystallization has been used to fractionate solutions in specialized applications, the technique has never been adopted on a large scale.

In some applications, freezing can perform a separation with a 75 to 90% reduction of the energy required by conventional distillation—acetic acid and acrylic acid refining are two examples.

Freeze crystallization also shows significant energy reduction in many applications for which evaporators are commonly employed. Here, the savings are not as spectacular as when replacing distillation, because most large evaporators employ energy-conserving multiple effects. Energy consumption of paper-plant black-liquor evaporators could be reduced by a half by using freezing, as could the energy used in the concentration of citrus juices.

Batch vs. continuous processes

In the past, most chemical process industry freeze separation processes have been batch operations. They have been labor intensive, inefficient (compared with their potential), and limited to low-throughput operations. These shortcomings have been tolerated because the process has been the only one that would work at all.

But, continuous process equipment is now on the market—equipment that makes available more of the inherent capabilities of the technology. Such equipment will be discussed later.

Advantages and disadvantages

Freeze separation is not the answer to every separation problem. Under some circumstances, either capital cost or energy consumption can be excessive.

Freeze separation is a relatively new process; although many of the equipment components are well developed, process and application work with them is still limited. Development work is needed on some of the simplified, less-costly freeze cycles—even some of the cycles that have been proposed and built are not fully developed.

However, experience with new applications in the chemical process industries will indicate where the present processes need improvement. Enough work has been done to warrant consideration of freezing for any separation process.

Some background

The history of modern freeze-separation processes dates only from the early 1950s, when the first continuous equipment was developed. Two very diverse routes were involved.

In the petroleum industry, Phillips Petroleum and then several others developed indirect-contact, scraped-surface heat exchange processes for the separation of xylene isomers. Although these developments were tested in the chemical and food industries, they proved to be too-expensive alternatives. So the xylene application remained the only large-scale user of that type of process.

At the same time, a number of direct-contact freeze separation processes were under development for seawater desalting. This was the first of the separation unit operations that received large-scale development funding from the (now defunct) Office of Saline Water; this funding was indicative of the process's potential for low energy consumption.

These desalting processes operated either at the triple point of water (vacuum freezing using either an absorption cycle or a vapor-compression cycle) or with a secondary refrigerant (both butane and halocarbon refrigerants were used). All of these processes were developed

Originally published May 7, 1979

to pilot-plant scales of 10,000 to 100,000 gal/d. The plants were large by most process-application standards, but too small for water-supply applications. The direct-contact processes are inherently more efficient than the indirect, but are also (generally) more complex.

Several scraped-surface processes have been commercialized in recent years and have been applied in selected applications where the energy savings can justify this expensive type of heat exchanger. Now in advanced stages of development for industrial applications are freeze processes that use either direct-contact heat transfer, or scraperless exchanger surfaces.

The basis of freeze separation

All freeze separation processes are based on the difference in component concentrations between solid and liquid phases that are in equilibrium. This can be most easily understood by referring to Fig. 1a. As a solution (say, at Point A) is cooled, there will be some temperature at which a solid crystalline phase begins to appear in the liquid phase (Point B).

Usually only one component in the solution crystallizes, and that crystal is pure. This permits operation in a single theoretical stage—as contrasted to the incremental separation and multistage processes needed in vapor-liquid separations.

The energy efficiency of the freeze separation process results both from this single-stage capability and from the lower latent heats associated with the solid-liquid phase change; crystallization latent heats are one-half to one-tenth those of vaporization.

At the initial freezing temperature, only a small amount of crystal will be formed. As crystals of a component are formed, the concentration of that component remaining in the solution is decreased. This causes the crystallization temperature of the remaining liquid to drop minutely, so that a lower operating temperature is needed to effect further crystallization. Thus, higher conversions to the crystal phase require successively lower temperatures, as shown by the operating line BCD.

In a binary mixture, a point is eventually reached where both components crystallize simultaneously; this is called the eutectic, and is shown as Point D. At the eutectic, the concentrations of the solid and liquid phase remain constant. Removal of more heat converts more of the liquid phase to solid but without any change in temperature.

Individual crystals of the two components are pure. That is, two separate kinds of crystals are formed, rather than one crystal that incorporates the molecules of both components. In some cases, it is possible to fractionate these solid phases. This will be discussed briefly later.

In multicomponent solutions, the occurrence of a dual-precipitation point may not result in constant operating temperature. For example, when freezing the gas-liquor waste streams from coal-conversion processes, three separate solids are formed before a relatively constant operating temperature is achieved. Water and ammonium bicarbonate crystals are formed at progressively lower temperatures; finally, phenol begins to precipitate. In multicomponent solutions, the continued

Freeze-cycle descriptions

Freeze separation processes are usually identified by describing the freezer-refrigeration cycles in the process. (Although the other components used in the process are necessary, they can usually be interchanged indiscriminately, so they do not provide any unique character to the cycle.) The freeze processes are divided according to the freezing mechanism, and subdivided by the type of refrigeration cycle.

Indirect-contact cycles—These processes all transfer heat (from the freezing solution) through a heat-transfer surface. A closed-cycle refrigeration system (one in which the refrigerant never contacts the process fluid) is used.

- Scraped surface. When the driving force for crystal growth comes through a heat-transfer surface, crystallization will start on that surface. Crystals then build up, cutting down heat-transfer. The conventional method in freeze processing has been to use a scraped surface, to remove the crystal buildup.
- Non-scraped surface. New methods are under development to eliminate buildup, and thus the very expensive scraped-surface heat exchanger.

Direct-contact, secondary-refrigerant cycles—Here, a synthetic organic refrigerant (halocarbon), or a volatile hydrocarbon (such as butane) is injected, as a liquid, into the process liquid in the freezer. Since the pressure of the system is less than the vapor pressure of the refrigerant, the refrigerant boils up, removing heat from the process liquid, thereby causing crystallization. A mechanical compressor is used to raise the pressure of the refrigerant to a point at which the refrigerant can be condensed by either cooling water or melting crystals.

- Subdivisions of this process are made by naming the refrigerant used, and by the way the refrigerant is mixed with the process fluid.

Direct-contact, triple-point processes—The triple point of a substance is that temperature and pressure at which vapor, liquid and solid are all in equilibrium. Removal of 1 lb of vapor from the liquid will not lower the temperature, but will crystallize 2 to 7 lb of the remaining solution. The processes are further subdivided according to the method by which the vapors are removed.

- Vacuum-freeze, vapor-compression. The vapor phase is removed by a mechanical compressor that compresses the vapor sufficiently to permit it to condense either directly on pure crystals or through a heat-transfer surface.
- Absorption-freeze, vapor compression. In this process, the vapors are absorbed in a material that has a vapor pressure below the triple point. The absorbent must be regenerated (i.e., the absorbed material removed) to keep the vapor pressure sufficiently low; here, a conventional refrigeration cycle (vapor compression), uses the condenser (heat discharge) portion of the cycle to provide the heat to drive off the absorbed material.
- Vacuum-freeze, ejector-absorption. Here, vapor is removed by two mechanisms—an absorption cycle and a low-pressure steam ejector. The steam that drives the ejector is also used to regenerate the absorbent. The ejector acts as a (thermal) compressor to raise the quality of the removed vapor so that the vapor can be condensed.

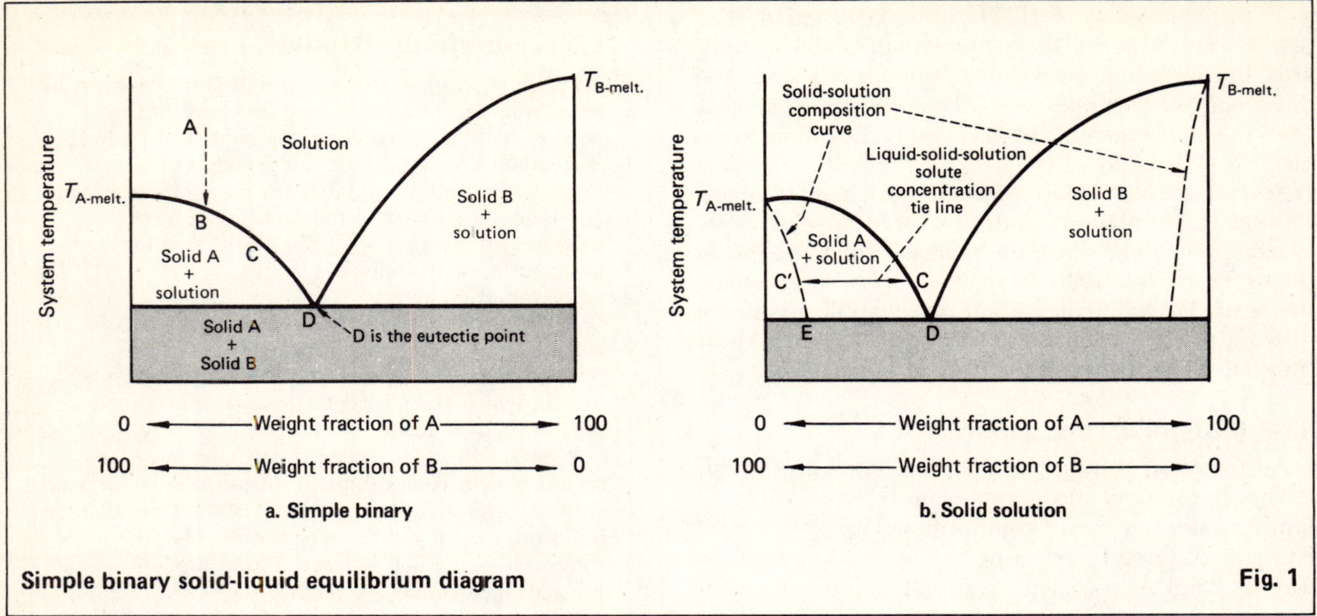

Simple binary solid-liquid equilibrium diagram **Fig. 1**

concentration of impurities in the liquid phase depresses the freezing point of the eutectic composition by an amount dependent on the concentration of the impurity and the specific contaminant.

In some circumstances, two or more compounds can exist in the same crystal. Such crystals are called solid solutions and are common in metallurgy. (Alloy properties are a result of careful control of the solid solution concentrations.) However, such solutions are very rare in chemical and petroleum applications.

Since the liquid and solid phases in equilibrium with each other contain different concentrations of impurity, the crystal can be purified in a multistage crystallization-melting operation called zone melting [*1,2*]. This is the basis for: the multistage freeze-separation process developed by Union Carbide-Australia; the Brodie Purifier [*3*]; and a device recently announced in Japan by Kureha Chemical Co. This multistage capability results in greater process complexity and equipment expense; higher energy consumption is a tradeoff to minimize the added expense of the scraped-surface heat exchangers.

Components of freeze process systems

A continuous freezing process, such as that shown schematically in Fig. 2, consists of at least the following modules:

A *freezer,* where sufficient heat is removed from the process fluid to crystallize up to 15% of the mass.

A *crystal purifier,* where the crystal is separated from the unfrozen concentrate and washed with a few percent of the melted product to remove any adhering concentrate from the surface of the crystal. The separated concentrate is recycled to the freezer to provide any desired recovery.

A *heat pump* that takes heat out of the freezer and transfers it either to cooling water or to the melting crystal that is removed from the purification section.

A *feed heat-exchanger,* usually employed to precool the feed by using the cold product and concentrate, thus reducing the load on the freezer.

Each of these operations can be performed with different types of equipment. At least six processes, using significantly different equipment, have been devised and built in the size range of 50,000 to 500,000 lb/d of crystal removed. (Only one of these processes has ever been widely used commercially.) A summary of the six processes is included in Table I.

The major pieces of equipment that can be used for each unit operation are summarized in Table II, as is the status of the operation. (Another class of freeze separation processes—the hydrate processes—has been studied for water desalting, but they have limited application in the chemical process industries and are not discussed here.)

Solid-liquid equilibria data

The method of predicting or experimentally determining solid-liquid equilibrium is not important so long as it provides the accuracy needed for the intended use; preliminary analyses require much less precise data than would a final design. But, somehow, the data must be obtained because they are the starting point for any analysis or design of a freeze separation process.

Experimental procedures

Accurate and reliable knowledge of the phase equilibria for multicomponent solutions requires experimental data that usually do not exist and are difficult to obtain. Several techniques can be used by the engineer to obtain data from actual samples of the material to be processed:

■ A *differential scanning calorimeter* can be used to obtain both initial freezing points and heat-capacity/latent-heat data. In some cases, this device can also predict high-recovery operations and eutectic formation.

■ A *laboratory ice bath* will predict initial freezing temperatures of a solution—manipulation of the composition of numerous solutions provides data for constructing an equilibrium model for the system throughout the concentrations of interest.

■ A freeze-separation *pilot plant* having adequate flexibility can operate over a variety of product-recovery ranges to give data for a particular application.

These methods are cumbersome, time-consuming and expensive. They should be used only when a decision to use a freeze-separation process has already been made, and accurate design data are required. For preliminary evaluations, there are predictive techniques that yield sufficiently accurate data.

Predictive techniques

Predictive techniques vary in their accuracy, and even the best methods lose accuracy around the eutectic point. Rather than going into solid-liquid equilibrium thermodynamics, we will review the various existing predictive techniques and refer the reader to the literature for more-comprehensive treatment.

Theoretical equilibrium—Freezing points of ideal solutions can be predicted from theoretical thermodynamics. In an ideal solution, the solubility of a solid (or the crystallization point of the dissolved material) is independent of the properties of the solvent and is given by the equation [5]:

$$\ln(1/x_2) = \frac{h_{fus}}{RT}\left(1 - \frac{T}{T_t}\right)$$

where x_2 = mol fraction of solute; h_{fus} = molar heat of fusion of the solute; T_t = triple-point temperature of the solute; and T = operating temperature for crystallization to occur at concentration x_2.

Real binary systems—The ideal equilibrium equation assumes ideal solid and liquid fugacities. In nonideal systems, the liquid fugacity is altered by an experimentally determined activity coefficient, γ, and the equilibrium equation then becomes:

$$\ln(1/\gamma\, x_2) = \frac{h_{fus}}{RT}\left(1 - \frac{T}{T_t}\right)$$

where R = ideal gas low constant.

Activity coefficients in solid-liquid systems are sparse but they do exist for many binary vapor-liquid equilibrium systems. Temperature/activity-coefficient relationships can be projected to the low temperatures associated with solid-liquid equilibria, and used with the above equation to predict binary system behavior.

Real multicomponent systems—The exponential increase in the possible number of multicomponent systems, with increasing numbers of components, requires that binary data be generated and used to predict vapor-liquid equilibria. Various equations have been proposed for combining binary data into multicomponent predictions. Some of the methods are discussed in Refs. [6,7]. This approach can also be used for solid-liquid equilibria predictions. Prausnitz and coworkers [8–10] have developed a technique using a functional-group-contribution method. This technique, originally developed for vapor-liquid and liquid-liquid equilibria, has recently been expanded to predict solid-liquid equilibrium as well [11,12].

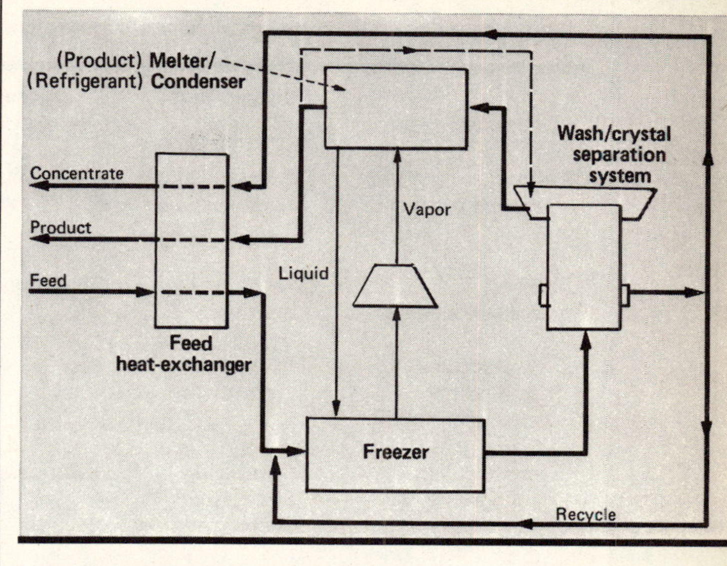

Freeze separation process schematic Fig. 2

Process considerations

A wide variety of equipment is available to the process engineer as a result of the numerous options for each of the major unit-operations involved in a complete freeze-separation system.

At this stage in the commercialization of freeze-separation processes, no equipment supplier is likely to guarantee (nor is any user likely to purchase) a system without doing some pilot-plant work to demonstrate the abilities of the process.

A general procedure for evaluating new applications is discussed and illustrated later in this article. The procedure will yield total power requirements and capital costs for any prospective application. These are the data from which a process engineer can derive the economic incentive to use a freeze-separation process in place of conventional techniques.

Certain process variables associated with a prospective application can indicate the ability of freeze separation to work in a given situation.

Since reduced power-consumption provides the economic incentive for the process, anything that affects energy requirements is an important factor in the evaluation. The energy needed for a freezing separation is used to remove heat from the process solution. This heat is then pumped to a higher level, by the refrigeration system, so that it can be rejected to either cooling water or one of the process streams (e.g., to melt the crystal). Factors that affect power consumption include:

- Freezing temperature.
- Melting temperature of the pure crystal.
- Cooling-water temperature.
- Heat of fusion of the crystal.
- Heat-capacity ratio (C_p/C_v) of the refrigerant.

Usually, if the freezing temperature is not more than 50°F lower than the crystal melt temperature or cool-

Various freeze crystallization systems, applications and unit sizes **Table I**

Freezer type	Refrigeration cycle	Crystal separation	Washing	Application	Size, lb/d
Indirect, scraped surface	Closed cycle, fluorocarbon	Wash column	Yes	Xylene separation (other organic isomer separations as well)	50,000
Indirect, scraped surface	Closed cycle, ammonia	Centrifuge	No	Citrus juice concentrate	
Direct-contact, triple-point vapor compression	Triple-point compression	Wash column	Yes	Seawater	750,000
Direct-contact triple-point	Triple-point absorption cycle	Wash column	Yes	Seawater	50,000
Direct-contact, secondary refrigerant	Vapor-compression, fluorocarbon	Wash column	Yes	Seawater	600,000
Direct-contact, secondary refrigerant	Vapor-compression, butane	Wash column	Yes	Seawater	20,000

ing-water temperature, the energy consumption is very low (0.05 kWh/lb of crystal, or less). Temperature differences above 100°F result in large energy consumption (0.10 to 0.25 kWh/lb), and are justifiable only when the conventional process is unusually energy consumptive (2,000 to 5,000 Btu/lb of purified material). The 100,000 gal/d, vacuum-freeze vapor-compression seawater desalting plant, built by Colt Industries for the Office of Saline Water, operated at 45 kWh/1,000 gal, which equals 0.0053 kWh/lb.

A multicomponent eutectic (several crystals forming simultaneously) or the occurrence of a solid solution can require a multistage freezing process. In the case of the solid solution, a Brodie Purifier type of device can provide the multistage operation needed for fractionation. For eutectic operations, pilot-plant work has shown that the two crystals can be effectively separated if one is lighter than the process liquid, and the other heavier.

Gravity separations of two crystals can be made quite effectively, if the crystal growth conditions are controlled and the specific gravities differ by 5% or more.

Freezer/heat-pump systems

The only freeze-separation processes that have been demonstrated on large-scale equipment for non-seawater applications use indirect scraped-surface freezers. The scraped-surface heat exchanger adds significantly to the cost of the freezing process. Indirect freezers that do not require scraped surfaces have been the subject of past research. Development of this concept is now going on in several companies.

Triple-point freezers, using both compressors and absorption cycles, have been demonstrated in seawater pilot plants. The freezing apparatus is well developed, but the refrigeration cycles present major problems in certain applications. With a vapor-compression cycle, the volumes to be handled at the triple-point are relatively large, and the adiabatic head requirements often dictate a multistage compressor. Fans, centrifugals and blowers are the only compressors that meet the process requirements, but they all have limited adiabatic-heat capabilities.

The absorption cycle will work only if all volatile materials are absorbed. If a nonabsorbable volatile material is present in the process fluid, it will build up in the vapor space and inhibit the mass transfer between the process fluid and the absorbent.

Secondary-refrigerant freezers overcome many of the problems of the triple-point process. The refrigerant can be selected to yield any desired pressure at the freezing conditions. A smaller compressor (lower inlet volume) results, but the compressor savings must be weighed against the added complexity introduced by the refrigerant.

Refrigerants are always soluble (to some degree) in the process fluids, so both the product and concentrate must be stripped. Usually, this added complexity is justified only in larger plants, where the economy of scale favors the simple and low-cost design of the direct-contact freezer.

Crystal separation/purification

The *wash column* has proven a very effective device for both separating the crystals produced in the freezer and purifying them. For a slurry to remain pumpable, the upper limit on crystal content cannot be over 15 to 20%, so the column has to remove a great deal of liquid. To

Check list of freeze crystallization processes that have been tried on pilot plant or commercial scale **Table II**

Freezers/refrigerators								Crystal separation		Melting			
Indirect contact		Triple-point			Secondary refrigerant								
Scraped surface	Non-scraped surface	Vapor compression	Steam ejector	Absorption	Butane	R-114	R-C318	Centrifuge	Wash column	Direct	Indirect	Process name	Status
✓									✓		✓	Brodie purifier Phillips crystal-lizer Kureha crystal purifier	Six installations worldwide, all on organic separations Several xylene separation installations Two installations on organics
✓								✓			✓	Orange juice	Tested on >10,000-lb/d equipment during mid-1960s. Fruit-solids losses in centrifuge. Project abandoned
		✓							✓	✓		Vacuum freeze, vapor-compression	100,000-gal/d plant for Office of Saline Water (OSW) in late 1960s by Colt Industries
			✓	✓					✓	✓		Vacuum freeze, ejector-absorption	6,000-gal/d bench-scale plant built for OSW in late 1960s by Colt Industries
				✓					✓		✓	Absorption freeze, vapor-compression	25,000-gal/d seawater pilot plant now being built for Office of Water Research and Technology (OWRT) by Concentration Specialists
					✓				✓		✓	Immiscible refrig-erant freezing process	10,000-gal/d seawater bench-scale unit built for process development by U.K. Atomic Energy Authority
					✓				✓	✓		Secondary refrig-erant freezing process	15,000-gal/d seawater plant built for OSW in late 1960s by Carrier Corp.
						✓			✓		✓	Crystallex	75,000-gal/d seawater plant built for OSW in early 1970s
							✓		✓		✓	Direct-freeze sep-aration process	Bench-scale tests for OSW in mid-1960s by Avco

work effectively, the column should receive a stream of relatively uniform and large (>150 μm) crystals. The wash column is unable to separate two different crystals unless one of them is more dense than the process fluid, or unless it is so small (<25 μm) that it can be washed out.

Centrifuges have been used for crystal separation in some xylene fractionation systems, and in a developmental program on citrus-juice concentration. A centrifuge can remove most of the brine from the crystal mass, but it lacks the wash column's abilities to counterwash the crystals with melted product. In the citrus application, this drawback led to excessive loss of fruit solids.

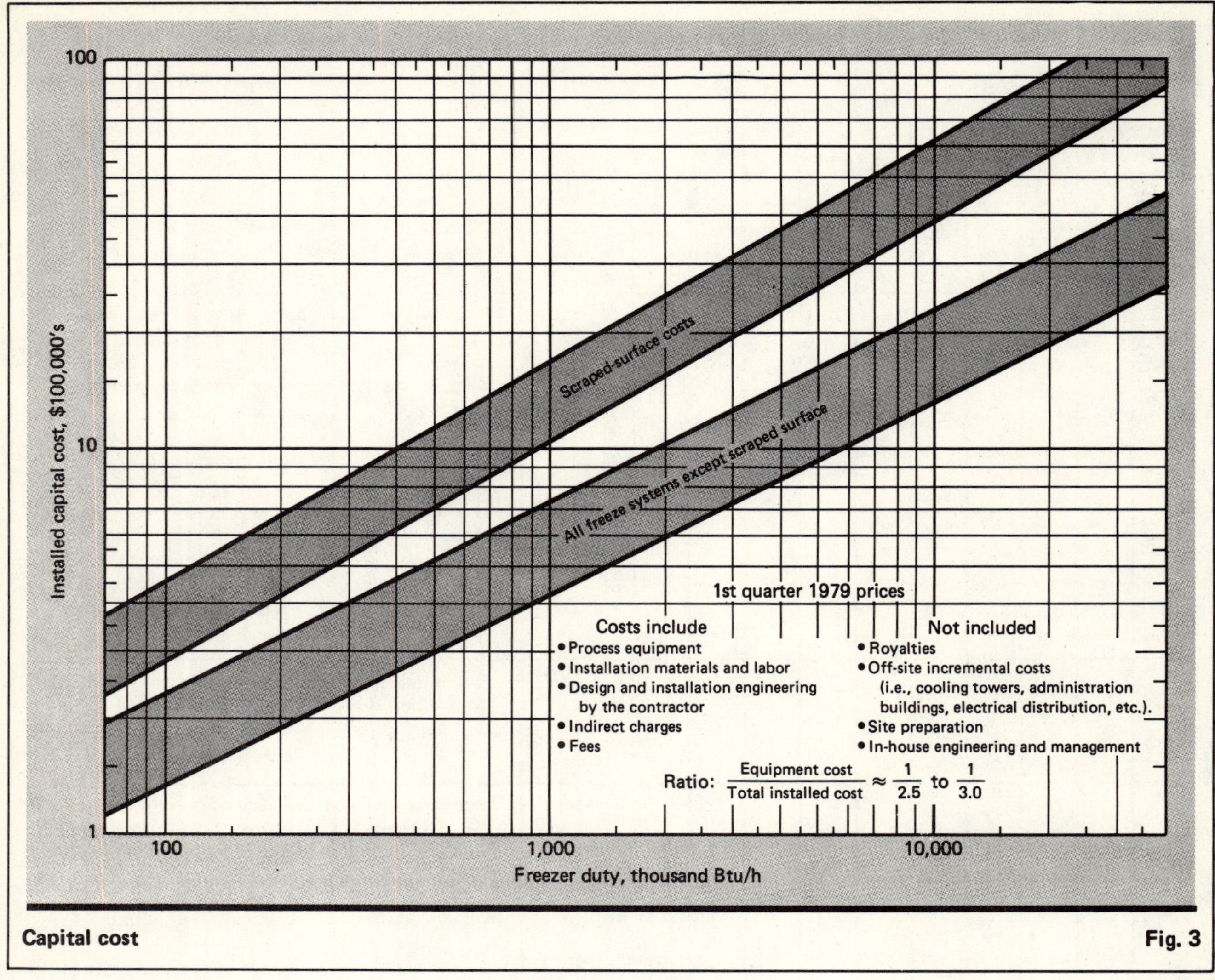

Capital cost **Fig. 3**

Filtration has proven less effective for crystal separation. It also cannot be used for washing the crystals. And, the screens or filter-cloths have a history of "freezing up" (brine freezing in the openings of the weave).

For separating eutectic crystals, it may be necessary to use the centrifuge, followed by reslurrying of the crystals in a wash column. When one crystal is heavier than the solution and the other is lighter, the separation can be made hydraulically; when they are both lighter (or heavier), this cannot be done. Scroll centrifuges can make fractional separations based on relatively small differences in specific gravities—as demonstrated when the precipitates from lime softening operations are separated for purging impurities from the calcium salts before recalcination.

Melter/condenser

As explained earlier, much of the refrigerant vapor can be condensed at low temperatures by melting the crystal products. The device for doing this can be either a direct-contact or shell-and-tube one. Direct-contact condensers are used only with direct-contact freezers; in an indirect cycle, the refrigerant remains isolated, to simplify the refrigeration system and ensure reliability.

Under some circumstances, an indirect melter-condenser may be mandatory—i.e., when product purity is essential, or refrigerant losses due to solubility are excessive.

Solid-liquid equilibria analyses

For preliminary analyses, crude estimates can be made in a number of ways. A binary solution is the simplest system to work with, and it can often be assumed in the preliminary work. The International Critical Tables [*13*] have data in various sections, often expressed as freezing point depressions, that approximate more-complex solutions.

Theoretical calculation of the temperature-solubility relationship can be used in organic solutions to predict both temperatures and eutectic compositions—this relationship becomes less accurate as the difference in molecules increases. Solution properties of electrolytes in water can be predicted by the colligative properties; the necessary details are available in most physical chemistry texts.

If your company has the computer program FLOWTRAN, or a similar simulations program with physical property generation, it can be used to provide the needed data.

Before final design or any major commitment is

made, laboratory analyses with real (not simulated) solutions should always be made.

Process definition

Once the temperature/composition relationship has been defined, a process can be formulated to make the separation. Here are a few generalizations to note:

■ First, determine the total recovery; if sufficient recovery can be achieved before a eutectic is reached, the system will be much simpler. Eutectic separations are best done in a separate stage dedicated to that purpose. Separation of the two crystals must be accomplished so that the product-rich stream can be recycled for purification, and the other material purged from the system. This separation can be nearly ideal—as occurs when the specific gravities of the crystals are above and below that of the solution—or as low as only a few percent upgrading of the product concentration.

■ If the final crystallization stage is more than 40°F lower than the initial freezing point, a second stage that is designed to carry half of the freezing heat load is usually economically justified by virtue of energy reduction. In small systems (<500,000-Btu/h freezer heat load), a single-stage process is most economical.

■ Three-stage systems are seldom justified on energy savings alone, unless the system is very large (>25-million-Btu/h freezer heat load). If a second eutectic point is reached, and it is desirable to separate the second and third crystals from each other, then a third stage should be added, so as to produce as much as possible of the second crystal (in Stage 2) independently of the third (produced only in Stage 3).

■ At this point in the evaluation, it is not necessary to stipulate the freeze cycle beyond differentiating between scraped-surface and nonscraped-surface types. The scraped-surface cycle can be substantially more expensive than the others.

■ If a solid solution occurs, one of the multistage fractional-crystallization processes must be used, unless the solid solution concentration of impurities is low enough not to cause a problem.

Mass and energy balances

Once the number of stages and the operating conditions in these stages have been defined, a total mass balance can be constructed around the system. Process heat loads on the freezers can be calculated from the heat of fusion of the crystallized material, together with the sensible heat load in cooling the process fluid between the inlet and discharge temperatures of that stage. In calculating the heat load, the following assumptions can be used for simplification:

■ The full feed stream is cooled to the discharge temperature, and then the crystal is precipitated at the low temperature.

■ Pumps and mechanical agitators add energy, ultimately in the form of heat. Full-shaft horsepowers are converted ideally at 3,413 Btu per kWh, or 2,545 Btu/(hp)(h). In small, multistage systems (i.e., <100,000 Btu/h), add 25% to the freezer heat load for mechanical heat additions. For large multistage systems (>500,000 Btu/h total process heat load), add 15%, and for large single-stage systems, 10%.

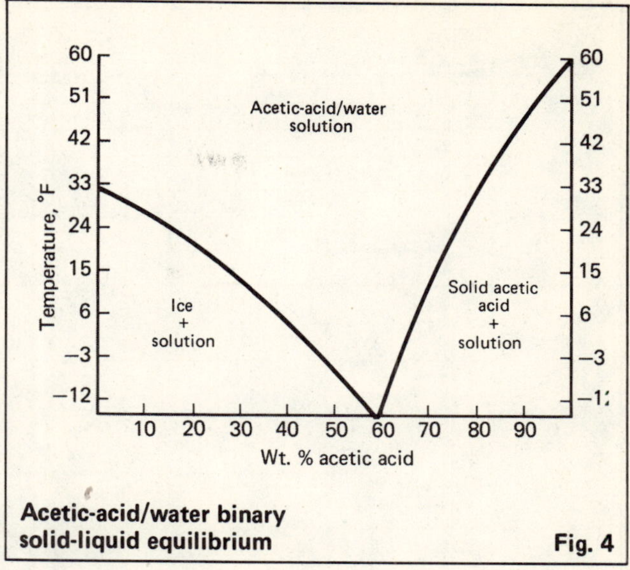

Acetic-acid/water binary solid-liquid equilibrium **Fig. 4**

■ Ambient heat loads also have to be removed by the refrigeration system, and are seen by the system as process heat load. This is usually about 10% of this load.

■ Take the sum of these three heat loads to estimate the capital cost of the system, by using Fig. 3. No further process or design calculations are necessary for a preliminary evaluation. For intermediate levels of sophistication, the individual components must be designed. Operating and maintenance costs can be estimated from the energy balance carried out as above.

Component sizing

The individual pieces of process equipment can be sized by using criteria available in the literature, by getting performance estimates from equipment suppliers, or by applying performance data from similar equipment with which the engineer is familiar.

Cost estimates

Crude capital costs can be estimated by using Fig. 3. In a multistage process, the heat load on each stage should be estimated, and the cost per stage then estimated and totalled to give the system cost. These costs are on an installed basis and include: equipment, installation materials and labor, engineering (by the contractor only), and indirect charges. Not included are: offsite incremental costs (e.g., electrical distribution, cooling towers, boilers), site preparation, in-house engineering, management or administrative costs, and royalties.

The direct-contact processes and the nonscraped-surface indirect-contact processes have capital costs that are relatively the same, and have therefore been lumped together in one band on the cost curve. The costs of the scraped-surface system reflect systems that have been designed for energy conservation rather than economic optimization. Local conditions and energy costs will affect these capital costs.

The driving forces (ΔT's) used in determining the costs have generally been 10 to 20°F; depending on process variables, cost minimums can require ΔT's of 40°F or more. Since scraped-surface devices are the

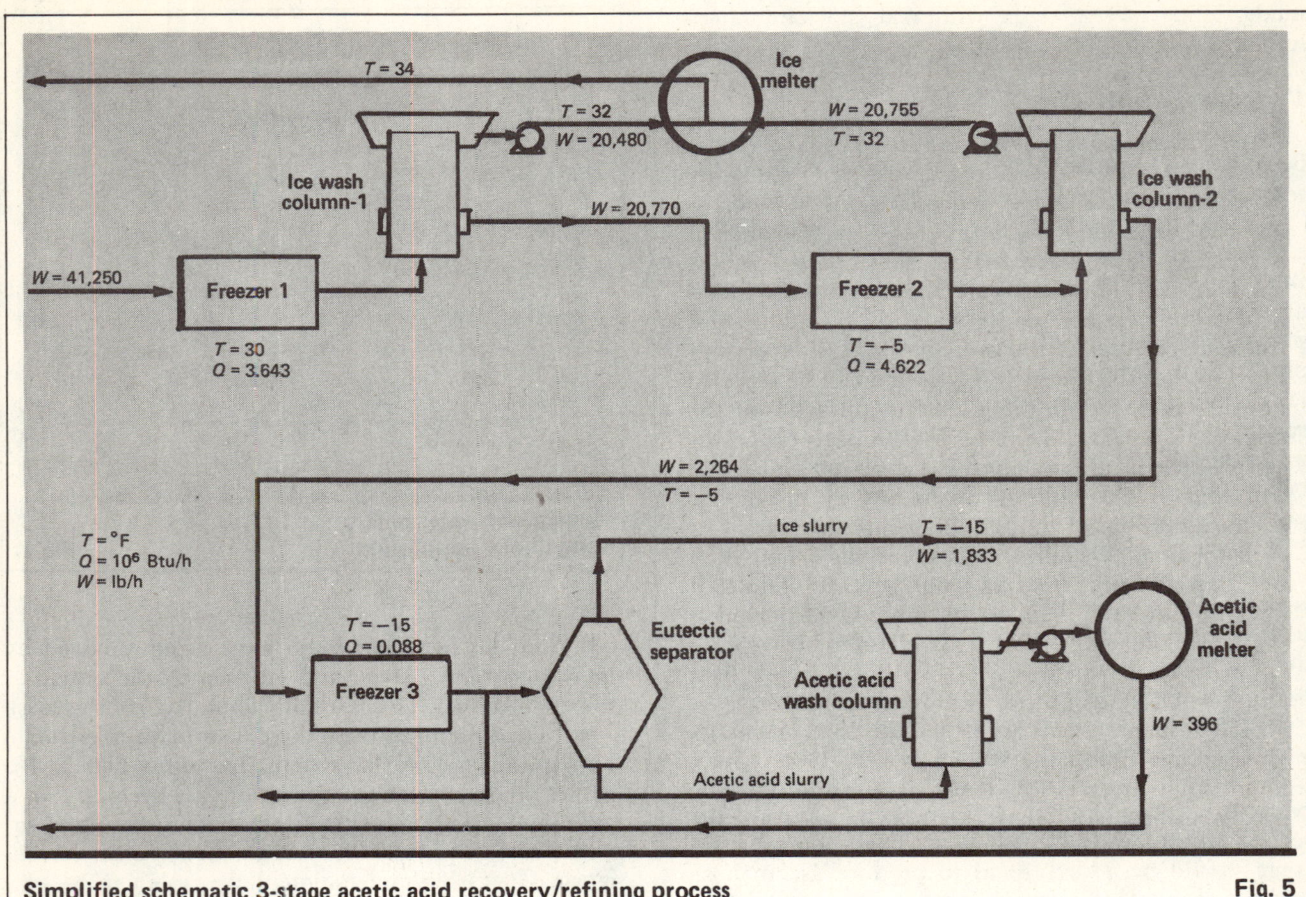

Simplified schematic 3-stage acetic acid recovery/refining process **Fig. 5**

biggest costs in those systems, that design philosophy can reduce capital costs significantly.

The major operations and maintenance costs are for energy and operating/maintenance personnel. Maintenance costs for crystallization systems are the best basis for this estimate, if the user has experience in that area.

Otherwise, a value of 6 to 9% of the capital cost can be used to estimate the yearly maintenance budget. The process requires only one operator per shift if a manpower pool for maintenance assistance is available. The process is usually heavily instrumented, and large plants should have an instrument mechanic assigned full time (one shift per day).

Energy costs can be estimated from the energy balance. Refrigeration-system power requirements will depend on the operating temperatures and the ΔT's assumed. (See the following example for details on this calculation. Note that, in the example, the refrigeration system is a cascade arrangement to minimize energy consumption.)

Cooling-water requirements are needed for compressor intercoolers and the final refrigerant condenser that is used to reject heat from the system. (All ambient heat loads, mechanical additions, inefficiencies, and exothermic concentration effects must be rejected from the system by compressing the refrigerant to a high enough temperature to transfer that heat to cooling water.)

Sample calculation

Application

In the chemical process industries, weak aqueous acetic acid streams are typical of many wastes. They are

Energy balance for sample problem on acetic acid concentration **Table III**

	Physical data								Operating temperatures		Heat loads, Btu/h X10^-6				
	Water				Acetic acid						Sensible		Latent		
Stage	In, lb/h	Crystallized, lb/h	C_p	λ_{fus}	In, lb/h	Crystallized, lb/h	C_p	λ_{fus}	T_{in}	T_{out}	Water	HAc	Water	HAc	Total
1	41,250	20,480	1.0	144	416	–	0.52	–	30	28	0.0825	0.0004	2.9491	–	3.0356
2	20,770	20,480	1.0	144	416	–	0.52	–	28	–15	0.8931	0.0093	2.9491	–	3.8515
3	290	275	1.0	144	416	396	0.52	84	–15	–16	0.0003	0.0002	0.0396	0.0333	0.0734

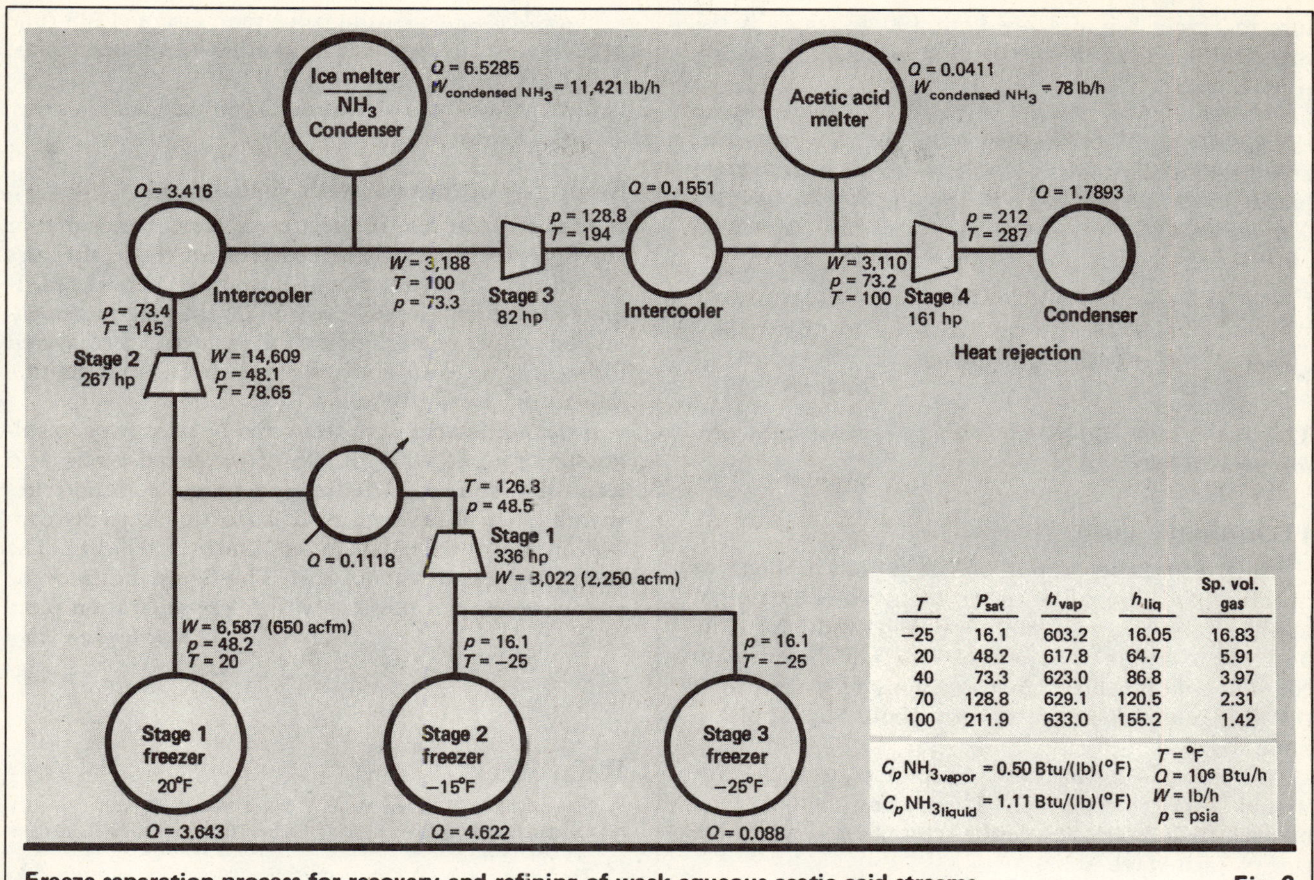

T	P_{sat}	h_{vap}	h_{liq}	Sp. vol. gas
−25	16.1	603.2	16.05	16.83
20	48.2	617.8	64.7	5.91
40	73.3	623.0	86.8	3.97
70	128.8	629.1	120.5	2.31
100	211.9	633.0	155.2	1.42

Freeze separation process for recovery and refining of weak aqueous acetic acid streams **Fig. 6**

produced from such operations as fiber spinning and partial-oxidation processes.

Freeze separation is uniquely able to recover the organic components in these wastes, by crystallizing and separating ice. In the case of acetic acid, no other process is capable of economic recovery.

Recovery of the acid has an added advantage because it would otherwise incur high waste-treatment costs.

These waste streams typically have 1 to 5% organics. In some cases, the organic fraction is predominantly acetic acid. In partial-oxidation wastes, the stream is usually predominantly acetic, with two or three other acids present at 10 to 20% of the weight of acetic acid. For this example, we will consider a stream containing 1% of acetic acid only.

Solid-liquid equilibria

The water/acetic-acid binary is well studied, and data are available in the literature—these are summarized in Fig. 4. Note that the eutectic point is at 59% acetic acid and −16°F. If the stream contains no (or few) impurities, all of the acid can be recovered.

Process definition

The basis for this calculation is: 1 million lb/d of feed (or approximately 83 gal/min); 1% acetic acid; 95% recovery of the acid.

Since the initial freezing point is 30°F, and the eutectic point is at −16°F, the dehydration step will be done in two stages, with a third stage operating at the eutectic to crystallize acetic acid and water simultaneously. The two crystals can be separated very effectively by gravity. The ice (sp. gr. = 0.92) floats in the solution (sp. gr. = 1.03), while the acetic acid crystals (sp. gr. = 1.05) settle to the bottom. The first two stages are designed for equal ice removal. The process schematic is shown in Fig. 5, and the energy balance in Table III.

The refrigeration system is cascaded, as shown in Fig. 6. For simplicity, we will assume an indirect process with an ammonia refrigeration cycle. The 20% added heat load for pumps and ambient loading is included in the values shown next to each freezer.

If capital costs are all that is required, this information is sufficient for estimating. However, a complete evaluation requires some estimate of the operating costs as well. Major operating costs are: power, manpower and maintenance. Power estimates require some equipment sizing.

Component sizing

Compressors—Heat loads and operating conditions in the compressors are shown in Fig. 6. The suction pressures of the first two compressor stages correspond to the saturated temperatures of the freezers. The suction and discharge pressures for Stages 3 and 4 correspond to the pressures at which the refrigerant will condense in the ice and acetic acid melters—in this case, condensing at 40°F when melting ice (32°F), and at 70°F when melting acetic acid (61°F). Note that the suction temperature to Stages 3 and 4 are both 100°F, which is the temperature out of the intercoolers (using 80°F cooling water).

Mechanical heat sources—Agitation and pumping horsepower can be estimated using the 10% heat-load factor suggested earlier. Conversion from the equivalent heat load (Btu/h) is 2,545 Btu/(hp)(h), and an assumed 0.92 motor efficiency. The pump horsepower calculated on this basis is:

Stage 1	$Q = 0.3036 \times 10^6$ Btu/h	hp = 130
Stage 2	$Q = 0.3852 \times 10^6$ Btu/h	hp = 165
Stage 3	$Q = 0.0073 \times 10^6$ Btu/h	hp = 3
		Total hp = 298

This is the only component sizing necessary in a preliminary design.

Preliminary costs

When a freeze separation process is built in stages, as in this application, all of the economies-of-scale do not apply. However, a refrigeration system would normally be built in stages with intercoolers. Combined wash columns would achieve good economy of scale so that the economic advantage of size would still apply to some degree.

As a conservative measure of capital costs, each stage should be priced separately. Here, using the heat load on each freezer, and the capital cost chart (Fig. 3), the cost of each stage is:

Stage	High estimate, $	Low estimate, $	Average, $
1	1,500,000	900,000	1,200,000
2	1,700,000	1,000,000	1,350,000
3	210,000	125,000	170,000
			Total 2,720,000

Equipment costs for this system would be $700,000 to $1 million. Note that the midpoint costs from the band in Fig. 3 were used; the band represents a +35/−15% range that results from the uncertainty of the application and process details at this point.

All of these data are meaningless unless they can be converted into costs and benefits. Methods for an absolute analysis vary between companies, but the basic data are easily developed at this point.

Benefits—Acetic acid value at $0.24/lb. 9,500 lb/d × 365 d/yr × 0.94 time onstream × 0.24/lb = $780,000/yr.

Biological treatment credit—1.067 lb COD/lb acetic × 0.6 lb BOD/lb COD × 10,000 lb acetic/d × 365 d/yr × 0.94 onstream × $0.12/lb BOD = $265,000/yr. (Note, COD = chemical oxygen demand, BOD = biochemical oxygen demand.)

If a biological treatment system is already installed, only the operating portion of this $0.12/lb BOD would be saved. Biological treatment costs are divided approximately equally between amortization and operating costs.

Costs: Power—Refrigeration, 850 hp; pumps, 200 hp; total, 1,050 hp. 1,050 hp × 0.7457 kW/hp × 24 h/d × 365 d/yr × 0.94 onstream × $0.03/kWh = $195,000/yr.

Costs: Manpower—Assume one man assigned full time at $8/h with 50% add-ons for overhead and supervision. 8,760 h/yr × $12/h = $105,000/yr.

Costs: Maintenance—Assume 0.060 of capital costs = $165,000/yr.

Savings compared with distillation

Amortization and indirect costs are not included in this analysis. If the acid is recovered for resale, the sales and distribution costs would have to be considered. If an organization already exists to do this, the incremental costs might be negligible. If the material is recovered for recycle, the costs above these operating and maintenance costs would be minor.

If the acetic acid were recovered by azeotropic distillation, about 125,000 Btu/lb of recovered acetic acid would be required. The freezing system consumes less than 2 kWh/lb product. At 35% fuel-to-electricity conversion, this is the equivalent of under 20,000 Btu. The energy reduction is about 85%. This figure indicates the energy reductions possible with freeze-separation processes.

R. V. Hughson, Editor

References

1. Pfann, William G., "Zone Melting," Wiley, New York, 1958.
2. Cronan, Calvin S., Zone Refining Excites Hopes for Purity Gain, *Chem. Eng.*, Apr. 20, 1959, pp. 80–82.
3. The Brodie Purifier: A Breakthrough in Fractional Crystallization, Sales Literature of the C. W. Nofsinger Co., Kansas City, Mo.
4. Anon., (Chementator), *Chem. Eng.*, Nov. 6, 1978, p. 67.
5. Prausnitz, J. M., "Molecular Thermodynamics of Fluid Phase Equilibria," Prentice-Hall, Englewood Cliffs, N.J. 1979.
6. Chien, H. H. Y., and Null, H. R., Generalized Multicomponent Equation for Activity Coefficient Calculation, *J. AIChE* (*Am. Inst. Chem. Eng.*), Vol. 18, No. 6, p. 1,177 (1972).
7. Null, H. R., "Phase Equilibria in Process Design," Wiley, New York, 1970.
8. Abrams, D. S., and Prausnitz, J. M., Statistical Thermodynamics of Liquid Mixtures: A New Expression for the Excess Gibbs Energy of Partly or Completely Miscible Systems, *J. AIChE* (*Am. Inst. Chem. Eng.*), Vol. 21, No. 1, p. 116 (1975).
9. Fendenslund, A., others, Group Contribution Estimation of Activity Coefficients in Non-Ideal Liquid Mixtures, *J. AIChE* (*Am. Inst. Chem. Eng.*), Vol. 21, No. 6, p. 1,086 (1975).
10. Fendenslund, A., others, data supplement to Ref. 9 (above), private communication.
11. Gmehling, J., others, Solid-Liquid Equilibria Using UNIFAC, paper presented at 71st Annual Meeting, AIChE, Miami Beach, Fla., Nov. 1978.
12. Rasmussen, P., others, Prediction of Phase Equilibria Using UNIFAC Group-Contribution: A Progress Report, paper presented at 86th National Meeting, AIChE, Houston, Tex., Apr. 1979.
13. Washburn, Edward W., ed., "International Critical Tables," McGraw-Hill, New York, 1926–1930.

The author

James A. Heist is Vice-President and Director of Systems Development, Concentration Specialists, Inc., Andover, MA 01810, (617) 470-0650, where he is responsible for process design for new applications of advanced separation processes. He holds a B.S. in chemical and petroleum refining engineering from Colorado School of Mines, Golden, Colo., is a member of AIChE, the Water Pollution Control Federation and the National Water Supply Improvement Assn. He is a registered professional engineer in the State of California.

INDEX